Lecture Notes in Artificial Intelligence 5678

Edited by R. Goebel, J. Siekmann, and W. Wahlster

Subseries of Lecture Notes in Computer Science

Ronghuai Huang Qiang Yang Jian Pei
João Gama Xiaofeng Meng Xue Li (Eds.)

Advanced Data Mining and Applications

5th International Conference, ADMA 2009
Beijing, China, August 17-19, 2009
Proceedings

Springer

Volume Editors

Ronghuai Huang
Beijing Normal University, R&D Center for Knowledge Engineering
100875, Beijing, China, E-mail: huangrh@bnu.edu.cn

Qiang Yang
The Hong Kong University of Science and Technology, Hong Kong, China
E-mail: qyang@cse.ust.hk

Jian Pei
Simon Fraser University, School of Computing Science
V5A 1S6 Burnaby BC, Canada, E-mail: jpei@cs.sfu.ca

João Gama
University of Porto, Faculty of Economics, 4200-465 Porto, Portugal
E-mail: jgama@fep.up.pt

Xiaofeng Meng
Renmin University, School of Information, 100872 Beijing, China
E-mail: xfmeng2006@gmail.com

Xue Li
The University of Queensland, School of Information Technology
and Electrical Engineering, 4072 St. Lucia, Queensland, Australia
E-mail: xueli@itee.uq.edu.au

Library of Congress Control Number: 2009931218

CR Subject Classification (1998): H.2.8, H.3.3, I.5.3, I.2.4, I.2.6, K.3.2

LNCS Sublibrary: SL 7 – Artificial Intelligence

ISSN	0302-9743
ISBN-10	3-642-03347-4 Springer Berlin Heidelberg New York
ISBN-13	978-3-642-03347-6 Springer Berlin Heidelberg New York

springer.com

Printed in Germany

Typesetting: Camera-ready by author, data conversion by Scientific Publishing Services, Chennai, India
Printed on acid-free paper SPIN: 12727162 06/3180 5 4 3 2 1 0

Preface

This volume contains the proceedings of the International Conference on Advanced Data Mining and Applications (ADMA 2009), held in Beijing, China, during August 17–19, 2009. We are pleased to have a very strong program. Acceptance into the conference proceedings was extremely competitive. From the 322 submissions from 27 countries and regions, the Program Committee selected 34 full papers and 47 short papers for presentation at the conference and inclusion in the proceedings. The contributed papers cover a wide range of data mining topics and a diverse spectrum of interesting applications. The Program Committee worked very hard to select these papers through a rigorous review process and extensive discussion, and finally composed a diverse and exciting program for ADMA 2009.

An important feature of the main program was the truly outstanding keynote speakers program. Edward Y. Chang, Director of Research, Google China, gave a talk titled "Confucius and 'Its' Intelligent Disciples". Being right in the forefront of data mining applications to the world's largest knowledge and data base, the Web, Dr. Chang described how Google's Knowledge Search product help to improve the scalability of machine learning for Web-scale applications. Charles X. Ling, a seasoned researcher in data mining from the University of Western Ontario, Canada, talked about his innovative applications of data mining and artificial intelligence to gifted child education. His talk "From Machine Learning to Child Learning" generated much interest among data miners alike. Daniel S. Yeung, who is a Chair Professor in the School of Computer Science and Engineering, South China University of Technology, in Guangzhou, China, talked about his research insights on sensitivity-based generalization error for supervised learning problems and feature selection. As the President of the IEEE Systems, Man and Cybernetics (SMC) Society and a Fellow of the IEEE, Professor Yeung shared much insight on the integration of data mining theory with practice. To highlight the importance of data mining practice, Longbing Cao from the University of Technology, Sydney Australia, talked about "Data Mining in Financial Markets", a timely and important topic that can only be understood well with his vast experience in financial data mining applications.

ADMA 2009 continued the success of the ADMA conference series, which is one of the leading international forums for data mining researchers, practitioners, developers and users to exchange edge-cutting ideas, techniques, and experience.

The Organizing Committee and the Program Committee thank all those who submitted their papers to the conference. We also thank the external reviewers, listed separately, who helped in the review process.

As Conference Co-chairs and Program Committee Co-chairs, we would like to thank the members of the Program Committee for their hard work and for bearing with the primitive conference management tool. Particularly, we would like to extend our special thanks to Yushun Li, Na Zhai, and Bin Jiang who made significant contributions to the review process and the conference organization. Their efforts were essential to the quality of the conference.

We hope that all the participants of ADMA 2009 take the opportunity to exchange exciting research ideas and explore the beautiful city of Beijing.

Qiang Yang
Ronghuai Huang
Jian Pei
João Gama
Xiaofeng Meng

Organization

ADMA2009 was organized by Beijing Normal University, China.

Steering Committee Chair

Xue Li	University of Queensland (UQ), Australia

General Co-chairs

Ronghuai Huang	Beijing Normal University, China
Qiang Yang	Hong Kong University of Science and Technology, China

Program Co-chairs

Jian Pei	Simon Fraser University, Canada
João Gama	University of Porto, Portugal
Xiaofeng Meng	Renmin University of China, China

Local Organization Co-chairs

Guangzuo Cui	Beijing Normal University, China
Yushun Li	Beijing Normal University, China
Fusheng Yu	Beijing Normal University, China

Publicity Co-chairs

Ying Zhou	Beijing Normal University, China

Regional Organization Co-chairs

Xiaochun Cheng	Middlesex University in London, UK
Soon-Joo Hyun	Information and Communications University, Korea
Ah-Hwee Tan	Nanyang Technological University, Singapore

Finance Chair

Ping Li	Beijing Normal University, China

Registration Chair

Lanqin Zheng	Beijing Normal University, China

Web Master

Jiangjian Ma	Beijing Normal University , China
Ying Yuan	Beijing Normal University , China

Program Committee

Hassan Abolhassani	Sharif University of Technology, Iran
Reda Alhajj	University of Calgary, Alberta, Canada
José del Campo Ávila	University of Malaga, Spain
James Bailey	University of Melbourne, Australia
Petr Berka	University of Economics, Czech Republic
Michael R. Berthold	University of Konstanz, Germany
Fernando Berzal	University of Granada, Spain
Rongfang Bie	Beijing Normal University, China
Rui Camacho	University of Porto, Portugal
André de Carvalho	University of Sao Paulo at Sao Carlos, SP, Brazil
Nick Cercone	York University, Canada
Jian Chen	Southern China University of Technology, China
Yu Chen	Sichuan University, China
Xiaochun Cheng	Middlesex University, UK
Frans Coenen	University of Liverpool, UK
Bruno Cremilleux	Universite de Caen, France
Guangzuo Cui	Beijing Normal University, China
Kevin Curran	Ulster University, UK
Alfredo Cuzzocrea	Institute of High Performance Computing and Networking - Italian National Research Council and University of Calabria, Italy
Xiangjun Dong	Shandong Institute of Light Industry, China
ZhaoYang Dong	The University of Queensland, Australia

Xiaoyong Du	Renmin University, China
Mohamad El-hajj	University of Alberta, Canada
Floriana Esposito	University of Bari, Italy
Yi Feng	Zhejiang Gongshang University, China
Raymond Yun Fu	University of Illinois at Urbana Champaign (UIUC), USA
João Gama	University of Porto, Portugal
Dragan Gamberger	Rudjer Boskovic Institute, Croatia
Jean-Gabriel Ganascia	University Pierre et Marie Curie (Paris VI), France
Hong Gao	Harbin Institute of Technology, China
Junbin Gao	Charles Sturt University, Australia
Peter Geczy	National Institute of Advanced Industrial Science and Technology (AIST), Japan
Raúl Giráldez	Pablo de Olavide University, Seville , Spain
Christophe Giraud-Carrier	Brigham Young University, USA
Bing Guo	SiChuan University, China
Ming Hua	Simon Fraser University, Canada
Jimmy Huang	York University, Canada
Tan Ah Hwee	Nanyang Technological University, Singapore
Iñaki Inza	University of the Basque Country, Spain
Ping Jiang	The University of Bradford, UK
Shengyi Jiang	GuangDong University of Foreign Studies, China
Alípio Jorge	University of Porto, Portugal
Rajkumar Kannan	Bishop Heber College, India
Dimitrios Katsaros	University of Thessaly, Greece
Mehmet Kaya	Firat University, Elazig, Turkey
Adam Krzyzak	Concordia University, Canada
Andrew Kusiak	University of Iowa, USA
Charles Li	Sichuan University, China
Gang Li	Deakin University, Melbourne Campus, Australia
Guohe Li	University of Petroleum, China
Jing Li	Sheffield University, UK
Xiaoli Li	Institute for Infocomm Research, Singapore
Xuelong Li	University of London, UK
Yingshu Li	Georgia State University, USA
Zhanhuai Li	Northwestern Polytechnical University, China
Jing Liu	Xidian University, China

Ricardo Vilalta	University of Houston, USA
Paul Vitanyi	CWI, The Netherlands
Guoren Wang	Northeastern University, China
Huiqiong Wang	City University of Hong Kong, China
Shuliang Wang	Wuhan University, China
Wei Wang	Fudan University, China
Hau San Wong	City University of Hong Kong, China
Dash Wu	University of Toronto, Canada
Qingxiang Wu	Ulster University, UK
Zhipeng Xie	Fudan University, China
Bingru Yang	Beijing Science and Technology University, China
Jingtao Yao	University of Regina, Canada
Fusheng Yu	Beijing Normal University, China
Jeffrey Xu Yu	The Chinese University of Hong Kong, China
Ras Zbyszek	University of North Carolina – Charlotte, USA
Sarah Zelikovitz	College of Staten Island of CUNY, USA
Jianzhou Zhang	Sichuan University, China
Shichao Zhang	University of Technology, Sydney, Australia
Tianhao Zhang	UPenn, USA
Yang Zhang	Northwest A&F University, China
Aoying Zhou	East China Normal University, China
Huiyu Zhou	Brunel University, UK
Mingquan Zhou	Beijing Normal University, China
Shuigeng Zhou	Fudan University, China
Xiaofang Zhou	University of Queensland (UQ), Australia
Zhi-Hua Zhou	Nanjing University, China
Zhanli Zhu	Xian Shiyou University, China

External Reviewers

Alexandra Carvalho
Alexandre Francisco
Ana Cachopo
Andre Rossi
Anna Lisa Gentile
Kelvin Ran Cheng
Kevin Kai Zheng
Laurent Gillard
Linhao Xu
Liwei Wang

Asadullah Shaikh
Bingru Yang
Bruno Feres de Souza
Cataldo Musto
Claudia Antunes
Dino Ienco
Dominique Gay
Elena Roglia
Felix Qing Xie
Frédéric Flouvat
François Rioult
Geoffrey Macintyre
George Tzanis
Ghim-Eng Yap
Henning Koehler
Huangliang Sun
Huining Qiu
Ioannis Katakis
Ioannis Partalas
Jia Rong
Jia Zhu
Jinlong Wang
João M. Moreira
Jonathan Ortigosa
Jose Luis Flores
Jun He
Keien Liu
Kein Xie
Longjiang Guo
Marc Plantevit
Marcio Basgalupp
Marcos A. Domingues
Max Pereira
Micky Eunjun Chin
Mona Alkhattabi
Murilo Naldi
Nattapon Harnsamut
Nicola Stokes
Nuno Escudeiro
Nuno Fonseca
Rajkumar Gaire
Ruggero Pensa
Sara Madeira
Shuo Shang
Sophia Alim
Sophie Aubin
Thierry Charnois
Xiaoshi Yin
Xiaoyan Liu
Xiaoyuan Wang
Xing Jiang
Xuan-Hong Dang
Yiping Ke
Yongli Ren
Zaiben Chen
Zhipeng Cai

Sponsoring Institutions

National Science Foundation of China, China
School of Educational Technology, Beijing Normal University, China

Table of Contents

Keynotes

Regular Papers

Short Papers

Confucius and "Its" Intelligent Disciples

Edward Y. Chang

Director of Research, Google China
http://infolab.stanford.edu/~echang/

Abstract. Confucius is a great teacher in ancient China. His theories and principles were effectively spread throughout China by his disciples. Confucius is the product code name of Google Knowledge Search product, which is built at Google Beijing lab by my team. In this talk, I present Knowledge Search key disciples, which are machine learning subroutines that generates labels for questions, that matches existing answers to a question, that evaluates quality of answers, that ranks users based on their contributions, that distills high-quality answers for search engines to index, etc. I will also present the scalable machine learning services that we built to make these disciples effective and efficient.

R. Huang et al. (Eds.): ADMA 2009, LNAI 5678, p. 1, 2009.

From Machine Learning to Child Learning

Charles Ling

Department of Computer Science, University of Western Ontario, Canada
http://www.csd.uwo.ca/faculty/cling

Abstract. Machine Learning endeavors to make computers learn and improve themselves over time. It is originated from analyzing human learning, and is now maturing as computers can learn more effectively than human for many specific tasks, such as adaptive expert systems and data mining. The effective and fruitful research in machine learning can now be used to improve our thinking and learning, especially for our children. In this talk, I will discuss my efforts in using machine learning (and AI) for child education in Canada and China. In early 2009, I hosted a TV series （天才孩子家家有） in a major talk show （湖湘讲堂） in China. The impact of such work in China and around the world can be huge.

R. Huang et al. (Eds.): ADMA 2009, LNAI 5678, p. 2 , 2009.

Sensitivity Based Generalization Error for Supervised Learning Problems with Application in Feature Selection

Daniel S. Yeung

Professor, the School of Computer Science and Engineering,
South China University of Technology, Guangzhou, Guangdong

Abstract. Generalization error model provides a theoretical support for a classifier's performance in terms of prediction accuracy. However, existing models give very loose error bounds.

This explains why classification systems generally rely on experimental validation for their claims on prediction accuracy. In this talk we will revisit this problem and explore the idea of developing a new generalization error model based on the assumption that only prediction accuracy on unseen points in a neighbourhood of a training point will be considered, since it will be unreasonable to require a classifier to accurately predict unseen points "far away" from training samples. The new error model makes use of the concept of sensitivity measure for multiplayer feedforward neural networks (Multilayer Perceptrons or Radial Basis Function Neural Networks). The new model will be applied to the feature reduction problem for RBFNN classifiers. A number of experimental results using datasets such as the UCI, the 99 KDD Cup, and text categorization, will be presented.

R. Huang et al. (Eds.): ADMA 2009, LNAI 5678, p. 3, 2009.

Data Mining in Financial Markets

Longbing Cao

Associate Professor, the University of Technology,
Sydney (UTS), Australia

Abstract. The ongoing global financial recession has dramatically affected public confidence and market development. An example is the market manipulation schemes hidden in capital markets, which have caused losses in billions of dollars, dramatically damaging public confidence and contributing to the global financial and credit crisis. While most investors lost during market falls, for instance, sophisticated speculators can manipulate markets to make money by illegally using a variety of maneuvering techniques such as wash sales. With financial globalization, manipulators are becoming increasingly imaginative and professional, employing creative tactics such as using many nominee accounts at different broker-dealers. However, regulators currently are short on effective technology to promptly identify abnormal trading behavior related to complex manipulation schemes. As a result, shareholders are complaining that too few market manipulators were being caught. In this talk, I will discuss issues related to this topic, present case studies and lessons learned in identifying abnormal trading behavior in capital markets. I will discuss the use of data mining techniques in this area such as activity mining, combined mining, adaptive mining and domain-driven data mining.

R. Huang et al. (Eds.): ADMA 2009, LNAI 5678, p. 4, 2009.

Cluster Analysis Based on the Central Tendency Deviation Principle

Julien Ah-Pine

Xerox Research Centre Europe
6 chemin de Maupertuis
38240 Meylan, France
julien.ah-pine@xrce.xerox.com

Abstract. Our main goal is to introduce three clustering functions based on the central tendency deviation principle. According to this approach, we consider to cluster two objects together providing that their similarity is above a threshold. However, how to set this threshold ? This paper gives some insights regarding this issue by extending some clustering functions designed for categorical data to the more general case of real continuous data. In order to approximately solve the corresponding clustering problems, we also propose a clustering algorithm. The latter has a linear complexity in the number of objects and doesn't require a pre-defined number of clusters. Then, our secondary purpose is to introduce a new experimental protocol for comparing different clustering techniques. Our approach uses four evaluation criteria and an aggregation rule for combining the latter. Finally, using fifteen data-sets and this experimental protocol, we show the benefits of the introduced cluster analysis methods.

Keywords: Cluster analysis, clustering functions, clustering algorithm, categorical and real continuous data, experimental protocol.

1 Introduction

Clustering is one of the main tasks in data analysis and in data mining fields and has many applications in real-world problems. Given a set of N objects $\mathbb{O} = \{o^1, \ldots, o^N\}$, described by a set of P features $\mathbb{D} = \{D^1, \ldots, D^P\}$, the clustering problem consists in finding homogeneous groups of these objects. However, clustering is a NP-hard problem and one has to use heuristics for processing large data-sets. Reviews of such heuristics can be found in [1,2,3]. With respect to the usual taxonomy of clustering methods [1], the present paper aims at contributing to the family of hard partitional techniques. As it was underlined in [3,4], one important factor in that case, is the partitioning function that a clustering algorithm attempts to optimize. Hence, the contributions of this paper are the following ones.

R. Huang et al. (Eds.): ADMA 2009, LNAI 5678, pp. 5–18, 2009.

First, we introduce three clustering functions that have the following general formulation:

$$F(S,\mu,X) = \sum_{i,i'=1}^{N} (S_{ii'} - \mu_{ii'})X_{ii'} \ . \tag{1}$$

where: S is a similarity matrix; μ is a matrix of central tendency measures of similarities given for each pair of objects; and X is a relational matrix which general term, $X_{ii'}$, equals 1 if o^i and $o^{i'}$ are in the same cluster; 0 otherwise. X is similar to an adjacency matrix, yet, as it must be a partition, it has to satisfy the following linear constraints [5,6], $\forall i, i', i'' = 1, \ldots, N$:

$$\begin{array}{ll} X_{ii} = 1 \ . & \text{(reflexivity)} \\ X_{ii'} - X_{i'i} = 0 \ . & \text{(symmetry)} \\ X_{ii'} + X_{i'i''} - X_{ii''} \leq 1 \ . & \text{(transitivity)} \end{array} \tag{2}$$

Following the relational analysis method (RA in the following) in cluster analysis [6,7], the related clustering problems that we want to solve can be formally stated as: $\max_X F(S,\mu,X)$ with respect to the linear constraints[1] given in (2).

Regarding (1), our main point of interest concerns the variable μ. By maximizing $F(S,\mu,X)$, we highly consider to group o^i and $o^{i'}$ together if their similarity $S_{ii'}$ is greater than $\mu_{ii'}$. However, how to set $\mu_{ii'}$? On which basis should we compute $\mu_{ii'}$? Several existing clustering methods are based on the same kind of objective function [9,6]. However, most of them consider $\mu_{ii'}$ as a constant parameter which can be set by the user. *On the contrary, in this paper, we define new clustering functions for which the quantities $\mu_{ii'}$ are data-dependent and are interpreted as central tendency measures of pairwise similarities.*

Then, in order to rapidly find an approximate solution X, we also introduce a clustering algorithm that amounts to a local search based technique. This approach is quite similar to the leader algorithm proposed in [9] and is also studied in the RA context in [6,2]. These techniques interesting as they don't require to fix the number of clusters.

The secondary purpose of this work, is to suggest a new experimental protocol for assessing different cluster analysis methods. Indeed, in most papers covering clustering techniques, only one or two assessment measures are used. *In this work, we propose to take into account four different evaluation criteria and a score aggregation method in our experimental protocol.* Hence, we propose to use this approach for benchmarking the introduced clustering methods.

The rest of this paper is organized as follows. In section 2, we introduce the central tendency deviation principle and define three new clustering functions. Next, in section 3, we detail the associated clustering algorithm. Then, we present our experimental protocol and the obtained results, in section 4. Finally, we give some conclusions and future work in section 5.

---

[1] Note that we can use integer linear programming to solve this clustering problem [5,7] but only for very small data-sets. Indeed, as it was already mentioned, this optimization problem is a NP-hard one [8].

2 Maximal Association Criteria and the Central Tendency Deviation Principle

In this section, we introduce three clustering functions for partitioning continuous numerical data. Firstly, we recall the maximal association approach in the RA framework [7]. In this context, association measures are used in their relational representations in order to design objective functions for clustering categorical data. Secondly, we extend these clustering functions to the more general case of continuous data. In that perspective, we underline the central tendency deviation principle.

2.1 Contingency and Relational Representations of Association Measures

Contingency table analysis aims at measuring the association between categorical variables. Let V^k and V^l be two categorical variables with category sets $\{D^k_s; s = 1, \dots, p_k\}$ and $\{D^l_t; t = 1, \dots, p_l\}$ (p_k being the number of category of V^k). Their associated contingency table denoted n of size $(p_k \times p_l)$, is defined as follows: n_{st} = Number of objects in both categories D^k_s and D^l_t. Then, it exists many association criteria based on the contingency table [10,11]. We are particularly interested in the three following ones:

- The Belson measure (denoted B) introduced in [12] which is related to the well-known χ^2 measure. The former is actually a non-weighted version of the latter.
- The squared independence deviation measure (denoted E) which was introduced in [11] and studied in [13] for measuring similarities between categorical variables.
- The Jordan measure (denoted J), which is a coefficient based upon [14] but which was formally introduced in [11].

We recall below, the contingency representation of these criteria[2].

$$B(V^k, V^l) = \sum_{s=1}^{p_k} \sum_{t=1}^{p_l} \left(n_{st} - \frac{n_{s.} n_{.t}}{N} \right)^2 .$$

$$E(V^k, V^l) = \sum_{s,t} n_{st}^2 - \frac{\sum_s n_{s.}^2 \sum_t n_{.t}^2}{N^2} .$$

$$J(V^k, V^l) = \frac{1}{N} \sum_{s,t} \left(n_{st} \left(n_{st} - \frac{n_{s.} n_{.t}}{N} \right) \right) .$$

where $n_{s.} = \sum_{t=1}^{p_l} n_{st}$.

In the context of the RA approach in contingency table analysis [11,15,7], we can equivalently express the previous criteria using relational matrices. For V^k,

[2] Notice that all of them are null when the variables V^k and V^l are statistically independent.

its relational matrix is denoted C^k and its general term $C^k_{ii'}$ equals 1 if o^i and $o^{i'}$ are in the same category; 0 otherwise. Then, one can observe the following identities between contingency and relational representations [16,15]:

$$\sum_{s,t} n_{st}^2 = \sum_{i,i'} C^k_{ii'} C^l_{ii'} \ ; \ \sum_{s} n_{s.}^2 = C^k_{..} \ ; \ \sum_{s,t} n_{st} n_{s.} n_{.t} = \sum_{i,i'} \frac{C^k_{i.} + C^k_{.i'}}{2} C^l_{ii'} \ . \tag{3}$$

where $\sum_i C^k_{ii'} = C^k_{.i'}$ and $\sum_{i,i'} C^k_{ii'} = C^k_{..}$.

Consequently, we obtain the following relational representations of the studied coefficients [15]:

$$B(C^k, C^l) = \sum_{i=1}^{N} \sum_{i'=1}^{N} \left(C^k_{ii'} - \frac{C^k_{i.} + C^k_{.i'}}{N} + \frac{C^k_{..}}{N^2} \right) C^l_{ii'} \ . \tag{4}$$

$$E(C^k, C^l) = \sum_{i,i'} \left(C^k_{ii'} - \frac{C^k_{..}}{N^2} \right) C^l_{ii'} \ . \tag{5}$$

$$J(C^k, C^l) = \frac{1}{N} \sum_{i,i'} \left(C^k_{ii'} - \frac{C^k_{i.}}{N} \right) C^l_{ii'} \ . \tag{6}$$

The relational formulation of association measures is of interest for analyzing the differences between such coefficients [17]. In our context, this formalism allows to define clustering functions for categorical data. We recall and extend this aspect in the following subsection.

2.2 Maximal Association Criteria and Their Extensions

Let suppose M categorical variables $V^k; k = 1, \ldots, M$ and let Δ represent one of the three studied association measures (4), (5) or (6). Then, the maximal association problem introduced in [7], can be formally stated as follows:

$$\max_X \frac{1}{M} \sum_{k=1}^{M} \Delta(C^k, X) \ .$$

where X is a relational matrix which satisfies (2).

In other words, we want to find the consensus partition X that maximizes the mean average association with all categorical variables $C^k; k = 1, \ldots, M$. This amounts to a clustering model for categorical data or consensus clustering [18]. Thanks to the relational representation, we have the following property [7]:

$$\frac{1}{M} \sum_{k=1}^{M} \Delta(C^k, X) = \frac{1}{M} \Delta \left(\sum_{k=1}^{M} C^k, X \right) = \frac{1}{M} \Delta(\mathbf{C}, X) \ . \tag{7}$$

where $\mathbf{C} = \sum_{k=1}^{M} C^k$ and $\mathbf{C}_{ii'}$ is the number of agreements between o^i and $o^{i'}$ (the number of variables for which both objects are in the same category).

Therefore, it appears that the mean average of the association between X and the relational matrices $C^k; k = 1, \dots, M$, is equivalent to the association between the former and an aggregated relational matrix[3], $\mathbf{C}$.

The summation property (7) is the basis of the extension of the studied objective functions, that we are going to introduce. Indeed, let represent categorical data as an indicator matrix T of size $(N \times P)$ where $P = \sum_{k=1}^{M} p_k$. In that representation, we consider all categories of all nominal variables. Accordingly, we denote $\mathbb{D} = \{D^1, \dots, D^P\}$ the set of all categories. Then, the general term of T, T_{ij}, equals 1 if o^i is in category D^j; 0 otherwise. In that representation, any object o^i is a binary vector of size $(P \times 1)$ and the following observation becomes straightforward: $\mathbf{C}_{ii'} = \sum_{j=1}^{P} T_{ij} T_{i'j} = \langle o^i, o^{i'} \rangle$, where $\langle ., . \rangle$ is the euclidean dot product. *Given this geometrical view, we can consider the extension of the clustering functions recalled previously to the more general case of continuous numerical features. Indeed, we simply interpret* $\mathbf{C}_{ii'}$ *as a dot product between vectors* o^i *and* $o^{i'}$ *and we symbolically replace any occurrence of* $\mathbf{C}$ *with a generic notation* S*. Henceforth, we assume that* S *is a Gram matrix (also called similarity matrix) and the vectors* $o^i; i = 1, \dots, N$*, are represented in a continuous feature space* T *of dimension* P*.*

2.3 The Central Tendency Deviation Principle

According to the previous subsection, the objective functions that we want to study are the following ones[4].

$$B(S, X) = \sum_{i,i'} \left(S_{ii'} - \left(\frac{S_{i.}}{N} + \frac{S_{i'.}}{N} - \frac{S_{..}}{N^2} \right) \right) X_{ii'} . \tag{8}$$

$$E(S, X) = \sum_{i,i'} \left(S_{ii'} - \frac{S_{..}}{N^2} \right) X_{ii'} . \tag{9}$$

$$J(S, X) = \sum_{i,i'} \left(S_{ii'} - \frac{1}{2} \left(\frac{S_{i.}}{N} + \frac{S_{i'.}}{N} \right) \right) X_{ii'} . \tag{10}$$

where $S_{i.} = S_{.i} = \sum_{i'} S_{ii'}$ (since S is symmetric) and $S_{..} = \sum_{i,i'} S_{ii'}$.

Given the previous equations, we can draw out the central tendency deviation principle. Indeed, one can observe that *all objective functions are based on a comparison between the similarity of two objects and a central tendency measure.*

In the case of B defined in (8), the transformation from $S_{ii'}$ to $\left(S_{ii'} - \frac{S_{i.}}{N} - \frac{S_{i'.}}{N} + \frac{S_{..}}{N^2} \right)$ is a geometrical one and is known as the Torgerson transformation [19]. Let $g = \frac{1}{N} \sum_{i=1}^{N} o^i$ be the mean vector. Then, we have: $\left(S_{ii'} - \frac{S_{i.}}{N} - \frac{S_{i'.}}{N} + \frac{S_{..}}{N^2} \right) = \langle o^i - g, o^{i'} - g \rangle$. For the Belson function, the *objects* o^i *and* $o^{i'}$ *could be clustered together providing that their dot product, centered with respect to the mean vector* g*, is positive.*

[3] Also called the collective Condorcet matrix in the RA approach [5].

[4] For the Jordan function, we drop the factor $\frac{1}{N}$.

Regarding E given in (9), the central tendency is the mean average over all pairwise similarities, $\frac{S_{..}}{N^2}$. This approach is also a global one as it considers all (pairs of) objects. In that case, *o^i and $o^{i'}$ are more likely to be in the same cluster if their own similarity is above the mean average of all similarities.*

Unlike the previous cases, the function J introduced in (10), is based on a local central tendency approach. For the Jordan function, *o^i and $o^{i'}$ have more chances to be grouped together if their similarity is greater than the arithmetic mean of the mean average of their similarity distributions $\frac{S_{i.}}{N}$ and $\frac{S_{i'.}}{N}$.*

However, one special case to consider is when the data are already centered. Indeed, if $S_{ii'} = \langle o^i - g, o^{i'} - g \rangle$, all three clustering functions become equivalent as $\frac{S_{i.}}{N} = \frac{S_{i'.}}{N} = \frac{S_{..}}{N} = 0$. Despite this point, we propose a version of the clustering functions that combines two kinds of central tendency approaches.

Following the previous observation and the Belson function, we first center the data. This leads to similarities $S_{ii'}$ that are either positive or negative. Next, *we focus on positive similarities only.* Indeed, the latter are related to pairs of vectors whose cosine index is positive which indicates that they are rather similar. Thus, let $\mathbb{S}^+$ be the set of pairs of objects having positive similarities: $\mathbb{S}^+ = \{(o^i, o^{i'}) \in \mathbb{O}^2 : S_{ii'} \geq 0\}$. Then, we compute the central tendency measures related to the clustering criteria, on the basis of pairs belonging to $\mathbb{S}^+$. More concretely, below are the clustering functions that we propose to define:

$$B^+(S,X) = \sum_{i,i'} \left(S_{ii'} - \left(\frac{S^+_{i.}}{N^+_{i.}} + \frac{S^+_{i'.}}{N^+_{i'.}} - \frac{S^+_{..}}{N^+_{..}} \right) \right) X_{ii'} \ . \tag{11}$$

$$E^+(S,X) = \sum_{i,i'} \left(S_{ii'} - \frac{S^+_{..}}{N^+_{..}} \right) X_{ii'} \ . \tag{12}$$

$$J^+(S,X) = \sum_{i,i'} \left(S_{ii'} - \frac{1}{2} \left(\frac{S^+_{i.}}{N^+_{i.}} + \frac{S^+_{i'.}}{N^+_{i'.}} \right) \right) X_{ii'} \ . \tag{13}$$

with, $\forall i = 1, \ldots, N$: $S^+_{i.} = \sum_{i':(o^i,o^{i'}) \in \mathbb{S}^+} S_{ii'}$; $S^+_{..} = \sum_{i,i':(o^i,o^{i'}) \in \mathbb{S}^+} S_{ii'}$; $N^+_{i.} = \#\{o^{i'} \in \mathbb{O} : (o^i, o^{i'}) \in \mathbb{S}^+\}$ and $N^+_{..} = \#\mathbb{S}^+$ ($\#$ being the cardinal).

Intuitively, this two-step approach allows to *obtain more compact clusters since pairs of objects that are finally clustered together must have very high similarities compared to central tendency measures based upon the most relevant*

Table 1. Clustering functions and their central tendency measures

Clus. func.	Central tendency measures
$B^+(S,X) = F(S, \mu^{B^+}, X)$	$\mu^{B^+}_{ii'} = \frac{S^+_{i.}}{N^+_{i.}} + \frac{S^+_{i'.}}{N^+_{i'.}} - \frac{S^+_{..}}{N^+_{..}}$
$E^+(S,X) = F(S, \mu^{E^+}, X)$	$\mu^{E^+}_{ii'} = \frac{S^+_{..}}{N^+_{..}}$
$J^+(S,X) = F(S, \mu^{J^+}, X)$	$\mu^{J^+}_{ii'} = \frac{1}{2} \left(\frac{S^+_{i.}}{N^+_{i.}} + \frac{S^+_{i'.}}{N^+_{i'.}} \right)$

similarities which are the positive ones and which actually correspond to the nearest neighbors of objects.

To sum up, we give in Table 1, the parameters μ of the respective clustering functions. This table uses the notations provided in equation (1). Henceforth, we are interested in the following clustering problems: $\max_X \Delta^+(S, X)$ with respect to the constraints in (2), where Δ^+ is one of the three functions in Table 1.

3 A Clustering Algorithm Based on Transfer Operations

The clustering problems that we have formally presented as integer linear programs, can be optimally solved but for very small data-sets. In this paper, we target large data-sets and our purpose is to introduce a clustering algorithm that allows to rapidly find an approximate solution X to these clustering problems. Regarding the notations, let $U = \{u_1, \ldots, u_q\}$ be the partition solution represented by the relational matrix X (q being the number of clusters of U). Hence, $u_l; l = 1, \ldots, q$, represents the set of objects that are within this cluster. As we have mentioned in section 1, the core of this heuristic is not new. Actually it can be seen as a prototype-based leader algorithm but employing similarities instead of euclidean distances. The basic ideas of this algorithm were already suggested in [9,6,2] for example. After introducing the mechanism of such an algorithm, we recall its particular weakness and propose a simple solution to overcome it.

3.1 The Basic Algorithm and Its Scalable Version

Given U, a current partition of the objects, we want to find out if the transfer of an object o^i to any other clusters of U distinct from its current cluster, can increase the clustering function value. To this end, we introduce the following contribution measures. In the equations below, u_i symbolically represents the current cluster of o^i; u_l an existing cluster but different from u_i and $\{o^i\}$ the singleton constituted of o^i.

$$cont_{u_i}(o^i) = 2 \sum_{i':o^{i'} \in u_i} (S_{ii'} - \mu_{ii'}) - (S_{ii} - \mu_{ii}) \,. \tag{14}$$

$$cont_{u_l}(o^i) = 2 \sum_{i':o^{i'} \in u_l} (S_{ii'} - \mu_{ii'}) + (S_{ii} - \mu_{ii}) \quad \forall u_l \neq u_i \,. \tag{15}$$

$$cont_{\{o^i\}}(o^i) = (S_{ii} - \mu_{ii}) \,. \tag{16}$$

With respect to the objective function (1), and all other things being equal, one can observe that the quantities[5] measures given in (14), (15) and (16), allow to decide whether object o^i should: stay in its current cluster; be transferred into another cluster; or generate a new cluster.

[5] Notice that S and μ are symmetric in our context. In the case of $cont_{u_i}(o^i)$ for example, the non-symmetric formulation would be: $\sum_{i':o^{i'} \in u_i} (S_{ii'} - \mu_{ii'}) + \sum_{i':o^{i'} \in u_i} (S_{i'i} - \mu_{i'i}) - (S_{ii} - \mu_{ii})$.

Regarding (15), let $cont_{u_{l*}}(o^i)$ be the maximum contribution of o^i to an existing cluster (distinct from its current cluster). Then, in order to maximize the objective function (1), one can observe that the algorithm should:

- create a new cluster if $cont_{\{o^i\}}(o^i)$ is greater than $cont_{u_i}(o^i)$ and $cont_{u_{l*}}(o^i)$,
- transfer o^i to u_{l^*} if $cont_{u_{l*}}(o^i)$ is greater than $cont_{u_i}(o^i)$ and $cont_{\{o^i\}}(o^i)$,
- do nothing in all remaining cases.

Given this basic operation, the algorithm processes all objects and continue until a stopping criterion is full-filled. Typically, we fix a maximal number of iterations over all objects (denoted *nbitr* in the following).

Since we improve the clustering function value at each operation, this algorithm converges to a local optimum.

In order to have an efficient implementation of the algorithm, we need to compute the contribution quantities (14) and (15) efficiently. In the following, we start by discussing the computation of the central tendencies $\mu_{ii'}$. In a second time, we present an efficient way for computing the contribution quantities (14) and (15) using prototypes.

According to Table 1, we need to compute the following statistics to determine $\mu_{ii'}$: the $(N \times 1)$ vectors of general terms $S^+_{i.}$ and $N^+_{i.}$ and/or the scalars $S^+_{..}$ and $N^+_{..}$. All these statistics will be referred as the "components of μ". They are computed before applying the clustering heuristic and are considered as inputs of the algorithm. The computation cost of these statistics is $O(N^2 \times P)$. However, we only need one iteration to calculate them. Moreover, we can compute these different vectors incrementally.

Next, if we consider the formulation of the contributions given in (14) and (15) in terms of S, the computation cost of the algorithm is of order $O(N^2 \times P \times nbitr)$. When N is high, this implementation is costly. Hopefully, in our context, *we can reduce the complexity cost of these quantities.* Let us recall that, we are given the feature matrix T of size $(N \times P)$ as input. Furthermore, let us assume that the space dimension is much lower[6] than the number of objects, $P << N$. Then, since $S = T \cdot T'$, we can use the linearity properties of the dot products in order to quickly compute the contributions (14) and (15) by using prototypes. First, one can observe that:

$$\sum_{i':o^{i'} \in u_l} S_{ii'} = \sum_{i':o^{i'} \in u_l} \langle o^i, o^{i'} \rangle = \langle o^i, h^l \rangle \quad \text{where} \quad h^l = \sum_{i':o^{i'} \in u_l} o^{i'} . \tag{17}$$

h^l is the non-weighted mean vector of size $(P \times 1)$ representing the cluster u_l. Hence, using $h^l; l = 1, \ldots, q$, as prototypes allows to reduce the computation cost of $\sum_{i':o^{i'} \in u_l} S_{ii'}$ from $O(\#u_l \times P)$ to $O(P)$. Second, the computation of the aggregated central tendencies measures, $\sum_{i':o^{i'} \in u_l} \mu_{ii'}$, can also be reduced by keeping up to date the vector ν of size $(q \times 1)$ of general term:

$$\nu_l = \sum_{i':o^{i'} \in u_l} \frac{S^+_{i'.}}{N^+_{i'.}} . \tag{18}$$

6 For high-dimensional space we can assume a pre-processing step that reduces the dimension of the feature space.

Table 2. Clustering functions and aggregated central tendency measures of the contribution of object o^i to a cluster

Clus. func.	Aggregated central tendency measures
$B^+(S,X)$	$\sum_{i':o^{i'}\in u_l} \mu_{ii'}^{B+} = \#u_l \frac{S_i^+}{N_i^+} + \nu_l - \#u_l \frac{S^+}{N^2}$
$E^+(S,X)$	$\sum_{i':o^{i'}\in u_l} \mu_{ii'}^{E+} = \#u_l \frac{S^+}{N^2}$
$J^+(S,X)$	$\sum_{i':o^{i'}\in u_l} \mu_{ii'}^{J+} = \frac{1}{2}\left(\#u_l \frac{S_i^+}{N_i^+} + \nu_l\right)$

Using ν, we can reduce the computation cost of $\sum_{i':o^{i'}\in u_l} \frac{S_{i'}^+}{N_{i'}^+}$, that is involved in the calculation of $\sum_{i':o^{i'}\in u_l} \mu_{ii'}$, from $O(\#u_l)$ to $O(1)$. Accordingly, we give in Table 2, the aggregated central tendency measures of the contribution of o^i to a cluster u_l, for the different clustering functions.

To sum up, if we pre-compute the components of μ and use the prototypes $h^l; l = 1,\ldots,q$, and ν then we can reduce the computation cost of the clustering algorithm to $O(N \times q \times P \times nbitr)$. In the meantime, the memory cost is kept to $O(N \times P)$. *These results are quite satisfying as they make the computation cost and memory cost of such an algorithm comparable to the popular k-means method with respect to the number of objects to be clustered.* We finally give in Algorithm 1, the pseudo-code of the proposed clustering algorithm.

Algorithm 1. Transfer based heuristic

Require: $nbitr$ = number of iterations; T = the feature matrix; μ = the central tendency components.
Take the first object o^i as the first element of the first cluster u_1:
$q \leftarrow 1$ where q is the current number of cluster
Update h^1 and ν_1
for $itr = 1$ to $nbitr$ **do**
  for $i = 1$ to N **do**
    if o^i hasn't been assigned a cluster yet **then**
      Set $cont_{u_i}(o^i) \leftarrow -\infty$ and compute $cont_{\{o^i\}}(o^i)$ using (16)
    else
      Compute $cont_{u_i}(o^i)$ using (14), (17), Table 2 and compute $cont_{\{o^i\}}(o^i)$ using (16)
    end if
    for u_l in the set of already constituted clusters **do**
      Compute $cont_{u_l}(o^i)$ using (15), (17), Table 2
    end for
    Find u_{l^*} the cluster which has the highest contribution with object o^i
    if $cont_{\{o^i\}}(o^i) > cont_{u_{l^*}}(o^i)$ and $cont_{\{o^i\}}(o^i) > cont_{u_i}(o^i)$ **then**
      Create a new cluster $u_{l'}$ whose first element is o^i:
      $q \leftarrow q + 1$
      Update $h^{l'}$ and $\nu_{l'}$ and update h^i and ν_i
    else
      if $cont_{u_{l^*}}(o^i) > cont_{\{o^i\}}(o^i)$ and $cont_{u_{l^*}}(o^i) > cont_{u_i}(o^i)$ **then**
        Transfer object o^i to cluster u_{l^*}:
        Update h^{l^*} and ν_{l^*} and update h^i and ν_i
        if the cluster u_i where was taken o^i is empty **then**
          $q \leftarrow q - 1$
        end if
      end if
    end if
  end for
end for

3.2 Setting the Scanning Order of Objects

One important issue related to Algorithm 1 is its dependency regarding the scanning order of the objects to cluster. To tackle this problem we propose to use one of the component of the central tendency μ that we have introduced beforehand. More precisely, *we propose to scan the objects according to the increasing order of* $(N_{1.}^+, \ldots, N_{N.}^+)$. For object o^i, $N_{i.}^+$ represents the number of objects with which it has a positive similarity (assuming centered data). Accordingly, we first process the less positively connected objects. This approach can be seen as a strategy for finding small and stable clusters rapidly. To some extent, it also can be viewed as a way for eliminating noise. Indeed, if we choose the decreasing order, the most positively connected objects will be processed first and they will bring in their clusters noisy objects.

4 Experiments

In this section, we introduce an experimental protocol that aims to compare different clustering techniques. This protocol takes into account four different evaluation criteria. The latter can be seen as four different "point of views" when ranking the clustering techniques. Therefore, we use an aggregation rule for combining the different rankings into one global ranking. In our experiments, we take the k-means algorithm as the baseline. We compare the results provided by the latter to the ones given by the clustering heuristic defined in Algorithm 1 and associated to the clustering functions mentioned in Table 1. We used fifteen data-sets of the UCI Machine Learning repository [20]. The results that we obtained show improvements over the k-means procedure.

4.1 Experimental Protocol

We propose to use four evaluation criteria defined in the literature for assessing the clustering algorithms' results. Usually, only one or two assessment coefficients are used. In this paper, we argue that the more evaluation criteria we use in an experimental protocol, the more robust the conclusions we can draw out from the latter.

We assume that for a given data-set, we have at our disposal, the true label of all objects. Accordingly, we denote by $V = \{v_1, \ldots, v_k\}$ the true partition of the data-set. In that case, $v_m; m = 1, \ldots, k$, is called a class.

Then, the evaluation criteria we propose to use are the following ones: the entropy measure [4], the Jaccard index [21,22], the adjusted Rand Index [23] and the Janson-Vegelius coefficient [24,15,22]. They are formally defined below where X and C are the respective relational matrices (see subsection 2.1) of U and V:

$$Ent(U,V) = \sum_{l=1}^{q} \frac{\#u_l}{N} \left(\frac{-1}{\log(k)} \left(\sum_{m=1}^{k} \frac{\#(u_l \cap v_m)}{\#u_l} \log(\frac{\#(u_l \cap v_m)}{\#u_l}) \right) \right) .$$

$$Jac(X,C) = \frac{\sum_{i,i'=1}^{N} C_{ii'} X_{ii'} - N}{\sum_{i,i'=1}^{N} (C_{ii'} + X_{ii'} - C_{ii'} X_{ii'}) - N} .$$

$$AR(X,C) = \frac{N^2 \sum_{i,i'=1}^{N} C_{ii'} X_{ii'} - \sum_{i,i'=1}^{N} C_{ii'} \sum_{i,i'=1}^{N} X_{ii'}}{\frac{N^2}{2} \left(\sum_{i,i'=1}^{N} C_{ii'} + \sum_{i,i'=1}^{N} X_{ii'} \right) - \sum_{i,i'=1}^{N} C_{ii'} \sum_{i,i'=1}^{N} X_{ii'}} .$$

$$JV(X,C) = \frac{\sum_{i,i'=1}^{N} (C_{ii'} - \frac{1}{k}) \left(X_{ii'} - \frac{1}{q} \right)}{\sqrt{\sum_{i,i'=1}^{N} (C_{ii'} - \frac{1}{k}) \sum_{i,i'=1}^{N} \left(X_{ii'} - \frac{1}{q} \right)}} .$$

Each of these four coefficients allows to rank the different clustering functions. Except for the entropy measure, the higher the score, the better the clustering output U. Since we want to use several data-sets in the experiments, we have as many rankings as pairs in (evaluation criteria × data-sets). Consequently, we need to combine all these rankings. To this end, we propose to use Borda's rank aggregation rule. In that case, let assume that we need to aggregate r rankings of c clustering techniques' result. Then, for each ranking, Borda's rule consists in scoring the best result with $c-1$, the second one with $c-2$ and so on until the last one which has a null score. Then, the final ranking of the clustering techniques is obtained by summing the r different scores distributions. The best method is the one which has the highest aggregated score.

As for each clustering method, we have to aggregate rankings with respect to two dimensions (evaluation criteria × data-sets), we apply the previous aggregation method in a two-step framework. Given a data-set, we start by aggregating the ranks provided by the different evaluation criteria. Thus, for each data-set, we obtain a first aggregated ranking of the clustering techniques. Then, we aggregate these consolidated rankings given by each data-set which are now seen as criteria. This second aggregated score distribution provides us with the global ranking that allows to compare the clustering techniques from a global viewpoint.

4.2 Experiments Settings

We report in Table 3, the description of the fifteen data-sets of the UCI Machine Learning repository [20], that we included in our experiments. These data-sets concern several domains with distinct numbers of objects, features and classes. *Our aim is to have a global evaluation of the introduced cluster analysis models rather than a specific evaluation restricted to a certain kind of data-sets.* For each data-set we centered[7] and standardized the feature matrix. We also normalized each object's vector in order to have a unit norm. The resulting feature matrix T is the input of all clustering techniques. Concerning the clustering functions

[7] Following the comments given in subsection 2.3.

Table 3. Data-sets description

Name	iris	sonar	glass	ecoli	liv-dis	ionos	wdbc	synt-cont	veh-silh	yeast	mfeat-fou	img-seg	abalo	pag-blo	land sat
Nb. obj.	150	208	214	336	345	351	569	600	846	1484	2000	2310	4177	5473	6435
Nb. feat.	4	60	10	7	6	34	32	60	18	8	76	18	8	10	36
Nb. clas.	3	2	6	8	2	2	2	6	4	8	10	7	28	5	6

we have introduced in this paper, this amounts to take the Gram matrix S as the cosine similarity matrix between centered vectors. Besides, when applying Algorithm 1, we set $nbitr = 10$.

We take the k-means algorithm as the baseline since this technique is pretty closed in spirit to our proposal (transfer operations) and since it was shown that it provides relevant results compared to other clustering techniques [25]. In our experiments, we set k as the number of true classes for the k-means algorithm while keeping free this parameter for Algorithm 1. Moreover, for each data-set, we launched 5 different runs with random seeds for the k-means method and took the mean average of the evaluation criteria measures.

4.3 Experiments Results and Discussions

We report in Table 4, the experiments results (in %) we obtained for each triple (evaluation measure × data-set × clustering algorithms). For each pair (evaluation measure × data-sets) we put in bold the score of the best method. We also give in the bottom lines of Table 4, the number of clusters found by the different methods. According to Table 4, one general comment we can make is that *the evaluation measures do not necessarily agree about their rankings.* Given a data-set, it happens that one clustering method is ranked first for one evaluation measure and last for another one. Besides, *some assessment measures are quite dependent on the number of clusters found.* This is typically the case for the entropy measure (better for larger number of clusters) and the Jaccard index (better for smaller number of clusters). *These observations justify our approach that supports the use of several evaluation criteria in order to have many "opinions".* Accordingly, if we apply the experimental protocol we have described in subsection 4.1, we find the following global ranking: $J^+ \succ E^+ \succ B^+ \sim k$-means. As we can see, cluster analysis based on the central tendency deviation principle can outperform the k-means technique. Besides, the Jordan criterion J^+ seems to be the most consensual clustering approach. Interestingly, it is not the one that was most often ranked first. This indicates that the J^+ objective function is quite robust compared to the other approaches and with respect to the wide variety of data-sets we tested. However, it is worth noticing that regarding higher dimensional data-sets such as synt-cont or mfeat-fou, the B^+ function seems to perform better. Regarding the number of clusters found, E^+ gives, on average, the highest number of clusters; next comes B^+ (except for the iris and abalone data-sets); and then J^+. The number of clusters found using our proposals are distinct from the original number of classes. For most of the data-sets, the

Table 4. Results according to the four evaluation criteria and number of clusters found

Ent	iris	sonar	glass	ecoli	liv-dis	ionos	wdbc	synt-cont	veh-silh	yeast	mfeat-fou	img-seg	abalo	pag-blo	land sat
B^+	35.7	80.5	50.4	26.7	**94.1**	45.1	**17.4**	**17.0**	80.0	53.1	44.6	44.1	61.8	21.5	38.6
E^+	**30.6**	75.5	**43.4**	**23.0**	94.7	**33.6**	18.1	25.8	**78.8**	52.1	**42.5**	49.8	64.5	**16.0**	37.4
J^+	34.1	**72.3**	45.1	23.6	94.6	36.6	18.8	27.3	82.8	52.4	43.8	**41.4**	66.0	17.9	**36.5**
k-m	36.6	97.9	55.4	24.3	98.0	82.1	43.8	26.8	89.4	**50.3**	51.0	43.2	**59.3**	20.1	36.8
Jac															
B^+	36.7	12.4	22.7	32.1	11.8	23.8	28.5	**57.4**	14.4	16.2	**31.6**	36.0	10.7	23.2	**48.7**
E^+	54.3	12.2	**28.1**	**52.8**	13.2	23.9	34.5	43.0	17.1	16.2	26.1	31.2	11.5	24.9	44.1
J^+	42.0	10.8	27.8	48.6	11.4	24.7	30.7	44.5	17.0	16.2	30.1	38.4	**11.6**	23.5	47.3
k-m	**57.1**	**36.4**	27.7	37.6	**44.0**	**42.9**	**73.6**	51.8	**18.3**	**19.4**	27.6	**39.2**	5.5	**39.9**	44.5
AR															
B^+	36.4	**4.1**	22.3	37.2	**1.0**	18.0	26.2	**67.7**	8.0	14.3	**42.6**	44.9	**7.0**	3.6	**57.3**
E^+	57.8	3.0	26.2	**59.6**	0.1	18.5	32.2	51.0	8.8	14.2	35.7	37.5	5.9	7.4	51.8
J^+	45.4	3.9	**26.7**	55.4	0.8	**21.1**	28.5	52.8	**8.9**	14.3	40.7	**47.8**	6.1	5.6	55.5
k-m	**58.7**	1.9	22.1	43.6	-0.6	17.0	**66.8**	61.2	7.9	**18.0**	36.8	47.6	4.9	**10.7**	53.0
JV															
B^+	33.2	**6.3**	24.1	41.1	**1.8**	26.5	37.5	**67.5**	8.2	18.1	**42.3**	44.3	15.7	26.2	**57.0**
E^+	55.0	4.5	**31.8**	**64.5**	0.6	27.3	42.0	46.3	8.3	20.3	34.9	36.2	**18.0**	**40.0**	49.6
J^+	46.5	**6.3**	31.0	60.3	1.5	**31.8**	39.5	50.4	**8.5**	19.6	39.9	47.0	17.3	32.2	54.2
k-m	**58.7**	1.9	27.2	46.6	0.8	17.1	**67.0**	61.6	7.9	**21.5**	36.9	**48.2**	6.2	37.6	53.1
Nb. clus.															
B^+	24	11	7	9	10	13	8	7	9	12	16	11	146	9	10
E^+	7	22	15	15	13	48	21	66	21	19	42	16	69	20	49
J^+	6	17	9	12	11	17	15	14	13	16	27	11	19	12	16
k-m	3	2	6	8	2	2	2	6	4	8	10	7	28	5	6

former is greater than the latter. This suggests that classes can contain several homogeneous subclasses that is interesting to find out in a knowledge discovery context.

5 Conclusion and Future Work

We have introduced three clustering functions for numerical data based on the central tendency deviation concept. These partitioning criteria can be seen as extensions of maximal association measures that were defined for clustering categorical data. We have presented an algorithm that approximately solves the clustering problems we have proposed. Moreover, we have defined a robust experimental protocol that involves four different assessment measures and an aggregation rule. We tested our clustering functions using fifteen data-sets and have showed that our proposals can perform better than the k-means algorithm.

In our future work, we intend to further analyze the geometrical properties of the clustering functions we have introduced in order to potentially apply them in more specific contexts such as high dimensional data or clusters with different shapes for example.

References

1. Jain, A.K., Murty, M.N., Flynn, P.J.: Data clustering: a review. ACM Comput. Surv. 31(3), 264–323 (1999)
2. Grabmeier, J., Rudolph, A.: Techniques of cluster algorithms in data mining. Data Min. Knowl. Discov. 6(4), 303–360 (2002)
3. Xu, R., Wunsch, D.I.: Survey of clustering algorithms. IEEE Transactions on Neural Networks 16(3), 645–678 (2005)
4. Zhao, Y., Karypis, G.: Criterion functions for document clustering: Experiments and analysis, Technical Report TR01-40, University of Minnesota (2001)
5. Michaud, P., Marcotorchino, J.F.: Modèles d'optimisation en analyse des données relationnelles. Mathématiques et Sciences Humaines 67, 7–38 (1979)
6. Marcotorchino, J.F., Michaud, P.: Heuristic approach of the similarity aggregation problem. Methods of operations research 43, 395–404 (1981)
7. Marcotorchino, J.F.: Cross association measures and optimal clustering. In: Proc. of the Computational Statistics conference, pp. 188–194 (1986)
8. Wakabayashi, Y.: The complexity of computing medians of relations. Resenhas IME-USP 3, 323–349 (1998)
9. Hartigan, J.: Clustering Algorithms. John Wiley and Sons, Chichester (1975)
10. Goodman, L., Kruskal, W.: Measures of association for cross classification. Journal of The American Statistical Association 49, 732–764 (1954)
11. Marcotorchino, J.F.: Utilisation des comparaisons par paires en statistique des contingences partie I. Technical Report F069. IBM (1984)
12. Belson, W.: Matching and prediction on the principle of biological classification. Applied statistics 7, 65–75 (1959)
13. Abdallah, H., Saporta, G.: Classification d'un ensemble de variables qualitatives. Revue de statistique appliquée 46, 5–26 (1998)
14. Jordan, C.: Les coefficients d'intensité relative de korosy. Revue de la société hongroise de statistique 5, 332–345 (1927)
15. Marcotorchino, J.F.: Utilisation des comparaisons par paires en statistique des contingences partie II. Technical Report F071. IBM (1984)
16. Kendall, M.G.: Rank correlation methods. Griffin, Londres (1970)
17. Ah-Pine, J., Marcotorchino, J.F.: Statistical, geometrical and logical independences between categorical variables. In: Proceedings of ASMDA 2007 (2007)
18. Lebbah, M., Bennani, Y., Benhadda, H.: Relational analysis for consensus clustering from multiple partitions. In: Proceedings of ICMLA 2008, pp. 218–223 (2008)
19. Torgerson, W.: Multidimensional scaling: I theory and method. Psychometrika 17, 401–419 (1952)
20. Asuncion, A., Newman, D.: UCI machine learning repository (2007)
21. Jaccard, P.: The distribution of the flora in the alpine zone. The New Phytologist 11, 37–50 (1912)
22. Youness, G., Saporta, G.: Some measures of agreement between close partitions. Student 51, 1–12 (2004)
23. Hubert, L., Arabie, P.: Comparing partitions. Journal of classification 2, 193–218 (1985)
24. Janson, S., Vegelius, J.: The J-index as a measure of association for nominal scale response agreement. Applied psychological measurement 6, 111–121 (1982)
25. Jain, A.K., Topchy, A., Law, M.H.C., Buhmann, J.M.: Landscape of clustering algorithms. In: Proceedings of ICPR 2004, vol. 1, pp. 260–263 (2004)

A Parallel Hierarchical Agglomerative Clustering Technique for Billingual Corpora Based on Reduced Terms with Automatic Weight Optimization

Rayner Alfred*

Universiti Malaysia Sabah,
Center for Artificial Intelligence, Locked Bag 2073,
88999 Kota Kinabalu, Sabah, Malaysia
ralfred@ums.edu.my

Abstract. Multilingual corpora are becoming an essential resource for work in multilingual natural language processing. The aim of this paper is to investigate the effects of applying a clustering technique to parallel multilingual texts. It is interesting to look at the differences of the cluster mappings and the tree structures of the clusters. The effect of reducing the set of terms considered in clustering parallel corpora is also studied. After that, a genetic-based algorithm is applied to optimize the weights of terms considered in clustering the texts to classify unseen examples of documents. Specifically, the aim of this work is to introduce the tools necessary for this task and display a set of experimental results and issues which have become apparent.

Keywords: Parallel Clustering, Hierarchical Agglomerative Clustering, Billingual Corpora, Natural Language Processing.

1 Introduction

Effective and efficient document clustering algorithms play an important role in providing intuitive navigation and browsing mechanisms by categorising large amounts of information into a small number of meaningful clusters. In particular, clustering algorithms that build illustrative and meaningful hierarchies out of large document collections are ideal tools for their interactive visualisation and exploration, as they provide data-views that are consistent, predictable and contain multiple levels of granularity.

There has been a lot of research on clustering text documents [1,2,3,4]. However, there are few experiments that examine the impacts of clustering bilingual parallel corpora, possibly due to the problem of the availability of large corpora in translation, i.e. parallel corpora [5]. Fortunately, we have obtained a large

* Head of Machine Learning Research Group, Center for Artificial Intelligence, Universiti Malaysia Sabah, Kota Kinabalu, Sabah, Malaysia.

R. Huang et al. (Eds.): ADMA 2009, LNAI 5678, pp. 19–30, 2009.

collection of over 20,000 pairs of *English-Bulgarian* documents that form our bilingual parallel corpus.

Compared to a clustering algorithm based on a single language, applying clustering to the same documents in two languages can be attractive for several reasons. Firstly, clustering in one language can be used as a source of annotation to verify the clusters produced for the other language. Secondly, combining results for the two languages may help to eliminate some language-specific bias, e.g., related to the use of homonyms, resulting in classes of better quality. Finally, the alignment between pairs of clustered documents can be used to extract words from each language and can further be used for other applications, such as *cross-linguistic information retrieval* (*CLIR*) [6,7].

It is also interesting to examine the impact on clustering when we reduce the set of terms considered in the clustering process to the set of the most descriptive terms taken from each cluster. Using the reduced set of terms can be attractive for several reasons. Firstly, clustering a corpus based on a set of reduced terms can speed up the process. Secondly, with the reduced set of terms, we can attempt to use a genetic algorithm to tune the weights of terms to users' needs, and subsequently classify unseen examples of documents.

There are three main parts of this paper. In the first part, we describe the task of clustering parallel corpora of English-Bulgarian texts in Section 2. The two main areas examined in this paper are English-Bulgarian cluster mapping and English versus Bulgarian tree structures. In the second part, we describe the task of clustering the English texts using the reduced set of terms in Section 3 and compare the results with the previous results of clustering English texts using the complete set of terms. Finally, in the last part, we describe the application of a genetic-based algorithm to optimise the weights of terms considered in clustering the English texts in Section 4. We provides the experimental results in Section 5 and this paper is concluded in Section 6.

2 Clustering Parallel Corpora

In the first part of this work, there are two parallel corpora (*News Briefs* and *Features*), each in two different languages, English and Bulgarian. In both corpora, each English document E corresponds to a Bulgarian document B with the same content. In Table 1, it is worth noting that the Bulgarian texts have a higher number of terms after stemming and stopword removal.

Table 1. Statistics of Document News Briefs and Features

Category (Num Docs)	Language	Total Words	Average Words	Different Terms
News briefs	English	279,758	152	8,456
(1835)	Bulgarian	288,784	157	15,396
Features	English	936,795	431	16,866
(2172)	Bulgarian	934,955	430	30,309

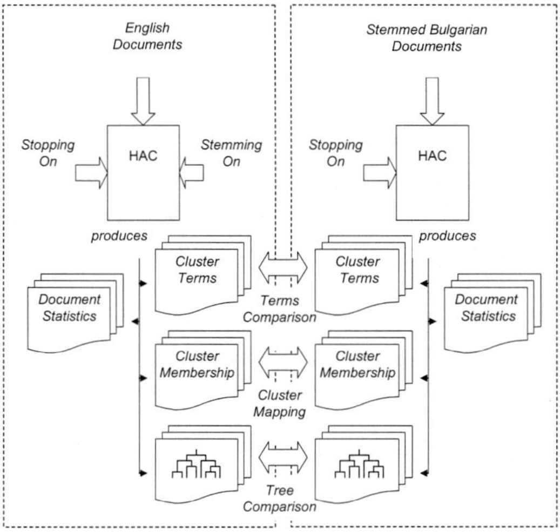

Fig. 1. Experimental set up for parallel clustering task

The process of stemming English corpora is relatively simple due to the low inflectional variability of English. However, for morphologically richer languages, such as Bulgarian, where the impact of stemming is potentially greater, the process of building an accurate algorithm becomes a more challenging task. In this experiment, the Bulgarian texts are stemmed by the BulStem algorithm [8]. English documents are stemmed by a simple affix removal algorithm, Porter Stemmer [9]. Figure 1 illustrates the experimental design set up for the first part of this work. The documents in each language are clustered separately according to their categories (*News Briefs* or *Features*) using a hierarchical agglomerative clustering (HAC) technique. The output of each run consists of two elements: the cluster members and the cluster tree for each set of documents. A detailed comparison of the results for the two languages looking at each of these elements is illustrated in the Experimental Results section.

3 Clustering Documents with a Set of Reduced Terms

In the second part of this work, after clustering the English texts, the terms that characterise the clusters are examined and n terms from each cluster are extracted into another set of terms used for re-clustering the English document again. Previously, there are 10 clusters generated ($k = 10$), in which there will be at most ($n \times k$) number of terms selected, due to the fact that the same term may appear in more than one cluster, where n is the number of terms extracted from each cluster and k is the number of clusters formed. In this stage, the English documents are re-clustered again based on these ($n \times k$) selected terms and the results are compared with the previous clustering results.

4 A Semi-supervised Clustering Technique Based on Reduced Terms

The last part of this work uses a corpus where documents are labelled with their target cluster ID. Clustering is then combined with a genetic algorithm optimising the weight of the terms so that clustering matches the annotation provided as closely as possible. There are two possible reasons for such an approach. Firstly, one can use the clusters provided for some of the documents in language X as a cluster membership annotation for the same documents in language Y. The additional tuning that Genetic Algorithm provides could help cluster the rest of the documents in language Y in a way that resembles more closely the result expected if the translation to language X was used. Secondly, experts such as professional reviewers often produce clusters that are different from the ones generated in an automated way. One can hope that some of their expertise can be captured in the way some of the term weights are modified, and reused subsequently when new documents from the same domain are added for clustering.

The representation of the problem in the Genetic Algorithm setting can be described as follows. A population of X strings of length m is randomly generated, where m is the number of terms (e.g. cardinality of reduced set of terms). These X strings represent the scale of adjustment that will be performed to the values of inverse-document-frequency (IDF) of the corpus and they are generated with values ranging from 1 to 3 uniformly distributed within $[1, m]$. Each string represents a subset of (S_1, S_2, S_3,..., S_{m-1}, S_m). When S_i is 1, the IDF value for the ith term will not be adjusted; otherwise, it will be adjusted by multiplying the IDF value for the ith term by 2 or 3, depending on the value of S_i. In this work, the scale's values of 0 and greater than 3 are not used, as these values may reduce the number of terms further as it adjusts the tf-idf values for a particular term to become 0 or too large. For example, given the tf and idf values shown in Figures 2 and 3 and a string of chromosomes generated shown in Figure 4, the new adjusted values for tf-idf for all terms will be $tf \times idf \times scale$.

The computation of the fitness function has two parts: Cluster dispersion (Eq. 1) and cluster purity (Eq. 3). In order to get clusters of better quality, the DBI measure [10] needs to be minimised, where $S_c(Q_k)$ and $S_c(Q_l)$ denote the *within-cluster* distances for clusters k and l and d_{ce} is the *centroid-linkage*

tf	T_1	T_2	T_3	T_4	T_5	T_6	T_7	...	T_m
D_1	1	0	1	1	0	3	4	...	1
D_2	0	2	3	0	4	1	0	...	0
D_3	1	4	5	0	0	0	0	...	0
...	...	...	...	...	...	...	...	...	...
D_n	1	0	0	0	0	2	1	...	2

Fig. 2. A sample of *term-frequency* (tf) table for n documents and m terms

Term	T_1	T_2	T_3	T_4	T_5	T_6	T_7	...	T_m
idf	0.93	1.63	2.32	4.22	0.67	0.67	1.63	...	3.43

Fig. 3. A sample of *inverse-document-frequency* (*idf*) table for n documents and m terms

Term	T_1	T_2	T_3	T_4	T_5	T_6	T_7	...	T_m
scale	1	1	3	2	1	2	2	...	3

Fig. 4. A string of chromosomes representing the scales of adjustment for m terms

distance that reflects the distance between the centres of two clusters. On the other hand, in order to group the same type of objects together in a cluster, the impurity function, as defined in Eq. 2, needs to be minimised, where n is the number of class, P_{k_c} is the number of points belonging to cth class in cluster k and N_k is the total number of points in cluster k. As a result, the fitness function (OFF) that needs to be maximised will be as defined in Eq. 4.

$$DBI = \frac{1}{C}\sum_{k=1}^{C} max_{l\neq k}\left\{\frac{S_c\left(Q_k\right)+S_c\left(Q_l\right)}{d_{ce}\left(Q_k,Q_l\right)}\right\} \tag{1}$$

$$GiniC_k = 1-\sum_{c=1}^{n}\left(\frac{P_{k_c}}{N_k}\right)^2 \tag{2}$$

$$Impurity = \frac{\sum_{k=1}^{K} T_{C_k}\cdot GiniC_k}{N} \tag{3}$$

$$OFF = \frac{1}{DBI}+\frac{1}{Impurity} \tag{4}$$

5 Experimental Results

5.1 Mapping of English-Bulgarian Cluster Membership

In the first part of this work, every cluster in English (EN) is paired with the Bulgarian (BG) cluster with which it shares the most documents. In this case, there will be a *one-to-one* mapping between the EN and BG clusters. The same is repeated in the direction of BG to EN mapping. Two precision values for each pair are then calculated: the precision of the English-Bulgarian mapping (EBM), $Precision_{EBM}$, and that of the Bulgarian-English mapping (BEM), $Precision_{BEM}$. The precision of the English-Bulgarian mapping, $Precision_{EBM}$, is defined in Eq. 5, where $C(E)$ is the clustered English documents and $C(B)$ is the clustered Bulgarian documents.

$$Precision_{EBM}(C(E),C(B)) = \frac{|C(E)\cap C(B)|}{|C(E)|} \tag{5}$$

Similarly, the precision of the Bulgarian-English mapping, $Precision_{BEM}$, can be defined in a similar manner as defined in Eq. 6. Thus, *purity* can be defined as the percentage of correctly clustered documents, given in Eq. 7, where $P(C(E), C(B))$ is the best precision obtained for $C(E)$ and $C(B)$ and C_{ALL} is the set of clusters formed and $C(E)$ belongs to C_{ALL}.

$$Precision_{BEM}(C(B), C(E)) = \frac{|C(B) \cap C(E)|}{|C(B)|} \tag{6}$$

$$purity = \sum_{C(E) \in C_{ALL}} \frac{|C(E)|}{|D|} \cdot P(C(E), C(B)) \tag{7}$$

Figures 5 and 6 show the precisions for the EBM and BEM for the cluster pairings obtained with $k = 10$ (number of clusters) for each of the two domains, *News Briefs* and *Features*. The X axis label indicates the ID of the cluster whose nearest match in the other language is sought, while the Y axis indicates the precision of the best match found, based on one-to-one mapping between them. For example, in Figure 5, EN cluster 7 is best matched with BG cluster 6 with the EBM mapping precision equal to 58.7% and BEM precision equal to 76.1%.

A final point of interest is the extent to which the mapping EBM matches BEM. Table 2 shows that alignment between the two sets of clusters is 100% when $k = (5, 10, 15, 20, 40)$ for both domains, *News Briefs* and *Features*. However, as the number of clusters increases, there are more clusters that are unaligned between the mappings. This is probably due to the fact that Bulgarian documents have a greater number of distinct terms. As the Bulgarian language has more word forms to express the same concepts as English phrases, this may affect the computation of weights for the terms during the clustering process.

It is also possible to study the purity of the mappings. Table 3 indicates the purity of the English-Bulgarian document mapping for various values of k. This measure has only been based on the proportion of clusters that have been aligned, so it is possible to have a case with high purity, but a relatively low number of aligned pairs.

Table 2. Degree of Purity for Cluster Mapping for English-Bulgarian Documents

Category	k=5	k=10	k=15	k=20	k=40
News briefs	0.82	0.63	0.67	0.65	0.59
Features	0.85	0.77	0.64	0.61	0.54

Table 3. Percentage Cluster Alignment

Category	k=5	k=10	k=15	k=20	k=40
News briefs	100.0%	100.0%	85.0%	85.0%	82.5%
Features	100.0%	100.0%	90.0%	90.0%	80.0%

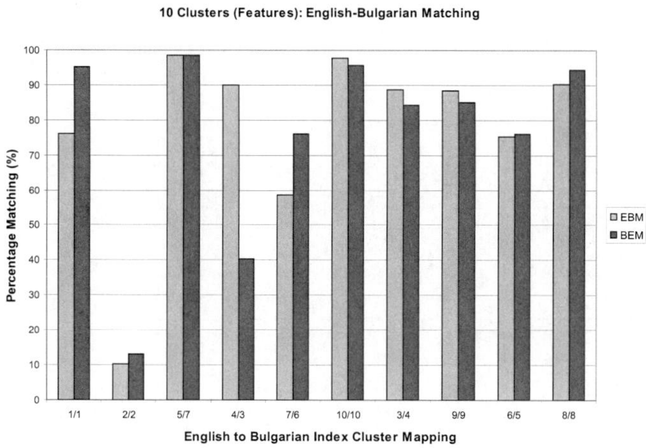

Fig. 5. Ten clusters, Features corpus

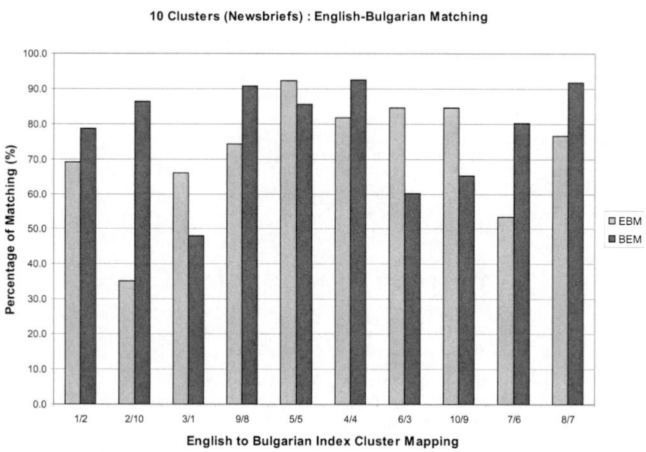

Fig. 6. Ten clusters, News Briefs corpus

5.2 Comparison of HAC Tree Structure

The cluster trees obtained for each language are reduced to a predefined number of clusters ($k = 10$) and then the best match is found for each of those clusters in both directions (EBM, BEM). A Bulgarian cluster BG is only paired with an English cluster EN if they are each other's best match, that is, $BG \Rightarrow BEM \Rightarrow EN$ and $EN \Rightarrow EBM \Rightarrow BG$.

The pair of cluster trees obtained for each corpus are compared by first aligning the clusters produced, and then plotting the corresponding tree for each language. Figures 7 and 8 show that when $k = 10$, all clusters can be paired, and the tree structures for both the English and Bulgarian documents are identical

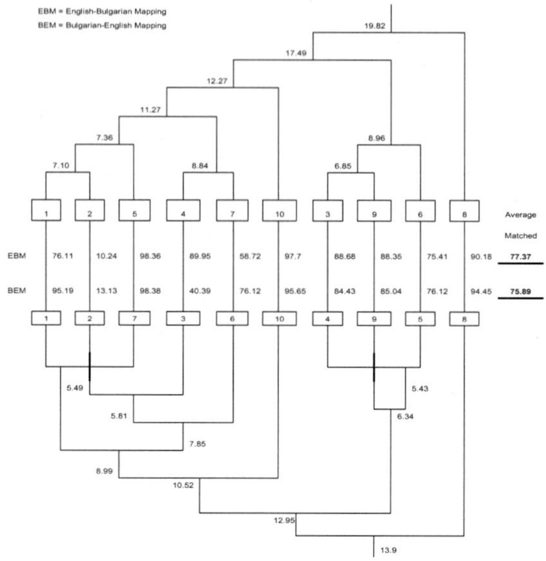

Fig. 7. HAC Structure Tree: Ten clusters, Features corpus

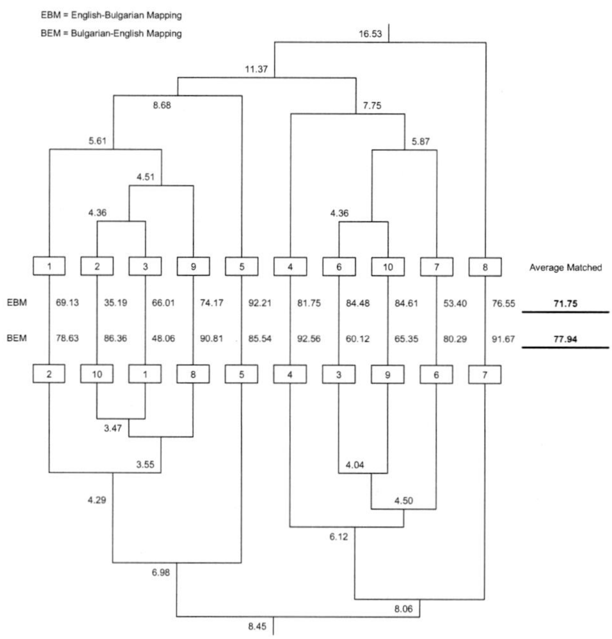

Fig. 8. HAC Structure Tree: Ten clusters, News Briefs corpus

(although distances between clusters may vary). However, there are unpaired clusters in both trees when $k = 20$ [11], and after the matched pairs are aligned, it is clear that the two trees are different. It is hypothesised that this may be a result of the higher number of stems produced by the Bulgarian stemmer, which demotes the importance of terms that would correspond to a single stem in English.

5.3 Clustering Based on a Set of Reduced Terms

Having seen in the previous experiment [11] that the most representative words for each cluster are similar for each language, an interesting question is whether clustering using only these words improves the overall accuracy of alignment between the clusters in the two languages. The intuition behind this is that, as the words characterising each cluster are so similar, removing most of the other words from consideration may be more akin to filtering noise from the documents than to losing information.

The clustering is rerun as before, but with only a subset of terms used for the clustering. That is to say, before the $TF\text{-}IDF$ weights for each document are calculated, the documents are filtered to remove all but $(n \times k)$ of the terms from them. These $(n \times k)$ terms are determined by first obtaining k clusters for each language, and then extracting the top n terms which best characterise each cluster, with the total number of terms equal to at most $(n \times k)$. In this experiment, we use $k = 10$ as the number of clusters and also $n = 10$ and $n = 50$ for the number of terms extracted from each cluster. Four new sets of clusters are thus created, one for each language and number of terms considered $(10 \times 10, 50 \times 10)$. The results in the four cases are compared to each other, and to the sets of previously obtained sets of clusters for which the full set of terms was used.

The results of comparing clusters in English and Bulgarian are shown in Table 4. These clearly indicate that as the number of terms used in either language falls, the number of aligned pairs of clusters also decreases. While term reduction in either language decreases the matching between the clusters, the effect is fairly minimal for English and far more pronounced for Bulgarian.

In order to seek to explain this difference between the languages, it is possible to repeat the process of aligning and calculating purity, but using pairs of clusters from the same language, based on datasets with different levels of term reduction. The results of this are summarised in Table 5. This table demonstrates that, for both languages, as the number of terms considered decreases, the clusters formed

Table 4. Number of aligned clusters and their purity (%) for reduced term clustering ($k = 10$)

	Bulgarian Terms		
English Terms	All	500	100
All	10 (74.9%)	4 (54.2%)	3 (53.0%)
500	9 (72.9%)	4 (46.0%)	3 (51.5%)
100	10 (70.3%)	4 (60.1%)	3 (75.5%)

Table 5. Number of aligned clusters and their purity (%) for reduced term datasets against the unreduced datset ($k = 10$)

	Number of Terms	
Languages	500	100
English	10 (80.1%)	9 (74.2%)
Bulgarian	4 (53.0%)	3 (53.0%)

deviate further and further from those for the unreduced documents. While the deviation for English is quite low (and may indeed be related to the noise reduction sought), for Bulgarian reducing the number of terms radically alters the clusters formed. As with the earlier experiments, the high morphological variability of Bulgarian compared to English may again be the cause of the results observed.

5.4 Semi-supervised Clustering Based on Genetic Algorithm

It has been shown that when the language is English, one can reduce the number of terms used without a great loss in performance. This could help reduce the search space and achieve a speed up when the term weights used by a clustering algorithm are fine-tuned by machine learning (e.g. a genetic algorithm) to obtain a tree of clusters in one language that more closely matches the tree for the other language.

Table 6 shows the number of aligned clusters and their purity (%) for the document clustering based on two sets of terms, namely:

1. The unreduced terms.
2. The reduced terms.

The documents are clustered based on a set of reduced terms with two different settings, one without terms weights adjustment (*Non-GA*) and one with GA-based terms weight adjustment (*GA*). In table 6, the purity of the clusters increases when the GA-based term weight adjustment is used.

Table 6. Number of aligned clusters and their purity for reduced term datasets against the unreduced dataset without weights adjustment and with GA-based weight adjustment ($k = 10$)

	Clusters Alligned		Purity	
Number of Terms	Non-GA	GA	Non-GA	GA
10	10	10	80.1%	81.0%
50	9	10	74.2%	82.0%

6 Conclusion

This paper has presented the effect of applying a clustering technique to parallel multilingual texts, which involves studies related to the differences of the cluster mappings and tree structure of the English and Bulgarian cluster texts. In this paper, we have also studied the effects of reducing the set of terms used to cluster English and Bulgarian texts. Finally, the results of the clustering task were obtained by applying a genetic-based algorithm to optimise the weights of terms considered to cluster the texts.

This work has presented the idea of using hierarchical agglomerative clustering on a bilingual parallel corpus. The aim has been to illustrate this technique and provide mathematical measures which can be utilised to quantify the similarity between the clusters in each language. The differences in both clusters and trees (dendrograms) have been analysed. It can be concluded that with a smaller number of clusters, k, all the clusters from English texts can be mapped into clusters of Bulgarian texts with a high degree of purity. In contrast, with a larger number of clusters, fewer clusters from English texts can be mapped into the clusters of Bulgarian texts with a lower degree of purity. In addition, the tree structures for both the English and Bulgarian texts are quite similar when k is reasonably small (and identical for $k \leq 10$).

A common factor in all the aspects of parallel clustering studied was the importance that may be attached to the higher degree of inflection in Bulgarian. From the very beginning, the significantly lower degree of compression that resulted from stemming Bulgarian was noted. This implies that there were a larger number of Bulgarian words which expressed the same meaning, but which were not identified as such. It is likely that this is one of the factors responsible for decreasing the alignment between the clusters for larger values of k.

To summarise, the results of the clustering of documents in each of two languages with quite different morphological properties are compared: English, which has a very modest range of inflections, as opposed to Bulgarian with its wealth of verbal, adjectival and nominal word forms. (This difference was additionally emphasised by the fact that the Bulgarian stemmer used produced results which were not entirely consistent in the choice between removing the inflectional or derivational ending.) The clusters produced and the underlying tree structures were compared.

In this work, the bilingual English-Bulgarian corpus has also been clustered based on a set of reduced terms, and the application of a genetic algorithm to tune the weights of terms considered in the clustering process has been shown. As most of the top terms seemed to represent the same concepts in the two languages, the possibility of restricting the number of terms used to a much smaller set than the original one was considered as a way of making the results more robust with respect to differences between languages and speeding up clustering.

Reducing the number terms alone resulted in a slight decline in performance (a drop of up to 10% in the clusters paired and 4.6% lower cluster purity) when reducing the list of English terms, and a catastrophic decline when this is done for Bulgarian in the cases of 100 and 500 terms studied. When the genetic-based algorithm is applied to the reduced set of terms to tune the weights of the terms (a maximum of 500 terms) to be considered in the clustering process, the result actually showed an increase in the purity of the clusters, as shown in Table 6.

Success here would also encourage other possible applications, such as training the algorithm on a hand-clustered set of documents, and subsequently applying it to a superset, including unseen documents, incorporating in this way expert knowledge about the domain in the clustering algorithm.

References

1. Steinbach, M., Karypis, G., Kumar, V.: A Comparison of Document Clustering Techniques. Technical Report 00-34, University of Minnesota
2. Zhao, Y., Karypis, G.: Evaluation of Hierarchical Clustering Algorithms for Document Datasets. ACM Press, New York (2002)
3. Moore, J., Han, E., Boley, D., Gini, M., Gross, R., Hastings, K., Karypis, G., Kumar, V., Mobasher, B.: Web Page Categorisation and Feature Selection using Association Rule and Principal Component Clustering. In: 7th Workshop on Information Technologies and Systems (1997)
4. Zamir, O., Etzioni, O.: Web Document Clustering: A Feasibility Demonstration. In: Research and Development in Information Retrieval, pp. 46–54 (1998)
5. Romaric, B.M.: Multilingual Document Clusters Discovery. In: RIAO, pp. 116–125 (2004)
6. Kikui, G., Hayashi, Y., Suzaki, S.: Cross-lingual Information Retrieval on the WWW. In: Multilinguality in Software Engineering: The AI Contribution (1996)
7. Xu, J., Weischedel, R.: Cross-lingual Information Retrieval Using Hidden Markov Models. In: The Proceedings of the Joint SIGDAT Conference on Empirical Methods in Natural Language Processing and Very Large Corpora (EMNLP/VLC 2000) (2000)
8. Nakov, P.: BulStem: Design and Evaluation of Inflectional Stemmer for Bulgarian. In: Proceedings of Workshop on Balkan Language Resources and Tools (2003)
9. Porter, M.F.: An Algorithm for Suffix Stripping. Program 14(3), 130–137 (1980)
10. Davies, D.L., Bouldin, D.W.: A Cluster Separation Measure. IEEE Trans. Pattern Analysis and Machine Intelligence, 224–227 (1979)
11. Alfred, R., Paskaleva, E., Kazakov, D., Bartlett, M.: HAC For Cross-language Information Retrieval. International Journal of Translation 19(1), 139–162

Automatically Identifying Tag Types

Kerstin Bischoff, Claudiu S. Firan, Cristina Kadar, Wolfgang Nejdl, and Raluca Paiu

L3S Research Center / Leibniz Universität Hannover
Appelstrasse 9a
30167 Hannover, Germany
{bischoff,firan,nejdl,paiu}@L3S.de, cristina.kadar@gmail.com

Abstract. Web 2.0 applications such as *Delicious*, *Flickr* or *Last.fm* have recently become extremely popular and as a result, a large amount of semantically rich metadata produced by users becomes available and exploitable. Tag information can be used for many purposes (*e.g.* user profiling, recommendations, clustering *etc.*), though the benefit of tags for search is by far the most discussed usage. Tag types differ largely across systems and previous studies showed that, while some tag type categories might be useful for some particular users when searching, they may not bring any benefit to others. The present paper proposes an approach which utilizes rule-based as well as model-based methods, in order to automatically identify exactly these different types of tags. We compare the automatic tag classification produced by our algorithms against a ground truth data set, consisting of manual tag type assignments produced by human raters. Experimental results show that our methods can identify tag types with high accuracy, thus enabling further improvement of systems making use of social tags.

Keywords: collaborative tagging, classification, tag types, social media.

1 Introduction

Collaborative tagging as a flexible means for information organization and sharing has become highly popular in recent years. By assigning freely selectable words to bookmarked Web pages (*Delicious*), to music (*Last.fm*) or pictures (*Flickr*) users generate a huge amount of semantically rich metadata. Consequently, several well known tagging systems have been acquired by search engine companies to exploit this additional information during search. Especially for multimedia resources, accurate annotations are extremely useful, as these additional textual descriptions can be used to support multimedia retrieval. Prior studies, which started to investigate users' motivations for tagging and the resulting nature of such user provided annotations, discovered that both motivations for tagging, as well as the types of assigned tags differ quite a lot across systems. However, not all tags are equally useful for search. For example, a user might tag a picture on *Flickr* with some of the things depicted on it, like "flowers", "sun", "nature", or with the associated location ("London") and time ("2008"). Since such tags are factual in nature, *i.e.* they are verifiable at least by common sense, they are potentially relevant to all other users searching for pictures *e.g.* from this location. However, to provide some more context for sharing her images with friends, she may also add more subjective, contextual tags like "awesome" or "post-graduate trip", or she may refer to

R. Huang et al. (Eds.): ADMA 2009, LNAI 5678, pp. 31–42, 2009.

herself by using the annotation "my friends". Assuming a certain amount of interpersonal agreement, subjective tags may still be useful for some users. For the majority of users, the tag "awesome" for example, may be an indicator of the quality of the picture, but not for people disagreeing with popular opinion. Self reference tags on the other hand are so highly personal that another person may not understand the tag at all or associate something different with it (*e.g.* her own post-graduate trip to Asia). Thus, personal tags are not applicable to other users of the system, except from the user herself and maybe some of her friends. Still, for estimating similarity between resources or users search engines and recommendation algorithms exploiting user generated annotations but not differentiating types of tags and their interpersonal value incorporate all (frequent) tags and thus introduce noise. Being able to distinguish between the types of tags associated to resources would thus be highly beneficial for search engines and recommendation algorithms to best support users in their information needs. Besides, tag classes enable building enhanced navigation tools. While currently the user faces a potentially infinite, unordered tag space, tag classes would allow for browsing pictures, web sites or music by the different informational facets of the associated tags.

In this paper we tackle exactly this aspect presenting an approach to automatically identify tag types. We rely on a tag type taxonomy introduced in [1] and analyzed with respect to the potential of the tag classes for search. Our approach is applicable to any tagging system and is not bound to one particular resource type.

2 Related Work

Recent scientific work has started examining tagging behaviors, tag types and automatic tag classification, many studies focusing on one specific collaborative tagging system.

2.1 Tagging Motivations and Types of Tags

Analyses of collaborative tagging systems indicate that incentives for tagging are quite manifold and so are the kinds of tags used. According to [2], organizational motivations for enhanced information access and sharing are predominant, though also social motivations can be encountered, such as opinion expression, attraction of attention, self-presentation [2,3]. Which of those incentives is most characteristic for a particular system seems to vary, depending on tagging rights, tagging support, aggregation model, *etc.* – all influencing why certain kinds of tags are used. [3] and [4] indicate that in free-for-all tagging systems like *Last.fm*, opinion expression, self-presentation, activism and performance tags become frequent, while in self-tagging systems like *Flickr* or *Delicious* users tag almost exclusively for their own benefit of enhanced information organization.

Despite the different motivations and behaviors, stable structures do emerge in collaborative tagging systems [3,5,6]. The evolving patterns follow a scale-free power law distribution, indicating convergence of the vocabulary to a set of very frequent words, coexisting with a long tail of rarely used terms [5,6]. Studying the evolution of tagging vocabularies in the MovieLens system, [7] use controlled experiments with varying system features to prove how such design decisions heavily influence the convergence process within a group, *i.e.* the proportions "Factual", "Subjective" and "Personal" tags will have. According to these results, being able to display automatically identified "Factual" tags only would lead to even more factual and interpersonally useful tags. Similarly, in their paper on collaborative tag suggestions, [8] introduce a taxonomy of five classes: Content, Context, Attribute, Subjective and Organizational tags.

[1] introduce an empirically verified tag type taxonomy comprising eight categories (Topic, Time, Location, Type, Author/Owner, Opinions/Quality, Usage context, Self reference) that is applicable to any tagging system, not bound to any particular resource type. Besides establishing type distributions for *Last.fm*, *Delicious* and *Flickr*, the authors discuss the potential of the different identified categories for supporting search. A complementing query log analysis showed that *e.g.* highly personal self-reference tags are indeed not used in querying a web search engine. Similarly, subjective usage context and opinions are rarely queried for, nor judged very useful for searching public web pages. Only for music these queries play an important role with people often searching for "wedding songs" or "party music". Here, interpersonal agreement seems higher due to the restricted domain and, probably, shared culture. In the present paper we will make use of this taxonomy and focus on automatically classifying tags accordingly.

2.2 Automatic Classification of Tags

So far, there have been only few studies trying to automatically categorize user tags. However, they all focus solely on specific domains and make no statements about the generalizability of their approaches to other areas apart from the original ones. Focusing on the domain of pictures, [9] try to extract event and place semantics from tags assigned to *Flickr* photos - making use of location (geographic coordinates) and time metadata (timestamp: upload or capture time) associated with the pictures. The proposed approach relies on burst analysis: tags referring to event names are expected to exhibit high usage patterns over short time periods (also periodically, *e.g.* "Christmas"), while tags related to locations show these patterns in the spatial dimension.

In [10], different tag categories used by users to annotate their pictures in *Flickr* are analyzed automatically. Using the WordNet lexical database the authors are able to classify 52% of their sample tags into the WordNet categories: Location (28%), Artefact/Object (28%), Person/Group (28%), Action/Event (28%), Time (28%) or Other (27%). However, tag classification is not the main focus of the paper, the authors being rather interested in recommending tags to users for supporting them in the annotation process. The authors of [11] map *Flickr* tags to anchor text in Wikipedia articles which are themselves categorized into WordNet semantic categories. Thus the semantic class can be inferred – improving the classifiable portion of *Flickr* tags by 115% (compared to using WordNet only). Given a set of *Delicious* bookmarks and tags assigned by users, [12] investigate the predictability of social tags for individual bookmarks. The proposed classification algorithms make use of the page's textual content, anchor text, surrounding hosts, as well as other tags already applied to the URL. This way, most tags seem to be easily predictable, page text providing the superior attributes for classification.

Thus, existing approaches often focus on predicting certain tag types only and they do so within one particular tagging system. Some techniques are restricted to the system used, *e.g.* as they require additional metadata [9] or assume content to be textual [12]. In contrast to previous work, we present a general approach to tag type classification applying our algorithms on collections containing different kinds of resources.

3 Tag Type Taxonomy

For automatically classifying user generated tags according to their functions into types, we chose the tag type taxonomy presented in [1]. This scheme was build upon the classification presented by [3] and adapted to be applicable for various types of resources (music, Web pages and pictures). It is fine grained enough for distinguishing different tag functions and the associated interpersonal value of the corresponding tag types and, more important, it was tested for its reliability. Table 1 shows the eight classes with corresponding example tags, found in the three systems *Last.fm*, *Delicious* and *Flickr*.

Table 1. Tag type taxonomy with examples of the used tagging systems (from [1])

Nr	Category	*Delicious*	*Last.fm*	*Flickr*
1	Topic	*webdesign, linux*	*love, revolution*	*people, flowers*
2	Time	*daily, current*	*80s, baroque*	*2005, july*
3	Location	*slovakia, newcastle*	*england, african*	*toronto, kingscross*
4	Type	*movies,mp3*	*pop, acoustic*	*portrait, 50mm*
5	Author/Owner	*wired, alanmoore*	*the beatles, wax trax*	*wright*
6	Opinions/Qualities	*annoying, funny*	*great lyrics, yum*	*scary, bright*
7	Usage context	*review.later,travelling*	*workout, study*	*vacation, honeymoon*
8	Self reference	*wishlist, mymemo*	*albums i own, seen live*	*me, 100views*

Topic is probably the most obvious way to describe an arbitrary resource, as it describes what a tagged item is about. For music, topic was defined to include theme (*e.g.* "love"), title and lyrics. The Topic of a picture refers to any object or person (*e.g.* "clowns") displayed, for web sites, it is associated with the title or the main subject (*e.g.* "data mining"). Tags in the **Time** category add contextual information about hour, day, month, year, season, or other time related modifiers. It may tell the time when a picture was taken (*e.g.* "2004"), a song was recorded (*e.g.* "80s"), a Web page was written or its subject event took place (*e.g.* "November 4"). Similarly, **Location** adds additional information, telling us about the country or the city, elements of the natural landscape, sights, nationality or place of origin. It can be the place where a concert took place (*e.g.* "Woodstock"), where a picture was taken (*e.g.* "San Francisco") or a location in a Web page (*e.g.* "USA"). Tags can also specify the **Type** of a resource – *i.e.* what something is. In general it refers to types of files (*e.g.* "pdf"), media (*e.g.* "movie") or Web pages (*e.g.* "blog"). For music this category contains tags specifying the music genre (*e.g.* "hip-hop"), as well as instrumentation (*e.g.* "piano"). For images it includes photo types (*e.g.* "portrait") as well as photographic techniques (*e.g.* "macro"). Yet another way to organize resources is by identifying the **Author/Owner** of a resource. It specifies who created the resource: the singer or the band name for songs, the name of a blogger or a photographer. It can also refer to the owner of the resource: a music label, a news agency or a company. Other tags like "depressing", "funny" or "sexy" contain subjective comments on the characteristics or on the quality of a resource. Users make use of such **Opinion** tags to express opinions either for social motivations or just for simplifying personal retrieval. **Usage context** tags suggest what to use a resource for, the context in which the resource was collected or the task for which it is used. These tags, although subjective, may still be a good basis for recommendations to other users. They can refer for example to a piece of music suitable to "wake up", a text "toRead" or a URL useful for "jobsearch". Last, the **Self reference** category contains highly personal tags, only meaningful and helpful for the tagger herself. Typical examples are "my favorite song", "home" *e.g.* referring to the start page of a site or "my friend" to indicate the presence of the user's friend on some picture.

Resources usually have more than one single tag associated with them, the tags falling into several categories. Our methodology focuses on automatically classifying all these tags from three different data sets into the eight functional categories.

4 Data Sets

For our experiments we used data from some well known collaborative tagging systems covering three different types of resources: music files, general Web pages, and pictures. For the later experiments the raw tags are used, *i.e.* no lemmatizing is applied.

We used an extensive subset of *Last.fm* pages corresponding to tags, music tracks and user profiles fetched in May 2007. We obtained information about a total number of 317,058 tracks and their associated attributes, including track and artist name, as well as tags for these tracks plus their corresponding usage frequencies. Starting from the most popular tags, we found a number of 21,177 different tags, which are used on *Last.fm* for tagging tracks, artists or albums. For each tag we extracted the number of times each tag has been used as well as the number of users who used the tag.

The *Delicious* data for our analysis was kindly provided by research partners. This data was collected in July 2005 by gathering a first set of nearly 6,900 users and 700 tags from the start page of *Delicious*. These were used to download more data in a recursive manner. Additional users and resources were collected by monitoring the *Delicious* start page. A list of several thousand usernames was collected and used for accessing the first 10,000 resources each user had tagged. The resulting collection comprises 323,294 unique tags associated with 2,507,688 bookmarks.

For analyzing *Flickr* tags, we took advantage of data crawled by our research partners during January 2004 and December 2005. The crawling was done by starting with some initial tags from the most popular ones and then expanding the crawl based on these tags. We used a small portion of the first 100,000 pictures crawled, associated with 32,378 unique tags assigned with different frequencies.

5 Automatic Tag Type Classification

For automatically classifying *Last.fm*, *Flickr* and *Delicious* tags, we propose two basic approaches to categorization, depending on the category that needs to be identified. Though some tags could be assigned to more than one functional type, we will categorize each tag according to its most popular type, mainly to make evaluation of automatic classification, *i.e.* human assessment of a ground truth data set feasible. For this, we use both straight forward matching rules against regular expressions and table look-ups in predefined lists, as well as more complex model-based machine learning algorithms.

5.1 Rule-Based Methods

Five of the eight tag type categories can be identified by using simple rules, implemented as regular expressions, or table look-ups in predefined lists.

Time. Spotting time-tags is done with the help of both several date/time regular expressions and by using lists of weekdays, seasons, holiday names, *etc.* The same predefined lists where used for all three systems. This approach can easily capture most time tags – since time vocabulary of the predominately English tags is rather restricted. Less trivial approaches, like detecting time related tags as bursts over short time periods [9], on the other hand, require time related metadata (*e.g.* upload) that is not present in all tagging systems. In total we used 19 complex regular expressions containing also 106 predefined time-related expressions (*e.g.* "May", "Thanksgiving", "monthly").

Location. For identifying location tags in *Last.fm*, *Flickr* and *Delicious*, we made use of the extensive knowledge provided by available geographic thesauri. From GATE[1], an open source tool for Natural Language Processing, a total of 31,903 unique English, German, French and Romanian location related words were gathered. These terms comprised various types of locations: countries (with abbreviations), cities, sites, regions, oceans, rivers, airports, mountains, *etc.* For *Delicious*, the list needed to be slightly adapted by manually excluding some (about 120) extremely common words (*e.g.* "java", "nice", "church") in order to assure better accuracy[2].

Type. Since the Type category, denoting what kind of resource is tagged, is system/ resource dependent, separate lists were used for the three systems. A list of 851 music genres gathered from AllMusic[3] was used in order to identify type tags in the *Last.fm* data set. This inventory of genres is highly popular and also used in ID3 tags of MP3 files. As music is only a (small) part of resources tagged in *Delicious*, we gathered a list of 83 English and German general media and file format terms, *e.g. document, pdf, foto, mpg* or *blog, messenger*. For *Flickr*, the type or genre list (45 items) covers besides file formats also picture types (like *portrait* or *panoramic*), photographic techniques (like *close-up* or *macro*) or camera-related words (like *megapixel, shutter*).

Author/Owner. From the information available on *Last.fm*, *i.e.* the tracks collected, a huge catalogue of artist names resulted, against which candidate tags were matched to identify whether a tag names the author or owner of a resource. In case of *Delicious* with its wide variety of Web pages bookmarked, finding the author or owner is not trivial. Since processing of a page's content and possibly extraction of named entities seems a costly procedure, we made use of an inexpensive heuristic assuming domain owners/authors to appear in a Web page's URL. With the help of regular expressions, we checked whether the potential owner or the author of the resource appears inside the corresponding URL (`http://xyz.author.com`). For *Flickr*, classifying tags into the Author/Owner category was not possible, as pictures are mostly personal and no user-related information was included in our data set.

[1] `http://gate.ac.uk/`

[2] Can be automated *e.g.* by filtering words whose most popular WordNet synset is not a location.

[3] `http://www.allmusic.com/`

Self reference. For identifying self reference tags from *Last.fm*, we created an initial list of 28 keywords, containing references to the tagger herself in different languages (*e.g.* "my", "ich" or "mia") and her preferences (*e.g.* "favo(u)rites", "love it", "listened"). For *Delicious* we adapted the list slightly to include also structural elements of a Web site (like "homepage", "login" or "sonstiges") that do not appear in the music tagging portal. Finally, for *Flickr* the list was adapted to include some personal background references, like "home" or "friends".

The rule-based methods are run over all tags to be classified; the remaining, unclassified tags are then used for training the classifiers. This filtering simplifies the subsequent task of learning to discriminate topic, opinion and usage context tags.

5.2 Model-Based Methods

Since building a reasonably comprehensive register of topics, usage contexts or opinion expressions is due to practically inexhaustible lists impossible, model-based machine learning techniques are necessary for identifying these kinds of tags. To find the Topic, Usage context and Opinion tags, different binary classifiers were trained to decide, based on given tag features, whether a tag belongs to the respective tag class or not. Here, we used classifiers available in the machine learning library Weka[4].

Classification Features. For all three systems *Last.fm*, *Delicious* and *Flickr*, we extracted the same features to be fed into the binary classifiers: Number of users or tag frequency respectively, Number of words, Number of characters, Part of speech, and Semantic category membership.

Number of users is an external attribute directly associated to each tag, measuring prevalence in the tagging community, and thus indicating a tag's popularity, relevance and saliency. For *Flickr*, we used the absolute usage *frequency* since our data does not contain the necessary user-tag tuples and it can be considered to be an equally useful, though different, indicator of popularity. Since it has been suggested that, often highly subjective opinion tags in *Last.fm*– like "lesser known yet streamable artists" – exhibit both a higher *number of words* and *number of characters* [4], we used these intrinsic tag features as well for training our classifiers. Similarly, many of these opinion tags are adjectives while topic tags are mostly nouns [3]. Thus, we included *part of speech* as additional feature. For determining word class, we employ the lexical database WordNet 2.1[5]. In form of a derived tree of hypernyms for a given word sense, WordNet also provides valuable information about the semantics of a tag. The three top level categories extracted from here complete our tag feature set. For *Last.fm* with its multi-word tags, we collected the latter two features for each word in the tag, *i.e.* we matched all terms individually if the phrase as a whole did not have a WordNet entry.

Sense Disambiguation and Substitution. For exploiting tag information like part of speech and WordNet category during machine learning, choosing the right meaning of a tag, for example "rock", is critical. Since statistical or rule-based part-of-speech

[4] `http://www.cs.waikato.ac.nz/ml/weka/`

[5] `http://wordnet.princeton.edu/`

tagging can not be applied for the one-word tags found in *Delicious* and *Flickr*, we decided to make use of the rich semantic information provided implicitly through tag co-occurrences. For the *Last.fm* and *Delicious* sample tags, we extracted all co-occurring tags with the corresponding frequencies. To narrow down potential relations, we computed second order co-occurrence. For all sample tags, we determined similarity with all other tags by calculating pairwise the cosine similarity over vectors of their top 1000 co-occurring tags. A very high similarity should indicate that two tags are almost synonymous because they are so frequently used in the same context (*i.e.* co-tags) – the two tags themselves rarely appearing together directly [13]. Given an ambiguous tag, we now search for the newly identified similar tags in the definitions, examples and synset words in WordNet. If this does not decide for one meaning, then by default the sense returned by WordNet as most popular is chosen. Since some tags are not found in WordNet at all, we make further use of similar tags by taking the most similar one having a match in WordNet as a substitute for the original tag. Due to missing coccurrence relationships for *Flickr*, neither disambiguation nor substitution could be applied.

To build models from the features listed that enable finding Topic, Usage context and Opinion tags from our sample tags, we experimented with various machine learning algorithms Weka offers: Naïve Bayes, Support Vector Machines, C4.5 Decision Trees, *etc.* For each, we moreover used different combinations of the basic features described. As the Weka J48 implementation of C4.5 yielded the best results, only the results obtained with this classifier are presented in the following section on evaluation results.

6 Results and Discussion

6.1 Ground Truth and Evaluation

For evaluating the proposed algorithms, we built a ground truth set containing sample tags from each system that were manually classified into one of the eight categories. To make manual tag categorization feasible a subset of 700 tags per system was assessed. Thus, we intellectually analyzed 2,100 tags in total. The samples per system included the top 300 tags, 200 tags starting from 70% of probability density, and 200 tags beginning from 90% – prior work suggests that different parts of the power law curve exhibit distinct patterns [5]. Clearly, such classification schemes only represent one possible way of categorizing things. Quite a few tags are ambiguous due to homonymy, or depending on the intended usage for a particular resource they can fall into more than one category. We based our decision on the most popular resource(s) tagged. On a subset of 225 tags we achieved a good and substantial inter-rater reliability for this scheme and method – a Cohen's Kappa value of κ 0.7. In general, it was often necessary to check co-occurring tags and associated resources to clarify tag meaning (see also [1]).

For measuring the performance of our tag type classification algorithms we use classification accuracy. For the model-based methods we perform a 10-fold cross-validation on the samples, and for the rule-based method we compute the accuracy by determining the number of true/false positives/negatives. Table 2 summarizes results for all systems and classes. It shows the best performing features, the achieved accuracy, precision and recall, and the percentage of tags (*i.e.* the sample of 700 tags) belonging to a

Table 2. Best results for rule-based and model-based methods. *(Features: POS=part of speech, C=WordNet categories, F=tag frequency, N=number of words and characters, RegEx=regular expressions, List=list lookup).*

	Class	Features	Accuracy	P	R	% Man.	% Auto.
Delicious	Topic	POS,C	81.46	83.89	96.38	67.14	76.00
	Time	RegEx,List	100.00	100.00	100.00	0.86	0.86
	Loc.	List	97.71	70.37	70.37	3.86	3.86
	Type	List	93.71	66.67	42.86	8.00	5.14
	Author	RegEx	70.20	9.85	38.57	6.29	2.14
	Opinion	N,POS,C	93.40	0.00	0.00	5.14	0.00
	Usage	POS,C	89.66	0.00	0.00	7.86	0.14
	Self ref.	List	99.00	33.33	16.67	0.86	0.29
	Unknown						*11.57*
Flickr	Topic	F,POS,C	79.39	84.62	88.82	46.07	45.92
	Time	RegEx,List	98.86	93.10	81.82	4.72	4.15
	Loc.	List	86.70	76.88	72.68	26.18	21.89
	Type	List	95.99	84.62	29.73	5.29	1.72
	Author	N/A				0.14	
	Opinion	N,POS,C	93.21	82.86	55.77	7.44	5.87
	Usage	N,POS,C	85.48	39.53	32.08	7.58	4.58
	Self ref.	List	97.85	100.00	16.67	2.58	0.43
	Unknown						*15.45*
Last.fm	Topic	F,N	90.32	0.00	0.00	2.43	0.00
	Time	RegEx,List	99.14	66.67	66.67	1.29	1.29
	Loc.	List	97.43	87.04	81.03	8.29	7.71
	Type	List	77.14	91.60	60.89	51.14	33.71
	Author	List	88.65	58.67	43.56	8.14	3.29
	Opinion	F,N,POS,C	74.73	79.84	83.06	17.71	18.43
	Usage	POS,C	79.57	62.96	37.78	6.43	5.29
	Self ref.	List	98.71	92.59	78.13	4.57	3.71
	Unknown						*26.57*

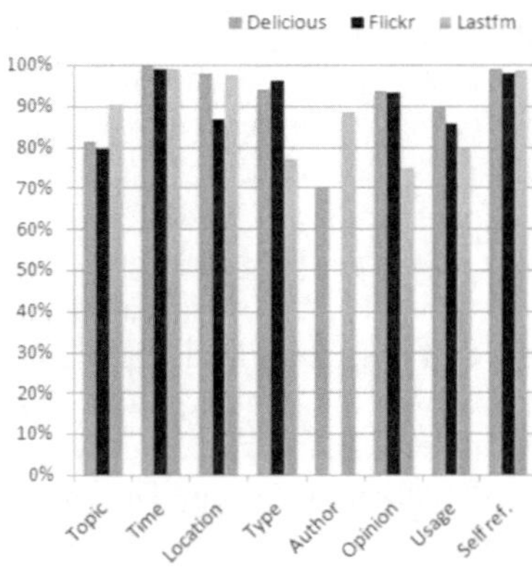

Fig. 1. Accuracy per class and system

certain category: both the real, manual value ("Man.") and the predicted, automatic value ("Auto."). A graphic representation of the accuracies is given in Figure 1.

6.2 Performance of Rule-Based Methods

The regular expressions and table look-ups performed very well in predicting the five categories Time, Location, Type, Author/Owner and Self-reference. With about 98% accuracy, performance was especially satisfactory for the highly standardized Time tags as well as for Self reference tags. However, accuracy is considerably lower for Type in *Last.fm*. This is mainly due to the used lists not being exhaustive enough. For example, the list of genres did not contain all potential sub-genres, newly emerging mixed styles or simply spelling variants and abbreviations. Its quota decreased progressively with the "difficulty" of the data set, *i.e.* the less frequent and more idiosyncratic the tags became. Since such handcrafted lists are never complete, automatic extension of the initial set should be achieved by expanding it with similar tags, *e.g.* based on second order co-occurrence. Similarly, our artist database did not contain all naming variants for a band or a singer and it had wrong entries in the artist's rubric. Allowing for partial matching, on the other hand, adds noise and results in predicting a much larger proportion of tags to denote Author/Owner than in the ground truth. For *Delicious* similarly, the regular expressions-based method just found a portion of the tags of interest, while more rules *e.g.* including named entity recognition would probably lead to many false positives. Last but not least, system specific design choices influence the accuracy of regular expressions for *Delicious* and *Flickr*. Since space characters in tags are not allowed here, compound names are written together (like "dead.sea", "sanfrancisco", "seattlepubliclibrary") and some location names may range over multiple tags (*e.g.* "new" and "york").

6.3 Performance of Model-Based Methods

The C4.5 decision tree yielded extremely good results for tag classification into Topic, Opinion and Usage tags. From the different intrinsic and extrinsic tag attributes used as features, part of speech and the semantic category in WordNet were present for all best performing classifiers, except for Topic in *Last.fm*. Here, number of users and number of words and characters alone achieved the best results. The number of words and characters obviously helped identifying Opinion tags in all three systems as well as Usage tags in *Flickr*. However, as a consequence of the relatively small training set of 700 tags as well as the highly unbalanced 'natural' distribution of tags over the three categories, robustness needs to be improved. Although in training the classifiers the set of positive and negative examples have been balanced, some classes had very few postive examples to learn from. For *Delicious* and *Flickr* the rate of false negatives is very high for the rare Opinion and Usage tags. Thus, none (for *Delicious*) or only part of the true Opinion and Usage tags are found. In contrast, almost all true Topic tags are correctly identified, but at the same time the number of predicted Topic tags overestimates the real proportion in the ground truth for *Delicious* and *Flickr*. The opposite happens for *Last.fm*. The classifiers learn well to reject non-Topic and non-Usage tags, but they also miss more than half of the true positives. Thus, our classifiers reinforce the tendencies to focus on one particular tag type depending on the system.

Nevertheless, the average accuracy is good, lying between 82% and 88%. As shown in Table 2 and Figure 2, except for Opinion in *Delicious* and Topic in *Last.fm*, the machine learning algorithms perform well in predicting tag type shares per system correctly. For example, the Opinion classifier matches 18.43% of the tags in *Last.fm*, compared to 17.71% by human rating.

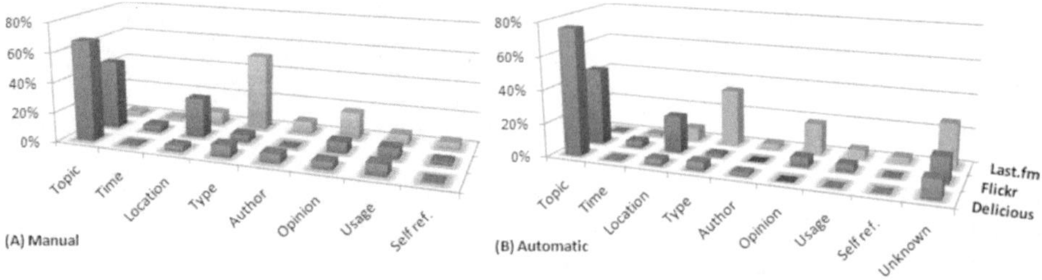

Fig. 2. Tag distribution per tagging systems: (A) manual assignment, (B) automatic assignment

6.4 Word Sense Disambiguation

Exploiting similar tags, extracted by computing second order co-occurrence, during learning improves classification performance on average by only 2% for *Last.fm*, while there is no noticeable difference for *Delicious*. Although using this method some meaningful disambiguations can be performed and a considerable part of tags not directly found in WordNet can be substituted, it does not have a big influence on classification accuracy. Some positive examples for similar tags for *Delicious* capturing synonyms, translations or simply singular/plural variations would be: "flats" and "Home.Rental", "Daily.News" and "noticias" or "technique" and "techniques". For *Last.fm*, we could find pairs like "relaxing" and "calm", "so beautiful" or "feelgood tracks". Though quite some of the similar tags found seem not to be synonymous, the strategy proved successful for disambiguation as (almost) synonymous and even strongly related words usually explain the meaning of a word. For example in the case of *Last.fm*, tags like "rock" or "pop" were correctly disambiguated and the musical meaning was chosen.

6.5 Overall Results

The linear average of all accuracies is 89.93%, while a more meaningful average, weighted by the real (*i.e.* manual) percentages of tags for each class, is 83.32%. This measure accounts for the different occurrence frequencies of the distinct tag types in the ground truth data. The weighted average values per system are: *Delicious*- 83.93%, *Flickr*- 85.07%, *Last.fm*- 81.08%. As initially shown in [1] for a smaller sample, tag class distributions vary significantly across the different systems. We observe that vocabulary and tag distribution depend on the resource domain, *e.g.* images and Web pages can refer to any topic, whereas music tracks are more restricted in content, thus leading to a more restraint and focused set of top tags. The most numerous category for *Delicious* and *Flickr* is Topic, while for *Last.fm* Type is predominant, followed by Opinion. A portion of tags could not be classified with reasonably confidence, the percentage for the "Unknown" tag type varying between 12% and 27%. Our methods overestimate the occurrences of Topic tags for *Delicious* at the expense of Opinion tags. Similary, not all Type and Author tags could be identified for *Last.fm*. Apart from this, our methods predict comparable class shares as the human raters in the overall distribution (Figure 2).

7 Conclusions and Future Work

Tag usage is rapidly increasing in community Web sites, providing potentially interesting information to improve user profiling, recommendations, clustering and especially search. It has been shown that some tag types are more useful for certain tasks than others. This paper extended previous work by building upon a verified tag classification scheme consisting of eight classes, which we use to automatically classify tags from three different tagging systems, *Last.fm*, *Delicious* and *Flickr*. We introduced two types of methods for achieving this goal – rule-based, relying on regular expressions and predefined lists, as well as model-based methods, employing machine learning techniques. Experimental results of an evaluation against a ground truth of 2,100 manually classified sample tags show that our methods can identify tag types with 80-90% accuracy on average, thus enabling further improvement of systems exploiting social tags.

For future work, first, multi-label classification is planned to better reflect the sometimes ambiguous nature of tags. Allowing for the prediction of multiple types per tag will render unnecessary a decision mechanism for choosing the right class from those predicted by independent binary classifiers. We also like to exploit resource features like title/description for web pages, lyrics for songs, or attributes extracted by content-based methods to learn a tags' type based on the concrete resource tagged. In addition, we intend to extend the model-based methods to enable machine learning of some categories now identified by rules or look-ups.

Acknowledgments. We are greatly thankful to our partners University Koblenz/Landau and the Tagora project for providing the *Flickr* data set and from the Knowledge and Data Engineering/Bibsonomy at the University of Kassel for providing the *Delicious* data set. This work was partially supported by the PHAROS project funded by the European Commission under the 6th Framework Programme (IST Contract No. 045035).

References

1. Bischoff, K., Firan, C.S., Nejdl, W., Paiu, R.: Can all tags be used for search? In: CIKM 2008, pp. 193–202. ACM, New York (2008)
2. Marlow, C., Naaman, M., Boyd, D., Davis, M.: Ht06, tagging paper, taxonomy, flickr, academic article, to read. In: HYPERTEXT 2006, pp. 31–40. ACM, New York (2006)
3. Golder, S.A., Huberman, B.A.: Usage patterns of collaborative tagging systems. Journal of Information Science 32(2), 198–208 (2006)
4. Zollers, A.: Emerging motivations for tagging: Expression, performance, and activism. In: WWW Workshop on Tagging and Metadata for Social Information Organization (2007)
5. Halpin, H., Robu, V., Shepherd, H.: The complex dynamics of collaborative tagging. In: WWW 2007, pp. 211–220. ACM, New York (2007)
6. Hotho, A., Jäschke, R., Schmitz, C., Stumme, G.: Information retrieval in folksonomies: Search and ranking. In: Sure, Y., Domingue, J. (eds.) ESWC 2006. LNCS, vol. 4011, pp. 411–426. Springer, Heidelberg (2006)
7. Sen, S., Lam, S.K., Rashid, A.M., Cosley, D., Frankowski, D., Osterhouse, J., Harper, F.M., Riedl, J.: Tagging, communities, vocabulary, evolution. In: CSCW, pp. 181–190. ACM, New York (2006)
8. Xu, Z., Fu, Y., Mao, J., Su, D.: Towards the semantic web: Collaborative tag suggestions. In: WWW Workshop on Collaborative Web Tagging (2006)
9. Rattenbury, T., Good, N., Naaman, M.: Towards automatic extraction of event and place semantics from flickr tags. In: SIGIR 2007, pp. 103–110. ACM, New York (2007)
10. Sigurbjörnsson, B., van Zwol, R.: Flickr tag recommendation based on collective knowledge. In: WWW 2008, pp. 327–336. ACM, New York (2008)
11. Overell, S., Sigurbjörnsson, B., van Zwol, R.: Classifying tags using open content resources. In: WSDM 2009, pp. 64–73. ACM, New York (2009)
12. Heymann, P., Ramage, D., Garcia-Molina, H.: Social tag prediction. In: SIGIR 2008, pp. 531–538. ACM, New York (2008)
13. Cattuto, C., Benz, D., Hotho, A., Stumme, G.: Semantic grounding of tag relatedness in social bookmarking systems. In: Sheth, A.P., Staab, S., Dean, M., Paolucci, M., Maynard, D., Finin, T., Thirunarayan, K. (eds.) ISWC 2008. LNCS, vol. 5318, pp. 615–631. Springer, Heidelberg (2008)

Social Knowledge-Driven Music Hit Prediction

Kerstin Bischoff, Claudiu S. Firan, Mihai Georgescu,
Wolfgang Nejdl, and Raluca Paiu

L3S Research Center / Leibniz Universität Hannover
Appelstrasse 9a
30167 Hannover, Germany
{bischoff,firan,georgescu,nejdl,paiu}@L3S.de

Abstract. What makes a song to a chart hit? Many people are trying to find the answer to this question. Previous attempts to identify hit songs have mostly focused on the intrinsic characteristics of the songs, such as lyrics and audio features. As social networks become more and more popular and some specialize on certain topics, information about users' music tastes becomes available and easy to exploit. In the present paper we introduce a new method for predicting the potential of music tracks for becoming hits, which instead of relying on intrinsic characteristics of the tracks directly uses data mined from a music social network and the relationships between tracks, artists and albums. We evaluate the performance of our algorithms through a set of experiments and the results indicate good accuracy in correctly identifying music hits, as well as significant improvement over existing approaches.

Keywords: collaborative tagging, classification, hit songs, social media.

1 Introduction

The benefits of being able to predict which songs are likely to become hits is various and is of big interest for both music industry and artists, as well as for listeners. In their attempt to release only profitable music, producers may want to have an indication of the potential of the music songs they will work with. Artists can profit from the results of such techniques by identifying the most suitable markets for their songs, music lovers' niches and by choosing the best channels and targets. Last but not least, normal music listeners can enjoy good music as a benefit of accurate hit predictions on a daily basis – radio stations can use such methods in order to improve their program by playing only songs which are highly likely hits.

Most previous attempts to identify hit songs have focused on intrinsic characteristics of songs, such as lyrics and audio features. In the prevailing view it is all about musical quality, so the task is to reveal the audience's preferences about music – *e.g.* by finding the similarity to what they liked before. However, it is often neglected that people are not independently deciding on what they like, but rather they like what they think other people may also like [1]. Despite 'intrinsic quality' success seems also to depend on the already known or

R. Huang et al. (Eds.): ADMA 2009, LNAI 5678, pp. 43–54, 2009.

assumed popularity, *i.e.* we find a rich get richer effect (a.k.a. preferential attachment or cumulative advantage). Thus, subjective opinions of a few early-arriving individuals account for hit potential as well.

As social networks become more and more popular and some specialize on certain topics, information about users' music tastes becomes available and easy to exploit. The wisdom of the crowds has become a famous notion of the collective intelligence manifesting itself in such collaborative tagging systems. Most important in our case, these networks set and identify trends and hot topics. In the present paper we propose a method for predicting the success of music tracks by exploiting social interactions and annotations, without relying on any intrinsic characteristics of the tracks. We predict the potential of music tracks for becoming hits by directly using data mined from a music social network (*Last.fm*) and the relationship between tracks, artists and albums. The social annotations and interactions enable both measuring similarity (*i.e.* intrinsic quality) of songs and finding those critical early-stage effects of cumulative advantage. Our approach requires only the social data corresponding to a track's first week life in *Last.fm* (*i.e.* the track is released *only* to the *Last.fm* audience), in order to be able to make good predictions about its potential and future evolution[1].

2 Related Work

Below we discuss the most relevant existing work, structured according to the two main directions along which we develop our methodology for predicting music hits.

2.1 Music Hits Prediction

Some previous work focused on automatic prediction of hit songs: in [2], the authors explore the automatic separation of hits from non-hits by extracting both acoustic and lyrics information from songs using standard classifiers on these features. Experiments showed that the lyrics-based features were slightly more useful than the acoustic features in correctly identifying hit songs. As ground truth data the authors made use of the Oz Net Music Chart Trivia Page[2]. This set is somewhat limited as it only contains top-1 hits in US, UK and Australia and the corpus used in the experiments was quite small – 1700 songs. In our approach we use a larger corpus and a much richer ground truth data set – the *Billboard.com* charts. Besides, our algorithms do not rely on lyrics or acoustic information but exploit social network data.

[3] focus on a complementary dimension: given the first weeks' sales data, the authors try to predict how long albums will stay in the charts. They also analyze whether a new album's position in the charts can be predicted for a certain week in the future. One of the most prominent commercial products

---

[1] *Last.fm* offers to artists the possibility to upload their own music to the portal (`http://www.last.fm/uploadmusic?accountType=artist`).

[2] `http://www.onmc.iinet.net.au/trivia/hitlist.htm`

for music hit prediction HSS[3] employs Spectral Deconvolution for analyzing the underlying patterns in music songs, *i.e.* it isolates patterns such as harmony, tempo, pitch, beat, and rhythm. Users of this service can upload a song, the system then analyzes it and compares it against existing chart hits from its database. The drawback of this system is that by using low-level features only, it cannot correctly predict the success of completely new types of music.

[4] claim that the popularity of a track cannot be learned by exploiting state-of-the-art machine learning (see also [1]). The authors conducted experiments contrasting the learnability of various human annotations from different types of feature sets (low-level audio features and 16 human annotated attributes like genre, instruments, mood or popularity). The results show that while some subjective attributes can be learned reasonably well, popularity is not predictable beyond-random – indicating that classification features commonly used may not be informative enough for this task. We investigate whether user generated interaction and (meta)data can serve as the missing link.

2.2 Social Media and Collaborative Tagging

Social media data provides ground for a wide range of applications. In [5], the authors make use of social media data for identifying high-quality content inside the Yahoo! Answers portal. For the community question-answering domain, they introduce a general classification framework for combining the evidence from different sources of information. Their proposed algorithms prove the ability to separate high-quality items from the rest with an accuracy close to that of humans. Our algorithms have a similar goal, though applied to a different domain – music. Though, it does not use tags, to a certain extent, the work of [6] is similar to ours: the authors analyze the potential of blog posts to influence future sales ranks of books on *Amazon.com*. The authors showed that simple predictors based on blog mentions around a product can be effective in predicting spikes in sales ranks.

Especially user generated tags have been extensively exploited: for building user profiles, for improving (personalized) information retrieval, results' clustering, classification or ontology building. Focusing on trend detection, in [7] the authors propose a measure for discovering topic-specific trends within folksonomies such as *Delicious*. Based on a differential adaptation of Google's PageRank algorithm, changes in popularity of tags, users, or resources within a given interval are determined. Similarly, [8] measure the interestingness of *Flickr* tags by applying a TFxIDF like score for different intervals in time.

For music, [9] found that *Last.fm* tags define a low-dimensional semantic space which - especially at the track level highly organized by artist and genre - is able to effectively capture sensible attributes as well as music similarity. We use this valuable folksonomy information for predicting music hits. To our best knowledge, user tags have not been used so far to infer hit potential of songs.

[3] `http://www.hitsongscience.com`

3 Data Sets

3.1 *Last.fm*

The method we propose for predicting music hits relies on external social information extracted from the popular music portal, *Last.fm*, a UK-based Internet radio and music community website, founded in 2002 and now owned by CBS Interactive. Statistics of the site claim 21 million users in more than 200 countries are streaming their personalized radio stations provided by *Last.fm*.

One of the most popular features of *Last.fm* user profiling is the weekly generation and archiving of detailed personal music charts and statistics. Users have several different charts available, including Top Artist, Top Tracks and Top Albums. Each of these charts is based on the actual number of times people listened to the track, album or artist. Similar global charts are also available and these are created based on the total number of individual listeners. Another important feature of *Last.fm* and crucial for our algorithms is the support for user-end tagging or labeling of artists, album and tracks. Thus, *Last.fm* creates a site-wide folksonomy of music. Users can browse musical content via tags or even listen to tag radios. Tags can fall into many categories, starting from genre, mood, artist characteristics and ending with users' personal impressions (for a detailed description of the existing types of *Last.fm* tags see [10]).

We collected 317,058 tracks and their associated attributes, such as artist and song name, number of times the tracks have been listened to on *Last.fm*, the name of the albums featuring the tracks, as well as tags that have been assigned to the songs. Additionally, the crawl contained information about 12,193 *Last.fm* users, all of them having listened to at least 50 songs and having used at least 10 tags. We started from the initial set of 12,193 crawled users and for all of them we downloaded all their available weekly charts. For this task we made use of the Audioscrobbler[4] web services. As not all of the 12,193 users from our initial set have been active since May 2007, we could gather charts for only 10,128 of them. A weekly chart consists of a list of songs that the user has listened during that particular week. The weekly charts we could gather span over 164 weeks and our final data collection consisted of 210,350 tracks, performed by 37,585 unique artists. 193,523 unique tags are associated with the tracks, 163,483 of these tags occurring as well along with artists.

3.2 Billboard Charts

For being able to asses the quality of our predictions, we also needed a good ground-truth data set. The most suitable for our purposes was the data exposed by *Billboard.com*. Billboard is a weekly American magazine devoted to the music industry, which maintains several internationally recognized music charts that track the most popular songs and albums in various categories on a weekly basis[5]. The charts, based on sales numbers and radio airplays are released as

[4] `http://www.audioscrobbler.net`

[5] `http://www.billboard.biz/bbbiz/index.jsp`

.html pages and represent the top tracks of the previous week. Every chart has associated a name, an issue date, and stores information about the success of the songs in form of rank, artist name and album/track name. Moreover, each chart entry has a previous week rank, as well as a highest rank field – *i.e.* the highest Billboard position ever reached by that song.

There are 70 different charts available for singles and 57 different ones for albums and a detailed list can be found at `http://www.billboard.biz/bbbiz/charts/currentalbum.jsp` and `http://www.billboard.biz/bbbiz/charts/currentsingles.jsp`, for albums and singles, respectively. We collect all these Billboard charts and aggregate the information, the resulting charts thus spanning over a range of almost 50 years, namely between August 1958 and April 2008. In total, the aggregated Billboard single chart contained 1,563,615 entries, 68,382 of them being unique songs. With respect to albums, the aggregated chart had 1,200,156 entries and among those, only 49,961 proved to be different albums.

The final set of tracks and the corresponding information around these tracks (artist, album, Billboard rank, *etc.*) is represented by the intersection of the set of unique tracks gathered from the *Last.fm* users' weekly charts and the set of tracks included in the Billboard charts. This intersection resulted in 50,555 unique music songs, on which we will perform our experiments.

4 Predicting Music Hits

We make use of the social information around the tracks, which we gather from the popular music portal *Last.fm*. This information is processed and transformed into a list of features, which is fed to a classifier for training it to discover potentially successful songs. The approach we propose relies on the following assumptions:

- The initial popularity (*i.e.* the popularity among listeners after only one week after the upload) of a track is indicative of its future success.
- Previous albums of the same artist have a direct influence on the future success of the songs.
- The popularity of other tracks produced by the same artist and included on the albums we consider has also an impact on the future success of a song.
- Popularity of the artist performing a track, in general, has a direct influence on the future success of new songs.

With these hypotheses fitting perfectly to the principles of preferential attachment/cumulative advantage, we now proceed describing the details of our music hit prediction algorithm based on social media data.

4.1 Feature Selection

The features used for training the classifiers are chosen such that the assumptions listed above are supported. It is thus natural to build a model where the main

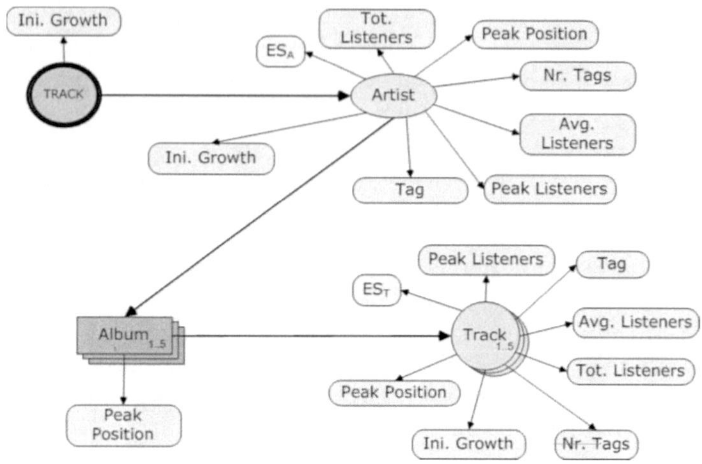

Fig. 1. Features used for training the classifiers

entities correspond to the interpreting *Artist*, previous popular *Albums* of the same artist and *Tracks* included on the albums considered. Moreover, each of these entities has associated a set of attributes, which are as well taken as input features for our classifiers. In Figure 1 we present the complete set of features considered.

All entities and their associated attributes related to a particular track, for which we would like to predict whether it will be a hit or not, form a tree having the $TRACK$ as root. Each of the features can be reached by starting from the root of the feature-tree and following the corresponding branches. We now discuss in detail the main feature entities and their associated attributes composing the feature tree.

Artist-Relevant Features. Artists as the performers of the songs we make predictions for are likely to have an influence on their hit potential. Usually, artist entities have associated a set of tags assigned by *Last.fm* users – we consider the top 5 most used tags, $Tag_{1..5}$. In case the artist does not yet have 5 tags, we exploit as many as available. Besides, we also include the total number of tags available for an artist, $Nr.\ Tags$, as well as its overall number of listeners, $Tot.\ Listeners$. $Ini.\ Growth$ represents the number of listeners for this artist during the first week it appeared in *Last.fm* charts (Note that these are *Last.fm* user charts.). The higher this number, the bigger the probability that this artist is quite popular and his future songs will become hits with a high probability. The $Peak\ Listeners$ feature measures the maximum number of listeners in one week over all weeks and all *Last.fm* user charts. With $Avg.\ Listeners$ we capture the average number of listeners over all *Last.fm* user charts and the value is computed as:

$$Avg.\ Listeners = \frac{Tot.\ Listeners}{\#weeks\ in\ Last.fm\ user\ charts} \quad (1)$$

The *Peak Position* represents the highest Billboard position this artist reached so far (for new artists, this value will not be known).

An *Artist* is directly connected to an *Album*-entity, since the performer might have produced several albums already. Thus, we also include as features the artist's top-5 albums, or as many as currently available.

Track-Relevant Features. In the model presented in Figure 1, containing the features used for training the classifiers, the *Track* entity occurs twice. It is important to distinguish between the two different instances: $TRACK$ represents the song for which we aim to automatically predict whether it will belong to the class "HIT" or "NHIT", which is also the root of the tree resulted from the complete set of features. Beside this, we also consider top-5 tracks, $Track_{1..5}$, appearing on the albums we include as feature for the given $TRACK$.

For $TRACK$, the track for which we want to make the predictions, only the *Ini. Growth* feature is considered (the maximum number of *Last.fm* listeners after the first week this song appeared on the *Last.fm* portal). The artist of the track, *Artist* represents an entity directly connected to $TRACK$ and for the case that the song has more authors (*e.g.* Madonna featuring Justin Timberlake) we consider only the first artist.

For $Track_{1..5}$, the tracks associated to other albums of the same artist, we include as feature the overall number of listeners on *Last.fm*, *Tot. Listeners*, as a strong indicator of its popularity. The tags given by the *Last.fm* users to a track are as well good popularity indicators. We consider the top-5 tags, $Tag_{1..5}$, or like in the case of the artists, if there are less than 5 tags, we consider as many as available. *Peak Position*, *Avg. Listeners*, *Peak Listeners* and *Ini. Growth* have the same meaning as the corresponding artist-related features.

Album-Relevant Features. Similar to the case of artist-like entities, albums also have associated a series of features: their popularity can be measured based on the highest position reached in Billboard, the *Peak Position* feature. Since for some artists the previously released albums can be quite many, we include only the top-5 albums which reached positions in the Billboard charts. Besides, from each album we also consider the top-5 Billboard listed tracks, $Track_{i_{1..5}}$.

Additional Features. In addition to the direct features discussed above, we also extract some implicit features for the artist and track entities. We associate ES-Entity Scores features, as a combination of the entities' Billboard top reached position and their HITS [11] scores – computed by applying HITS on a graph using artists, tracks and tags as nodes. Given an artist A, a track T and a tag TG, we create links as follows:

- From A to T, if track T is played by artist A;
- From T to TG, if track T has been tagged with tag TG;
- From A to TG, if artist A has been tagged with tag TG.

On the resulted graph we apply the HITS algorithm and compute the corresponding hub and authority scores. We present below the formulas for computing the HITS scores for artists, HS_A and tracks entities, HS_T:

$$HS_{A|T} = \begin{cases} 0, \text{ if } hubS_{A|T} == 0 \wedge authS_{A|T} == 0; \\ authS_{A|T}, \text{ if } hubS_{A|T} == 0 \wedge authS_{A|T}! = 0; \\ hubS_{A|T}, \text{ if } hubS_{A|T}! = 0 \wedge authS_{A|T} == 0; \\ authS_{A|T} \cdot hubS_{A|T}, \text{ otherwise.} \end{cases} \quad (2)$$

$hubS_{A|T}$ and $authS_{A|T}$ represent the hub- and authority scores of the artist and track (represented as subscripts A or T respectively).

The final Entity Scores (ES) will be based both on the outcome of calculating the HITS scores and the corresponding best positioning in any Billboard charts ever. This score will give an estimation of the popularity of certain artists and tracks in relation to the tags used, between themselves and in the opinion of a recognized authority in the domain, as the Billboard charts are. The formula for computing ES_A and ES_T, the entity scores for artists and tracks is given below:

$$ES_{A|T} = \begin{cases} \frac{1}{1000} \cdot HS_{A|T}, \text{ if } PeakPos_{A|T} \text{ is missing;} \\ \\ \frac{1}{PeakPos_{A|T}} \cdot HS_{A|T}, \text{ otherwise.} \end{cases} \quad (3)$$

$PeakPos_{A|T}$ represents the best reached position by the artist or track in all considered Billboard charts. If these entities do not occur in any of the charts (they never got that successful as to be included in the music tops), we consider a large number (1000) to substitute their missing Billboard rank. The inverse of this number or of the best Billboard position is considered for the computation of the final Entity Score (see Equation 3). The resulting $ES_{A|T}$ scores for artists and tracks will be used as features for our music hits prediction algorithm. They will be attached to the corresponding entities depicted in the feature graph from Figure 1.

4.2 Music Hit Prediction Algorithm

The core of our music hit prediction method is a classifier trained on the *Billboard.com* ground truth and using as features social media data extracted from *Last.fm* or inferred from it. We experiment with a number of different classifiers (Support Vector Machines, Naïve Bayes, Bayesian Networks and Decision Trees) and for building the classifiers we use the corresponding implementations available in the open source machine learning library Weka[6] [12]. Given the four hypotheses mentioned above, the classifiers learn a model from a training set of data. Once the model is learned, it can be applied to any unseen data from *Last.fm* and predict whether the corresponding songs have the potential of becoming hits or not.

The set of songs described in Section 3 is split into two partitions: one partition for training and one for testing the classifiers. We train classifiers for several rank ranges, such that the partitioning of the data satisfies the following: For hit class $1 - 1$, we consider as hit songs only those tracks which have reached top-1 in Billboard charts. All other songs starting with the second position in Billboard are considered non-hits. Similarly, other hit rank ranges are considered: $1 - 3$

[6] `http://www.cs.waikato.ac.nz/~ml/weka`

(*i.e.* tracks which have reached top-3 Billboard positions are regarded hits, while the rest, starting from position 4, are non-hits), $1-5$, $1-10$, $1-20$, $1-30$, $1-40$ and $1-50$. The number of hit and non-hit instances is approximately the same for all classifiers. We select as many songs as available from the rank ranges considered as hits. For non-hits, we randomly pick about the same number of songs from the set of music tracks with Billboard positions greater than the right margin of the hit class or from the set of tracks not appearing at all in the Billboard charts (*i.e.* "clear" non-hits). We summarize in Table 1 the resulting number of instances for each of the hits' rank ranges.

Each classifier is trained and tested on the total set of instances (both hits and non-hits), corresponding to each of the hit class ranges. For the songs in the training set, we build the set of corresponding features according to the attributes attached to the main entities (artist, albums, tracks) as depicted in Figure 1. The classifier is trained on the resulting set of features and a model is learned from it. After this step, the model is applied to all songs from the test data and a prediction is made.

5 Experiments and Results

For measuring the performance of our prediction algorithm we use the following metrics:

- Accuracy (Acc) – Statistical measure of how well the classifier performs overall;
- Precision (P) – Probability for items labeled as class C of indeed belonging to C;
- Recall (R) – Probability of all items belonging to class C of being labeled C;
- F1-measure (F1) – Weighted harmonic mean of Precision and Recall;
- Area under ROC (AUC) – Probability that a randomly chosen example not belonging to C will have a smaller estimated probability of belonging to C than a randomly chosen example indeed belonging to C.

We experimented with several multi-class classifiers: Support Vector Machines, Naïve Bayes, Decision Trees and Bayesian Networks with 1 or 2 parents, but given the space limitations only the best results are presented – this was the case of Bayesian Networks with 2 parents. In Table 1 we also present the averaged results of the 10-fold cross validation tests.

As observed from Table 1, the best results are obtained for the classifier built for detecting top-1 music hits. For this case, we obtain a value of 0.883 for the AUC measure, 0.788 precision and 0.858 recall for hits, while the overall accuracy is 81.31%. In [2] the authors reported AUC values of 0.69 for the best performing classifiers, trained to recognize top-1 hits from charts in Unites States, UK and Australia. Having similar data sets' sizes and song sets with no bias on any particular music genre (though the tracks might be different), our results for class $1-1$ are comparable with the ones reported by [2]. Our approach performs better, providing $\approx 28\%$ improvement in terms of AUC values over the

Table 1. Classifiers' evaluation for predicting Hits/Non-Hits, considering different rank intervals for the hit-classes

Hits' Range	#Hits	#Non-Hits	Acc[%]	Hits				Non-Hits			
				P	R	F1	AUC	P	R	F1	AUC
1 – 1	2,335	2,331	81.31	0.788	0.858	0.821	0.883	0.844	0.768	0.804	0.883
1 – 3	3,607	3,594	79.73	0.768	0.852	0.808	0.875	0.833	0.742	0.785	0.875
1 – 5	4,354	4,339	79.57	0.765	0.854	0.807	0.871	0.834	0.737	0.783	0.87
1 – 10	5,553	5,515	79.24	0.771	0.835	0.801	0.857	0.818	0.75	0.783	0.856
1 – 20	7,016	6,913	75.84	0.804	0.688	0.741	0.848	0.724	0.83	0.773	0.848
1 – 30	8,035	7,897	75.87	0.808	0.684	0.741	0.85	0.722	0.835	0.774	0.85
1 – 40	8,744	8,538	75.28	0.802	0.679	0.735	0.843	0.716	0.829	0.768	0.843
1 – 50	9,024	8,807	75.19	0.803	0.676	0.734	0.84	0.714	0.83	0.768	0.84

methods described [2]. It has been argued that AUC values between 0.5 – 0.7 are an indicator of low accuracy, while AUC values between 0.7 – 0.9 indicate good accuracy [13].

For all other classifiers, the results present as well characteristics which indicate good classification accuracy. In terms of AUC values, the performance is a bit worse than for the very restrictive case of hits taken only from top-1 Billboard charts (class $1-1$). The main reason for this is the fact that as we increase the rank range for what we call hits, the tracks begin to have a more heterogeneous set of features making it more difficult for classifiers to distinguish the correct hits from the rest of the songs. However, as we increase the interval ranges, precision improves in the detriment of recall, the best value being achieved for hit predictions from the interval $1-30$.

For the scenarios we consider, precision is actually more important than recall: a music label would be interested in promoting as far as possible only those music tracks which definitely have the potential of becoming hits; most radio stations try to play only music tracks which are already popular and on their way to top positions in the music charts. The main advantage of relying on such an approach is the fact that they can easily identify new and fresh sounds after just one week of letting the song "in the hands" of the *Last.fm* users.

In addition to the experiments described above, we also tested the accuracy of the built classifiers on a concrete scenario: we created a set of 100 songs, all having reached position-7 in Billboard, as their best rank. The resulted set of tracks was afterwards used for testing all classifiers (the set of 100 rank-7 songs was removed from all training sets of all classifiers). In Figure 2 we present the average probabilities for the 100 rank-7 tracks as assigned by the different classifiers and indicating the likelihood of the tracks to belong to the particular hit range class. The thick line at the 50% average probability corresponds to random class assignment. We observe that classifiers corresponding to classes $1-1$, $1-3$ and $1-5$ all have probabilities below the threshold, which is perfectly correct since all tested tracks have rank position 7. Starting with the classifier for the range $1-10$, the average probabilities are showing the track position to be included in the respective intervals.

Regarding the features we used for classification, we investigated which features were especially valuable for classification. We analyzed Information Gain and

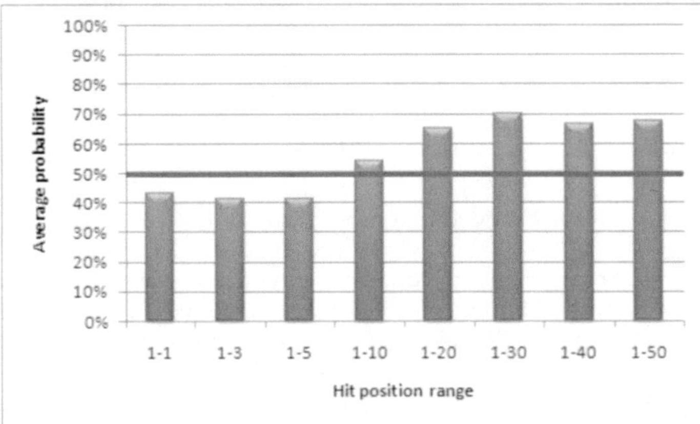

Fig. 2. Classification probability for chart position 7 averaged over 100 songs

Chi Square values for all features and found out that entity scores for artists and tracks ($ES_{A|T}$, see Equation 3) were particularly useful, ES_T being the top feature. Thus, the (prior) popularity of the artist and his earlier tracks and albums – measured as embeddedness in the *Last.fm* graph (artist-tracks-tags)– is the most distinct indicator of success for new songs. We clearly find a rich-get-richer effect: Once artists are popular chances for subsequent success are high.

6 Conclusions and Future Work

Previous attempts to identify music hits relied entirely on lyrics or audio information for clustering or classifying song corporas. By using data from a Web 2.0 music site, our approach adds a new dimension to this kind of research. Our algorithms exploit social annotations and interactions in *Last.fm* that enable both measuring intrinsic similarity of songs and finding critical early-stage effects of cumulative advantage for tracks assumed to be popular. In order to be able to make accurate predictions about evolution and hit-potential of songs, it only requires those tracks to be inside the portal for one week. The large scale experiments we performed indicate good classification accuracy for our method and compared with previous comparable work we achieve $\approx$ 28% improvement in terms of AUC. The applications of our algorithm are manifold: record companies, radio stations, the artists themselves and last but not least, the users.

As future work we plan to experiment with an extended set of input features for the classifiers, including besides social attributes also audio and lyrics information. Since marketing is known to have a great impact on the future success of songs, we intend to study other online information sources, such as advertisements, blogs, or forums, which could as well give strong indications of a songs' hit potential and possibly underlying mechanisms of preferential attachment.

Acknowledgments. This work was partially supported by the PHAROS project funded by the European Commission under the 6th Framework Programme (IST Contract No. 045035).

References

1. Watts, D.J.: Is justin timberlake a product of cumulative advantage? New York Times, April 15 (2007)
2. Dhanaraj, R., Logan, B.: Automatic prediction of hit songs. In: 6th International Conference on Music Information Retrieval (ISMIR 2005), pp. 488–491 (2005)
3. Chon, S.H., Slaney, M., Berger, J.: Predicting success from music sales data: a statistical and adaptive approach. In: 1st ACM workshop on Audio and music computing multimedia (AMCMM 2006), pp. 83–88. ACM, New York (2006)
4. Pachet, F., Roy, P.: Hit song science is not yet a science. In: 9th International Conference on Music Information Retrieval (ISMIR 2008) (2008)
5. Agichtein, E., Castillo, C., Donato, D., Gionis, A., Mishne, G.: Finding high-quality content in social media. In: 1st ACM International Conference on Web Search and Data Mining (WSDM 2008), pp. 183–194. ACM, New York (2008)
6. Gruhl, D., Guha, R., Kumar, R., Novak, J., Tomkins, A.: The predictive power of online chatter. In: 11th ACM SIGKDD international conference on Knowledge discovery in data mining (KDD 2005), pp. 78–87. ACM, New York (2005)
7. Hotho, A., Jäschke, R., Schmitz, C., Stumme, G.: Trend detection in folksonomies. In: Avrithis, Y., Kompatsiaris, Y., Staab, S., O'Connor, N.E. (eds.) SAMT 2006. LNCS, vol. 4306, pp. 56–70. Springer, Heidelberg (2006)
8. Dubinko, M., Kumar, R., Magnani, J., Novak, J., Raghavan, P., Tomkins, A.: Visualizing tags over time. In: 15th international conference on World Wide Web (WWW 2006), pp. 193–202. ACM, New York (2006)
9. Levy, M., Sandler, M.: A semantic space for music derived from social tags. In: 8th International Conference on Music Information Retrieval (ISMIR 2007), pp. 411–416 (2007)
10. Bischoff, K., Firan, C.S., Nejdl, W., Paiu, R.: Can all tags be used for search? In: 17th ACM Conference on Information and Knowledge Management (CIKM 2008), pp. 193–202. ACM, New York (2008)
11. Kleinberg, J.: Authoritative sources in a hyperlinked environment. Journal of the Association for Computing Machinery 46(5), 604–632 (1999)
12. Witten, I.H., Frank, E.: Data Mining: Practical Machine Learning Tools and Techniques, 2nd edn. Morgan Kaufmann Series in Data Management Systems. Morgan Kaufmann, San Francisco (2005)
13. Fischer, J.E., Bachmann, L.M., Jaeschke, R.: A readers' guide to the interpretation of diagnostic test properties: clinical example of sepsis. Intensive Care Medicine Journal 29(7), 1043–1051 (2003)

Closed Non Derivable Data Cubes Based on Non Derivable Minimal Generators

Hanen Brahmi[1], Tarek Hamrouni[1,2], Riadh Ben Messaoud[3], and Sadok Ben Yahia[1]

[1] Faculty of Sciences of Tunis, Tunisia
{tarek.hamrouni,sadok.benyahia}@fst.rnu.tn
[2] CRIL-CNRS, Lille Nord University, Lens, France
hamrouni@cril.univ-artois.fr
[3] Faculty of Economic and Management Sciences of Nabeul, Tunisia
riadh.benmessaoud@fsegn.rnu.tn

Abstract. It is well recognized that data cubes often produce huge outputs. Several efforts were devoted to this problem through closed cubes, where cells preserving aggregation semantics are losslessly reduced to one cell. In this paper, we introduce the concept of *closed non derivable data cube*, denoted $\mathcal{CND}$-$\mathcal{C}$ube, which generalizes the notion of bi dimensional frequent closed non derivable patterns to the multidimensional context. We propose a novel algorithm to mine $\mathcal{CND}$-$\mathcal{C}$ube from multidimensional databases considering three anti-monotone constraints, namely "*to be frequent*", "*to be non derivable*" and "*to be minimal generator*". Experiments show that our proposal provides the smallest representation of a data cube and thus is the most efficient for saving storage space.

Keywords: Data warehouse, data cube, closed pattern, non derivable pattern, minimal generator.

1 Introduction

Since the 90s, the *Data Warehouse* architecture has gained widespread acceptance as an integrating data solution. In this architecture, data coming from multiple external sources are extracted, filtered, merged, and stored in a central repository. The content of a data warehouse is analyzed by *Online Analytical Processing* (OLAP) applications in order to discover trends, patterns of behaviors, and anomalies as well as to find hidden dependencies between data [1]. The outcomes of these analyses are then used to support various business decisions.

In order to efficiently answer OLAP queries, a widely adopted solution is to compute and materialize data cubes [2]. For example, given a relation "Car-Sale", a set of dimensions *DIM* = {"Model", "Year", "Color"}, a measure "Sales", and an aggregation function *SUM*, the *CUBE* operator [2] is expressed as follows:

```
SELECT Model, Year, Color, SUM (Sales)
FROM Car-Sale
GROUP BY CUBE Model, Year, Color.
```

As illustrated in Table 1, the query achieves all possible group-by operations according to all combinations of attributes belonging to *DIM*. Then, we obtain the so called

R. Huang et al. (Eds.): ADMA 2009, LNAI 5678, pp. 55–66, 2009.

Table 1. Data cube example using the relation "Car-Sale"

Model	Year	Color	Sales
Ford	1991	Blue	7
Chevy	1990	Red	5
Ford	1990	Blue	99
Ford	1990	Red	64

Model	Year	Color	Sales
Chevy	1990	Red	5
Chevy	1990	All	5
Chevy	All	Red	5
Chevy	All	All	5
Ford	1990	Blue	99
Ford	1990	Red	64
Ford	1990	All	163
Ford	1991	Blue	7
Ford	1991	All	7
Ford	All	Blue	106
Ford	All	Red	64
All	1990	Blue	99
All	1990	Red	69
All	1991	Blue	7
Ford	All	All	170
All	1990	All	168
All	1991	All	7
All	All	Blue	106
All	All	Red	69
All	All	All	175

Data Cube. It is obvious that computing data cubes is a combinatory problem. In fact, the size of a cube exponentially increases according to the number of dimensions. Furthermore, the problem worsens since we deal with large data sets. For instance, Ross and Srivastava exemplify the problem by achieving a full data cube encompassing more than 210 million of tuples from an input relation having 1 million of tuples [3]. In general, the problem is due to two reasons: the exponential number of dimensional combinations to be dealt with and, the number of attributes per dimension. In addition, data cubes are generally sparse [3], thus scarce value combinations are likely to be numerous and, when computing an entire data cube, each exception must be preserved. In such a context, two streams of approaches seems to be possible: (i) approaches favoring the efficiency of OLAP queries despite of the storage space, and (ii) those privileging optimal data representations instead of enhancing query performances.

In this paper, we investigate another way of tackling the problem. First, we introduce the concept of *closed non derivable cube* and prove that the latter greatly reduces the size of a data cube. Then, we propose an algorithm to efficiently compute the closed non derivable cubes. Through extensive carried out experiments on benchmarks and real-world datasets, we show the effectiveness of our proposal on both runtime performances and information lossless reduction of storage space.

The remainder of the paper is organized as follows. We scrutinize, in Section 2, the related work. Our proposal is detailed in Section 3. We define the concepts of our representation and introduce the CLOSENDMG algorithm in Section 4. We also relate the encouraging results of the carried out experiments in Section 5. Finally we conclude by resuming the strengths of our contribution and sketching future research issues.

2 Related Work

Approaches addressing the issue of data cube computation and storage attempt to reduce at least one of the quoted drawbacks. The BUC [4] and HCUBING [5] algorithms enforce anti-monotone constraints and partially compute data cubes to reduce both execution time and disk storage requirements. The underlying argument is that OLAP users are only interested in general trends. In the same way, other methods use the statistical structure of data to compute density distributions and give approximate answers to OLAP queries [6]. These approaches are efficient since they reduce the execution time and compress the space storage. However, they do not efficiently answer OLAP queries.

Another category of approaches, called "*information lossless*", aims at finding the best compromise between efficiency of OLAP queries and the requirements of storage without discarding any possible – even unfrequent – queries. The key idea consists in pre-computing and storing frequently used aggregates while preserving all data needed to compute results of an unforeseen query. These approaches mostly rely on the materialization of views. The following three methods also fit in the information lossless trend, and are based on data mining algorithms.

- The **Quotient Cube** is a summarizing structure for a data cube that preserves its semantics [7]. It can be efficiently constructed and achieves a significant reduction of the cube size. The key idea behind a quotient cube is to create a summary by carefully partitioning the set of cells of a cube into equivalent classes while keeping the cube roll-up and drill-down semantics and lattice structure. Moreover, Casali *et al.* proved that the quotient cube can be computed by the application of frequent closed itemset mining algorithms, *e.g.*, the CLOSE algorithm [8].
- The **Closed Cube** represents a size-reduced representation of a data cube when compared to the quotient cube [9]. It only consists of closed cells. A cell, say *c*, is a closed cell if there is no cell, *d*, such that *d* is a specialization (descendant) of *c*, and *d* has the same measure value as *c*. Casali *et al.* proved that the closed cube can be computed using frequent closed itemsets mining algorithms [8].
- The **Representative Slice Mining (*RSM*) approach** is a three-phase based approach that exploits 2D frequent closed itemset mining algorithms to mine frequent closed cubes [10]. The basic idea is: (i) to transform a 3D dataset into 2D datasets; (ii) mine the 2D datasets using an existing 2D frequent closed itemset mining algorithm; and (iii) prune away any frequent cubes that are not closed.

Table 2 summarizes the surveyed approaches dedicated to an information lossless data cube reduction. Approaches fitting in the information lossless trend using data mining algorithms attempt to reduce storage space. Due to its usability and importance, reducing the storage space of a data cube still a thriving and a compelling issue. In this respect, the main thrust of this paper is to propose a new information lossless concise representation called *Closed Non Derivable Cube*, $\mathcal{CND}$-Cube, which can be seen as an extension of the closed non derivable patterns to the multidimensional search space. The main idea behind our approach comes from the conclusion drawn from the Data Mining community that focused on the lossless reduction of frequent itemsets. Even though blamed to miss underling semantics, (closed) non derivable itemsets have been

Table 2. Approaches of data cube reduction fitting in the information lossless trend

Approaches seeking an exact and concise representation based on data mining algorithms		**Approaches based on a compromise between storage space and OLAP queries efficiency**	
Approaches	Algorithms	Approaches	Algorithms
Quotient Cube [7] (Lakshmanan *et al.*, 2002)	CLOSE [8]	CURE for Cubes [11] (Morfonios *et al.*, 2006)	CURE
Closed Cube [9] (Casali *et al.*, 2003)	CLOSE	Condensed Cube [12] (Wang *et al.*, 2002)	BST
RSM [10] (Ji *et al.*, 2006)	CLOSE	Dwarf Cube [13] (Sismanis *et al.*, 2002)	STA

shown to present the best lossless compactness rates. In this respect, we attempt to mine the $\mathcal{CND}$-$\mathcal{C}$ube that permits to obtain the smallest multidimensional data representation in order to efficiently save the storage space. To build up the $\mathcal{CND}$-$\mathcal{C}$ube, we introduce a novel algorithm called CLOSENDMG (*closed non derivable itemsets based on minimal generators*).

3 $\mathcal{CND}$-$\mathcal{C}$ube: A Novel Data Cube Approach

The $\mathcal{CND}$-$\mathcal{C}$ube fully and exactly captures all the information enclosed in a data cube. Moreover, we apply a simple mechanism which significantly reduces the size of aggregates that have to be stored. Our aim is to compute the smallest representation of a data cube when compared to the pioneering approaches of the literature. In this respect, Casali *et al.* [9,14] proved that there is a lattice isomorphism between the Closed Cube and the Galois lattice (concept lattice) computed from a database relation *R*. Such an isomorphism is attractive since it allows the use of data mining algorithms. It is also proved to be efficient, to compute concise representations of a data cube. For example, the computation of *Quotient Cube*, *Closed Cube*, and *RSM* representations is based on data mining algorithms. On the other hand, the approach of Casali *et al.* is based on the Birkhoff theorem [15] to bridge a concept lattice to a closed Cube lattice.

In our approach, we also use this isomorphism and the Birkhoff theorem in order to apply our CLOSENDMG algorithm to compute the $\mathcal{CND}$-$\mathcal{C}$ube. More precisely, starting from a database relation *R*, we look for extracting closed non derivable patterns by computing the closures of non derivable minimal generators. Then, based on this isomorphism, we use the Birkhoff theorem to obtain the $\mathcal{CND}$-$\mathcal{C}$ube. In order to do so, we propose to use our CLOSENDMG algorithm. This latter operates in two steps: The first step extracts patterns fulfilling three anti-monotone constraints, namely "*to be frequent*", "*to be non derivable*" and "*to be minimal generator*". Whereas, the second one computes closures of non derivable minimal generators.

4 Computation of the $\mathcal{CND}$-$\mathcal{C}$ube

We start this section by presenting the key settings that will be of use in the remainder.

4.1 Formal Background

Closed Itemset. One condensed representation of itemsets is based on the concept of closure [8].

Definition 1. *The closure γ of an itemset X is the maximal superset of X having the same support value as that of X.*

Example 1. According to the relation "Car−Sale" shown by Table 1, the set of closed itemsets is as follows: { ($\emptyset$: 4), ('Ford': 3), ('1990': 3), ('Blue', 'Ford': 2), ('1990', 'Red': 2), ('Ford', '1990': 2), ('Ford', '1991', 'Blue': 1), ('Chevy', '1990', 'Red': 1), ('Ford', '1990', 'Blue': 1), ('Ford', '1990', 'Red': 1)}.

Minimal Generator. The concept of minimal generator [16] is defined as follows.

Definition 2. *An itemset $g \subseteq \mathcal{I}$ is said to be a minimal generator of a closed itemset f iff $\gamma(g) = f$ and $\nexists\, g_1 \subset g$ such that $\gamma(g_1) = f$. For a user−defined support threshold MinSup, the set of frequent minimal generators includes all generators that are frequent.*

Example 2. According to the relation "Car-Sale" shown by Table 1, the set of minimal generators is: {('Ford': 3), ('Chevy': 1), ('1991': 1),('1990': 3), ('Blue': 2), ('Red': 2), ('Ford', '1991': 1), ('Ford', 'Red': 1), ('1991', 'Red': 1)}.

Non Derivable Itemset. The collection of non-derivable frequent itemsets, denoted $\mathcal{NDI}$, is a lossless representation of frequent itemsets based on the inclusion-exclusion principle [17,18].

Definition 3. *Let X be an itemset and Y a subset of X. If $|X \backslash Y|$ is odd, then the corresponding deduction rule for an upper bound of Supp(X) is:*

$$Supp(X) \leq \sum_{Y \subseteq I \subset X} (-1)^{|X \backslash I| + 1}\, Supp(I)$$

If $|X \backslash Y|$ is even, the sense of the inequality is inverted and the deduction rule gives a lower bound instead of an upper bound of the support of X. Given all subsets of X, and their supports, we obtain a set of upper and lower bounds for X. In the case where the smallest upper bound equals the highest lower bound, the support of X is exactly derived. Such an itemset is called *derivable*. In the remainder, the lower and upper bounds of the support of an itemset X will respectively be denoted $X.l$ and $X.u$.

Example 3. According to the relation "Car-Sale" shown by Table 1, the set of non derivable itemsets is: {('1991': 1), ('Chevy': 1), ('Red': 2), ('Blue': 2), ('Ford': 3), ('1990': 3), ('1991', 'Blue': 1), ('1991', 'Ford': 1), ('Chevy', 'Red': 1), ('Chevy', '1990': 1), ('Ford', 'Blue': 2), ('1990', 'Blue': 1), ('Ford', 'Red': 1), ('1990', 'Red': 2), ('Ford', '1990': 2)}.

Closed Non Derivable Itemset. The set of frequent closed non-derivable itemsets, denoted $\mathcal{CNDI}$, has been introduced by Muhonen and Toivonen [19].

Definition 4. *Let $\mathcal{NDI}$ be the collection of frequent non-derivable itemsets. The set of frequent closed non-derivable itemsets is as follows: $\mathcal{CNDI} = \{\gamma(X) \mid X \in \mathcal{NDI}\}$.*

Example 4. According to the relation "Car-Sale" shown by example 1, the set of closed non derivable itemsets is { (∅: 4), ('Ford': 3), ('1990': 3), ('Blue', 'Ford': 2), ('1990', 'Red': 2), ('Ford', '1990': 2), ('Ford', '1991', 'Blue': 1), ('Chevy', '1990', 'Red': 1), ('Ford', '1990', 'Blue': 1), ('Ford', '1990', 'Red': 1)}.

4.2 Link between CNDI and Minimal Generators

The computation of frequent closed non derivable itemsets can be optimized if we use minimal generators. Indeed, each closed non-derivable itemset can easily be shown to be the closure of at least a non-derivable minimal generator. "Non-derivable minimal generator" is both a "non-derivable" and "minimal generator" itemset. Hence, instead of computing the whole set of frequent non derivable itemsets for which the associated closures must be computed as did by the authors in [19], we can only use the set of frequent non derivable minimal generators. To get out the set of frequent non-derivable minimal generators, a modification of algorithms dedicated to frequent non-derivable itemset mining has to be performed. Its main aim is to only retain the itemsets fulfilling the minimal generator constraint among the set of frequent non-derivable itemsets. The introduction of minimal generators within NDI and FIRM[1] algorithms will hence optimize both the candidate generation and closure computation steps. Indeed, the number of frequent non-derivable minimal generators is lower than that of non derivable itemsets.

4.3 CLOSENDMG Algorithm

In this subsection, we introduce a novel algorithm intended to compute the $\mathcal{CND}$-$\mathcal{C}$ube.

Non Derivable Minimal Generator. The main idea is to only retain the itemsets fulfilling the minimal generator constraint among the set of frequent non-derivable itemsets.

Definition 5. *Given an itemset $I \subseteq \mathcal{I}$, the set of $\mathcal{MG}$-$\mathcal{NDI}$ is defined as follows: $\mathcal{MG}\text{-}\mathcal{NDI} = \{I \subseteq \mathcal{I} \mid I.l \neq I.u \text{ and } I \text{ is a } \mathcal{MG}\}$.*

We can conclude the following theorem about the cardinality of the set of $\mathcal{MG}$-$\mathcal{NDI}$:

Theorem 1. *The cardinality of the set of non-derivable minimal generator itemsets $\mathcal{MG}$-$\mathcal{NDI}$ is always smaller than or equal to the cardinality of the set of non-derivable itemsets $\mathcal{NDI}$, i.e., $\mid \mathcal{MG}\text{-}\mathcal{NDI}| \leq \mid \mathcal{NDI}|$.*

Proof. According to definition 5, the set of $\mathcal{MG}$-$\mathcal{NDI}$ retains only the itemsets that fulfill the minimal generator constraint among the set of frequent non-derivable itemsets, we trivially have $|\mathcal{MG}\text{-}\mathcal{NDI}| \leq |\mathcal{NDI}|$.

Example 5. The set of non derivable minimal generators, generated from the relation "Car-Sale" is: {('1991': 1), ('Chevy': 1), ('Red': 2), ('Blue': 2), ('Ford': 3), ('1990': 3), ('1990', 'Blue': 1), ('Ford', 'Red': 1), ('Ford', '1990': 2)}.

[1] The FIRM algorithm [19] mines closed non derivable patterns.

Table 3. List of used notations in the CLOSENDMG algorithm

$\mathcal{MGC}_k$	: The set of non derivable minimal generator candidates of size k.
$\mathcal{FMG}_k$	: The set of frequent non derivable minimal generators of size k.
$\mathcal{G}en$	: The set of non derivable minimal generators of size k from which we generate non derivable minimal generator candidates of size k+1.
$\mathcal{P}re\mathcal{C}_{k+1}$	: Frequent non derivable minimal generator itemsets of size k+1.
$\mathcal{MG}$-$\mathcal{NDI}$	: The set of frequent non derivable minimal generator itemsets.
$\mathcal{MG}$-$\mathcal{CNDI}$	: The set of frequent closed non derivable minimal generator itemsets generated using the CLOSENDMG algorithm with their respective supports.
Estimated-Supp	: The *estimated support* is a pruning strategy introduced in the TITANIC algorithm [20] such that, for a generator candidate itemset *g* having a size k, if the minimum of the supports of the subsets of *g* of size (k-1) is different from the real support of *g*, then *g* is a minimal generator.

Closed Non Derivable Minimal Generators. To compute the set of closed non-derivable minimal generators, we introduce the following definition.

Definition 6. *Given* $\mathcal{MG}$-$\mathcal{NDI}$ *a set of frequent non-derivable minimal generators. The set of frequent closed non-derivable minimal generators is* $\mathcal{MG}$-$\mathcal{CNDI}$= $\{\gamma(X)|$ $X \in \mathcal{MG}$-$\mathcal{NDI}\}$.

The definition of $\mathcal{MG}$-$\mathcal{CNDI}$ straightforwardly gives CLOSENDMG whose pseudo-code is shown by Algorithm 1. The used notations are summarized in Table 3.

Likewise the comparison presented by Muhonen and Toivonen in [19] between the $\mathcal{CNDI}$ and the $\mathcal{NDI}$ sets, we compare in the following theorem the cardinality of $\mathcal{MG}$-$\mathcal{CNDI}$ *vs.* that of $\mathcal{MG}$-$\mathcal{NDI}$.

Theorem 2. *The cardinality of the set of closed non-derivable minimal generators* $\mathcal{MG}$-$\mathcal{CNDI}$ *is smaller than or equal to the cardinality of the set of non-derivable minimal generators* $\mathcal{MG}$-$\mathcal{NDI}$*:* $|\mathcal{MG}$-$\mathcal{CNDI}| \leq |\mathcal{MG}$-$\mathcal{NDI}|$.

Proof. According to Definition 6, the operator γ gives exactly one closure for each itemset. On the other hand, many itemsets can be mapped to the same closure. We thus have $|\mathcal{MG}$-$\mathcal{CNDI}| \leq |\mathcal{MG}$-$\mathcal{NDI}|$.

Example 6. According to the relation "Car-Sale" shown by example 1, the set of closed non derivable itemsets based on minimal generators is as follows: {(∅: 4), ('Ford': 3), ('1990': 3), ('Blue', 'Ford': 2), ('1990', 'Red': 2), ('Ford', '1990': 2), ('Ford', '1991', 'Blue': 1), ('Chevy', '1990', 'Red': 1), ('Ford', '1990', 'Blue': 1), ('Ford', '1990', 'Red': 1)}.

In this paper, we attempt to mine $\mathcal{CND}$-$\mathcal{C}$ube that delivers "closed non derivable" relationships among dimensions. Indeed, the first step, *i.e.*, in the computation phase, performs the extraction of non-derivable minimal generators (lines 2–23). The main idea behind their extraction is to ensure an efficient computation that reduces the run-time requirements. As a result, we obtain a data cube consisting of only non-derivable minimal generators. In addition, the second step allows to compress the latter by finding the closed non-derivable minimal generators in a data cube relation (lines 24–25).

Algorithm 1. CLOSENDMG($\mathcal{D}$, *MinSup*)

Input. A dataset $\mathcal{D}$ and a support threshold *MinSup*
Output. The collection $\mathcal{MG}$-$\mathcal{CND}$ of closed non-derivable itemsets based on minimal generators

```
Begin
    k := 1; MG-NDI := ∅;
    MG-CNDI := ∅;
    MGC_1 = {{i}| i ∈ I};
    Foreach i ∈ MGC_1 do
        i.l := 0; i.u := |D|;
    While MGC_k not empty do
        Count the estimated support of each candidate in one pass over D;
        Estimated-Supp(MGC_k)=Min(Supp(subsets(MGC_k)));
        FMG_k := {I ∈ MGC_k | Supp(I) ≠ Estimated-Supp(I) and Supp(I) ≥ MinSup};
        MG-NDI := MG-NDI ∪ FMG_k; Gen := ∅;
        Foreach I ∈ FMG_k do
            If (Supp(I) ≠ I.l) and (Supp(I) ≠ I.u) then
            Gen := Gen ∪ I
        PreC_{k+1} := Apriori-Gen( Gen)
        MGC_{k+1} := ∅;
        Foreach J ∈ PreC_{k+1} do
            Count the upper bound and lower bound of J;
            If l≠u then
            If Supp(J)≠ Estimated-Supp(J) then J.l := l ;
            J.u := u ;
            FMG_{k+1} := FMG_{k+1} ∪ {J}
        k := k+1;
    forall I ∈ MG-NDI do
        MG-CNDI ⟵ MG-CNDI ∪ {γ(I)}
    return MG-CNDI
End
```

The obtained $\mathcal{CND}$-$\mathcal{C}$ube exactly captures all the information enclosed in a data cube. Moreover, our representation provides a simple mechanism that significantly reduces the size of aggregates to be stored.

5 Experimental Results

We compare our approach with the pioneering approaches falling within the information lossless trend, namely, *Quotient Cube* and *Closed Cube*[2]. All experiments were carried out on a PC equipped with a 3GHz Pentium IV and 2GB of main memory running under Linux Fedora Core 6.

[2] The closed Cube was extracted thanks to the CLOSE algorithm [8].

Table 4. The considered datasets at a glance

Datasets	Attributes	Tuples
COVTYPE	54	581012
SEP85L	7871	1015367
MUSHROOM	119	8124
CHESS	75	3196
RETAIL	16470	88162
T10I4D100K	1000	100000

During the carried out experimentation, we used two dense benchmarks datasets: CHESS, MUSHROOM, two sparse benchmarks datasets: RETAIL, T10I4-D100K[3], and two real datasets used in the data cube context: COVTYPE[4], SEP85L[5]. Table 4 sketches dataset characteristics used during our experiments. In the last column, the size, in KB, of the dataset is reported.

Through these experiments, we have a twofold aim: first, we have to stress on comparing the computation time obtained by CLOSENDMG *vs.* that FIRM[6] to compute the $\mathcal{CND}$-$\mathcal{C}$ube. Second, put the focus on the assessment of the compactness in storage terms of our approach *vs.* that proposed by the related approaches of the literature.

Performance aspect. Figure 1 plots the runtime needed to generate the $\mathcal{CND}$-$\mathcal{C}$ube for the considered datasets, using the algorithms CLOSENDMG and FIRM. Clearly, in efficiency terms, the algorithm CLOSENDMG largely outperforms the FIRM algorithm especially for dense and real datasets. Indeed, the gap between both curves tends to become wider as far as the *MinSup* values decreases.

Storage reduction aspect. In this respect, we denote by "full cube", a data cube without any compression, *i.e.*, it is a non-reduced one, generated using the "CUBE" operator as illustrated in Table 1 for the relation example "Car-Sale". In the following, we have to compare the size of the $\mathcal{CND}$-$\mathcal{C}$ube to be stored *vs.* the size of, respectively, full cube, closed and quotient cubes. Table 5, presents the space on the disk in (KB) of need to store these data cube representations.

The datasets require too much main memory ($>$ 4GB) when computing the data cube representations, *i.e.*, $\mathcal{CND}$-$\mathcal{C}$ube, Quotient Cube and the Closed Cube, with a minimum threshold equal to 1 (all the possible patterns). Thus, we were obliged to set a minimum threshold for each dataset that make us able to extract the data cube representation. As highlighted by Table 5, the carried out experiments show that the introduced representation $\mathcal{CND}$-$\mathcal{C}$ube provides an important reduction of storage space on the disk when compared to the data cube, *Quotient Cube* and *Closed Cube*.

Considering the three concise representations (Closed Cube, Quotient Cube and $\mathcal{CND}$-$\mathcal{C}$ube), we conclude that the best compression rates of a full data cube are obtained for dense and real datasets, *i.e.*, CHESS, MUSHROOM, COVTYPE and SEP85L.

[3] Available at: *http://fimi.cs.helsinki.fi/data/.*

[4] Available at: *http://ftp.ics.uci.edu/pub/machine-learning-databases/covtype.*

[5] Available at: *http://cdiac.esd.ornl.gov/cdiac/ndps/ndp026b.html.*

[6] Available at: *http://www.cs.helsinki.fi/u/jomuhone/.*

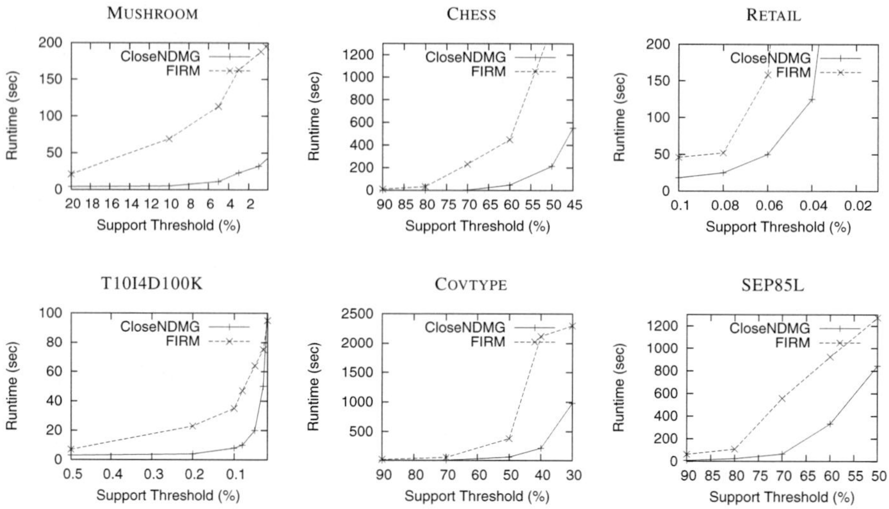

Fig. 1. Mining time of $\mathcal{CND}$-$\mathcal{C}$ubes using the FIRM and CLOSENDMG algorithms

Table 5. Size of the data cubes generated (KB)

	Full cube	$\mathcal{CND}$-**Cube**	Closed Cube	Quotient Cube
MUSHROOM	10147	**1578**	2972	4021
CHESS	15104	**1009**	2386	2500
COVETYPE	20825	**1428**	5410	6900
SEP85L	32912	**3827**	5925	7383
T10I4D100K	15398	**9590**	10987	12890
RETAIL	13025	**10523**	11913	11986

A smaller compression rates are given with sparse data, *i.e.*, RETAIL and T10I4D100K datasets.

The space percentage of our concise representation is smaller than the classical data cube storage space. For COVTYPE and SEP85L datasets, our condensed representation requires, respectively, 6.85% and 11.62% of the space needed to fully store the data cube. Compared to the *Quotient Cube* and *Closed Cube*, our rates are smaller. For example, the *Closed Cube* requires, respectively, 25.37% and 18%, of the space needed to store a full data cube of COVTYPE and SEP85L data sets. The *Quotient Cube* rates obtained for the two latter datasets are, respectively, 33.13% and 22.43%. We conclude that for real datasets the compression is greater when using $\mathcal{CND}$-$\mathcal{C}$ube *vs.* both the *Closed Cube* and the *Quotient Cube*. The compression rates obtained for both dense datasets, *i.e.* MUSHROOM and CHESS, by the $\mathcal{CND}$-$\mathcal{C}$ube are also significant and by far greater than those obtained respectively by the *Quotient Cube* and the *Closed Cube*. As expected, the compression rates are nevertheless much more modest for sparse data sets, *i.e.* RETAIL and T10I4D100K, *i.e.*, 62.28% and 90.57% for respectively T10I4D100K and RETAIL datasets. Interestingly enough, the obtained rates for sparse contexts outperform those obtained by the other representations for the same datasets.

6 Conclusion and Perspectives

In this paper, we focused on the lossless information approaches using data mining algorithms to tackle the mentioned above challenges, *i.e.*, costly execution time of the data cube computation as well as a large storage space on the disk. Thus, we introduced a closed cube called $\mathcal{CND}$-$\mathcal{C}$ube based on an efficient mining algorithm called CLOSENDMG. The carried out experimental results showed the effectiveness of the introduced approach and highlighted that the $\mathcal{CND}$-$\mathcal{C}$ube outperforms the pioneering approaches in information lossless reduction. Future issues for the present work mainly concern: (1) the consideration of the dimension hierarchies in the very same spirit as Cure for Cubes [11], (2) the study of the extraction of "generic multidimensional" association rules based on the $\mathcal{CND}$-$\mathcal{C}$ube [21].

References

1. Chaudhuri, S., Dayal, U.: An overview of data warehousing and OLAP technology. SIGMOD Record 26(1), 65–74 (1997)
2. Gray, J., Chaudhuri, S., Bosworth, A., Layman, A., Reichart, D., Venkatrao, M.: Data cube: A relational aggregation operator generalizing group-by, cross-tab, and sub totals. Data Mining and Knowledge Discovery 1(1), 29–53 (1997)
3. Ross, K., Srivastava, D.: Fast computation of sparse data cubes. In: Proceedings of the 23rd International Conference on Very Large Databases (VLDB 1997), Athens, Greece, pp. 116–125 (1997)
4. Beyer, K., Ramakrishnan, R.: Bottom-up computation of sparse and iceberg cubes. In: Proceedings of the 1999 ACM-SIGMOD International Conference on Management of Data (SIGMOD 1999), Philadelphia, Pennsylvania, USA, pp. 359–370 (1999)
5. Han, J., Pei, J., Dong, G., Wang, K.: Efficient computation of iceberg cubes with complex measures. In: Proceedings of the International Conference on Management of Data (SIGMOD 2001), Santa Barbara, California, USA, pp. 441–448 (2001)
6. Pedersen, T., Jensen, C., Dyreson, C.: Supporting imprecision in multidimensional databases using granularities. In: Proceedings of the 11th International Conference on Scientific and Statistical Database Management (SSDBM 1999), Cleveland, Ohio, USA, pp. 90–101 (1999)
7. Lakshmanan, L., Pei, J., Han, J.: Quotient cube: How to summarize the semantics of a data cube. In: Proceedings of the 28th International Conference on Very Large Databases (VLDB 2002), Hong Kong, China, pp. 778–789 (2002)
8. Pasquier, N., Bastide, Y., Taouil, R., Lakhal, L.: Efficient mining of association rules using closed itemset lattices. Journal of Information Systems 24(1), 25–46 (1999)
9. Casali, A., Cicchetti, R., Lakhal, L.: Closed cubes lattices. Annals of Information Systems 3, 145–165 (2009); Special Issue on New Trends in Data Warehousing and Data Analysis
10. Ji, L., Tan, K.L., Tung, A.K.H.: Mining frequent closed cubes in 3D datasets. In: Proceedings of the 32nd International Conference on Very Large Data Bases (VLDB 2006), Seoul, Korea, pp. 811–822 (2006)
11. Morfonios, K., Ioannidis, Y.E.: Cure for cubes: Cubing using a ROLAP engine. In: Proceedings of the 32nd International Conference on Very Large Data Bases, Seoul, Korea, pp. 379–390 (2006)
12. Wang, W., Lu, H., Feng, J., Yu, J.: Condensed cube: An effective approach to reducing data cube size. In: Proceedings of the 18th International Conference on Data Engineering (ICDE 2002), San Jose, USA, pp. 213–222 (2002)

13. Sismanis, Y., Deligiannakis, A., Roussopoulos, N., Kotidis, Y.: DWARF: shrinking the petacube. In: Proceedings of the 2002 ACM-SIGMOD International Conference on Management of Data (SIGMOD 2002), Madison, USA, pp. 464–475 (2002)
14. Casali, A., Nedjar, S., Cicchetti, R., Lakhal, L., Novelli, N.: Lossless reduction of datacubes using partitions. International Journal of Data Warehousing and Mining (IJDWM) 4(1), 18–35 (2009)
15. Ganter, B., Wille, R.: Formal Concept Analysis. Springer, Heidelberg (1999)
16. Bastide, Y., Pasquier, N., Taouil, R., Stumme, G., Lakhal, L.: Mining minimal non-redundant association rules using frequent closed itemsets. In: Palamidessi, C., Moniz Pereira, L., Lloyd, J.W., Dahl, V., Furbach, U., Kerber, M., Lau, K.-K., Sagiv, Y., Stuckey, P.J. (eds.) CL 2000. LNCS, vol. 1861, pp. 972–986. Springer, Heidelberg (2000)
17. Calders, T., Goethals, B.: Non-derivable itemset mining. Data Mining and Knowledge Discovery 14(1), 171–206 (2007)
18. Gunopulos, D., Khardon, R., Mannila, H., Toivonen, H.: Data mining, hypergraph transversals, and machine learning. In: Proc. of the 16th ACM Symp. on Principles of Database Systems (PODS), Tuscon (1997)
19. Muhonen, J., Toivonen, H.: Closed non-derivable itemsets. In: Fürnkranz, J., Scheffer, T., Spiliopoulou, M. (eds.) PKDD 2006. LNCS, vol. 4213, pp. 601–608. Springer, Heidelberg (2006)
20. Stumme, G., Taouil, R., Bastide, Y., Pasquier, N., Lakhal, L.: Computing Iceberg concept lattices with TITANIC. Data and Knowledge Engineering 42(2), 189–222 (2002)
21. Messaoud, R.B., Rabaséda, S.L., Boussaid, O., Missaoui, R.: Enhanced mining of association rules from data cubes. In: Proceedings of the 9th ACM International workshop on Data warehousing and OLAP, Arlington, Virginia, USA, pp. 11–18 (2006)

Indexing the Function: An Efficient Algorithm for Multi-dimensional Search with Expensive Distance Functions

Hanxiong Chen[1], Jianquan Liu[1], Kazutaka Furuse[1], Jeffrey Xu Yu[2], and Nobuo Ohbo[1]

[1] Computer Science, University of Tsukuba, 1-1-1 Tennodai, Ibaraki, 305-8577, Japan
{chx,ljq,furuse,ohbo}@dblab.is.tsukuba.ac.jp

[2] Systems Engineering & Engineering Management, Chinese University of HongKong, China
yu@se.cuhk.edu.hk

Abstract. Indexing structures based on space partitioning are powerless because of the well-known "curse of dimensionality". Linear scan of the data with approximation is more efficient in high dimensional similarity search. However, approaches so far concentrated on reducing I/O, ignored the computation cost. For an expensive distance function such as L_p norm with fractional p, the computation cost becomes the bottleneck. We propose a new technique to address expensive distance functions by "indexing the function" by pre-computing some key values of the function once. Then, the values are used to develop the upper/lower bounds of the distance between each data and the query vector. The technique is extremely efficient since it avoids most of the distance function computations; moreover, it does not spend any extra storage because no index is constructed and stored. The efficiency is confirmed by cost analyses, as well as experiments on synthetic and real data.

Keywords: Similarity search, high dimensional space, function index.

1 Introduction

Implementing efficient similarity search mechanisms for high dimensional data sets is one of the important research topic in the field of data engineering, and has been well studied in recent years[1][2][3][4][5][6]. Difficulties of this topic mainly arise from the well-known property called "curse of dimensionality". In high dimensional spaces, it is observed that hypersphere covering results of a nearest neighbor query tends to have huge radius. Because of this property, space partitioning and data partitioning techniques become worse than simple sequential scan[7]. For this reason, recent research attention is mainly paid to improve the performance of sequential scan. Such techniques include VA-file (or vector approximation file) and its variations[5][6][8][9].

Another research topic of the similarity search is what distance metric should be used[10][11][12][13]. For vector data sets, L_p norm (Minkowski metric) is generally used in various applications. The parameter p refers to the degree of power, and mainly used are $p = 1$ (Manhattan metric) and $p = 2$ (Euclid metric). In [10], it is argued that the value of p is sensitive to meaningfulness in high dimensional spaces, and smaller value of p is preferable. It is also mentioned that fractional values less than 1 are rather

R. Huang et al. (Eds.): ADMA 2009, LNAI 5678, pp. 67–78, 2009.

effective than the cases of $p = 1$ and $p = 2$. However, using fractional p makes the distance calculation costly, and it considerably decreases performance of the similarity search. This is because we need time consuming numerical computations for calculating pth powers.

In this paper, we propose an efficient technique for the similarity search with L_p norm of fractional p. The motivation comes from the observation on our first experiment (Figure 7) that the computation cost of the functions affects the performance as significant as the I/O cost. Our approach is based on the idea of "indexing the costly distance function". Some approximated values of pth powers which can be used to calculate distance between vectors are computed in advance, and they are used to obtain upper and lower bound in the course of similarity search procedure. The technique is extremely efficient since we do not have to perform numerical computations for every search. We present the efficiency of the proposed technique by both analysis and experiments performed on synthetic and real data sets.

This paper is organized as follows. In the next section, we explain the motivation of our approach with an example. Following this, Section 3 describes the proposed method and its performance analysis. In Section 4, we present empirical results. Section 5 provides summary and conclusions.

2 Motivation by Example

We first explain how the bounds are assembled by Figure 1. For simplicity, suppose that the space is two dimensions and normalized (that is, $[0, 1)^2$) and that the distance function is L_p with a fractional p ($0 < p < 1$). Because the calculation of $(\cdot)^p$ is expensive, our mission is to avoid such calculation as much as possible.

When interval [0, 1) is divided into 10 equal blocks by points $t/10$, preparing $c[t] = (t/10)^p$, $t = 0, 1, \ldots, 10$ takes 9 times of calculation of p-power ($c[0] = 0$, $c[10] = 1$ are known without real calculation). These $c[t]$'s are enough to construct upper and lower bounds for efficient nearest search for *all* vectors.

$c[\cdot]$ pre-computes the values of the "knots" of the grid net, for instance, $L_p^p(q, p1) = c[2] + c[2]$. Each vector falls in one of the rectangles bounded by 4 corner knots, whose nearest one and farthest one to q represent the lower bound and the upper bound of the distance between v and q, respectively. For vector v of Figure 1 c), the bounds are $L_p(q, p1)$ and $L_p(q, p2)$ because $L_p(q, p1) \le L_p(q, v) < L_p(q, p2)$. We would like to emphasize that, the grid net in the figure is drawn only for understanding. It means neither the partition of the real space, nor the coordinate system. Hence, as in Figure 1 a) and b), it does not mean that the query point q locates in the origin.

In Figure 1 c), taking a data vector $v = (v.x,\ v.y)$, suppose the task is to find whether $L_p(q, v) \le r$ holds. Before computing the exact distance $L_p(q, v) = \sqrt[p]{|q.x - v.x|^p + |q.y - v.y|^p}$, in some cases we can give the answer with less computation. Denote $|q.x - v.x|$ and $|q.y - v.y|$ shortly by δ_x and δ_y, respectively, obviously $0 \le \delta_x < 1$. So there exists a $t_x \in \{0, 1, \ldots, 9\}$ such that

$\frac{t_x}{10} \le \delta_x < \frac{t_x+1}{10}$ and thus $c[t_x] \le \delta_x^p < c[t_x + 1]$. For the same reason, there exists a t_y such that $c[t_y] \le \delta_y^p < c[t_y + 1]$. Consequently,

$$l_b \equiv c[t_x] + c[t_y] \le L_p^p(q, v) = \delta_x^p + \delta_y^p < c[t_x + 1] + c[t_y + 1] \equiv u_b$$

This tells that l_b and u_b are a lower bound and an upper bound of the distance between q and v, respectively. In other words, if $l_b > r^p$ then we know that the real distance will never less than r^p hence the vector can be safely discarded. On the other hand, if $u_b \leq r^p$ is satisfied then v is added to the answer set immediately. Noting that both the upper bound and the lower bound can be simply assembled by looking up the pre-computed array $c[\cdot]$, most vectors are judged without the expensive p-power calculation.

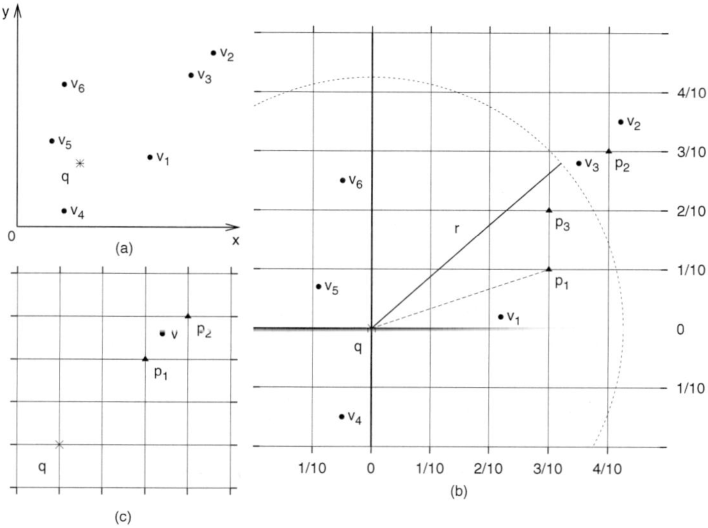

Fig. 1. A 2-dim. example of function index. a) the data space and an arbitrary query vector. b) "cover" the data space with a virtual grid net centered at q. and c) focus on one vector to see how bounds are assembled.

The image of the range query and k Nearest Neighbors (k-NN) query processing is illustrated in Figure 2. In a), v_1 is included into the answer set automatically because even its upper bound to q ($L_p^p(q, p1) = c[3] + c[1]$) is less than r. On the other hand, v_2 is discarded because its lower bound ($L_p^p(q, p2) = c[4] + c[3]$) has already exceeded r. Unfortunately, whether v_3 satisfies the range condition is unknown and the real distance must be examined. Nevertheless, more detailed partition of [0, 1) to larger $c[\cdot]$ can reduce such v_3 area easily. Because $c[\cdot]$ is not stored, the total extra cost to double the partition is double of computation of $c[\cdot]$ and the in-memory array of it. VA-file, though good for some k-NN, can never good for range query.

In b), suppose that database V consists of eight vectors $v_1, v_2 \ldots, v_8$ being accessed in the order of their subscriptions and the query is to find 2-NN of q from V. The k-NN is process in two phases. The first phase excludes those vectors which have no possibility to be k-NN. In filtering only the bounds of the distance between each vector and q are necessary. The second phase refines the candidate of the previous phase and give answer of k-NN.

To process 2-NN, firstly, the lower bounds of v_1, v_2 ($l_b(v_1)$ and $l_b(v_2)$) are obtained by looking up $c[\cdot]$ as above. Meanwhile, v_1, v_2 are taken as the first two candidates.

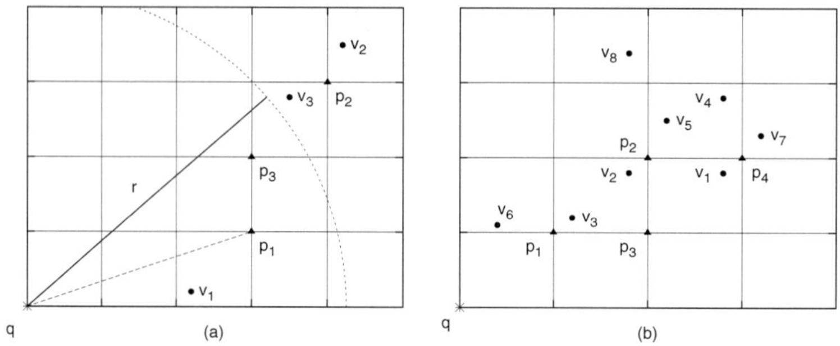

Fig. 2. Examples of a) Range query and b) 2-NN query

Their upper bounds and lower bounds are sorted in ascending order, respectively. Now that there are k (here, 2) candidates found, ideally newly scanned approximation can be discarded or can replace existing ones.

When v_3 is encountered, its lower bound $l_b(v_3)(\equiv L_p^p(q, p1))$ is compared with the existing larger (hence v_1's) upper bound $u_b(v_1)(\equiv L_p^p(q, p4))$. Because the former one is smaller, v_3 is added to the candidates because we are not sure currently which of v_3 and v_1 has a smaller real distance to q. On the other hand, the next encountered v_4 is discarded because $l_b(v_4) \geq max\{u_b(v_2), u_b(v_3)\}$. Therefore the real distance $L_p(q, v_4)$ between v_4 and q can never less than that of either v_2 or v_3. Since there are already exist at least two other nearer vectors to q than v_4, v_4 will never have chance to be 2-NN of q.

Similarly, v_5, v_7 and v_8 are discarded. On the contrary, v_6 remains because its lower bound is less than that of v_5.

So, after the filtering phase, v_1, v_2, v_3, v_6 are left as candidates. Then in the refinement phase, the real distance between v_i and q is examined for $i = 1, 2, 3, 6$. This examination finally decides that v_3 and v_6 are the answers to 2-NN of q.

3 Indexing the Expensive Function

In Table 1 we first give the notations that will be used throughout this paper, though some have been used in the previous section.

Let V and q be as in Table 1, r be a real number and k be a natural number. A range query is to find those vectors of V within distance r from q, with respect to L_p. A k-NN query aims at finding the k nearest vectors of V of q with respect to L_p.

A range query is formally expressed by finding the answer set V_r where

$$V_r = \{v | v \in V, L_p(q, v) \leq r\}$$

On the other hand, the answer set V_k of a k-NN query satisfies all the following conditions all together. The first condition says that there are at least k vectors in the answer set. By the second condition, V_k becomes least than k vectors if the farthest vector(s) are removed. This happens when several vectors have exactly same distance from q.

$$\forall v \in V_k \forall u \in (V - V_k), L_p(q, v) < L_p(q, u) \text{ and}$$
$$|V_k| \geq k \wedge |V_k - arg\ max_{v \in V_k}\{L_p(q, v)\}| < k$$

Table 1. Notations and Basic Definitions

V, D	vector database, $V \subseteq [0, 1)^D$, D: number of dimensions
N	the number of vectors in V, that is $N = \|V\|$
d	subscription range over dimensions, $d \in \{1, 2, \ldots, D\}$
v_i	ith data vector, $v_i \in V$, sometimes be omitted by v
$v.x_d$	v's coordinate value of dth dimension. $v.x_d \in [0,\ 1)$
q	a query vector, $q \in [0, 1)^D$
δ_d	the difference between q and v on dth dimension, $\|q.x_d - v.x_d\|$
L_p	p-norm distance function. $L_p(q, v) = \sqrt[p]{\sum_{d=1}^{D} \delta_d^p}$
$l_b(v), u_b(v)$	lower & upper bounds of v, respectively: $l_b(v) \leq L_p^p(q, v) < u_b(v)$
B	parameter: the number of *knot*s dividing [0,1)

Since the exponential is computationally expensive, it is meaningless to compute each element δ_d^p of the arbitrary norm L_p by its original definition. We generalize the description of the processing mentioned in Section 2 and develop the following efficient solution.

The Range Query Algorithm. First of all, Function *get-bound* is commonly used by both range query algorithm and k-NN algorithm so we isolate it alone for readability. The action of this function is clear and needs no explanation.

Our algorithm for processing range queries is very simple. As in Algorithm Rang-query (Fig. 4), it scans each vector, assembling its bounds by calling Function *get-bound*, then comparing them with the given range r. Naturally, instead of comparing $L_p(q, v)$ with r for all v of V, it is much more cheaper to compare $L_p^p(q, v)$ with r^p because $L_p^p(q, v)$ is assemble from $c[\cdot]$ straightforwardly. This is reflected by line 2 in the algorithm. Line 5 decides those vectors surely be answer while line 7 excludes those vectors which never have possibility to be answer. If either of the above failed, then we have to examine the real distance in line 9.

The k-NN Query Algorithm. The algorithm for processing k-NN queries is in filtering phase and refinement phase. The essential difference comparing with VA-file is that we do not aim at the reduction of I/O cost but computation cost. Since our algorithm needs not to depend on a certain index, we can also process a query flexibly by changing B dynamically. We introduce heap structures in both phases. For readability, we assume that the heaps are sorted in ascending order.

The filtering phase is based on the idea that $\forall v_1, v_2 \in V$, $l_b(v_1) \geq u_b(v_2) \Rightarrow L_p(q, v_1) \geq L_p(q, v_2)$. Generally, if we have k candidates in hand and the largest upper bound among them is $u_b(v_k)$, then a new encountered vector v can be safely discarded the moment we found $l_b(v) \geq u_b(v_k)$. On the other hand, if any lower bound of the exist candidates is large than the upper bound of this v, then we can discard that candidate by replacing it by v. Algorithm kNN-filtering (Fig. 5) describes this processing formally. Before all, $c[t]$ is obtained by $B - 1$ p-power calculations. Then the first k encountered vectors along with their lower bounds sorted in ascending order are add to candidate heap $Cand$. The first k upper bounds are also inserted to heap H_u(of fix size k, and sorted). This guarantees that there at least k candidates (line 3-7). Then each later coming vector is compared to the largest upper bound found so far ($H_u[k]$ and line 10).

Function get-bound(v, q, $c[B]$)
$v = (v.x_1, \ldots, v.x_D)$: $v \in V$,
$q = (q.x_1, \ldots, q.x_D)$: the query vector

begin
  $t_d \leftarrow \lfloor |q.x_d - v.x_d| \times B \rfloor$,
    $d = 1, 2, \ldots, D$.
  $l_b(v) \leftarrow \sum_{d=1}^{D} c[t_d]$
  $u_b(v) \leftarrow \sum_{d=1}^{D} c[t_d + 1]$
  return l_b, u_b
end

Fig. 3. Computation of the Upper/Lower Bounds

Algorithm Range-query
Input: V, q, B as in Table 1, and r.
Output: Answer set V_r

begin
$c[t] \leftarrow (t/B)^p$ for $t = 1, 2, \ldots, B-1$.
$r \leftarrow r^p$; $V_r \leftarrow \emptyset$;
foreach $v \in V$
  $(l_b(v), u_b(v)) \leftarrow$ get-bound(v, q, c)
  if $u_b(v) < r$ **then**
    $V_r \leftarrow V_r \cup \{v\}$
    else if $l_b(v) > r$ **then**
      discard v; **break**
    else if $L_p^p(q, v) < r$ **then** // real dist.
      $V_r \leftarrow V_r \cup \{v\}$
  end if
end foreach
end

Fig. 4. Range Query

The vector is added to $Cand$ only if its lower bound does not exceed $H_u[k]$. Adding a vector to $Cand$ also causes the replacement of H_u (line 12), which *tightens* the upper bounds contiguously.

In the algorithms, we hid some details to make the description simple and more readable. In real world, the heap H_u is implemented as fix size k. Then for example in line 6, before an insertion to H_u is really carried out, the kth value is removed unconditionally. Overwriting the kth position without checking whether it is empty makes the execution more efficient. Moreover, actually it is not necessary that H_u is sorted. Being a heap, that is, the largest element is found in root, is enough for H_u.

Usually several times more vectors than $|V_k|$ remain after the filtering phase, so it is necessary to examine the real distance. Algorithm kNN-refinement (Fig. 6) provides a sophisticated method to do so. For a similar reason as in the previous phase, to guarantees k vectors, the first k candidates in $Cand$ is added to answer set V_k unconditionally (line 3-7). Noting again that V_k is sorted by $L_p^p(q, v_i)$, instead of simply examining all the remaining vectors in $Cand$ and updating V_k, a technique is developed to accelerate the termination. Using the fact that real distance ($= u_b = u_l$) is the tightest bound, the condition in line 10 terminates the algorithm any time a lower bound $l_b(v_i)$ ($i > k$) is found exceeds the k'th distance $L_p^p(q, v_k)$ of $V_k[k]$. Because $Cand$ was sorted in ascending order of l_b (in the filtering phase), vectors (say, v_j) after v_i has larger lower bounds than $l_b(v_i)$, hence their real distance to q can never be under $L_p(q, v_k)$. It is clear that such v_j can be safely discarded by expressed the relationship formally as follows.

$$L_p^p(q, v_j) > l_b(v_j) \geq l_b(v_i) > L_p^p(q, v_k)$$

Algorithm kNN-filtering
Input: Vector set V and query vector q, B and k.
Output: Candidate set and bounds $Cand$

begin
create heaps $Cand$ and H_u
$c[t] \leftarrow (t/B)^p$ for $t = 1, 2, \ldots, B-1$.
foreach $v \in \{v_1, v_2, \ldots, v_k\}$
  $(l_b(v), u_b(v)) \leftarrow$ get-bound(v, q, c)
  insert $(v, l_b(v))$ into $Cand$ // $l_b(v)$ as sorting key
  insert $u_b(v)$ into H_u // H_u is sorted
end foreach
foreach $v \in V \backslash \{v_1, v_2, \ldots, v_k\}$
  $(l_b(v), u_b(v)) \leftarrow$ get-bound($v, q, c[B]$)
  if $l_b(v) \leq H_u[k]$ **then**
    insert $(v, l_b(v))$ into $Cand$
    insert $u_b(v)$ into H_u
  end if
end foreach
return $Cand$
end

Fig. 5. Filtering phase for k-NN Query

Algorithm kNN-refinement
Input: V, q, B, k.
Output: Answer set V_k (in heap structure).

begin
$Cand \leftarrow$ **Call** kNN-filtering
$V_k \leftarrow \emptyset$
for $i = 1, 2, \ldots, k$
  $(v, l_b(v)) \leftarrow Cand[i]$
  compute $L_p^p(q, v)$ //the real distance
  insert $(v, L_p^p(q, v))$ into V_k
end for
for ($i = k + 1; i \leq$ size-of($Cand$); i++)
  $(v, l_b(v)) \leftarrow Cand[i]$
  if $l_b(v) > L_p^p(q, v_k)$ of $V_k[k]$ **then break**
  else insert $(v, L_p^p(q, v))$ into V_k
end for
return $V_k[i], i = 1, 2 \ldots, k$
end

Fig. 6. Refinement phase for k-NN Query

3.1 Efficiency Analysis

There are various factors that influence the performance. First of all the data set. Either for data index or for our *function index*, the efficiency of index-based query processing depends on the dimensions of the data space, the number of points in the database, the data distribution, and even on the correlation of dimensions. *function index* is especially affected by the complexity of the functions.

Our algorithms reduce the computation of p-power to ignorable multiply and summation for those vectors judged by the bounds. In other words, the efficiency of the algorithm appears in the filtering effect; the more vectors filtered out by bounds, the more efficient it is. We analyze the filtering effect in both phases. The yardstick used here is one computational cost of $(\cdot)^p$, denoted by C_p.

The filtering effect strongly correlates with the tightness of the bounds, that is $(u_b - l_b)$. By Function get-bound (Fig. 3),

$$u_b - l_b = \sum_{d=1}^{D}(c[t_d + 1] - c[t_d]) = \sum_{d=1}^{D}((\frac{t_d + 1}{B})^p - (\frac{t_d}{B})^p)$$

In each dimension, $(u_b - l_b)$ is projected to a band of width $1/B$, and vectors falls in such bands cannot be judged at the point of time and are kept for further examination in the next phase. Number of such vectors is in proportion to the areas of the bands. Apparently, our purpose is to reduce such vectors to as few as possible.

Range Queries. In the best case when a vector is close to q, $\delta_d = |q.x_d - v.x_d|$ is close to the original point and the area is 1. Imagining the 2-dimension case, the worst case arises when the vector is far away from q and the areas will be the rim of the square, $B^2 - (B-1)^2$. Obviously the average estimation is $(B/2)^2 - ((B/2)-1)^2$. The analysis is extended easily to general D dimension where the best, average, and worst estimations are 1, $(B/2)^D - ((B/2)-1)^D$ and $B^D - (B-1)^D$, respectively.

When B is large enough, $B^D - (B-1)^D$ can be simplified to DB since $B^D - (B-1)^D = B^D - (B^D + DB^{D-1} + D(D-1)B^{D-2} + \ldots \sim DB^{D-1}$ and similarly $(B/2)^D - ((B/2)-1)^D \sim DB^{D-1}/2^D$. Consequently, the best average and worst areas in proportion to the data space $[0,1)^D$ is $1/B^D$, $D - 1/2^D B$, and D/B Consequently.It is worth noting that the worst case happens only in the case when q is the given as the vertexes (all $q.x_d$'s are binary 0 or 1.). and all the data concentrate to opposite angle of q, that is, all $v.x_d$'s are 1 or 0. A simple example is when q is the original point, then all $N = |V|$ data vectors concentrate to a single point $(1, 1, \ldots, 1)$.

Based on the above discussion, the number of vectors have to be examine the real distance is between 1 and D/B.

***k*-NN Queries.** In the fist phase, computation of $(\cdot)^p$ is necessary for $c[\cdot]$. Therefor the cost is $B \times C_p$.To estimate the number of real distance computation in the second phase, we need to estimate the candidates left after the phase (size of H_l in Algorithm kNN-filtering (Fig. 5)). Imaging from two extreme cased, it can be known that the number strongly depends on the distribution of the vectors: when at least k vectors falls in the same block of q, then it is the number of all vectors in this block; when more than $N - k$ vectors distribute on the *surface* of $[0,1)^D$, then the number may as large as N!

As in the experiments, for uniform distributions, this number is usually several ten times large than k. In the second phase of Algorithm kNN-refinement (Fig. 6), with the technique that accelerates the termination, the number of vectors need actual computation is reduced to several times of k

4 Experimental Results

To confirm the efficiency of the proposed technique, we performed an experimental evaluation and compared it to simple sequential scan. All experimental results presented in this section are performed on a Intel-based computer system running under Linux. CPU is Intel(R) Xeon (TM) 2.80 GHz and the amount of main memory is 3.2GB. Programs are implemented in the C++ language. We used two kinds of data sets for the experiments: a synthetic data set and 4 real data sets. The synthetic one is the set of uniformly distributed vectors. The real data set is Corel Image Features taken from UCI KDD Archive[1]. As mentioned below, proposed technique is consistently effective for all the data sets. Parameters used in the experiments and their default values are given in Table 2.

Table 2. Parameters Used in the Experiments

symbol	default value	description
N	1048576	the number of vectors in data set
k	100	the number of similar vectors to be searched
D	64	dimensionality
p	0.9	the degree of power
B	2048	the number of "knots"

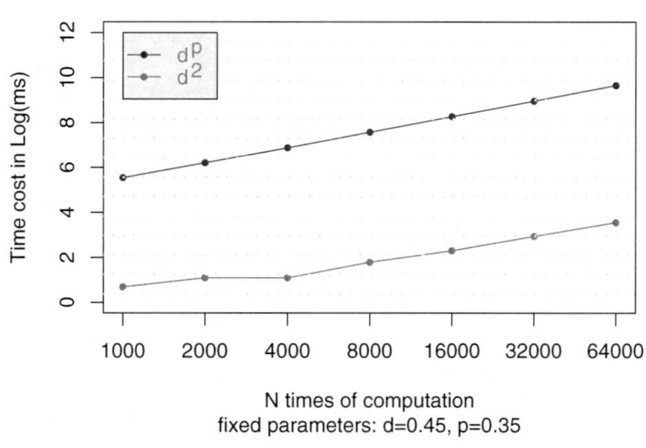

Fig. 7. Cost about d^p and d^2

[1] http://kdd.ics.uci.edu/databases/CorelFeatures/CorelFeatures.data.html

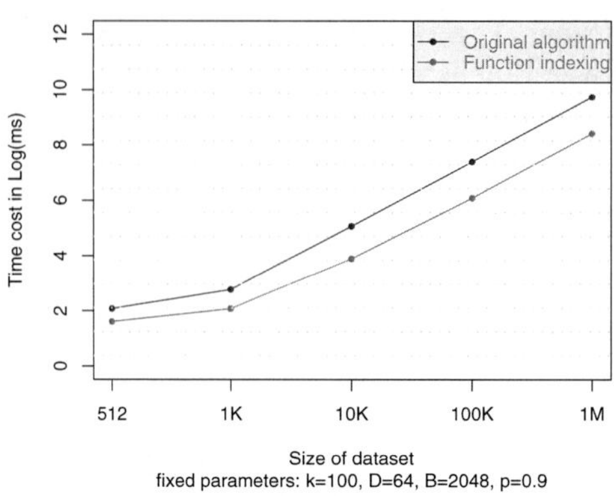

Fig. 8. Cost about different size of synthetic dataset

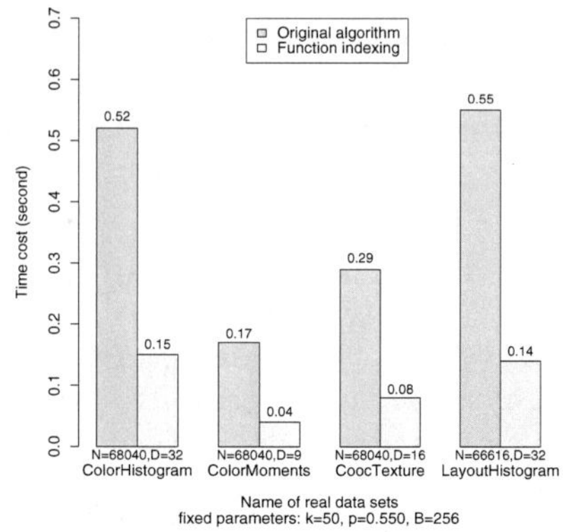

Fig. 9. Comparison on real dataset

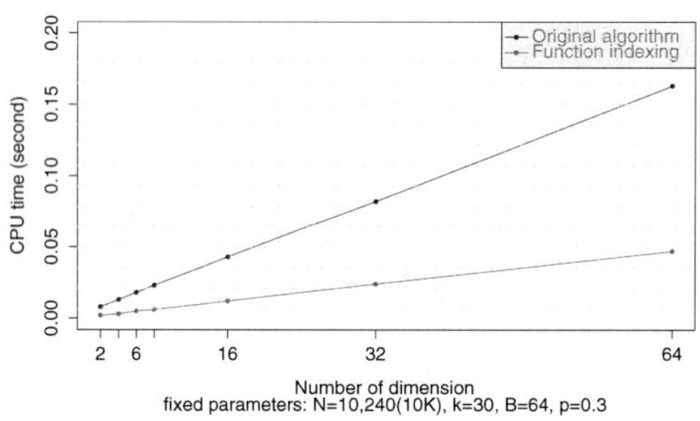

Fig. 10. (a) Cost about different D

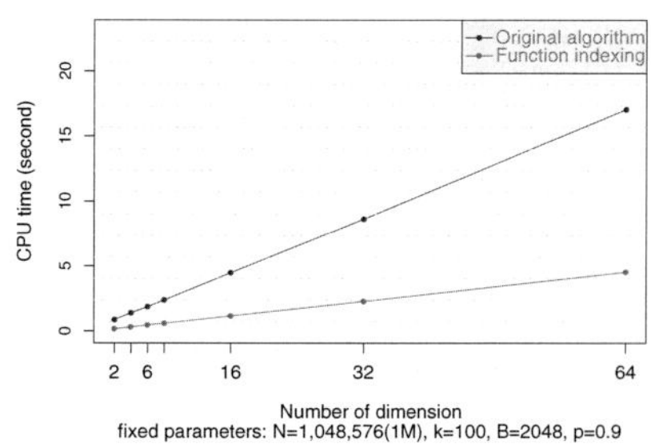

Fig. 11. (b) Cost about different D

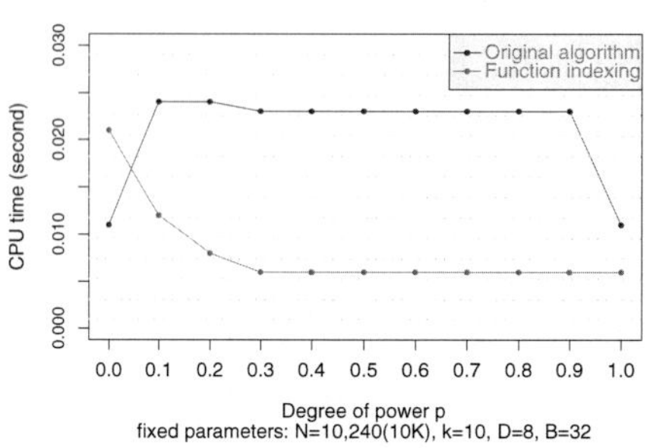

Fig. 12. (a) Cost about different p

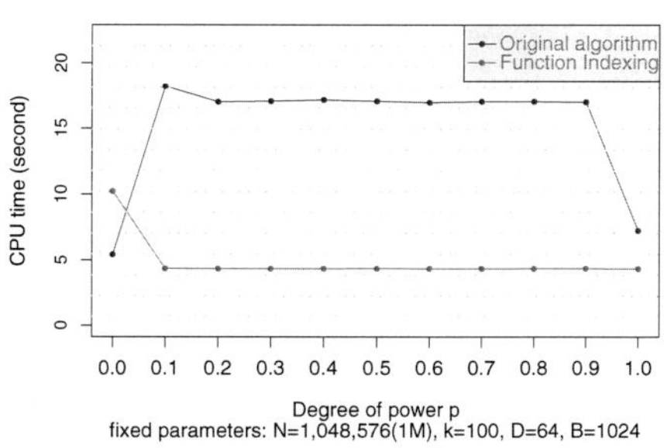

Fig. 13. (b) Cost about different p

First of all, we present the cost for computing pth powers is dominant when p is fractional. Figure 7 shows the total time elapsed to perform similarity search (including file I/Os) and total time of the distance computations elapsed during the search. As shown in this figure, the cost of the distance computations is significant portion while the cost of I/Os is relatively small. In remaining experiments, we compare the proposed technique and simple sequential scan in elapsed time to perform similarity search (excluding file I/Os).

Figure 8 and Figure 9 show the scalability of the proposed technique on synthetic and real data sets, respectively. As expected, the proposed method is fairly successful in reducing computational cost. The cost is linear to the size of data sets, since our approach is a variation of sequential scan.

In the next experiment, we evaluated performance by varying dimensionality. Results are shown in Figure 10 and Figure 11. In both results, we can observe that the elapsed time is linear to the dimensionality. This is because the number of computations of pth powers is linear to the dimensionality. We can also confirm that the proposed method is stably effective.

Figure 12 and Figure 13 is the results when we varying the value of p. We can see that the computation of pth powers is quite costly when p is fractional. Since the proposed method computes approximated values of pth powers in advance, and there is no need to compute p powers during the search, the performance is much better, independing to the value of p.

5 Conclusions

For a long time it is believed that I/O cost dominates the performance of almost any kind of searching. Efforts are thus put on developing of data index to reduce I/O while processing searching. We figured out in this paper that the computation cost for multi-dimensional searches with expensive distance functions is also a dominative factor. To reduce such computation cost, we developed an efficient filtering algorithm based on the new technique called *function indexing*. We also designed range query algorithm as well as k-NNquery algorithm based on this technique. Analyses on the filtering effect of the algorithms show that most of the distance computation can be avoided. As a general technique, it is also widely applicable to any kind of applications with multi-dimensional searches. Experiments on real and synthetic data confirm the efficiency.

To extend the technique, we are planning to combine it with other indexes, such as VA-file. We will also investigate the effect on non-uniform data and dimension-wise distance functions.

References

1. Berchtold, S., Böhm, C., Keim, D., Kriegel, H.P.: The X-tree: An index structure for high-dimensional data. In: Proceedings of 26th International Conference on Very Large Data Bases, pp. 28–39 (1996)
2. Berchtold, S., Ertl, B., Keim, D.A., Kriegel, H.P., Seidl, T.: Fast nearest neighbor search in high-dimensional space. In: Proceedings of the 14th International Conference on Data Engineering, pp. 209–218 (1998)
3. Berchtold, S., Keim, D., Kriegel, H.P.: The pyramid-technique: Towards breaking the curse of dimensional data spaces. In: Proceedings of the 1998 ACM SIGMOD International Conference on Management of Data, pp. 142–153 (1998)
4. Böhm, C., Berchtold, S., Keim, D.A.: Searching in high-dimensional spaces: Index structures for improving the performance of multimedia databases. ACM Computing Surveys 33(3), 322–373 (2001)

5. Ferhatosmanoglu, H., Tuncel, E., Agrawal, D., Abbadi, A.E.: Vector approximation based indexing for non-uniform high dimensional data sets. In: Proceedings of the ACM International Conference on Information and Knowledge Management, pp. 202–209 (2000)
6. Weber, R., Schek, H.J., Blott, S.: A quantitative analysis and performance study for similarity-search methods in high-dimensional spaces. In: Proceedings of 24th International Conference on Very Large Data Bases, pp. 194–205 (1998)
7. Berchtold, S., Böhm, C., Keim, D., Kriegel, H.P.: A cost model for nearest neighbor search in high-dimensional data space. In: ACM PODS Symposium on Principles of Database Systems, pp. 78–86 (1997)
8. An, J., Chen, H., Furuse, K., Ohbo, N.: Cva-file: An index structure for high-dimensional datasets. Knowledge and Information Systems Journal 7(3), 337–357 (2005)
9. Chen, H., An, J., Furuse, K., Ohbo, N.: C^2VA:trim high dimensional indexes. In: Meng, X., Su, J., Wang, Y. (eds.) WAIM 2002. LNCS, vol. 2419, pp. 303–315. Springer, Heidelberg (2002)
10. Aggarwal, C., Hinneburg, A., Keim, D.A.: On the surprising behavior of distance metrics in high dimensional spaces. In: Van den Bussche, J., Vianu, V. (eds.) ICDT 2001. LNCS, vol. 1973, pp. 420–434. Springer, Heidelberg (2000)
11. Beyer, K.S., Goldstein, J., Ramakrishnan, R., Shaft, U.: When is "nearest neighbor" meaningful. In: Proceedings of the 7th Int. Conf. on Database Theory, pp. 217–235 (1999)
12. Hinneburg, A., Agrawal, D., Keim, D.A.: What is the nearest neighbor in high dimensional spaces? In: Proceedings of the 26th VLDB Conference, pp. 506–515 (2000)
13. Yi, B., Faloutsos, C.: Fast time sequence indexing for arbitrary L_p norms. In: Proceedings of 26th International Conference on Very Large Data Bases, pp. 385–394 (2000)

Anti-germ Performance Prediction for Detergents Based on Elman Network on Small Data Sets

Anqi Cui, Hua Xu, and Peifa Jia

State Key Laboratory on Intelligent Technology and Systems,
Tsinghua National Laboratory for Information Science and Technology,
Department of Computer Science and Technology,
Tsinghua University, 100084, Beijing
caq08@mails.tsinghua.edu.cn

Abstract. Anti-germ performance test is critical in the production of detergents. However, actual biochemical tests are often costly and time consuming. In this paper, we present a neural network based model to predict the performance. The model made it much faster and cost less than doing actual biochemical tests. We also present preprocessing methods that can reduce data conflicts while keeping the monotonicity on small data sets. This model performs well though the training data sets are small. Its input is the actual value of key ingredients, which is not widely used in solving biochemical problems. The results of experiments are generated on the base of two detergent products for two types of bacteria, and appear reasonable according to natural rules. The prediction results show a high precision and fitting with the monotonicity rule mostly. Experts in biochemical area also give good evaluations to the proposed model.

Keywords: Anti-germ performance prediction, Artificial neural networks, Monotonicity rule, Pre-processing methods.

1 Introduction

Anti-germ performance test of key ingredients is one of the critical aspects in detergent manufacturing. This work enables product researchers to adjust the amount of key ingredients to achieve a minimum requirement or to try different combinations of ingredients in the product solution. However, actual biochemical tests are often costly and time-consuming. Therefore, a rapid and low cost prediction model is needed in detergent manufacturing.

Key ingredients take effects on the anti-germ performance by their biochemical structure. Previous work on mathematical modeling mainly focuses on numerical fitting methods[1]. However, this method can only deal with simple models such as linear, logarithmic, or exponential fitting, which cannot fully express the natural rules.

R. Huang et al. (Eds.): ADMA 2009, LNAI 5678, pp. 79–90, 2009.

Artificial neural network (ANN) is one of the well known prediction models. The feature of dealing with non-linear problems makes ANN popular in biochemical simulations. Neural network is often used in predicting performances according to some macroscopical factors (e.g. temperature, humidity, etc.)[2]. This can be intuitive but to some extent ignores certain biochemical information of those key ingredients. To the opposite, some work is based on QSAR (Quantitative structure-activity relationship)[3], which focuses on molecular structures. This method is detailed but reduces the flexibility when ingredients are changed.

Another issue is that the performance of ANN depends on its structure and the training data (training data sometimes affects more significantly). The ANN itself cannot remove all the noises in the training data, thus the accuracy between calculated and actual performance is reduced. Increasing the amount of training data can reduce some noises' effect. However, as biochemical tests require quite some time and efforts, the number of experimental data is often too limited to meet the training requirements of the model.

Our Contribution. In order to overcome the disadvantages of prediction models, we designed an ANN model with specialized data preprocessing methods. The neural network is chosen for its non-linear and easy-to-extend feature, and a flexible number of inputs for predicting more ingredients' performance. As training data is quite limited, the preprocessing methods remove some conflicts in order to improve the prediction results. A monotonicity rule is applied in the treatment as well, which raises the consistency between the prediction results and the natural rules.

The paper is organized as the following: In the next section we introduce some related work in this field. The detailed problem and algorithm is presented in Section 3. The experimental results and evaluations are shown in Section 4. The conclusion and future work is introduced in Section 5.

2 Related Work

Artificial neural network is often used in predicting anti-germ performances or shelf life of products. This kind of work mainly chooses some macroscopic features as inputs of neural network model. Shelf life prediction in the work [2] chooses humidity, temperature, sorption properties and so on. These factors are easy to collect, but their influence to shelf life is complicated and empirical. Other similar works choose number of bacteria, white blood cell count and so on to predict aminoglycoside response[4], or protein, pH, etc. to predict urinary tract infection[6]. These works don't consider the effect related to the amount of the key ingredients, and the choice of factors can be empirical.

Another type of neural network applications is QSAR studies. This method uses molecular descriptors to generate the input from chemical structure to neural network. The descriptors include atom count, molecular weight, total energy and other molecule-level factors. QSAR is widely used in drug analysis for its accurate analysis of compounds. Anti-germ activity prediction can also use this method such as [3,5]. However, this method always deals with a single or a particular family of chemical compounds. It's not easy to use this method to predict different type of chemical ingredients together, and is also hard to extend more ingredients.

Mathematical models can be used to estimate the shelf life of food products[1]. This method works as a fitting method, which totally ignored the chemical effects of key ingredients.

As a conclusion, neural network works better to predict such anti-germ performances. But using the amount (exact values) of ingredients to predict the performances is not widely used. The reason may be lack of experimental data (as training data) or data conflicts. A method that solves this problem is introduced in the following sections.

3 Design of Model

In this section, we'll first describe the main problem, then introduce the data preprocessing methods and the design of the corresponding network.

3.1 Problem Statement

Anti-germ performance problem is to predict the reduction of germs when the amount of key ingredients is given. A typical method for testing the anti-germ efficacy works as the following:

1. Determine the amount of key ingredients and prepare the solution with certain amount.
2. Put the detergent solution into a prepared medium of concerned bacteria.
3. Use some methods to simulate a washing process.
4. Extract the liquid after washing and count the number of bacteria.
5. Calculate the log reduction of bacteria number as the final result.

A traditional method[7] usually takes a long time from step 2 to step 4. Instead, the proposed model will generate the final result in step 5 after given the amount of ingredients in step 1.

Existing experimental data consists of the amount of each key ingredient and the log reduction value of the concerned bacteria. For example, if the amount of each ingredient is $l_1, l_2, \cdots, l_n$, respectively, and the log reduction value of bacteria is b, this group of data can be represented as the following:

$$(l_1, l_2, \cdots, l_n) : b$$

The problem can be described in this way: If a set of data (formatted above) is given, how can we predict the performance b for a new group of $(l_1, l_2, \cdots, l_n)$?

This problem seems to be a very simple mathematical fitting problem. However, the conflicts in experimental data affect a lot. The next sub-section will introduce this problem.

3.2 Preprocessing Methods

As errors always occur in actual experiments, the experimental data is not as ideal as we expected. For example, the log reduction value has a range of 0 to 5 and over 5, which means the actual reduction of bacteria counts is 10^0 to 10^5 and $> 10^5$. The error of this value is ± 0.5 log, which is quite big within such a small range.

Generally, there are two types of data conflicts:

(1) Inconsistency with each other
Same amount of ingredients may result in different performances. This is common in actual experiments. It can be expressed in a formal way:

For t groups of data:

$$(l_{11}, l_{12}, \cdots, l_{1n}) : b_1$$
$$(l_{21}, l_{22}, \cdots, l_{2n}) : b_2$$
$$\cdots$$
$$(l_{t1}, l_{t2}, \cdots, l_{tn}) : b_t$$

If $\forall i \in \{1, 2, \cdots, t\}, (l_{i1}, l_{i2}, \cdots, l_{in})$'s are equal (the corresponding amount of ingredients equals), but some b's are different from others, there's a conflict in these groups.

To solve this conflict, simply take the average value $\bar{b} = \frac{1}{t}\sum_{i=1}^{t} b_i$ as the performance, and replace all the t groups of data with only one group:

$$(l_{11}, l_{12}, \cdots, l_{1n}) : \bar{b}$$

(2) Inconsistency with monotonicity rule
The monotonicity rule is an important prior knowledge in the prediction of anti-germ performance. The performance follows this natural principle: *The more amount of ingredients we add, the better efficacy it shows towards the bacteria.* Considering the errors in the experiments, the performance should *at least keep flat*. More formally, for t groups of data:

$$(l_{11}, l_{12}, \cdots, l_{1n}) : b_1$$
$$(l_{21}, l_{22}, \cdots, l_{2n}) : b_2$$
$$\cdots$$
$$(l_{t1}, l_{t2}, \cdots, l_{tn}) : b_t$$

If $l_{1i} \leq l_{2i} \leq \cdots \leq l_{ti}$ for all i's, but $b_1 > b_j (\forall j \in \{2, 3, \cdots, t\})$, there's a monotonicity conflict in these groups.

For this problem, still take the average value $\bar{b}$ as the performance. The only difference is to keep all those groups – just replace b_i's with $\bar{b}$.

After the above two steps, the conflicts in experimental data are reduced largely. Thus, they can be used as the training data of the neural network model, which will be discussed in the next subsection.

3.3 The Neural Network Model

There are many types of artificial neural networks. In this paper, we use one of the recurrent neural networks, Elman network. Elman network is first mentioned in Elman's work[8]. A typical two-layer Elman network[9] is shown in Fig. 1.

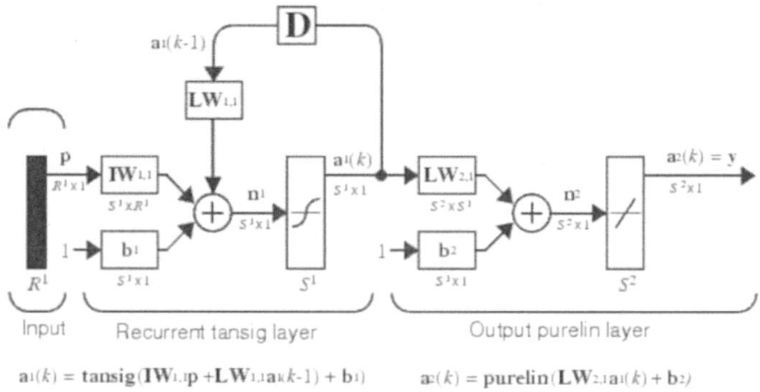

Fig. 1. A two-layer Elman network architecture[9]

Some of the network parameters are determined by experiments. The results will be shown in the next section.

Generally speaking, neural network is good at dealing with a large amount of training data. However, we have only 20 to 40 groups of data. The following section will show that the network model works well as we expected.

4 Experiments and Evaluation

The experimental data comes from actual experiments made in a company which mainly manufactures daily use chemical products. Two detergent products are tested in the experiments. Product 1 has three key ingredients and product 2 has five. Both series of experiments are tested towards two types of bacteria. The numbers of data for each kind of experiments are listed in Table 1.

In the preprocessing methods mentioned before, some data may be removed if it conflicts with others. The numbers of data after the preprocessing step are listed in Table 2. As we can see in the table, the remaining input data is quite limited.

Table 1. The No. of experimental data

Detergent	Bacteria1	Bacteria2
Product1	28	35
Product2	28	23

Table 2. The No. of preprocessed data

Detergent	Bacteria1	Bacteria2
Product1	26	33
Product2	18	16

This section contains the following: First, the structure of the network is determined by experiments. Then the model is evaluated from the following aspects: The necessity of data preprocessing methods, the precision of the model and the fact that the model fits the monotonicity rule. We also consulted some experts on biochemistry area for the efficacy of the model.

Note that the training algorithm uses random initial values for the Elman neural network, which causes the parameters (e.g. weight values, bias values, etc.) different each time. Therefore, the results shown in the following subsections are not the unique ones.

4.1 The Structure of Network

According to the problem, we use a two-layer Elman network model. However, in order to avoid the overfitting problem caused by a too complicated model, the number of hidden neurons needs to be determined by actual experiments.

The training function for the network updates weight and bias values using Levenberg-Marquardt optimization. The self-adaptive learning function is the gradient descent with momentum weight and bias learning function. The performance function measures the performance of the network according to the mean of squared errors. The transfer function of each neuron is the traditional Tan-Sigmoid transfer function (Hyperbolic tangent sigmoid transfer function, i.e. $\text{tansig}(n) = \frac{2}{1+\exp(-2*n)} - 1$). Actually, these functions are set according to MathWorks' Matlab (R2007b) Neural Network Toolbox's default settings.

The experimental data is divided into three parts: training, validation and test. The number of the data in these parts is 60%, 20% and 20% of the total amount, respectively. The data are normalized within the range of [-1, 1]. The number of iterations for training is 300. We set the number of hidden neurons to 5, 10 and 15, and do training for these three models. The training of each model is repeated for 20 times. Each time the parts of data are divided randomly. As long as the training is finished, we calculate the error of the test set. Then we compare the average of the twenty errors. The results are listed in Table 3.

As shown in the table, we conclude that for most of the time, 10 hidden neurons' model works better than that of 5 or 15. Therefore, we use 10 hidden neurons in our model.

Table 3. Average errors of models using different number of hidden neurons

Product	Bacteria	No. of hidden neurons		
		5	10	15
1	1	0.1654	0.1296	0.2317
1	2	0.0998	0.0868	0.0638
2	1	0.0450	0.0260	0.0289
2	2	0.0885	0.0851	0.1106

4.2 Necessity of Preprocessing Methods

In order to show that the preprocessing methods are important for removing conflicts in training data, we test the performance of network with raw data and preprocessed data and make a comparison.

With raw experimental data in Table 1, we make a similar operation for testing average errors as in the previous section, i.e. training the network for 20 times and calculate the average errors for each group. The only difference is that the structure of network (the number of hidden neurons) is certain (10). The errors are shown in Table 4.

Table 4. Average testing errors with raw data

Detergent	Bacteria1	Bacteria2
Product1	0.2676	0.1199
Product2	0.0741	0.1112

A clearer comparison with the second column (10 hidden neurons) in Table 3 is shown in Table 5.

Table 5. Comparison of errors with raw and preprocessed data

Product	Bacteria	Err. of raw	Err. of preprocessed
1	1	0.2676	0.1296
1	2	0.1199	0.0868
2	1	0.0741	0.0260
2	2	0.1112	0.0851

It's quite clear to see that the preprocessing methods significantly raise the precision of the network.

4.3 Precision of Training and Testing

In order to show the effectiveness of the preprocessing methods and the model, we test the precision of the model. Fig. 2 shows the training results of each product and bacteria.

In each of the four sub-figures in Fig. 2, the x-axis is the normalized performance value from actual experimental data, while the y-axis is the normalized predicted performance value. Therefore, a data point should fall around the line $y = x$ to meet the accuracy between actual data and prediction.

The data points are denoted in three symbols. "+" points are data that are used for training, while "x" and "o" points are for validating and testing, respectively. From the figure, we see there's no significant difference between the performances of three sets.

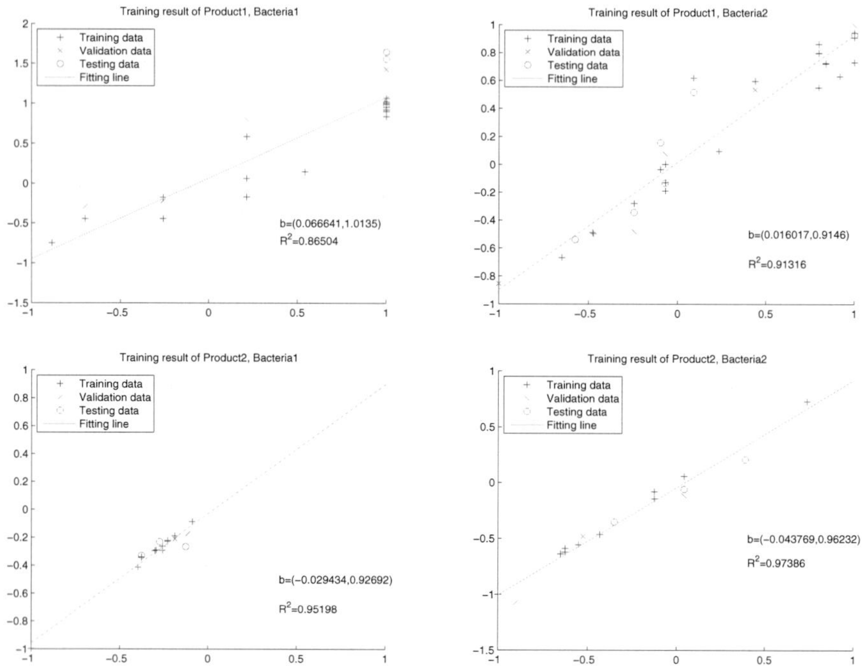

Fig. 2. Training results of products and bacteria

The green line in the figure is the fitting line of data points by linear regression. The figure also shows the coefficients and R^2 statistic value of the regression. In the texts, "$b = (b_0, b_1)$" means the equation of the line is $y = b_0 + b_1 x$. The fitting results show that the slope of the line and R^2 value are all nearly one, which indicates that the prediction of data is reasonable. Notice that the precisions of each sub-figure are not quite the same. This is because each of the experimental data is different, and the quality of those data may affect a great deal on the experiments.

As existing actual data is limited and is not ideal enough, we used all the data to train new models without dividing them into training, validating and testing sets. These new models will be used in predicting new inputs (whose performances are unknown), or in other words, in practical applications. Therefore, the monotonicity performance will also be validated with this "no division" model.

4.4 Monotonicity Rule

As mentioned before, the monotonicity rule is that *the more amount of ingredients a solution has, the better (or at least not worse) efficacy it works against the bacteria.*

To show the monotonicity results, we generate inputs that cover the whole input space, i.e. $[-1,1]^n$ (where [-1,1] is a normalized dimension). Then we predict the performances for all the input data. As there are more than two dimensions in the input (more than two key ingredients), we select two of the ingredients to draw a surface. The variations of other ingredients are drawn separately.

The results for product 1 are shown in Fig. 3. In order to show the performances clearly, the range of performance is limited to [-1, 1]. So points upper than 1 or lower than -1 are set to be 1 or -1.

In each of the surfaces, the cross points of the mesh's border are predicted performances. The colors in the meshes also denote the value of performance, where lower values are darker, higher values are lighter. Consider a single surface in a sub-figure, We see that the performance increases as the amount of Ingredient 1 or 2 grows.

The amount of Ingredient 3 also changes in a sub-figure. However, we didn't generate a dense variation of Ingredient 3 in order not to make the figure too crowded. In each of the sub-figures, we just draw five values of Ingredient 3 (in normalized value): -1, -0.5, 0, 0.5, 1. The values are in ascent order, and the corresponding surfaces are from lower to upper in the figures. This shows that the performance increases as the amount of Ingredient 3 grows.

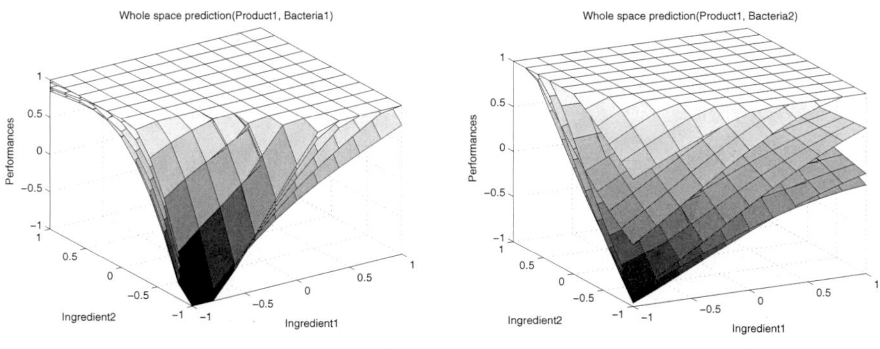

Fig. 3. Monotonicity results for Product 1

For Product 2, since there are five key ingredients, we tested the dimensions separately. Fig. 4 and Fig. 5 show the results of Product 2. The sub-figures from left to right show Ingredient 1 and 2, 2 and 3, 4 and 5 respectively. The other fixed amounts of ingredients are all -1.

Product 2's monotonicity results are not as good as Product 1's. The most important reason is that the number of inputs is five, more than that of Product 1. Thus, the instance space is much larger than Product 1's. To make things worse, the actual experimental data of Product 2 is no more than Product 1's (see Table 1); after the preprocessing step, the effective data is much less (even less than a half, see Table 2). Therefore, the results appear worse. Even so, we see some ingredients fit the monotonicity rule. For example, in Bacteria 1, Ingredient 1 and Ingredient 2's increasing will bring the growth of performance; in

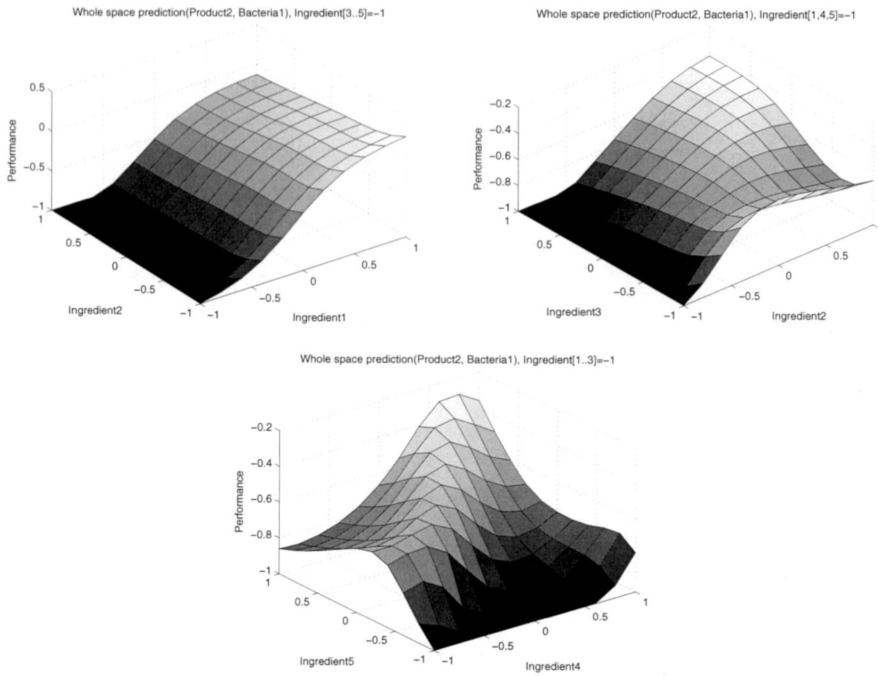

Fig. 4. Monotonicity results for Product 2, Bacteria 1

Bacteria 2, Ingredient 1, 4, and 5's increasing will also cause the performance to increase. The overall trend of ingredients appears reasonable.

4.5 Experts' Evaluation

When we finished creating the model, two scientists were invited to test the model. They generated some input data and tested the performance using the model. Then they evaluated the results according to their biochemical experiences. The evaluation results are shown in Table 6.

In most cases, all prediction results are reasonable. Only for Product 2 Bacteria 2, 12 of the results are not good enough. However, these data's amounts of ingredients are far away from existing experimental data. They keep the monotonicity, but the actual values should be higher. In general, this problem can be solved by adding training data in certain space.

Table 6. Evaluation results (Good results/Total number)

Detergent	Bacteria1	Bacteria2
Product1	89/89	89/89
Product2	62/62	50/62

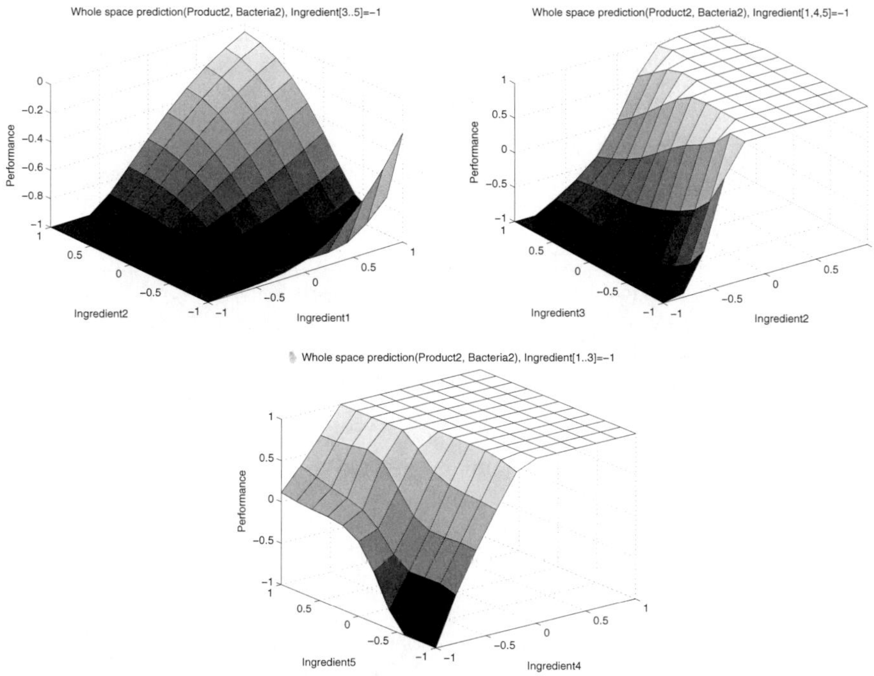

Fig. 5. Monotonicity results for Product 2, Bacteria 2

5 Conclusion

In this paper, we developed a neural network model and preprocessing methods that help predict the anti-germ performance for detergents. This model helps a great deal in industrial production, which reduces much of the time-consuming and costing biochemical experiments. As the number of actual experimental data for training is often limited, The preprocessing methods can reduce data conflicts while keeping the monotonicity.

The model's input is the amount of key ingredients. This kind of input makes the model extensible and flexible for other problems.

The experiments are made upon two detergent products, each of which contains two types of bacteria. As a flexible algorithm, the model (with preprocessing methods) appears reasonable towards the data. Even though the actual data is quite limited (about 10-30 in each group), the prediction results show a high precision and fitting with the monotonicity rule. Therefore, the algorithm can be widely used in practical problems.

Acknowledgments. The work in this paper is supported by National Natural Science Foundation of China (Grant No: 60405011, 60575057, 60875073).

References

1. Azanha, A.B., Faria, J.A.F.: Use of Mathematical Models for Estimating the Shelf-life of Cornflakes in Flexible Packaging. Packaging Technology and Science 18, 171–178 (2005)
2. Siripatrawan, U., Jantawat, P.: A Novel Method for Shelf Life Prediction of a Packaged Moisture Sensitive Snack Using Multilayer Perceptron Neural Network. Expert Systems with Applications 34, 1562–1567 (2008)
3. Buciński, A., Socha, A., Wnuk, M., et al.: Artificial Neural Networks in Prediction of Antifungal Activity of a Series of Pyridine Derivatives Against Candida Albicans. Journal of Microbiological Methods 76, 25–29 (2009)
4. Yamamura, S., Kawada, K., Takehira, R., et al.: Prediction of Aminoglycoside Response Against Methicillin-Resistant Staphylococcus Aureus Infection in Burn Patients by Artificial Neural Network Modeling. Biomedicine & Pharmacotherapy 62, 53–58 (2008)
5. Zou, C., Zhou, L.: QSAR Study of Oxazolidinone Antibacterial Agents Using Artificial Neural Networks. Molecular Simulation 33, 517–530 (2007)
6. Heckerling, P.S., Canaris, G.J., Flach, S.D., et al.: Predictors of Urinary Tract Infection Based on Artificial Neural Networks and Genetic Algorithms. International Journal of Medical Informatics 76, 289–296 (2007)
7. Petrocci, A.M., Clarke, P.: Proposed Test Method for Antimicrobial Laundry Additives. Journal of the Association of Official Analytical Chemists 52, 836–842 (1969)
8. Elman, J.L.: Finding Structure in Time. Cognitive Science 14, 179–211 (1990)
9. Demuth, H., Beale, M., Hagan, M.: Neural Network Toolbox 5 User's Guide. The MathWorks, Inc. (2007)

A Neighborhood Search Method for Link-Based Tag Clustering

Jianwei Cui[1,2], Pei Li[1,2], Hongyan Liu[3], Jun He[1,2], and Xiaoyong Du[1,2]

[1] Key Labs of Data Engineering and Knowledge Engineering, Ministry of Education, China
[2] School of Information, Renmin University of China
100872 Beijing
{cjwruc,lp,hejun,duyong}@ruc.edu.cn
[3] Department of Management Science and Engineering, Tsinghua University,
100084 Beijing
hyliu@tsinghua.edu.cn

Abstract. Recently tagging has been a flexible and important way to share and categorize web resources. However, ambiguity and large quantities of tags restrict its value for resource sharing and navigation. Tag clustering could help alleviate these problems by gathering relevant tags. In this paper, we introduce a link-based method to measure the relevance between tags based on random walk on graphs. We also propose a new clustering method which could address several challenges in tag clustering. The experimental results based on del.icio.us show that our methods achieve good accuracy and acceptable performance on tag clustering.

Keywords: Tag clustering, tag relevance, tags, Web2.0.

1 Introduction

In Web2.0 system, tags are widely used to annotate resources and share contents. Although tag is an import part of Web2.0 systems, there are two problems in tag applications. The first problem is that different users may use different words to tag the same resource.

Table 1 lists the tagging information of a resource about Sakai system. Sakai is a famous open-source e-learning system based on Web2.0. If educators own this resource, they tend to tag it with "education", "classes" etc, based on their background. On the other hand, web technicians tend to use "web2.0", "j2ee" to express their views of the resource. Therefore, if educators share the resource and tag as "classes", technicians search for "j2ee" will not find it. This will affect the resource sharing in such a tagging system.

The second problem is that the large quantity of tags makes the exploration of the tag space and discovery of the resource difficult. Tags could help the traditional search engine because users can discover their interested resources by exploring tag space, but this advantage will be discounted when tags become numerous, because it's hard for users to determine which tag could be useful for them facing so many

R. Huang et al. (Eds.): ADMA 2009, LNAI 5678, pp. 91–103, 2009.

Table 1. Tags of Sakai

Resource	Tags of Educators	Tags of Technician
Sakai system	education, classes, teaching, learning	web2.0, opensource, freeware, j2ee

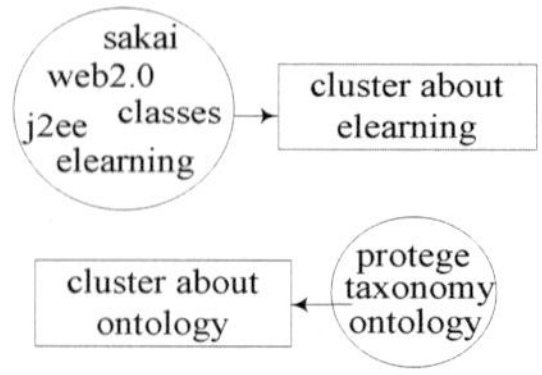

Fig. 1. Tag Cluster

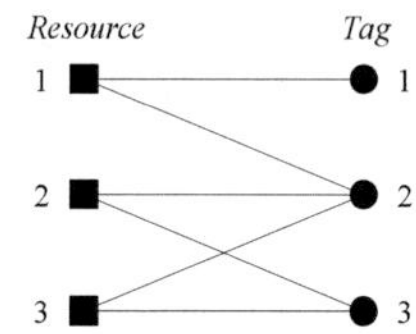

Fig. 2. Tag-Resource Bipartite Graph

tags. Although there have been great visualization technologies, such as *tagclouds*[2], it is also difficult to find resource in such *tagclouds*[3], because tags are not grouped and there is too much information when exploring the *tagclouds.*

Tag clustering could alleviate these problems by grouping relevant tags. Relevant tags are those tags which describe the same topic or tend to annotate same resources. Figure 1 shows the effectiveness of tag clustering. Relevant tags such as "classes", "web2.0", "j2ee" are grouped together, which could help resource sharing because user can find their desired resources through not only their own tags but also a group of relevant tags. Tag clustering could also help users navigate the tag space because users needn't face the whole set of tags. For example, users interested in "elearning" don't need to focus on tags about "ontology".

For tag clustering, the first task is to assess the relevance of tags. Tag relevance could be measured from several aspects. From the content aspect, similarity of tag content combined with some external semantic dictionary could help find the relationship between tags. However this method suffers when new words and concepts not contained in the dictionary are used [1]. From the annotation aspect, the number of co-occurrence is used to measure the tag similarity [3]. Co-occurrence means two tags are used to describe the same resource. However, co-occurrence is not enough to measure tag relevance accurately because two tags may be relevant although they don't describe the same resource. Based on this observation, in this paper, we propose a link-based method to analyze the relevance between tags. As Figure 2 shows, tags and resources are viewed as nodes in a bipartite graph, and edges in the graph represent the annotate information from tags to resources. We not only consider the similarity between tags, but also the similarity between resources. For example, *Tag* 1 and *Tag* 3 don't annotate the same resource, but they are relevant according to our methods because *Resource* 1 (annotated by *Tag* 1) and *Resource* 2 (annotated by *Tag* 3) are related and both annotated by *Tag* 2. We calculate the relevance between tags in the bipartite graph by propagating the relevance between tags and resources with a certain transition probability.

We can do clustering after getting the relevance between tags. There are some challenges to do this. First, because the number of tags is large, the clustering algorithm must have an acceptable time complexity when dealing with thousands of tags. Second, the number of clusters is unknown. Therefore, the algorithm must determine the cluster number automatically. Third, the result of clustering should not depend on data input order, since the meaning of a tag is determined, it should be grouped into a same cluster no matter which input order is used. In this paper we introduce a clustering algorithm to address these challenges. Our algorithm finds a cluster by searching its neighborhood and expands it by adding the most relevant tag to it gradually. Experiments conducted on del.icio.us [12] demonstrate our algorithm has a good accuracy and an acceptable performance in tag clustering.

The rest of this paper is organized as follows. Section 2 presents related works. In section 3, we introduce our method to assess the relevance of tags. Section 4 introduces our clustering method. The experiment results are shown in Section 5. We present the conclusion in Section 6.

2 Related Works

The work of assessing the relevance between tags could be summarized from three aspects. Firstly, from the aspect of tag content, [4] evaluated the similarity of two tags by their contents and information from lexical or terminological resources such as Leo Dictionary, Wordnet, Google and Wikipedia. Secondly, from the similarity of resources, [7] measured tags similarity by comparing the similarity of documents which they annotate. The similarity of documents could be calculated through VSM (Vector Space Model). Thirdly, from the annotation information, [1] viewed the resources as a vector-space, in which a vector can be created for every tag, and each element in the vector represents the frequency with which the tag was used in corresponding resource. Then the cosine similarity is used to measure the tag similarity. [3] mentioned the concept of tag co-occurrence and used the number of co-occurrence between tags to measure their similarity. For the tag-resource bipartite graph respect, relevance between tags is actually the relevance between nodes in the graph. SimRank [8] provided a definition for similarity between nodes in a link graph. The similarity of two nodes is the sum of first meeting possibility at different steps when two random surfers start out from two nodes.

For tag clustering, [1] shows there are many underlying and emergent structures in tagging systems, which could help for tag organization, and there are three main methods for data clustering. The first is partition-based method. [5] introduced *PAM*(Partition Around Mediods) a *k-mediods* algorithm to partition data into k clusters, but the value of k must be given by user. [3] mentioned the limitation of tagging system and explained how tag clustering could help to cure these limitations. They proposed a partitioning-based clustering method, which uses spectral bisection to split data into two clusters recursively, and modularity function to decide whether the partition is acceptable. [1] introduced a divisive method upon a graph, in which tags correspond to nodes and relevance values between tags are edge weights. Then they divide the graph into unconnected sub-graphs by removing the lowest weighted edges. But methods proposed in [1, 3] are not effective enough when dealing with large datasets. The second is hierarchy method. *BIRCH* [13] is a hierarchy clustering algorithm which merges two nearest clusters in every step, but it is also hard to determine at which step the clustering should stop. The third is density method.

DBSCAN [6] is a density-based algorithm which generates clusters from core points to their density-reachable points. It has the $O(n \cdot logn)$ time complexity after doing some preprocessing work of dataset, but the clustering result may change with different data input orders.

3 The Assessment of Tag Relevance

In this section, we introduce our method to assess the relevance between tags. Our method is based on the Random Walk Theory, so in the first part of this section, we briefly introduce the Random Walk Theory and SimRank[8]— which is an algorithm used to assess the similarity between nodes in a directed graph which bases on Random Walk Theory, and then we will introduce our method.

3.1 Random Walk on Graphs

Random walk on graphs is a special case of Markov Chain [9]. According to the random walk theory, given a graph G(V, E), assuming there is a random surfer standing on node a, he has an identical possibility to walk to each neighbor of node a, and no possibility to any other nodes in this graph. If this graph is weighted, the possibility will be adjusted according to the edge weights. Usually, researchers use the term "transition probability" to describe the probability when transferring between nodes, and "transition matrix" for the matrix form of these transition probabilities.

There are lots of approaches and applications based on random walk on graphs. The most commonly mentioned work is PageRank [10], which assumes a random surfer makes random transitions between web pages in a Web Graph. Another well-known algorithm is SimRank [8], which is an iteratively reinforced similarity measuring method in a link graph. The similarity between node a and b is regarded as the first-meeting probability of two random surfers who start from these two nodes respectively. If $a = b$, the similarity $S(a, b)$ is defined to be 1; Otherwise, an iterative process is introduced, as shown in Equation (1).

$$S_k(a,b) = \frac{d}{|I(a)||I(b)|} \cdot \sum_{i=1}^{|I(a)|} \sum_{j=1}^{|I(b)|} S_{k-1}(I_i(a), I_j(b)) \tag{1}$$

where d is a decay constant ($0<d<1$) and $I(a)$ is a set of neighbors of node a, and analogously for $I(b)$. Moreover, Equation (1) can be viewed from the perspective of transition matrix (we use T to represent it) and an equivalent description can be given in Equation(2).

$$S_k = d \cdot T S_{k-1} T^T + M \tag{2}$$

where T^T is the transpose of T and M is a correction matrix making every element on diagonal of S_k to be 1. Equation (2) can be also found in [11].

The similarity obtained by SimRank ranges from 0 to 1. When studying SimRank algorithm from the viewpoint of random walk, we learn that

$$S_k = \sum_{i=1}^{k} P_i \tag{3}$$

where P_i is the first-meeting probability of two random surfers by means of i steps. It means S_k is the summation of all first-meeting probabilities by means of random walk steps no more than k. This perception is very important to the assessment of tag relevance.

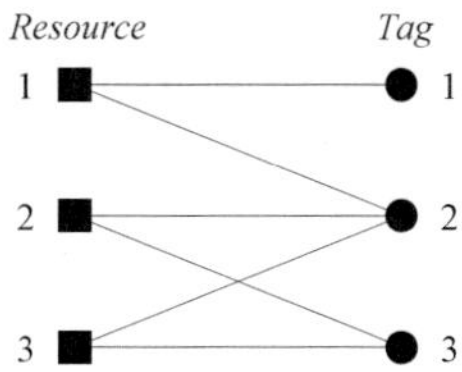

Fig. 3. A Resource-Tag example

R_0 = an identity matrix;
$P_1 = R_1 = d \cdot T_{T\to R} R_0 (T_{T\to R})^T + M$;
while P_k is above a threshold
    using Equation (6) to obtain P_{k+1};
    $k = k+1$;
return R = $\sum P_k$ where k is an odd number.

Fig. 4. The Computation of Tag Relevance

3.2 The Assessment of Tag Relevance Based on Random Walk

In this paper, the relevance degree between two tags a and b, $R(a, b)$ is defined as the probability of co-occurrence. If $a = b$, we define $R(a, b) = 1$ reasonably. This definition well captures the intuition of tag relevance based on link analysis, which means the more possibly these two tags occur in the same annotation, the larger relevance these two tags have. Two kinds of co-occurrences are distinguished here, the direct co-occurrence and indirect co-occurrence. If two tags occur in the same annotation, we call it direct co-occurrence. For example, in Figure 3, *Tag* 1 and *Tag* 2 occur in the same annotation to *Resource* 1. For indirect co-occurrence, two tags are not occurred in the same annotation but in similar annotations. For example, *Tag* 1 and 3 do not occur in the same annotation, but since *Resource* 1, 2 and 3 are related, *Tag* 1 and 3 also indirectly co-occur to some extent. We use iterative computation to assess the indirect co-occurrence probability.

Based on random walk theory, the co-occurrence probability of *Tag a* and *b* can be equivalent to the first-meeting possibility of two random surfers starting from *Tag a* and *b* respectively and meeting on some resource. Note that, since we need to obtain the co-occurrence probability in annotation, only the first-meeting on resources needs to be calculated. Remember SimRank calculates the first-meeting probability on every node, which differs from our definition. In our measurement, it is apparent that two random surfers starting from tags cannot meet on resources by even steps. To be specific, the relevance of *Tag a* and *b* by ($2k$-1) steps can be described in the following equation(4).

$$R_{2k-1} = \sum_{i=1}^{k} P_{2i-1} \tag{4}$$

Compared with SimRank, our relevance measurement only calculates the first-meeting probability by odd steps. Since the relationship between resources and tags can be formalized as a bipartite graph, we use $T_{T\to R}$ to describe the transition probability from tags to resources, and analogously for $T_{R\to T}$. For initial stage, $R(a, b) = 0$ if $a \neq b$, so relevance matrix R_k is an identity matrix when $k = 0$. (Readers may need to refer to Bipartite SimRank in [8] to understand the setting here.)

For direct co-occurrence probability, it is described by P_1 and computed by

$$R_1 = P_1 = d \cdot T_{T\to R} R_0 (T_{T\to R})^T + M \tag{5}$$

where M is a correction matrix making every element on diagonal of R_1 to be 1.

For indirect co-occurrence probability, an iterative computation is performed. Since we have two transition matrices to describe two different transfer directions, given P_k, we have

$$P_{k+1} = \begin{cases} d \cdot T_{T\to R} P_k (T_{T\to R})^T + M' & when\ \ k\ is\ even \\ d \cdot T_{R\to T} P_k (T_{R\to T})^T + M' & when\ \ k\ is\ odd \end{cases} \tag{6}$$

where M' is a correction matrix making every element on diagonal of P_{k+1} to be 0. Because $d<1$, P_k approaches 0 when k increases. So when every element of P_k is below a threshold, the iterative computation will end. The major step of this algorithm to compute tag relevance is shown in Figure 4.

4 Tag Clustering

4.1 The Clustering Method

Before describing our algorithm, we give some definitions. We use $R(a, b)$ to denote the relevance value between two different tags a and b.

Definition 1. If t is a tag and C is a cluster, $R(t, C)$ is the relevance value between t and C which is defined as follows:

$$R(t,C) = \begin{cases} \sum_{a\in C} R(t,a)/n & t \notin C \\ \sum_{\substack{a\in C \\ t\neq a}} R(t,a)/(n-1) & t \in C \end{cases} \tag{7}$$

where n is the size of cluster C. The neighborhood of a tag or a cluster is an important concept for our algorithm. The neighborhood of a tag t is an area in which the tags are relevant enough to t, analogously for a cluster. From this intuition, we give following definitions:

Definition 2. If t is a tag, $Nbh(t, d)$ is a set of tags, in which any tag a satisfies $a \neq t$ and $R(t, a) \geq d$ where d is a relevance threshold.

Definition 3. If C is a cluster, $Nbh(C, d)$ is a set of tags, in which any tag t satisfies $t \notin C$ and $R(t, C) \geq d$ where d is a relevance threshold.

Bases on the definition above, we give our Tag Clustering Algorithm, *TagClus*

```
Algorithm TagClus
Input: a set T of tags and relevance threshold d;
Output: a set of clusters Θ;
Initialization: Θ=Φ, cluster=NIL, scan tags to get Nbh(t, d) for each tag t;
Main Procedure:
 1. while T≠Φ do
 2.     if cluster=NIL then
 3.         (a, b)=arg max_{a,b∈T} R(a,b);
 4.         if R(a, b) ≥d then
 5.             cluster=new cluster({a, b}); remove a, b from T;
 6.         else
 7.             Θ=Θ∪{new clusters(T)}; break;
 8.     else
 9.         if Nbh(cluster, d)≠NIL
10.             t= arg max_{t∈Nbh(cluseter,d)} R(t,cluster); cluster=cluster∪t; update Nbh(cluster, d);
11.         else
12.             Θ=Θ∪ cluster ; cluster =NIL;
13. return Θ;
```

Fig. 5. The Algorithm of Tag Clustering

In algorithm *TagClus*, every cluster in Θ consists of a set of tags. Firstly, we need one scan of tags to initialize *Nbh*(*t*, *d*) for every tag *t*. We start clustering by selecting a pair of most relevant tags(step 3 in Figure 5) to create a *cluster* and initialize *Nbh*(*cluster*, *d*). For the current *cluster*, if *Nbh*(*cluster*, *d*) is not empty, search *Nbh*(*cluster*, *d*) to find a tag *t* which has the maximum relevance value to the cluster, *R*(*t*, *cluster*), then add *t* to *cluster*, update *Nbh*(*cluster*, *d*) and remove *t* from *T*. Otherwise, we find a pair of tags *a* and *b*, which have the maximum relevance in the rest of tags. If $R(a ,b) \geq d$, we remove *a* and *b* from *T* and use *a* and *b* to create a new cluster, or else it means that the relevance value between any tags in *T* is smaller than *d*, so we create a cluster for each tag in *T* and add these clusters to Θ.

Our method depends on a parameter *d*. It is easy to infer that the larger value of *d* will lead a smaller number of clusters and the structure of cluster is looser because tags in cluster tends to be less relevant to each other. We will discuss how to determine the appropriate value of *d* in the next section.

Analyzing algorithm *TagClus*, we can infer that the cluster results don't depend on the input order of tags, because when we add a tag to *cluster*, we always choose the most relevant tag to it, and when we decide to create a new cluster, we always choose a pair of most relevant tags. We can also infer when a new cluster is created, no tag will be added to clusters created before, because we always choose tag in *Nbg*(*cluster*, *d*), and only create a new cluster when *Nbg*(*cluster*, *d*) becomes empty. Because of this, we only need to maintain the current *cluster*. Clusters created before will not change and don't need to store in memory, which will help to save memory and we can output a cluster once we start to create a new cluster.

4.2 The Time Complexity Analysis of *TagClus*

The most time consuming step of *TagClus* is the updating of *Nbh(cluster, d)* (step 10 in Figure 5). All the relevance values between tags in *T* should be updated as a new tag added, this will cause a $O(n^2)$(n is the number of tags) time complexity although we do this incrementally according to Definition 1. To solve this problem, we use some pruning techniques.

When we add a tag *t* to the current cluster *C*, we don't update the relevance values between all the rest tags and *C*. Instead we only update tags in $Nbh(t, d) \cup Nbh(C, d)$. Tags not in $Nbh(t, d) \cup Nbh(C, d)$ are far enough to *t* and *C*, so they have very small possibility to be a member of *Nbh(C, d)* when *t* is added. Figure 6 shows the situation when we add a new tag *t* to the current cluster *C*, where *t*, *r*, *a*, *b*, *x*, *y*, *z* are tags. When adding the most relevant tag *t* to *C*, we only update the relevance values between any tag in {*a*, *b*(in *Nbh(C, d)*), *r*(in *Nbh(t, d)*)} and *C*, and don't care *x* , *y*, *z* at this step. When we update the *Nbh(C, d)*, tags in *Nbh(C, d)* which are far to *t* may be filtered out because their relevance values to *C* may become smaller than *d* as *t* is added, and tags in *Nbh(t, d)* may be added to *Nbh(C, d)* because the insertion of *t* may increase their relevance to *C*. So the size of *Nbh(C, d)* may change all the time, and when *Nbh(C, d)* become empty, we create a new cluster.

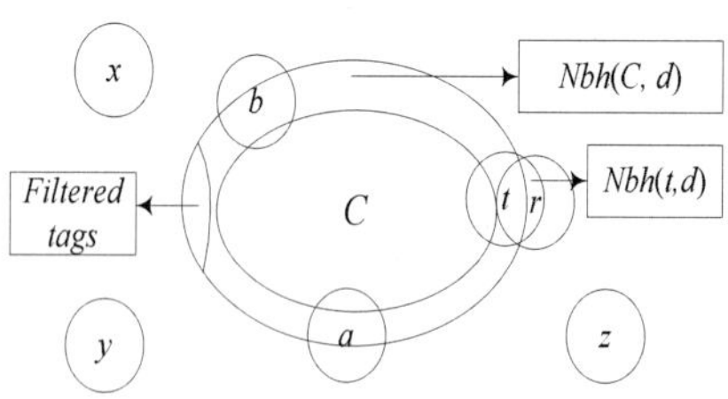

Fig. 6. Neighborhood Search

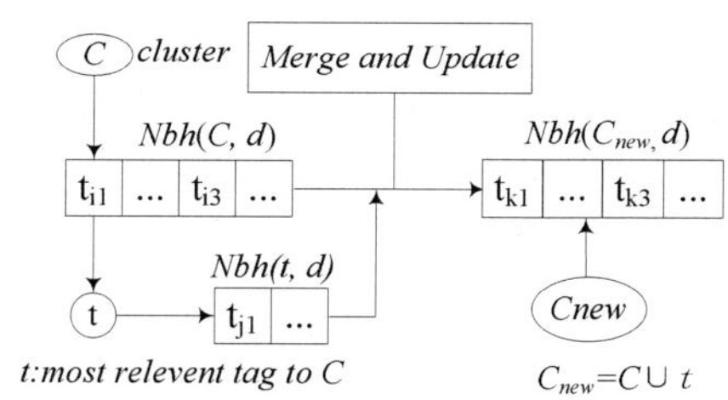

Fig. 7. The Update Process of *Nbh(C, d)*

Figure 7 shows the update process of *Nbh(C, d)*, in which $t_{i1}...t_{in}$ are tags in *Nbh(C, d)*, $t_{j1}...t_{jm}$ are tags in *Nbh(t, d)*, and C_{new} is the new cluster contains *C* and *t*. Tags in *Nbh(C, d)* and *Nbh(t, d)* are stored in the same order. We merge *Nbh(C, d)* and *Nbh(t,d)* to get $Nbh(C_{new}, d)$. At the process of merge, we update the $R(a, C_{new})$ where *a* is in $Nbh(t, d) \cup Nbh(C, d)$, and filter tags which don't satisfy $R(a, C_{new}) \geq d$. The most relevant tag to the C_{new} can also be selected in the merge process.

Based on the analysis above, we define a value *p* as follows:

$$p = (\sum_{i=1}^{k} (\max\ size(Nbh(C_i, d))) \cdot size(C_i)) \Big/ \sum_{i=1}^{k} size(C_i) \tag{8}$$

where size($Nbh(C_i, d)$) is number of tags in $Nbh(C_i, d)$, analogously for $size(C_i)$ for cluster C_i. *p* is the average update times when a new tag *t* is added to current cluster *C*, so we can conclude that *TagClus* has a time complexity $O(p \cdot n)$.

5 Experimental Study

5.1 Clustering Results

Our experiment is conducted on del.icio.us, which is a famous bookmark tagging system. Users in del.icio.us can tag their own or others' shared bookmarks freely. Del.icio.us has an extremely large amount of data. We obtain a subset of it. Firstly, we choose a group of users who used tags "java", "j2ee" and "web", get their recent shared bookmarks and filter bookmarks which were tagged less than 6 times. Then we get tagging information for each of these bookmarks. After that, we perform a cleaning step. We remove tags which only contain stop words according to the stop-words list taken from SMART Retrieval System[15], merge tags which stem leftover words by the Porter Stemmer algorithm[14], and filter tags used less than 11 times. Table 2 shows the statistic of our dataset.

Table 2. Dataset Statistics

Items	Before Cleaning	After Cleaning
Number of Resources	1190	1190
Number of Tags	55931	4710
Number of Tagging	6089449	5985089

Table 3. Clustering Result

clusters	Tags in cluster
size=12	applications.information, java6, java-docu, shortcut:ja, progtoolbar, javase, ns, 1.6, j2se, 6, jdk6, j2se6
size=10	classification,ontologie,protégé knowledge_management,ontologia, taxonomy,ontology,stanford,protégé
size=1	knowledge/opensouce/pojo/design

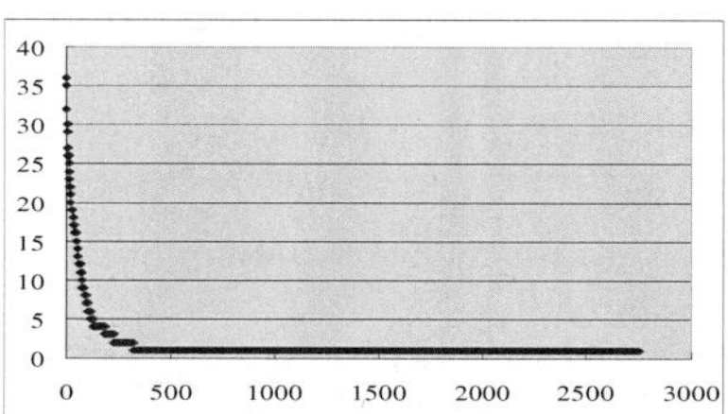

Fig. 8. Cluster Size Distribution

Table 4. Clustering Size Distribution

Item	number
Singleton clusters	2426
Clusters size=2	97
Clusters 2<size<=4	97
Clusters 4<size<=16	99
Clusters Size>16	42
Largest cluster size	35

Table 3 gives some examples of the clustering result. The first row shows a cluster which refers to the topic "java" and the second row is about "ontology". The last row gives some examples of singleton clusters (contain only one tag) in which tags are separated by slash. Figure 8 shows the cluster size distribution of the result. Clusters were sorted according to descent order of their size. The x-axis is the cluster id of each cluster and the y-axis is the size of cluster. We can see the number of clusters increase rapidly as the size of cluster decreases. Table 4 shows the number of clusters in different intervals. The last row shows the largest cluster contains 35 tags. The first row shows singleton clusters contain about half of the total tags of our dataset. By analyzing tags in singleton clusters, we find that these tags tend to describe some general concepts. They are relevant to many tags. However, relevance values between these tags and other clusters are not high enough to absorb them into these clusters.

5.2 The Effectiveness of Tag Clustering

In this section, we introduce two ways to evaluate the effectiveness of the clustering result. Silhouette coefficient [5] (denoted as SC) is a famous coefficient to evaluate the clustering result. [5] shows that clusters whose SC values greater than 0.5 could be viewed as reasonable clusters.

From Table 5, we can see that about 40% of the clusters have SC greater than 0.5 and about 65% of clusters have SC greater than 0.4. Therefore we can draw the conclusion that about 2/3 of clusters generated by algorithm *TagClus* are about reasonable.

Table 5. SC Distribution, d=0.5

SC Interval	Percentage
SC>=0.5	39%
0.45<=SC<0.5	13%
0.40<=SC<0.45	13%
0.35<=SC<0.4	10%
SC<0.35	25%

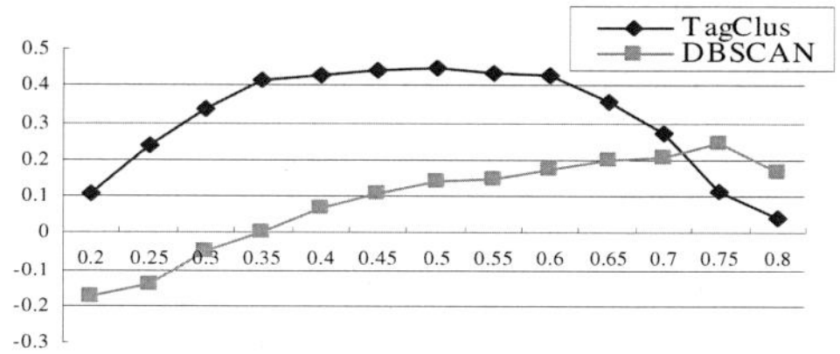

Fig. 9. SC Distribution of *TagClus* and *DBSCAN*

DBSCAN[6] is a density based clustering method which could find any shape clusters. *DBSCAN* could run under a parameter $noiserate \in [0,1]$, and *TagClus* depends on a parameter $d \in [0,1]$, so we compare the average SC value of clusters between *TagClus* and *DBSCAN* at parameter space [0.2,0.8]. The result is illustrated in Figure 9 where the x-axis is the parameter and the y-axis is the SC value. We can see *TagClus* reaches its best effect when *d* is about 0.5 with the SC value is about 0.44, and *DBSCAN* performs best when *noiserate* is about 0.75 with SC value is about 0.24, and at most part of the parameter space, *TagClus* performs better than *DBSCAN* in terms of SC. Figure 9 also shows *TagClus* tend to perform best when $d \in [0.45,0.55]$, and experiment results on different size of datasets also shows the same phenomenon. Therefore, we can use *d*=0.5 for our algorithm at different situations.

We have another way to evaluate our clustering result. Tags in the same cluster should be more relevant than those in different clusters. If we search two tags in the same cluster by search engine, we expect to get more pages. We input two tags *a* and *b* to the search engine, use three coefficients of Equation (9) to evaluate the relevance of two tags. *page*(*a*) is the number of pages contain *a* and *page*(*a*, *b*) is the number of pages contain both *a* and *b*. We choose 100 pairs of tags intra the same cluster, in different clusters and randomly respectively, and use Google to get the average value of three coefficients for each situation. Table 6 show that tags in the same clusters have the highest values for all three coefficients (*d*=0.5). We also use this method to compare *TagClus* (*d*=0.5) with *DBSCAN* (*noiserate* =0.75) for tags in the same clusters. Table 7 shows our method also performs better for all three coefficients.

$$\begin{cases} Avg(a,b) = page(a,b) / avg\{page(a), page(b)\} \\ Max(a,b) = page(a,b) / \max\{page(a), page(b)\} \\ Min(a,b) = page(a,b) / \min\{page(a), page(b)\} \end{cases} \quad (9)$$

Table 6. Clustering Results of *TagClus*

Tags	Avg	Max	Min
Intra cluster	0.00895	0.00597	0.10784
Inter cluster	0.00449	0.00305	0.01807
Random	0.00509	0.00351	0.02787

Table 7. *TagClus* compared with *DBSCAN*

Intra cluster	Avg	Max	Min
TagClus	0.00895	0.00597	0.10784
DBSCAN	0.00630	0.00416	0.08321

5.3 Time Complexity of Clustering

Figure 10 shows the run time of *TagClus* at different size of datasets where the x-axis is data size and the y-axis is run time by millisecond. The runtime curve is higher than $O(n{\cdot}logn)$ but far lower than $O(n^2)$.

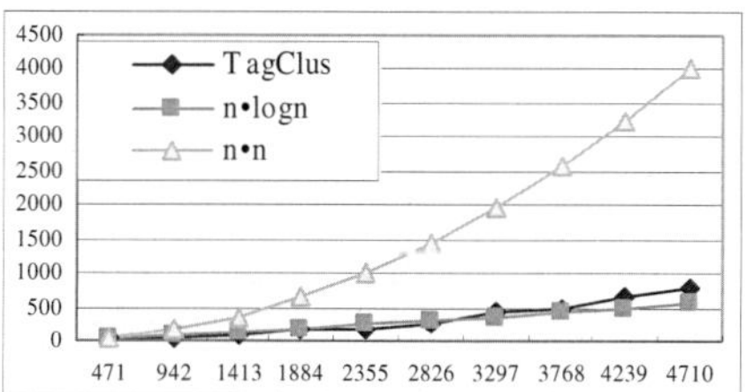

Fig. 10. Run Time Distribution

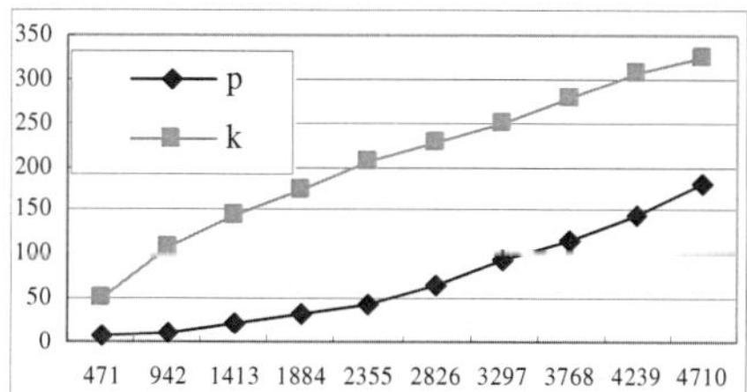

Fig. 11. Distribution of *p* and *k*

In section 4.2, we prove the time complexity of *TagClus* is $O(p{\cdot}n)$. Figure 11 shows the increase of *p* and *k* (not count the singleton clusters) at different size of data. We can see *p* is always less than *k*, so *TagClus* performs better than $O(n{\cdot}k)$.

5.4 The Membership Degree of Tags

$$m(t,C) = \mathrm{R}(t,C) / \sum_{i=1}^{k} R(t,C_i) \tag{10}$$

There are many singleton clusters in the results, where a cluster contains only one tag. However, the tag in a singleton cluster may be also relevant to other clusters, and tags in non-singleton clusters may also have this feature. The membership degree $m(t, C)$ is the possibility of a tag t belong to a cluster C. We use Equation (10) to calculate $m(t, C)$. For a tag, we sort the membership degree descendently and draw its curve. Figure 12 is for a tag in a singleton cluster where the x-axis represents the clusters and the y-axis represents the membership degree. We can see its membership degree is relatively high in some clusters, analogously for Figure 13, which shows a tag in a non-singleton cluster whose size is 9. To determine the number of clusters which are relevant enough to a tag, we use the method of proposed in [3]. We draw the first derivation and second derivation of the membership degree curve. Find the first peak of the first derivation (that is when the second derivation goes from positive to negative) from tail and check the whether peak is relatively high. If two conditions are both satisfied, this point is cut-off (red circles in Figure 13 and Figure 14), clusters before this cut-off are relevant enough to the tag.

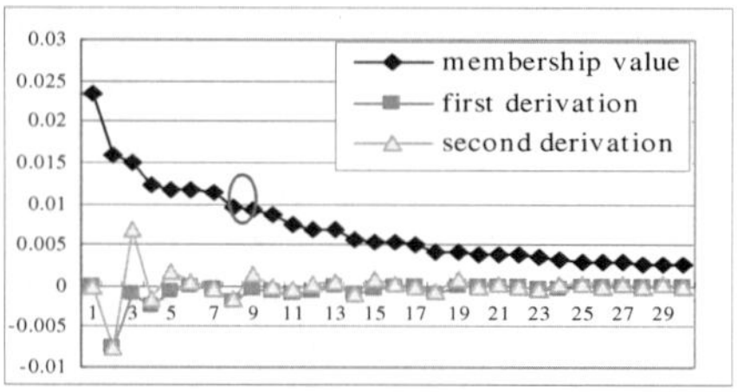

Fig. 12. Membership Degree of a Tag in Singleton Cluster

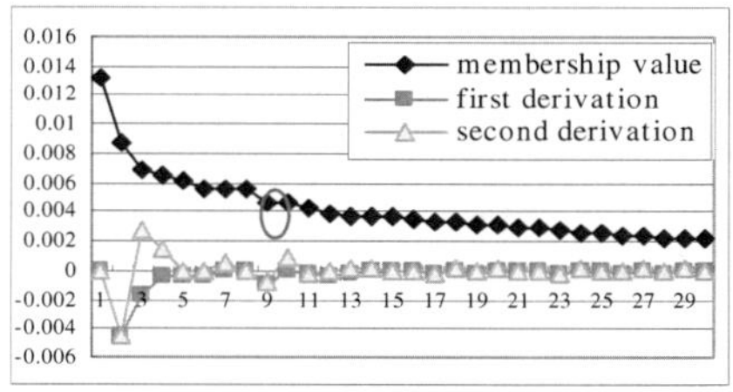

Fig. 13. Membership Degree of a Tag in Non-Singleton Cluster

6 Conclusion

In this paper, we propose to use tag clustering to alleviate the problems in tag application. Tag clustering could help for sharing contents, discovering resources and navigating tag space at Web2.0 system as we analyzed.

We use link-based method to measure the relevance between tags. Our method considers both the relevance between tags and the relevance between resources. We also introduce a new clustering algorithm, *TagClus* which could deal with thousand of tags efficiently, automatically determine the number of clusters and generate the result insensitive to the data input order. The experiments on del.icio.us show that our methods achieve a good result and an acceptable performance in tag clustering.

References

1. Simpson, E.: Clustering Tags in Enterprise and Web Folksonomies. Technical report, HP Labs (2008)
2. Newzingo: Your Map to Google News, `http://www.newzingo.com`
3. Grigory, B., Philipp, K., Frank, S: Automated Tag Clustering: Improving search and exploration in the tag space. WWW (2006)
4. Celine, V.D., Martin, H., Katharina, S.: Folksontology: An integrated approach for turning folksomomies into ontology. SemNet, 57–70 (2007)
5. Leonard, K., Peter, J.R.: Finding Groups in Data: an Introduction to Cluster Analysis. Wiley Interscience, Hoboken (1990)
6. Martin, E., Hans-Peter, K., Jorg, S., Xiaowei, X.: A Density-Based Algorithm for Discovering Clusters in Large Spatial Databases with Noise. In: SIGKDD 1996 (1996)
7. Christopher, H.B., Nancy, M.: Improved Annotation of the Blogopshere via Autotagging and Hierarchical Clustering. WWW (2006)
8. Glen, J., Jennifer, W.: SimRank: A measure of structural-context similarity. In: SIGKDD, pp. 538–543 (2002)
9. Kallenberg, O.: Foundations of Modern Probability. Springer, New York (1997)
10. Page, L., Brin, S., Motwani, R., Winograd, T.: The PageRank citation ranking: Bringing order to the Web. Technical report, Stanford University Database Group (1998)

11. Pei, L., Zhixu, L., Li, H., Jun, H., Xiaoyong, D.: Using Link-Based Content Analysis to Measure Document Similarity Effectively. APWeb/WAIM, 455–467 (2009)
12. Del.icio.us, http://delicious.com
13. Tian, Z., Raghu, R., Miron, L.: BIRCH: An Efficient Data Clustering Method for very Large Databases. In: SIGMOD, pp. 103–114 (1996)
14. Porter, M.: An algorithm for suffix stripping. Program 14(3), 130–137 (1980), http://www.tartarus.org/~martin/PorterStemmer
15. The stop-words list, http://members.unine.ch/jacques.savoy/clef/englishST.txt

Mining the Structure and Evolution of the Airport Network of China over the Past Twenty Years

Zhengbin Dong[1], Wenjie Wu[2,3], Xiujun Ma[1,*], Kunqing Xie[1], and Fengjun Jin[2]

[1] Key Laboratory of Machine Perception, Ministry of Education, Peking University, 100871 Beijing
{dongzhengbin,maxj,kunqing}@cis.pku.edu.cn
[2] Institute of Geographic Sciences and Natural Resources Research, Chinese Academy of Sciences, 100101 Beijing
{wuwj.07s,jinfj}@igsnrr.ac.cn
[3] Graduate School of the Chinese Academy of Sciences, 100101 Beijing

Abstract. In this paper we study the Airport Network of China (ANC), which represents China's domestic civil aviation infrastructure, as a complex network. We mine the structure and evolution of ANC over the past twenty years by using the real aviation data in the year of 1984, 1993 and 2006. The main contributions can be summarized as three-fold: *firstly*, we analyze ANC by using the complex network analysis method and find that ANC is a typical small world network with high clustering coefficient and small diameter; *secondly*, we find that the evolution of ANC over the past twenty years meets the densification law and shrinking/stabilizing diameter law; *lastly*, some interesting patterns of airports in ANC are found by the visual data mining, such as Circle Pattern, Province Capital Pattern and Star Pattern.

Keywords: Airport Network of China, ANC, network structure, network evolution, network diameter, degree distribution, clustering coefficient, betweeness centrality.

1 Introduction

Transportation infrastructures are of crucial importance to the development of a country and are important indicators of its economic growth. They form the backbone of tourism industry, support movement of goods and people across the country, thereby driving the national economy [1]. Roadways, railways and airways are the major means of transport in China, although contribution of airways is small compared to that of the other two. The civil aviation in China has been developed very fast since the Reform and Opening of China in 1980s. There are great changes in the structure of civil aviation in China. Understanding of the civil aviation system and its changes over the past twenty years is important for reasons of policy, administration and efficiency.

* Corresponding author.

R. Huang et al. (Eds.): ADMA 2009, LNAI 5678, pp. 104–115, 2009.

Complex Network Analysis is a novel method for mining the network data. During the past few years, complex network analysis has been used to study many real-life complex systems. Examples include the Internet, the World Wide Web, email networks, peer-to-peer networks, scientific co-authorship networks [2], human sexual network [3], mobile social network [4] and etc. These researches have shown some ubiquitous properties about the real-life social networks: the small-world effect, the power-law and heavy-tails distributions, the scale-free network, the small diameter and etc.

There have been some researches that focus on the transportation systems, including the railways [5] and the airways [1, 6-10]. The World-wide Airport Network (WAN), as the global airways system, has been studied from many different aspects, such as the properties of topological structure, community structure, traffic dynamics and modeling methods. Paper [6] studies the structure properties of WAN and concludes that WAN is a small-world network. The same result is found in paper [7] that WAN is a scale-free small-world network. It is also found that in WAN the most connected cities are not necessarily the most central because of the multi-community structure of WAN. Moreover, the community of WAN has been detected and the results show that community structure cannot be explained solely based on geographical constraints and that geopolitical considerations have to be taken into account. Beyond the topological properties, WAN has been studied [9] as a complex weighted network, where the weight is the traffic flow amount – strength of interactions between the cities. The correlations among weighted quantities and the topological structure of WAN are investigated for the first time. A model with geo-political constraints is proposed [10] to explain the evolution and growth of WAN. Beyond the study of the global airport network, there are also some researches on the regional airport network. For example, the paper [1] studies the Airport Network of India (ANI), which represents India's domestic civil aviation infrastructure. It is found that ANI is a small-world network characterized by a truncated power-law degree distribution and has a signature of hierarchy. The Airport Network of China (ANC), a network much smaller than WAN, is also analyzed [8] for its topology and traffic dynamics. Its topology was found to be having small-world network features and a two-regime power-law degree distribution.

The Airport Network of China (ANC) is a crucial part of WAN and its detail structure and properties may have its own features. The evolution of ANC is also very important and useful for understanding the growth of economic in China. Moreover, the knowledge about the structure and evolution of ANC can be used to modify the current policy and improve the efficiency. So in this paper we mine the structure properties and the evolution of ANC by using the real-life aviation data in 1984, 1993 and 2006. The data is from the Civil Aviation Administration of China and the duration is about twenty years. The main contributions of this paper can be summarized as three-fold: *firstly*, we find that ANC is a typical small world network with small diameter and high clustering coefficient; *secondly*, we find that the evolution process of ANC over the past twenty years meets the densification law and shrinking/stabilizing diameter, i.e., the airlines between the cities growth much fast than the growth of the airport cities and the diameter of ANC shrinks year by year; *lastly*, we investigate some important city in detail and find some very interesting patterns of them in ANC, such as Star Pattern of Urumqi or Kunming.

The paper is organized as follows: in the next section, we describe the dataset and ANC. In section 3, we analyze the global structure and the evolution of ANC over past twenty years. In section 4, we analyze ANC from the city level by the visual mining method and investigate an anomaly in ANC. We finally summarize our work and discuss the future research directions in section 5.

2 Data Modeling for Airport Network of China (ANC)

The Airport Network of China (ANC) comprises domestic airports of China and airlines connecting them. There is traffic flow on each airline. In this paper we use an undirected binary graph to represent ANC without considering the traffic flow. Let an undirected binary graph be $G = \{(V, E) \mid V$ is a set of nodes, E is a set of edges. $E \subseteq V \times V$, an edge $e = (i, j)$ connects two nodes i and j and $i, j \in V$, $e \in E\}$. In ANC, the nodes of the network represent the airports and the edges between the pairs of nodes represent the airlines between the cities.

Table 1. The aviation data of ANC in 1984, 1993 and 2006

Year	Airport Number	Airline Number	Passenger Traffic (10^4)
1984	60	156	391
1993	82	422	3385
2006	91	471	15968

In this paper we choose the aviation data of 1984, 1993 and 2006 to analyze the structure and evolution of ANC. The data (Table 1) is from the Civil Aviation Administration of China.

3 Mining the Global Structure and Evolution of ANC

In this section we will use some important network metrics to analyze the global structure and evolution of ANC over the past twenty years.

3.1 Degree Distribution

The degree of a node v in a network, represented as $d(v)$, is the number of connections or edges the node has to other nodes. Let $N(v) = \{u \mid (v, u) \in E$ and $v, u \in V\}$, which is a set of the neighbor nodes of v in the graph G. so $d(v)$ is the size of set $N(v)$. The degree distribution $p(k)$ of a network is then defined to be the fraction of nodes in the network with degree k. Thus if there are n nodes in total in a network and n_k of them have degree k, we have $p(k) = n_k/n$.

The degree distribution is very important in studying both real networks, such as the Internet and social networks, and the theoretical networks. The simplest model of network is the ER random model [11] introduced by Erdos and Renyi. The degree distribution of the network generated by the ER random model follows Poisson distribution but it is found that many real networks follow the heavy-tail distribution such as power-law [2-3]: $p(k) \sim k^{-r}$, where r is a constant whose value is typically in the range $2<r<3$. The networks that follow power-law degree distribution are called scale-free networks [12].

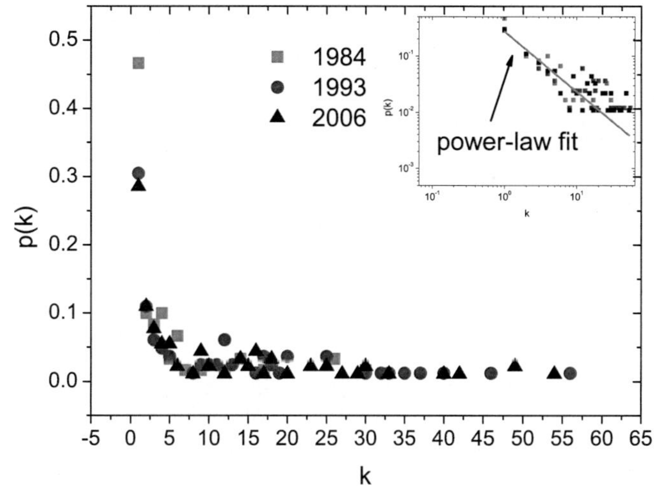

Fig. 1. The degree distribution of ANC in 1984, 1993 and 2006 and the inserted figure is the log-log plot. The green line is the fitting curve by the power-law function.

The degree distribution of ANC in 1984, 1993 and 2006 is plotted in Fig. 1. We can conclude that the degree distribution fits the heavy-tail distribution: most nodes have lower degrees but few nodes, such as Beijing, Shanghai and etc, have very higher degrees. The city with higher degree is called "hub" in ANC, which connects the node with lower degree. The inserted figure is the log-log plot of the degree distribution and from it we can find that the degree distribution does not follow the power-law like other airport networks [7, 8]. Power-law is a line on log-log plot [13] but the curve of ANC in log-log scale is not a line at all (see the green line in figure), so ANC is not a typical scale-free network.

The average degrees of ANC in 1994, 1993 and 2006 are 4.47, 10.1 and 9.95 separately. The growth of average degree indicates the great amount of the new airlines among the airports and the denser of ANC over the past twenty years.

3.2 Shortest Path and Diameter

A path in a network is defined as a sequence of nodes $(n_1, \ldots, n_k)$ such that from each of its nodes there is an edge to the successor node. The path length is the number of edges in its node sequence. A shortest path between two nodes, i and j, is a minimal length path between them. A shortest path between two nodes is referred to as a *geodesic*. The distance between i and j, noted as $d(i, j)$, is the length of its shortest path.

The diameter of the network is the length of the longest shortest path, which is important because it quantifies how far apart the farthest two nodes in the graph are.

In Table 2 we give the average shortest path length and the diameter of ANC in 1984, 1993 and 2006. The length of average shortest path and diameter become shorter and shorter over the past twenty years, which indicates that ANC is becoming very dense. As a result, the efficiency of ANC is improved because there is no need to take several flights from one place to another by air in ANC.

Table 2. The average shortest path length and diameter of ANC in 1984, 1993 and 2006

Year	Airport Number	Average Shortest Path Length	Diameter
1984	60	2.5	5
1993	82	2.15	4
2006	91	2.22	4

3.3 Clustering Coefficient

The clustering coefficient of a vertex in a network quantifies how close the vertex and its neighbors are to being a clique (complete graph). This measure is first introduced by Duncan J. Watts and Steven Strogatz in 1998 [14] to determine whether a network is a small-world network.

The clustering coefficient of node v, noted as C_v, measures the extent of the interconnectivity between the neighbors of node v and is the ratio of the number of edges between the nodes in the direct neighborhood to the number of edges that could possibly exist among them, C_v can be defined as:

$$C_v = \frac{2\left|\bigcup_{i,j\in N(v)} e(i,j)\right|}{d(v)(d(v)-1)} : e(i,j)\in E \tag{1}$$

where $d(v)$ is the degree of node v and $N(v)$ is the set of the neighbor nodes of v, which have been defined in section 3.1.

After the clustering coefficient of a node is defined, then we give the definition of clustering coefficient of a network, which is the average of the clustering coefficients of all nodes in the graph:

$$\overline{C} = \frac{1}{n}\sum_{i=1}^{n} C_i \tag{2}$$

The clustering coefficient distribution $p(C)$ of a network, like the degree disturbing, is defined to be the fraction of nodes in the network with clustering coefficient C. Thus if there are n nodes in total in a network and n_k of them have clustering coefficient C, we have $p(C) = n_k/n$. The average clustering coefficient of degree k, noted as $C(k)$, is defined as the average value of clustering coefficient of nodes with degree k.

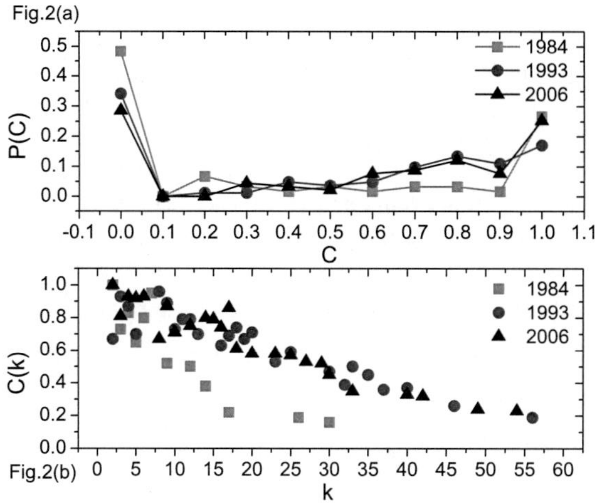

Fig. 2. There are two figures: (a) the clustering coefficient distribution of ANC in 1984, 1993 and 2006; (b) the average clustering coefficient versus degree of ANC in 1984, 1993 and 2006

In Fig. 2(a) the distribution of clustering coefficient of ANC in 1984, 1993 and 2006 is plotted. We can get two conclusions that: 1) most value of clustering coefficient of node is zero (the percent of each years is more than 0.3) because of the great number of degree one in ANC. The clustering coefficient of node with degree one is zero; 2) the distribution does not fit the right-skewed law like other real-life network, such as scientific co-authorship networks [2] and mobile social network [4], which fits the left-skewed distribution. In other networks with right-skewed distribution, most nodes have small clustering coefficient and few nodes have large value, but in ANC the number of nodes which value is grater than 0.9 is relatively higher. The reason for this interesting result is that ANC is very dense as a whole and the airports connect each other very close while other networks are sparse as a whole but very dense in some local region.

The average clustering coefficient versus degree of ANC in 1984, 1993 and 2006 is plotted in Fig. 2(b). The figure indicates that the node with lower degree have higher values of clustering coefficient while the node with higher degree have lower value. The value of $C(k)$ decays from 1.0 to lower than 0.2. The reason for this result is that in ANC the high degree nodes, i.e., the hub airports, connect other low degree nodes. For example, Beijing, the capital of China, connects almost all other cities in China, so its degree is high and clustering coefficient is low.

The network clustering coefficient of ANC in 1984, 1993 and 2006 are 0.38, 0.48 and 0.54 separately, which also indicates like other metrics that ANC is becoming very dense over the past twenty years.

The small-world network is the network with two properties: 1) a small average shortest path length and 2) a large clustering coefficient, which is proposed by Duncan J. Watts and Steven Strogatz in 1998 [14]. In ANC the diameter is very small compared to its size (see section 3.2) and the clustering coefficient is much larger than the random network, so we can conclude that ANC is a typical small-world network like other airport network, such as WAN [6] and ANI [1].

3.4 Betweenness Centrality

Centrality is a core concept for the analysis of social networks, and betweenness is one of the most prominent measures of centrality. It was introduced independently by Anthonisse (1971) [15] and Freeman (1977) [16], and measures the degree to which a vertex is in a position of brokerage by summing up the fractions of shortest paths between other pairs of vertices that pass through it. Betweenness is therefore classified as a measure of mediation in Borgatti and Everett (2006) [17].

The formal definition of betweenness centrality in a network is as below [18]: denote by $\sigma(s,t)$ the number of shortest paths (sometimes referred as geodesics) from s to t and $\sigma(s,t \mid v)$ be the number of shortest number from s to t passing through some vertex v other than s, t. If $s = t$, let $\sigma(s,t) = 1$, and if $v \in \{s,t\}$, let $\sigma(s,t \mid v) = 0$. Then the betweenness $c_B(v)$ of a vertex v can be defined to be:

$$c_B(v) = \sum_{s,t \in V} \frac{\sigma(s,t \mid v)}{\sigma(s,t)} \tag{3}$$

where $\frac{0}{0} = 0$ by convention. The measure is therefore usually interpreted as the degree to which a vertex has control over pair-wise connections between other vertices, based on the assumption that the importance of connections is equally divided among all shortest paths for each pair.

In airport network the betweeness is a indicator of "central" of a vertex and the vertex with high betweenness is very important from the angle of flight change because it is on the position of brokerage between other pairs of vertices.

We use $\overline{c_B}(k)$ to represent the average value of betweenness of the vertices with the same degree k. In order to compare the distribution of $\overline{c_B}(k)$ in ANC in three years, we scale the value by dividing the maximum value in each year and then the scaled values of $\overline{c_B}(k)$ are among 0 and 1. In Fig.3 we plot the distribution of $\overline{c_B}(k)$ in 1984, 1993 and 2006.

We find that the $\overline{c_B}(k)$ distributions of ANC after scaling follow the exponential distributions and the fit curves in Fig.3 are plotted by using the exponential function as below:

$$y = y_0 + Ae^{R_0 x} \tag{4}$$

The exponential distribution of $\overline{c_B}(k)$ means that there is a strong relation between degree and betweenness: the higher the degree the higher value of betweenness. In addition, in ANC the most-connected node (with highest degree) is the most-central node (with highest betweeness), which is different with the result of WAN [7]. In WAN, the most connected cities are not necessarily the most central. The reason for this interesting pattern is that WAN has multi-community structure while ANC does not have this structure. The node connecting different communities will have higher betweeness. However in ANC, all the nodes are connected closely and form only one large group, which can also be proved by the abnormal left-skewed distribution of clustering coefficient in section 3.3. The central of the community of ANC is the capital of China, i.e., Beijing. As a result, it has the highest degree and betweenness.

It is clear that there is an obvious anomalous point (see the green arrow) of $\overline{c_B}(k)$ in 2006, which represents a very important pattern in ANC and we will discuss it in detail in section 4.

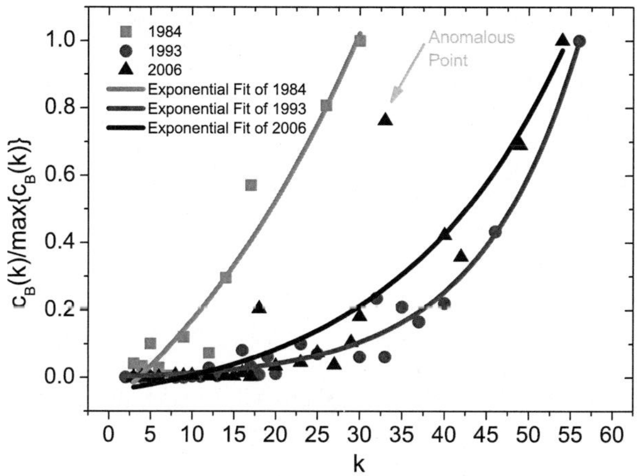

Fig. 3. The $\overline{c_B}(k)$ distribution of ANC in 1984, 1993 and 2006 and the curves are the fit results by using exponential function

3.5 Evolution of ANC

We have analyzed the global structure of ANC in section from 3.1 to 3.4. We also get some knowledge about the evolution of ANC over the past twenty years: the number of airports and the airlines are increasing year by year and the whole network of ANC is becoming denser. But what is the relation between the number of nodes and the number of edges overtime and what is the law of diameter change?

The conventional wisdom or intuition of these two questions is that: 1) constant average degree, i.e., the number of edges grows linearly with the number of nodes; 2) slowly growing diameter, i.e., as the network grows the distances between nodes grow. However the real-life network does not follow these two laws, J. Leskovec (2005) [19] find: that 1) networks are denser over time and the number of edges grows faster than the number of nodes – average degree is increasing, which follow the power-law. This result is called densification power law; 2) the diameter actually exhibits a gradual decrease as the network grows, which is called shrinking/stabilizes diameter law.

In ANC, the evolution law over past twenty years obviously follows the densification power law and shrinking/stabilizes diameter law, which can be proved by the increase of average degree in section 3.1 and the decrease of diameter in section 3.2.

4 City Level Pattern Detecting of ANC

In this section, we will analyze some important cities of ANC and find some interesting patterns in node level by using the visualization of ANC. First, we will visualize ANC of three years and discuss some obvious patterns that can be observed. Then we discuss the anomalous point at section 3.4.

4.1 Visualization of ANC

Visualization is an important tool to analyze the properties of network when the size of network is relative small. In this paper we use Pajek [20] to visualize ANC, which is a very famous and powerful network analysis software.

The visualization of ANC in 1984, 1993 and 2006 is plotted in Fig. 4, where the red nodes represent the provincial capital cities and the blue nodes represent the non-provincial capital cities. The size of the node represents the degree of the node, which is scaled by the maximum value of each year.

There are some very interesting patterns can be observed from the Fig. 4, we discuss three examples.

Circle Pattern. We can see that there is no obvious community structure in ANC, while ANC is connected as a whole and from a circle. The centers of the circle are some major cities, for example, Beijing, Shanghai, Guangzhou and etc. These centers connect very close at the central of circle and other cities connect center cities at the surrounding. The more close to the center the more connections of the city.

Province Capital Pattern. This pattern reflects the differences of some province capital and non-province capital cities. In the year of 1984, Chongqing is not yet a province capital city, but we can see from Fig. 4(a) that it is very close to the center of ANC and its degree is relative high. However, Tianjin and Haikou are on the fringe of ANC in 1984 although they are province capital cities all the time. The similar examples can be found in 1993 (Shenzhen VS Lhasa) and 2006 (Qingdao VS Lhasa), so what the reason for this pattern? This pattern indicates that the development of airport and airline has not much relation with the geopolitical. Some cities that are very important from the view of geopolitical are not very important from the view of network of airport and vice versa. We think that the major factor for the importance of airport is the economy of the city. If the economy of a city is very good, it will have the great demand of traffic flow by air because of the great growth of people and amount of exchange goods, then the city becomes an important airport in ANC. Shenzhen is a good example to illustrate our point. In 1984 Shenzhen is still a small village and there is no airport at all. In 2006, Shenzhen become a very major airport because of the development of economy since the Reform and Opening in 1980s. Although Shenzhen is very close to another major city in geography, Guangzhou, but it is nothing to prevent the development of Shenzhen airport, which indicates that the economy factor is beyond the geography factor.

Star Pattern. This pattern can be observed obviously in 1984 and 2006. We will discuss this pattern in detail in next section.

(a) Year of 1984

(b) Year of 1993

(c) Year of 2006

Fig. 4. The visualization of ANC in 1984, 1993 and 2006

4.2 Anomaly of Betweeness Centrality

In section 3.4 we find an anomalous point of betweenness in 2006, this anomalous point indicates a very important pattern in ANC that will be discussed in this section. We analyze the value of betweenness in 2006 and find that the anomalous point is Kunming. The degree of Kunming in 2006 is 33 and it is not very high compared to the maximum value of 54 of Beijing, but the betwwenness value is very high in 2006. As a result, it produces an obvious anomalous point in Fig. 3 in section 3.4.

What is the pattern of Kunming in 2006 with mediate degree but high betweenness? We can see it from the Fig. 4(c) that Kunming connected many nodes with one degree, such as Zhaotong, Lincang, Simao, Baoshan and etc. These nodes with one degree connect to other nodes in ANC through Kunming and so the betweeness of Kunming is much higher according to the definition of betweenness. We call this pattern Star Pattern. There is another obvious Star Pattern in 2006, i.e., Urumqi. The reason for Urumqi is not an anomalous point is that the average value of betweenness with the same degree hides the high value of Urumqi. The Star Pattern can also be found in other years, for example, Xi'an and Guangzhou in 1984 (see Fig. 4(a)), Beijing in 1993 (see Fig. 4(b)).

The next question is that what is the mechanism of forming the Star Pattern? We think there are two factors to form this pattern. One is the development of economy and another is the geography of some cites. Let us take Urumqi as an example. In 1993 there is no Star Pattern for Urumqi but there is in 2006. We can observe that in 2006, Urumqi connects some small cities, such as Kurle, Kashi and Aksu, which are not in 1993. That is the result of the development of economy of these small cities, i.e., they need to connect to other airports in ANC for the economy reason. But why they connect Urumqi but not Beijing or other cities? That is the result of geography. Urumqi is the capital of Xinjiang and the nearest major city for these small cities. So the Star Pattern is the result of development of economy and the limit of geography. The Star Pattern make the major city like Urumqi be the center of the regional area.

5 Conclusion

In this paper we mine the structure and evolution of Airport Network of China (ANC) over the past twenty years by using the complex network analysis method. We find that ANC is a typical small-world network with high clustering coefficient and small diameter although the degree distribution does not follow power-law like other airport networks. And the evolution of ANC follows the densification law and shrinking/stabilizing diameter law. In addition, we find some very interesting patterns in ANC with the help of visual mining, such as the Star Pattern, Circle Pattern and etc.

In the future we will continue to study the properties of ANC from the view of complex network and pay more attentions to the weight analysis of ANC.

Acknowledgments

The author is grateful to Prof. Fengjun Jing and Master Wenjie Wu of Institute of Geographic Sciences and Natural Resources Research, CAS, for help in obtaining the aviation data. The author is thankful to Dr. Xiujun Ma and Master Wenjie Wu for some useful discussions.

This work is supported by the National Natural Science Foundation of China (NSFC) under Grant No.40801046 & 60874082 and the Key Project of National Natural Science Foundation of China (NSFC) under Grant No.40635026.

References

1. Bagler, G.: Analysis of the Airport Network of India as a complex weighted network. arXiv:cond-mat/0409773 (2004)
2. Newman, M.E.J.: Who is the best connected scientist? A study of scientific co-authorship networks. Phys., Rev. E64, 06131–06132 (2001)
3. Fredrik, L., Christofer, R.E., Luis, A.N.A., et al.: The web of human sexual contacts. Nature 411, 907–908 (2001)
4. Dong, Z.B., Song, G.J., Xie, K.Q., Wang, J.Y.: An experimental study of large-scale mobile social network. In: Proc. of WWW 2009, Madrid, Spain, April 20-24 (2009)
5. Sen, P., Dasupta, S., Chatterjee, A., et al.: Small-world properties of the Indian railway network. Phys., Rev. E67, 036106 (2003)
6. Amaral, L.A.N., Scala, A., Barthelemy, M., Stanley, H.E.: Classes of small-world networks. Proc. Natl. Acad. Sci (USA) 97, 11149 (2000)
7. Guimera, R., Mossa, S., Turtschi, A., Amaral, L.A.N.: The worldwide air transportation network: anomalous centrality, community structure, and cities' global roles. Proc. Natl. Acad. Sci. (USA) 102, 7794 (2005)
8. Li, W., Cai, X.: Statistical analysis of airport network of China. Phys. Rev. E69, 046106 (2004)
9. Barrat, A., Barthelemy, M., Pastor-Satorras, R., et al.: The architecture of complex weighted networks. Proc. Natl. Acad. Sci. (USA) 101, 3747 (2004)
10. Guimera, R., Amaral, L.A.N.: Modeling the world-wide airport network. Eur. Phys. J.B 38, 381 (2004)
11. Erdos, P., Renyi, A.: On random graphs. Publ. Math. (Debrecen) 6, 290 (1959)
12. Albert, R., Barabasi, A.L.: Statistical mechanics of complex networks. Rev. Mod. Phys. 74, 47–97 (2002)
13. Clauset, A., Shalizi, C.R., Newman, M.E.J.: Power-law distributions in empirical data. arXiv:076.1062v1 (2007)
14. Watts, D.J., Steven, S.: Collective dynamics of 'small-world' networks. Nature 393, 440–442 (1998)
15. Anthonisse, J.M.: The rush in a directed graph. Tech. Rep. BN 9/71 (1971)
16. Freeman, L.C.: A set of measures of centrality based upon betweenness. Sociometry 40, 35–41 (1977)
17. Borgatti, S.P., Everett, M.G.: A graph-theoretic perspective on centrality. Social Networks 28, 466–484 (2006)
18. Brandes, U.: On variants of shortest-path betweenness centrality and their generic computation. Social Networks 30(2), 36–45 (2008)
19. Leskovec, J., Kleinber, J., Faloutsos, C.: Graphs over time: densification laws, shrinking diameters and possible explanaionts. In: ACM KDD (2005)
20. Batagelj, V., Mavar, A.: Pajek: Program for large network analysis. Connections (1998)

Mining Class Contrast Functions by Gene Expression Programming*

Lei Duan, Changjie Tang, Liang Tang, Tianqing Zhang, and Jie Zuo

School of Computer Science, Sichuan University,
610065 Chengdu, Sichuan
{leiduan,cjtang}@scu.edu.cn

Abstract. Finding functions whose accuracies change significantly between two classes is an interesting work. In this paper, this kind of functions is defined as class contrast functions. As Gene Expression Programming (GEP) can discover essential relations from data and express them mathematically, it is desirable to apply GEP to mining such class contrast functions from data. The main contributions of this paper include: (1) proposing a new data mining task – class contrast function mining, (2) designing a GEP based method to find class contrast functions, (3) presenting several strategies for finding multiple class contrast functions in data, (4) giving an extensive performance study on both synthetic and real world datasets. The experimental results show that the proposed methods are effective. Several class contrast functions are discovered from the real world datasets. Some potential works on class contrast function mining are discussed based on the experimental results.

Keywords: Gene Expression Programming, Contrast Mining, Data Mining.

1 Introduction

Discovering mathematic models that can precisely describe the underlying relationships and be easily understood from observed data is helpful for scientists to know better on the unknown. Given a set containing several class samples, it is interesting to find some models or relationships that exist in one class while not in others. For example, the height of an ordinary person equals to his/her arm span length. However, for most basketball players and swimmers, their arm span lengths are greater than their heights. This kind of relationships differs from the regression models which do not take class information into consideration.

In this paper, we propose a new data mining task called class contrast function mining. The class contrast function has following characteristics:

- Each variable in class contrast function is an attribute in the sample set.
- A class contrast function has high accuracy in one class but not in other classes.

* This work was supported by the National Natural Science Foundation of China under grant No. 60773169, and the 11th Five Years Key Programs for Sci. &Tech. Development of China under grant No. 2006BAI05A01.

R. Huang et al. (Eds.): ADMA 2009, LNAI 5678, pp. 116–127, 2009.

Definition 1 (Contrast Ratio). Given a dataset *D* that contains two class samples (c_1 and c_2). Let *f* be a function. Suppose $Err(f, c_1)$ and $Err(f, c_2)$ are average errors of *f* in c_1 and c_2, respectively. If $Err(f, c_1) \leq Err(f, c_2)$, then the contrast ratio of *f*, $CR(f)$:

$$CR(f) = Err(f, c_1) \,/\, Err(f, c_2)$$

And $Err(f, c_1)$ is called tiny error, $Err(f, c_2)$ is called coarse error.

Definition 2 (Class Contrast Function). Let $CR(f)$ be the contrast ratio of *f* on dataset *D*. Given an user defined parameter ε, $0 < \varepsilon < 1.0$. If $CR(f) < \varepsilon$, then *f* is a class contrast function on *D*.

Example 1. Given a sample set with two classes (c_1 and c_2), the samples are list as follows. Given a function $f(a_1, a_2, a_3, a_4)$: $a_1 + a_2 + a_3 - a_4 = 0$, then the average absolute errors of *f* in c_1 and c_2 are 0.3333 and 4.0 respectively. Let $CR(f)$ be 0.25. Then *f* is a class contrast function of c_1.

Table 1. A sample set with two classes (c_1 and c_2)

	a_1	a_2	a_3	a_4	a_5	Class
s_1	2	4	4	9	5	c_1
s_2	1	1	2	4	9	c_1
s_3	4	3	1	8	3	c_1
s_4	3	4	3	8	1	c_2
s_5	4	5	2	6	8	c_2
s_6	6	2	4	7	9	c_2

From Definition 1 and Definition 2, we can see that a good class contrast function should have small contrast ratio. In other words, the tiny error should be small and the coarse error should be large. However, small contrast ratio may not mean small tiny error. So, in our work we define a contrast threshold *t* that the tiny error of the class contrast function discovered must be less than it. In other words, if the tiny error of a discovered class contrast function is greater than contrast threshold, this function may be meaningless (big error in each class).

Intuitively, the concept of class contrast function is similar to the concept of emerging patterns (EPs) which was proposed by Dong and Li [1]. Due to wide applications of EPs, many high performance algorithms on discovering EPs have been proposed [2-8]. However, class contrast function is not as the same as EP, since EPs are set of items whose support changes significantly between the two classes. Algorithms of emerging patterns discovery cannot be applied to datasets with numeric attributes directly, unless some discretization method is adopt. However, information may be lost in the process of converting numeric values to items.

To the best of our knowledge, there is no previous work on finding function relationships whose accuracies are different between classes have been done. Generally, finding the class contrast functions has following challenges.

- How to determine the form of class contrast functions to be discovered. In the real world applications, there is no priori knowledge on function form, variables, and parameters of the function. Even we are not sure whether there exists class contrast function in data or not.

- How to find class contrast functions from high dimensional dataset?
- How to find all class contrast functions those exist in the given dataset?

Traditional regression methods can be used to discover functions from dataset. However, such methods need user to define some hypothesis. Gene Expression Programming (GEP) [9, 10], which is the newest development of Genetic Algorithms (GA) and Genetic Programming (GP), has strong calculation power due to the special individual structure. GEP can evolve functions with little priori knowledge. And it is unnecessary to define the function form in GEP. GEP can select function variables automatically from all given attributes. GEP has been widely used in data mining [11-16]. Section 2 introduces the preliminary knowledge of GEP.

The main contributions of this work include: (1) proposing a new data mining task – class contrast function mining, (2) designing a GEP based method to find class contrast functions, (3) presenting several strategies for finding multiple class contrast functions in datasets, (4) giving an extensive performance study on proposed methods and discussing some potential work on class contrast functions.

The rest of this paper is organized as follows. Section 2 introduces related works. Section 3 presents the main ideas used by our algorithms and the algorithms. Section 4 reports an experimental evaluation of the algorithms. Section 5 discusses future works, and concluding remarks.

2 Related Work

2.1 Emerging Patterns

Emerging Patterns (EPs) are contrast items between two classes of data whose support changes significantly between the two classes [1]. Specially, pattern which just occurs in some samples of one class is called jumping EP [1]. Since the first EP mining algorithm was proposed in [1], several methods had been designed, including: Constraint-based approach [3], Tree-based approach [4, 5], projection based algorithm [6], ZBDD based method [7], and Equivalence Classes based method [8]. The complexity of finding emerging patterns is MAX SNP-hard [17].

EPs can be found in many real world datasets. Since EPs have high discrimination power [17], many EP based classification methods have been proposed, such as CAEP [18], DeEPs [19, 20], Jumping EP based method [21]. By using EPs, the discriminating power of low support EPs, together with high support ones and multi-feature conditions are taken into consideration when building a classifier. The research results show that EP based classification methods often out perform state of the art classifiers, including C4.5 and SVM [17].

2.2 The Basic Concepts and Terminology Definitions of GEP

The basic steps of using GEP to seek the optimal solution are the same as those of GA and GP [9]. The main players in GEP are: the chromosome and the expression tree. The chromosome is a linear, symbolic string of fixed length, while the expression tree contains the genetic information of the chromosome. A chromosome consists of one or more genes. Each gene is divided into a head and a tail. The head contains symbols that represent functions or terminals, whereas the tail contains only terminals.

In GEP, the length of a gene and the number of genes composed in a chromosome are fixed. However, each gene can code for an expression tree of different sizes and shapes. The valid part of GEP genes can be got by parsing the expression tree from left to right and from top to bottom. Since the structural organization of GEP genes is flexible, any modification made in the chromosome can generate a valid expression tree. So all programs evolved by GEP are syntactically correct.

Based on the natural selection principle, GEP operates iteratively evolving a population of chromosomes, encoding candidate solutions, through genetic operators, such as selection, crossover, and mutation, to find an optimum solution. The details of GEP implementation can be referred in [6]. Other than C. Ferreira's researches [9-12], GEP has been widely used in data mining research fields, such as, symbolic regression [13], classification [14, 15], and time series analysis [16].

3 Class Contrast Function Mining

3.1 Fitness Function Design in GEP

The GEP algorithm begins with the random generation of a set of chromosomes, which is called the initial population. Then the fitness of each individual is evaluated according to fitness function. The individuals are then selected according to fitness to reproduce with modification, leaving progeny with new characteristics. The individuals of this new generation are subjected to the same evolution process: expression of the genomes, confrontation of the selection environment, and reproduction with modification. This procedure is repeated until a satisfactory solution is found, or a predetermined number of generations is reached. Then evolution stops and the best-so-far solution is returned [9, 10].

The fitness function in GEP determines the evolution direction of candidate solutions. As stated before, we prefer to find class contrast function that not only has small contrast ratio but also small tiny error. Given a GEP individual *g*, the fitness of g is calculated as follows.

$$fitness(g) = \begin{cases} 1/sErr \times t \times 0.5 & sErr > t \\ (1 - sErr/cErr) \times 0.5 + 0.5 & sErr \le t \end{cases} \tag{1}$$

where *sErr* is the tiny error, *cErr* is the coarse error and *t* is contrast threshold.

There are two phrases of evaluating fitness of a GEP individual *g*,

- The tiny error is greater than the predefined value. In this phrase, the fitness is evaluated by the tiny error. The fitness range is (0, 0.5].
- The tiny error is not greater than the predefined value. In this phrase, the fitness is evaluated by both tiny error and coarse error. The fitness range is [0.5, 1.0].

Based on Equation (1), the GEP individual whose tiny error is small and contrast ratio is small will get high fitness value. Since individuals with higher fitness values will get larger opportunities to survive and evolve, using Equation (1) in GEP can generate desirable results. In our work, we are interested in finding functions that has high accuracy in one class but not in other classes. This is the main difference between our work and other function discovery works. Algorithm 1 describes the pseudo code of evaluating GEP individuals to evolve class contrast functions.

Algorithm 1: GEP_Evaluate(Pop, D1, D2, t)
Input: (1) A set of evolving GEP individuals: Pop; (2) A set of samples that belong to class *c*: D1; (3) A set of samples that do not belong to class *c*: D2; (4) A user defined threshold: t.
Output: the individual with the highest fitness: bestIndividual.

```
begin
    1. bestFit ← 0
    2. bestIndividual ← NULL
    3. For each individual ind in Pop
    4.   Err1 ← Min(getAvgErr(ind, D1), getAvgErr(ind, D2))
    5.   Err2 ← Max(getAvgErr(ind, D1), getAvgErr(ind, D2))
    6.     if Err1 > t
    7.        Fit ← 1/Err1 * t * 0.5
    8.     else
    9.        Fit ← (1 - Err1/Err2) * 0.5 + 0.5
   10.     if bestFit < Fit
   11.        bestFit ← Fit
   12.        bestIndividual ← ind
   13. return bestIndividual
end.
```

Algorithm 1 states the process of the evaluation in GEP. In Step 4 and 5, Function getAvgErr() returns the average errors of current individual *ind* in dataset D1 and D2, respectively. From Step 6 to 9, fitness is evaluated according to Equation (1). The time complexity of Algorithm 1 is $O(m*n)$, where m is the number of GEP individuals in population, and n is the number of samples in D1 and D2.

There are two stop conditions for GEP evolving class contrast function:

- A function whose contrast ratio is greater than the predefined value is found.
- The number of generations equals to the predetermined value.

If no class contrast function is found by GEP, we have two choices: increasing the predefined generation number or restarting GEP once more. If no satisfactory function is found after several independent GEP runs, we stop the searching and conclude there may be no class contrast function exists in the given dataset.

3.2 Finding Class Contrast Function in High Dimensional Dataset

In Subsection 3.1, we describe the basic steps of applying GEP to finding class contrast functions. In a naïve way, we take all attribute values of samples as GEP terminals. However, this naïve way may be inefficient when the dataset is high dimensional. The reason lies that in GEP when the number of terminals is large, the evolution efficiency is low. As a result, it is undesirable to apply GEP to evolving class contrast functions in high dimensional dataset directly.

As a more challenging problem in this work, we investigate how to select a part of original attributes of dataset as GEP terminals. Naturally, the selected attributes should contribute to generating good class contrast functions. Intuitively, attributes whose values differ greatly in different classes are preferable. Since class distribution information is available for each attribute in class contrast function mining, we adopt information gain based method to select GEP terminals from dataset. In addition, we take the correlation information of selected attributes into consideration.

The details of calculating information gain are presented in [22, 23]. For each attribute in the dataset, we calculate its information gain. And we sort all attributes in information gain descending order. Suppose the size of GEP terminal set is k. The simple way is fetching attributes with top k information gains. However, some correlations may exist in attributes. So, we select attributes that have high information gains and low correlations among them. Let v_1 and v_2 be two attribute vectors. The correlation between them, denoted as $Cor(v_1, v_2)$, is calculated in following way.

$$Cor(v_1, v_2) = v_1 \bullet v_2 / (\|v_1\| \, \|v_2\|) \tag{2}$$

Suppose V is the set of selected attribute vectors. For each unselected attribute vector v, we calculate its correlation with each selected attribute. And take the maximum as the correlation between v and V. The correlation score between v and V, denoted as $CorScore(v, V)$, is defined as follows.

$$CorScore(v, V) = 1 - \mathrm{Max}\{Cor(v, v') \mid v' \in V\} \tag{3}$$

In Equation (3), we take the maximal value, since want to select the attribute that has the minimal correlation value with selected attributes. Suppose $Gain(v)$ is the information gain of attribute v, we define the contrast score, denoted as $ConScore(v)$, as follows.

$$ConScore(v) = Gain(v) * CorScore(v, V) \tag{4}$$

Given a high dimensional dataset, suppose we want to select k attributes as GEP terminals. First, we select the attribute with the highest information gain. Then we selected another $k - 1$ attributes based on Equation (4) one by one.

3.3 Strategies of Searching Multiple Class Contrast Functions

The next problem is how to discover multiple class contrast functions which may exist in the dataset. It is a challenging work since we have no idea about the form of the next class contrast function. Even we are not sure whether another class contrast function exists. Moreover, if we apply GEP to the dataset again, the same class contrast function which has been discovered may be found again. So, we should avoid this situation. We design three strategies for searching multiple class contrast functions in three different cases.

- Keep the dataset unchanged, and record each discovered class contrast function. If the current GEP individual is a class contrast function discovered already, its fitness is assigned as the minimum value.
- For each discovered class contrast function, attributes in it are removed from the original dataset before finding the next class contrast function.
- For each discovered class contrast function, the attribute in the function and has the highest information gain is removed from the original dataset before finding the next class contrast function.

The first strategy takes all attributes as GEP terminals each time, so it is suitable for the case that the number of dimensions of the dataset is small. The second strategy takes different attributes as GEP terminals each time, so it is suitable for the case that the number of dimensions of the dataset is large. And the third strategy is suitable for the case the number of dimensions of the dataset is neither large nor small.

4 Performance Evaluation

To evaluate the performance of our method for mining class contrast functions from data, we implement all proposed algorithms and GEP algorithm in Java. The experiments are performed on an Intel Pentium Dual 1.80 GHz (2 Cores) PC with 2G memory running Windows XP operating system.

Table 1 lists the parameters of GEP in our experiments. Moreover, to improve the function discovery ability of GEP, the constant set is {0.1, 0.2, 0.3, 0.4, 0.5, 0.6, 0.7, 0.8, 0.9, 2, 4, 5, 6, 7, 8, 9} in our experiments. Please refer to [10] for the detail usages of these parameters.

Table 2. Default parameters for GEP algorithm

parameter	value	parameter	value
Population size	500	One-point recombination rate	0.4
Number of generations	200	Two-point recombination rate	0.2
Linking function	addition	Gene recombination rate	0.1
Function set	{+, −, *, / }	IS transposition rate	0.1
Number of genes	3	IS elements length	1, 2, 3
Gene head size	7	RIS transposition rate	0.1
Selection operator	tournament	RIS elements length	1, 2, 3
Mutation rate	0.3	Gene transposition rate	0.1

4.1 Experiments on Synthetic Datasets

First, we generate some synthetic datasets that contain linear relation or non-linear relation to validate the effectiveness of the algorithms for mining class contrast function. For each synthetic dataset, there are 100 samples belong to c_1 class, and 100 samples belong to c_2 class. All values in each synthetic dataset are generated randomly under an even distribution, and the value range is [-200, 200].

Let $D_1 = \{X_1, X_2, X_3, X_4, X_5\}$ be a synthetic dataset. Suppose all samples of c_1 in D_1 satisfy linear relation f_1: $2X_1 + X_2 - X_3 = 0$. That is, f_1 is a class contrast function of c_1 in D_1. Let $D_2 = \{X_1, X_2, X_3, X_4, X_5\}$ be a synthetic dataset. Suppose all samples of c_1 in D_2 satisfy non-linear relation f_2: $X_1 * X_2 + X_3 - X_5 = 0$. That is, f_2 is a class contrast function of c_1 in D_2.

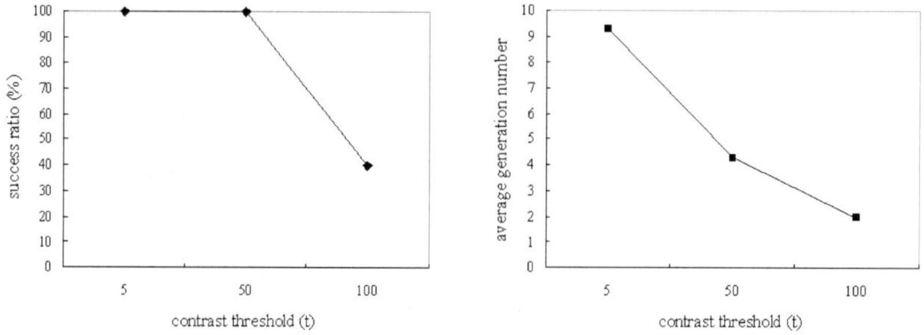

Fig. 1. The success ratios and average generation numbers of finding f_1 in D_1

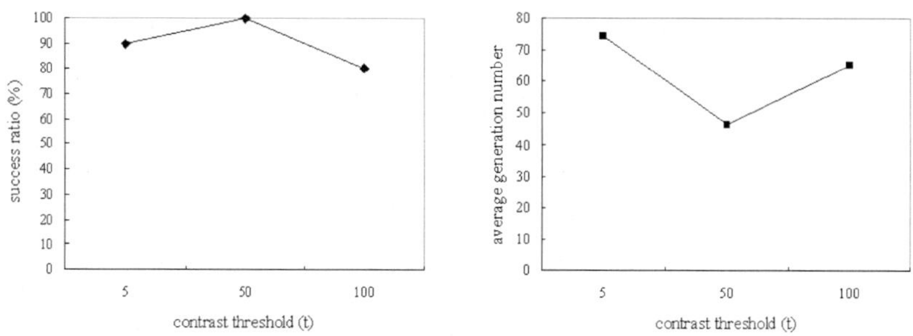

Fig. 2. The success ratios and average generation numbers of finding f_2 in D_2

We apply our method to discovering f_1 and f_2 from D_1 and D_2 respectively. Three different contrast thresholds (5, 50 and 100) are set in the experiment. For each contrast threshold, we run the method 20 times independently, and record the GEP generation number when the predefined function is discovered. The success ratio is the percent of the times of finding the predefined function compared to the total running times. Figure 1 and 2 respectively illustrates the success ratios and average generation numbers of finding f_1, f_2 in D_1 and D_2 under different contrast thresholds.

From Figure 1 and Figure 2, we can see that when the contrast threshold is 50, our method can find the predefined class contrast functions (f_1 and f_2) efficiently. The success ratios are 100% in both D_1 and D_2. The average generation number in each case is 4.3 and 46.2, respectively. Moreover, we can see that when the contrast threshold is 5, the average generation number is larger compared with when the contrast threshold equals to 50. The reason lies that if the contrast threshold is small, the first phrase of fitness evaluation ($tErr > t$) will be tough for GEP individuals, most potential good individuals may be eliminated. So the evolution process is decreased. When the contrast threshold is 100, the success ratio is smaller compared with when the contrast threshold equals to 50. The reason lies that if the contrast threshold is large, the first phrase of fitness evaluation ($tErr > t$) cannot filter some bad GEP individuals out. Individuals which have big errors in either class but small contrast ratio may be selected as the best result. As a result, the success ratio is decreased.

4.2 Experiments on Real World Datasets

Next, we apply our method to some microarray datasets, which are downloaded from Kent Ridge Bio-medical Dataset (http://datam.i2r.a-star.edu.sg/datasets/krbd). The characteristics of microarray dataset include: high dimension, numeric attributes, etc. Table 2 lists the characteristics of the microarray data test in our experiments.

Table 3. Data characteristics of 3 microarray datasets

Dataset	# samples in class 1	# samples in class 2	# attributes
Breast Cancer	44 (non-relapse)	34 (relapse)	24481
Central Nervous	21 (survivors)	39 (failures)	7129
Colon Cancer	22 (normal)	40 (cancer)	2000

Table 4. The experimental results on Breast Cancer data subset

	$t = 0.08$	$t = 0.085$	$t = 0.09$
Success ratio	0%	80%	100%
Avg. contrast ratio	\	0.176728	0.006567
Best contrast ratio	\	0.094032	0.006150
Avg. tiny error	\	0.083988	0.086885
Best tiny error	\	0.083578	0.084548
Avg. coarse error	\	0.475237	13.22988
Best coarse error	\	0.888820	13.74682
Best class contrast function	\	x_{13620}*(x_{376}-0.4)-x_{19967} =0	(x_{7813}+0.3)*(0.6-0.8)-x_{21943}*x_{376}* 0.7*x_{376}-x_{19967}=0

As stated previously, the evolution efficiency of GEP is low when the terminal set size is large. For each microarray dataset, we calculate the contrast score of each attribute based on the method introduced in Subsection 3.2. We fetch 10 attributes with highest contrast scores, and find class contrast functions from them. Specifically, for Breast Cancer dataset, the index set of selected attribute is {376, 7813, 8781, 13620, 6326, 21943, 18424, 726, 7508, 19967}[1]. For Central Nervous dataset, the index set of selected attribute is {7015, 5527, 2473, 2141, 843, 3419, 3774, 4605, 2088, 10}. And for Colon Cancer dataset, the index set of selected attribute is {1670, 248, 1041, 1292, 142, 1410, 1327, 1324, 1771, 896}.

Subsection 3.3 introduces three strategies for finding multiple class contrast functions. In this experiment, we adopt the first strategy to find more class contrast functions. For each dataset, we apply our proposed method 20 times independently to discover class contrast functions. If a class contrast function whose fitness is greater than 0.5 is found, this run is marked as a success. In this case, the tiny error of the discovered function is greater than the contrast threshold. The success ratio is the percent of the number of success runs compared with the total running times. The number of generations is set as 500. Three different contrast thresholds (t) are set in experiments. We choose these values so that different success ratios can be got which is helpful for us to analyze the experimental results. Table 3 to Table 5 list the experimental results on these three data subsets.

From Table 3, we can see that in Colon Cancer data subset it is failed to find class contrast function whose tiny error is less than 0.08. The reasons may include: first, there is no function satisfies this constraint; Second, larger generation number for GEP should be set; Third, more functions should be added into GEP's function set. The success ratio can be increased by setting larger contrast threshold. But the tiny error of the best individual may be increased. It is worth to note that the difference between the average tiny error and the best tiny error is small. Since the fitness of each individual depends on its contrast ratio, some individuals are chosen due to the large coarse error. So, determining suitable contrast threshold is necessary and important. Similar conclusions can be got from Table 4 and Table 5. In Table 5, when the contrast threshold is 180 or 190, we get the same class contrast function several times. So the average values equal to the values of the best one.

[1] The index of the first attribute in the original dataset is 0. Here, we list the indexes of selected attributes in contrast score descending order.

Table 5. The experimental results on Central Nervous data subset

	$t = 85$	$t = 90$	$t = 95$
Success ratio	40%	70%	100%
Avg. contrast ratio	0.147789	0.121467	0.106082
Best contrast ratio	0.132213	0.095922	0.095922
Avg. tiny error	82.62214	83.45789	88.88393
Best tiny error	82.82429	87.26190	87.26190
Avg. coarse error	559.0565	687.0856	837.8832
Best coarse error	626.4439	909.7179	909.7179
Best class contrast function	$(0.3+x_{3419}+x_{5527}-x_{4605})*0.3-x_{10} = 0$	$x_{3419}-x_{4605}-2.0*x_{10} = 0$	$x_{3419}-x_{4605}-2.0*x_{10} = 0$

Table 6. The experimental results on Colon Cancer data subset

	$t = 180$	$t = 190$	$t = 200$
Success ratio	10%	25%	100%
Avg. contrast ratio	0.282965	0.389621	0.199963
Best contrast ratio	0.282965	0.389621	0.124137
Avg. tiny error	177.8966	183.0485	194.3224
Best tiny error	177.8966	183.0485	179.1640
Avg. coarse error	628.6869	469.8115	971.7921
Best coarse error	628.6869	469.8115	1443.278
Best class contrast function	$x_{1292}*(x_{248}+0.7)/(x_{1292}+0.4+x_{1041})-x_{896}=0$	$(x_{142}+x_{248})*0.2-x_{896}=0$	$x_{248}/((0.3+x_{1771}/5.0)*0.1)-x_{896}=0$

Furthermore, we can adopt the second or the third strategy described in Subsection 3.3 to find other class contrast functions in these datasets. The basic process is similar to adopting the first strategy.

5 Discussions and Conclusions

Finding the difference between different classes is an important data mining task. Some concepts on contrast mining have been proposed. For example, Emerging Patterns are item sets whose support changes significantly between two classes. However, methods for finding EPs cannot be applied to numeric dataset directly.

In this paper, we propose a new data mining task called class contrast function mining. Shortly, class contrast functions are functions whose accuracies change significantly between two classes. Class contrast function mining is a challenging work. It is related to regression, contrast mining, classification, etc. Initially, we design a GEP based method to discover class contrast functions, and apply it to both synthetic and real world datasets. The experimental results show that our proposed methods are effective. Several class contrast functions are discovered from the real world datasets. We also get some conclusions by analyzing the experimental results.

There are many works worth to be deeply analyzed in the future. For example, in our experiments we demonstrate that the contrast threshold is important. How to design a method that can adjust the contrast threshold self-adaptively is a desirable future work. Moreover, we will consider how to find all class contrast functions.

References

1. Dong, G., Li, J.: Efficient Mining of Emerging Patterns: Discovering Trends and Differences. In: Proc. of KDD 1999, pp. 43–52 (1999)
2. Dong, G., Li, J.: Mining Border Descriptions of Emerging Patterns from Dataset Pairs. Knowl. Inf. Syst. 8(2), 178–202 (2005)
3. Zhang, X., Dong, G., Ramamohanarao, K.: Exploring Constraints to Efficiently Mine Emerging Patterns from Large High-dimensional Datasets. In: Proc. of KDD 2000, pp. 310–314 (2000)
4. Bailey, J., Manoukian, T., Ramamohanarao, K.: Fast Algorithms for Mining Emerging Patterns. In: Elomaa, T., Mannila, H., Toivonen, H. (eds.) PKDD 2002. LNCS, vol. 2431, pp. 39–50. Springer, Heidelberg (2002)
5. Fan, H., Ramamohanarao, K.: An Efficient Single-Scan Algorithm for Mining Essential Jumping Emerging Patterns for Classification. In: Chen, M.-S., Yu, P.S., Liu, B. (eds.) PAKDD 2002. LNCS, vol. 2336, pp. 456–462. Springer, Heidelberg (2002)
6. Bailey, J., Manoukian, T., Ramamohanarao, K.: A Fast Algorithm for Computing Hypergraph Transversals and its Application in Mining Emerging Patterns. In: Proc. of ICDM 2003, pp. 485–488 (2003)
7. Loekito, E., Bailey, J.: Fast Mining of High Dimensional Expressive Contrast Patterns Using Zero-suppressed Binary Decision Diagrams. In: Proc. of KDD 2006, pp. 307–316 (2006)
8. Li, J., Liu, G., Wong, L.: Mining Statistically Important Equivalence Classes and Delta-Discriminative Emerging Patterns. In: Proc. of KDD 2007, pp. 430–439 (2007)
9. Ferreira, C.: Gene Expression Programming: A New Adaptive Algorithm for Solving Problems. Complex Systems 13(2), 87–129 (2001)
10. Ferreira, C.: Gene Expression Programming: Mathematical Modeling by an Artificial Intelligence. Angra do Heroismo, Portugal (2002)
11. Ferreira, C.: Discovery of the Boolean Functions to the Best Density-Classification Rules Using Gene Expression Programming. In: Proc of the 4th EuroGP, pp. 51–60 (2002)
12. Ferreira, C.: Mutation, Transposition, and recombination: An analysis of the evolutionary Dynamics. In: 4th Int'l Workshop on Frontiers in Evolutionary Algorithms, Research Triangle Park, North Carolina, USA, pp. 614–617 (2002)
13. Lopes, H.S., Weinert, W.R.: EGIPSYS: An Enhanced Gene Expression Programming Approach for Symbolic Regression Problems. Int'l Journal of Applied Mathematics and Computer Science 14(3), 375–384 (2004)
14. Zhou, C., Xiao, W., Tirpak, T.M., Nelson, P.C.: Evolution Accurate and Compact Classification Rules with Gene Expression Programming. IEEE Transactions on Evolutionary Computation 7(6), 519–531 (2003)
15. Duan, L., Tang, C., Zhang, T., et al.: Distance Guided Classification with Gene Expression Programming. In: Li, X., Zaïane, O.R., Li, Z.-h. (eds.) ADMA 2006. LNCS, vol. 4093, pp. 239–246. Springer, Heidelberg (2006)
16. Zuo, J., Tang, C., Li, C., et al.: Time Series Prediction based on Gene Expression Programming. In: Li, Q., Wang, G., Feng, L. (eds.) WAIM 2004. LNCS, vol. 3129, pp. 55–64. Springer, Heidelberg (2004)
17. Bailey, J., Dong, G.: Contrast Data Mining: Methods and Applications. In: Tutorial at 2007 IEEE ICDM (2007)
18. Dong, G., Zhang, X., Wong, L., Li, J.: CAEP: Classification by Aggregating Emerging Patterns. Discovery Science, 30–42 (1999)

19. Li, J., Dong, G., Ramamohanarao, K.: Instance-Based Classification by Emerging Patterns. In: Zighed, D.A., Komorowski, J., Żytkow, J.M. (eds.) PKDD 2000. LNCS, vol. 1910, pp. 191–200. Springer, Heidelberg (2000)
20. Li, J., Dong, G., Ramamohanarao, K., Wong, L.: DeEPs: A New Instance-Based Lazy Discovery and Classification System. Machine Learning 54(2), 99–124 (2004)
21. Li, J., Dong, G., Ramamohanarao, K.: Making Use of the Most Expressive Jumping Emerging Patterns for Classification. In: Terano, T., Chen, A.L.P. (eds.) PAKDD 2000. LNCS, vol. 1805, pp. 220–232. Springer, Heidelberg (2000)
22. Dougherty, J., Kohavi, R., Sahami, M.: Supervised and Unsupervised Discretization of Continuous Features. In: Proc. of the 20th ICML, pp. 194–202 (1995)
23. Fayyad, U., Irani, K.: Multi-interval Discretization of Continuous-valued Attributes for Classification Learning. In: Proc. of the 13th IJCAI, pp. 1022–1029 (1993)

McSOM: Minimal Coloring of Self-Organizing Map

Haytham Elghazel, Khalid Benabdeslem, and Hamamache Kheddouci

University of Lyon, F-69622 Lyon, France, University of Lyon 1, Villeurbanne, LIESP, EA4125
{elghazel,kbenabde,hkheddou}@bat710.univ-lyon1.fr

Abstract. A *Self-Organizing Map* (SOM) is an artificial neural network tool that is trained using unsupervised learning to produce a low-dimensional representation of the input space, called a map. This map is generally the subject of a clustering analysis step which aims to partition the referents vectors (map neurons) in compact and well-separated groups. In this paper, we consider the problem of clustering self-organizing map using a modified graph minimal coloring algorithm. Unlike the traditional clustering SOM techniques, using k-means or hierarchical classification, our approach has the advantage to provide a partition of self-organizing map by simultaneously using dissimilarities and neighborhood relations provided by SOM. Experimental results on benchmark data sets demonstrate that the proposed clustering algorithm is able to cluster data in a better way than classical ones and indicates the effectiveness of SOM to offer real benefits for the original minimal coloring clustering approach.

Keywords: Self-organizing map, clustering, minimal coloring, neighborhood relations.

1 Introduction

Clustering can be considered as the most important *unsupervised learning* problem which deals with finding a *structure* in a collection of unlabeled data. To this end, it conducts a process of organizing objects into groups whose members are similar in some way and dissimilar to those of other groups. Existing clustering algorithms can be generally categorized into five classes: partitioning, hierarchical, density-based, grid-based and model-based [1].

In this paper, we focus on the problem of *model-based clustering* using *self-organizing map* (SOM) [2]. This technique, proposed by Kohonen, is a powerful tool in visualization and analysis of high-dimensional data which has been widely used in in various application domains such as pattern recognition, biological modeling, Web analysis, information retrieval, and many other domains.

SOM is considered as a set of neurons arranged in a a *low-dimensional structure* such that there are *neighborhood relations* among the neurons. After training, each neuron is attached to a feature vector of the same dimension as the

R. Huang et al. (Eds.): ADMA 2009, LNAI 5678, pp. 128–139, 2009.

input space. Based on these feature vectors, SOM is usually a subject of clustering step in a two-level approach, which aims to cluster input data based on the SOM. The main idea is that the first level is to train data by the SOM and the second level is to cluster data based on the SOM. Several attempts have been made to cluster self-organizing map generally using classical clustering methods, such as *k-means* or *hierarchical clustering* algorithms [3,4]. However, these approaches fail to incorporate the important property of topological neighborhood relations offered by SOM. Motivated by this, in this paper, we propose a novel and efficient approach for clustering SOM, McSOM (for *Minimal coloring of Self-Organizing Map*).McSOM is an extension of the *minimal graph coloring based clustering technique*. The main advantage of our proposal is its ability to take into account the topological relations provided by SOM. Two problems are therefore to be considered: 1) how to incorporate as much as useful the *neighborhood informations* offered by SOM in the second-level to clustering SOM; 2) which neighborhood order is able to deliver satisfactory results.

McSOM is evaluated against benchmark data sets and the results of this study demonstrate that the proposed clustering algorithm is able to cluster data in a better way than classical clustering algorithms of SOM and indicates the effectiveness of SOM to offer real benefits (*clustering quality* and *runtime*) for the original minimal graph coloring based clustering approach.

2 Minimal Coloring of Graphs

In this section, we provide some background on the minimal graph coloring based clustering framework that was proposed in [5].

When the dissimilarities among all pairs of data $\{w_1, w_2, \ldots, w_n\}$ (in our case w_i is a *p-dimensional referent vector* corresponding to the *SOM neuron* i, $w_i = \{w_i^1, w_i^2, \ldots, w_i^p\}$) are specified, these can be summarized as a weighed dissimilarity matrix $\mathcal{D}$ in which each element $\mathcal{D}(w_i, w_j)$ stores the corresponding dissimilarity. Based on $\mathcal{D}$, the data can also be conceived as a weighted linkage graph $G = (\boldsymbol{V}, \boldsymbol{E})$, where $\boldsymbol{V} = \{v_1, v_2, \ldots, v_m\}$ is the vertex set which corresponds to the data (*i.e. SOM neurons*, v_i for w_i), and $\boldsymbol{E} = \boldsymbol{V} \times \boldsymbol{V}$ is the edge set which corresponds to a pair of vertices (v_i, v_j) weighted by the dissimilaritiy $\mathcal{D}(w_i, w_j)$. A widely adopted definition of optimal clustering is a partitioning that *minimizes* dissimilarities within and *maximizes* dissimilarities between clusters. These two conditions amount to saying that edges between two vertices within one cluster should be small weighted (denoting *high similarity*), and those between vertices from two clusters should be large weighted (*weak similarity*). The clustering problem can be formulated as a *graph theoretical problem.*

In that framework, the clustering step consists in finding *combinatorial structures* within the dissimilarity graph. Today the most widespread graph-theoretic clustering approaches consist in finding *complete subgraphs* derived from an input *threshold graph.* These structures are based on *highly connected graph components* and are generally considered to provide highly homogenous groups. Several works exploring this idea are reported in the literature [6]. Hansen and Delattre

[5] reduced the partitioning problem into k classes with a minimal diameter[1], to the minimal coloring problem of a *superior threshold graph* in which vertices correspond to objects and edges correspond to dissimilarities between two elements which is higher than a given threshold value θ chosen among the dissimilarity table $\mathcal{D}$. In other words, $G_{>\theta}$ is given by $\boldsymbol{V} = \{v_1, v_2, \ldots, v_m\}$ as vertex set and $\{(v_i, v_j) | \mathcal{D}(w_i, w_j) > \theta\}$ as edge set. The goal is to divide the vertex set $\boldsymbol{V}$ into a partition $\boldsymbol{P_k} = \{C_1, C_2, \ldots, C_k\}$ where for $\forall C_i, C_j \in \boldsymbol{P_k}, C_i \cap C_j = \phi \; for \; i \neq j$ (when the number of clusters k is not predefined). The notation of $\boldsymbol{P_k}$ is used to both represent a set of clusters as well as a set of colors.

In such *superior threshold graph* $G_{>\theta}$, the *minimal coloring* is NP-*complete* and consists to determine the minimum number of colors (clusters) needed to color the vertices of the graph such that no two adjacent vertices (dissimilar in the sense of threshold θ) have the same color (*proper coloring*). A variety of approximations and search algorithms have been developed to solve the minimal graph coloring problem in a reasonable amount of time. A number of simplest graph coloring algorithms that have been proposed and analyzed in the literature follow the *paradigm of sequential coloring* as described in [7]. In the sequential coloring paradigm a strategy for ordering of the vertices is firstly prescribed. The top vertex is put in color class number one. The remaining vertices are considered in order, and each is placed in the first color class for which it has no adjacencies with the vertices already assigned to the class. If no such class exists, then a new class is created. Several different schemes have been used for the initial ordering.

The simplest and well-known graph minimal coloring algorithm is the *Largest First* (LF) one developped by *Welsh and Powell* in [8]. This algorithm, easy to implement and fast, sorts the vertices by decreasing degree. Althought this algorithm (when adopted in cluster analysis as in Hansen and Delattre approach [5]) tends to build a partition of the data set with effectively compact clusters, it don't give any importance to the *cluster-separation*. This is generally due to the selection strategy of vertices with *same degree* to affect them to one color class in the building of the *minimal coloring* of *threshold graph*. Indeed, given a threshold graph $G_{>\theta}$, by changing the order of such treated vertices, many coloring functions become possibles and valids. The main problem is to find the appropriate coloring which is constrained to maximize the *intracluster homogeneity* and the *intercluster separation* of the returned partition $\boldsymbol{P_k}$.

3 Graphical Clustering of SOM

This section is devoted to discuss our modification to the minimal coloring approach for clustering *self-organizing map* by considering the *SOM neighborhood relations*. The main idea of our McSOM approach is to use the SOM neighborhood relations to constrain the construction of the *threshold graph* (indispensable

[1] The *diameter* of one cluster is the largest dissimilarity between two objects belonging to the same cluster.

for the coloring algorithm) and the possible vertex selections during the building of the minimal coloring of this graph.

As mentioned above, the *minimal coloring based clustering approach* requires a *non complete edge-weighted graph* $G_{>\theta} = (\boldsymbol{V}, \boldsymbol{E}_{>\boldsymbol{\theta}})$ to return a partition $\boldsymbol{P_k}$ of $\mathcal{W} = \{w_1, w_2, \ldots, w_m\}$ neurons (referent vectors) set. In order to incorporate the SOM neighborhood relations into the clustering problem, our first modification concerns the construction of this non-complete graph that will be presented to the *LF algorithm*. This *non complete edge-weighted graph* is now given by $G_{>\theta,\alpha} = (\boldsymbol{V}, \boldsymbol{E}_{>\boldsymbol{\theta},\boldsymbol{\alpha}})$, where:

- $\boldsymbol{V} = \{v_1, v_2, \ldots, v_m\}$ is the vertex set which corresponds to the data (*i.e. SOM neurons*, v_i for w_i).
- $\boldsymbol{E}_{>\boldsymbol{\theta},\boldsymbol{\alpha}}$ is the edge set, where for each two vertices v_i and v_j in $\boldsymbol{V}$, the edge $(v_i, v_j) \in \boldsymbol{E}_{>\boldsymbol{\theta},\boldsymbol{\alpha}}$ *iff* $\mathcal{D}(w_i, w_j) > \theta$ *or* $\mathcal{SN}(w_i, w_j) > \alpha$. $\mathcal{D}$ is the dissimilarity matrix and $\mathcal{SN}$ is the *SOM rectangular-neighborhood order matrix*. For exemple, two neurons considered from the red rectangle in figure 1 have a *SOM rectangular-neighborhood order* equal to 1.
- θ is the *dissimilarity* threshold.
- α is the *SOM rectangular-neighborhood order* threshold.

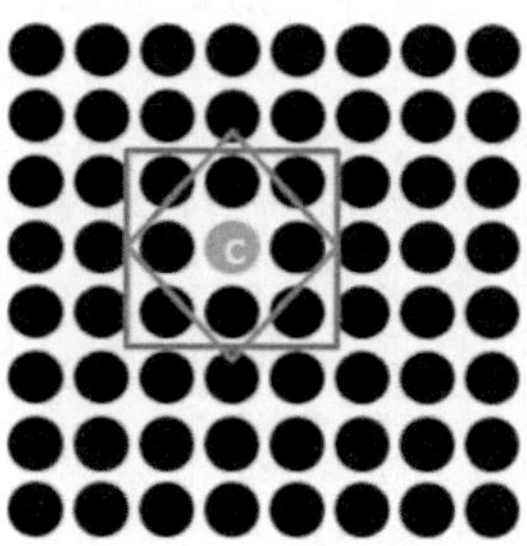

Fig. 1. Two dimensional topological map with 1-neighborhood of a neuron c. Rectangular (red) with 8 neighbors and diamond (blue) with 4 neighbors.

This proposal offers the possibility to perform the *minimal coloring based clustering approach* multiple runs, each of them increasing the value of the *SOM rectangular-neighborhood order threshold* α. Once all threshold values passed, the algorithm provides the *optimal neighborhood order* (corresponding to one threshold value α_o) which is able to deliver satisfactory results of clustering SOM. Consequently, this can provide a solution to the second problem highlighted in Section 1.

The data to be clustered (*i.e.* SOM neurons) are now depicted by a non-complete edge-weighted graph $G_{>\theta,\alpha}$. Additional modifications of the *minimal coloring based clustering approach* are considrerd in order to improve the quality of clustering by incorporating the topological relations offred by SOM. The changes concern now the *Largest First* (LF) algorithm of *Welsh and Powell* used for coloring the graph $G_{>\theta,\alpha}$. In fact, after sorting the vertices of $G_{>\theta,\alpha}$ by decreasing degree, the *LF algrithm* of *Welsh and Powell* starts from the vertex of

$\boldsymbol{V}$ which has the maximum degree Δ. The algorithm colors this vertex using the color one. Then, algorithm tries to color the remaining vertices (by respecting the decreasing order of their degree) according to the following principle: each vertex is placed in the first color class for which it has no neighbors. If no such color exists, then a new color is created for it.

The main problem is to find the appropriate vertex to color it when there is a choice between many vertices with the same degree. For an illustration purpose, suppose that we have two adjacent vertices v_i and v_j having the same degree and no neighbors in one color c. Therefore, if v_i is selected for coloring, it can be assigned to color c which will not be possible after for v_j, and vice versa. We note the reliance of the coloring based clustering result to the selection manner for such vertices. This choice is constrained to maximize the *intracluster homogeneity* and then the *intercluster dissimilarity* of the returned partition $\boldsymbol{P_k}$. As a solution, we propose the following strategy: when one vertex v_i with degree d is selected for coloring and the first color c different from those of its neighborhood is found, the vertices not yet colored, having the same degree d and without any neighbor in c, will be simultanously considered for coloring. So the vertex whose dissimilarity with c is minimal will be the first to color with c and the remaining vertices will be considered later. For our McSOM approach, this dissimilarity relies only on the neighborhood relations between neurons (*c.f.* figure 1 for the neighborhhod of order 1). Hence, the distance between the vertex v_i and a color c is given by the average of *SOM rectangular-neighborhood order* between w_i related to v_i and the neurons related to the vertices colored with c.

$$d(v_i, c) = \frac{1}{|c|} \sum_{j=1}^{|c|} \mathcal{SN}(w_i, w_j) \tag{1}$$

where $|.|$ denote the cardinality.

Our idea is summarized in the procedure *McSOM_coloring*(). For that, we need to introduce the following notations and routines:

- $degree(v_i)$: the degree of the vertex v_i in $G_{>\theta,\alpha}$. It is the number of neighbors of v_i.
- Δ: the maximum degree of $G_{>\theta,\alpha}$. It is also the maximal number of colors available for coloring this graph.
- $c(v_i)$: the color (integer value) of the vertex v_i in $G_{>\theta,\alpha}$.
- $\boldsymbol{N}(v_i)$: the neighborhood of vertex v_i in $G_{>\theta,\alpha}$.
- $\boldsymbol{N_c}(v_i)$: the neighborhood colors of vertex v_i.
- $ncolors$: the number of colors in the graph $G_{>\theta,\alpha}$. Initially, since vertices of G are not yet colored, $ncolors$ is equal to 0.
- $Dequeue(v_i, \boldsymbol{V})$ is the method which removes the vertex v_i from the vertex set $\boldsymbol{V}$ when it is colored.
- $Update(d(v_j, c))$ is the method which updates the dissimilarity value between the vertex v_j and the color c when a vertex v_i is recently affected to c. In other words, the coloring of v_i with c causes to change the dissimilarity values $d(v_j, c)$ for each vertex v_j of $G_{>\theta,\alpha}$. Furthermore, although naive calculation

Algorithm 1. Procedure *McSOM_coloring*()

Require: $G_{>\theta,\alpha} = (\boldsymbol{V}, \boldsymbol{E}_{>\theta,\alpha})$; *//A graph with a set of vertices and a set of edges.*

Select the vertex v from $\boldsymbol{V}$ with the highest degree;
$c(v) := 1$;
$Dequeue(v, \boldsymbol{V})$;
for each vertex $v_j \in \boldsymbol{V}$ **do**
    $Update(d(v_j, c(v)))$
end for
$ncolors := 1$;
repeat
    Select v from $\boldsymbol{V}$;
    $c := min\{h|\ 1 \leq h \leq ncolors, h \notin \boldsymbol{N_c(v)}\}$;
    if $(c \leq ncolors)$ **then**
        $\boldsymbol{M} := \boldsymbol{V} \setminus \boldsymbol{N}(v)$;
        $\boldsymbol{H} := \{v_h | v_h \in \boldsymbol{M}\ and\ degree(v_h) = degree(v)\ and\ c \notin \boldsymbol{N_c(v_h)}\}$;
        $\boldsymbol{H} := \boldsymbol{H} \cup \{v\}$;
        $v_i := argmin_{v_h \in \boldsymbol{H}}(d(v_h, c))$;
        $c(v_i) := c$;
        $Dequeue(v_i, \boldsymbol{V})$;
    else
        $c(v) := c$;
        $ncolors := ncolors + 1$;
        $Dequeue(v, \boldsymbol{V})$;
    end if
    for each vertex $v_j \in \boldsymbol{V}$ **do**
        $Update(d(v_j, c))$
    end for
until $(\boldsymbol{V} = \phi)$

of $d(v_j, c)$ takes $O(m)$ (*i.e.* m is the number of SOM neurons), it can also be reduced to $O(1)$ using the $Update(d(v_j, c))$ routine based on its old value as defined in the following equation:

$$d(v_j, c) = \frac{|c|\, d^{old}(v_j, c) + \mathcal{D}(w_j, w_i)}{|c| + 1} \tag{2}$$

where w_i is the neuron related to vertex v_i recently affected to the color c.

Proposition 1. *The Procedure McSOM_coloring() runs in* $O(m^2)$.

Proof. Procedure *McSOM_coloring*() starts from the vertex of $\boldsymbol{V}$ which has the maximum degree Δ. The procedure colors this vertex using the color one. Then, algorithm tries to color the remaining vertices (at most m SOM neurons) according to the following principle: the neighborhood colors (at most Δ colors) of each vertex v_i is checked in order to find the first possible color c for it. If such color exists, each vertex v_h, with the same degree of v_i and for which c has no adjacencies with it (at most m), is selected and for all possible vertices,

Table 1. Characteristics of used data sets

Data sets	n	p	*#labels*
Pima	768	8	2
Two-Diamonds	800	2	2
Heart Disease	303	13	2
Yeast	1484	8	10
Rings	1000	3	2
Engytime	4096	2	2
Haberman	306	3	2

including v_i, the appropriate one is chosen to be placed in c. Otherwise, if no such color exists, then a new color c is created for v_i. Futhermore, in all cases, the dissimilarities between all vertices in $\boldsymbol{V}$ (at most m) and the selected color c are updated ($O(1)$). Therefore Procedure *McSOM_coloring*() uses at most $(m \times (\Delta + m + m \times 1))$ instructions, and the complexity is $O(m^2)$. □

Proposition 2. *The two-level approach using self-organizing map and our proposed McSOM approach reduces considerably the runtime of our clustering approach.*

Proof. Consider the clustering of n instances ($z_i \in R^p, i = 1, 2, ..., n$) using the McSOM approach, an extension of the *minimal coloring based clustering method.* This approach generates the minimal coloring of any graph $G_{>\theta,\alpha}$ in $O(n^2)$ (*c.f.* Proposition 1). When the data set (n instances) is first clustered using *self-organizing map*, McSOM is then applied, during a second step, for clustering a set of m neurons returned by SOM (*i.e.* $\mathcal{W} = \{w_1, w_2, \ldots, w_m\}$ where $m = 5\sqrt{n})^2$. Therefore, McSOM uses at most $((5\sqrt{n})^2)$ instructions. The clustering using a two-level approach (*self-organizing map* and *modified minimal coloring approach i.e.* McSOM) allows to reduce the complexity of this latter to $O(n)$. □

4 Experiments on Benchmark Data Sets

In this section, we illustrate our algorithm's performance on several relevant benchmark data sets, viz., *Pima*, *Two-Diamonds*, *Heart Disease*, *Yeast*, *Rings*, *Engytime* and *Haberman* [9] (*c.f.* Table 1). For each trained SOM, McSOM, *Orig-m-col* (the *Original minimal coloring based clustering approach without any neighborhood information* in [5]), *Agglomerative Hierarchical* (*AHC* in the form of *Ward-based* approach) and *k-means* clustering SOM algorithms were applied. A comparison is made based on the quality of the partitioning obtained from them. We remind that the Euclidian distance is applied to define the dissimilarity level $\mathcal{D}$ between two p-dimensional referent vectors. Moreover, for each *SOM*

[2] Heuristic proposed by T. Kohonen for automatically providing the number of neurons in the map.

rectangular-neighborhood order threshold (choosed between 1 and the number of SOM row's *nrows*), McSOM is performed multiple runs, each of them increasing the value of the dissimilarity threshold θ. Once all neighborhood and dissimilarity threshold values passed, the algorithm provides the optimal partitioning which maximizes *Generalized Dunn's* quality index [10]. This index is designed according some simulation studies. On the other hand, *AHC* and *k-means* approaches have been performed for several partitions over SOM neurons and for each of them the optimal partition was returned using *Davies-Bouldin* quality index [10]. For an interesting assess of the results gained with the different clustering approaches, the following performance indices are used:

- Two *dissimilarity-based validity* schemes called *Davies-Bouldin* and *Generalized Dunn's* indices [10]. Considered as *internal criteria*, they give an idea about the compromise between the *intracluster similarity* (tightness) and *intercluster dissimilarity* (separation) offered by the returned partition. Basically we would like to ***minimizes*** the *Davies-Bouldin* index and to ***maximize*** the *Dunn's generalized index* to achieve high quality clustering.
- Two *statistical-matching* schemes called *Purity* and *Adjusted Rand index* [11] which concern the *clustering accuracy*. In our case, the used UCI data sets include *class information* (*label*) for each data instance. These labels are available for evaluation purposes but not visible to the clustering algorithm. Indeed, evaluation is based on these two schemes in order to compare clustering results against *external criteria*. *Purity* is a simple evaluation measure which can be expressed as the percentage of elements of the assigned label in a cluster. To compute this measure, each cluster is assigned to the label which is most frequent in the cluster, and then the accuracy of this assignment is measured by counting the number of correctly assigned instances and dividing by n (the total number of instances). On the other hand, the partition obtained using one clustering approach will be assessed by considering relations upon instances. Recommended by Milligan and Cooper [12], on the basis of an extensive empirical study, we adopt the *Adjusted Rand index* proposed by Hubert and Arabie [11] to assess the *degree of agreement* between two partitions (the one obtained with the clustering algorithm (*clusters*) and the correct predefined one (*labels*)). When comparing two clustering algorithms, the one that produces the greater *Adjusted Rand index* and *Purity* should be preferred since the partition correctly identifies the underlying classes in the data set.

We report here our experiments using seven relevant benchmark data sets (*c.f.* Table 1) chosen from UCI database [9]. Tables 2 and 3 provide the clustering results according to *Generalized Dunn's* and *Davies-Bouldin* indices. Both $Dunn_G$ and DB measures indicate better clustering for all partitions generated by the proposed McSOM approach, except for *Engytime data set* where we obtain identical results with the *original minimal coloring based clustering approach* (*Orig-m-col*). The clusters given from McSOM are thus *compact* and *well-separated*. This confirms the pertinence of the *graph minimal-coloring technique* associated with the *neighborhood informations* provided by SOM (1) to

Table 2. Evaluation of clustering SOM approaches on Generalized Dunn's index

Data sets	*k-means*	*AHC*	*Orig-m-col*	*McSOM*
Pima	0.8499(4)	1.0458(3)	1.0617(3)	**1.2697(2)**
Two-Diamonds	1.9678(2)	1.9753(2)	1.1833(3)	**1.9754(2)**
Heart Disease	0.9537(3)	0.9457(3)	0.9351(3)	**1.0109(2)**
Yeast	0.9254(6)	0.8861(6)	0.8094(4)	**0.9263(6)**
Rings	0.9639(10)	1.3313(10)	1.1175(4)	**1.3565(2)**
Engytime	0.9466(10)	0.7588(10)	1.1551(3)	**1.1551(3)**
Haberman	0.9447(5)	0.8258(9)	1.1601(3)	**1.2440(2)**

Table 3. Evaluation of clustering SOM approaches on Davies-Bouldin index

Data sets	*k-means*	*AHC*	*Orig-m-col*	*McSOM*
Pima	1.8147(4)	1.7892(3)	1.7378(3)	**1.5665(2)**
Two-Diamonds	1.0093(2)	1.0088(2)	1.2682(3)	**1.0082(2)**
Heart Disease	1.9826(3)	1.9847(3)	1.9796(3)	**1.9563(2)**
Yeast	1.4368(6)	1.4458(6)	1.4061(4)	**1.3361(6)**
Rings	0.9682(10)	0.8435(10)	1.2224(4)	**0.8042(2)**
Engytime	1.1573(10)	1.1594(10)	1.1190(3)	**1.1190(3)**
Haberman	1.6409(5)	1.7226(9)	1.6998(3)	**1.5678(2)**

Table 4. Evaluation of clustering SOM approaches on Adjusted Rand index rate (%)

Data sets	*k-means*	*AHC*	*Orig-m-col*	*McSOM*
Pima	34.56(4)	35.20(3)	28.98(3)	**35.50(2)**
Two-Diamonds	98.68(2)	99.34(2)	86.67(3)	**99.67(2)**
Heart Disease	50.29(3)	50.66(3)	50.50(3)	**51.49(2)**
Yeast	66.52(6)	66.69(6)	55.55(4)	**66.97(6)**
Rings	56.93(10)	58.63(10)	47.92(4)	**59.19(2)**
Engytime	57.02(10)	57.07(10)	57.53(3)	**57.53(4)**
Haberman	46.71(5)	44.17(9)	46.54(3)	**49.97(2)**

offer a compromise between the *intercluster separation* and the *intracluster homogeneity*, and (2) to improve the quality of clustering obtained by the *original minimal coloring based clustering approach* (*without any neighborhood information*). By considereing the topologial oganization offered by SOM, McSOM offers a partition of data with effectively *well-separated* clusters and then gives a solution to the problem of *weak cluster-distinctness* of *original minimal coloring approach*.

Tables 4 and 5 list the clustering results according to *Adjusted Rand index* and *Purity rate*. Additionally, the *purity rates* provided by our approach, for *Pima, Two-Diamonds, Rings* and *Engytime* data sets, are compared with those returned from a two recently proposed clustering SOM techniques [13] (*c.f.* Table 6). They are *AT-Neigh-W* (Hierarchical clustering based on *artificial ant* and a new dissimilarity measure which take into account the topological order of referent vectors) and *AHC-Neigh-W* (Agglomerative Hierarchical Clustering approach based on the same dissimilarity measure)[3]. According to the *Adjusted Rand index* measures, we observe that McSOM provides always the highest values, except for *Engytime data set* where we obtain identical results with the *original minimal coloring based clustering approach* (*Orig-m-col*). Considering the *purity rate* measures, McSOM provides *generally* better clustering results, except for:

- *Two-Diamonds* and *Pima*, where McSOM provides the highest purity value than all approaches except the *AT-Neigh-W* one. However, there is a natural explanation. These results are expected since *AT-Neigh-W* provides a large number of clusters (7 for *Two-Diamonds* and 5 for *Pima*) than McSOM (2 for *Two-Diamonds* and 2 for *Pima*). Indeed, high purity is easy to achieve when the number of clusters is large - in particular, purity is 100% if each instances gets its own cluster (this is the main problem of the *purity rate scheme*). Moreover, we note that McSOM identifies the same number of clusters as for the *correct predefined partition* (labels) which is respectively 2 for *Two-Diamonds* and 2 for *Pima*.
- *Rings* and *Engytime*, where we observe that *k-means, AHC, AT-Neigh-W*, and *AHC-Neigh-W* achieve the best purity rates. These results are also expected since the number of clusters returned from these approaches is also greater than the one provided by McSOM.
- *Haberman* data set. Althought *k-means* and *AHC* offer slightly better purity rate than McSOM, the latter provides the same number of clusters as for the *correct predefined partition* which is much smaller than the one returned from *k-means* and *AHC* (5 for *k-means*, 9 for *AHC* and 2 for McSOM).

Consequently, it is observed that McSOM *generally* achieves the highest *Adjusted Rand* and acceptable *purity rates* with the smallest number of clusters (generally the same as for the *correct predefined partition*). It can be concluded that McSOM generates *meaningful* clusters than the other clustering SOM approaches. On the other hand, by looking the results provided by the *original minimal coloring approach*, we deduce that incorporating *neighborhood informations* (offered by SOM) in our approach increases the *clustering accuracy* related to the previously discussed measures.

Notice. Throughout the experiments reported above, the optimal partition indicates that a *SOM rectangular-neighborhood* of order between $\frac{nrows}{2}$ and $nrows$[4] (but generally close to $\frac{nrows}{2}$) gives satisfactory results. This confirm the idea that there is a *SOM rectangular-neighborhood order* which is well-suited to partition SOM neurons in compact and well-separated clusters.

[3] We note that the results of *AT-Neigh-W* and *AHC-Neigh-W* algorithms are given from [13] and not be reproduced in this paper.

[4] *nrows* is the number of SOM row's.

Table 5. Evaluation of clustering SOM approaches on Purity rate (%)

Data sets	*k-means*	*AHC*	*Orig-m-col*	*McSOM*
Pima	66.28(4)	66.28(3)	66.02(3)	**67.63(2)**
Two-Diamonds	99.50(2)	99.75(2)	95.25(3)	**99.86(2)**
Heart Disease	56.11(3)	59.08(3)	55.78(3)	**59.19(2)**
Yeast	43.40(6)	40.84(6)	38.07(4)	**45.70(6)**
Rings	95.70(10)	100(10)	70.8(4)	**78.5(2)**
Engytime	94.46(10)	94.48(10)	70.87(3)	**70.87(3)**
Haberman	75.16(5)	75.49(9)	72.32(3)	**73.53(2)**

Table 6. Additional comparisons on Purity with other clustering SOM approaches

Data sets	*AT-Neigh-W*	*AHC-Neigh-W*
Pima	72.4(5)	65.10(2)
Two-Diamonds	100(7)	96.88(5)
Rings	81.5(5)	100(11)
Engytime	88.04(7)	93.90(5)

5 Conclusion

In this work we have proposed McSOM, a new *graph minimal coloring approach* for clustering *self-organizing map*. The purpose of this approach is to improve the existing minimal coloring based clustering approach using the topological organization (offered by SOM) to find a cluster partition of self-organizing map where internal cluster cohesion and separation among clusters are simultaneously effective. Improvements concern the whole of the original minimal coloring approach. The SOM topological relations are used to constrain not only the construction of the weighted linkage graph of *referent vectors* but also the possible vertex selections during the building of the minimal coloring of this graph. We have implemented, performed experiments, and compared our method to other clustering SOM approaches. We have shown significant improvements in *clustering quality* and *time complexity* as demonstrated by the results obtained over seven UCI data sets, in the form of *internal* and *external criteria*. It is concluded that combining minimal coloring based clustering and *topological organization* provided by SOM achieves better performances than either in isolation.

The present work can be extended in many directions: (1) we are leading additional experiments on a larger real data set, and (2) producing summary information of obtained clusters in order to evaluate the influance of *SOM neighborhood informations* on the results, to name a few.

References

1. Jain, A., Murty, M., Flynn, P.J.: Data clustering: A review. ACM Computing Surveys 31, 264–323 (1999)
2. Kohonen, T.: Self-organizing Maps, vol. 30. Springer, Heidelberg (2001)

3. Vesanto, J., Alhoniemi, E.: Clustering of the self-organizing map. IEEE Transactions on Neural Networks 11(3), 586–600 (2000)
4. Wu, S., Chow, T.W.S.: Self-organizing-map based clustering using a local clustering validity index. Neural Processing Letters 17(3), 253–271 (2003)
5. Hansen, P., Delattre, M.: Complete-link cluster analysis by graph coloring. Journal of the American Statistical Association 73, 397–403 (1978)
6. Matula, D.W., Beck, L.L.: Smallest-last ordering and clusterings and graph coloring algorithms. Journal of the ACM 30(3), 417–427 (1983)
7. Matula, D.W., Marble, G., Issacson, J.D.: Graph coloring algorithms, pp. 109–122 (1972)
8. Welsh, D.J.A., Powell, M.B.: An upper bound for the chromatic number of a graph and its application to timetabling problems. Computer Journal 10(1), 85–87 (1967)
9. Blake, C., Merz, C.: Uci repository of machine learning databases (1998)
10. Kalyani, M., Sushmita, M.: Clustering and its validation in a symbolic framework. Pattern Recognition Letters 24(14), 2367–2376 (2003)
11. Hubert, L., Arabie, P.: Comparing partitions. Journal of Classification 2, 193–218 (1985)
12. Milligan, G.W., Cooper, M.C.: A study of comparability of external criteria for hierarchical cluster analysis. Multivariate Behavioral Research 21(4), 441–458 (1986)
13. Azzag, H., Lebbah, M.: Clustering of self-organizing map. In: European Symposium on Artificial Neural Networks (ESANN 2008), pp. 209–214 (2008)

Chinese Blog Clustering by Hidden Sentiment Factors*

Shi Feng[2], Daling Wang[1,2], Ge Yu[1,2], Chao Yang[2], and Nan Yang[2]

[1] Key Laboratory of Medical Image Computing, Northeastern University, Ministry of Education
[2] College of Information Science and Engineering, Northeastern University, 110004 Shenyang, Liaoning
fengshi@ise.neu.edu.cn, {dlwang,yuge}@mail.neu.edu.cn

Abstract. In the Web age, blogs have become the major platform for people to express their opinions and sentiments. The traditional blog clustering methods usually group blogs by keywords, stories or timelines, which do not consider opinions and emotions expressed in the articles. In this paper, a novel method based on Probabilistic Latent Semantic Analysis (PLSA) is presented to model the hidden emotion factors and an emotion-oriented clustering approach is proposed according to the sentiment similarities between Chinese blogs. Extensive experiments were conducted on real world blog datasets with different topics and the results show that our approach can cluster Chinese blogs into sentiment coherent groups to allow for better organization and easy navigation.

1 Introduction

Most recently, Weblogs (also referred to as blogs) have become a very popular type of media on the Web. Blogs are often online diaries published and maintained by individual users (bloggers), reporting on the bloggers' daily activities and feelings. The contents of the blogs include commentaries or discussions on a particular subject, ranging from mainstream topics (e.g., food, music, products, politics, etc.), to highly personal interests [11].

Currently, there are a great number of blogs on the Internet. According to the reports from CNNIC (China Internet Network Information Center), by December 2008, the number of Chinese blog writers has researched 162 million, and it is reaching half of the total Internet users in China [4]. As the number of blogs and bloggers increases dramatically, developing an effective way to collect, monitor and analyze information on blogs can provide users key insights on 'public opinion' on a variety of topics, such as commercial products, political views, controversial events, etc. There are some previous literatures about blog content based analysis [2] [5]. Different kinds of tools are also being provided to help users retrieve, organize and analyze the blogs. Several commercial blog search engines and blog tagging systems [6] [25] have been published on the Web.

* This work is supported by National Natural Science Foundation of China (No. 60573090, 60703068, 60673139) and the National High-Tech Development Program (2008AA01Z146).

R. Huang et al. (Eds.): ADMA 2009, LNAI 5678, pp. 140–151, 2009.

The vast majority of blog contents are about bloggers' opinions, feelings and emotions [16]. Most of the previous studies usually try to cluster blogs by their keywords and underlying topics. However, during the clustering approach, little attention is paid to the sentiments, which are very important feature of the blog documents. Moreover, there may be thousands of blog entries with various opinions on one certain topic, and it is really difficult for people to collect, analyze and extract the public opinions on this topic. For example, there is a burst of blogs writing about the famous Chinese 110 meter hurdler Liu Xiang's withdrawing from Beijing Olympics Games in late August, 2008. Traditional topic-oriented blog search engines, such as Google Blog Search [6], can generate story coherent clusters for Liu's events. However, these tools could not provide users with a summarization or guideline about people's attitudes and reactions for Liu's behavior in blogs. From the discussions above, we can see that there are still some obstacles and limitations for people to make better understanding of bloggers' opinions and sentiments in blogs.

In this paper, we propose a novel method to cluster Chinese blogs by hidden sentiment factors. Usually, bloggers express their feelings and emotions in a much complex way, but most previous work on sentiment mining simply classify blogs into positive and negative categories, and do not provide a comprehensive understanding of the sentiments reflected in blogs. Therefore, we employ a Probabilistic Latent Semantic Analysis (PLSA) based approach to model the hidden sentiment factors, and cluster blogs by the underlying emotions embedded in them. We develop a summarization algorithm to label each cluster with sentiment key words, and a ranking algorithm is proposed to reorganize blog entries by their contribution to the hidden sentiment factors in each cluster. By this sentiment clustering method, we can make Chinese blogs into groups to allow for better organization and easy navigation.

The rest of the paper is organized as follows. Section 2 introduces the related work on blog mining and sentiment analysis. Section 3 models Chinese blogs using hidden sentiment factors. Section 4 describes the Chinese blog sentiment clustering algorithm based on sentiment similarity. Section 5 provides experimental results on real world datasets. Finally we present concluding remarks and future work in Section 6.

2 Related Work

There are two types of previous literatures relevant to our work. One is about blog mining, and the other is about sentiment analysis.

2.1 Blog Mining

Blogs have recently attracted a lot of interest from both computer researchers and sociologists. One direction of blog mining focuses on analyzing the contents of blogs. Glance et al. [5] gave a temporal analysis on blog contents and proposed a method to discover trends across blogs. In [19], the similarity between two blogs was calculated at the topic level, and Shen et al. presented the approach to find the latent friends who shared the similar topic distribution in their blogs.

Some literatures have been published on blog clustering. Qamra et al. [17] proposed a Content-Community-Time model that can leverage the content of entries, their timestamps, and the community structure of the blogs, to automatically discover story clusters. Bansal et al. [1] have observed that given a specific topic or event, a set of keywords will be correlated. They presented efficient algorithms to identify keyword clusters in large collections of blog posts for specific temporal intervals.

Our work is quite different from the previous studies on blogs. Most of existing work focuses on developing topic-based clustering methods or conducting content analysis for blogs. We propose a novel method to group Chinese blogs into sentiment clusters, which could facilitate public opinion monitoring for governments and business organizations.

2.2 Sentiment Analysis

Sentiment analysis is the main task of opinion mining, and most of existing work focuses on determining the sentiment orientations of documents, sentences and words. In document level sentiment analysis, documents are classified into positive and negative according to the overall sentiment expressed in them [18] [21]. However, the emotions that bloggers want to express are much more complex. Therefore, it would be too simplistic to classify the document into just positive or negative categories. In [12], a PLSA based method was used to model the hidden sentiment in the blogs, and an autoregressive sentiment-aware model was presented to predict movie box office. Lu et al. [13] used semi-supervised PLSA to solve the problem of opinion integration.

Different from the traditional classification approaches for sentiment analysis, in this paper, we propose a PLSA based sentiment clustering method for Chinese blogs. An interactive sentiment clustering method on movie reviews has been proposed in [3]. Users need to participate in the clustering approach and the clustering results are highly relevant to the users' experiences. Besides that, the emotions expressed in blogs are more complex than in movie reviews. Our method can model the multifaceted nature of sentiments and group the blogs according to sentiments they contain.

3 Modeling the Hidden Sentiment Factors in Chinese Blogs

3.1 Sentiment Lexicon Acquisition and Preparation

Previous clustering methods usually focus on the underlying topics and stories in each blogs. Vector Space Model (VSM) is used to represent the blog entries and each element in the vector is the weight of word's frequency in blog entries. Since our intention is to cluster blogs by their embedded sentiments, we employ Chinese sentiment lexicon to give the blogs a new sentiment-bearing representation.

There are some previous literatures on building sentiment dictionaries. We obtained the Chinese sentiment lexicon NTUSD used by Ku et al. [10], which contains 2,812 positive words and 8,276 negative words in Chinese. We also collect the data from Hownet Sentiment Dictionary (Hownet for short), which contains 4,566 positive words and 4,370 negative words in Chinese [8].

3.2 Chinese Blogs Preprocessing

The preprocessing approach can be described in the following steps.

(1) Given a Chinese blog *b*, firstly we segment the sentences into words by using Chinese text processing tools. We add sentiment dictionary into the user defined

vocabulary of the tools, so the precision of segment can be improved. We also tag the words with the part-of-speech information for next steps.

(2) We choose the appropriate words for sentiment analysis. Since not all kinds of the words are good emotion indicators, we need not to extract all words as candidate sentiment words. Here words with part-of-speech adjective, verb, proper noun, adverb and conjunction are selected for further processing steps.

(3) We pick out the words in the sentiment dictionary, so finally we get the sentiment words set $W = \{w_1, w_2, \ldots, w_M\}$ in the blog b.

(4) The frequencies of the words in W are counted, and the blog b is represented as feature vector. So given a blog set $B = \{b_1, b_2, \ldots, b_N\}$, we can use a $N \times M$ matrix $A = (f(b_i,w_j))_{N \times M}$ to describe the blogs, and $f(b_i,w_j)$ is the occurring frequency of sentiment word w_j in the blog b_i.

After preprocessing, the Chinese blogs can be represented as sentiment matrix A and each element denotes a sentiment word and its frequency in one blog.

3.3 A PLSA Based Approach for Modeling Hidden Sentiment Factors

The traditional methods of sentiment analysis usually focus on classifying the documents and sentences by their emotion orientation. Different from the previous work, we attempt to find the sentiment similarities between blog articles and cluster the blogs by their Hidden Sentiment Factors (HSF).

Probabilistic latent semantic analysis (PLSA) [7] and its extensions [14] have recently been applied to many text mining problems with promising results. The PLSA model can be utilized to identify the hidden semantic relationships among general co-occurrence activities. The bloggers' opinions and sentiments, which are usually written in natural languages, are often expressed in complex ways. Here we consider the blog as being generated under the influence of a number of hidden sentiment factors and each hidden factor may reflect one specific aspect of the sentiments embedded in the blog.

Given a blog set B, B is generated from a number of hidden sentiment factors $Z = \{z_1, z_2, \ldots, z_k\}$. Suppose $P(b)$ denotes the probability of picking a blog document b from B. According to probability theory, we have the following formula:

$$P(w,b) = P(b)P(w \mid b) \tag{1}$$

We model the embedded emotions as hidden sentiment factor $Z = \{z_1, z_2, \ldots, z_k\}$ and the probability $P(w|b)$ can be rewritten by latent variables z as:

$$P(w,b) = P(b)\sum_z P(w \mid z)P(z \mid b) \tag{2}$$

Where $P(w|z)$ represents the probability of choosing a word w from the sentiment word set W; $P(z|b)$ denotes the probability of picking a hidden sentiment factor z from Z. Applying Bayes rule, we can transform Formula (2) as follows:

$$P(w,b) = \sum_z P(w \mid z)P(b)P(z \mid b) = \sum_z P(w \mid z)P(z)P(b \mid z) \tag{3}$$

The EM algorithm is used to estimate the parameters in PLSA model. The probability that a word w in a blog b is explained by the latent sentiment corresponding to z is estimated during the E-step as:

$$P(z \mid b, w) = \frac{P(z)P(b \mid z)P(w \mid z)}{\sum_{z'} P(z')P(b \mid z')P(w \mid z')} \tag{4}$$

And the M-step consists of:

$$P(w \mid z) = \frac{\sum_{d} f(b, w)P(z \mid b, w)}{\sum_{b,w'} f(b, w')P(z \mid b, w')} \tag{5}$$

$$P(b \mid z) = \frac{\sum_{w} f(b, w)P(z \mid b, w)}{\sum_{d',w} f(b', w)P(z \mid b', w)} \tag{6}$$

$$P(z) = \frac{\sum_{b,w} f(b, w)P(z \mid b, w)}{\sum_{b,w} f(b, w)} \tag{7}$$

After several iteration steps, the algorithm converges when a local optimal solution is achieved. Using the Bayes rule, we can compute the posterior probability $P(z|b)$ as follows:

$$P(z \mid b) = \frac{P(b \mid z)P(z)}{\sum_{z} P(b \mid z)P(z)} \tag{8}$$

The result $P(z|b)$ in Formula (8) represents how much a hidden sentiment factor z "contributes" to the blog b and the hidden factor probability set $\{P(z|b)|z\in Z\}$ can reflect the embedded sentiments in blog b. In the next section, the sentiment similarity between blogs can be computed based on the hidden factor probabilities, and a blog sentiment clustering method is proposed.

4 Clustering Chinese Blogs by Hidden Sentiment Factors

4.1 Blog Sentiment Similarity Measurement

The sentiments that people expressed in blogs are really complex, and multi-emotions can coexist in just one blog. For example, in one blog article, the blogger can feel sad and regretful for Liu Xiang's injury, and he or she may also express hopeful altitudes and encourage Liu to survive this injury and recover as soon as possible.

Various words can be used for bloggers to express one state of emotion. For example, the words *glad*, *joyful*, *pleased* and *delightful* can express a state of happiness; the words *sore*, *sorrowful*, *woeful* can express a state of sadness. Therefore, when computing similarity, the traditional Vector Space Model will be faced with sparseness problem. The Fig.1 shows the VSM representation of two blogs. $Blog_1 = \{w_1,w_2,w_5,w_6,w_9,w_{12},w_{16},w_{17},w_{20},w_{22},w_{23}\}$ and $Blog_2 = \{w_3,w_7,w_8,w_{11},w_{15},w_{18},w_{19},w_{21}\}$. There is no overlap between two vectors. However, the two blogs may be similar from the emotion state point of view, that is to say, the words in $Blog_1$ an $Blog_2$ can reflect the same sentiment meanings. Using Formulas (1) to (8), we can model bloggers' emotions by hidden sentiment factors. As can been seen from Fig.2, the HSF has built a emotion state layer for the blog datasets, and the sentiment similarity can be calculated even if there are no overlap sentiment words between two blog articles.

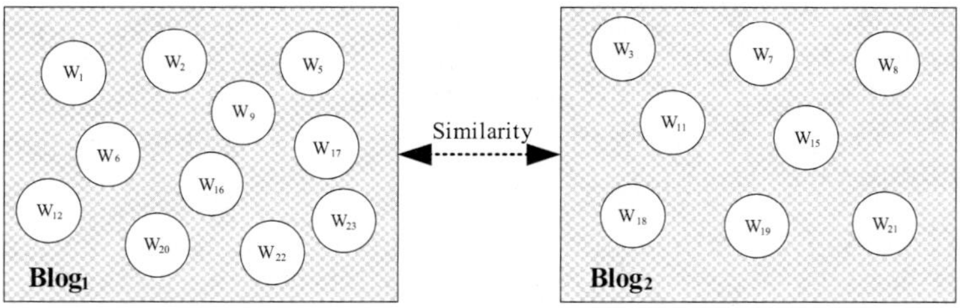

Fig. 1. The similarity measurement by Vector Space Model

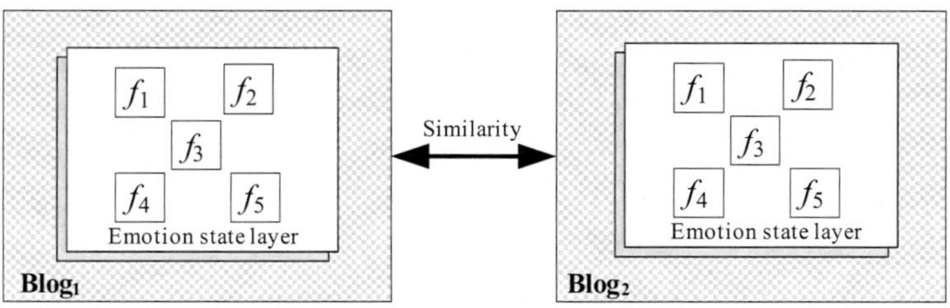

Fig. 2. The similarity measurement by Hidden Sentiment Factors

From the discussion above, we know that the set $\{P(z|b)|z\in Z\}$ can conceptually represent the probability distribution over the hidden sentiment factor space for blog set B. Suppose N represents the number of blogs, k represents the number of hidden sentiment factors, then we can build a $N\times k$ matrix $D=(p(z_j|b_i))_{N\times k}$ to reflect the relationship between blogs and latent sentiment factors, and $p(z_j|b_i)$ denotes the probability of the z_j hidden sentiment factor in blog b_i. The distance between two blog vectors can represent the emotion similarity between them. Therefore we define sentiment similarity between blogs by applying classic cosine similarity as:

$$SentiSim(\vec{b_i},\vec{b_j}) = \frac{\vec{b_i}\bullet\vec{b_j}}{\|\vec{b_i}\|\times\|\vec{b_j}\|} \tag{9}$$

where $\vec{b_i}\bullet\vec{b_j} = \sum_{m=1}^{k} p(z_m|b_i)p(z_m|b_j)$, $\|\vec{b_i}\| = \sqrt{\sum_{l=1}^{k} p(z_l|b_i)^2}$

4.2 Chinese Blogs Sentiment Clustering and Label Words Extraction

The *SentiSim* function in Formula (9) provides us a new way to measure the similarity between Chinese blogs at emotion state level. Based on *SentiSim*, we can employ the existing clustering methods to do sentiment clustering tasks. In this paper, we cluster Chinese blog entries based on K-Means algorithm, which has already been proven to be an effective text clustering method. Several concrete problems during the clustering approach are discussed as follows.

The number of clustering result categories. Given the parameter k denotes the number of hidden sentiment factors in blog dataset and k' denotes the number of clustering result categories in K-Means algorithm. If we know the number of result categories in advance, we set k' to be the number of predefined categories. However, when the number of categories is not defined beforehand, it is difficult to predict the parameter k'. In this paper, we set k' equal the number of hidden sentiment factors if there's no predefined categories.

The sentiment labels. Our intention is to make users to get better known the overall sentiments that bloggers express in the blog dataset. For each cluster C_i, suppose the number of blogs in C_i is N_i, the center of C_i can be represented by:

$$\begin{aligned} Center_i &= \{\sum_{n=1}^{N_i} p(k_1 \mid b_n)/N_i, \sum_{n=1}^{N_i} p(k_2 \mid b_n)/N_i, ..., \sum_{n=1}^{N_i} p(k_k \mid b_n)/N_i\} \\ &= \{\overline{k_1}, \overline{k_2}, ..., \overline{k_k}\} \end{aligned} \tag{10}$$

The dominant HSF θ_{ci} is:

$$\theta_{ci} = \arg\max(\overline{k_1}, \overline{k_2}, ..., \overline{k_k}) \tag{11}$$

The labels of C_i are the words that can reflect the key sentiments in C_i and contribute little to other clusters:

$$Rank(w) = p(w \mid \theta_{ci}) - \frac{1}{k-1} \sum_{C \neq C_i} p(w \mid \theta_{cn}) \tag{12}$$

We extract the top five sentiment words by $Rank(w)$:

$$L(C_i) = \{w_a, w_b, w_c, w_d, w_e \text{ where } Rank(w_a) \geq Rank(w_b) \geq Rank(w_c) \geq Rank(w_d) \geq Rank(w_e) > \text{others}\} \tag{13}$$

Sentiment rank the blogs. When blog dataset is large, it is a time consuming task for people to read all of the articles to get the overall sentiments expressed in blogs. We rank the blogs in cluster C_i by $p(b|\theta_{ci})$. θ_{ci} is the dominant HSF of C_i, therefore, if $p(b|\theta_{ci})$ has bigger value, it indicates the blog is more relevant to the main sentiments expressed in C_i. By ranking the blogs, the most topical blogs can be accessed easily and convenient for the users to quickly get the public opinion in the given blog dataset.

Using Chinese Blog Sentiment Clustering algorithm (CBSC), we can group the blogs into k' clusters and the elements in each cluster reflect similar emotions.

5 Experiments

Our experiment is conducted on a commodity PC with Windows XP, Core2 Duo CPU and 4GB RAM. Since there are no standard dataset on Chinese blogs to do the work, in this experiment we collected the blogs from the MSN Spaces [15], which are in Chinese.

Determining the orientation of movie reviews is very common and foundational work for sentiment analysis. So firstly we collect blogs about reviews on Stephen Chow's movie "*CJ7*" (Long River 7). Totally 658 blog entries are crawled, and our unsupervised sentiment clustering results are compared to annotators'.

Liu Xiang's withdrawing from Olympic Games is the hottest topic in China, and people are willing to publish their opinions about this matter on blogs. We collect the blog entries about Liu Xiang since August 18th 2008 when he quitted the Games, and these blog articles reflect people's various attitudes and feelings toward Liu's behavior. We use Google Blog Search [6] as topic retrieval tools to find the relevant blog entries and total 1,578 articles are crawled for further analysis.

Not all of the crawled blog articles contain the authors' emotions. Three people are asked to annotate the dataset by the following rules:

(1) The first two people tag the blogs as "Positive", "Negative" and "Neutral";
(2) The first two people tag the blogs without authors' sentiments as "Irrelevant";
(3) If there's a disagreement between the first two people, the third people determine the finally category.

By this annotating method, we classify "*CJ7*" blogs into three parts (Positive, Negative, Neutral) and finally we pick 100 blogs for each category to conduct the clustering experiments. Different from the blogs on movies and commercial products, people express their sentiments on controversial topics are more complex. The blogs contain the emotions such as compliment, blessing, encouragement are classified into "Positive" category; the blogs express a state of emotion such as condemnation, disappointment, sadness are classified into "Negative" category. The blogs express authors' opinions, which are hard to determine orientation, are classified into "Neutral" category. Finally, we pick out 300 blog entries on Liu's injury for each category.

Unlike English and Spanish, there is no delimiter to mark word boundaries and no explicit definition of words in Chinese languages. So the preprocessing steps need to segment Chinese text into unique word tokens. ICTCLAS [9] is a Chinese lexical analysis system, which is able to make Chinese word segmentation and part-of-speech tagging with about 98% precision. The result words with part-of-speech adjective, verb, proper noun, adverb and conjunction are selected for further calculations of sentiment scores and finally a sentiment word matrix $A = (f(b_i,w_j))_{N\times M}$ to describe the blog entries of each dataset.

Firstly, we evaluate our algorithm on predefined category clustering task. We use clustering purity [20] as a quality measure of predefined categories blog sentiment clustering. For category r, N_r denotes the number of elements in r and N_r^i denotes the number of elements in clustering result i, which belong to r, therefore we get the purity for category S_r:

$$P(S_r) = \frac{1}{N_r} \max(N_r^i) \quad (14)$$

The entire clustering purity is defined as:

$$P(B) = \sum_{r=1}^{k'} \frac{n_r}{n} P(S_r) \tag{15}$$

The clustering purities for PLSA-based algorithm are shown in Fig.3. The X-axis is the parameter k during the PLSA iteration steps, namely the number of hidden sentiment factors. Fig.3 shows that the clustering purity is relevant to the number of hidden sentiment factors. Fig.3(a) depicts the clustering purity for the dataset on the movie "*CJ7*" and we can get the best clustering purity 41.5% when k=4. Fig.3(b) shows that using NTUSD, we can get the best purity 39.3% for "*Liu Xiang's injury*" dataset when k=5. When k grows bigger, the purities decrease gradually.

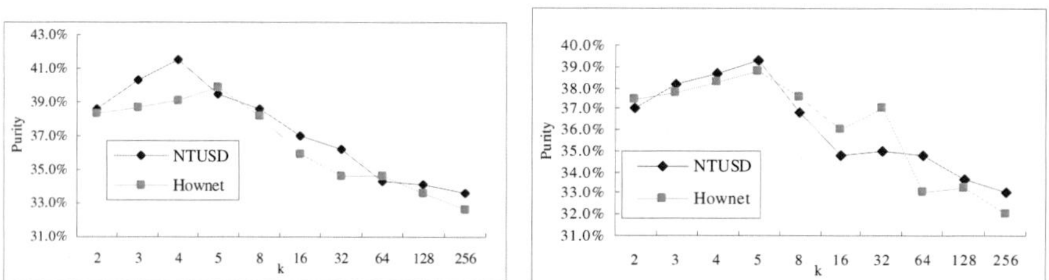

(a) Blogs about the movie "*CJ7*" (b) Blogs about "*Liu Xiang's injury*"

Fig. 3. The purity of PLSA-based blog sentiment clustering method

Table 1 and Table 2 summarize our observations for different clustering models.

Table 1. The clustering purities for "*CJ7*" blogs based on different models

	NTUSD	Hownet
VSM	36.6%	34.6%
LSA	40.9%	38.6%
CBSC	41.5%	39.9%

Table 2. The clustering purities for "*Liu Xiang's injury*" blogs based on different models

	NTUSD	Hownet
VSM	34%	32.1%
LSA	35%	34.8%
CBSC	39.3%	38.7%

We predefine result categories k'=3 for VSM and LSA model. We can see from above two tables that when using NTUSD and proposed CBSC method, we can get best clustering results. And we get better clustering purities on "*CJ7*" dataset than on "*Liu Xiang's injury*" dataset, however, our CBSC method achieve less improvement on "*CJ7*" dataset than on "*Liu*" dataset. It can be seen from Table 2 that blog sentiment clustering is really a hard problem, and this may because that the emotions expressed on controversial topics are much more complex than on movie and product reviews. Table 2 shows that in sentiment clustering task, CBSC-based method could achieve 15.5% improvement than VSM method and 12% improvement than LSA method.

Since the embedded emotions in given dataset are complex, we attempt to cluster the blogs into more categories, by this way we can extract more detailed and topical sentiments in the dataset. NTUSD is employed as sentiment lexicon. For "*CJ7*" blogs, we set hidden sentiment factor to be 4 and set the hidden sentiment factor to be 5 for "*Liu Xiang's injury*" blogs. Finally, we get the following clustering results, as shown in Fig.4, Table 3 and Table 4.

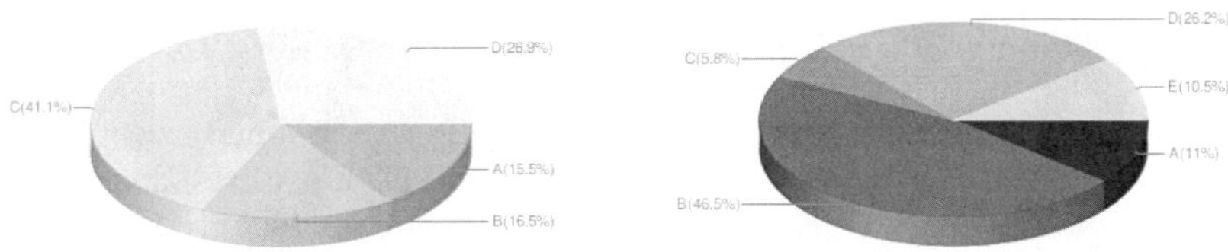

(a) Blogs about the movie "*CJ7*" (b) Blogs about "*Liu Xiang's injury*"

Fig. 4. The clustering results for CBSC algorithm

Table 3. Cluster labels extracted from each cluster of "CJ7"

Clusters	Key Sentiment Words
A	Hope, recommend, ability Believe, hardship
B	weep, familiar, touched deserved, dream
C	disappointed, prepare, exaggerated interest, bored
D	cute, real, finished humor, comedic

Table 4. Cluster labels extracted from each cluster of "*Liu Xiang's injury*"

Clusters	Key Sentiment Words
A	pathetic, monopolize, commiserative finished, regretful
B	development, bad, eliminate cheat, abuse
C	fair, perfect, ego fortunate, hope
D	patriotic, logic, courtesy stunning, contention
E	attention, battlefield, fearful condemnation, obstacle

Fig.4(a) shows that the "*CJ7*" blog dataset has been partitioned into 4 clusters. We use the Formula (12) to extract cluster labels in each cluster and the corresponding cluster labels are shown in Table 3. We can see from Fig.4(b) that, our method has clustered the "*Liu Xiang's injury*" dataset into 5 groups and the largest group B has about 46.5% of all blog articles and the smallest group C has about 5.8% of all blog articles. We can see from Table 3 and Table 4 that cluster labels provide a very brief summarization of the sentiments in each cluster. The blog ranking algorithm is applied in each cluster. Volunteers are asked to explore in the clustering result data space. A survey is done among the volunteers and people response that in most cases, the top *N* articles in each group reflect similar emotions and the clustering results can help them to make better understanding of bloggers' opinions about Liu Xiang's withdrawing Olympic Games. However, as *N* grows bigger, the sentiments in blogs in the lower position are less relevant to the corresponding cluster labels.

From the above experiments, we can see that sentiment clustering is a hard problem for blog data. This is because that people usually express multi-facet sentiments in blogs on controversial topics. Grouping the blogs into just three categories (Positive, Negative, Neutral) may lose many detailed emotions embedded in blogs. The proposed clustering algorithm based on hidden sentiment factors can reflect the complexity of emotions and extracted cluster labels are effective navigation guidelines for users to quickly get the public opinions on given topics.

6 Conclusion and Future Work

The traditional blog clustering methods usually partition blogs by topics or keywords. In this paper, we propose an emotion-oriented clustering algorithm to group Chinese blog articles according to hidden sentiment factors. Experimental results demonstrate that we can generate correct clusters by their embedded emotions and extract the sentiment labels for each cluster, thus could help users to quickly get bloggers' topical sentiments on given topics.

Further research directions include finding a way to extract the common opinions in each sentiment cluster. And also, the product reviews are useful data source for both individual customers and business companies. We will use sentiment clustering method to analyze people's opinions about certain product and help latent customers and company leaders to make decisions.

References

1. Bansal, N., Chiang, F., Koudas, N., Tompa, F.: Seeking Stable Clusters in the Blogosphere. In: 33rd International Conference on Very Large Data Bases, pp. 390–398 (2007)
2. Bar-Ilan, J.: An Outsider's View on "Topic-oriented" Blogging. In: 13th International Conference on World Wide Web Alternate Papers Track, pp. 28–34 (2004)
3. Bekkerman, R., Raghavan, H., Allan, J., Eguchi, K.: Interactive Clustering of Text Collections According to a User-Specified Criterion. In: 20th International Joint Conference on Artificial Intelligence, pp. 684–689 (2007)
4. China Internet Network Information Center (CNNIC), `http://www.cnnic.cn/en/index`

5. Glance, N., Hurst, M., Tornkiyo, T.: Blogpulse: Automated Trend Discovery for Weblogs. In: WWW 2004 Workshop on the Weblogging Ecosystem (2004)
6. Google Blog Search, http://blogsearch.google.com
7. Hofmann, T.: Probabilistic Latent Semantic Indexing. In: 22nd International ACM SIGIR Conference on Research and Development in Information Retrieval, pp. 50–57 (1999)
8. HowNet, http://www.keenage.com/html/e_index.html
9. ICTCLAS, http://www.ictclas.org
10. Ku, L., Chen, H.: Mining Opinions from the Web: Beyond Relevance Retrieval. Journal of American Society for Information Science and Technology 58(12), 1838–1850 (2007)
11. Kumar, R., Novak, J., Raghavan, P., Tomkins, A.: Structure and Evolution of Blogspace. Commun. ACM 47(12), 35–39 (2004)
12. Liu, Y., Huang, X., An, A., Yu, X.: ARSA: a Sentiment-aware Model for Predicting Sales Performance Using Blogs. In: 30th International ACM SIGIR Conference on Research and Development in Information Retrieval, pp. 607–614 (2007)
13. Lu, Y., Zhai, C.: Opinion Integration through Semi-supervised Topic Modeling. In: 17th International Conference on World Wide Web, pp. 121–130 (2008)
14. Mei, Q., Zhai, C.: A Mixture Model for Contextual Text Mining. In: Twelfth ACM SIGKDD International Conference on Knowledge Discovery and Data Mining, pp. 649–655 (2006)
15. MSN Spaces, http://home.services.spaces.live.com
16. Ni, X., Xue, G., Ling, X., Yu, Y., Yang, Q.: Exploring in the Weblog Space by Detecting Informative and Affective Articles. In: 16th International Conference on World Wide Web, pp. 281–290 (2007)
17. Qamra, A., Tseng, B., Chang, E.: Mining Blog Stories Using Community Based and Temporal Clustering. In: Thirteen ACM Conference on Information and Knowledge Management, pp. 390–398 (2004)
18. Pang, B., Lee, L., Vaithyanathan, S.: Thumbs up? Sentiment Classification Using Machine Learning Techniques. In: 2002 Conference on Empirical Methods in Natural Language Processing, pp. 79–86 (2002)
19. Shen, D., Sun, J., Yang, Q., Chen, Z.: Latent Friend Mining from Blog Data. In: 6th IEEE International Conference on Data Mining, pp. 552–561 (2006)
20. Steinbach, M., Karypis, G., Kumar, V.: A Comparison of Document Clustering Techniques. In: Sixth ACM SIGKDD International Conference on Knowledge Discovery and Data Mining (2006)
21. Turney, P.: Thumbs Up or Thumbs Down? Semantic Orientation Applied to Unsupervised Classification of Reviews. In: 40th Annual Meeting of the Association for Computational Linguistics, pp. 417–424 (2002)

Semi Supervised Image Spam Hunter: A Regularized Discriminant EM Approach

Yan Gao[1], Ming Yang[2], and Alok Choudhary[1]

[1] Dept. of EECS, Northwestern University,
Evanston, IL, USA
[2] NEC Laboratories America,
Cupertino, CA, USA
{ygao@cs,choudhar@eecs}.northwestern.edu, myang@sv.nec-labs.com

Abstract. Image spam is a new trend in the family of email spams. The new image spams employ a variety of image processing technologies to create random noises. In this paper, we propose a semi-supervised approach, regularized discriminant EM algorithm (RDEM), to detect image spam emails, which leverages small amount of labeled data and large amount of unlabeled data for identifying spams and training a classification model simultaneously. Compared with fully supervised learning algorithms, the semi-supervised learning algorithm is more suitedin adversary classification problems, because the spammers are actively protecting their work by constantly making changes to circumvent the spam detection. It makes the cost too high for fully supervised learning to frequently collect sufficient labeled data for training. Experimental results demonstrate that our approach achieves 91.66% high detection rate with less than 2.96% false positive rate, meanwhile it significantly reduces the labeling cost.

1 Introduction

Spam is e-mail that is both unsolicited by the recipient and sent in nearly identical form to numerous recipients. Research reveals that 96.5% of incoming e-mails received by businesses were spam by June 2008 [1], and spam management costs U.S. businesses more than $70 billion annually [2]. As of 2007, image spam accounted for about 30% of all spam [3]. Image spam has become a more and more deteriorating issue in recent years [4].

Most current content-based spam filtering tools treat conventional email spam detection as a text classification problem, utilizing machine learning algorithms such as neural networks, support vector machine (SVM) and naïve Bayesian classifiers to learn spam characteristics [5–10]. These text-based anti-spam approaches achieved outstanding accuracy and have been widely used. In response, spammers have adopted a number of countermeasures to circumvent these text-based filters. Embedding spam messages into images, usually called "image spam", is one of the most recent and popular spam construction techniques.

R. Huang et al. (Eds.): ADMA 2009, LNAI 5678, pp. 152–164, 2009.

Fig. 1. Sample spam images

Typically the image spam contains the same types of information advertised in traditional text-based spams, while similar techniques from CAPTCHA (Completely Automated Public Turing Test to Tell Computers and Humans Apart) are applied to manipulate the texts in the image. These techniques include adding speckles and dots in the image background, randomly changing image file names, varying borders, randomly inserting subject lines, and rotating the image slightly. Figure 1 shows some examples. The consequence is an almost infinite number of image-based spams that contain random variants of similar spam content. This kind of spam images is typically attached to or embedded in text with randomly generated good words or content lifted from famous literature. Through this, image spam has successfully bypassed text-based spam filters and presented a new challenge for spam researchers.

In the early stage, there are several organizations and companies working on filtering image-based spam using Optical Character Recognition (OCR) techniques [11, 12]. SpamAssassin (SA) [13] pulls words out of the images and uses the traditional text-based methods to filter spams. This strategy was unavoidably defeated by the appearance of CAPTCHA. Therefore, it is an urgent need to develop a fully automatic image content based spam detection system. Several recent research works are targeting on it, such as the image spam hunter proposed by Gao et al. [14], Dredze et al's fast image spam classifier [15], and near duplicate image spam detection [16, 17]. Most of them leverage supervised machine learning algorithms to build a classifier for filtering spam images [14, 15] by using image-based features.

However, in an adversary classification problem [18] like spam detection, it is not sufficient to just train a classifier once. The reason resides in the fact that any machine learning algorithms are estimating models based on the data statistics, and the assumption is that the statistics of the data used for training are similar to the data statistics in testing. However, spammers are always trying to counteract them by adapting their spamming algorithms to produce image spam emails with feature statistics different from what the anti-spam algorithms have been trained upon. Therefore, the anti-spam algorithms may need to be re-trained from time to time to capture the adversary changes of spam statistics.

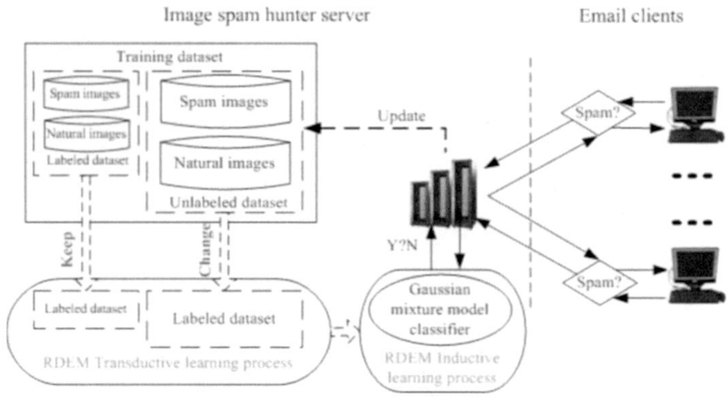

Fig. 2. Prototype system diagram

Furthermore, collecting sufficient labeled spam images for re-training the classifiers is a very labor intensive and costly task. Therefore, it is not desirable, if not possible, to label a large amount of images attached in emails for re-training the classifiers each time, especially when it has to be done very frequently to keep the pace with the spammers. In order to avoid such high cost of labeling large amount of data, a semi-supervised learning scheme [19] is a more efficient choice, where we can leverage small amount of labeled image data and large amount of unlabeled data for training the classifiers.

In this paper, we propose a regularized discriminant EM algorithm (RDEM) to perform semi-supervised spam detection. RDEM improves discriminant EM (DEM) algorithm by leveraging semi-supervised discriminant analysis. In particular, we regularize the cost function of multiple discriminant analysis (MDA) [20] in DEM with a Laplacian graph embedding penalty. Inherited from the DEM, our approach performs both *transductive* learning and inductive learning. We test the proposed approach on an image spam dataset collected from Jan 2006 to Mar 2009, which contains both positive spam images collected from our email server, and negative natural images randomly downloaded from Flickr.com and other websites by performing image search using Microsoft Live Search. Our approach achieves 2.96% false positive rates with 91.66% true positive rates with less than 10% labeled data on average. Comparison results with previous literature demonstrate the advantages of our proposed approach.

2 System Framework of RDEM Image Spam Hunter

In this section, we describe a semi-supervised Image Spam Hunter system prototype to differentiate spam images from normal image attachments. Figure 2 shows the system diagram. We first randomly choose and label a small percentage of spam images as the positive samples and general photos as the negative samples to form the labeled training dataset. The unlabeled training dataset is randomly chosen from the mixed pool of spam images and normal photos. There is no need for clustering the spam images and normal photos into groups in our prototype system, because the Gaussian mixture model (GMM) in our algorithm is able to deal with the multi-class categorization problem automatically.

The RDEM algorithm, which will be further detailed in Section 3.3, is then applied to the training dataset, which includes both small amount of labeled training data and large amount of unlabeled training data, to build a model for distinguishing the image spams from good emails with image attachments. The unlabeled training data are labeled in this process, which is the *transductive* learning part of the proposed RDEM algorithm. A Gaussian mixture model is induced simultaneously in a discriminative embedding space, which could be further used to classify new data. This is the *inductive* learning part of the RDEM algorithm. Because of this joint transductive and inductive learning process, our proposed semi-supervised image spam hunter is robust to the random variations that exist in current spam images, and easy to adapt to the new changes for image spams in terms of the low labeling cost.

It is worth noting that our semi-supervised spam hunter also fits well as a helpful component running at the beginning of other supervised anti-spam systems to boost the small amount of labeled data. Once enough labeled data is generated through the component, a fully supervised classifier could be further trained for automated spam detection. In a sense, the proposed semi-supervised spam detection scheme could also be functioned as a bootstrap system for a fully supervised spam hunter such as the one proposed by Gao et al. [14].

3 Regularized Discriminant EM Algorithm

We improve the discriminant EM (DEM) [21] algorithm for semi-supervised learning, which introduces a graph Laplacian penalty [22, 23] to the discriminant step of the DEM algorithm. We call it regularized DEM algorithm (RDEM). In the rest of this section, we first introduce the classical unsupervised EM algorithm [24], then present the details of the DEM algorithm [21] and RDEM algorithm, respectively.

3.1 EM Algorithm

EM [24] algorithm is an iterative method to perform maximum likelihood parameter estimation with unobserved latent variables in a probabilistic model. Formally, let $D = \{(x_i, z_i)\}_{i=1}^N$ where $x_i \in R^n$ is the observed data, z_i is the unobserved data, and θ is the parameter vector which characterizes the probabilistic model of D. Denote $Z = \{z_i\}_{i=1}^N$, $X = \{X_i\}_{i=1}^N$, and the log likelihood function by $L(X, Z, \theta)$. In our formulation, we assume that the data model is a Gaussian mixture model (GMM) of k components, therefore $\theta = \{(\omega_j, \mu_j, \Sigma_j)\}_{j=1}^k$, where ω_j, μ_j and Σ_j are the mixture probability, mean, and covariance matrix of the j-th Gaussian component $G(x|\omega_j, \mu_j, \Sigma_j)$, respectively. Furthermore, we define $z_i = \{z_{ij}\}_{j=1}^k$ where $0 \le z_{ij} \le 1$ represents how likely data point x_i belongs to the j-th Gaussian component. Let θ^{t-1} be the estimated parameter at the iteration $t-1$ of the EM algorithm, at iteration t, the EM algorithm runs the following two steps to estimate the Gaussian mixture model:

E-Step: Calculate the expected value of $L(X,Z,\theta)$ w.r.t. $p(z|x,\theta^{t-1})$, given the current θ^{t-1}, i.e., $Q(\theta|\theta^{t-1}) = E(L(X,Z,\theta)|x,\theta^{t-1})$. We have

$$z_{ij}^t = \frac{\omega_j^{t-1} G\left(x_i \middle| \omega_j^{t-1}, \mu_j^{t-1}, \Sigma_j^{t-1}\right)}{\sum_k \omega_k^{t-1} G(x_i|\omega_k^{t-1}, \mu_k^{t-1}, \Sigma_k^{t-1})}, \tag{1}$$

$$Q(\theta|\theta^{t-1}) = \sum_{i=1}^{N}\sum_{j=1}^{k} z_{ij}^t [\log \omega_j^{t-1} - \frac{1}{2}\log|\Sigma_j^{t-1}| - \frac{1}{2}(x_i - \mu_j)^T \sum_j^{t-1^{-1}} (x_i - \mu_j) - \frac{n}{2}\log 2\pi]. \tag{2}$$

M-Step: Find the parameter θ^t such that $\theta^t = \arg\max_\theta Q(\theta|\theta^{t-1})$. We have

$$\omega_j^t = \frac{1}{n}\sum_i z_{ij}^t, \mu_j^t = \frac{\sum_i z_{ij}^t x_i}{\sum_i z_{ij}^t}, \tag{3}$$

$$\Sigma_j^t = \frac{\sum_i z_{ij}^t (x_i - \mu_j^t)(x_i - \mu_j^t)^T}{\sum_i z_{ij}^t}. \tag{4}$$

Estimating GMM using EM is a popular approach for unsupervised clustering of data. The EM iterations are guaranteed to find a local optimal estimation.

3.2 Discriminant EM Algorithm

Discriminant EM [21] (DEM) is a semi-supervised extension of the original EM algorithm. It assumes that the data can be categorized into c different classes, and the structure of the data can be captured by a GMM with c components in a $c-1$ dimensional discriminant embedding space. Let $D = \{(x_i, l_i)\}_{i=1}^{N_1}$ be the set of labeled data, where $l_i \in \{1,2,\cdots,c\}$ is the label of the data $x_i \in R^n$. Let $U = \{(u_i, z_i)\}_{i=1}^{N_2}$ be the set of unlabeled data, where $z_i = \{z_{ij}\}_{j=1}^{c}$ is the unknown soft labels of $u_i \epsilon R^n$. Moreover, let $\widetilde{x_\iota}$ and $\widetilde{u_\iota}$ be the projection of x_i and u_i in the $c-1$ dimensional discriminant embedding represented by an n × (c − 1) projection matrix W, i.e., $\widetilde{x_\iota} = W^T x_i$ and $\widetilde{u_\iota} = W^T u_i$. Also let $(\widetilde{\omega_\iota}, \widetilde{\mu_\iota}, \widetilde{\Sigma_\iota})$ be the parameters of the i-th Gaussian component $G(x|\widetilde{\omega_J}, \widetilde{\mu_J}, \widetilde{\Sigma_J})$ of the GMM in the embedding space. The DEM algorithm is composed of the following three steps:

E-Step: Estimate the probabilities of the class labels for each unlabeled data u_i, i.e.,

$$z_{ij}^t = \frac{\widetilde{\omega}_j^{t-1} G\left(\tilde{u}_i^{t-1} \middle| \widetilde{\omega}_j^{t-1}, \tilde{\mu}_j^{t-1}, \widetilde{\Sigma}_j^{t-1}\right)}{\sum_k \widetilde{\omega}_k^{t-1} G\left(\tilde{u}_i^{t-1} \middle| \widetilde{\omega}_k^{t-1}, \tilde{\mu}_k^{t-1}, \widetilde{\Sigma}_k^{t-1}\right)}. \tag{5}$$

D-Step: Perform multiple discriminant analysis [20] based on the labeled data D and soft labeled data U, by solving the following optimization problem to identify the optimal embedding W^t, i.e.,

$$w^t = arg\max_w \frac{w^T S_b w}{w^T S_w w}, \tag{6}$$

where

$$S_b = \sum_{j=1}^{C} (m - m_j)(m - m_j)^T, \tag{7}$$

$$S_w = \sum_{i=1}^{N_1} (x_i - m_{l_i})(x_i - m_{l_i})^T + \sum_{i=1}^{N_2} \sum_{j=1}^{C} z_{ij} (u_i - m_j)(u_i - m_j)^T, \tag{8}$$

$$m = \frac{1}{N_1 + N_2} \left(\sum_{i=1}^{N_1} x_i + \sum_{i=1}^{N_2} u_i \right), \tag{9}$$

$$m_j = \frac{1}{\sum_{i=1}^{N_1} \delta(l_i, j) + \sum_{i=1}^{N_2} z_{ij}} \left(\sum_{i=1}^{N_1} \delta(l_i, j) x_i + \sum_{i=1}^{N_2} z_{ij} u_i \right), \tag{10}$$

and $\delta(l_i, j)$ is the Dirac delta function which takes value one when $l_i$equals j and zero otherwise. $W^t$is composed of the eigen vectors corresponding to the largest $C - 1$ eigen values of the generalize eigen system $S_b w = \lambda S_w w$. Then both the labeled and unlabeled data are projected into the embedding, i.e.,

$$\tilde{x}_i^t = W^{t^T} x_i, \tilde{u}_i^t = W^{t^T} u_i. \tag{11}$$

M-Step: Estimate the optimal parameters of the GMM in the embedding space, i.e.,

$$\tilde{\omega}_j^t = \frac{1}{N_1 + N_2} \left(\sum_{i=1}^{N_1} \delta(l_i, j) + \sum_{i=1}^{N_2} z_{ij}^t \right), \tag{12}$$

$$\tilde{\mu}_j^t = \frac{\sum_{i=1}^{N_1} \delta(l_i, j)\, \tilde{x}_i^t + \sum_{i=1}^{N_2} z_{ij}^t\, \tilde{u}_i^t}{\sum_{i=1}^{N_1} \delta(l_i, j) + \sum_{i=1}^{N_2} z_{ij}^t}, \tag{13}$$

$$\Sigma_j^t = \frac{\sum_{i=1}^{N_1} \delta(l_i, j)\, (\tilde{x}_i^t - \mu_j^t)(\tilde{x}_i^t - \mu_j^t)^T + \sum_{i=1}^{N_2} z_{ij}^t\, (\tilde{u}_i^t - \mu_j^t)(\tilde{u}_i^t - \mu_j^t)^T}{\sum_{i=1}^{N_1} \delta(l_i, j) + \sum_{i=1}^{N_2} z_{ij}^t}. \tag{14}$$

These three steps are iterated until convergence. As we have already discussed, although DEM itself is a semi-supervised algorithm, the D-step is a purely supervised step. This is not desirable because it fully trusts the labels estimated from the E-step. We proceed to replace it with a semi-supervised discriminant analysis algorithm.

3.3 Regularized Discriminant EM Algorithm

Cai et al. [25] and Yang et al. [26] propose a semi-supervised discriminant analysis algorithm to leverage both labeled and unlabeled data to identify a discriminant embedding for classification. Following the common principle of learning from unlabeled data, which is to respect the structure of the data, semi-supervised discriminant analysis introduces a graph Lapacian regularization term into multiple discriminant analysis, based on the regularized discriminant analysis framework proposed by Friedman [27]. The intuition of applying the graph Laplacian regularization is that in a classification problem, data points which are close to one another are more likely to be categorized in the same class. More formally, for the unlabeled data set U, let $U = [u_1, u_2, \cdots, u_{N_2}]$ be the $n \times N_2$data matrix, we define

$$s_{kl} = \begin{cases} 1, & u_k \in N_p(u_l) \| u_l \in N_p(u_k) \\ 0, & otherwise \end{cases}, \quad (15)$$

where $N_p(u)$ indicates the p-nearest neighbors of the data point u. Let $S = [s_{kl}]$, and $D = diag[d_{kk}]$ where $d_{kk} = \sum_{l=1}^{N_2} s_{kl}$. Both are $N_2 \times N_2$ matrices. S defines a p-nearest neighbor graph. Following previous work on spectral cluster [22, 23], the graph Laplacian is naturally defined as

$$J(w) = \sum_{k=1}^{N_2} \sum_{l=1}^{N_2} s_{kl} (w^T u_k - w^T u_l)^2 \quad (16)$$

$$= 2 \sum_{k=1}^{N_2} w^T u_k d_{kk} u_k^T w - 2 \sum_{k=1}^{N_2} \sum_{l=1}^{N_2} w^T u_k s_{kl} u_l^T w \quad (17)$$

$$= 2w^T U(D - S) U^T w = 2w^T U L U^T w, \quad (18)$$

there L is the Laplacian matrix [22, 23]. It is clear minimizing $J(w)$ with respect to w would result in that data close to one another would be also close to one another in the embedding space. Following the regularized discriminant analysis [27], we introduce this graph Laplacian [22] regularization term into the multiple discriminant analysis cost function (i.e., Equation 6), i.e.,

$$w^t = \arg\max_w \frac{w^T S_b w}{w^T S_w w + \beta w^T U L U^T w}, \quad (19)$$

where $S_b$and $S_w$are defined in Equation 7-10, and β is a control parameter to balance between the supervised term and unsupervised term, respectively. In the D-Step, we shall replace Equation 6 with Equation 19 and perform a semi-supervised discriminant analysis. We denote $S_w' = S_w + \beta U L U^T$. Because of the graph Laplacian regularization term, W is composed of the C eigenvectors corresponding to the C largest eigen values in the generalized eigen system $S_b w = \lambda S_w' w$. Keep the other two steps unchanged in the DEM algorithm discussed in the previous subsection, we propose an improved DEM algorithm. Named after the regularized discriminant analysis, we call it regularized DEM algorithm (RDEM). One thing we should notice that there is a small difference between Equation 19 and the formulation proposed by Cai et al. [25], because our formulation also takes the soft labels of the unlabeled data into consideration. It is clear that when β= 0, Equation 19 degenerates to Equation 6. In a sense, DEM is a special case of the regularized DEM algorithm.

Inherited from the DEM, our proposed approach performs both *transductive* learning and *inductive* learning. It performs transductive learning since the unlabeled data will be labeled after the training process by the maximum a posteriori estimation. Meanwhile, the induced GMM model in the discriminative embedding space can be used straightforwardly for classifying new data samples.

4 Image Features

We adopt an effective set of 23 image statistics [28–30] integrating color, texture, shape, and appearance properties for image spam detection.

Color Statistics: We build a 10^3 dimensional color histogram in the joint RGB space, i.e., each color band is quantized into 10 different levels. The entropy of this joint RGB histogram is the first statistics we adopted. We also build an individual 100 dimensional histogram for each of the RGB band, 5 statistics are calculated from each of these three histograms, including the discreteness, mean, variance, skewness, and kurtosis. The discreteness is defined as the summation of all the absolute differences between any two consecutive bins. The other four are all standard statistics. Hence we adopt 16 color statistics in total.

Texture Statistics: We employ the local binary pattern (LBP) [31] to analyze the texture statistics. A 59 dimensional texture histogram is extracted. It is composed of 58 bins for all the different uniform local binary patterns, i.e., those with at most two 0∼1 transitions in the 8-bit stream, plus an additional bin which accounts for all other non-uniform local binary patterns. The entropy of the LBP histogram is adopted as one feature. This adds in 1 texture statistics.

Shape Statistics: To account for the shape information of the visual objects, we build a 40×8 = 320 dimensional magnitude-orientation histogram for the image gradient. The difference between the energies in the lower frequency band and the higher frequency band are used as 1 feature. The entropy of the histogram is another feature. We further run a Canny edge detector [32] and the total number of edges and the average length of the edges are adopted as two statistics. These produce 4 shape statistics in total.

Appearance Statistics: We build the spatial correlogram [33] of the grey level pixels within a 1-neighborhood. The average skewness of the histograms formed from each slice of the correlogram is utilized as one feature. Another feature is the average variance ratio of all slices, where the variance ratio is defined as the ratio between the variance of the slice and the radius of the symmetric range over the mean of the slice that accounts for 60% of the total counts of the slice. These add up to 2 appearance statistics.

5 Experiments

Three quantities, i.e., *recognition accuracy, true positive rate, and false positive rate*, are adopted to compare the different approaches. The recognition accuracy stands for the overall classification accuracy of both spam and non-spam images. The true

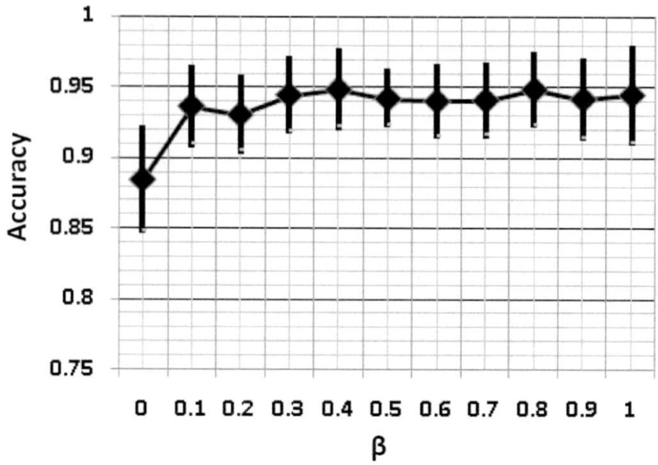

Fig. 3. The recognition accuracy of the RDEM algorithm with different setting of β. Each bar presents the average recognition accuracy and the standard deviation over 20 random label/unlabel splits. Note $\beta = 0$ is equivalent to run the original DEM algorithm.

positive rate represents the portion of the spam images being classified as spams, while the false positive rate indicates the portion of the non-spam images being classified as spam. We have to remark that most often a spam detection system would prefer to work with low false positive rate, and very few missing detections of spams are acceptable.

5.1 Data Collection

We collected two sets of images to evaluate our semi-supervised spam hunter system: normal images and spam images. We collected 1190 spam images from real spam emails received by 10 graduate students in our department between Jan 2006 and Feb 2009. These images were extracted from the original spam emails and converted to jpeg format from bmp, gif and png formats. Since we anticipate the statistics of normal images will be similar to those photo images found in social networking sites and image search results from popular search engines, we collected 1760 normal images by either randomly downloading images from Flickr.com or fetching the images from other websites from the search results on Microsoft Live Image Search (http://www.live.com/?scope=images).

5.2 Comparison to DEM

To simulate the real application scenarios, we randomly sample 10% (small portion) of the images from the data we collected to represent the labeled data, and the rest are regarded as the unlabeled data. We call one such random sample as 1 split. Since β in Equation 19 controls the impact of the graph Laplacian regularization in the RDEM algorithm, we first explore the impact of it. The recognition accuracy is calculated based on the maximum a posteriori label estimate.

Figure 3 presents the recognition accuracy with different settings of β. We test 10 different settings to vary β from 0.1 to 1.0 with a stride of 0.1. Each black marker presents the average recognition accuracy over 20 random splits. The bar overlayed on each marker presents standard deviation of the recognition accuracy over 20 splits. As we can clearly observe, the recognition accuracy of RDEM is not that sensitive to the setting of β. For $\beta \neq 0$, the average detection accuracies are all above 90%. Indeed, the marker corresponding to $\beta = 0$ is exactly the recognition accuracy of the DEM algorithm. It shows that RDEM is superior to DEM with all the different settings of β.

In the experiments, we use 3 Gaussian components to model the spam images, and another 3 Gaussian components to model ordinary images. Since $\beta = 0.8$ presents the best recognition accuracy of 94.84% for RDEM, we use it for all the other comparisons. We further compare RDEM and DEM in terms of the average true positive rate and false positive rate in Table 1. As we can see the RDEM outperforms the original DEM algorithm significantly, i.e., it achieves both higher true positive rate and lower false positive rate than the DEM. It is also worth noting that the detection results of the RDEM algorithm are also more consistent, i.e., the standard deviation of the detection rates from it are smaller. This manifests that RDEM is statistically more stable.

Table 1. Comparison of RDEM with DEM algorithm.

Method	Ave. True Positive	Ave. False Positive
DEM	85.38%±5.20%	9.69%±7.00%
RDEM($\beta = 0.8$)	91.66%±2.33%	2.96%±1.45%

5.3 Comparison to Supervised Learning Methods

Table 2 shows the comparison of the accuracy of our RDEM method against two popular supervised learning methods, the Boosting tree [34] and SVM [35], with different amount of labeled data. The Boosting tree [34] was leveraged by Gao et al. [14] to detect the spam images, and SVM [35] has demonstrated to be the optimal classifier in many applications. We can observe that RDEM demonstrates consistent performance gain over the Boosting tree and SVM with either 1%, 5% or 10% of labeled data. For example, RDEM can still achieve 88.40% true positive rate with a false positive rate of 5.61%, given the labeled data only accounts for 1% of the total data in our data collection (i.e., 12 spam images and 18 normal photos).

As we also observe, when the number of labeled data is small, the variances of the true positive rates of both the Boosting tree and the SVM are much higher than that of the RDEM algorithm. This is quite understandable since for strong supervised learning algorithms such as boosting tree and SVM, lacking of labeled training data would make the learning process very brittle and unstable. Hence they show very high variances. Our preliminary results show a very good cost performance of RDEM. The small number of labeled data in the training stage is extremely valuable for the real client-side email spam detection system, as it avoids annoying the end users by the tedious task of labeling a lot of image spams, and provides the spam detection system a good start.

Table 2. Comparison of the RDEM against the Boosting tree and SVM methods with different amounts of labeled data

Method	Ave. True Positive Rate		
	1.0%	5.0%	10.0%
RDEM ($\beta = 0.8$)	88.40%±3.19%	90.89%±3.57%	91.66%±2.33%
Boosting tree	67.09%±38.92%	72.99%±36.64%	86.87%±5.71%
SVM	19.39%±33.74%	51.59%±25.21%	68.51%±17.48%
Method	Ave. False Positive Rate		
	1.0%	5.0%	10.0%
RDEM ($\beta = 0.8$)	5.61%±4.09%	3.61%±2.82%	2.96%±1.45%
Boosting tree	4.85%±4.15%	5.06%±3.94%	3.44%±1.94%
SVM	12.65%±29.72%	9.80%±13.48%	9.25%±9.60%

6 Conclusion and Future Work

We proposed a semi-supervised system prototype based on a regularized discriminant EM algorithm to detect the spam images attached in emails. The proposed method employs a small amount of labeled data and extracts efficient image features to perform both transductive and inductive learning to detect the spam images, and achieves promising preliminary results. Future research will be focusing on further improving the computational efficiency of the RDEM algorithm, and exploring more discriminative image features.

References

1. Sophos Plc: `http://www.sophos.com/pressoffice/news/articles/2008/07/dirtydozjul08.html`
2. Necleus Research: `http://nucleusresearch.com/research/notes-and-reports/spamthe-repeat-offender/`
3. McAfee: `http://www.avertlabs.com/research/blog/index.php/2007/05/25/arespammers-giving-up-on-image-spam/`
4. Hayati, P., Potdar, V.: Evaluation of spam detection and prevention frameworks for email and image spam a state of art. In: Proc. Conf. on Information Integration and Web-based Application and Services, Linz, Austria (November 2008)
5. Sahami, M., Dumais, S., Heckerman, D., Horvitz, E.: A bayesian approach to filtering junk e-mail. In: Proc. AAAI Workshop on Learning for Text Categorization, Madison, Wisconsin (July 1998)
6. Drucker, H., Wu, D., Vapnik, V.N.: Support vector machines for spam categorization. IEEE Transactions on Neural Networks 10, 1048–1054 (1999)
7. Carreras, X., Salgado, J.G.: Boosting trees for anti-spam email filtering. In: Proc. the 4th International Conference on Recent Advances in Natural Language Processing, Tzigov Chark, BG, pp. 58–64 (2001)
8. Boykin, P.O., Roychowdhury, V.P.: Leveraging social networks to fight spam. Computer 38(4), 61–68 (2005)

9. Blosser, J., Josephsen, D.: Scalable centralized bayesian spam mitigation withbogofilter. In: USENIX LISA (2004)
10. Li, K., Zhong, Z.: Fast statistical spam filter by approximate classifications. In: ACM SIGMETRICS, pp. 347–358 (2006)
11. Fumera, G., Pillai, I., Rolir, F.: Spam filtering based on the analysis of text information embedded into images. Journal of Machine Learning Research 6, 2699–2720 (2006)
12. Biggio, B., Fumera, G., Pillai, I., Roli, F.: Image spam filtering using visual information. In: ICIAP (2007)
13. SpamAssassin: `http://spamassassin.apache.org`
14. Gao, Y., Yang, M., Zhao, X., Pardo, B., Wu, Y., Pappas, T., Choudhary, A.: Imagespam hunter. In: Proc. of the 33rd IEEE International Conference on Acoustics, Speech, and Signal Processing, Las Vegas, NV, USA (April 2008)
15. Dredze, M., Gevaryahu, R., Elias-Bachrach, A.: Learning fast classifiers for imagespam. In: Proc. the 4th Conference on Email and Anti-Spam (CEAS), California, USA (August 2007)
16. Mehta, B., Nangia, S., Gupta, M., Nejdl, W.: Detecting image spam using visual features and near duplicate detection. In: Proc. the 17th International World Wide Web Conference, Beijing, China (April 2008)
17. Wang, Z., Josephson, W., Lv, Q., Charikar, M., Li, K.: Filtering image spam with near-duplicate detection. In: Proc. the 4th Conference on Email and Anti-Spam (CEAS), California, USA (August 2007)
18. Dalvi, N., Domingos, P., Mausam, S.S., Verma, D.: Adversarial classification. In: Tenth ACM SIGKDD international conference on Knowledge discovery and data mining, pp. 99–108 (2004)
19. Zhu, X.: Semi-supervised learning literature survey. Technical Report 1530, Computer Sciences, University of Wisconsin-Madison (2005)
20. Fisher, R.A.: The use of multiple measurements in taxonomic problems. Annals of Eugenics 7, 179–188 (1936)
21. Wu, Y., Tian, Q., Huang, T.S.: Discriminant-em algorithm with application to image retrieval. In: Proc. IEEE Conf. on Computer Vision and Pattern Recognition, Hilton Head Island, SC, USA, June 2000, vol. I (2000)
22. He, X., Niyogi, P.: Locality preserving projections. In: Thrun, S., Saul, L., Schölkopf, B. (eds.) Advances in Neural Information Processing Systems 16. MIT Press, Cambridge (2004)
23. He, X., Yan, S., Hu, Y., Niyogi, P., Zhang, H.: Face recognition using laplacianfaces. IEEE Transaction on Pattern Analysis and Machine Intelligence 27(3), 328–340 (2005)
24. Dempster, A.P., Laird, N.M., Rubin, D.B.: Maximum likelihood from incomplete data via the em algorithm. Journal of the Royal Statistical Society. Series B (Methodological) 39(1), 1–38 (1977)
25. Cai, D., He, X., Han, J.: Semi-supervised discriminant analysis. In: Proc. the 11th IEEE International Conference on Computer Vision (ICCV), Rio de Janeiro, Brazil (October 2007)
26. Yang, J., Yan, S., Huang, T.: Ubiquitously supervised subspace learning. IEEE Transactions on Image Processing 18(2), 241–249 (2009)
27. Friedman, J.H.: Regularized discriminant analysis. Journal of the American Statistical Association 84(405), 165–175 (1989)
28. Ng, T.T., Chang, S.F.: Classifying photographic and photorealistic computer graphic images using natural image statistics. Technical report, Columbia University (October 2004)

29. Ng, T.T., Chang, S.F., Hsu, Y.F., Xie, L., Tsui, M.P.: Physics-motivated features for distinguishing photographic images and computer graphics. In: ACM Multimedia, Singapore (November 2005)
30. Ng, T.T., Chang, S.F., Tsui, M.P.: Lessons learned from online classification of photo-realistic computer graphics and photographs. In: IEEE Workshop on Signal Processing Applications for Public Security and Forensics (SAFE) (April 2007)
31. Mäenpä, T.: The local binary pattern approach to texture analysis extensions and applications. Ph.D thesis, Infotech Oulu, University of Oulu, Oulu, Finland (August 2003)
32. Canny, J.: A computational approach to edge detection. IEEE Trans. Pattern Anal. Mach. Intell. 8(6), 679–698 (1986)
33. Huang, J., Kumar, S.R., Mitra, M., Zhu, W.J., Zabih, R.: Image indexing using color correlograms. In: IEEE Computer Society Conference on Computer Vision and Pattern Recognition. IEEE Computer Society, Los Alamitos (1997)
34. Tu, Z.: Probabilistic boosting-tree: Learning discriminative models for classification, recognition, and clustering. In: Tenth IEEE International Conference on Computer Vision (2005)
35. Cortes, C., Vapnik, V.: Support-vector networks. Machine Learning 20(3), 273–297 (1995)

Collaborative Filtering Recommendation Algorithm Using Dynamic Similar Neighbor Probability

Chuangguang Huang[1], Jian Yin[1], Jing Wang[1], and Lirong Zheng[2]

[1] Department of Computer Science, Sun Yat-sen University, 510275 Guangzhou, Guangdong
{hchuangg,issjyin}@mail.sysu.edu.cn, jingyun_wj@163.com
[2] Guangdong Institute of Medical Information, 510180 Guangzhou, Guangdong
zhenglir@hotmail.com

Abstract. In this paper, we focus on how to overcome several limitations in the traditional research of collaborative filtering(CF). We present a novel CF recommendation algorithm, named DSNP(Dynamic Similar Neighbor Probability). This algorithm improves the neighbors' similarities computations of both users and items to choose the neighbors dynamically as the recommendation sets. How to select the confident subsets which are the most effective neighbors to the target object, it is the first stage. A major innovation is by defining a dynamic neighbor probability over the trustworthy subsets. Moreover, we define a prediction algorithm that combines the advantages of dynamic neighbor coefficient with the user-based CF and the item-based CF algorithms. Experimental results show that the algorithm can achieve consistently better prediction accuracy than traditional CF algorithms, and effectively leverage the result between user-based CF and item-based CF. Furthermore, the algorithm can alleviate the dataset sparsity problem.

Keywords: Dynamic Similar Neighbor Probability, Collaborative Filtering, Recommendation System, Similarity Criterion, Trustworthy Subset.

1 Introduction

Over the years, the researches of recommendation systems discuss several different technologies, mainly as follows: memory-based CF recommendation [1],[2], clustering-model recommendation[3[,[4], content-based CF recommendation[5],[6], ontology-based model recommendation[7], etc. Because of requiring so much time to train and upgrade the model, model-based methods cannot handle with growing data range as collaborative filtering approaches. But with the continuous dramatic increase in the number of users and items, it is anticipated that, for scalability in practically, CF recommendation method is used more and more widely.

The recent techniques in CF research by collecting the neighbors from the user-related or item-related objects, which used to alleviate several challenges of recommender systems, such as: data sparsity [10], cold start[11]and scalability[7]. However, the traditional CF research still has some limitations:

R. Huang et al. (Eds.): ADMA 2009, LNAI 5678, pp. 165–174, 2009.

(1) In CF research, many methods [1,8] use kNN(k-Nearest Neighbors) to select the recommender objects for the unknown ratings. But sometimes, k may not be the optimum.
(2) Traditional CF methods adopt the ranking by the similarity to select the recommender objects for the target one[4, 9]. The higher similarity object, the more possibility to be selected. However, the higher similarity object maybe be probably reflected by the less common interest ratings.
(3) In the past, the research in CF usually only consider one of the user-based and the item-based method, or overestimate the influence of one from the other [1,9]. Before we make some prediction about the unknown ratings, the analysis to the user-based and item-based method is required. However, the researchers less learn the relation between the both methods. We should know how to combine both methods dynamically to improve the quality of the recommender systems.

The main contribution of this paper, we focus on how to overcome several limitations in the research area of collaborative filtering(CF). The remainder of the paper is arranged as follows. In section 2, we introduce the definition of the problem and the related work in traditional CF research. Section 3 describes the main contribution of our algorithms step by step and discusses the time complexity. The experimental results on the MovieLens dataset and some discussions are shown in Section 4. Finally, we conclude our work in section 5.

2 Problem Definition and General Approach

In the typical recommender system, there is a list of s users and a list of t items. Every user can express his/her opinions about the list of items. The goal of a collaborative filtering algorithm is to suggest new items or to predict the utility of a certain item for a particular user based on the user's previous likings and the opinions of other like-minded users.

Table 1. R(s×t)User-item ratings matrix

	I_1	...	I_j	...	I_t
U_1	$R_{1,1}$	...	$R_{1,j}$	...	$R_{1,t}$
...					
U_a	$R_{a,1}$	...	$R_{a,j=}?$	...	$R_{a,t}$
...					
U_s	$R_{s,1}$	...	$R_{s,j}$	...	$R_{s,t}$

In the matrix, there is a list of s users $U = \{U_1, U_2, ..., U_s\}$ in the rows, while the cols represent the items $I = \{I_1, I_2, ..., I_t\}$. The rating of the user U_a to item I_j is $R_{a,j}$, which denotes the user interest.

2.1 Similarity Measure

There are a number of different ways to compute the similarity between users, such as: cosine-based similarity, Pearson correlation coefficient and some other revised similarity.

(1) Cosine-based similarity

$$\text{Sim}(U_a, U_b) = \text{cosine}(\overrightarrow{R_a}, \overrightarrow{R_b}) = \frac{\sum_{k=1}^{t} R_{a,k} \times R_{b,k}}{\sqrt{\sum_{k=1}^{t} (R_{a,k})^2} \times \sqrt{\sum_{k=1}^{t} (R_{b,k})^2}} \tag{1}$$

(2) Pearson correlation coefficient

In order to fix the different rating bias in different users, using Pearson correlation coefficient(PCC) method to cut off the average rating of the user. Let the set of items which rated by both user a and b is denoted by I', which $I' = (I_{U_a} \cap I_{U_b})$.

$$\text{Sim}(U_a, U_b) = \frac{\sum_{i_k \in I'} (R_{a,k} - \overline{R_a}) \times (R_{b,k} - \overline{R_b})}{\sqrt{\sum_{i_k \in I'} (R_{a,k} - \overline{R_a})^2} \times \sqrt{\sum_{i_k \in I'} (R_{b,k} - \overline{R_b})^2}} \tag{2}$$

(3) Revised similarity method

The user similarity value is based on the items they both rated. If the number of co-rated items is too small, the result will be not sound enough. To shrink this effect, Herlocker [12,13] proposed to use the following modified similarity computation equation. From Herlocker's research, Ma[2] proposed to set the threshold value γ to revise the similarity.

$$\text{Sim}'(U_a, U_b) = \frac{\min(|I'|, \gamma)}{\gamma} \times \text{Sim}(U_a, U_b) \tag{3}$$

2.2 Traditional kNN-Based CF Algorithm

Through the similarity measure, select the k nearest neighbors(kNN) of user a which have rated the item x, noted as $S_k(U_a)$. ($|S_k(U_a)| = k$). Using user-based CF(UBCF) to predict the un-known rating $R_{a,j}$.

$$R_{a,j} = \overline{R}_a + \frac{\sum_{U_x \in S_k(U_a)} Sim'(U_a, U_x) \times (R_{x,j} - \overline{R}_x)}{\sum_{U_x \in S_k(U_a)} Sim'(U_a, U_x)} \tag{4}$$

$\overline{R}_a$ and $\overline{R}_x$ individually expressed the average ratings of user a and x. Finally, $R_{a,j}$ is predicted by formula(4). The traditional CF algorithm usually adopt kNN-based to predict the unknown ratings.

3 CF Recommendation Algorithm Using Dynamic Similar Neighbor Probability

Traditional CF algorithms basically calculate the relationships among the objects of the users set or the items set independently, and predict the feature of the target object. These researches usually only consider one-side of the user-based and the item-based method, or overestimate the influence of some one from the other. For example, if we want to predict the rating, we not only use user-based CF(UBCF) to select the neighbors of the target user who have rated the target item, but also use item-based CF(IBCF) to select the neighbors of target item which have been rated by the target user. However, if the target user have shown his interest to so many items, but less users have rated the target item, UBCF methods will get lower accuracy than IBCF methods because IBCF methods will be more efficient to get the neighbors of the target item. In the practice, we should learn how to effectively leverage the result between UBCF and IBCF.

3.1 Dynamically Similar Neighbors Selection

The weakness of k nearest neighbors algorithm for large, datasets diversity led us to explore alternative neighbor selection technique. Our approach attempted to select the dynamically similar neighbors (DSN) for the target. We introduce two thresholds μ and ν, the one is used for the similarities among the users, the other is for the similarities among the items. For example, If the similarity between the neighbor and the target user is larger than μ, then this neighbor is selected as the recommender user. A set of recommend users $S(U_a)$ can be defined as (5), the set of recommend items $S(I_j)$ can be formulated as (6).

$$S(U_a) = \left\{ U_x \middle| Sim'(U_a, U_x) > \mu, a \neq x \right\} \tag{5}$$

$$S(I_j) = \left\{ I_y \middle| Sim'(I_j, I_y) > \nu, j \neq y \right\} \tag{6}$$

According the equation (5) and (6), we calculate $m = \left| S(U_a) \right|$ and $n = \left| S(I_j) \right|$.

3.2 Definition of the Confidence Subset

During the process of recommender objects selection, the similarity is the main standard guideline. Traditional CF methods use the ranking of the similarity to select the recommender objects for the target rating[4,8,12]. The higher similarity object, the more possibility to be selected. However, the higher similarity object maybe be probably reflected by the less common interest ratings, the accuracy of the prediction will be decreased. If we judge the neighbors only by their similarity, it maybe partly low confident. So, we not only consider the similarity larger than the threshold, but consider the number of common interest ratings within these two objects. We introduce two thresholds ε and γ, which are applied as the thresholds of the common ratings. We define $S'(U_a)$ for user a as the confident subset of the recommend set $S(U_a)$ in (7), and define $S'(I_j)$ for item j as the confident subset of the recommend set $S(I_j)$ in (8).

$$S'(U_a) = \left\{ U_x \middle| Sim'(U_a, U_x) > \mu \wedge \left| I_{U_a} \cap I_{U_x} \right| > \varepsilon, a \neq x \right\} \tag{7}$$

$$S'(I_j) = \left\{ I_y \middle| Sim'(I_j, I_y) > \nu \wedge \left| U_{I_j} \cap U_{I_y} \right| > \gamma, j \neq y \right\} \tag{8}$$

We calculate the number of the two subsets: $m' = \left| S'(U_a) \right|$ and $n' = \left| S'(I_j) \right|$.

3.3 Leverage the Neighbors Sets

Using UBCF algorithm predicts the unknown rating by the ratings of the similar users, while IBCF algorithm predicts the unknown rating by the ratings of the similar items. However, predicting the unknown rating only using UBCF algorithm or only using IBCF algorithm will loss some important information of the ratings matrix. Even if we use the average of the two results or set a experimental weight, it can't improve the accuracy of the predictions. We propose to dynamically integrate UBCF and IBCF algorithms, and take advantage of users' correlations and items' correlations in the ratings matrix. Hence, our approach focuses the confidence of the two different algorithms results to define the dynamic neighbor coefficient, which is used in the CF algorithm to produce a new prediction result. Which algorithm prediction has higher confidence, it will be appropriate the more proportion of the new prediction result.

For example, we want to predict the ratings of $R_{a,j}$=?. It can get the confidence values of UBCF and IBCF by (7),(8). Respectively, $m' = |S'(U_a)|$ and $n' = |S'(I_j)|$. The values of m' and n' depend on the neighbors of the unknown ratings, they will change at different target. So we can't figure them before prediction.

We introduce the coefficient λ which is used to determine how the prediction relies on UBCF prediction, while $1-\lambda$ is used to determine the proportion of the IBCF prediction. They are defined as:

$$\Pr(S'(U_a)) = \frac{\phi \times m'}{\phi \times m' + n'}, \Pr(S'(I_j)) = \frac{n'}{\phi \times m' + n'}, (\phi \geq 0)$$

The optimal parameter ϕ can be trained through the training process. We will illustrate it in the experiment.

3.4 CF Based on Dynamic Similar Neighbor Probability

We combine the parameter and coefficient defined above into the CF algorithm. Using CF algorithm using DSNP(Dynamic Similar Neighbor Probability), which integrates the UBCF and IBCF dynamically to create a new prediction. The recommender formula is divided into two situations as below.

If $m' > 0$ or $n' > 0$, it means some neighbors of the unknown ratings are confident enough to make a recommendation , we prior to choose (9) to recommend:

$$
\begin{aligned}
R_{a,j} &= \Pr(S'(Ua))\times(\overline{R}_a + \frac{\sum_{U_x\in S(U_a)} Sim'(U_a,U_x)\times(R_{x,j}-\overline{R}_x)}{\sum_{U_x\in S(U_a)} Sim'(U_a,U_x)}) \\
&+ \Pr(S'(Ij))\times(\overline{R}_j + \frac{\sum_{I_y\in S(I_j)} Sim'(I_j,I_y)\times(R_{a,y}-\overline{R}_y)}{\sum_{I_y\in S(I_j)} Sim'(I_j,I_y)}) \\
&= (\frac{\phi\times m'}{\phi\times m'+n'})\times(\overline{R}_a + \frac{\sum_{U_x\in S(U_a)} Sim'(U_a,U_x)\times(R_{x,j}-\overline{R}_x)}{\sum_{U_x\in S(U_a)} Sim'(U_a,U_x)}) \\
&+ (\frac{n'}{\phi\times m'+n'})\times(\overline{R}_j + \frac{\sum_{I_y\in S(I_j)} Sim'(I_j,I_y)\times(R_{a,y}-\overline{R}_y)}{\sum_{I_y\in S(I_j)} Sim'(I_j,I_y)})
\end{aligned} \tag{9}
$$

$\overline{R}_a$ and $\overline{R}_x$ represent the average rating of user a and user x in ratings matrix respectively. $\overline{R}_j$ and $\overline{R}_y$ represent the average rating of item j and item y. Supposing the similar users set of user a is null(m' =0), while the similar items set of item j is not null(n' >0). As (9), the result completely depends on collaborative filtering by the similar neighbors of item j. In opposition, we can get the final result by the UBCF algorithm.

3.5 Algorithms and Performance Analysis

According to the DSN method mentioned in 3.1 and 3.2, our algorithm based dynamically similar neighbors selection. This algorithm automatically searches the recommender sets and confident subsets for the unknown ratings to complete the prediction task.

Algorithm 1. *Search_DSNObj* (Search the *DSN* set of the target object)

Input: Ratings matrix $R(s\times t)$, the user U_a , item I_j , threshold $\mu,\nu,\varepsilon,\gamma$

Step 1: In ratings matrix $R(s\times t)$, we use the revised similarity method to calculate the similarity matrix of the users and items respectively, and store in two matrixes: $Arr\_UserSim(s\times s)$ and $Arr\_ItemSim(t\times t)$.

Step 2: Considering the similarity matrix $Arr\_UserSim(s\times s)$. Every user *x*, which has the similarity value $\mathrm{Sim}'(U_a,U_x)$ larger than μ , is selected into $S(U_a)$. If the common ratings number of user *a* and user *x* larger than ε , as $|I_{U_a}\cap I_{U_x}|>\varepsilon$, then we put user *x* into $S'(U_a)$.

Step 3: Considering the similarity matrix $Arr\_ItemSim(t\times t)$. Every item *x*, which has the similarity value $\mathrm{Sim}'(I_j,I_x)$ larger than ν , is selected into $S(I_j)$. If the number of the users co-rated item *j* and item *x* larger than γ , as $|U_{I_j}\cap U_{I_x}|>\gamma$, then we put user *x* into $S'(I_j)$.

Output: The *DSN* set $S(U_a)$ and $S'(U_a)$ of U_a , the *DSN* set $S(I_j)$ and $S'(I_j)$ of I_j

In DSNP algorithm, through step 1 and 2 to get the similar neighbors of the user and item. Our *Search_DSNObj* algorithm sets the thresholds to limit the similar neighbors in a acceptable scale, usually the number of the similar neighbors set is a integer. So the time complexity is $2\times O(1)$. Step 4 and 5 is to provide the prediction rating, which need to loop the formula by the number of the neighbors set. So the time complexity is $2\times O(1)$. Therefore, DSNP algorithm complexity is $4\times O(1)$.

Algorithm 2. CF recommendation algorithm using Dynamic Similar Neighbor Probability(DSNP)

Input: The target user *a* and item *j*, coordinator parameters ϕ

Step 1: In the similar neighbor set $S(U_a)$ of the user *a*, calculate $m=|S(U_a)|$, while the number of confidence neighbors set is $m'=|S'(U_a)|$.

Step 2: In the similar neighbor set $S(I_j)$ of the item *j*, calculate $n=|S(I_j)|$, while the number of confidence neighbors set is $n'=|S'(I_j)|$.

Step 3: select the appropriate coordinator parameter ϕ,

Step 4: If $m'>0$ or $n'>0$, as formula (9), return the prediction rating $R_{a,j}$.

Step 5: Else if m>0 or n >0, using m and n replace m' and n', as formula (9), return the prediction rating $R_{a,j}$.

Output: The prediction rating of $R_{a,j}$

---

4 Experiment

Through experiments, we expect to evaluate the recommendation quality of our algorithms, and to answer the following questions: 1) How about the coordinator parameter will impact on the result of the prediction algorithm, can it get a better prediction result? 2) The comparison of the DSNP CF algorithm and other CF algorithms. Which algorithm can get the better prediction accuracy?

4.1 Dataset

We used real dataset from the MovieLens recommender system[14]. The density of the ratings matrix is 6.3%. It shows in this dataset the ratings matrix is a very sparse. As usual experiments, we randomly select 300 users from the dataset Firstly. Then, we use 80% of the data was used as training set and 20% of the data was used as test set. The thresholds for the experiments are empirically set as follows: $\mu=\nu=0.2$, $\varepsilon=30$, $\gamma=20$.

4.2 Evaluating the Quality

There are several types of measures for evaluating the quality of a recommender system. Evaluate the accuracy of a system by comparing the numerical recommendation scores against the actual user ratings for the user-item pairs in the test dataset. Mean-Absolute-Error (MAE) between real ratings and predictions is a widely used measurement. The lower the MAE, the more accurately the recommendation algorithm predicts user ratings.

4.3 Experiment with Different Parameter ϕ

The size of parameter ϕ has significant impact on the prediction quality. This experiment is designed to set different value of coordinator parameter ϕ, and to compare with the UBCF and IBCF algorithm. Through the comparison, our purpose is to study how to choose the appropriate parameter ϕ to reconcile the DSNP recommendation algorithm between the UBCF and IBCF algorithm.

In this experiment, the x-axis presents the value of parameter ϕ, the y-axis presents the MAE of the test dataset. Because the result of UBCF algorithm is better than the result of IBCF algorithm, hence, the parameter ϕ will start from 1 and gradually increase. Through the experiments of the 300 users are shown in figure 1 , we can see the results of DSNP are not as good as the results of UBCF, when the parameter ϕ is too small.

During the parameter ϕ increasing from 1 to 20, the MAE values of the experiments step down significantly, which are lower than the other algorithms. When the value of parameter ϕ increases from 20 to 200, as the trend analysis, the value of $\lambda = \frac{\phi \times m'}{\phi \times m' + n'}$ will increase towards 1. It means the final prediction will more reliable to the result of UBCF algorithm. We can obtain the optimal value of parameter ϕ in the experiment as example. When the parameter ϕ is near to 20, the MAE value is to a minimum. Hence, we select ϕ =20 as an optimum value for the subsequent experiment.

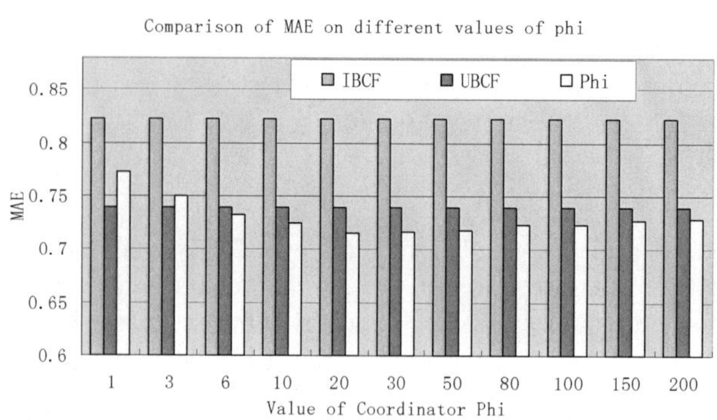

Fig. 1. Comparison of MAE on the different values of phi

4.4 Comparison Traditional CF and DSNP Algorithms

The purpose of this study is to experimentally determine the prediction accuracy of the DSNP. We compare them to the traditional CF algorithms and the EMDP (Effective Missing Data Prediction) algorithm, which is the relatively industry-leading CF research[2]. The x-axis in the experiment figure is the number of similar neighbors of the target item; the y-axis presents the MAE of the testing dataset.

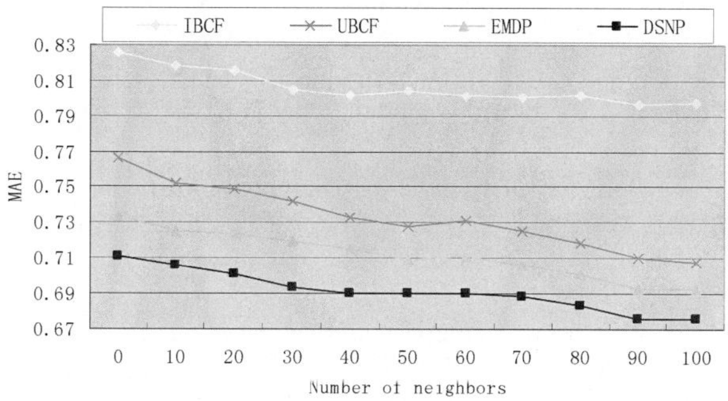

Fig. 2. Comparison of traditional CF and DSNP algorithm

In the experiment, comparing with the traditional UBCF, IBCF algorithm and the recent EMDP algorithm, we can learn that the DSNP algorithm can consistently get a lower MAE and provide better quality of predictions. When the neighbors of the target item increase, the MAEs step down together, the qualities of recommendation are improved. The more neighbors in the dataset, the more significant results will be get.

5 Conclusion

Recommendation system is a hot research in the very wide application areas. The contribution in this paper, we overcome the several limitations in CF research and present a novel CF recommendation algorithm using DSNP (Dynamic Similar Neighbor Probability). Through the similarities computations of both user-based and item-based, we can dynamically choose the neighbors, which as the trustworthy sets. Using these sets to select the confident subsets, which are the most effective objects to the target object. The prediction algorithm DSNP combines the advantages of trustworthy subset for recommendation. Experimental results show that the algorithm can consistently achieve better prediction accuracy than traditional CF algorithms, and effectively leverage the result between user-based CF and item-based CF. Furthermore, the algorithm can alleviate the dataset sparsity problem.

Acknowledgments. This work is supported by the National Natural Science Foundation of China (60773198, 60703111), Natural Science Foundation of Guangdong Province (06104916, 8151027501000021), Research Foundation of Science and Technology Plan Project in Guangdong Province (2008B050100040), Program for New Century Excellent Talents in University of China(NCET-06-0727).

References

1. Sarwar, B., Karypis, G., Konstan, J., Riedl, J.: Item-Based collaborative filtering recommendation algorithms. In: Proc. of the 10th World Wide Web Conf., pp. 285–295. ACM Press, New York (2001)
2. Ma, H., King, I., Lyu, M.R.: Effective Missing Data Prediction for Collaborative Filtering. In: Proc. of the 30th annual international ACM SIGIR conference, pp. 39–46 (2007)
3. Kohrs, A., Merialdo, B.: Clustering for collaborative filtering applications. In: Proceedings of computational intelligence for modelling, control & automation. IOS Press, Vienna (1999)
4. Xue, G.-R., Lin, C., Yang, Q., Xi, W., Zeng, H.-J., Yu, Y., Chen, Z.: Scalable collaborative filtering using cluster-based smoothing. In: Proc. of the 28th annual international ACM SIGIR, pp. 114–121 (2005)
5. Hofmann, T.: Collaborative filtering via gaussian probabilistic latent semantic analysis. In: Proc. of SIGIR, pp. 259–266 (2003)
6. Hofmann, T.: Latent semantic models for collaborative filtering. ACM Trans. Inf. Syst. 22(1), 89–115 (2004)
7. Vincent, S. -Z.: Boi Faltings: Using Hierarchical Clustering for Learning the Ontologies used in Recommendation Systems. In: KDD, pp. 599–608 (2007)
8. Chen, M.-C., Chen, L.-S., Hsu, F.-H., Hsu, Y., Chou, H.-Y.: HPRS: A Profitability based Recommender System. In: Industrial Engineering and Engineering Management, pp. 219–223. IEEE, Los Alamitos (2007)
9. Lee, H.-C., Lee, S.-J., Chung, Y.-J.: A Study on the Improved Collaborative Filtering Algorithm for Recommender System. In: Fifth International Conference on Software Engineering Research, Management and Applications, pp. 297–304 (2007)
10. Sarwar, B.M., Karypis, G., Konstan, J.A., Riedl, J.: Application of Dimensionality Reduction in Recommender System–A Case Study. In: ACM WebKDD Web Mining for E-Commerce Workshop, pp. 82–90 (2000)
11. Massa, P., Avesani, P.: Trust-aware collaborative filtering for recommender systems, pp. 492–508. Springer, Heidelberg (2004)
12. Herlocker, J., Konstan, J.A., Riedl, J.: An empirical analysis of design choices in neighborhood-based collaborative filtering algorithms. Information Retrieval 5, 287–310 (2002)
13. McLaughlin, M.R., Herlocker, J.L.: A collaborative filtering algorithm and evaluation metric that accurately model the user experience. In: Proc. of SIGIR, pp. 329–336 (2004)
14. GroupLens Research, `http://www.grouplens.org`

Calculating Similarity Efficiently in a Small World

Xu Jia[1,2], Yuanzhe Cai[1,2], Hongyan Liu[3], Jun He[1,2], and Xiaoyong Du[1,2]

[1] Key Labs of Data Engineering and Knowledge Engineering, Ministry of Education, Beijing
[2] Department of Computer Science, Renmin University of China, 100872 Beijing
{jiaxu,yzcai,hejun,duyong}@ruc.edu.cn
[3] Department of Management Science and Engineering, Tsinghua University, 100084 Beijing
liuhy@sem.tsinghua.edu.cn

Abstract. *SimRank* is a well-known algorithm for similarity calculation based on link analysis. However, it suffers from high computational cost. It has been shown that the world web graph is a "small world graph". In this paper, we observe that for this kind of small world graph, node pairs whose similarity scores are zero after first several iterations will remain zero in the final output. Based on this observation, we proposed a novel algorithm called *SW-SimRank* to speed up similarity calculation by avoiding recalculating those unreachable pairs' similarity scores. Our experimental results on web datasets showed the efficiency of our approach. The larger the proportion of unreachable pairs is in the relationship graph, the more improvement the *SW-SimRank* algorithm will achieve. In addition, *SW-SimRank* can be integrated with other *SimRank* acceleration methods.

Keywords: Similarity, Simrank, Linkage Mining.

1 Introduction

Searching similar web pages for a query is a significant task for the search engine. Recently, there are a variety of link-based similarity measures using the hyperlinks in the Web. One of the best well-known algorithms in similarity calculation based on the linkage analysis is *SimRank* [1], which has higher accuracy than other algorithms [2] [3] [4]. To speed up this algorithm is very important for two reasons. First, calculating *SimRank* quickly is necessary to reduce the lag time from the time the web graph is already calculated to the time this graph changed. Second, this algorithm suffers from high computational cost, which is quadratic complexity. Despite the small size of web graph (containing about 5000 web pages), this computation can take more than 10 hours. Therefore, these two reasons intensify the need for faster methods to calculate similarity using the basic idea of *SimRank*.

In our research, we have an observation: the world web graph is not a connected-graph. In other words, it is impossible to discover the path between each block on the World Wide Web. Thus during the whole iteration processes, the *SimRank* scores of

R. Huang et al. (Eds.): ADMA 2009, LNAI 5678, pp. 175–187, 2009.

these unreachable pairs are always 0. Therefore, in order to decrease the time cost, we can search these pairs and don't recalculate these similarity values in the following steps of iteration.

Based on this observation, we proposed a novel algorithm called *SW-SimRank* to speed up similarity calculation by avoiding recalculating those unreachable pairs' similarity scores. In our experiment, this algorithm speeds up the performance of *SimRank* by about 10% on the web datasets.

The main contributions of this paper are as follows:

- Based on most pairs of vertices in web graphs connected by short even-step paths, we develop a new algorithm, *SW-SimRank*, which improves the performance of *SimRank* by about 10%. This method can also be integrated with other algorithms [5][6][7][8] to speed up the similarity calculation.
- We prove the convergence of the *SW-SimRank* algorithm by theoretical study.
- We evaluate the method on real datasets to confirm their applicability in practice.

This paper is organized as follows. We introduce the related work in section 2 and define the graph models in section 3. Preliminaries are presented in section 4, and the *SW-SimRank* algorithm is described in section 5. Our performance study is reported in section 6, and finally this study is concluded in section 7.

2 Related Work

We categorize existing works related to two classes: clustering based on link analysis and small world theory.

Clustering based on link analysis: The earliest research work for similarity calculation based on link analysis focuses on the citation patterns of scientific papers. The most common measures are *co-citation* [2] and *co-coupling* [3]. *Co-citation* means if two documents are often cited together by other documents, these two documents may have the same topic. *Co-coupling* means that if two papers cite many papers in common, they may focus on the same topic. In [4], Amsler proposed to fuse bibliographic *co-citation* and *co-coupling* measures to determine the similarity between documents. However, all these methods compute the similarity only by considering their immediate neighbors. *SimRank* [1] is proposed to consider the entire graph to determine the similarity between two nodes. But this method has a high time complexity, which limits its use to large datasets. Thus, there are some methods proposed to improve the performance of *SimRank*. In the same paper of *SimRank*, authors also proposed a *Pruning-SimRank* [1] algorithm which computes similarities by small scope of relationship graph around these two nodes. Another algorithm called *Fingerprint-SimRank* [5] pre-computes several steps of random walk path from each object and use these steps to calculate the similarity between objects. Although *Fingerprint-SimRank* improves the computation performance, this algorithm has a high space complexity. In addition, Xiaoxin Yin et al [6] proposed a hierarchical structure—*SimTree* to represent the similarity between objects and develop an effective similarity calculation algorithm—*LinkClus*. However, in that paper, there is no sound theoretical proof of the convergence for that algorithm. Thus, it is difficult to decide in which iteration *LinkClus* will gain the best accuracy. Dmitry Lizorkin et al[7] presents a technique to estimate the accuracy of computing *SimRank* iteratively and some optimization techniques that improve the computational complexity of the iterative algorithm from $O(n^4)$ to $O(n^3)$ in the worse situation. All these techniques can combine with our method to improve the performance of *SimRank.*

Small world theory: In 1960s, Stanley Mailgram discovered the famous phenomena that a letter passed from people to people is able to reach a target individual by about only six steps. This experiment first reveals the small-world theory, "most pairs in most network is connected by a short path through the graph." The more rigorous math work[9] is published by Pool and Kochen. Adamic[10] analyzed the world wide web and showed that web is also a small world graph, i.e., the sites are highly clustered but the path length is not long.

3 Problem Definition

Our research work focuses on the relationship graph. The definition of this graph is given as follows:

A relationship graph $G(V, E)$ is a directed graph, where vertices in V represent objects in a particular domain and edges in E describe the relationship between nodes. Furthermore, we use $I(v)$ represent all the neighbors of a node v . $|I(v)|$ is the number of neighbor nodes of v. For example, Fig.1 is a relationship graph that describes the web page relationship from Cornell computer science department. In this graph, a directed edge <p, q> means from page p to page q corresponds to a reference (hyperlink) from page p to page q. Our research work focuses on how to cluster the similar pages on the web graph. For this web relationship graph, these web pages from the web site of department of computer science in Texas, Cornell, Washington, and Wisconsin have been manually classified into seven domains, course, department, faculty, project, staff, student, and others. We want to cluster these pages automatically into these seven fields.

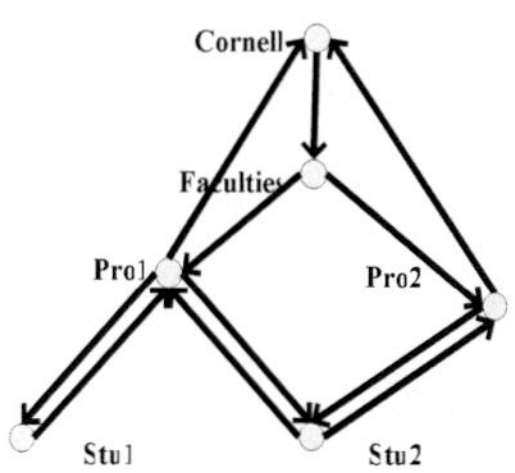

(a)

Name	URL
Cornell	http:\\www.cs.cornell.edu
Faculties	http:\\www.cs.cornell.edu\People\faculty\index.htm
Pro1	http:\\www.cs.cornell.edu\~professor1
Pro2	http:\\www.cs.cornell.edu\~professor2
Stu1	http:\\www.cs.cornell.edu\~stu1
Stu2	http:\\www.cs.cornell.edu\~stu2

(b)

Fig. 1. Web page relationship graph

4 Review of *SimRank*

In this section, we introduce the *SimRank* algorithm.

4.1 Introduction of *SimRank*

SimRank [1], a classical linkage-based similarity calculation algorithm, measures similarity of two objects based on the principle, "two objects are similar if they link to similar objects". The recursive similarity computational equation is as follows:

$$\begin{cases} 1 & (V_a = V_b) \\ s(V_a, V_b) = \dfrac{c}{|I(V_a)||I(V_b)|} \displaystyle\sum_{i=1}^{|I(V_a)|} \sum_{j=1}^{|I(V_b)|} s(I_i(V_a), I_j(V_b)) & (V_a \neq V_b) \end{cases} \quad (1)$$

Where c is a decay value and $c \in [0, 1]$, $I(V_a)$ is the set of neighbor nodes of V_a and $I_i(V_a)$ is i^{th} neighbor node for V_a. $|I(V_a)|$ is the number of neighbors of V_a.

From another point of view, $s(V_a, V_b)$ can be considered as how soon and how many times two surfers, when they walk starting with V_a and V_b respectively and travel randomly on the graph G. Thus, based on the *expected f-meeting distance* theory [1], Glen Jeh al et gives another definition of *SimRank* score $s(V_a, V_b)$.

$$s(V_a, V_b) = \sum_{t:(V_a, V_b) \sim\sim> (V_x, V_x)} P[t] c^{l(t)} \quad (2)$$

Where c called the decay factor, is a constant in (0,1). $t = <w_1, w_2, \ldots, w_k>$ is the travel path from a to b and $l(t) = k/2$, the number of steps starting from the beginning position V_a and V_b. $P[t]$ of travelling path t is $\prod_{i=1}^{k-1} \frac{1}{|I(w_i)|}$, which represents the probability of a surfer traveling along this path.

In order to explain formula (2), we calculate the similarity between Pro1's web page and Pro2's web page. Two surfers starting from Pro1 and Pro2 respectively walk one step. Then, they may meet at Cornell node and the travelling path is <Pro1, Cornell, Pro2>. The probability of a surfer starting form Pro1 to Cornell is 1/3, because there are three web pages connected to node Pro1. Similarly, the probability from Pro2 to Cornell is 1/2. Thus, in this path t, $P[t] = 1/3 \times 1/2 = 1/6$. We set c = 0.8 and the number of step is 1. Thus the *SimRank* score on this path is $1/6 \times (0.8)^1 = 0.133$. Because there are only two paths between Pro1 and Pro2, such as < Pro1, Cornell, Pro2> and < Pro1, Stu2, Pro2>, thus s(Pro1, Pro2) is equal to 0.267 at the first iteration. If walking two steps, there is another path <Pro1, Stu1, Pro1, Stu2, Pro2> (i.e., Pro1→Stu1→Pro1, Pro2→Stu2→Pro1), the *SimRank* score on this path is $(1/3 \times 1 \times 1/2 \times 1/2) \times 0.8^2 = 0.0533$. *SimRank* will search all paths and sum them. s(Pro1,Pro2) = 0.267 + 0.0533+ …= 0.404.

In this example, we can understand that *SimRank* score is considered as how "close" to a usual "source" of similarity.

The major steps of *SimRank* algorithm are as follows:

Algorithm 1. *SimRank*

Input: Decay Factor c, Transfer Probability Matrix T (the probability of moving from state i to state j in one step), Tolerance Factor ε (under normal case, ε = 0.001)

Output: Similarity Matrix s_k

```
k←1;
s_0 ←identity matrix;
while(Max(|s_k(V_a, V_b) - s_{k-1}(V_a, V_b)| / |s_{k-1}(V_a, V_b)|) > ε )
    k←k+1;
    s_{k-1} ←s_k;
    for each element s_k(V_a,V_b)
```

$$s_k(V_a, V_b) = c \cdot \sum_{i=1}^{|I(V_a)|} \sum_{j=1}^{|I(V_b)|} T_{V_a V_i} \cdot T_{V_b V_j} \cdot s_{k-1}(V_i, V_j)$$

```
    end for
end While
return s_k
```

4.2 Definition of *SimRank* Convergence

In paper [1], the author proposed that the *SimRank* has a rapid convergence, with relative similarity stabilizing in five iterations. However, some pairs' *SimRank* scores still have huge changes after five iterations in our experiments. We define the convergence factor d as follows:

$$d = max\ (|s_{k+1}(V_a, V_b) - s_k(V_a, V_b)| \ / \ |s_k(V_a, V_b)|) \tag{3}$$

We refer[11] to setting then say when $d < 10^{-3}$, *SimRank* scores are convergent. We calculate the convergence factor d for four web datasets as shown in table 1. As can be seen from table 1, these *SimRank* scores converge to fixed values after about 40 steps of iteration in these four real world datasets.

Table 1. Statistics of factor d about Web Dataset convergence

iteration / Web DataSets	1	2	3	4	5	6	…	Final Iteration
Wisconsin	20	114.9	207.9	59.7	33.11	4.44	…	*44 iter.*
Washington	48.9	181.6	593.6	633.8	250.9	29.7	…	*46 iter.*
Texas	47.4	313.6	352.1	49.2	4.1	1.8	…	*40 iter.*
Cornell	48.8	260.5	260.5	140.4	5.46	2.10	…	*42 iter.*

5 *SW-SimRank* Algorithm

In this section, we first introduce our observation of *SimRank* calculation on real world datasets. Then, based on our observation, we propose the *SW-SimRank* algorithm. Finally, we analyze *SW-SimRank* and discuss reachable threshold, r, on the web graph.

5.1 *SW-SimRank* Intuition

Ten years ago, "bow tie theory"[12] shatters one myth about the Web. In fact, the Web is less connected than previously thought. As shown in Fig.2, the web network can be divided into four different regions: 1)disconnected pages, 2)the origination, 3)the termination and 4)the core. All these form a whole almost like the big bow tie. We can see that the proportion of disconnected pages is very large for the whole web graph. We also analyze the four web datasets and observe that these four web site graphs are not connected. Table 2 shows that Texas web site contains 34 different regions, Cornell web site 52 regions, Washington 120 regions, Wisconsin 87 regions. According to the *expected f-meeting distance theory*, because it is impossible to search the path between two nodes in different regions, *SimRank* scores of these pairs are always 0 during the whole process of iterations. Thus, if we can find these pairs and avoid recalculating these *SimRank* scores between these nodes in the iterations, that will greatly reduce the time cost for *SimRank* algorithm.

Fig.3. shows that the degree of web site appears to have the power-law degree distributions. In our observation on real web graphs, these power law graphs are usually

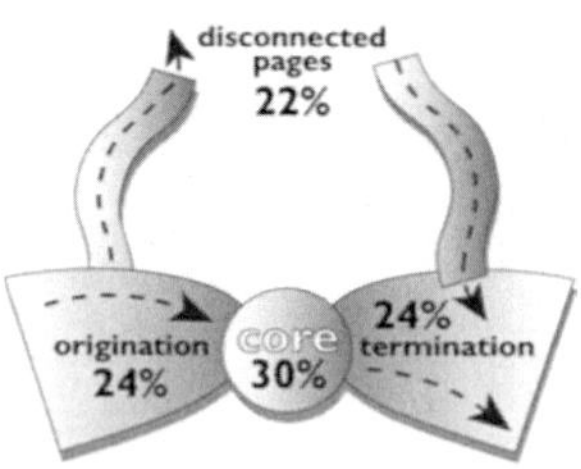

Fig. 2. Bow-Tie Theory

Table 2. Partition of Four Datasets

Dataset	T.	C.	Wa.	Wi.
Regions#(n_r)	34	52	120	87
β	0.04	0.06	0.100	0.068

Note: $\beta = n_r/n$, describing the proportion of regions in the network. Where n is the number of vertices.

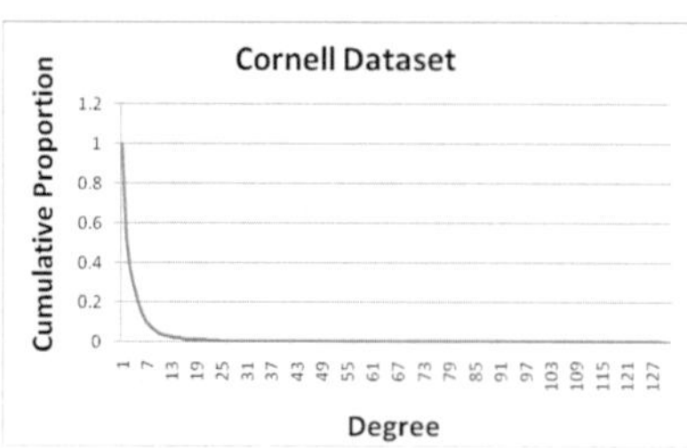

Fig. 3. Cornell Dataset-Degree Distribution. (X-aixs represents the degree of graph, Y-aixs discribes the cumulative probability distribution of degrees.)

Table 3. Statistics of Four Datasets

Data Set	Vertices#(n)	Edges#(s)	α
Texas	827	2667	0.008
Cornell	867	2691	0.007
Washington	1205	3299	0.005
Wisconsin	1263	5305	0.007

Note: $\alpha = s / (n*(n-1)/2)$, describing the density of graph.

Table 4. Harmonic Mean Distance (l_1) for Four Datasets

Texas	Cornell	Washington	Wisconsin
4	4.76	5.9	4.68

Note: $l_1^{-1} = \frac{1}{n(n+1)/2} \sum_{i \geq j} d_{ij}^{-1}$ [9], describing the average distance between two vertices in the relationship graph. Where, d_{ij} is the distance between two vertices, when there is no path in the graph between i and j, $d_{ij} = +\infty$ and, consistently $d_{ij}^{-1} = 0$.

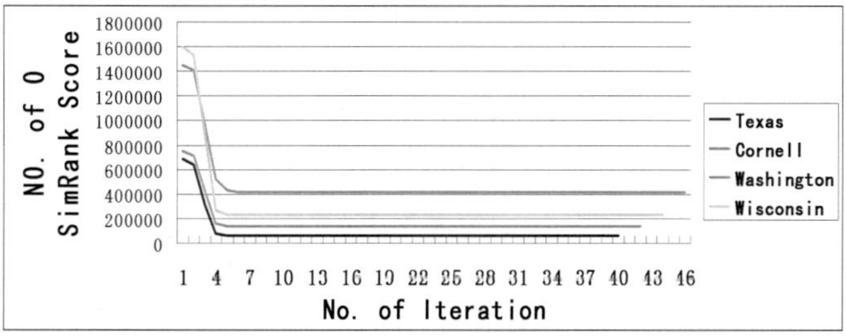

Fig. 4. Num. 0 SimRank score VS No. of Iteration

sparse and the path length between them is very short. In table 3, we can obverse that α of each graph is very small, less than 0.008, which indicates that all of these real world graphs are sparse graphs. However, though the edges of these web graphs are quite few, the average length reachable path in these graphs is extremely short. In table 4, we can observe that average path length for these four web graph is about 5, which means that we can browse other five pages to visit these pages from any connected pages in the web site.

Another interesting phenomenon also exists in the process of *SimRank* computation. In Fig. 4, we can observe that after only about 5~6 steps of iteration, if those pairs' *SimRank* scores are 0, they will keep 0 during the whole process of iteration. Thus, based on this phenomenon, we don't need to recalculate these 0-value pairs after a special number of iterations.

5.2 *SW-SimRank*

Above-mentioned observation of 0-value distribution suggests that the computational time of *SimRank* algorithm can be reduced by avoiding unnecessary computation. In particular, if these pairs' *SimRank* scores remain zero after several steps of iteration, we don't need to recalculate them in the following iterations. Based on this observation, we propose a novel algorithm called *SW-SimRank.* The major steps are shown as follows:

Algorithm 2. ***SW-Simrank***

Input: Decay Factor c, Transfer Probability Matrix T, Tolerance Factor ε
Output: Similarity Matrix s

```
k←1;
s_0←identity matrix;
flagmatrix ←0;  // 0: need calculate similarity further;
                // 1: need not calculate similarity further;
while(Max(|s_k(V_a, V_b) - s_{k-1}(V_a, V_b)| / |s_{k-1}(V_a, V_b)|) > ε )
    k←k+1;
    s_{k-1} ←s_k;
    for each element s_k(V_a,V_b)
        if(flagmatrix(V_a,V_b) == 0)
            s_k(V_a,V_b) = c · Σ_{i=1}^{|I(V_a)|} Σ_{j=1}^{|I(V_b)|} T_{V_aV_i} · T_{V_bV_j} · s_{k-1}(V_i,V_j) ;
    end for
    if(k = r)   // r: Reachable Threshold
        for each element s_k(V_a,V_b)
            If(|s_k(V_a, V_b) == 0)
                flagmatrix(V_a,V_b) = 1;
        end for
end while
return s_k
```

After r steps of iteration, we mark the pair whose *SimRank* score is 0 in the similarity matrix and don't recalculate these pairs during the following steps of iteration. Our method's time and space complexity is the same as *SimRank*. In next section, we will discuss how to determine r for *SW-SimRank*.

5.3 Analysis of *SW-SimRank* Algorithm

For the above introduction of *SimRank*, the process of node pair similarity calculation is the process to search the even-step paths between two nodes. In the kth step of iteration, *SimRank* algorithm searches for $2k$-step paths for each pair. For example, we calculate the *SimRank* score in Fig.5(a) and list their *SimRank* values on the right. We can find that in the first iteration, 2-step paths, $<V_a, V_c, V_b>$ and $<V_a, V_d, V_b>$, have been searched and similarity score in this iteration is $(0.5\times0.5 + 0.5\times0.5) \times 0.8^1 = 0.4$. In the second iteration, 4-step paths, $<V_a, V_c, V_a, V_d, V_b>$, $<V_b, V_d, V_b, V_c, V_a>$, $<V_a, V_d, V_a, V_c, V_b>$, $<V_a, V_d, V_b, V_c, V_b>$are observeed and added to the original similarity score. In the end, *SimRank* score will converge. Based on the small world theory, even though these graphs are extremely sparse, the reachable distance is very short. Because *SimRank* searches the even-length paths, we analyze the reachable even-step path between two nodes. In table 5, the average reachable even-path length (l_2) is a little longer than the average reachable path length (l_1) but still very short, about 6. These statistics data also mean that on the web graph, most pairs of pages seem to be connected by a short even-length path through the network. The largest reachable length of these even-paths (max-l_2) is also not very long, about 8 to14, which is approximately 2 times of l_2. That

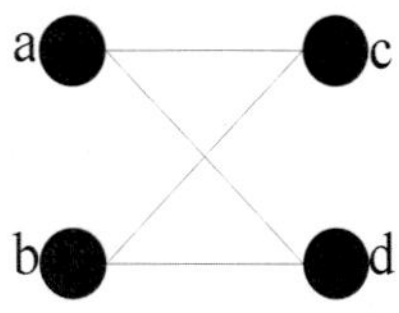

(a)

Iter. / Val.	1	2	3	4	5	…	+∞
s(a,b)	0.4	0.56	0.62	0.65	0.66	…	0.67
s(c,d)	0.4	0.56	0.62	0.65	0.66	…	0.67

(b)

Fig. 5. SimRank Calculation

Table 5. Path Statistics of Web Datasets

Iteration / Web DataSet	Texas	Cornell	Washington	Wisconsin
l_1	4	4.76	5.9	4.68
max-l_1	10	11	16	12
l_2	4.673	5.4624	6.628	5.3
max- l_2	8	10	14	12
r	5	6	7	6

Note: l_1 is the same as table 1. $l_2^{-1} = \frac{1}{n(n+1)/2}\sum_{i \ge j} d_{2ij}^{-1}$ describe the average distance between two vertices in the relationship graph. Where, d_{2ij} is the distance of even-length path between two vertices. *Max*- l_2 is the max length of even-length path between pairs.

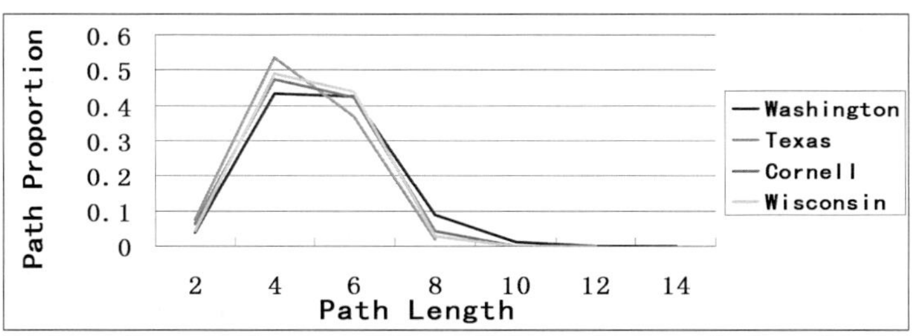

Fig. 6. Even-Length Path Proportion VS Path Length

means if we can't find an even-length path whose length is less than or equal to max-l_2 between these nodes, it is impossible to observe any even-path between these nodes. Meanwhile, in the process of *SimRank* calculation, in the kth iteration, 2k-length paths are searched. Thus, the reachable threshold, r, is 1/2(max-l_2).

Fig. 6 discribes the disribution of the even-length paths for these four datasets. On the observation, these distributions are approximately symmetric, but for a fast decaying tail. The Symmetry of distribution graph means the average reachable even-length path(l_2) is about half of max-l_2 and the fast decaying tails shows that though the real length of max-l_2 is slightly longer than two times of l_2, the proportion of these longer paths is very less. Thus, the experiential function to predict r shows as follows.

$$r = \lceil l_2 \rceil; \tag{4}$$

6 Empirical Study

In this section, we want to evaluate the performance of the *SW-SimRank*. We focus on testing the accuracy, the computational cost and the factors influencing the performance of our algorithm *SW-SimRank*.

6.1 Experiment Setting

Our experiments focus on website datasets.

Web Site Dataset: We use the CMU four university datasets [13]. These datasets contain web pages from computer departments of four universities: Cornell, Texas, Washington and Wisconsin. Web pages in these datasets have been manually divided into seven classes, student, faculty, staff, department, course, project and others. These classes will be used as the standard to evaluate the accuracy of our algorithm.

When we test the accuracy of these methods, we take PAM [14], a k-medoids clustering approach, to cluster objects based on similarity score calculated by these methods. We randomly select the initial centroids for 50 times when doing clustering. We compare the clustering result with the real class and choose the most accurate results among the 50 results. This evaluation method will also be used by [6].

All of our experiments are conducted on the PC with a 3.0GHz Intel Core 2 Duo Processor, 4GB memory, under windows XP Professional SP2 environment. All of our programs are written in java.

6.2 Accuracy of *SW-Simrank* Algorithm

For accuracy experiments, we compare *SW-SimRank* with the *SimRank* algorithm on the website datasets. We set reachable threshold *r* by the equation (4). The result is shown in Fig.7.

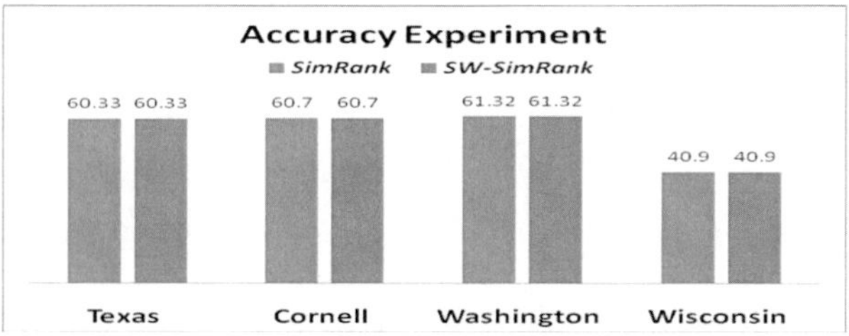

Fig. 7. Accuracy on Website Dataset

In Fig.7, we can see that the accuracy of *SW-SimRank* is the same as *SimRank*. In table 5, we can observe that in *r* steps of iteration, all of the reachable even-paths have been searched. Thus, *SW-SimRank* doesn't lose any accuracy on the web datasets of four universities.

6.3 Computation Cost of *SW-Simrank* Algorithm

The computation cost for all these algorithms depends on two aspects: the time for each step of iteration and the number of iterations.

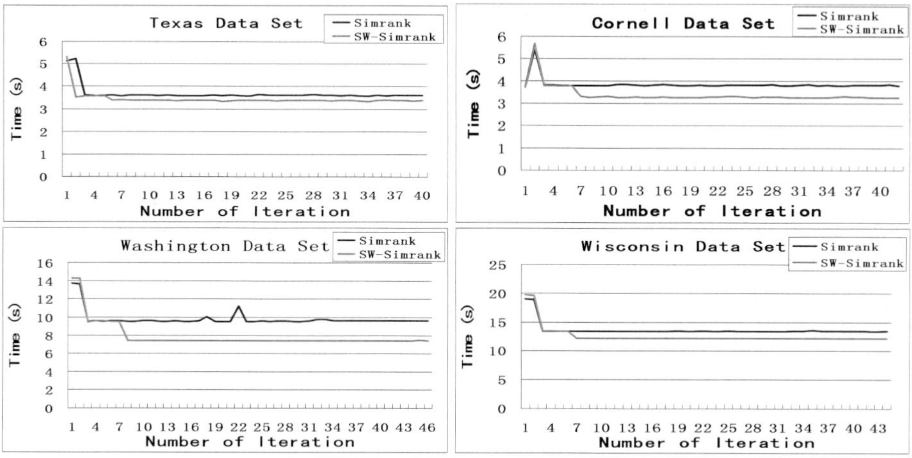

Fig. 8. Iteration VS Time(s)

Table 6. Total Time VS Algorithm

Dataset / Alg.	Texas	Cornell	Washington	Wisconsin
SimRank	147.9s	162.0s	453.1s	602.7s
SW-SimRank	***138.2s***	***143.2s***	***366.0 s***	***557.3 s***

Fig.8 shows time cost at each step of iteration. We can observe that after r steps of iteration, the time cost for *SW-SimRank* drops a lot, lower than the cost of original *SimRank* algorithm. Table 6 discribes the total time of these two methods, we can see that our method speeds up the orginal *SimRank* by about 10%.

6.4 The Factor Affecting *SW-SimRank*

Fig.9. shows the performance improvement proportion of these four datasets and table 7 describes the proportion of these unreachable pairs on these web graphs. In Fig.9, *SW-SimRank* takes the biggest performance improvement on Washington dataset and the

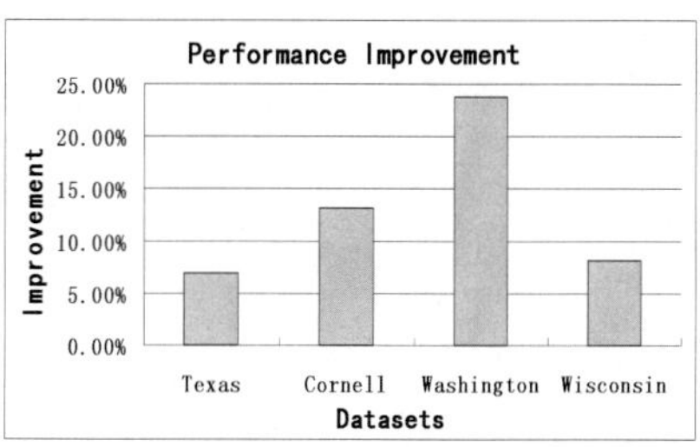

Fig. 9. Performance Improvement Proportion for four datasets

Table 7. URPP for each Datasets

Dataset	T.	C.	Wa.	Wi.
URPP	0.09	0.19	0.29	0.15

Note: URPP means the proportion of unreachable pairs. The larger the URPP, the more zeros in the similarity matrix.

unreachable pairs in Washington graphs are also the most. Thus, the higher proportion of unreachable pairs is in the relationship graph, the higher proportion of performance improvement is for *SW-SimRank*. Therefore, the worst situation is that if the application graph is a connected graph, the time cost of *SW-SimRank* is the same as *SimRank*.

7 Conclusion

In this paper, we proposed a novel algorithm called *SW-SimRank* to speed up similarity calculation by avoiding recalculating the unreachable pairs' similarity scores. This work is based on our observation that node pairs' similarity scores which are zero after first several rounds of iterations will remain zero in the final output. Our experimental results showed that the new algorithm achieves higher efficiency than the original *SimRank* algorithm. Meanwhile, we proved the convergence of the *SW-SimRank* algorithm in the appendix.

Acknowledgments

This work was supported in part by the National Nature Science Foundation of China under Grant No.70871068, 70621061, 70890083 and 60873017.

References

1. Jeh, G., Widom, J.: SimRank: A measure of structural-context similarity. In: SIGKDD, pp. 538–543 (2002)
2. Small, H.: Co-citation in the scientific literature: A new measure of the relationship between two documents. Journal of the American Society for Information Science 24(4), 265–269 (1973)
3. Kessler, M.M.: Bibliographic coupling between scientific papers. American Documentation 14(1), 10–25 (1963)
4. Amsler, R.: Applications of citation-based automatic classification. Linguistic Research Center (1972)
5. Fogaras, D., Racz, B.: Scaling link-based similarity search. In: WWW, pp. 641–650 (2005)
6. Yin, X.X., Han, J.W., Yu, P.S.: LinkClus: Efficient Clustering via Heterogeneous Semantic Links. In: VLDB, pp. 427–438 (2006)
7. Dmitry, L., Pavel, V., Maxim, G., Denis, T.: Accuracy Estimate and Optimization Techniques for SimRank Computation. In: VLDB, pp. 422–433 (2008)
8. Xi, W., Fox, E.A., Zhang, B., Cheng, Z.: SimFusion: Measuring Similarity Using Unified Relationship Matrix. In: SIGIR, pp. 130–137 (2005)
9. Pool, I., Kochen, M.: Contacts and influence, Social Network (1978)
10. Lada, A.A.: The Small World Web. In: Abiteboul, S., Vercoustre, A.-M. (eds.) ECDL 1999, vol. 1696, p. 443. Springer, Heidelberg (1999)
11. Langville, A.N., Meyer, C.D.: Deeper Inside PageRank. Internet Mathematics 1(3), 335–400 (2004)
12. Broder, A., Kumar, R., Maghoul, F., Raghavan, P., Rajagopalan, S., Stata, R., Tomkins, A., Wiener, J.: Graph Structure in the Web. In: WWW (2000)
13. CMU four university data set, `http://www.cs.cmu.edu/afs/cs/project/theo-20/www/data/`
14. Han, J.W., Kamber, M.: Data Mining Concepts and Techniques. Morgan Kaufmann Publishers, San Francisco (2001)

Appendix

Proof of convergence of the *SW-SimRank*

Lemma A: Let $s(V_a,V_b)$ be the similarity value calculated by *SimRank*,

$$s(V_a,V_b) = \sum_{t:(V_a,V_b)\sim\sim>(V_x,V_x)} P[t]c^{l(t)} = c\sum_{i=1}^{n_1} P[t] + c^2\sum_{i=1}^{n_2} P[t] + \ldots + c^k\sum_{i=1}^{n_k} P[t] + \ldots$$

Let $s_{sw}(V_a,V_b)$ be the similarity value calculated by *SW-SimRank*,

$$S_{sw}(V_a,V_b) = \sum_{t:(V_a,V_b)\sim\sim>(V_x,V_x)} P[t]c^{l(t)} = c\sum_{i=1}^{m_1} P[t] + c^2\sum_{i=1}^{m_2} P[t] + c^k\sum_{i=1}^{m_k} P[t] + \ldots$$

Let $u_k = c^k\sum_{i=1}^{n_k} P[t]$, and $v_k = c^k\sum_{i=1}^{m_k} P[t]$, Then $0 \le v_k \le u_k$.

Proof: There are three situations for the kth step of iteration for *SW-SimRank.*

Case 1: In k^{th} iteration, *SW-SimRank* don't search even path between V_a and V_b, but will search the even path in the following iteration. Thus, $0=v_k<u_k$.

Case 2: In k^{th} iteration, *SW-SimRank* don't search even path between V_a and V_b, and will not search the even path in the following iteration. Thus, $0=v_k=u_k$.

Case 3: In k^{th} iteration, *SW-SimRank* searches all of the paths between V_a and V_b. Thus, $0\le v_k=u_k$.

In sum, $0\le v_k \le u_k$. ■

Lemma B: $s_{swk}(V_a,V_b) \le s_k(V_a,V_b) \le c\dfrac{1-c^k}{1-c}$

Proof: According to Lemma A, it's easy to get $s_{swk}(V_a,V_b) \le s_k(V_a,V_b)$.

$$s_k(V_a,V_b) = c\sum_{i=1}^{n_1} P[t] + c^2\sum_{i=1}^{n_2} P[t] + \ldots + c^k\sum_{i=1}^{n_k} P[t] + \ldots$$

Let 1 equal $max(\sum_{i=1}^{n_1} P[t], \sum_{i=1}^{n_2} P[t], \ldots, c^k\sum_{i=1}^{n_k} P[t], \ldots)$

then $s_k(V_a,V_b) \le c + c^2 + \ldots + c^k + \ldots = c\dfrac{1-c^k}{1-c}$

Thus, $s_{swk}(V_a,V_b) \le s_k(V_a,V_b) \le c\dfrac{1-c^k}{1-c}$. ■

Theorem: $s_{sw}(V_a,V_b)$ will converge to a fixed value.

Proof: $s_{sw}(V_a,V_b) = \lim_{k->+\infty} s_{swk}(V_a,V_b)$, $s(V_a,V_b) = \lim_{k->+\infty} s_k(V_a,V_b)$

$$\lim_{k->+\infty} s_{swk}(V_a,V_b) \le \lim_{k->+\infty} s_k(V_a,V_b) \le \lim_{k->+\infty} c\frac{1-c^k}{1-c} = \frac{c}{1-c}$$

Therefore, $s_{sw}(V_a,V_b)$ has a upper bound.

In Lemma A, $0\le v_k$, thus $s_{sw}(V_a,V_b)$ is the positive term series.

Thus, *we have proved the convergence of the SW-SimRank.* ■

A Framework for Multi-Objective Clustering and Its Application to Co-location Mining

Rachsuda Jiamthapthaksin, Christoph F. Eick, and Ricardo Vilalta

Computer Science Department, University of Houston,
77204-3010 Houston, TX, USA
{rachsuda,ceick,vilalta}@cs.uh.edu

Abstract. The goal of multi-objective clustering (MOC) is to decompose a dataset into similar groups maximizing multiple objectives in parallel. In this paper, we provide a methodology, architecture and algorithms that, based on a large set of objectives, derive interesting clusters regarding two or more of those objectives. The proposed architecture relies on clustering algorithms that support plug-in fitness functions and on multi-run clustering in which clustering algorithms are run multiple times maximizing different subsets of objectives that are captured in compound fitness functions. MOC provides search engine type capabilities to users, enabling them to query a large set of clusters with respect to different objectives and thresholds. We evaluate the proposed MOC framework in a case study that centers on spatial co-location mining; the goal is to identify regions in which high levels of Arsenic concentrations are co-located with high concentrations of other chemicals in the Texas water supply.

Keywords: Multi-objective clustering, knowledge discovery, spatial data mining, co-location mining, clustering with plug-in fitness functions.

1 Introduction

The goal of clustering is to group similar objects into clusters while separating dissimilar objects. Although clustering is a very popular data mining technique which has been used for over 40 years, its objectives and how to evaluate different clustering results is still subject to a lot of controversy. For example, Saha et al. observe "*evidently one reason for the difficulty of clustering is that for many data sets no unambiguous partitioning of the data exists, or can be established, even by humans*" [1]. Moreover, matters are even worse for applications that seek of non-homogeneous groupings, because "*virtually all existing clustering algorithms assume a homogeneous clustering criterion over the entire feature space … because its intrinsic criterion may not fit well with the data distribution in the entire feature space*" [2]; furthermore, "*by focusing on just one aspect of cluster quality, most clustering algorithms … are not robust to variations in cluster shape, size, dimensionality and other characteristics*" [3]. Finally, in specific application domains, users seek for clusters with similar extrinsic characteristics; their definitions of "interestingness" of clusters are usually different from those used in typical clustering algorithms. Consequently, clusters obtained by typical clustering algorithms frequently do not capture what users are really looking for.

R. Huang et al. (Eds.): ADMA 2009, LNAI 5678, pp. 188–199, 2009.

The goal of this paper is to alleviate the problems that have been identified in the previous paragraph; in particular, a novel framework for multi-objective clustering is introduced and evaluated. The goal of multi-objective clustering is to decompose a dataset into similar groups maximizing multiple objectives in parallel. Multi-objective clustering can be viewed as a special case of multi-objective optimization which aims to simultaneously optimize trade-off among multiple objectives under certain constraints. In this paper, we provide a methodology, architecture and algorithms that, based on a large set of objectives, derive clusters that are interesting with respect to two or more of those objectives. The proposed architecture relies on clustering algorithms that support plug-in fitness functions and on multi-run clustering in which clustering algorithms are run multiple times maximizing different subsets of objectives that are captured in compound fitness functions. Using different combinations of single objective functions a cluster repository is created; final clusterings are created by querying the repository based on user preferences.

In the following, we discuss an example to better illustrate what we are trying to accomplish. Let us assume we like to assist a travel agent in finding different, interesting places (spatial clusters) with respect to a large set of objectives his customers are interested in, such as quality and price of hotels, quality of hiking, desirable weather patterns, acceptable crime rates, etc. This set of objectives Q will be captured in reward functions $reward_q$ each of which corresponds to a single objective q; moreover, for each objective $q \in Q$ a minimum satisfaction threshold θ_q is provided. Our proposed framework identifies the "best" places for our travel agent for which two or more objectives[1] in Q are satisfied. These places will be stored in a cluster repository. However, if places are overlapping certain dominance relations should be enforced and only non-dominated places will be stored in the repository. For example, if there is a region in South Texas that satisfies objectives $\{A,B\}$ and a slightly different region satisfying objectives $\{A,B,C\}$, then only the later region will be reported; similarly, if two overlapping regions satisfy the same objectives, but the rewards of one region are higher, only that region should be reported. As we will further discuss in section 2, dominance relations between clusters can be quite complicated, because the scope of a cluster has to be considered. For example, a cluster in South Texas is disjoint in scope from a cluster in West Texas; consequently, no dominance relation exists between these two regions. Finally, dealing with clusters with highly overlapping scope when creating the final clustering poses another challenge: for example, if we have two highly similar clusters, one satisfying objectives $\{A,B\}$ and the other satisfying objectives $\{B,C\}$: should we include both clusters in the final results; or—if we decided not to do so—how should we select the cluster to be included in the final clustering? In summary, the ultimate vision of this work is to develop a system that, given a large set of objectives, automatically identifies the "best" clusters that satisfy a large number of objectives; moreover, different clusters will usually serve quite different objectives.

An idea employed by our approach is to run a clustering algorithm with a plug-in compound fitness function multiple times to generate clusters that maximize sets of objectives captured in the compound fitness function. Our clustering approach to cope

---

[1] A cluster x is potentially interesting with respect to objective q, if its reward value is larger than q's satisfaction threshold: $reward_q(x) > \theta_q$.

with multi-objective problem is similar to the ensemble clustering approach as it is based on the key hypothesis that better clustering results can be obtained by combining clusters that originate from multiple runs of clustering algorithms [4]. However, our approach is an incremental approach that collects and refines clusters on the fly, and the search for alternative clusters takes into consideration what clusters already have been generated, rewarding novelty.

The rest of our paper is organized as follows: Section 2 proposes an architecture and algorithms for multi-objective clustering. Section 3 demonstrates the proposed system for a co-location mining case study. Section 4 discusses related work and Section 5 summarizes our findings.

2 An Architecture and Algorithms for Multi-run Clustering

A main goal of multi-objective clustering (MOC) is to find individual clusters that are good with respect to multiple objectives; due to the nature of MOC only clusters that are good with respect to at least two objectives are reported. In the remainder of this section we focus on a specific architecture and algorithms for MOC. The following features are of major importance for the proposed architecture: the use of clustering algorithms that support plug-in fitness/reward functions, the capability to create compound fitness functions to instruct clustering algorithms to seek for clusters that are good with respect to multiple objectives, the use of a repository *M* that stores clusters that are potentially interesting, the use of a multi-objective dominance relation to determine what clusters should be stored in *M*, and the availability of a cluster summarization tool that creates final clusterings based on user preferences from the clusters in the repository *M*.

2.1 Building Blocks for Multi-Objective Clustering

A. Clustering algorithms that support plug-in fitness and reward functions

In this paper, we assume that $Q=\{q_1, q_2, \ldots, q_z\}$ is the set of objectives that multi-objective clustering maximizes. For each objective $q \in Q$ a reward function $Reward_q$ has to be provided that measures to which extent the objective q is satisfied—higher rewards mean better clusters with respect to objective q. Moreover, reward thresholds $\theta_{q1},\ldots,\theta_{qz}$ are associated with each reward function. If $Reward_q(x)>\theta_q$ holds, we say that "*cluster x satisfies objective q*". In general, the goal of MOC is to seek for clusters that satisfy a large number of objectives in Q, but rarely all objectives in Q; different clusters usually serve different objectives.

Our approach employs clustering algorithms that support plug-in fitness function; given a dataset $O=\{o_1,\ldots,o_n\}$, the algorithm seeks for a clustering $X=\{x_1,\ldots,x_k\}$ that maximizes a plug-in fitness function q:

$$q(X) = \sum_{i=1}^{k} Reward_q(x_i)\,. \tag{1}$$

subject to: $x_i \cap x_j = \emptyset$ for $i \neq j$, $x_i \subseteq O$ for $i = 1, \ldots, k$ and $\bigcup_{i=1}^{k} x_i \subseteq O$.

A family of clustering algorithms that support such fitness functions (CLEVER [5], SCMRG [6], and MOSAIC [7]) has been designed and implemented in our past research. Our approach measures the quality of a clustering $X=\{x_1,..,x_k\}$ as the sum of rewards obtained for each cluster x_i (i=1,…,k) using the reward function $Reward_q$. Additionally, reward functions are used in our multi-objective clustering approach to determine dominance and when creating the final clustering. Reward functions typically correspond to popular cluster evaluation measures, such as entropy or compactness.

B. The role of the cluster repository *M*

Our approach runs clustering algorithms multiple times with the same or different reward/fitness functions and stores the potentially interesting clusters in a cluster list *M*. Each time a new clustering *X* is obtained, *M* is updated; some clusters in *X* might be inserted into *M*, and some clusters in *M* might have to be deleted due to the arrival of better clusters in *X*. Only non-dominated, multi-objective clusters are stored in *M*. We will define what clusters *M* can contain more formally next.

Definition 1. *x* is a multi-objective cluster with respect to a set of objectives *Q*

$$MO\_Cluster(x,Q) \Leftrightarrow \exists q \in Q\ \exists q' \in Q (q \neq q' \wedge Reward_q(x) \geq \theta_q \wedge Reward_{q'}(x) \geq \theta_{q'}) . \quad (2)$$

Definition 2. *x* dominates *y* with respect to *Q*

$$Dominates(x,y,Q) \Leftrightarrow \forall q \in Q\ ((Reward_q(x) \geq Reward_q(y) \vee Reward_q(x) < \theta_q \wedge Reward_q(y) < \theta_q) \wedge Similarity(x,y) \geq \theta_{sim}) . \quad (3)$$

Definition 2 introduces dominance relations between clusters. It is also important to observe that if *x* and *y* are both bad clusters with respect to a single objective *q*, the rewards associated with objective *q* are not used to determine dominance between *x* and *y*; in general, we only compare *x* and *y* based on those objectives that at least one of them satisfies. Moreover, the above definition assumes that clusters *x* and *y* have an agreement in their scope to make them comparable; a user-defined similarity threshold θ_{sim} has to be provided for this purpose. In our current work, similarity between two clusters *x* and *y* is assessed (|*c*| returns the cardinality of set *c*) as follows:

$$Similarity(x,y)=|x \cap y|/|x \cup y| . \quad (4)$$

It takes the ratio of the number of common objects between *x* and *y* over the total number of objects in *x* and *y*.

In the following, we use the symbol '$\rangle$' to express dominance relationships between clusters:

$$x \rangle y \Leftrightarrow Dominates(x,y,Q) . \quad (5)$$

In general, *M* should only store non-dominated clusters, and algorithms that update *M* should not violate this constraint; that is:

$$m \in M \Rightarrow \sim\exists m' \in M\ m' \rangle m . \quad (6)$$

2.2 The Proposed MOC Framework

The architecture of the MOC system that we propose is depicted in Fig. 1; it consists of 4 main components: a clustering algorithm, storage unit, goal-driven fitness function generator and cluster summarization unit. MOC is performed as follows: First, the goal-driven fitness function generator (*FG*) selects a new fitness function for the clustering algorithm (*CA*), which generates a new clustering *X* in the second step. Third, the storage unit (*SU*) updates its cluster list *M* using the clusters in *X*. The algorithm iterates over these three steps until a large number of clusters has been obtained. Later, in the fourth step, the cluster summarization unit (*CU*) produces final clusters based on user preferences which are subsets of the clusters in *M*. Details of algorithms proposed to serve the individual tasks are given in the following discussions.

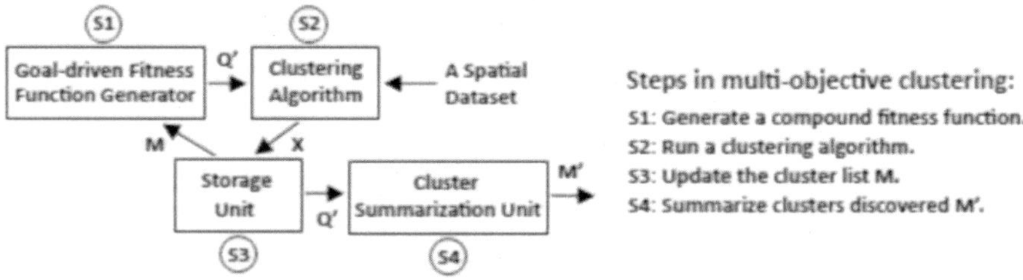

Fig. 1. An architecture of multi-objective clustering

Preprocessing step. The objective of this step is to obtain simple statistics for individual fitness functions for a given dataset. Results obtained from this step will be used to determine reward thresholds, $\theta_{q1},\ldots,\theta_{qz}$. Our current implementation uses percentiles to determine the satisfaction threshold for a particular reward function $Reward_q$. Alternatively, the thresholds could be acquired from a domain expert.

Step 1: Generate a compound fitness function *Q'*. *FG* selects a subset Q'($\subset Q$) of the objectives *Q* and creates a *compound fitness function* $q_{Q'}$ relying on a penalty function approach [8] that is defined as follows:

$$q_{Q'}(X) = \sum_{i=1}^{k} CmpReward(x_i) \ . \tag{7}$$

$$CmpReward(x) = \sum_{q \in Q'} (Reward_q(x) * Penalty(Q',x)) \, . \tag{8}$$

Penalty(*Q'*,*x*) is a penalty function that returns 1 if *x* satisfies all objectives in *Q'*, but returns a smaller number in the case that some objectives in *Q'* are not satisfied. In general, $q_{Q'}$ sums the rewards for all objectives $q \in Q'$; however, it gives more reward to clusters x_i that satisfy all objectives in order to motivate the *CA* to seek for multi-objective clusters. Our current implementation uses 2-objective compound fitness functions $q_{Q'}$ with *Q'*={*q*,*q'*} in conjunction with the following penalty function:

$$Penalty(\{q,q'\},x) = \begin{cases} 1 & Reward_q(x) \geq \theta_q \wedge Reward_{q'}(x) \geq \theta_{q'} \\ 0.5 & Otherwise\,. \end{cases} \tag{9}$$

Step 2: Run the *CA* with a compound fitness function $q_{Q'}$ to generate a clustering *X* maximizing *Q'*.

```
FOR ALL x∈X' DO
    IF ~MO_Cluster(x) ∨ ∃m∈M m⟩x THEN
        Discard x;
    ELSE
        Let D={m∈M | x⟩m};
        Insert x into M;
        Delete all clusters in D from M;
```

Fig. 2. Update_M_by_X algorithm

```
Let
DEDGE:={(c1,c2)|c1∈M ∧ c2∈M ∧ sim(c1,c2)>θ_rem
∧better(c2,c1)}
REMCAND:={c|∃d (c,d)∈DEDGE}
DOMINANT:={c|∃d (d,c)∈DEDGE ∧ c∉REMCAND}
REM:={c|∃d ((c,d)∈DEDGE ∧ d∈DOMINANT)}
Better(c1,c2)↔ ∀q∈Q̂, Reward_q(c1)>Reward_q(c2) ∨
                         (Reward_q(c1)=Reward_q(c2) ∧
clusterNumber(c1)>clusterNumber(c2))
Remark: Ties have to be broken so that DEDGE is always
a DAG; no cycles in
               DEDGE are allowed to occur.

Input: M, Q̂, θ_q  for each q∈Q̂
Output: M'⊆M

Remove {c∈M | ∃q∈Q̂ Reward_q(c)<θ_q}
               "Remove bad clusters with respect to Q̂."
Compute DEDGE from M;
Compute REMCAND;
Compute DOMINANT;
WHILE true DO
 {
  Compute REM;
  IF REM=∅ THEN EXIT ELSE M=M/REM;
  Update DEDGE by removing edges of deleted clusters in
REM;
  Update REMCAND based on DEDGE;
  Update DOMINANT based on DEDGE and REMCAND;
}
RETURN(M);
```

Fig. 3. MO-Dominance-guided Cluster Reduction algorithm (MO-DCR)

Step 3: Update the storage unit *M* using the obtained clustering *X*. To accomplish this task, we introduce the *Update_M_by_X* algorithm as specified in Fig. 2. The algorithm considers multi-objective dominance as discussed in Section 2.1. In a nutshell, the algorithm selectively inserts "good" clusters with respect to two or more objectives and makes sure that all clusters in *M* are non-dominated.

Step 4: Create a final clustering from *M*. The cluster summarization unit retrieves a subset of clusters *M'* from *M* based on user preferences. In this paper, we introduce an algorithm called *MO-Dominance-guided Cluster Reduction algorithm* (MO-DCR), whose pseudocode is given in Fig. 3. MO-DCR returns a subset of interesting clusters, based on the following two user-defined input parameters:

1. $\hat{Q} \subset Q$ and reward thresholds θ_q for each $q \in \hat{Q}$
2. A cluster removal similarly threshold θ_{rem}. (Basically, if two clusters have too much overlap, one is not included in the final clustering.)

The goal of the algorithm is to return a clustering that is good with respect to $\hat{Q}$ selected by a user, and to remove clusters that are highly overlapping. The algorithm iteratively performs two steps:

1. Identify multi-objective dominant clusters with respect to $\hat{Q}$, and
2. Remove dominated clusters which are in the θ_{rem}-neighborhood of a dominant cluster.

3 Experimental Results

In this section, we demonstrate the benefits of the proposed multi-objective clustering framework in a real world water pollution case study; the goal of the study is to obtain a better understanding of what causes high arsenic concentrations to occur. In this section, we report on experiments that use multi-objective clustering to identify regions in Texas where high level of Arsenic concentrations are in close proximity with high concentrations of other chemicals. For instance, the Rank3 region in Fig. 5b is found by the framework with an associated co-location pattern As↑Mo↑V↑B↑, indicating that high Arsenic concentrations are co-located with high Molybdenum, Vanadium and Boron concentrations in this region. TWDB has monitored water quality and collected the data for 105,814 wells in Texas over last 25 years. For this experiment we used a water well dataset called Arsenic_10_avg [9] which was created from a database provided by the Texas Water Development Board (TWDB) [10]. In this paper, we use a subset of this dataset containing 3 spatial attributes longitude, latitude and aquifer and the following 8 chemical concentrations for each water well: Arsenic (As), Molybdenum (Mo), Vanadium (V), Boron (B), Fluoride (F), Chloride (Cl^-), Sulfate (SO_4^{2-}) and Total Dissolved Solids (TDS).

We used CLEVER (CLustEring using representatives and Randomized hill climbing), introduced in [5], as the clustering algorithm in the experiments. In a nutshell, CLEVER is a prototype-based clustering algorithm that seeks for a clustering *X* maximizing a plug-in fitness function *q(X)*. A single-objective fitness function for regional co-location mining has been introduced in [5]. In the following, we will reformulate this problem as a multi-objective clustering problem.

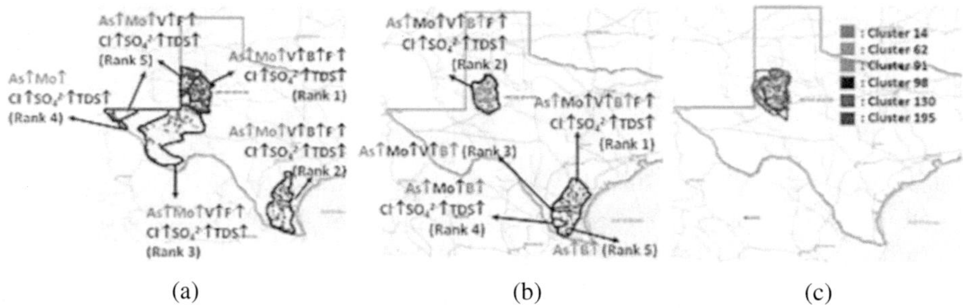

Fig. 5. Visualization of experimental results: (a) and (b) are the top 5 regions ordered by rewards using user-defined query {As↑,Mo↑} and {As↑,B↑}, respectively, and (c) is overlay of similar regions in the storage unit located in the Southern Ogallala aquifer

Let O be a dataset

$x \subseteq O$ be a cluster, called a region in the following
$o \in O$ be an object in the dataset O
$N=\{A_1,\ldots,A_m\}$ be the set of non-geo-referenced continuous attributes in the dataset O
$Q=\{A_1\uparrow,A_1\downarrow,\ldots,A_m\uparrow,A_m\downarrow\}$ be the set of possible base co-location patterns
$B \subseteq Q$ be a set of co-location patterns
z-score(A,o) be the z-score of object o's value of attribute A

$$z(A\uparrow,o)=\begin{cases} z\text{-}score(A,o) & if\ z\text{-}score(A,x)>0 \\ 0 & otherwise. \end{cases} \tag{10}$$

$$z(A\downarrow,o)=\begin{cases} -z\text{-}score(A,o) & if\ z\text{-}score(A,x)<0 \\ 0 & otherwise. \end{cases} \tag{11}$$

z(p,o) is called the *z-value* of base pattern $p \in Q$ for object o in the following. The interestingness of an object o with respect to a co-location set $B \subseteq Q$ is measured as the product of the z-values of the patterns in the set B. It is defined as follows:

$$i(B,o) = \prod_{p \in B} z(p,o)\,. \tag{12}$$

In general, the interestingness of a region can be straightforwardly computed by taking the average interestingness of the objects belonging to a region; however, using this approach some very large products might dominate interestingness computations; consequently, our additionally considers purity when computing region interestingness, where *purity*(B,x) denotes the *percentage of objects* $o \in c$ *for which* $i(B,o)>0$. The interestingness $\varphi(B,x)$ of a region x with respect to a co-location set B is measured as follows:

$$\varphi(B,x) = \frac{\sum_{o \in c} i(B,o)}{|x|} \times purity(B,x)^{\tau}\,. \tag{13}$$

The parameter $\tau \in [0,\infty)$ controls the importance attached to purity in interestingness computations; τ=0 implies that purity is ignored, and using larger value increases the importance of purity. Finally, the reward of the region x is computed as follows:

$$Reward_B(x) = \varphi(B,x) \times |x|^{\beta}. \tag{14}$$

where |x| denotes the number of objects in the region and β is a parameter that controls how much premium is put on the number of objects in a region. Finally, $Reward_B$ is used to define a plug-in fitness function q_B as follows (see also section 2):

$$q_B(X) = q_B(\{x_1,\ldots,x_k\}) = \sum_{i=1}^{k}\left(Reward_B(x_i)\right) . \tag{15}$$

In the experiment, we use 7 different objective functions: $q_{\{As\uparrow,Mo\uparrow\}}$, $q_{\{As\uparrow,V\uparrow\}}$, $q_{\{As\uparrow,B\uparrow\}}$, $q_{\{As\uparrow,F^-\uparrow\}}$, $q_{\{As\uparrow,Cl^-\uparrow\}}$, $q_{\{As\uparrow,SO4^{2-}\uparrow\}}$, $q_{\{As\uparrow,TDS\uparrow\}}$. We are basically interested in finding regions for which at least two of those functions are high; in other words, regions in which high arsenic concentrations are co-located with high concentrations of two or more other chemicals. We claim that the MOC approach for regional co-location mining is more powerful than our original approach [5] that uses a single-objective function that relies on maximum-valued patterns: $max_B(\varphi(B,x))$ is used to assess interestingness for a region—which ignores alternative patterns: for example, if *A* is co-located with *B*,*C* in one region and with *D*,*E* in another region and the two regions overlap, the original approach will not be able to report both regions.

In the following, we report results of an experiments in which MOC was used with the following parameter setting for the fitness function we introduced earlier: τ=1, β=1.5. In the MOC preprocessing step, we obtain the reward thresholds at 40 percentile based on the fitness distribution of the individual fitness functions: $\theta_{q\{As\uparrow,Mo\uparrow\}}$=0.59, $\theta_{q\{As\uparrow,V\uparrow\}}$=2.91, $\theta_{q\{As\uparrow,B\uparrow\}}$=0.27, $\theta_{q\{As\uparrow,F^-\uparrow\}}$=2.87, $\theta_{q\{As\uparrow,Cl^-\uparrow\}}$=0.53, $\theta_{q\{As\uparrow,SO4^{2-}\uparrow\}}$=1.35 and $\theta_{q\{As\uparrow,TDS\uparrow\}}$=1.51. After finishing iterative process in MOC which exploring all pairs of the 7 fitness functions, out of original generated 1,093 clusters, only 227 clusters were selectively stored in the repository.

Regarding the last step in MOC, we set up a user defined reward thresholds to create a final clustering to: $\theta_{q\{As\uparrow,Mo\uparrow\}}$=13, $\theta_{q\{As\uparrow,V\uparrow\}}$=15, $\theta_{q\{As\uparrow,B\uparrow\}}$=10, $\theta_{q\{As\uparrow,F^-\uparrow\}}$=25, $\theta_{q\{As\uparrow,Cl^-\uparrow\}}$=7, $\theta_{q\{As\uparrow,SO4^{2-}\uparrow\}}$=6, $\theta_{q\{As\uparrow,TDS\uparrow\}}$=8, and set the removal threshold θ_{rem}=0.1 to seek for nearly non-overlapping clusters. Examples of the top 5 regions and patterns with respect to two queries: $query_1$={As↑,Mo↑} and $query_2$={As↑,B↑} are shown in Fig. 5a and 5b, respectively. In general, query patterns are used to select and sort the obtained regions; that is, regions dissatisfying {As↑,Mo↑} would not be included in the result for $query_1$. The visualizations associate the chemicals whose co-location strengths are above the threshold with each visualized region. For instance, for the query {As↑,Mo↑}, all of the top 5 regions retrieved contain patterns whose length is 5 or more. It can be observed that the Rank1 region in Fig. 5a significantly overlaps with the Rank2 region in Fig. 5b and share the same co-location sets: different regions are reported to better serve the different interests expressed by the two queries. Moreover, as depicted in Fig. 5b MOC is also able to identify nested clusters (i.e. the regions ranked 3-5 are sub-regions of the Rank1 region), and particularly discriminate among companion elements, such as Vanadium (Rank3 region), or Chloride, Sulfate and Total Dissolved Solids (Rank4 region). In this particular case, the regions ranked 3-5 better serve specific objectives, whereas the Rank1 region satisfies a larger set of objectives; that is, there is no dominance relationship between the Rank1 region and the regions ranked 3-5. In general, in the experiment a large number of overlapping regions without any dominance relationships were identified in the Southern Ogallala aquifer, as depicted in Fig. 5c, which is a hotspot for arsenic pollution in Texas.

4 Related Work

Multi-objective clustering is considered a specific field of multi-objective optimization (MOO) whose goal is to simultaneously optimize trade-off between two or more objectives under certain constraints. According to our investigation, there are two approaches coping with multi-objective clustering: multi-objective evolutionary algorithms (MOEA) and dual clustering. MOEA have been widely used in MOO such as, NGSA-II [11] and PESA-II [12]. Their searching strategy, which automatically generates and preserves a set of diverse solutions, is desirable for this type of problem. In particular to multi-objective clustering, such MOEA are adapted to solve the multiple objectives clustering problem. The general idea is that by using clustering validation measures as the objective functions, the algorithm iteratively evolves clusters from one generation to another to improve quality as well as to explore diversity of cluster solutions. Handl and Knowles introduced VIENNA, an adaptive version of PESA-II EA incorporating specialized mutation and initialization procedures [3]. The algorithm employs two following internal measures to estimate clustering quality: overall deviation and connectivity. Such clustering quality measures have also been used in many other MOEA, e.g. MOCK [13] and MOCLE [14]. Finally, work by Molina et al. [15] employ scatter tabu search for non-linear multi-objective optimization which can potentially be utilized for multi-objective clustering. Dual clustering is another approach for multi-objective clustering [16]. It makes use of both clustering and classification algorithms to generate and to refine a clustering iteratively serving multiple objectives.

Our approach differs from those two approaches in that it seeks for good individual clusters maximizing multiple objectives that are integrated into a single clustering by a user-driven post-processing step. The proposed post processing algorithm operates like a search engine that allows users to query a large set of clusters with respect to different objectives and thresholds, obtaining a final clustering from the viewpoint of a single or a small set of objectives that are of particular interest for a user.

5 Conclusion

The paper centers on multi-objective clustering; in particular, we are interested in supporting applications in which a large number of diverse, sometimes contradictory objectives are given and the goal is to find clusters that satisfy large subsets of those objectives. Applications that need such capabilities include recommender systems, constraints satisfaction problems that involve a lot of soft constraints, complex design problems, and association analysis. Although the deployment of such systems is highly desirable, they are still quite far away from becoming commercial reality, because of the lack of useful research in this area.

The main contribution of this paper is to provide building blocks for the development of such systems. In particular, we proposed novel dominance relation for the case when we have a lot of objectives, but it is impossible to accomplish many of them. A second building block are clustering algorithms that support plug-in fitness functions, and the capability to construct compound fitness functions when a small set of objectives has to be satisfied. The third building block is a system architecture in which a large repository of clusters will be generated initially based on a given set of objectives, relying on multi-run clustering, dominance-relations, and compound fitness functions. The repository can be viewed as a meta-clustering with respect to all objectives investigated. Specific clusterings are generated by querying the repository based on particular user preferences and objective satisfaction thresholds. The fourth building block is the domain-driven nature of our approach in which users can express clustering criteria based on specific domain needs, and not based on highly generic, domain-independent objective functions which is the approach of most traditional clustering algorithms. Finally, we provided evidence based on a case study that multi-objective clustering approach is particularly useful for regional co-location mining, for which it is very difficult to formalize the problem using a single objective due to the large number of co-location patterns.

However, using the MOC approach creates new challenges for co-location mining: 1) a very large repository (containing more than 100,000 clusters) of highly overlapping, potentially interesting clusters has to be maintained and queried efficiently, and 2) coming up with sophisticated summarization strategies that extract clusters from the repository based on user preferences and possibly other user inputs is very challenging task. Our current summarization algorithm is only one of many possible solutions to this problem.

Acknowledgments. This research was supported in part by a grant from the Environmental Institute of Houston (EIH).

References

1. Saha, S., Bandyopadhyay, S.: A New Multiobjective Simulated Annealing Based Clustering Technique Using Stability And Symmetry. In: 19th International Conference on Pattern Recognition (2008)
2. Law, H.C.M., Topchy, A.P., Jain, A.K.: Multiobjective Data Clustering. In: IEEE Conputer Society Conference on Computer Vision and Pattern Recognition (2004)
3. Handl, J., Knowles, J.: Evolutionary Multiobjective Clustering. In: Yao, X., Burke, E.K., Lozano, J.A., Smith, J., Merelo-Guervós, J.J., Bullinaria, J.A., Rowe, J.E., Tiňo, P., Kabán, A., Schwefel, H.-P. (eds.) PPSN 2004. LNCS, vol. 3242, pp. 1081–1091. Springer, Heidelberg (2004)
4. Jiamthapthaksin, R., Eick, C.F., Rinsurongkawong, V.: An Architecture and Algorithms for Multi-Run Clustering. In: IEEE Computational Intelligence Symposium on Computational Intelligence and Data Mining (2009)
5. Eick, C.F., Parmar, R., Ding, W., Stepinki, T., Nicot, J.-P.: Finding Regional Co-location Patterns for Sets of Continuous Variables in Spatial Datasets. In: 16th ACM SIGSPATIAL International Conference on Advances in GIS (2008)

6. Eick, C.F., Vaezian, B., Jiang, D., Wang, J.: Discovery of Interesting Regions in Spatial Datasets Using Supervised Clustering. In: 10th European Conference on Principles and Practice of Knowledge Discovery in Databases (2006)
7. Choo, J., Jiamthapthaksin, R., Chen, C.-S., Celepcikay, O.C., Giusti, Eick, C.F.: MOSAIC: A Proximity Graph Approach to Agglomerative Clustering: In: 9th International Conference on Data Warehousing and Knowledge Discovery (2007)
8. Baeck, T., Fogel, D.B., Michalewicz, Z.: Penalty functions, Evolutionary computation 2. In: Advanced algorithms and operators. Institute of Physics Publishing, Philadelphia (2000)
9. Data Mining and Machine Learning Group website, University of Houston, Texas, http://www.tlc2.uh.edu/dmmlg/Datasets
10. Texas Water Development Board, http://www.twdb.state.tx.us/home/index.asp
11. Deb, K., Pratap, A., Agrawal, S., Meyarivan, T.: A fast and elitist multiobjective genetic algorithm: NSGA-II. J. Evolutionary Computation. 6, 182–197 (2002)
12. Corne, D.W., Jerram, N.R., Knowles, J.D., Oates, M.J.: PESA-II: Region-based Selection in Evolutionary Multiobjective Optimization. In: Genetic and Evolutionary Computation Conference, pp. 283–290 (2001)
13. Handl, J., Knowles, J.: An Evolutionary Approach to Multiobjective Clustering. J. Evolutionary Computation. 11, 56–57 (2007)
14. Faceli, K., de Carvalho, A.C.P.L.F., de Souto, M.C.P.: Multi-Objective Clustering Ensemble. J. Hybrid Intelligent Systems. 4, 145–146 (2007)
15. Molina, J., Laguna, M., Martí, R., Caballero, R.: SSPMO: A Scatter Search Procedure for Non-Linear Multiobjective Optimization. INFORMS J. Computing 19, 91–100 (2007)
16. Lin, C.-R., Liu, K.-H., Chen, M.-S.: Dual Clustering: Integrating Data Clustering over Optimization and Constraint Domains. J. Knowledge and Data Engineering 17 (2005)

Feature Selection in Marketing Applications

Stefan Lessmann and Stefan Voß

Institute of Information Systems, University of Hamburg, Von-Melle-Park 5, D-20146 Hamburg, Germany
lessmann@econ.uni-hamburg.de, stefan.voss@uni-hamburg.de

Abstract. The paper is concerned with marketing applications of classification analysis. Feature selection (FS) is crucial in this domain to avoid cognitive overload of decision makers through use of excessively large attribute sets. Whereas algorithms for feature ranking have received considerable attention within the literature, a clear strategy how a subset of attributes should be selected once a ranking has been obtained is yet missing. Consequently, three candidate FS procedures are presented and contrasted by means of empirical experimentation on real-world data. The results offer some guidance which approach should be employed in practical applications and identify promising avenues for future research.

Keywords: Marketing, Decision Support, Classification, Feature Selection, Support Vector Machines.

1 Introduction

The paper is concerned with the construction of predictive and parsimonious classification models to support decision making in marketing applications. Classification aims at approximating a functional relationship between explanatory factors (i.e., features) and a discrete target variable on the basis of empirical observations. The features characterize an object to be classified, whereas the target encodes its membership to a-priori known groups. In marketing contexts, the *objects* usually represent customers and the target variable encodes some behavioral trait. Exemplary applications include, e.g., the selection of responsive customers for direct-mailing campaigns or the identification of customers at the risk of churning (see, e.g., [13]).

A categorization of customers into groups suffices to solve a focal decision problem, e.g., whom to contact within a campaign. However, the more general data mining objective of deriving novel and actionable knowledge from data is not necessarily fulfilled, unless the prediction model reveals its internal functioning in a self-explanatory manner. In other words, it is essential that decision makers can understand the nature of the relationship that has been discerned from historical data and what factors govern customer behavior. This, in turn, facilitates adapting customer-centric business processes to, e.g., prevent future customer defection. Consequently, models are needed which are accurate in terms of the predictions they generate and interpretable with respect to their use of customer attributes.

R. Huang et al. (Eds.): ADMA 2009, LNAI 5678, pp. 200–208, 2009.

Tree- or rule-based classification models as well as linear classifiers like logistic regression are commonly considered comprehensible (see, e.g., [5]). Though, it is common that marketing datasets comprise a large number of attributes, which severely deteriorates the interpretability of a classification model [10]. To avoid cognitive overload of decision makers through employing an excessive number of features, parsimonious prediction models are required. This necessitates the use of FS to remove less informative attributes prior to building the (final) classification model.

FS is a popular topic in data mining and several approaches (e.g., forward or backward selection heuristics, evolutionary algorithms and meta-heuristics, etc.) have been proposed in the literature (see, e.g., [7]). The focus of respective work is commonly on efficiently searching the space of alternative feature subsets and providing a ranking of attributes according to their relevance for the prediction task. For example, such a ranking may reveal that using only the top-20% of attributes suffices to achieve 85% of the maximal accuracy. Whereas such information is of undoubted importance, it does not answer the question how many attributes should really be discarded. To achieve this, an auxillary selection criterion needs to be defined.

The objective of this work is to present results of a large empirical study that examines the effect of three different FS strategies on predictive accuracy and attribute set size to better understand the trade-off between highly accurate but complex and interpretable, parsimonious models within a marketing context. In that sense, the results serve as a first step towards the development of a general standard how to organize FS if an assessment of attribute importance is given by some feature ranking mechanism.

The paper is organized as follows: Section 2 describes the methodology for constructing classification models and computing attribute rankings, as well as the FS strategies considered in this paper. The datasets employed in the study as well as empirical results are presented in Section 3, before conclusions are drawn in Section 4.

2 Methodology

2.1 Classification with Support Vector Machines

Classification involves an automatic categorization of objects into a-priori defined classes. The objects are described by a set of attributes, which are assumed to affect class membership. However, the nature of the relationship between attribute values and class is unknown and has to be estimated from a sample of pre-classified examples $\{\boldsymbol{x}_i, y_i\}_{i=1}^{N}$, whereby $\boldsymbol{x} \in \Re^M$ represents an individual object and $y \in \{-1; +1\}$ its corresponding class label, which is assumed to be binary in this work. Thus, a classifier or classification model can be defined as a functional mapping from objects to classes: $f(\boldsymbol{x}) : \Re^M \mapsto \{-1; +1\}$.

The development of classification algorithms enjoys ongoing popularity in data mining, statistics and machine learning. Hence, a large number of candidate procedures are available for building classification models. The support vector machine (SVM) classifier is used throughout the paper. This choice is motivated by the following observations: 1) SVMs are linear classifiers like logistic regression and, as such, naturally enable assessing the contribution of individual attributes. 2) They have shown appealing performance in several classification benchmarks (see, e.g., [12,2]). 3) SVMs are commonly credited for their robustness towards large and possibly correlated feature sets (see, e.g., [18]). Therefore, they are able to distinguish between relevant and less informative attributes, which makes them a suitable candidate for FS [16]. 4) Very efficient learning algorithms for linear SVMs [9,8,3] are available to cope with the computational burden of constructing multiple models in a repetitive manner, as needed in (wrapper-based [11]) FS.

The SVM separates objects of opposite classes by means of a maximal margin hyperplane (MMH). That is, the plane is constructed to achieve a maximal distance between the nearest objects of adjacent classes. Let $\boldsymbol{w}$ and b denote the normal and intercept of a canonical hyperplane. Then, a MMH can be constructed by solving [18]:

$$\begin{aligned} \min_{w,\xi,b} \; & \|\boldsymbol{w}\| + \beta \sum_{i=1}^{N} \xi_i \\ s.t. \; & y_i(\boldsymbol{w} \cdot \boldsymbol{x} + b) \geq 1 - \xi_i \;\; \forall i = 1, \ldots, N \\ & \xi_i \geq 0 \qquad\qquad\qquad\quad \forall i = 1, \ldots, N \; . \end{aligned} \tag{1}$$

The slack variable ξ_i accounts for the fact that the data may not be linearly separable, whereas the tuning parameter β allows users to control the trade-off between having a large margin and few classification errors (i.e., objects on the wrong side of the separating plane or inside the margin with $\xi_i > 0$).

Let $\hat{\boldsymbol{w}}$ and $\hat{b}$ denote the optimal solution of (1). Then, the SVM classifier is given as:

$$f(\boldsymbol{x}) = sign(\hat{\boldsymbol{w}} \cdot \boldsymbol{x} + \hat{b}) \; . \tag{2}$$

Equation (2) exemplifies that linear classifiers are naturally interpretable in the sense that the influence of an individual attribute j on model predictions f is given by $\hat{w}_j$.

Note that SVMs can easily be extended to handle nonlinear classification problems by incorporating kernel functions [18]. However, such nonlinear SVMs are opaque in the sense that the influence of attributes on (class) predictions is concealed [15]. Therefore, attention is limited to the linear case (2) in this paper.

2.2 Feature Ranking with Recursive Feature Elimination

FS aims at identifying and discarding attributes that are of minor importance to the classifier, or possibly detrimental [7]. Furthermore, FS improves model comprehensibility since parsimonious models are easier to interpret [10].

In view of (2), attributes with small coefficient $\hat{w}_j$ have little impact on the classifier and can thus be discarded. This characteristic is exploited within the recursive feature elimination (RFE) algorithm [6]. RFE involves building a SVM with all attributes included, removing the attribute with minimal $|\hat{w}_j|$, and repetitively continuing this procedure until $M-1$ attributes are discarded. Adopting this approach, a ranking of attributes in terms of their coefficients in $\boldsymbol{w}$ and time of removal is produced.

RFE can be characterized as a recursive backward elimination procedure and requires constructing multiple SVMs with attribute sets of decreasing size. Although being expensive, re-training the SVM each time one attribute has been removed is important to assess the relevance of an attribute in relation with all other remaining attributes [6]. Consequently, the availability of fast training algorithms is essential to implement RFE in marketing applications, where a large number of attributes have to be processed. Therefore, the availability of highly efficient training algorithms to solve (1) [9,8,3] may be seen as a particular merit of the linear SVM classifier.

2.3 Alternative Feature Selection Criteria

A RFE-based ranking is valuable to appraise the relevance of individual attributes. However, RFE does not define the specific number of attributes to be removed [6]. Consequently, an auxiliary performance criterion is needed to utilize RFE for FS. This is demonstrated in Fig. 1, which displays the development of classification performance when recursively removing attributes according to RFE rankings. The experiment employs the US Census dataset [1], which comprises 48,842 examples with 17 attributes each. The data is randomly partitioned into a training set for model building and a test set for out-of-sample evaluation, repeating the sampling multiple times to obtain a robust estimate of the model's behaviour (see Section 3 for details).[1] Model performance is assessed in terms of the AUC criterion (area under a receiver-operating-characteristics curve), which ranges between 0,5 and 1,0 with higher values indicating higher accuracy (see, e.g., [4]). It has been suggested that the AUC is a particularly appropriate accuracy indicator in marketing applications [14]. Thus, it has been selected for this study to assess different classification models and FS strategies. Fig. 1 illustrates that using very few attributes suffices to classify this dataset with high accuracy. However, this will not generally be the case and a clear standard how to select the *right* number of attributes on the basis of a ranking like Fig. 1 is yet missing.

An intuitive approach to determine a particular feature subset (i.e. define which attributes should be discarded) is to use the feature set that yields highest accuracy. In fact, this strategy is commonly employed in other studies (see, e.g., [14]) and subsequently referred to as GFS (greedy FS). However, within the context considered here, GFS may not be ideal because it emphasizes only one FS objective (i.e., predictive accuracy), whereas model comprehensibility is not taken into account. In other words, focusing on accuracy alone may produce an overly complex model using (too) many features and, therefore, complicate its interpretation by decision makers. For example, GFS would discard only a single attribute in the case of US Census, whereas Fig. 1 indicates that using only the

[1] This is depicted by means of a box-plot in Fig. 1.

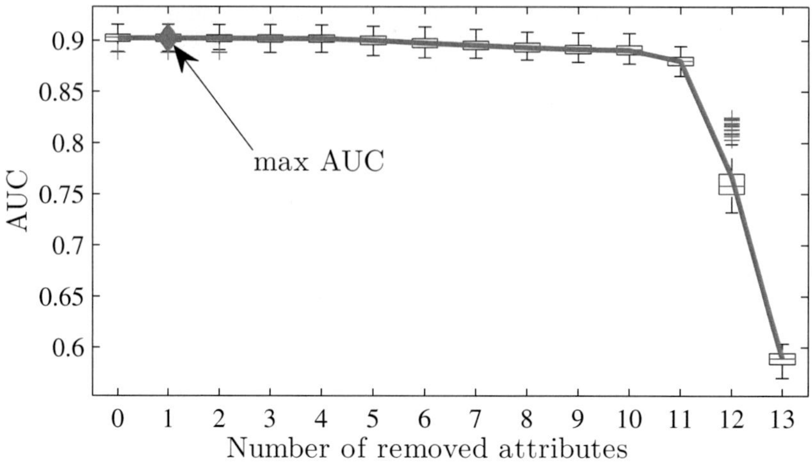

Fig. 1. Development of classification accuracy in terms of AUC during RFE

"best" four attributes would still achieve reasonable performance not far from the highest observed AUC.

To overcome the tendency of GFS to discard attributes (too) conservatively, an alternative approach could allow for a decrease in AUC up to a predefined threshold. Such a threshold-based FS (TFS) should help to further diminish the number of attributes, but requires an additional parameter (i.e., the threshold) to be defined. The choice of a suitable value should be governed by the financial loss incurred by losing (some) predictive accuracy. It seems reasonable that a rough estimate of this loss is available within customer-centric decision problems.

Third, the set of features may be augmented with randomly generated probe attributes. By definition, a random attribute will be uncorrelated with class membership. Therefore, all attributes that contain valuable information for predicting class membership should receive a higher rank than a random attribute within the RFE algorithm. If, on the other hand, a feature is ranked below a probe attribute by RFE, it is safe to assume that this attribute is not related to class and can thus be dismissed. This idea has been proposed in [17] and is subsequently referred to as RFS (random FS). A possible benefit of RFS stems from the fact that it avoids use of AUC. That is, the decision which attributes should be removed is based solely on the ranking. Consequently, some time may be saved by not assessing a classification model's accuracy on test data during RFE. On the other hand, the number of probe attributes to be incorporated into the dataset requires additional tuning to adapt RFS to a specific task. In the absence of a measure of predictive accuracy to guide the choice for a suitable number of probe attributes, determining this parameter may prove to be a complex endeavor.

3 Empirical Experiment

In oder to contrast the three FS strategies, an empirical study is conducted, which employs nine real-world datasets from different areas of customer-centric decision making like credit scoring, response modeling and churn prediction. The reader is referred to [14] for a detailed description of each task, whereas summary statistics for the datasets are provided in Table 1.

Each dataset is randomly partitioned into a training set (60%) and test set (40%). RFE is invoked to construct a series of SVM models on the training set, discarding the least important attribute (i.e., $\min(w_j)$), and assess the resulting model's performance on the test set. This procedure is repeated ten times and results are averaged to increase robustness and avoid possible bias due to a *lucky sample*. Hence, $10 \cdot (M-1)$ classifiers are built and assessed on each dataset. The algorithm of Keerthi and DeCoste [9] has been employed for SVM training.

The results of the comparison in terms of AUC and the number of features (M) for classifiers using all attributes as well as those which perform RFE-based FS by means of GFS, TFS, or RFS are given in Table 2. For the latter cases, the percentage decrease in accuracy and number of attributes is reported in the second row per dataset.

Table 2 illustrates the trade-off between parsimonious and (highly) accurate models. In particular, the performance of the SVM classifier with all attributes included defines an upper bound across all datasets. This confirms the robustness of SVMs towards large attribute sets. However, a key motivation for FS is to improve model comprehensibility, i.e., reduce the number of coefficients w_j that decision makers have to analyze. In this sense, the GFS criterion produces appealing results. It achieves reasonable reduction, especially on higher-dimensional datasets, and offers AUC results close to the full model. On the other hand, the number of attributes is further reduced by at least 74% when employing the TFS criterion. However, this comes at the cost of a decrease in accuracy of up to 6%. Clearly, TFS results depend upon the user-defined threshold, which has been set to 0.02 (i.e., features are removed until a model's performance falls below 98% of the full model's AUC). A 6% accuracy decrease may be acceptable in some applications, but prohibitive in others. A possible avenue for future research could thus comprise computing the full range of threshold values (i.e., 0 to 100%) and analyzing the resulting 'portfolio' of AUC scores and attribute set sizes. This would facilitate the construction of an *efficient frontier* of threshold settings, and, if decision makers are able to give a utility function

Table 1. Dataset characteristics

dataset	no. examples	no. attributes	prior class +1 [%]
AC	690	14	44.49
GC	1000	24	30.00
US Census	48,842	17	23.93
DMC 2000	38,890	96	5.87
DMC 2001	28,128	106	50.00
DMC 2002	20,000	101	10.00
DMC 2004	40,292	107	20.43
DMC 2005	50,000	119	5.80
DMC 2006	16,000	24	47.71

Table 2. Empirical results of the comparison of three FS strategies

Dataset	All		GFS		TFS		RFS	
	AUC	*M*	AUC	*M*	AUC	*M*	AUC	*M*
AC	0.92	14	0.92	8.0	0.92	3.7	0.92	11.3
			0.00	0.43	0.00	0.74	0.00	0.19
GC	0.79	24	0.78	15.4	0.74	2.9	0.77	17.3
			0.01	0.36	0.06	0.88	0.03	0.28
US Census	0.90	17	0.90	13.8	0.88	3.4	0.90	13.4
			0.00	0.19	0.02	0.80	0.00	0.21
DMC2000	0.81	96	0.81	48.5	0.77	12.4	0.81	64.3
			0.00	0.49	0.05	0.87	0.00	0.33
DMC2001	0.66	106	0.66	13.4	0.65	2.1	0.66	48.5
			0.00	0.87	0.02	0.98	0.00	0.54
DMC2002	0.66	101	0.65	33.9	0.65	9.6	0.65	66.2
			0.02	0.66	0.02	0.90	0.02	0.34
DMC2004	0.85	107	0.85	37.3	0.84	2.4	0.76	64.6
			0.00	0.65	0.01	0.98	0.11	0.40
DMC2005	0.67	119	0.67	71.9	0.64	24.5	0.66	82.0
			0.00	0.40	0.04	0.79	0.01	0.31
DMC2006	0.60	24	0.60	21.1	0.58	4.3	0.59	19.8
			0.00	0.12	0.03	0.82	0.02	0.18

that expresses their preferences regarding high accuracy and a small number of attributes, enable an *optimal* threshold to be identified.

Whereas, an unambiguous selection of GFS or TFS requires further analysis, Table 2 indicates that RFS is less suitable. That is, GFS usually achieves a higher reduction of attributes and smaller decrease in AUC. A minor advantage of RFS over GFS can be observed on *US Census* in the sense that it succeeds in discarding more features (4% c.f. 1%), while both strategies maintain the full model's AUC. In the case of *DMC 2006*, RFS removes more attributes than GFS, but at the cost of smaller AUC, whereas RFS is domiated by GFS on all other datasets. Consequently, the results provide strong evidence for RFS being a less suitable solution towards FS. However, it has to be noted that this result could - to some extent - be influenced by the use of SVMs for feature scoring. Alternative techniques may produce different attribute rankings and thereby affect the performance of FS strategies. In that sense, a replication of the observed results with other classifiers is another promising area for future research to shed light on the robustness of RFS, GFS and TFS with respect to the underlying feature ranking mechanism.

4 Conclusions

Marketing applications of classification analysis require accurate as well as parsimonious prediction models. Therefore, three candidate strategies for organizing FS on the basis of RFE-based attribute rankings have been contrasted by means of empirical experimentation on representative real-world data. The results indicate that RFS is a less suitable approach when used in conjunction with SVMs. On the contrary, the GFS solution has shown promising results, although it may suffer from an overly conservative selection mechanism. Consequently, it may be desirable to examine the behavior of TFS within the context of portfolio analysis to shed light upon the choice set of accuracy/complexity trade-offs decision makers are facing. In addition, the use of algorithms from the field of multi-criteria optimization may be a promising area for future research to address the two conflicting objectives in a unified framework.

References

1. Asuncion, A., Newman, D.J.: UCI machine learning repository (2007)
2. Baesens, B., Van Gestel, T., Viaene, S., Stepanova, M., Suykens, J., Vanthienen, J.: Benchmarking state-of-the-art classification algorithms for credit scoring. Journal of the Operational Research Society 54(6), 627–635 (2003)
3. Fan, R.-E., Chang, K.-W., Hsieh, C.-J., Wang, X.-R., Lin, C.-J.: Liblinear: A library for large linear classification. Journal of Machine Learning Research 9, 1871–1874 (2008)
4. Fawcett, T.: An introduction to ROC analysis. Pattern Recognition Letters 27(8), 861–874 (2006)
5. Friedman, J.H.: Recent advances in predictive (machine) learning. Journal of Classification 23(2), 175–197 (2006)
6. Guyon, I., Weston, J., Barnhill, S., Vapnik, V.: Gene selection for cancer classification using support vector machines. Machine Learning 46(1-3), 389–422 (2002)
7. Guyon, I.M., Elisseeff, A.: An introduction to variable and feature selection. Journal of Machine Learning Research 3, 1157–1182 (2003)
8. Joachims, T.: Training linear SVMs in linear time. In: Eliassi-Rad, T., Ungar, L.H., Craven, M., Gunopulos, D. (eds.) Proc. of the 12th ACM SIGKDD Intern. Conf. on Knowledge Discovery and Data Mining, Philadelphia, PA, USA, pp. 217–226. ACM Press, New York (2006)
9. Keerthi, S.S., DeCoste, D.: A modified finite newton method for fast solution of large scale linear SVMs. Journal of Machine Learning Research 6, 341–361 (2005)
10. Kim, Y.S., Street, W.N., Russell, G.J., Menczer, F.: Customer targeting: A neural network approach guided by genetic algorithms. Management Science 51(2), 264–276 (2005)
11. Kohavi, R., John, G.H.: Wrappers for feature subset selection. Artificial Intelligence 97(1-2), 273–324 (1997)
12. Lessmann, S., Baesens, B., Mues, C., Pietsch, S.: Benchmarking classification models for software defect prediction: A proposed framework and novel findings. IEEE Transactions on Software Engineering 34(4), 485–496 (2008)
13. Lessmann, S., Voß, S.: Supervised classification for decision support in customer relationship management. In: Bortfeldt, A., Homberger, J., Kopfer, H., Pankratz, G., Strangmeier, R. (eds.) Intelligent Decision Support, pp. 231–253. Gabler, Wiesbaden (2008)
14. Lessmann, S., Voß, S.: A reference model for customer-centric data mining with support vector machines. European Journal of Operational Research 199(2), 520–530 (2009)

15. Martens, D., Baesens, B., van Gestel, T., Vanthienen, J.: Comprehensible credit scoring models using rule extraction from support vector machines. European Journal of Operational Research 183(3), 1466–1476 (2007)
16. Sindhwani, V., Rakshit, S., Deodhar, D., Erdogmus, D., Principe, J., Niyogi, P.: Feature selection in MLPs and SVMs based on maximum output information. IEEE Transactions on Neural Networks 15(4), 937–948 (2004)
17. Stoppiglia, H., Dreyfus, G., Dubois, R., Oussar, Y.: Ranking a random feature for variable and feature selection. Journal of Machine Learning Research 3, 1399–1414 (2003)
18. Vapnik, V.N.: The Nature of Statistical Learning Theory. Springer, New York (1995)

Instance Selection by Border Sampling in Multi-class Domains

Guichong Li[1], Nathalie Japkowicz[1], Trevor J. Stocki[2], and R. Kurt Ungar[2]

[1] School of Information Technology and Engineering, University of Ottawa,
800 King Edward Ave,
Ottawa, ON, K1N 6N5, Canada
{jli136,nat}@site.uottawa.ca
[2] Radiation Protection Bureau, Health Canada,
Ottawa, ON, K1A1C1, Canada
{trevor_stocki,kurt_ungar}@hc-sc.gc.ca

Abstract. Instance selection is a pre-processing technique for machine learning and data mining. The main problem is that previous approaches still suffer from the difficulty to produce effective samples for training classifiers. In recent research, a new sampling technique, called Progressive Border Sampling (PBS), has been proposed to produce a small sample from the original labelled training set by identifying and augmenting border points. However, border sampling on multi-class domains is not a trivial issue. Training sets contain much redundancy and noise in practical applications. In this work, we discuss several issues related to PBS and show that PBS can be used to produce effective samples by removing redundancies and noise from training sets for training classifiers. We compare this new technique with previous instance selection techniques for learning classifiers, especially, for learning Naïve Bayes-like classifiers, on multi-class domains except for one binary case which was for a practical application.

Keywords: Instance Selection, Border Sampling, Multi-class Domains, Class Binarization method.

1 Introduction

It has been realized that the redundancies and noise in data hinder data mining and machine learning algorithms to achieve their goals [11]. Practitioners have made a lasting effort on developing effective pre-processing techniques in recent decades [8][13][17][20]. Instance selection (IS) is a pre-processing technique that selects a consistent subset of the original training set for a supervised learning [11][20]. As a result, IS brings in two benefits: reducing the learning cost with respect to computational cost and helping learners build successful classifiers.

However, the previously proposed IS techniques still suffer from ineffectiveness of the resulting samples for learning any classifier. For example, Condensed Nearest Neighbour rule (CNN) and Editing Nearest Neighbour rule (ENN) tend to be used for Instance-Based Learning (IBL) [2][20]. Instead of those issues for relieving the learning

R. Huang et al. (Eds.): ADMA 2009, LNAI 5678, pp. 209–221, 2009.

cost, in this paper, we emphasize the effectiveness of IS to help induction algorithms learn successful classifiers if a training set reduction becomes possible because mining an effective sample from the original training set, especially on a multi-class domain, is not a trivial issue.

Recently proposed Progressive Border Sampling (PBS) [9][10] can overcome the drawback encountered in previously proposed approaches by borrowing the idea of Progressive Sampling techniques (PS) [13]. PBS can produce a small sample from the original training set by identifying and augmenting the border points for training any classifier. In this paper, we discuss how to use PBS to produce effective samples by removing redundancies and noise on multi-class domains.

As we know, the previously proposed Repeated Nearest Neighbour rule (RENN) [20] can be used for removing noise by repeatedly applying ENN. The main problem of RENN is that the repeated process is subject to a loss of information. In this paper, we first improve RENN by incorporating the Progressive Learning (PL) technique of PS with ENN for algorithmic convergence. After the noise is removed by avoiding the loss of information, PBS identifies and augments border points by assuming a pairwise border sampling strategy on multi-class domains. We show that the new method by incorporating the strategies for noise on multi-class domains with PBS outperforms the previously proposed IS techniques for training Naïve Bayes-like classifiers such as Aggregating One-Dependence Estimators (AODE) [19], etc.

The remainder of this paper is organized as follows. In Section 2, we introduce the two related works to give some background. In Section 3, we discuss the method for border sampling on multi-class domains. We discuss a new strategy for removing noise, and then incorporate the new strategy with PBS for effective samples in Section 4. The experimental design and results are reported in Section 5. Finally, we draw our conclusion and suggest future work in Section 6.

2 Preliminary

We introduce the methodology related to instance selection on multi-class domains.

2.1 Instance Selection by Border Sampling

Instance selection techniques (IS) focus on selecting a consistent subset of the training set for Instance-Based Learning [11][20]. The Condensed Nearest Neighbour rule (CNN) [17][20], a pioneer of the IS, finds a minimally consistent subset S of the training sets T. Editing Nearest Neighbour rule (ENN) reduces training sets by removing noise, which cannot be correctly classified by their k nearest neighbours, e.g., $k = 3$ in this paper. Because the removal of a noisy data point might lead to a new source of noise, Repeated Editing Nearest Neighbour rule (RENN) repeatedly removes noisy data until no noise of this kind is found [20]. Further, a variant of Decremental Reduction Optimization Procedure 3 (DROP3) [20] and Iterative Case Filtering (ICF) [2], denoted as DROP3.1, can be used for removing redundant data points. DROP3.1 first executes ENN to remove noise from the original training set T, and sort the resulting instances S by distances to their nearest neighbours belonging to the other classes in S, and then remove redundant points, which can be classified by their k nearest neighbours, e.g., $k = 5$ in this paper, in S with a high probability p, e.g., $p \geq 0.8$ in this paper, without the redundant points.

On the other hand, border points potentially exist in a labelled dataset [9]. A full border consists of near borders and far borders, and it can be identified by a recent technique called Border Identification in Two Stages (BI_2) [9]. Because initial border points have high uncertainty, which is insufficient for adequate learning [5][12], Progressive Border Sampling (PBS) [9] has been proposed to augment border points for an effective sample the context of supervised learning by borrowing the basic idea behind Progressive Sampling technique (PS) [13], which progressively learns a small sample from the original training set with an acceptable accuracy by defining a sampling schedule and convergence condition [8][13].

2.2 Pairwise Naïve Bayes

Given a training set with a probability distribution P, Naïve Bayes assumes the probabilities of attributes $a_1,a_2,\ldots,a_n$ to be conditionally independent given the class $c_i \in C$ [5][12], and is given by

$$y_i = \arg\max_{c_i \in C} \prod_{j=1}^{n} P(a_j \mid c_i) P(c_i)$$

Because the conditional independence is not expected to be satisfied in practical applications [4], previous research has proposed Naïve Bayes-like classifiers for the enhancement of Naïve Bayes by relieving the restriction of conditional independence. Aggregating One-Dependence Estimators (AODE) [19] achieves higher accuracy by averaging over a constrained group of 1-dependence Naive-Bayes models built on a small space. AODE with Subsumption Resolution (AODEsr) [22] augments AODE by detecting the specialization-generalization relationship between two attribute values at classification time and deleting the generalization attribute value. Hidden Naïve Bayes (HNB) [21] constructs a hidden parent for each attribute. Weightily Averaged One-Dependence Estimators (WAODE) [7] weights the averaged 1-dependence classifiers by the conditional mutual information.

On the other hand, learning a Naïve Bayes is different from learning a Support Vector Machine (SVM) because SVM is originally designed as a binary classifier while other classifiers, e.g., Naïve Bayes and Decision Tree, are directly designed on either binary or multi-class domains. The class binarization methods [3][18], e.g., the one-against-one (oo) and one-against-all (oa), are used for enhancing binary classifiers on multi-class domains.

In general, the pairwise classification or the oo method transforms a multi-class domain with m class into m(m-1)/2 binary domains. Each binary domain consists of all examples from a pair of classes. A binary classifier is trained on each binary domain. For classification, an observation x is input to all binary classifiers, and the predictions of the binary classifiers are combined to yield a final prediction.

There is a theoretical discussion about the pairwise Naïve Bayes classifiers, which is related to the pairwise Bayes classifier [16]. A pairwise probabilistic classifier is trained on a binary domains consisting of all examples in either c_i or c_j, denoted as c_{ij}, to estimate probabilities $p_{ij} = P(c_i|x, c_{ij})$ and $p_{ji} = P(c_j|x,c_{ij}) = 1 - p_{ij}$ for voting. It has been shown that the resulting prediction from all binary classifiers by a linear combination of votes is equivalent to regular Bayes classification for class ranking.

The oa classification splits a multi-class domain into m binary domains consisting of one class c_i, i = 1…m, from all other classes, and train these binary classifiers using all examples of class c_i as positive examples and the examples of the union of all other classes $c_j = D - c_i$ as negative examples.

It has been realized that pairwise Naïve Bayes built on each pair of classes of a multi-class domain is reduced to a standard Naïve Bayes directly built on the multi-class domain [16]. Although the oa classification can be reduced to a regular Bayes classification, a Naïve Bayes classifier with the oa is not consistent with a regular Naïve Bayes because the related probability estimates are not equivalent [16].

3 Pairwise Border Sampling

We discuss two main issues related to border sampling on multi-class domains.

3.1 Class Binarization Method

Border sampling on multi-class domains is not a trivial issue. The previous class binarization methods for classification provide a direct venue for the border sampling on multi-class domains. As a result, two kinds of class binarization methods, i.e., one-against-one (oo) and one-against-all (oa), for border sampling on multi-class domains can be described as follows.

- **oo method**

It is also called the pairwise method. Border sampling with the oo strategy identifies the pairwise borders on each pair of classes. All obtained $c(c - 1)/2$, where c is the number of classes, pairwise borders are combined together by a simple union as the resulting sample.

- **oa method**

Border sampling with the oa strategy identifies individual borders b_i in each class by identifying a pairwise border b'_i between the class and the rest of classes such that b_i can be obtained by retaining border points in class i out of b'_i. All obtained individual borders b_i, i = 1,…, k are combined together by a simple union as the resulting border.

3.2 Naïve Bayes Validation

Initially identified border points have high uncertainty, which might be improper for sufficiently learning. Uncertainty can be overcome by progressively learning new border points on the remaining data obtained by removing the previously identified border points for an augmented border until this augmented border is sufficient for Bayesian learning [5].

The paiwise border sampling identifies and augments border points on each pair of classes by assuming the oo strategy. Heuristically, the augmentation on each pair can be validated by building a Naïve Bayes model and testing on the pair until the performance of the Naïve Bayes model does no longer ascend [9]. As a result, a pairwise Naïve Bayes are built from all pairs of classes. According to the early discussion, it is believed that this pairwise Naïve Bayes can be reduced to the standard Naïve Bayes built on the resulting sample.

However, a Naïve Bayes with the oa is not equivalent to a standard Naïve Bayes due to the probability estimation [16]. Moreover, because the oo is applied on each pair of classes, it requires less data access than the oa. As a result, the pairwise PBS is preferable to the PBS with the oa.

4 Instance Selection by Border Sampling in Multi-class Domains

Noise removal is an important issue for instance selection. In general, there are two kinds of methods: the Tomek Link based method and the RENN based method, for noise removal. A Tomek Link is a pair of adjunct data points belonging to different categories, and one in the pair is identified as a noise if it is farther from its nearest neighbour in the same category than the adjunct point in the Tomek Link [17][20].

As we know in Section 2.1, ENN is used for removing noise, which cannot be classified correctly by its nearest neighbours. RENN is a method to repeatedly remove noise of this kind by applying ENN. Therefore, this RENN based method for noise removal appears preferable to the Tomek Link based method because it has a more direct effect for classification than the latter. The main problem of RENN is that it suffers from the loss of information because some border points are also removed as noise while they are useful for training classifiers [20].

4.1 PENN

We propose a new algorithm for improving the original Repeatedly Editing Nearest Neighbour rule by assuming PL technique. The new algorithm is called Progressively

```
PENN algorithm
Input     D: a sample for training with c classes
Output    D′
begin
    D′ = D, oD = D, LCurve[k], k = 0..K(100),k = 1
    while(true)
        LCurve[k] = LearningNB(D′, D)
        if(LCurve[k] < LCurve[k-1])
           D′ = oD, break
        H_k = kNN(D′, 3), D″ = ∅, isFinished = true
        for(each p ∈ D′)
           if(H_k(p))
             D″ = D″ ∪ p
           else
             isFinished = false
        if(isFinished)
           break;
        oD = D′, D′ = D″, k++
    return D′
end
```

Fig. 1. PENN algorithm

Editing Nearest Neighbour (PENN), as shown in Figure 1. PENN has only input: D, which is the original training set. It outputs D′ as the reduced training set. The algorithm initializes its variables at Step 1. LCurve is used for describing the learning curve of Naïve Bayes. From Step 2 to Step 14, the algorithm progressively learns the resulting sample D′ by removing noise in the previously generated D′. A Naïve Bayes classifier is built on D′ and tested on the original data D during Step 3. If the learning curve descends at Step 4, the algorithmic convergence is detected, and the previously learned result oD is returned as D′ at Step 5. Otherwise, the algorithm builds a k-Nearest Neighbour classifier (kNN) on D′ with its parameter of 3 (the number of nearest neighbours) at Step 6, and the kNN model classifies each data point in D′ at Step 8. Actually, the algorithm from Step 6 to Step 11 corresponds to the original ENN. If all data are correctly classified, then the while loop exits at Step 13. Otherwise, the algorithm continues in the while loop.

4.2 PEBS Algorithm: A Hybrid of PENN and PBS

We can combine PENN and PBS for instance selection. The combination of PENN and PBS is a hybrid algorithm, called PEBS, as shown in Figure 2. First, PENN is used for removing noise. Second, PBS is used for removing redundancy. Arguably, PENN is not suggested to be invoked after PBS is invoked in PEBS because there is no chance to add new border points after noise is removed.

```
PEBS algorithm
Input      D: a sample for training with c classes
Output     B
begin
1    D = PENN(D), B = ∅;
2    C = getClassset(D), C = {C_i | i = 0, …, c}
3    for ∀i, j, where i < j, C_i ≠ ∅, and C_j ≠ ∅
4        B_ij = ∅, C'_i = C_i, C'_j = C_j; C_ij = C_i ∪ C_j
5        Acc[k] = 0, k = 0, 1, …, K, K = 100
6        while(true)
7            B'_ij = BI_2(C'_i, C'_j, B_ij)
8            B_ij = B_ij ∪ B'_ij,
9            C'_i = C'_i - B'_ij, C'_j = C'_j - B'_ij
10           Acc[k] = ValidateNBModel(B_ij, C_ij)
11           if(Acc[k] ≤ Acc[k-1])
12               B_ij = old; break;
13               continue
14           old = B_ij, k++
15       B = B ∪ B_ij
```

Fig. 2. PEBS: The hybrid of PENN and PBS

In Figure 2, PEBS applies pairwise border sampling on each pair of classes from Step 3 to Step 15. In the while loop from Step 6 to Step 14, PEBS identifies at Step 7, augments border points at Step 8, and validates a Naïve Bayes built at Step 10 on the current border points and tested on the pair of classes for convergence detection at Step 11. All augmented border points from all pairs of classes are unified together at Step 15 as a resulting sample.

The number of iterations of the while loop from Step 2 to Step 14 in PENN is expected to be bounded with a small number. However, PENN has a quadratic time complexity due to kNN is quadratic for classification [2][20]. PEBS is also quadratic due to PENN and the original PBS [9] although PBS can be scaled up [10].

5 Experiments

Our experimental design and results are reported as follows.

5.1 Datasets for Experiments

We conducted experiments on 10 benchmark multi-class datasets chosen from the UCIKDD repository [1] and one binary dataset obtained from a nuclear security application, as shown in Table 1, where the columns #attr, #ins, and #c are the number of attributes, instances, and classes in training sets, respectively; #PEN, #PEBS, #CNN, #ENN, #RENN, and #DROP3.1 are the sample sizes generated by PENN, PEBS, CNN, ENN, RENN, and DROP3.1, respectively; %PEN, %PEBS, and %RENN are the percents of #PEN, #PEBS, and #RENN to #ins, respectively.

For the application, a possible method of explosion detection for the Comprehensive nuclear-Test-Ban-Treaty [15] consists of monitoring the amount of radioxenon in the atmosphere by measuring and sampling the activity concentration of Xe-131m, Xe-133, Xe-133m, and Xe-135 [14]. Several samples are synthesized under different circumstances of nuclear explosions, and combined with various levels of normal concentration backgrounds so as to synthesize a binary training dataset, called XenonT2D1.

In our experiments, PEBS ran with the Radial-based function [12] as a similarity measure for computing the nearest neighbours. Several inductive algorithms are used for training Naïve Bayes-like classifiers on either the resulting samples generated by PEBS or the full training sets (Full), or those generated by previous approaches, i.e., CNN, ENN, RENN, and DROP3.1, which is implemented for experiments in this paper. The performances of these classifiers with respect to the Area under ROC curve (AUC) [6], based on averages obtained within 20 runs of the 10 cross validation, are used for comparison between PEBS and the other algorithms.

The software tools for Naïve Bayes and three Naïve Bayes-like learners: AODE, AODEsr, and HNB (WAODE is omitted due the limitation of space) are chosen from the Waikato Environment for Knowledge Analysis (Weka) [23]. The datasets have been pre-processed by using the ReplaceMissingValue tool in Weka for missing values and the unsupervised Discretize tool in Weka for discretizing continuous values. The classifiers are built with their default settings, with no loss of generality, e.g., NB with Maximum Likelihood estimator, and AODE with a frequencyLimit of 1, i.e., any attribute with values below this limit cannot be use as a parent, etc.

5.2 Experimental Results

Our initial results in Table 1 show that PEBS can produce much smaller samples, e.g., on average, 303 samples and 653 samples from Anneal and Hypothyroid, respectively, than other approaches, i.e., CNN, ENN, and RENN except DROP3.1, while it can retain most instances, e.g., in Vowel, if few redundancies can be found. The comparison between PENN and RENN is discussed later. On average, PEBS produces smaller samples than other approaches except for DROP3.1, which intends to produces the smallest samples among all approaches.

XenonT2D1 is a distinct case that the synthesized data contains much redundancy. PEBS can produce a much smaller sample from XenonT2D1 than other approaches while other approaches reduce a little redundancy except for DROP3.1.

We show the effectiveness of PEBS by comparing PEBS with CNN, ENN, RENN, and DROP3.1 for training the classifiers, i.e., NB, AODE, AODEsr, and HNB, as shown from Table 2 to Table 5. We use 'w' and 'l' to represent PEBS's wins and losses, respectively, against the corresponding methods in terms of the paired t-test (first) and Wilcoxon signed rank test (second) at significance levels of 0.05.

Table 1. The 11 datasets

Datasets	#attr	#ins	#c	#PEN	%PEN	#PEBS	%PEBS	#CNN	#ENN	#RENN	%RENN	#DROP3.1
Anneal	39	898	5	808	100	303	37	793	800	797	99	148
Audiology	70	226	24	203	100	164	80	179	154	140	69	88
Autos	26	205	6	185	100	161	87	179	155	143	78	107
Balance-s	5	625	3	528	94	459	82	458	500	499	89	306
Hypothyroid	30	3772	4	3395	100	653	19	3071	3185	3170	93	111
P-tumor	18	339	21	305	100	302	99	293	167	128	42	124
Soybean	36	683	18	615	100	541	88	519	582	573	93	162
Vehicle	19	846	4	720	95	697	91	753	624	592	78	336
Vowel	14	990	11	891	100	891	100	845	843	828	93	570
Zoo	18	101	7	88	97	39	43	71	88	87	96	23
XenonT2D1	5	640	2	572	99	26	5	578	567	567	98	30
Average		848		756	99	385	67	703	697	684	84	182

PEBS can help learn better NB and other three Naïve Bayes-like classifiers, as shown from Table 2 to Table 5, in most cases in terms of the paired t-test and Wilcoxon signed rank test as compared with Full, and other approaches. Especially, it is consistently superior to DROP3.1 in all cases for training classifiers.

The averaged AUC are shown at the bottoms of Table 2 to Table 5. We summarized the results for statistical test in Table 6. The results clearly show that PEBS consistently outperforms previously proposed instance selection approaches for training set reduction, and helps learn successful classifiers as compared with Full.

Table 2. Training NB

	PEBS	Full	CNN	ENN	RENN	DROP3.1
Anneal	0.9587	0.9601 -l	0.96 -l	0.9593 -l	0.9592	0.9501 -w
Audiology	0.6984	0.7012 -l	0.7002	0.6904 ww	0.6843 ww	0.6868 ww
Autos	0.9096	0.9119	0.9122	0.8736 ww	0.8602 ww	0.8712 ww
Balance-s	0.8942	0.8307 ww	0.8989	0.9074 -l	0.9075 -l	0.8442 ww
Hypothyroid	0.8995	0.8802 ww	0.8805 -w	0.8805 -w	0.7863 ww	0.8141 ww
P-tumor	0.7543	0.7544	0.7543	0.73 ww	0.7049 ww	0.7308 ww
Soybean	0.9983	0.9983 -w	0.9983	0.9981 -w	0.998 -w	0.9981
Vehicle	0.8109	0.8077 -w	0.8079 -w	0.8079 -w	0.7951 ww	0.7812 ww
Vowel	0.9591	0.9591	0.9574 -w	0.9493 ww	0.9416 ww	0.9572
Zoo	0.894	0.894	0.8917	0.8917	0.894	0.894
XenonT2D1	0.9873	0.9919	0.9919	0.9919	0.9919	0.955 ww
Average	0.8777	0.8698	0.8761	0.8688	0.8531	0.8528

Table 3. Training AODE

	PEBS	Full	CNN	ENN	RENN	DROP3.1
Anneal	0.9596	0.961	0.961	0.9602	0.9601	0.9515 -w
Audiology	0.6987	0.7015 -l	0.7008 -l	0.6907 ww	0.6844 ww	0.6872 ww
Autos	0.9326	0.9349	0.9352	0.8933 ww	0.8772 ww	0.8897 ww
Balance-s	0.8641	0.798 ww	0.8678	0.8877 -l	0.8856 -l	0.7699 ww
Hypothyroid	0.8952	0.8733 ww	0.8735 ww	0.8735 ww	0.7893 ww	0.8115 ww
P-tumor	0.7546	0.7547	0.7541	0.7305 ww	0.705 ww	0.7315 ww
Soybean	0.9986	0.9986	0.9985	0.9983 ww	0.9982 ww	0.9983
Vehicle	0.8994	0.9013 -l	0.9019 -l	0.9019 -l	0.877 ww	0.8615 ww
Vowel	0.994	0.994	0.9938	0.987 ww	0.9818 ww	0.9902 ww
Zoo	0.894	0.894	0.8917	0.8917	0.894	0.894
XenonT2D1	0.9878	0.9917	0.9917	0.9915	0.9915	0.9579 ww
Average	0.8891	0.8811	0.8878	0.8815	0.8653	0.8585

PENN is an improved method for noise removal by incorporating PL technique with ENN. As we can see in Table 1, PENN is not expected to reduce much noise from the original datasets. There are only four cases, i.e., Balance-s, Vehicle, Zoo, and XenonT2D1, where PENN can remove noise, which is less than that removed by RENN. We emphasize that PENN can guarantee few loss of information such that PEBS can produce effective samples for training classifiers as compared with Full and other instance selection approaches.

We compare PENN with RENN by training NB and other three Naïve Bayes-like classifiers on either the resulting samples generated by PENN and RENN or the full training sets (Full), as shown in Table 7, where the names of datasets are omitted, and

Table 4. Trianing AODEsr

	PEBS	Full	CNN	ENN	RENN	DROP3.1
Anneal	0.9647	0.9651	0.9651	0.9639 -w	0.9636 -w	0.9597 -w
Audiology	0.7082	0.7069	0.7075	0.6993 ww	0.6918 ww	0.6962 ww
Autos	0.9403	0.9419	0.9424 -l	0.8954 ww	0.8774 ww	0.8998 ww
Balance-s	0.8665	0.7073 ww	0.8691	0.8858 -l	0.8842 -l	0.7825 ww
Hypothyroid	0.9103	0.8916 -w	0.892 -w	0.892 -w	0.8048 ww	0.8525 ww
P-tumor	0.7576	0.758	0.7577	0.7305 ww	0.7044 ww	0.7343 ww
Soybean	0.9988	0.9989 -l	0.9989	0.9986	0.9986 -w	0.9987
Vehicle	0.8983	0.8979	0.8981	0.8981	0.873 ww	0.8714 ww
Vowel	0.9971	0.9971	0.9971	0.9929 ww	0.987 ww	0.9935 ww
Zoo	0.894	0.894	0.894	0.894	0.894	0.894
XenonT2D1	0.9891	0.9919	0.9919	0.9917	0.9917	0.9773
Average	0.8936	0.8759	0.8922	0.8851	0.8679	0.8683

Table 5. Training HNB

	PEBS	Full	CNN	ENN	RENN	DROP3.1
Anneal	0.9644	0.9641	0.9638	0.9635	0.9633	0.9583 -w
Audiology	0.7029	0.7044 -l	0.7029	0.6939 ww	0.6878 ww	0.6938 ww
Autos	0.9458	0.9451	0.945	0.8978 ww	0.8769 ww	0.8966 ww
Balance-s	0.8485	0.8808 -l	0.8536	0.8727 -l	0.8727 -l	0.7507 ww
Hypothyroid	0.9066	0.8864 -w	0.8848 -w	0.8848 -w	0.7842 ww	0.8448 ww
P-tumor	0.7557	0.7557	0.7556	0.727 ww	0.7016 ww	0.7273 ww
Soybean	0.999	0.999 -w	0.999	0.9988 -w	0.9987 ww	0.9988 -w
Vehicle	0.9075	0.9078	0.9077	0.9077	0.8794 ww	0.8742 ww
Vowel	0.9974	0.9974	0.9973	0.9931 ww	0.9861 ww	0.9939 ww
Zoo	0.894	0.894	0.894	0.894	0.8893	0.894
XenonT2D1	0.9816	0.9921	0.9921	0.9915	0.9915	0.977
Average	0.8922	0.8935	0.8904	0.8833	0.8640	0.8632

the rows correspond to the datasets in Table 1 in order without any confusion. The bottom row shows the average values.

As we can see, there is only case, i.e., Balance-s, where PENN is inferior to RENN for training NB and other three Naïve Bayes –like classifiers in terms of the paired t-test and Wilcoxon signed rank test. PENN is superior to RENN in all other cases by avoiding loss of information, and PENN consistently helps learn NB and other three Naïve Bayes-like classifiers without any loss of information as compared with Full.

Balance-s is also a case that PENN enhances PBS in PEBS. We conducted the related experiments in that PBS without PENN is inferior to other approaches for training Naïve Bayes and other three Naïve Bayes-like classifiers although it does not intend to degrade the performance of these classifiers built on the resulting sample as compared with learning on the original training set. In addition, the maximum tries of PEBS for pairwise border sampling is 16 on P-tumor case. Empirically, it is bound by a small number, as discussed in the previous research for PBS [9].

The results on XenonT2D1 surprise us that PEBS consistently outperforms other approaches for training successful classifiers by producing a much small sample.

Table 6. Summary of statistical tests

		Full	CNN	ENN	RENN	DROP3.1
Paired t-test	PEBS	5\39\0	1\43\0	18\26\0	26\18\0	29\15\0
Wilcoxon signed rank test	PEBS	10\27\7	6\34\4	25\13\6	29\11\4	34\10\0

Table 7. The comparison between PENN and RENN

NB			AODE			AODEsr			HNB		
PENN	Full	RENN	PENN	Full	RENN	PENN	Full	RENN	PENN	Full	RENN
0.9601	0.9601	0.9592 -w	0.961	0.961	0.9601 w	0.9651	0.9651	0.9636 -w	0.9641	0.9641	0.9633 -w
0.7012	0.7012	0.6843ww	0.7015	0.7015	0.6844ww	0.7069	0.7069	0.6918ww	0.7044	0.7044	0.6878ww
0.9119	0.9119	0.8602ww	0.9349	0.9349	0.8772ww	0.9419	0.9419	0.8774ww	0.9451	0.9451	0.8769ww
0.8797	0.8307-w	0.9075 -l	0.8421	0.798 ww	0.8856 ll	0.8034	0.7073ww	0.8842 ll	0.8724	0.8808	0.8727
0.8802	0.8802	0.7863ww	0.8733	0.8733	0.7893ww	0.8916	0.8916	0.8048ww	0.8864	0.8864	0.7842ww
0.7544	0.7544	0.7049ww	0.7547	0.7547	0.705 ww	0.758	0.758	0.7044ww	0.7557	0.7557	0.7016ww
0.9983	0.9983	0.998 -w	0.9986	0.9986	0.9982 -w	0.9989	0.9989	0.9986 -w	0.999	0.999	0.9987 -w
0.809	0.8077	0.7951ww	0.8976	0.9013	0.877 ww	0.8948	0.8979	0.873 ww	0.903	0.9078	0.8794ww
0.9591	0.9591	0.9416ww	0.994	0.994	0.9818ww	0.9971	0.9971	0.987 ww	0.9974	0.9974	0.9861ww
0.9919	0.9919	0.9919	0.9916	0.9917	0.9915	0.9917	0.9919	0.9917	0.9915	0.9921	0.9915
0.8846	0.8796	0.8629	0.8949	0.8909	0.8750	0.8949	0.8857	0.8777	0.9019	0.9033	0.8742

6 Conclusion and Future Work

Instance selection by PBS on multi-class domains is not a trivial issue. As a result, we argue that PBS prefers the pairwise border sampling to the one-against-all method on multi-class domains by borrowing class binarization methods for classification on multi-class domains. We show an improved PENN algorithm, which incorporates Progressive Learning (PL) technique with Editing Nearest Neighbour rule (ENN), for noise removal without any loss of information. Finally, we design a new hybrid method, called Progressively Editing Nearest Neighbour rule for Progressive Border Sampling (PEBS), for instance selection by incorporating PENN with PBS. PENN is used for noise removal first, and then PBS is used for removing redundancies.

The experimental results show that PEBS can produce much smaller samples than other instance selection approaches in some cases while it produces little larger samples than these approaches in other cases. On average, PBS can produce smaller samples than other approaches except DROP3.1. On the other hand, PEBS consistently outperforms other approaches to produce effective samples in all cases in terms of the paired t-test and in most cases in terms of the Wilcoxon signed rank test. Especially, PEBS consistently outperforms DROP3.1 in all cases. In addition, PENN is not expected to remove much noise as compared with RENN by avoiding loss of information. PENN produces a small sample consistent with the full training set by removing noise if possible. PENN outperforms RENN in most cases except for one case, where it is inferior to RENN. Especially, we show that PENN enhances PBS in the worse case as compared with the full training set.

PENN is not efficient due to its quadratic time complexity, and PEBS for border sampling is still subject to small failures in some case in terms of the Wilcoxon signed rank test. These drawbacks are expected to be overcome in future work.

References

1. Bay, S.D.: The UCI KDD archive, 1999 (1999), http://kdd.ics.uci.edu
2. Brighton, H., Mellish, C.: Advances in instance selection for instance-based learning algorithms. Data Mining Knowledge Discovery 6(2), 153–172 (2002)
3. Debnath, R., Takahide, N., Takahashi, H.: A decision based one-against-one method for multi-class support vector machine. Pattern Anal. Applic. 7, 164–175 (2004)
4. Domingos, P., Pazzani, M.: Beyond independence: Conditions for the optima-lity of the sample Bayesian classifier. Machine Learning 29, 103–130 (1997)
5. Duda, R.O., Hart, P.E.: Pattern Classification and Scene Analysis. A Wiley Interscience Publication, Chichester (1973)
6. Fawcett, T.: ROC graphs: Notes and practical considerations for researchers (2003), http://www.hpl.hp.com/-personal/TomFawcett/papers/index.html
7. Jiang, L., Zhang, H.: Weightily Averaged One-Dependence Estimators. In: Yang, Q., Webb, G. (eds.) PRICAI 2006. LNCS, vol. 4099, pp. 970–974. Springer, Heidelberg (2006)
8. John, G., Langley, P.: Static versus dynamic sampling for data mining. In: Proceedings of the Second International Conference on Knowledge Discovery and Data Mining, pp. 367–370. AAAI Press, Menlo Park (1996)
9. Li, G., Japkowicz, N., Stocki, T.J., Ungar, R.K.: Full Border Identification for reduction of training sets. In: Proceedings of the 21st Canadian Artificial Intelligence, Winsor, Canada, pp. 203–215 (2008)
10. Li, G., Japkowicz, N., Stocki, T.J., Ungar, R.K.: Border sampling through Markov chain Monte Carlo. In: Proceedings of the 2008 Eighth IEEE International Conference on Data Mining, Pisa, pp. 393–402 (2008)
11. Liu, H., Motoda, H.: On issues of instance selection. Data Mining and Knowledge Discovery 6, 115–130 (2002)
12. Mitchell, T.: Machine Learning. McGraw-Hill Companies, Inc., New York (1997)
13. Provost, F., Jensen, D., Oates, T.: Efficient Progressive Sampling. In: Proc. of the fifth ACM SIGKDD, San Diego, California, US, pp. 23–32 (1999)
14. Stocki, T.J., Blanchard, X., D'Amours, R., Ungar, R.K., Fontaine, J.P., Sohier, M., Bean, M., Taffary, T., Racine, J., Tracy, B.L., Brachet, G., Jean, M., Meyerhof, D.: Automated radioxenon monitoring for the comprehensive nuclear-test-ban treaty in two distinctive locations: Ottawa and Tahiti. J. Environ. Radioactivity 80, 305–326 (2005)
15. Sullivan, J.D.: The comprehensive test ban treaty. Physics Today 151, 23 (1998)

16. Sulzmann, J., Fürnkranz, J., Hüllermeier, E.: On Pairwise Naive Bayes Classifiers. In: Kok, J.N., Koronacki, J., Lopez de Mantaras, R., Matwin, S., Mladenič, D., Skowron, A. (eds.) ECML 2007. LNCS, vol. 4701, pp. 658–665. Springer, Heidelberg (2007)
17. Tomek, I.: Two modifications of CNN. IEEE Transactions on Systems, Man and Cybernetics 6(6), 769–772 (1976)
18. Tsujinishi, D., Koshiba, Y., Abe, S.: Why pairwise is better than one-against-all or all-at-once. In: Proceedings of IEEE International Conference on Neural Networks, vol. 1, pp. 693–698. IEEE Press, Los Alamitos (2004)
19. Webb, G.I., Boughton, J., Wang, Z.: Not So Naive Bayes: Aggregating One-Dependence Estimators. Machine Learning 58(1), 5–24 (2005)
20. Wilson, D.R., Martinez, T.R.: Reduction Techniques for Instance-Based Learning Algorithms. Machine Learning 38, 257–286 (2000)
21. Zhang, H., Jiang, L., Su, J.: Hidden Naive Bayes. In: Twentieth National Conference on Artificial Intelligence, pp. 919–924 (2005)
22. Zheng, F., Webb, G.I.: Efficient lazy elimination for averaged-one dependence estimators. In: Proc. 23th International Conference on Machine Learning (ICML 2006), pp. 1113–1120 (2006)
23. WEKA Software, v3.5.2. University of Waikato, `http://www.cs.waikato.ac.nz-/ml/weka/index_datasets.html`

Virus Propagation and Immunization Strategies in Email Networks

Jiming Liu[1,2,⋆], Chao Gao[1], and Ning Zhong[1,3]

[1] International WIC Institute, Beijing University of Technology, 100124 Beijing, China
[2] Department of Computer Science, Hong Kong Baptist University, Kowloon Tong, HK, China
jiming@comp.hkbu.edu.hk
[3] Department of Life Science and Informatics, Maebashi Institute of Technology, Japan

Abstract. This paper proposes a simulation model to characterize virus propagation in an email network and analyzes various defense strategies. We demonstrate the key factor which affects the efficiency of different strategies, and illustrate two phases of virus propagation. The results show that D-steps immunization is a feasible strategy in the case of finite resources and the efficiency of node-betweenness immunization is better than targeted immunization if we have the global information about a network.

1 Introduction

Recent research has shown that most real-world systems can be modeled as complex networks, where nodes represent individuals (e.g., computers, Web pages, email-box) and edges represent the connections among individuals (e.g., network links, hyper-links, relationships between two people) [1]. There are many research topics related to network-like environments [2]. Among them, how to control virus propagation in physical networks (e.g., trojan virus) and virtual networks (e.g., email worms) has become an interesting and challenging issue for scientists [1,3,4].

Reliable propagation models help researchers to further understand new virus attacks and/or new spreading mechanisms. At the same time, a reliable model provides a test-bed for developing or evaluating some new and/or improved security strategies for restraining virus propagation [5]. Furthermore, these models can be used to predict flaws in a global network infrastructure when exposed worm attacks [6]. Based on reliable models, researchers can design effective immunization strategies for preventing and controlling virus propagation in computer networks (e.g., worms). Generally speaking, there are two major issues in this research area: 1) how to exactly model the process of virus propagation in a complex network and 2) how to efficiently restrain virus propagation.

1.1 Propagation Models

Lloyd and May proposed an epidemic propagation model to depict the characteristics of propagation [1]. Some traditional epidemic models, e.g., SI [3], SIR [4], SIS [7], and

⋆ Corresponding author.

R. Huang et al. (Eds.): ADMA 2009, LNAI 5678, pp. 222–233, 2009.

SEIR [8], had been applied to simulate virus propagation and study system dynamic characteristics. These epidemic models were modeled as equations based on mean-filed theory, a type of black-box modeling approach, therefore they only provided a macroscopic understanding of the propagation. Some assumptions such as full mixing and equiprobable contacts are unreliable in the real world. And some microscopic interactive behaviors cannot be observed through the models. Meanwhile, these models overestimate the speed of propagation [9]. In order to overcome these shortfalls, Zou built an interactive email network to analyze worm propagation [9]. Two user behaviors, i.e., checking email-boxes and clicking suspected emails, were added into this model in order to examine the effects of these behaviors on the virus propagation. But there is a little information about how to restrain worm propagation in their model.

1.2 Immunization Strategies

Currently, one of the popular methods to effectively and efficiently restrain virus propagation is called network immunization, where a set of nodes in the network are immunized (or protected) and hence will not be infected by viruses any more. Pastor-Satorras studied the critical values of both random and targeted immunization [10]. Though targeted immunization gains the best result, it needs global information of a whole network. In order to overcome this shortfall, a local strategy called acquaintance immunization is proposed [11,12]. According to this strategy, the probability of selected nodes with a high degree is much higher than that of selected nodes with a low degree, because the nodes with higher degrees have more links in a scale-free network. Thus, the efficiency of acquaintance immunization is between those of random and targeted immunization. However, this strategy does not distinguish the differences among neighbors, and randomly selects some nodes and their direct neighbors. These issues seriously restrict the effectiveness of acquaintance immunization [13]. If we extend the selection range from direct neighbors to D-steps neighbors, the immunization problem can be translated into a graph covering problem, i.e., some initial "seed" nodes aim to cover the whole network within d steps [13,14]. So, acquaintance strategy can be seen as 1-step immunization.

An email network is a typical interactive social network where an edge between two users represents they have communicated before [9,15]. We take an interactive email model as an example to evaluate the efficiency of different immunization strategies, as well as the characteristics of propagation. Our model exactly exhibits the microscopic process of email worm propagation and further finds some hidden determining factors which affect propagation and immunization strategies, such as the power-law exponent and the average path length. Different from other models, here we focus on comparing the performances of the previous degree-based strategies and new betweenness-based strategies, replacing the critical value of epidemic in the network.

1.3 Organization

The remainder of this paper is organized as follows: Section 2 presents a model of interactive email network and some important research problems to be dealt with in this paper. Section 3 shows some experiments to compare the actual measurements of different strategies both in synthetic networks and Enron email network. Section 4 provides the analysis and discussions about the effect of power-law exponent on virus propagation and immunization strategies, and the performances of immunization strategies including cost, efficiency and robustness. Section 5 concludes the major results of this work.

2 Formal Definitions about Email Networks

We begin with a brief introduction to characterize the models of epidemic propagation and some well-known immunization strategies. We focus on the characteristics of virus propagation and the efficiency of immunization strategies. Our numerical simulation has two phases. Firstly, we establish a pre-existing email network, in which each node has some interactive behaviors, more details are shown in this section. And then, we deploy virus propagation into the network and study the epidemic behaviors when applying different immunization strategies, more details are shown in Section 3.

2.1 Structures of Email Networks

Structures of most popular networks follow power-law distributions [16] and the power-law exponents (α) are often between 2 and 3 [17]. Current research shows that an email network also follows heavy-tailed distribution [9,15]. So three synthetic power-law networks are generated based on the GLP algorithm [18] in which the exponent can be tuned. The three synthetic networks all have 1000 nodes with α =1.7, 2.7 and 3.7, respectively. In order to reflect more characteristics about a real-world network, we will also study Enron email network[1], which is built by Andrew Fiore and Jeff Heer. The degree statistics of these networks are shown in Table 2.

2.2 A General Model of Interactive Email Networks

We use a network-based environment to observe virus propagation. A network can be denoted as E=$\langle V, A\rangle$, where $V = \{v_1, v_2, ..., v_n\}$ is the set of nodes and $A = \{\langle v_i, v_j\rangle \mid 1 \le i, j \le n\}$ is the set of undirected links (if v_i in the hit-list of v_j, there is a link between v_i and v_j). The virus can propagate along edges in order to infect more nodes in a network.

In order to have a general definition, a node can be denoted as a tupe *<ID, Status, NodeLink*, $P_{behavior}$, B_{action},*VirusNum, NewVirus>*.

- ***ID***: the identifier of node, $v_i.ID = i$.
- ***Status***: the states of node, which includes four statuses:

$$v_i.Status = \begin{cases} healthy = 0, & if\ the\ node\ has\ no\ virus, \\ danger = 1, & if\ the\ node\ has\ virus\ but\ not\ infected, \\ infected = 2, & if\ the\ node\ has\ been\ infected, \\ immunized = 3, & if\ the\ node\ has\ been\ immunized. \end{cases}$$

[1] Http://bailando.sims.berkeley.edu/enron/enron.sql.gz

- ***NodeLink***: the information of its hit-list or adjacent neighbors, $v_i.NodeLink = \{\langle i,j\rangle \mid \langle i,j\rangle \in A\}$.
- $P_{behavior}$: the probability of node to perform different behaviors.
- B_{action}: the different behaviors.
- ***VirusNum***: the total number of new unchecked viruses before the next operation.
- ***NewVirus***: the number of new receiving viruses from its neighbors at each step.

In order to make our model as general as possible, we simulate two interactive behaviors according to Zou'model [9], i.e. the checking email intervals and the clicking email probabilities, which are all following the Gaussian-distribution. So the formula of $P_{behavior}$ in the tupe can be expressed as $P^1_{behavior} = ClickProb$ and $P^2_{behavior} = CheckTime$.

- ***ClickProb*** is the probability of user clicking suspected E-mail,

$$v_i.P^1_{behavior} = v_i.ClickProb = normalGenerator(\mu_p, \sigma_p) \sim N(\mu_p, \sigma_p^2).$$

- ***CheckRate*** is the probability of user checking E-mail,

$$v_i.CheckRate = normalGenerator(\mu_t, \sigma_t) \sim N(\mu_t, \sigma_t^2).$$

- ***CheckTime*** is the next checking E-mail time,

$$v_i.P^2_{behavior} = v_i.CheckTime = expGenerator(v_i.CheckRate).$$

B_{action} can be expressed as $B^1_{action} = receive\_email$, $B^2_{action} = send\_email$ and $B^3_{action} = update\_email$. If a user receives a virus email, the node will update its status by $v_i.Status \leftarrow danger$. If user opens an email which has virus attachment, the node will adjust its status by $v_i.Status \leftarrow infected$. At the same time, the node will send virus emails to all its friends according to hit-list. If a user is immunized, the node will update its status by $v_i.Status \leftarrow immunized$.

2.3 Research Issues in Email Networks

Based on the above email network, we will provide a detailed discussion of the key components in this paper by addressing the following issues step by step:

1) Process of virus propagation and some factors which will affect the efficiency of immunization strategy.
2) Effect of power-law exponent on virus spreading and immunization strategy.
3) Effect of node-betweenness and edge-betweenness on immunization.
4) Performances of different strategies, i.e., cost, efficiency and robustness.

3 Simulation of Immunization Strategies

In this section, we aim to compare the efficiency of different strategies. Specifically, we want to examine whether betweenness-based immunization strategies can restrain the worm propagation in email networks and which measure can be used to depict the efficiency of immunization strategies.

3.1 Some Assumptions and Measurements

Some assumptions in interactive email models are as follows:

- If a user opens an infected email, the node is infected and will send viruses to all its friends in its hit-list;
- When checking his/her mailbox, if a user does not click virus emails, we assume that the user deletes those suspected emails;
- If nodes are immunized, they will never send virus emails even if a user clicks an attachment.

At beginning, there are two methods to select infected nodes. One is the random infection, and the other is the malicious infection, i.e., the infected nodes with the maximal degree. Since the process of email worm propagation is stochastic, all numerical results are given in average values by simulating 100 times.

In order to evaluate the efficiency of immunization strategies and find a bridge between the local information and the global information, we test two statistic parameters: the sum of immunized degree (*SID*) which reflects the importance of the nodes in a network, and the average path length (*APL*) of a network which reflects the connectivity and transmission capacity of the network.

The sum of immunized degree can be expressed as $SID = \sum v_i.degree$, where $v_i.status = immunized$. And the average path length can be expressed:

$$APL = \frac{1}{\frac{1}{2}n(n-1)} \sum_{i>j} d_{ij}$$

Where d_{ij} is the shortest path between *i* and *j*. If there is no path between *i* and *j*, $d_{ij} \to \infty$. In order to facilitate the calculation, we use Harmonic Means of *APL*(APL^{-1}) to reflect the connectivity of a network:

$$APL^{-1} = \frac{1}{\frac{1}{2}n(n-1)} \sum_{i>j} d_{ij}^{-1}$$

if there is no path between *i* and *j*, $d_{ij} = 999$.

3.2 Simulation Results

We compare the efficiency of different immunization strategies, i.e., targeted and random strategy [10], acquaintance strategy(random and maximal neighbor) [11,12], D-steps strategy(D=2 and D=3) [13,14] (which are all introduced in section 1.2), and our proposed betweenness-based strategy(nodes and edges betweenness), in three synthetic networks and Enron email network when selecting different immunized proportion (5%, 10% and 30%). Two initial infected nodes are selected based on different infected models (random or malicious). Parts of numerical results are shown in Table 1. The results in other synthetic networks are coincident with NET1. The comparing charts of different strategies in Enron email network are shown in Fig.1.

From Table 1, we find that node-betweenness immunization has the best results (i.e. minimum final infected nodes, F) except immunization 5% nodes in NET2. So, if we

Table 1. NET1 is a synthetic network with $\alpha = 1.7$ in the case of random attack. NET2 is Enron email network in the case of malicious attack. There are two infected nodes at the initial situation. If there is no immunization, the final infected nodes of NET1 is 937, $APL^{-1} = 751.36(10^{-4})$, and NET2 is 1052, $APL^{-1} = 756.79(10^{-4})$. The total simulation time T=600.

		5%			10%			30%		
		F	SID	APL^{-1}	F	SID	APL^{-1}	F	SID	APL^{-1}
NET1	Targeted Nodes	679	**3697**	111.56	413	**4694**	57.88	7	**6666**	16.20
	Nodes	**653**	3680	**109.80**	**368**	4588	**52.31**	**3**	6418	**12.89**
	Edge	682	3619	116.73	398	4518	58.58	39	5829	21.47
	Random	879	414	658.05	803	845	570.01	563	2538	295.57
	Rand Neighbor	805	1828	399.11	673	2981	224.57	232	5339	39.99
	Max Neighbor	721	2897	203.75	556	4025	95.60	34	5946	23.28
	D-steps D=2	663	3625	118.19	456	4635	60.51	9	6617	16.6
	D-steps D=3	658	3690	112.82	431	4684	58.93	7	6664	16.18
NET2	Targeted	**396**	**1641**	401.28	300	**1992**	326.68	182	**2658**	176.2
	Nodes	441	1483	**358.83**	**239**	1780	**243.89**	**58**	2259	**39.42**
	Edge	964	80	635.73	914	147	540.10	255	1916	138.29
	Random	944	151	689.67	885	316	624.27	646	924	406.91
	Rand Neighbor	577	1092	491.86	458	1376	444.62	349	1697	403.65
	Max Neighbor	541	1154	474.25	437	1427	432.08	329	1797	383.85
	D-steps D=2	492	1417	412.29	346	1843	333.41	209	2591	190.95
	D-steps D=3	406	1580	412.53	342	1839	364.81	246	2375	226.83

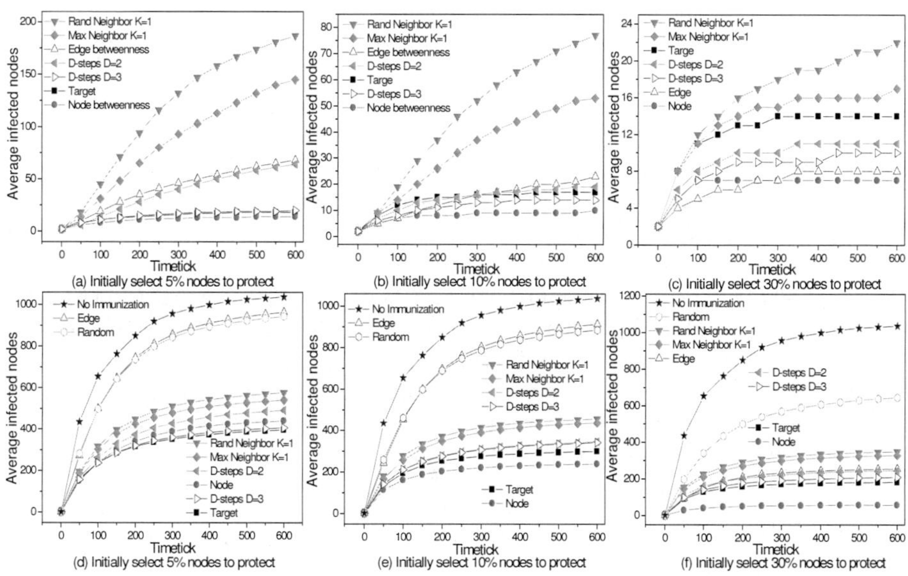

Fig. 1. Comparing different strategies in Enron email network. (a-c) correspond to random attacks and (d-e) are malicious attacks. In each plot, the curve labels are ranked by the total infected nodes from top to bottom.

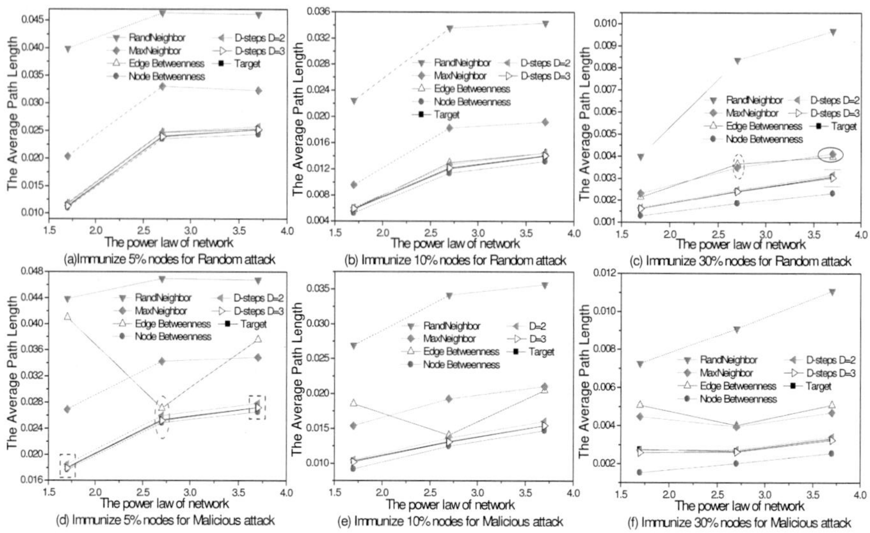

Fig. 2. Comparing different immunization strategies in three synthetic networks, (a-c) correspond to random attacks and (d-f) are malicious attacks

have the global information, node-betweenness immunization is the best strategy. The average degree of NET2 is $\langle k \rangle = 3$. It means that only a few nodes have large degree, others have low degree. In this star network, if nodes with the maximal degree are infected, they will spread in the network rapidly and the sum of final infected nodes will larger than other cases. So it does not illustrate that targeted strategy is better than node-betweenness strategy. On the contrary, the more proportion of immunized nodes, the better efficiency of node-betweenness immunization than that of targeted strategy.

As shown in Table 1, we obtain the maximal SID when we use targeted immunization. However, the final infected nodes are consistent with the Harmonic Means of APL, but not with the SID. Controlling virus epidemic does not depend on the degrees which are immunized, but the path length of the whole network, which can also explain why the efficiency of node-betweenness immunization is better than targeted immunization. Node-betweenness immunization selects nodes which are protected based on the average path length, but targeted immunization is based on the degree of nodes.

In order to obtain more detailed information, we compare the change of *APL* with different strategies in various synthetic networks. The comparing results are shown in Fig.2. Fig.3(a) and Fig.3(b) compare the changes of total infected nodes over time, which correspond to Fig.2(c) and Fig.2(d), respectively.

From the above experiments, we can obtain some conclusions:

(1) As shown in Table 1, *APL* can reflect the efficiency of immunization strategies. So, when we design a distributed immunization strategy, we will pay more attention to those nodes which have great impacts on *APL*.

(2) Node-betweenness immunization is more efficient than targeted immunization through comparing the number of final infected nodes.

(3) Power-law exponent (α) makes great impact on edge-betweenness immunization, but has little impact on others.

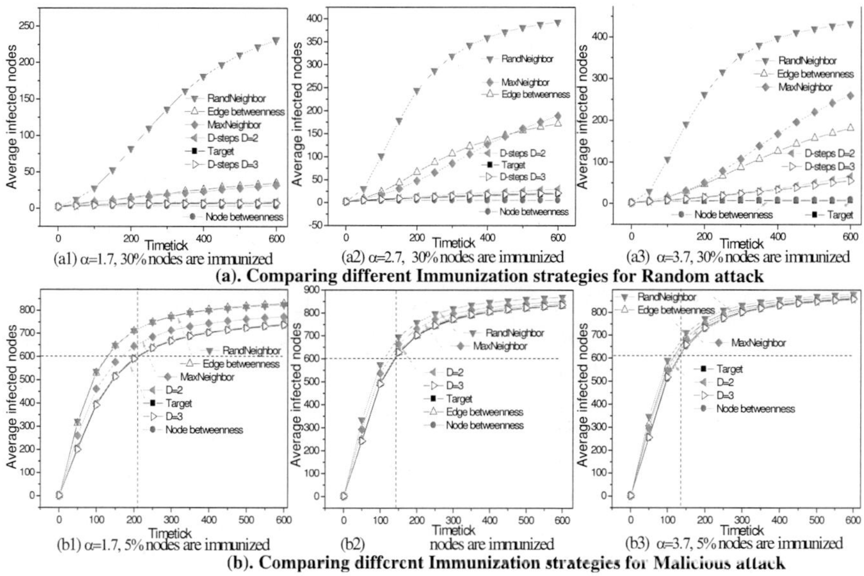

Fig. 3. Comparing different Immunization strategies for both random and malicious attack in synthetic networks

4 Analysis and Discussions

This section firstly analyzes the impact of power-law exponent on virus propagation, and then some comparing experiments are used to evaluate the performances (e.g., efficiency, cost and robustness) of the existing immunization strategies and betweenness-based strategies.

4.1 The Impact of Power-Law Exponent on Virus Propagation

Zou et al. compared virus propagation in synthetic networks with $\alpha = 1.7$ and $\alpha = 1.1475$ [9] and pointed out that worm initially propagation had two phases. However, they did not give more detailed illustrations on this result. In additionally, they did not compare the effect of power-law exponent on different immunization strategies during virus propagation. Table 2 presents more detailed degree statistics of different networks in order to illustrate the effect of power-law exponent on virus propagation and immunization strategies.

We firstly explore virus propagation in non-immunization networks. Figure 4(a) denotes the changes of average infected nodes over time and Fig.4(b) denotes the average degree of infected nodes at each tick. The results show that 1) the power-law exponent has no effect on the total infected nodes. The total infected nodes are related to the attack models, i.e., malicious attack is dangerous than random attack. 2) Because

Table 2. Degree statistics of both Enron email network and synthetic networks

the size of degree	> 80	80-60	60-50	50-30	30-20	< 20
$\alpha = 1.7$	7	16	7	29	38	903
$\alpha = 2.7$	0	4	5	45	34	912
$\alpha = 3.7$	0	2	7	34	60	897
Enron	9	2	1	6	9	1211

malicious attack initially infects highly connected nodes, the fall time of average infected degrees in malicious attack is less than that in random attack (T1<T2). The speed and range of infection will be amplified by these highly connected nodes in a network. In phase I, viruses propagate very quickly and infect more nodes in a network. However, in phase II, the curves of total infected nodes grow slowly (Fig.4(a)), because the viruses aim to infect those nodes that have small degrees (Fig.4(b)), and a node with fewer links is harder to be infected.

Figure 5 compares the effect of different immunization strategies on the average infected degree of nodes in different networks. The results show that 1) random immunization has no effect on restraining virus propagation because the curves of average infected degree are basically coincident with non-immunization circumstance. 2) Comparing (a,b,c) in Fig.5 as well as Fig.5(d,e,f), respectively, the peak value of average infected degree is the largest in the network with α=1.7 and the smallest in the network with α=3.7. This is because the network with lower exponent has higher connectedness nodes (the size of degree between 50 and 80) which are like amplifiers in the process of virus propagation. 3) The larger α, the more final infected nodes as well as the more duration time of virus propagation (T1<T2<T3). Because the larger α, the more APL^{-1}, it can increase the number of final infected nodes. And in a larger exponent network, virus needs more time to infect those nodes with middle or small size degree.

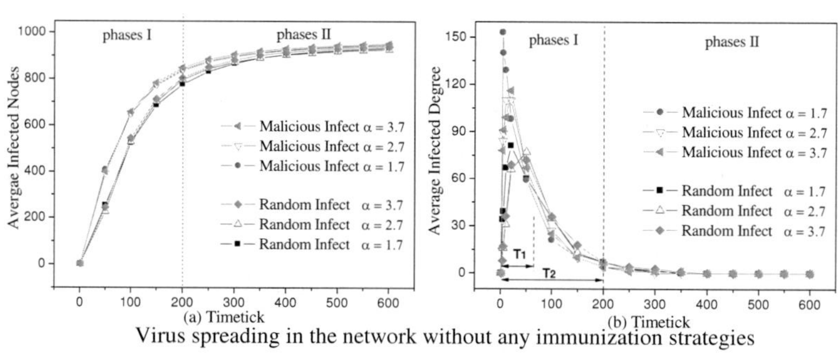

Fig. 4. Change of average infected nodes and average infected degree of nodes over time when virus propagation in different networks without applying any immunization strategies under different ways of attack circumstances

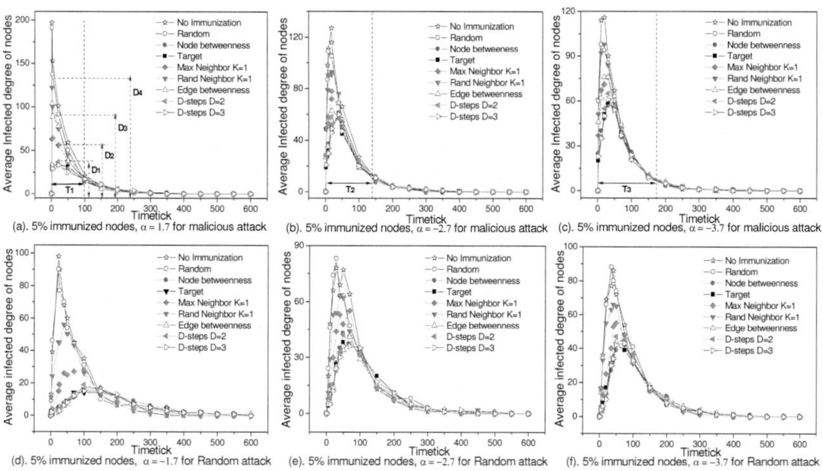

Fig. 5. Changes of average infected degree over time when the virus propagation in different networks which applying different immunization strategies. (a-c) correspond to malicious attack and (d-f) are random attack.

4.2 Evaluating Different Immunization Strategies

The structures of social networks change frequently, which can affect the efficiency of immunization strategies. To better evaluate different strategies, we define some measurements to compare them in detail.

• **Efficiency** can be defined as the total infected nodes when the propagation reaches to an equilibrium state. In Section 3, we assess different strategies in synthetic networks and Enron email network. The results show that node-betweenness immunization could obtain the best efficiency.

• **Cost** can be defined as how many nodes need to be immunized in order to achieve a given level of epidemic prevalence ρ. Generally, $\rho \rightarrow 0$. Now we define some parameters, f_c is the immunization critical value, ρ_0 is the infected density without any immunization strategy and ρ_f is the infected density with certain immunization strategy.

Figure 6 indicates the relationship between the reduced prevalence ρ_f/ρ_0 and the fraction of immunized nodes *f*. It shows that node-betweenness immunization can get the lowest prevalence by protecting the fewest nodes in a network. At the same time, the immunization cost increases with the increment of α, i.e., in order to achieve the epidemic prevalence $\rho \rightarrow 0$, node-betweenness immunization strategy needs 20%, 25% and 30% nodes to be immunized respectively in three synthetic networks.

• **Robustness** reflects the tolerance against the dynamic evolution of a network, i.e., the change of power-law exponents (α).

Figure 7 shows the relationship between the immunized threshold f_c and α. A low level of f_c with a small variation indicates that the immunization strategy is robust. The robustness is important when an immunization strategy is deployed into a scalable and dynamic network(e.g., P2P and email network). The advantage of D-steps

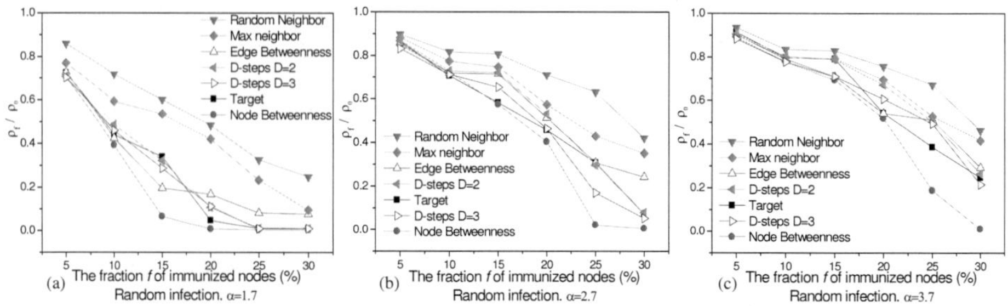

Fig. 6. Reduced epidemic prevalence ρ_f/ρ_0 as a function of the fraction f of immunized nodes

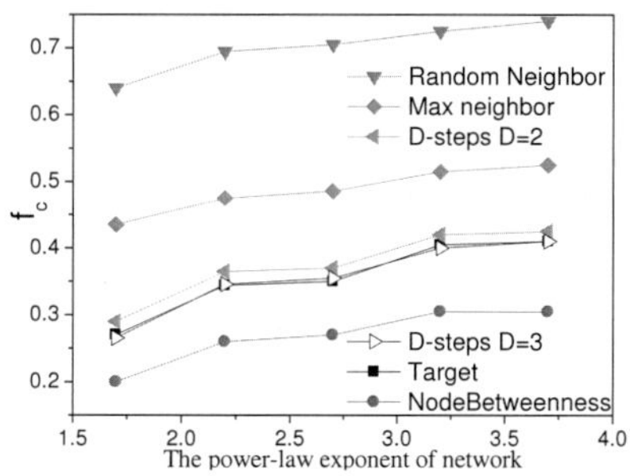

Fig. 7. Immunization critical value f_c as a function of power-law exponent

immunization is that it only needs local information. Fig.7 shows the robustness of D-steps immunization is close to that of targeted immunization. And the robustness of node-betweenness is the best.

5 Conclusion

In this paper, we have analyzed the process of virus propagation and compared the efficiency of the existing immunization strategies and betweenness-based immunization strategies in an interactive email model. The numerical results have shown that the key factor which affects the efficiency of immunization strategies is *APL*, not the total immunized degrees (*SID*).

Moreover, we have analyzed the impact of power-law exponent on virus propagation and immunization strategies in detail. Especially, we have presented a two-phases model of virus propagation by comparing the total infected nodes with the sum of infected degrees varied over time. Through the comparisons of the performances, we have found that D-steps immunization is a feasible strategy in the case of finite resources and node-betweenness immunization is better than targeted immunization if we have the global information about a network.

Acknowledgment

This work was supported in part by the National Natural Science Foundation of China Grant (60673015), The Doctoral Student Innovation Program of Beijing University of

Technology (bcx-2009-062), Open Foundation of Key Laboratory of Multimedia and Intelligent Software (Beijing University of Technology), Beijing, the Major State Basic Research Development Program of China (973 Program) (2003CB317001), and the Research Grant Council Central Allocation Fund of the Hong Kong SAR Government (HKBU 1/05C).

References

1. Lloyd, A.L., May, R.M.: How Viruses Spread Among Computers and People. Science 292(5520), 1316–1317 (2001)
2. Newman, M.E.J.: The Structure and Function of Complex Networks. SIAM Review 45(2), 167–256 (2003)
3. Pastor-Satorras, R., Vespignani, A.: Epidemic Spreading in Scale-Free Networks. Physical Review Letters 86(14), 3200–3203 (2001)
4. Moore, C., Newman, M.E.J.: Epidemics and Percolation in Small-world Network. Physical Review E 61(5), 5678–5682 (2000)
5. Whalley, I., Arnold, B., Chess, D., Morar, J., Segal, A., Swimmerr, M.: An Environment for Controlled Worm Replication and Analysis. Virus Bulletin, 1–20 (2000)
6. Serazzi, G., Zanero, S.: Computer Virus Propagation Models. In: Calzarossa, M.C., Gelenbe, E. (eds.) MASCOTS 2003. LNCS, vol. 2965, pp. 26–50. Springer, Heidelberg (2004)
7. Eguiluz, V.M., Klemm, K.: Epidemic Threshold in Structured Scale-Free Networks. Physical Review Letters 89(10), 108701 (2002)
8. May, S.R.: Enhanced: Simple Rules with Complex Dynamics. Science 287(5453), 601–602 (2000)
9. Zou, C.C., Towsley, D., Gong, W.: Modeling and Simulation Study of the Propagation and Defense of Internet E-mail Worms. IEEE Transaction on Dependable and Secure Computing 4(2), 105–118 (2007)
10. Pastor-Satorras, R., Vespignani, A.: Immunization of Complex Networks. Physical Review E 65(3), 36104 (2002)
11. Cohen, R., Havlin, S., Ben-Averaham, D.: Efficient Immunization Strategies for Computer Networks and Populations. Physical Review Letters 91(24), 0247901 (2003)
12. Gallos, L.K., Liljeros, F., Argyrakis, P., Bunde, A., Havlin, S.: Improving Immunization Strategies. Physical Review E 75(4), 045104 (2007)
13. Gomez-Gardenes, J., Echenique, P., Moreno, Y.: Immunization of Real Complex Communication Networks. European Physical Journal B 49(2), 259–264 (2002)
14. Echenique, P., Gomez-Gardenes, J., Moreno, Y., Vazquez, A.: Distance-d Covering Problem in Scale-Free Networks with Degree Correlation. Physical Review E 71(3), 035102 (2005)
15. Newman, M.E.J., Forrest, S., Balthrop, J.: Email Networks and the Spread of Computer Viruses. Physical Review E 66(3), 035101 (2002)
16. Strogatz, S.H.: Exploring Complex Networks. Nature 410(6825), 268–276 (2001)
17. Milo, R., Shen-Orr, S., Itzkovitz, S., Kashtan, N., Chklovskii, D., Alon, U.: Network Motifs Simple Building Blocks of Complex Networks. Science 298(5594), 824–827 (2002)
18. Bu, T., Towsley, D.: On Distinguishing Between Internet Power Law Topology Generators. In: Proceedings of the Twenty First Annual Joint Conference of the IEEE Computer and Communications Societies, pp. 638–647. IEEE Press, New York (2002)

Semi-supervised Discriminant Analysis Based on Dependence Estimation

Xiaoming Liu[1], J. Tang[1], Jun Liu[1], Zhilin Feng[2], and Zhaohui Wang[1]

[1] College of Computer Science and Technology, Wuhan University of Science and Technology, 430081 Wuhan, Hubei
[2] Zhijiang College, Zhejiang University of Technology, 310024 Hangzhou, Zhejiang
lxmspace@gmail.com

Abstract. Dimension reduction is very important for applications in data mining and machine learning. Dependence maximization based supervised feature extraction (SDMFE) is an effective dimension reduction method proposed recently. A shortcoming of SDMFE is that it can only use labeled data, and does not work well when labeled data are limited. However, in many applications, it is a common case. In this paper, we propose a novel feature extraction method, called Semi-Supervised Dependence Maximization Feature Extraction (SSDMFE), which can utilize simultaneously both labeled and unlabeled data to perform feature extraction. The labeled data are used to maximize the dependence and the unlabeled data are used as regulations with respect to the intrinsic geometric structure of the data. Experiments on several datasets are presented and the results demonstrate that SSDMFE achieves much higher classification accuracy than SDMFE when the amount of labeled data are limited.

Keywords: Semi-supervised learning, Dependence maximization, Dimension reduction.

1 Introduction

With the rapid accumulation of high-dimensional data such as digital images, videos and DNA microarray data, dimension reduction has been widely used in many applications, such as video classification, face recognition and protein structure prediction, etc., they can remove irrelevant and redundant features, increase learning accuracy and speed. A lot of dimension reduction methods have been proposed, such as Principal Component Analysis (PCA) [1], Linear Discriminant Analysis (LDA) [2, 3], Neighborhood Preserving Projection (NPP) [4, 5] and so on. These methods can be roughly classified into 2 classes, unsupervised methods, such as PCA and NPP, and supervised methods, such as LDA. Generally speaking, with the utilization of class label information, supervised methods can achieve better classification accuracies than unsupervised methods.

Most of the above methods are linear methods, kernel based nonlinear methods have been applied successfully in various data mining tasks such as clustering and classification [6]. They work by constructing kernel functions which implicitly map

R. Huang et al. (Eds.): ADMA 2009, LNAI 5678, pp. 234–245, 2009.

the data from the original data space to a high-dimensional feature space and compute the dot product in the feature space. Similar to dimension reduction methods in the original space, in many applications, the interesting patterns of the data may lie in a low-dimensional subspace of the kernel feature space. Thus, dimension reduction in the kernel space has received considerable attention in recent years [7-9]. Among them, supervised dimension reduction methods using the Hilbert-Schmidt Independence Criterion (HSIC) [9] have been proposed in the kernel space [10], we denote the method with SDMFE (Supervised Dependence Maximization Feature Extraction). Under HSIC, the optimal subspace kernel maximizes its dependence with the ideal kernel constructed from the class labels. The method has been shown to be very effective and the optimal subspace kernel can be obtained with eigenvalue decomposition and thus efficient [11]. One shortcoming of the method is that it requires all the data are labeled and cannot deal with unlabeled data. However, in real applications, it is common to have a data set with both unlabeled data and labeled data, and the size of labeled data is generally small due to the high cost to obtain them. The datasets of this kind present a serious challenge, the so-called "small labeled-sample problem", to supervised feature extraction, that is, when the labeled sample size is too small to carry sufficient information about the target concept, supervised feature extraction algorithms will fail because they either unintentionally remove many relevant features or select irrelevant features. Under the assumption that labeled and unlabeled data are sampled from the same population generated by target concept, it is expected to better estimate feature relevance using both labeled and unlabeled data. Learning from mixed labeled and unlabeled data, so called semi-supervised learning, has shown great improvement on many problems compared with the methods based only on labeled data [12, 13]. Although nonlinear methods are effective, they usual demand more computation time than linear methods. Linear methods are powerful tools for solving large-scale data-mining tasks with large number of features in the input space such as those arising in face recognition and the textual domain [13, 14]. In these problems, linear kernel can be used to obtain satisfying results without nonlinear kernel mapping.

In this paper, we propose a novel linear feature extraction algorithm-SSDMFE (Semi-Supervised DMFE), as a natural generalization of SDMFE to semi-supervised learning. Under the regularization framework, SSDMFE can utilize information in both labeled data and unlabeled data, and can be solved as an eigenvalue decomposition problem globally with a closed form solution. Experiments on face recognition data sets are presented. The results demonstrate that SSDMFE can significantly improve the classification accuracy of SDMFE method when labeled data are limited.

2 Background

Because our method is based on dependence maximization criterion, which is proposed in HSIC, we review the Hilbert-Schmidt independence criterion. Before this, we give a short introduction of general linear subspace learning.

2.1 Linear Subspace Learning

Several linear subspace learning methods have been proposed for dimension reduction, PCA and LDA are the most popular two methods among them. In a classification task, we are given a set of n centered training data $\{(x_1, y_1),\ldots,(x_n, y_n)\}$, where $x_i \in R^D$ and $y_i \in \{1,\ldots,k\}$ are the input and output respectively. In linear subspace learning, a low-dimensional subspace of the feature space is used to extract informative features. Let S be a d-dimensional subspace of the original D-dimensional space, and let $P = [p_1,\ldots, p_d] \in R^{D\times d}$ be the projection matrix, then a data point in low-dimensional space can be obtained as $z_i = P^T x_i$. Thus, the original kernel matrix $K =< X, X >= X^T X \in R^{n\times n}$ is transformed into

$$K_d =< P^T X, P^T X >= X^T PP^T X \tag{1}$$

where $X = [x_1,\ldots, x_n]$.

In PCA, data label information is discarded and the projection is obtained by maximizing $tr(K_d)$, which can be solved with eigen-value decomposition. In LDA, data label information is utilized and two scatter matrixes (between-class scatter matrix S_b and in-class scatter matrix S_w) are defined, and the criterion is to minimize the projected $S_w^l = P^T S_w P$ and maximize $S_b^l = P^T S_b P$. When S_w is not singular, the problem can be obtained with eigen-value decomposition, and when S_w is singular, several methods have been proposed to overcome the difficulty [3, 15]. In linear HSIC (see below and [9] for details) based method, the criterion is based on

$$P^* = \arg\max_P tr(HX^T PP^T XHL_y) \tag{2}$$

where $L_y = Y^T Y$ and Y is the class label matrix.

2.2 Hilbert-Schmidt Independence Criterion

Hilbert-Schmidt Independence Criterion (HSIC) is proposed recently for measuring the statistical dependences of random variables [9, 10] in the kernel space. Let F_x be a Reproducing Kernel Hilbert Space (RKHS) defined on the domain X associated with the kernel function $K_x : X \times X \to R$ and the mapping function $\phi_{K_x} : X \to F_x$, and let F_y be another RKHS defined on the domain Y with the kernel function $K_y : Y \times Y \to R$ and the mapping function $\psi_{K_y} : Y \to F_y$. Assume that $x \in X$ and $y \in Y$ can be drawn from some joint measure p_{xy} (probability distribution), then the cross-variance operator $C_{xy} : F_y \to F_x$ is defined as [9, 11]:

$$C_{xy} = E_{xy}[(\phi_{K_x}(x) - \mu_x) \otimes (\psi_{K_y}(y) - \mu_y)] \tag{3}$$

where $\otimes$ is the tensor product operator, $\mu_x = E[\phi_{K_x}(x)]$, and $\mu_y = E[\psi_{K_y}(y)]$ are the means in RKHS. Given separable RKHSs F_x, F_y and a joint measure p_{xy}, HSIC is defined as the squared Hilbert-Schmidt norm of the cross-covariance operator C_{xy} given by [9, 11]

$$HSIC(p_{xy}, F_x, F_y) \triangleq \left\| C_{xy} \right\|_{HS}^2 \tag{4}$$

In practice, given a finite set of data pairs $Z = \{(x_1, y_1), \ldots, (x_n, y_n)\} \in R^D \times R$ drawn independently from p_{xy}, the empirical estimate of HSIC is given by [9]

$$HSIC(Z, F_x, F_y) \triangleq (n-1)^{-2} tr(G_x H G_y H) \tag{5}$$

where $tr(\cdot)$ is the trace of a matrix, $G_x, G_y \in R^{n \times n}$ are the matrices of the inner product of instances in F_x and F_y which can also be considered as the kernel matrices of X and Y with the kernel functions $(G_x)_{ij} = K_x(x_i, x_j)$, $(G_y)_{ij} = K_y(y_i, y_j)$, and $H = I - ee^T / n$ is the center matrix, and e is the vector of all ones with length n. It can be seen that HSIC computes the traces of the product of two centered kernel matrices.

HSIC has several advantages for subspace learning [9], for example, it is an independence measure, unbiased, and can be computed efficiently. HSIC has been applied successfully in clustering [16], supervised feature selection, and extraction [10, 11] in kernel space. In this paper, motivated by applications such as face recognition and document classification[14], we apply HSIC in the original space instead of the kernel space, besides we extend its application to semi-supervised scenery.

3 Semi-supervised Subspace Learning with Dependence Maximization

3.1 Semi-supervised Learning via Dependence Maximization

We extend the subspace learning in Eqn(1) to semi-supervised learning via dependence maximization. Suppose we have l labeled centered examples $\mathbf{x}_1, \mathbf{x}_2, \cdots, \mathbf{x}_l \in R^D$ with class labels from k classes $\{1, \ldots, k\}$ and u unlabeled examples $\mathbf{x}_{l+1}, \mathbf{x}_{l+2}, \cdots, \mathbf{x}_{l+u} \in R^D$ with unknown class memberships. Without loss of generality, we assume the first l data are labeled. So there are a total of $n = l + u$ examples and usually $l \ll u$. When l is too small compared with the number of features, SDMFE may not perform well since the information it can use is very limited. To remedy this problem, we want to incorporate unlabeled data to improve the performance of SDMFE, at the same time; we also want to preserve the efficiency of the algorithm. More specifically, we need to find a set of projective vectors $P = [P_1, P_2, \cdots, P_d] \in R^{D \times d}$, such that the transformed low-dimensional representations $\mathbf{z}_i = P^T \mathbf{x}_i \in R^d$ can preserve the structure of the original data set. We propose to maximize a new objective function in the regularization framework as follows:

$$\max_{P \in R^{n \times l}} tr(HX^T PP^T XHL_y) + \lambda J(P)$$
$$s.t. \ P^T P = I_d \tag{6}$$

where $X = [x_1, \ldots, x_l, 0_{l+1}, \ldots, 0_n] \in R^{D \times n}$,

$$H = \begin{bmatrix} H_l & 0 \\ 0 & 0 \end{bmatrix}$$

$H_l = I - ee^T / l$ is the center matrix, λ is a control parameter, L_y is the ideal class label kernel of the labeled data, and $J(P)$ is a regularization term which can be used to obtain information from the unlabeled data. Note that the first term in equatin (6) only utilizes the information conveyed by the labeled data. There are many ways to formulate the regularization, such as Tikhonov regularization to control the complexity of the method and avoid the singularity problem, to minimize distance in reduced space of the data samples those are near in original space [17], to minimize distance between unlabeled data and their nearest neighbors (both labeled and unlabeled) [13].

We use an approach similar to [13], which utilizes a prior assumption of consistency. In this approach, for classification, it is assumed that different nearby points are likely to belong to the same class, and for dimension reduction, different nearby points will have similar low-dimensional representations. More specifically, given a set of examples $\{\mathbf{x}_i\}_{i=1}^n$, we use a p-nearest neighbor graph G to model the relationship between the nearby data points. We put an edge between nodes i and j if $\mathbf{x}_i$ and $\mathbf{x}_j$ are "close", i.e., $\mathbf{x}_i$ and $\mathbf{x}_j$ are among p nearest neighbors of each other.

Let the corresponding weight matrix be $\mathbf{S}$ defined by [5]

$$\mathbf{S}_{ij} = \begin{cases} \frac{1}{p}, \text{ if } \mathbf{x}_i \in N_p(\mathbf{x}_j) \text{ or } \mathbf{x}_j \in N_p(\mathbf{x}_i) \\ 0, \text{ otherwise} \end{cases} \tag{7}$$

where $N_p(\mathbf{x}_i)$ denotes the set of p nearest neighbors of $\mathbf{x}_i$. Note here, other weight definition can also be used, such as those defined in [13].

With our separation of labeled and unlabeled data, we now explain the utilization of geometry information by analyzing the neighbor information. For each labeled data point, its neighbor may have different configurations: 1) Only contains unlabeled data, 2) Contains data from the same labeled class (with or without unlabeled data), 3) Contains data from other classes (with or without unlabeled neighbors). Similarly, we can see that the neighbor structure of each unlabeled data x_{l+i} may have the following 3 configurations: 1) Only contains unlabeled data, 2) Contains data from one labeled class (with or without unlabeled data), 3) Contains data from 2 or more classes (with or without unlabeled neighbors). Our general assumption is that near data in original space should also be near in projected space, if near data contains different class labels, then the margin of the distance to different class should be large.

We can consider the first 2 configurations at the same time for labeled and unlabeled data at the same time. For configuration 1) and 2), we minimize the following:

$$\begin{aligned} J(P) &= \frac{1}{2}\sum_{\mathbf{x}_{l+i}} \sum_{\mathbf{x}_j \in N(\mathbf{x}_{l+i})} \left\| P^T\mathbf{x}_{l+i} - P^T\mathbf{x}_j \right\|^2 \mathbf{S}_{ij} \\ &= tr\left(P^T \left[\sum_{\mathbf{x}_{l+i}} \left(\sum_{\mathbf{x}_j \in N(\mathbf{x}_{l+i})} \mathbf{S}_{l+i,j} \right) \mathbf{x}_{l+i}\mathbf{x}_{l+i}^T - \sum_{\mathbf{x}_{l+i}} \sum_{\mathbf{x}_j \in N(\mathbf{x}_{l+i})} \mathbf{S}_{l+i,j}\mathbf{x}_{l+i}\mathbf{x}_j^T \right] P \right) \\ &= tr\left(P^T\mathbf{X}(\mathbf{D}-\mathbf{S})\mathbf{X}^T P \right) \\ &= tr\left(P^T\mathbf{XLX}^T P \right) \end{aligned} \tag{8}$$

where $\mathbf{D}$ is a diagonal matrix whose entries are the column sums of $\mathbf{S}$ and $\mathbf{L} = \mathbf{D} - \mathbf{S}$ is the Laplacian matrix of $\mathbf{S}$ [18]. Note here all $\mathbf{D}, \mathbf{R}$ and $\mathbf{L} \in R^{n\times n}$, with the first l rows and columns are zeros corresponding to the labeled data.

For the unlabeled data points with configuration 3), we simply ignore the regularization. This choice is based on two observations. Firstly, this configuration occurs rarely. Secondly, even it occurs, since these neighbors are usually also neighbors to each other, then dependence maximization criterion has been applied to them, and we assume the information dependence maximization used is more reliable. By ignoring this difficult regularization, our method can be expected to be more general. Our method is different from [17], while neighbors from different classes are also required to be neighbors in projected space with regularization, which may mix them up, and reduce the discriminant capability.

Combining Eqn. (8) into Eqn. (6), we can get the objective function for the optimization problem of our SSDMFE algorithm, which can be represented by a maximal problem with respect to P :

$$\begin{aligned} &\max_{P\in R^{D\times d}} tr(HX^T PP^T XHL_y) - \lambda tr(P^T XLX^T P) \\ &s.t.\ \mathrm{P}^T P = I_d \end{aligned} \tag{9}$$

where λ is a control parameter. For the first term, we have

$$\begin{aligned} tr(HX^T PP^T XHL_y) &= tr(PP^T XHL_y HX^T) \\ &= tr(P^T XHL_y HX^T P) \end{aligned}$$

Integrate the above equation into Eqn.(4), we have

$$\begin{aligned} &\max_{P\in R^{n\times l}} tr\left[P^T \left(XHL_y HX^T - \lambda XLX^T \right) P \right] \\ &s.t.\ P^T P = I_d \end{aligned} \tag{10}$$

Clearly, similar to PCA, the problem expressed by Equation (10) is a typical eigen-decomposition problem, which can be easily and efficiently solved by computing the eigenvectors of $XHL_y HX^T - \lambda XLX^T$ corresponding to the d largest eigenvalues. Since $XHL_y HX^T - \lambda XLX^T$ is symmetric, the eigenvalues are all real. If the optimal P^* has been obtained, the corresponding HSIC value is $\sum_{i=1}^{d} \lambda_i$. Since the eigenvalues reflect the contribution of the corresponding dimensions, we can control d by setting a threshold thr ($0 \le thr \le 1$) and then choosing the first d eigenvectors such as $\sum_{i=1}^{d} \lambda_i \ge thr \times (\sum_{i=1}^{D} \lambda_i)$.

The optimal P^* of Equation (10) is determined by kernel matrices $K(x)$ and L_y derived from the labeled data and the corresponding labels, respectively. Here for data space kernel we use the linear projection kernel, for label space kernel, we have several choices [11]. Let $y \in R^n$ be the label vector with the i-th entry $y_i \in \{1,\ldots,k\}$, and let $Y \in R^{k \times l}$ be the class indicator matrices defined as $Y(ij) = 1$ if $y_j = i$ and $Y(ij) = 0$ otherwise. We can use the label kernel matrix $L_y = K(y) = Y^T Y \in R^{n \times n}$.

3.2 SSDMFE Algorithm

The SSDMFE algorithm can be summarized as follows:

1. Construct the label space kernel of labeled L_y.

2. Compute the matrix $XHL_y HX^T$.

3. Construct the neighbor graph G with both labeled and unlabeled data, and calculate the regularization matrix $\mathbf{XLX}^T$ with Equation (8).

4. Solve the eigenvalue problem in Equation (9), select d eigenvectors corresponding to the largest eigenvalues.

5. Let $P = [P_1, P_2, \cdots, P_d]$. Data points can be embedded into the lower-dimensional space via the transform: $\mathbf{x} \to \mathbf{z} = P^T \mathbf{x}$.

4 Experiments

In this section, we describe the evaluation of the performance of the proposed SSDMFE algorithms in face recognition problems. We compared with another linear space method (PCA based eigenface method) and the original SDMFE method. Experiments were performed on PIE and YaleB face data sets.

4.1 PIE Data Set

We used the PIE face database [19] for the first set of experiments. The database contains 41,368 face images of 68 individuals, and each individual has 13 different poses, under 43 different illumination conditions and with 4 different expressions. We chose the frontal poses (C27) with varying lighting and illumination conditions for our experiments. There are about 49 images for each individual. Before the experiments, each image was clipped and resized to a resolution of 32×32 pixels. Some sample images are shown in Figure 1.

We used KNN for the final classification and used it to estimate the classification errors of each algorithm. KNN is a very simple yet effective classifier. With a proper distance metric to identify the nearest neighbors, KNN often yields competitive results to

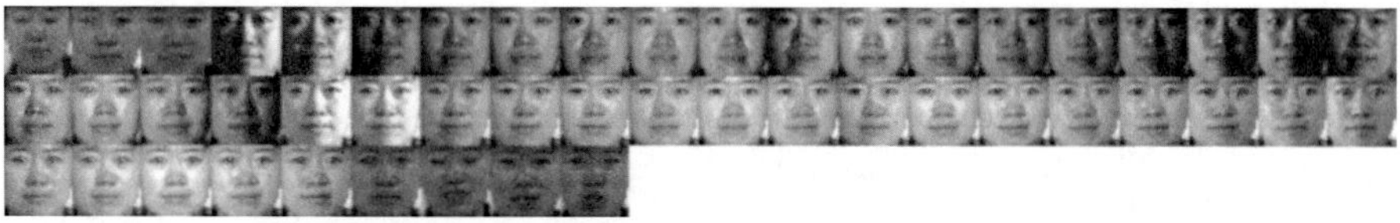

Fig. 1. Sample images of one subject in PIE face database

some advanced classification algorithms. The number of the nearest neighbors used in KNN was simply set as 1 for simplicity. For the neighbor graph construction, the number of neighbors need to be choosen, we used cross-validate to determine the number.

In the first experiment, 30 images were randomly selected for each person to form the training set and the rest to form the test set. From the 30 images for each person, one image was randomly selected and labeled while the other 29 images remained unlabeled. We performed 10 random splits and the averaged recognition rates of different methods evaluated on the unlabeled training data and the test data are reported in Table 1. The baseline method ignored the manifold structure of data and Eigenface did not utilize the labeled data, and hence the performances of these two methods were poor. On the other hand, both SDMFE and SSDMFE exploited the label information and hence obtained better results. It can be seen that SSDMFE presented better result than SDMFE, implying that unlabeled data can improve the accuracy of SDMFE when there are only very few labeled training samples.

A similar experiment but with 2 labeled images from the 30 training samples as training samples was performed. Table 2 reports the results of 10 averaged random splits. Similar results are observed that SSDMFE can achieve the best performance among all the methods. We also investigated the effect of parameter λ on the performance of SSDMFE. The recognition error rates on the unlabeled training data and test data are plotted in Fig.2. It can be seen that the performances do not change significantly when λ varies in the range of $[0.1,1.5]$. So it is not difficult to choose an appropriate value of λ.

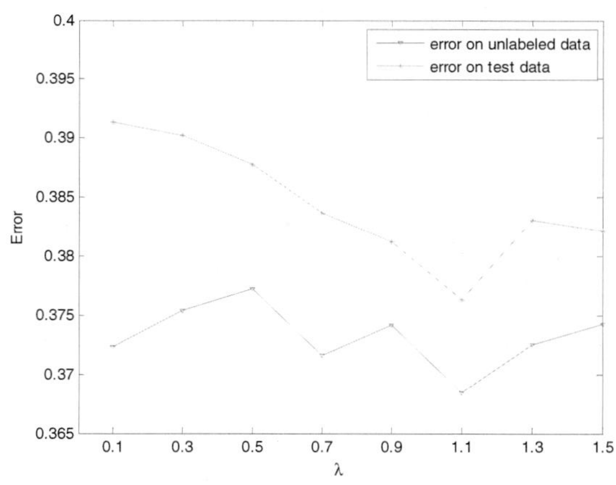

Fig. 2. Recognition error with different λ values on PIE data set

Table 1. Averaged recognition error rates on PIE when there are one labeled and 29 unlabeled examples

Method	Unlabeled error	Test error
Baseline	0.7835	0.7842
Eigenface	0.8104	0.8213
SDMFE	0.6302	0.6744
SSDMFE	0.5643	0.5672

Table 2. Averaged recognition error rates over 10 random splits on PIE when there are 2 labeled and 28 unlabeled examples

Method	Unlabeled error	Test error
Baseline	0.6232	0.2412
Eigenface	0.6424	0.6488
SDMFE	0.5127	0.5213
SSDMFE	0.3729	0.3876

4.2 Extended YaleB Face Database

The extended Yale Face Database B [20] contains images of human subjects under varying pose and illumination conditions. We used the cropped images and resized them to 32×32 pixels, the dataset has 38 individuals and around 64 near frontal images under different illuminiations per individual. Some sample images are shown in Fig.3.

Fig. 3. Sample images of one person in YaleB data set

Table 3. Averaged recognition error on YaleB data set with different t values. (a) $t = 2$; (b) $t = 5$.

Method	Unlabeled error	Test error
Baseline	0.8231	0.8343
Eigenface	0.8318	0.8136
SDMFE	0.7852	0.7682
SSDMFE	0.4438	0.4572

(a)

Method	Unlabeled error	Test error
Baseline	0.6824	0.6732
Eigenface	0.7124	0.7541
SDMFE	0.6638	0.5874
SSDMFE	0.2782	0.2644

(b)

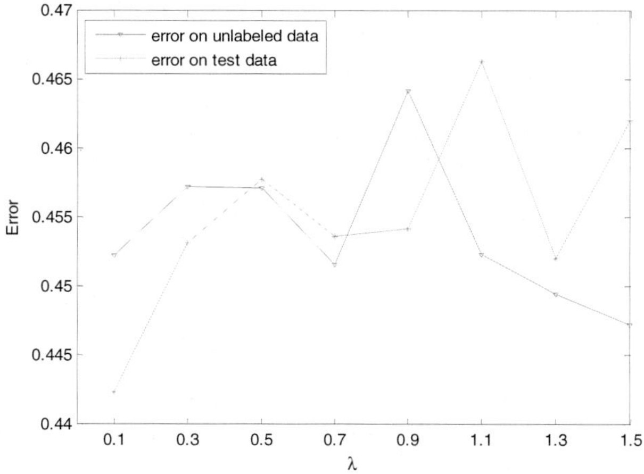

(a)

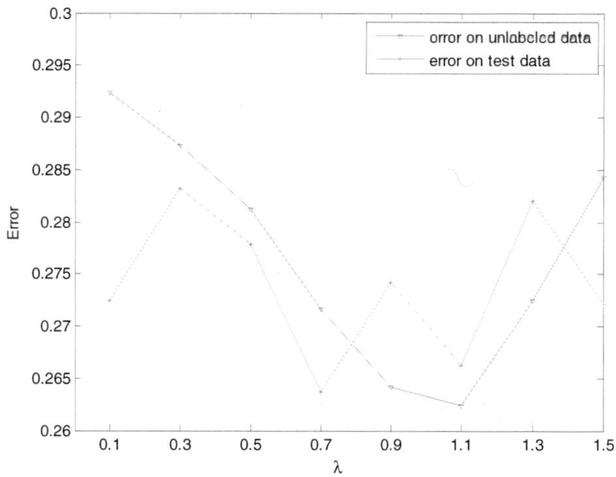

(b)

Fig. 4. Recognition error with different λ values on YaleB data set with different number t of labeled samples, (a) $t = 2$, (b) $t = 5$

We conducted two experiments on the YaleB database. For each subject, 10 images were randomly selected for training and the rest was used for the test set. Among the 10 training images, we randomly chose $t = 2$ or $t = 5$ images and labeled them. The two experiments correspond to different values of t. For each t, 10 randomly trials were performed and the averaged results are reported in Table 3. Similar results to PIE are observed. We can see that Eigenface is not a good choice for recognition on YaleB data set and SSDMFE achieved the best results among all the methods. SDMFE is better than Eigenface since it utilize the label discriminant information. The reason for SSDMFE achieving the best performance could be the utilization of discriminant information and local geometry information simultaneously, especially when only a few labeled training samples are presented.

As in the experiments on PIE database, we also investigated the effect of parameter λ in Eqn.(6) on the performance of SSDMFE. The recognition error rates on the unlabeled training and the test data are plotted in Fig.3 and Fig.4. We can see that when λ varies in the range $[0.1, 1.5]$, the performance of SSDMFE does not change much. So the value of parameter λ is not very hard to set.

5 Conclusion

In this paper, we have proposed a new dimension reduction method called Semi-Supervised Dependence Maximization based Feature Extraction (SSDMFE). It extended Supervised Dependence Maximization based Feature Extraction into semi-supervised environment, and can make use of both labeled and unlabeled data in learning a transformation to achieve dimension reduction. The selective utilization the local geometry information among labeled and unlabeled samples plays a crucial role in maximizing the discrimination ability of SSDMFE. Experimental results on two widely used face databases are very encouraging when compared with other related methods.

References

1. Jolliffe, I.: Principal component analysis. Springer, New York (2002)
2. Fisher, R.A.: The use of multiple measurements in taxonomic problems. Ann. of Eugenics 7, 179–188 (1936)
3. Liu, X., Wang, Z., Feng, Z., Tang, J.: A Pairwise Covariance-Preserving Projection Method for Dimension Reduction. In: Seventh IEEE International Conference on Data Mining, pp. 223–231 (2007)
4. He, X., Cai, D., Yan, S., Zhang, H.: Neighborhood preserving embedding. In: Tenth IEEE International Conference on Computer Vision, pp. 1208–1213 (2005)
5. Liu, X., Yin, J., Feng, Z., Dong, J., Wang, L.: Orthogonal Neighborhood Preserving Embedding for Face Recognition. In: IEEE International Conference on Image Processing, pp. 133–136 (2007)
6. Cristianini, N., Shawe-Taylor, J.: An introduction to support Vector Machines: and other kernel-based learning methods. Cambridge Univ. Pr., Cambridge (2000)
7. Mika, S., Ratsch, G., Muller, K.: A mathematical programming approach to the kernel fisher algorithm. Advances in neural information processing systems, 591–597 (2001)
8. Rosipal, R., Trejo, L.: Kernel partial least squares regression in reproducing kernel hilbert space. The Journal of Machine Learning Research 2, 97–123 (2002)
9. Gretton, A., Bousquet, O., Smola, A., Scholkopf, B.: Measuring statistical dependence with Hilbert-Schmidt norms. In: Jain, S., Simon, H.U., Tomita, E. (eds.) ALT 2005. LNCS, vol. 3734, pp. 63–77. Springer, Heidelberg (2005)
10. Smola, A., Gretton, A., Borgwardt, K., Bedo, J.: Supervised feature selection via dependence estimation. In: Proceedings of the 24th International Conference on Machine Learning, pp. 823–830. ACM, New York (2007)

11. Chen, J., Ji, S., Ceran, B., Li, Q., Wu, M., Ye, J.: Learning subspace kernels for classification. In: KDD, pp. 106–114 (2008)
12. Zhu, X.: Semi-supervised learning literature survey. Department of Computer Sciences, University of Wisconsin at Madison, Madison (2006)
13. Zhang, Y., Yeung, D.: Semi-supervised discriminant analysis using robust path-based similarity. In: CVPR, pp. 1–8 (2008)
14. Sindhwani, V., Selvaraj, S.: Large scale semi-supervised linear support vector machines. In: Proceedings of the 29th annual international ACM SIGIR conference on Research and development in information retrieval, pp. 477–484 (2006)
15. Chen, L., Liao, H., Ko, M., Lin, J., Yu, G.: A new LDA-based face recognition system which can solve the small sample size problem. Pattern recognition 33(10), 1713–1726 (2000)
16. Smola, A., Gretton, A., Borgwardt, K.: A dependence maximization view of clustering. In: Proceedings of the 24th International Conference on Machine Learning, pp. 815–822. ACM, New York (2007)
17. Cai, D., He, X., Han, J.: Semi-supervised discriminant analysis. In: Proceedings of the IEEE International Conference on Computer Vision, pp. 1–7 (2007)
18. Chung, F.: Spectral graph theory: American Mathematical Society (1997)
19. Sim, T., Baker, S., Bsat, M.: The CMU pose, illumination, and expression database. IEEE Transactions on Pattern Analysis and Machine Intelligence 25(12), 1615–1618 (2002)
20. Georghiades, A., Belhumeur, P., Kriegman, D.: From few to many: illumination cone models for face recognitionunder variable lighting and pose. IEEE Transactions on Pattern Analysis and Machine Intelligence 23(6), 643–660 (2001)

Nearest Neighbor Tour Circuit Encryption Algorithm Based Random Isomap Reduction

Wei Lu[1] and Zheng-an Yao[2]

[1] School of Information Technology of Beijing Normal University Zhuhai Campus, 519085 Zhuhai, Guangdong
[2] Mathematics Department, Zhongshan University, 510275 Guangzhou, Guangdong

Abstract. This paper presents nearest neighbor tour circuit encryption algorithm based random Isomap reduction. In order to be suited for privacy-preserving classification, we first alter the selection fashion of the parameters nearest neighbor number k and embedded space dimension d of Isomap reduction algorithm. Further we embed the tourists' sensitive attribution into random dimension space using random Isomap reduction, thus the sensitive attributes are encrypted and protected. Because the transformed space dimension d and nearest neighbor number k are both random, this algorithm is not easily be breached. In addition, Isomap can keep geodesic distance of two points of dataset, so the precision change of classification after encryption can be controlled in a small scope . The experiment show that if we select appropriate parameters, then nearest neighbors of every point may be completely consistent. The present algorithm can guarantee that the security and the precision both achieve the requirements.

Keywords: Random Isomap reduction; geodesic distance; nearest neighbor; sensitive information; encryption.

1 Introduction

Privacy-preserving data mining[1,2] techniques have already obtained widespread attention of computer domain, there are many article proposed for it. The general goal of the privacy-preserving data mining techniques is defined as to hide sensitive individual data values from the outside world or from unauthorized persons, and simultaneously preserve the underlying data patterns and semantics so that a valid and efficient decision model based on the distorted data can be constructed. In the best scenarios, this new decision model should be equivalent to or even better than the model suing the original data from the viewpoint of decision accuracy.

The proposed privacy preserving data mining methods are divided into two kinds. One is based on data conversion, and another is based on the cryptology. A large amount of research work has been devoted to this area, and resulted in such techniques as k-anonymity[3],data perturbation[4], and privacy preserving distributed data mining[5,6]. Data distortion is one of the most important parts

R. Huang et al. (Eds.): ADMA 2009, LNAI 5678, pp. 246–252, 2009.

in many privacy-preserving data mining tasks. The desired distortion methods must preserve data privacy, and at the same time, must keep the utility of the data after the distortion. The classical data distortion methods are based on the stochastic value perturbation.

Paper[1] proposes a method which adds random noise to the data to preserve the sensitive information while maintaining the distribution of the underlying data and hence the utility of data. Paper[7]proposes a random rotation technique to perturb data, in which the data matrix will be rotated randomly. Several methods have been proposed to reconstruct the original data from the perturbed one , such as spectral filtering[8], principal component analysis[9].

However, all these methods have one common shortcoming that if we try to preserve more privacy, we may have to loss more information. It seems contradictory to achieve the goal of preserving privacy and maintaining the utility of the data. In this paper, we present Isomap reduction[10]based privacy-preserving algorithm which ensure that the mining process will not violate privacy up to a certain degree of security. Isomap[10]is one new algorithm for manifold learning[11,12]. Isomap algorithm is usually used for dimensionality reduction and data compression. Given a set $x_1, x_2, \cdots x_n$ of n points in R^l , find a set of points $y_1, y_2, \cdots, y_n$ in $R^d (d < l)$ such that y_i "represents" x_i .

It is the first time that using Isomap reduction algorithm for privacy-preserving data mining. We embed the original data into random dimension space so that the sensitive information will be protected well. Isomap can keep geodesic distance of the dataset, so these mining methods based distance are not affected. We use the classification of tourists to verify the present algorithm. The experiment show that the present method can provide both enough protection and full keep of the data. In section 2, we present the nearest neighbor tour circuit encryption algorithm based random Isomap reduction; Section 3 is the experimental result and analysis; Finally in section 4, we give the conclusion.

2 Nearest Neighbor Tour Circuit Encryption Algorithm Based Random Isomap Reduction

2.1 Problem Description

For new tourist, the travel organization can estimate which class people it belongs based nearest neighbor method. So the organization can recommend appropriate tour circuit to the new tourist. But tourist must provide complete individual so that the travel organization can judge its class.

Usually tourist don't hope individual information be announced, for some sensitive information will need to obtain the security. Suppose the travel organization has clustered the tourists according to the historic record, moreover it has a most appropriate tour circuit to each kind of tourist, the data form is

$$Z = \{(x_1, y_1), \cdots, (x_n, y_n)\} = (X, y),$$

where $X = \begin{pmatrix} x_{11}x_{12}\cdots x_{1l} \\ \cdots \\ x_{n1}x_{n2}\cdots x_{nl} \end{pmatrix}$ is information data of tourists. There are n clustering tourists, and each class has l attributes such as average age and average income. $y = (line1, line2, \cdots, linen)^{\mathrm{T}}$ represents appropriate travel line for every class tourists.

For some sensitive information such as age, month income and holiday days, which must be transformed to hide its true value and not affecting the data mining.

2.2 Encryption Algorithm Based Random Isomap Reduction

The basic thought of the Isomap comes from the multi-dimensional scale (MDS)[13]. MDS is one non-supervised dimension reduction method. The basic thought of MDS is the distance of two points in lower embedding space is same of that in the original high space.

Isomap reduction algorithm is usually used for dimension reduction. Because Isomap reduction can keep inherent geodesic distance, so we can use it in sensitive data encryption algorithm. In the Isomap reduction algorithm, the parameter embedded space dimension d is obtained according to the computation. In order to apply the Isomap reduction algorithm in the encryption, we propose choosing the parameter with the stochastic way, thus enhancing the security of the algorithm.

Algorithm 1. Encryption Algorithm Based Random Isomap reduction

Input Dataset $X = \{x_1, x_2, \cdots, x_n\} \in R^l$

1: [Finding the Nearest Neighbor] Computing the nearest neighbor nodes of x_i. Generally we use k nearest neighbors (Parameter $k \in N$) or ε neighborhood.

2: [Constructing the weights graph] If nodes i and j are nearest neighbor points, put the weight of the edge as$d_x(i,j)$.

3: [Computing the distance] Computing the shortest distance of two points on graph, the obtained distance matrix is $D_G = \{d_G(i,j)\}$.

4: [Finding embedding space using MDS] Let

$$S = (S_{ij}) = (D_{ij}^2), H = (H_{ij}) = (\delta_{ij} - 1/N), \tau(D) = -HSH/2,$$

The lower embedding data is the characteristic vectors which are corresponding to $2, \cdots, d+1$ characteristic value of $\tau(D)$.

5: After the complete attribute of the dataset is encrypted using Isomap reduction, it can go step 1 once more, again carrying Isomap reduction encryption.

Output encrypted dataset X_1

2.3 Nearest Neighbor Tour Circuit Encryption Algorithm Based Random Isomap Reduction

Suppose the individual information of the new tourist (define as Alice below) is $x = (x^1, x^2, \cdots x^l)$. Alice needs the travel organization (define as Bob below) recommending an appropriate circuit for her. But Alice doesn't hope giving all primary information to Bob, so she can give the encrypted data using algorithm 1 to Bob.

Algorithm 2. Nearest neighbor tour circuit encryption algorithm based random Isomap reduction

Input. travel organization Bob input dataset (X, y); new tourist Alice input the new data x

1: Alice combines X and x producing

$$\tilde{X} = \begin{pmatrix} X \\ x \end{pmatrix}^T .$$

2: Alice uses algorithm 1 to encrypt $\tilde{X}$, embedding $\tilde{X}$ into a random dimension space. Suppose x turns into x^1 after encryption, and X to X^1 . In this step, Alice can select the values of two parameters k, d of Isomap reduction algorithm stochastically.

3: Alice transfers the encrypted data to Bob.

4: Bob use x^1 to find nearest neighbor in X^1 , finding what class Alice belongs. Suppose we find the ith class tourists ${x_i}^1$ that is near to Alice. Because Isomap can keep the inherent topology of the dataset, so the class that is nearest to Alice in map space is the same as that class found in original dataset. Alice belongs to ith class tourist.

Output Bob output the line i to Alice

For the encryption algorithm that Alice uses is random, Bob and other people are difficult in inferring original sensitive data x from encrypted data $\tilde{X}^1$.

2.4 Safety Analysis

The present algorithm is safe that can undergo simple attacks, for there are three random steps in encryption algorithm 1.

1)Selecting of the high dimension d is stochastic.
2)Selecting of the nearest neighbors'numbers k of each data x_i is stochastic.
3)Alice can use Isomap reduction algorithm to encrypt the dataset again and again.

According to these three stochasticicitys, each computation get a entire different random dimensional embedding manifold. Bob and other people are difficult to attack the encryption, so Alice's sensitive attributes can obtain good protection.

How to ensure the value of k, d is uncertain in manifold learning, but this paper changes it to stochastic selection. The present method not only settles the uncertain of the Isomap reduction algorithm, but also enhances the safety of the algorithm.

3 Experiment Result

3.1 Experiment Dataset

This section we give the experiment result of one actual dataset. For simplicity, suppose data is standard and numeric. The dataset only has 6 class tourists, and every record only has 3 attributes. The tour circuit is simple. One travel organization Bob has one dataset as table 1.

Table 1. Tour circuit sample dataset

No	Average age	Average income	Average holidays	The good tour circuit
1	30	3	15	South Korea → USA → Thailand
2	18	0	90	South Africa → Singapore → Malaysia → South Korea
3	25	1	8	Malaysia → Singapore → South Korea
4	55	13	25	USA → Hawaii → Singapore → Malaysia
5	40	2	14	Europe → Japan → Singapore
6	36	5.6	7	Japan → USA → Hawaii → Europe

The data of new tourist Alice is (age, income, holidays)=(38, 5.9, 8), and Alice needs Bob recommending appropriate tour circuit. Alice doesn't want to transfer her sensitive attributes such as age, income to Bob, so she can use algorithm 2 to encrypt the data. Table 2, table3, table4 are some sample of encrypted dataset.

Because in Algorithm the parameters k, d have infinite selectivities, Alice will have different encrypted dataset in every running time of algorithm 1. And Alice can also run encryption algorithm 1 arbitrarily. Bob and other people is difficult getting original dataset from encrypted dataset.

Alice sends the encrypted dataset to Bob. If Bob receive the embedded dataset 1 as table 2, he calculate the distances between the new data and initial six class tourist point are

11.0036 84.5943 13.9058 25.0337 5.8883 2.5702

So Bob thinks the new data of Alice is nearest to class 6, and she get the conclusion that Alice is most similar to 6th class tourist.

If Bob calculate the distances using original dataset as table 1, he will get the distances between Alice and six class

11.0186 84.6098 13.8928 25.0681 7.4303 2.2561

The class nearest to Alice is 6th class. Though the distances of original dataset and encrypted dataset are different, but the nearest data is same. So the encryption algorithm doesn't effect the result of the classification algorithm, which is

Table 2. Encrypted dataset 1 of tourist($k = 5, d = 2$)

Embedding attribute	class 1	2	3	4	5	6	new data
Attribute 1	7.4355	-68.4798	13.0776	3.5211	10.4644	17.9013	16.0798
Attribute 2	6.6378	2.2514	13.4072	-21.8262	-1.9423	1.6426	-0.1706

Table 3. Encrypted dataset 2 of tourist($k = 6, d = 2$)

Embedding attribute	class 1	2	3	4	5	6	new data
Attribute 1	-7.6002	68.2890	-13.2632	-3.6795	-10.6457	-16.8179	-16.2825
Attribute 2	6.5678	2.2560	13.3449	-21.8831	-2.0012	1.9314	-0.2157

Table 4. Encrypted dataset 3 of tourist($k = 6, d = 3$)

Embedding attribute	class 1	2	3	4	5	6	new data
Attribute 1	7.6002	-68.2890	13.2632	3.6795	10.6457	16.8179	16.2825
Attribute 2	6.5678	2.2560	13.3449	-21.8831	-2.0012	1.9314	-0.2157
Attribute 3	-0.5389	-0.0446	-0.5784	-0.9610	3.8540	-1.0855	-0.6457

decided by its character of keeping geodesic distance. Likewise, if Bob receives encrypted dataset such as table 3 and table 4, we will also get the nearest class tourist to Alice is 6. Bob can't get the original sensitive data of Alice, but also can commend the right tour circuit to Alice.

4 Conclusion

This paper presents nearest neighbor tour circuit encryption algorithm based random Isomap reduction. Isomap reduction algorithm is first used on privacy-preserving data mining. Because the Isomap can keep the geodesic distance, thus the nearest points that found on original dataset and encrypted dataset are same. And two parameters k and d of Isomap reduction are stochastically selected, so the embedded space's dimension is variable, the safe of the algorithm is assured. The experiment result show that the present method can not only give the user's sensitive attributes enough protection, but also gives the fit tour circuit commend.

For privacy-preserving data mining and manifold learning are all new theory which are far from maturation. It can further combine all encryption algorithm and will have further develop. For example, the analysis of selection of parameter k, d is very important, the selection of manifold algorithm, the generation of Isomap reduction and the choose of the nearest point, all these have upswing space. It is worth doing the further research to advance a better theory and algorithm.

Acknowledgments. This paper is supported by the scientific research foundation of Beijing Normal University Zhuhai Campus.

References

1. Agrawal, R., Srikant, R.: Privacy-Preserving Data mining. In: 2000 ACM SIGMOD International Conference on Management of Data, pp. 439–450. ACM Press, Dallas (2000)
2. Agrawal, S., Haritsa, J.R.: A Framework for High-Accuracy Privacy-Preserving Mining. In: 2005 IEEE International Conference on Data Engineer (ICDE), pp. 193–204. IEEE Press, Tokyo (2005)
3. Sweeney, L.: K-Anonymity: A Model for Protecting Privacy. International Journal on Uncertainty, fuzziness and Knowledge-based Systems 10(5), 557–570 (2002)
4. Xu, S., Zhang, J., Han, D., Wang, J.: Singular Value Decomposition Based Data Distortion Strategy for Privacy Distortion. Knowledge and Information System 10(3), 383–397 (2006)
5. Vaidya, J., Clifton, C.: Privacy Preserving K-means Clustering over Vertically Portioned Data. In: 9th ACM SIGKDD International Conference on Knowledge Discovery and Data Mining, pp. 206–215. ACM Press, Washington (2003)
6. Vaidya, J., Yu, H., Jiang, X.: Privacy Preserving SVM Classification. Knowledge and Information Systems 14, 161–178 (2007)
7. Chen, K., Liu, L.: A Random Rotation Perturbation Approach to Privacy Data Classification. In: 2005 IEEE International Conference on Data Mining (ICDM), pp. 589–592. IEEE Press, Houston (2005)
8. Kargupta, H., Datta, S., Wang, Q., Sivakumar, K.: Random Data Perturbation Techniques and Privacy Preserving Data Mining. Knowledge and Information Systems 7, 387–414 (2005)
9. Huang, Z., Du, W., Chen, B.: Deriving Private Information from Randomized Data. In: 2005 ACM SIGMOD International Conference on Management of Data, pp. 37–48. ACM Press, Baltimore (2005)
10. Tenenbaum, J.B., de Silva, V., Langford, J.C.: A Global Geometric Framework for Nonlinear Dimensionality Reduction. Science 290, 2319–2323 (2000)
11. Roweis, S.T., Saul, L.K.: Nonlinear Dimensionality Reduction by Locally Linear Embedding. Science 290, 2323–2326 (2000)
12. de Silva, V., Tenenbaum, J.B.: Global Versus Local Methods in Nonlinear Dimensionality Reduction. In: Advances in Neural Information Processing Systems 15 (NIPS 2002), pp. 705–712. MIT Press, Cambridge (2003)
13. Berger, M., Gostiaux, B.: Differential Geometry: Manifolds, Curves and Surfaces, GTM115. Springer, Heidelberg (1974)

Bayesian Multi-topic Microarray Analysis with Hyperparameter Reestimation

Tomonari Masada, Tsuyoshi Hamada, Yuichiro Shibata, and Kiyoshi Oguri

Nagasaki University, Bunkyo-machi 1-14,
852-8521 Nagasaki, Japan
masada@cis.nagasaki-u.ac.jp

Abstract. This paper provides a new method for multi-topic Bayesian analysis for microarray data. Our method achieves a further maximization of lower bounds in a marginalized variational Bayesian inference (MVB) for Latent Process Decomposition (LPD), which is an effective probabilistic model for microarray data. In our method, hyperparameters in LPD are updated by empirical Bayes point estimation. The experiments based on microarray data of realistically large size show efficiency of our hyperparameter reestimation technique.

1 Introduction

Latent Dirichlet allocation (LDA) [4], an epoch-making Bayesian multi-topic analysis method, finds its application in various research fields including natural language processing, information retrieval and image analysis [2][5][6][13][14]. We can also apply an LDA-like Bayesian multi-topic analysis to microarray data, where we regard samples as documents and genes as words. However, microarray data are given as a real matrix, not as a non-negative integer matrix. Therefore, researchers apply LDA after introducing Gaussian distributions in place of word multinomial distributions and provide an efficient probabilistic model, *Latent Process Decomposition (LPD)* [9], where topics in LDA are called *processes*.

As we can find Dirichlet prior distributions for word multinomials in LDA, we can find prior distributions for Gaussian distributions in LPD. To be precise, Gaussian priors are prepared for mean parameters, and Gamma priors are for precision parameters. However, as far as we know, there are still no reports on how we can reestimate *hyperparameters*, i.e., parameters of these prior distributions, and there are also no reports on whether we can improve microarray analysis by using hyperparameter reestimation. Therefore, in this paper, we provide a hyperparameter reestimation technique for LPD and show the results of experiments using microarray data of realistically large size.

Our method is based on a marginalized variational Bayesian inference (MVB) proposed by Ying et al. [15]. Marginalized variational Bayesian inference, alternatively

R. Huang et al. (Eds.): ADMA 2009, LNAI 5678, pp. 253–264, 2009.

called *collapsed* variational Bayesian inference [12], theoretically achieves better lower bounds than conventional variational Bayesian inferences [4][9]. In this paper, we propose a method for maximizing lower bounds further by reestimating hyperparameters, where we use empirical Bayes point estimates [4] as new values for hyperparameters. We denote our method by *MVB+*.

The experiments presented in [15] are only based on microarray data of small size. Therefore, we use microarray data of realistically large size, where the number of genes (features) ranges from 3,000 to 18,000. Our experiments using large microarray data will show that hyperparameter reestimation can achieve better results in lower bound maximization and also in sample clustering.

The rest of the paper is organized as follows. Section 2 describes MVB for LPD. In Section 3, we provide the details of MVB+. Section 4 presents the results of our experiments. Section 5 concludes the paper with future works.

2 Latent Process Decomposition (LPD)

Latent Process Decomposition (LPD) [9] can be regarded as a latent Dirichlet allocation (LDA) [4] re-designed for microarray data. LPD and LDA share a special feature, *topic multiplicity*. That is, both in LDA and in LPD, each document (sample) is modeled as a mixture of multiple topics (processes). With respect to this point, LPD and LDA are completely the same.

However, LPD is different from LDA with respect to observed data generation, because microarray data are always given as a matrix of real values. Therefore, LPD uses Gaussian distributions in place of word multinomial distributions in LDA. A generative description of LPD can be given as follows:

- Draw a Gaussian distribution $N(x;\mu_{gk},\lambda_{gk})$ for each pair of gene g and process k from prior distributions. To be precise, a mean parameter μ_{gk} is drawn from a Gaussian prior $N(\mu;\mu_0,\lambda_0)$ and a precision parameter λ_{gk} is drawn from a Gamma prior $\mathrm{Gam}(\lambda;a_0,b_0)$.
- In LDA, this part corresponds to a determination of word probability with respect to a specific topic by drawing a topic-wise word multinomial distribution from a corpus-wide Dirichlet prior.
- Draw a multinomial distribution $\mathrm{Mult}(z;\theta_d)$ for each sample d from a symmetric Dirichlet prior distribution $\mathrm{Dir}(\theta;\alpha)$.
- This part is completely the same with LDA by identifying samples in LPD with documents in LDA, and processes in LPD with topics in LDA.
- For an occurrence of gene g in sample d, draw a process z_{dg} from $\mathrm{Mult}(z;\theta_d)$, and then draw an observed real value x_{dg} from $N(x;\mu_{gz_{dg}},\lambda_{gz_{dg}})$.
- This part is similar to a topic drawing from $\mathrm{Mult}(z;\theta_d)$ followed by a word drawing from the word multinomial corresponding to the drawn topic in LDA. However, in case of LPD, observed data are real, and each gene occurs exactly once in each sample.

While LPD is described as a generative model for microarray data, it can be easily applied to other real matrix data.

The generative description shown above leads to the full joint distribution:

$$\begin{aligned}
&p(\mathbf{x},\mathbf{z},\theta,\mu,\lambda|\alpha,\mu_0,\lambda_0,a_0,b_0)\\
&=\prod_d p(\theta_d|\alpha)\prod_{g,k} p(\mu_{gk}|\mu_0,\lambda_0)\prod_{g,k} p(\lambda_{gk}|a_0,b_0)\prod_{d,g} p(z_{dg}|\theta_d)p(x_{dg}|\mu_{gz_{dg}},\lambda_{gz_{dg}})\\
&=\prod_d \frac{\Gamma(K\alpha)}{\Gamma(\alpha)^K}\prod_k \theta_{dk}^{\,n_{dk}+\alpha-1}\cdot\prod_{g,k}\sqrt{\frac{\lambda_0}{2\pi}}\exp\left\{-\frac{\lambda_0(\mu_{gk}-\mu_0)^2}{2}\right\}\\
&\cdot\prod_{g,k}\frac{b_0^{a_0}}{\Gamma(a_0)}\lambda_{gk}^{\,a_0-1}e^{-b_0\lambda_{gk}}\cdot\prod_{d,g}\sqrt{\frac{\lambda_{gz_{dg}}}{2\pi}}\exp\left\{-\frac{\lambda_{gz_{dg}}(x_{dg}-\mu_{gz_{dg}})^2}{2}\right\}
\end{aligned}\tag{1}$$

In this paper, we adopt a *marginalized variational Bayesian inference (MVB)* proposed in [15] for LPD and introduce hyperparameter reestimation to MVB, because MVB achieves better inference results than without marginalization as shown in [15]. An outline of MVB for LPD is given below.

We first marginalize process multinomial parameters θ_{dk} in Eq. (1) as follows:

$$\begin{aligned}
&p(\mathbf{x},\mathbf{z},\mu,\lambda|\alpha,\mu_0,\lambda_0,a_0,b_0)=\int p(\mathbf{x},\mathbf{z},\theta,\mu,\lambda|\alpha,\mu_0,\lambda_0,a_0,b_0)d\theta\\
&=\prod_d \frac{\Gamma(K\alpha)}{\Gamma(\alpha)^K}\frac{\prod_k\Gamma(n_{dk}+\alpha)}{\Gamma(\sum_k n_{dk}+K\alpha)}\cdot\prod_{g,k}\sqrt{\frac{\lambda_0}{2\pi}}\exp\left\{-\frac{\lambda_0(\mu_{gk}-\mu_0)^2}{2}\right\}\\
&\cdot\prod_{g,k}\frac{b_0^{a_0}}{\Gamma(a_0)}\lambda_{gk}^{\,a_0-1}e^{-b_0\lambda_{gk}}\cdot\prod_{d,g}\sqrt{\frac{\lambda_{gz_{dg}}}{2\pi}}\exp\left\{-\frac{\lambda_{gz_{dg}}(x_{dg}-\mu_{gz_{dg}})^2}{2}\right\},
\end{aligned}\tag{2}$$

where n_{dk} denotes the number of genes whose observed data in sample d are generated from a Gaussian distribution corresponding to process k. Our aim is to obtain a parameter values maximizing the log likelihood:

$$\log p(\mathbf{x}|\alpha,\mu_0,\lambda_0,a_0,b_0)=\log\int\sum_{\mathbf{z}} p(\mathbf{x},\mathbf{z},\mu,\lambda|\alpha,\mu_0,\lambda_0,a_0,b_0)d\mu d\lambda \tag{3}$$

The maximization of this log likelihood is intractable. Therefore, we use a variational method and introduce an approximated posterior $q(\mathbf{z},\mu,\lambda)$ to obtain a lower bound of the log likelihood via Jensen's inequality:

$$\log p(\mathbf{x}|\alpha,\mu_0,\lambda_0,a_0,b_0)\geq\log\int\sum_{\mathbf{z}} q(\mathbf{z},\mu,\lambda)\log\frac{p(\mathbf{x},\mathbf{z},\mu,\lambda|\alpha,\mu_0,\lambda_0,a_0,b_0)}{q(\mathbf{z},\mu,\lambda)}d\mu d\lambda \tag{4}$$

Further, we assume that $q(\mathbf{z},\mu,\lambda)$ can be factorized as $q(\mathbf{z})q(\mu)q(\lambda)$. For each of $q(\mathbf{z})$, $q(\mu)$, and $q(\lambda)$, we choose a multinomial distribution defined over processes $\mathrm{Mult}(z_{dg};\gamma_{dg})$, a Gaussian distribution $N(\mu_{gk};m_{gk},l_{gk})$, and a Gamma distribution $\mathrm{Gam}(\lambda_{gk};a_{gk},b_{gk})$, respectively. Among the parameters of the approximated posterior, multinomial parameters γ_{dgk} of $\mathrm{Mult}(z_{dg};\gamma_{dg})$ can be interpreted as showing how closely gene g in sample d relates to process k. This interpretation is important in applications.

After some calculations, we obtain a lower bound L of the log likelihood:

$$\begin{aligned}
L = &\sum_{\mathbf{z}} \prod_{d,g,k} \gamma_{dgk}{}^{\delta_{dgk}} \sum_{d,k} \log\Gamma(n_{dk}+\beta) - \frac{\lambda_0}{2}\sum_{g,k}\left\{\frac{1}{l_{gk}} + \left(m_{gk}-\mu_0\right)^2\right\} \\
&+\left(a_0 - a_{gk}\right)\sum_{g,k}\psi(a_{gk}) - b_0\sum_{g,k}\frac{a_{gk}}{b_{gk}} + \frac{1}{2}\sum_{d,g,k}\gamma_{dgk}\left\{\psi(a_{gk}) - \log b_{gk}\right\} \\
&-\frac{1}{2}\sum_{d,g,k}\gamma_{dgk}\frac{a_{gk}}{b_{gk}}\left\{\frac{1}{l_{gk}} + \left(x_{dg}-m_{gk}\right)^2\right\} - \sum_{d,g,k}\gamma_{dgk}\log\gamma_{dgk} - \frac{1}{2}\sum_{g,k}\log l_{gk} \\
&+\sum_{g,k}\log\Gamma(a_{gk}) - a_0\sum_{g,k}\log b_{gk} + \sum_{g,k} a_{gk} - \frac{DG\log 2\pi}{2} - DK\log\Gamma(a) \\
&+D\left\{\log\Gamma(K\alpha) - \log\Gamma(G+K\alpha)\right\} + GK\left\{\frac{\log\lambda_0 + 1}{2} + a_0\log b_0 - \log\Gamma(\alpha)\right\}
\end{aligned} \tag{5}$$

where δ_{dgk} is 1 when $z_{dg}=k$ and 0 otherwise, and $\psi(\cdot)$ means digamma function. Our aim is now to maximize L. The details of update formula derivation are referred to [15]. Here we only include the resulting formulas.

$$\begin{aligned}
\gamma_{dgk} \leftarrow &\log\left(\alpha + \sum_{d',g',k'}\gamma_{d'g'k'} - \gamma_{dgk}\right) - \frac{\sum_{d',g',k'}\gamma_{d'g'k'}(1-\gamma_{d'g'k'}) - \gamma_{dgk}(1-\gamma_{dgk})}{2(\alpha + \sum_{d',g',k'}\gamma_{d'g'k'} - \gamma_{dgk})^2} \\
&+\frac{1}{2}\left\{\psi(a_{gk}) - \log b_{gk}\right\} - \frac{a_{gk}}{2b_{gk}}\left\{\frac{1}{l_{gk}} + \left(x_{dg} - m_{gk}\right)^2\right\}
\end{aligned} \tag{6}$$

$$l_{gk} \leftarrow \frac{a_{gk}\sum_d \gamma_{dgk}}{b_{gk}} + \lambda_0 \tag{7}$$

$$m_{gk} \leftarrow \frac{1}{l_{gk}}\left\{\frac{a_{gk}\sum_d \gamma_{dgk}x_{dg}}{b_{gk}} + \mu_0\lambda_0\right\} \tag{8}$$

$$a_{gk} \leftarrow a_0 + \frac{1}{2}\sum_d \gamma_{dgk} \tag{9}$$

$$b_{gk} \leftarrow \frac{1}{2}\sum_d \gamma_{dgk}\left\{\frac{1}{l_{gk}} + \left(x_{dg} - m_{gk}\right)^2\right\} + b_0 \tag{10}$$

The lower bound L in Eq. (5) will also be utilized to estimate inference quality in the experiments presented in Section 4. However, the first term in Eq. (5) is intractable. [15] gives the following approximation:

$$\sum_{\mathbf{z}} \prod_{d,g,k} \gamma_{dgk}{}^{\delta_{dgk}} \sum_{d,k} \log\Gamma(n_{dk}+\alpha)$$
$$\approx DK\log\Gamma(\alpha) + \sum_{d,g,k} \gamma_{dgk} \left\{ \log\left(\alpha + \sum_{g'>g} \gamma_{dg'k}\right) - \frac{\sum_{g'>g} \gamma_{dg'k}(1-\gamma_{dg'k})}{2(\alpha + \sum_{g'>g} \gamma_{dg'k})^2} \right\} \tag{11}$$

3 MVB+

As shown in Eq. (5), , the lower bound L includes the following hyperparameters: α, μ_0, λ_0, a_0, b_0. We can maximize L by taking derivatives of L with respect to these hyperparameters. This idea is not pursued in previous researches [9][15]. Therefore, we show how to reestimate hyperparameters in this section. We denote MVB for LPD accompanied with hyperparameter reestimation by *MVB+*.

First, we consider the reestimation of α. However, the derivative of L with respect to α suggests a difficulty in obtaining an update efficient in execution time even after introducing an approximation in Eq. (11). Further, a marginalized variational Bayesian inference for LDA [12] uses a fixed value for α. Therefore, we do not update α in MVB+. We leave as an open problem deriving an efficient update for α.

Second, by taking the derivative of L with respect to μ_0, we can obtain a simple update formula:

$$\mu_0 \leftarrow \frac{\sum_{g,k} m_{gk}}{GK} \tag{12}$$

Third, the derivative of L with respect to λ_0 is

$$\frac{\partial L}{\partial \lambda_0} = -\frac{1}{2}\sum_{g,k}\left\{\frac{1}{l_{gk}} + \left(m_{gk} - \mu_0\right)^2\right\} + \frac{GK}{2\lambda_0} \tag{13}$$

While we can obtain an update $\lambda_0 \leftarrow GK \Big/ \sum_{g,k}\{1/l_{gk} + (m_{gk}-\mu_0)^2\}$, preliminary experiments reveal that this update often results in unstable numerical computations. Therefore, we do not reestimate λ_0.

Finally, we take the derivatives of L with respect to the rest two hyperparameters, a_0 and b_0. For b_0, a simple update formula

$$b_0 \leftarrow \frac{GKa_0}{\sum_g \sum_k a_{gk}/b_{gk}} \tag{14}$$

follows. However, some trick is required for a_0, because the derivative leads to:

$$\psi(a_0) \leftarrow \frac{\sum_g \sum_k \{\psi(a_{gk}) - \log b_{gk}\}}{GK} + \log b_0 \tag{15}$$

and digamma function should be inverted. We use a method in [8] for digamma function inversion. We reproduce the method for evaluating $\psi^{-1}(y)$ here.

1. If $y \geq -2.22$ then $x \leftarrow \exp(y) + 0.5$, else $x \leftarrow -1/(y - \psi(1))$.
2. Repeat the following until convergence: $x \leftarrow x - (\psi(x) - y)/\psi'(x)$.

We denote trigamma function by $\psi'(\cdot)$.

It should be noted that we can use the same formula Eq. (5) for computing L even when we reestimate hyperparameters. Therefore, we can compare inference quality of MVB+ with that of MVB by using L.

4 Experiments

4.1 Comparison Strategy

In this section, we present the results of our experiments to reveal how our hyperparameter reestimation works. We compare MVB+ with MVB by (i) lower bound L (see Eq. (5)) and also by (ii) quality of sample clusters. In comparing MVB+ with MVB by L, larger values are better, because our task is to maximize L as far as possible. On the other hand, when we compare MVB+ with MVB by sample clustering, we take the following strategy.

A clustering of samples is induced by determining a process k satisfying

$$k = \arg\max_{k'} \sum_{g} \gamma_{dgk'} \tag{16}$$

for each sample d, where γ_{dgk} for all d, g, k are obtained as a result of a sufficient number of update iterations in MVB+ or in MVB. We evaluate the quality of a clustering of samples by, first, computing precision and recall for each cluster. Then, for each pair of precision P and recall R, we compute an F-score as their harmonic mean, i.e., $F = 1 / (1/P + 1/R)$. We use the precision, recall, and F-score averaged over all clusters as an evaluation measure for each clustering results.

4.2 Datasets

Previous research [15] only use datasets of small size, where the number of genes ranges from 500 to 1,000. Therefore, in this paper, we use datasets of realistically large size, available at [1] and [16], whose specification is given in Table 1.

"Leukemia" dataset from [1], referred as LK in this paper, provides three labels ALL/MLL/AML as a prefix of each sample name. When we use these three labels as true cluster labels, both MVB+ and MVB give quite poor performance. Therefore, we compare MVB+ with MVB based on the binary clustering task (ALL,MLL)/AML after identifying ALL with MLL.[1]

[1] ALL: acute lymphoblastic leukemias, AML: acute myelogenous leukemias, MLL: lymphoblastic leukemias with mixed-lineage leukemia gene translocations [1].

For "Five types of breast cancer" dataset from [16], referred as D1 in this paper, we find a description telling that meaningfull classifier should try to distinguish labels A from B in [16]. However, LPD seems quite week in this task, because both MVB+ and MVB rarely give a cluster where the number of B samples is larger than that of A samples. In other words, almost all resulting clusters are dominated by A samples. Therefore, D1 dataset is used only for comparison based on lower bounds.

For "Three types of bladder cancer" dataset from [16], referred as D2, the prepared three labels T1/T2+/Ta are used as is.

"Healthy tissues" dataset from [16], referred as D3, has too many true cluster labels (35 labels) for only 103 samples. We faced difficulty in obtaining clustering results worthy to evaluate by precision, recall, and *F*-score. Therefore, D3 dataset is also used only for comparison based on lower bounds.

Table 1. Dataset specification

Dataset name (abbreviation)	# of samples	# of genes
Leukemia (LK) [1]	72	12582
Five types of breast cancer (D1) [16]	286	17816
Three types of bladder cancer (D2) [16]	40	3036
Healthy tissues (D3) [16]	103	10383

4.3 Implementation

We implemented MVB+ and MVB in C language and compile by `gcc -O3`. Inference computations are executed on a PC equipped with Intel Core2 Quad CPU Q9550 @ 2.83 GHz and 8GBytes memory. Digamma and trigamma functions are estimated from asymptotic series [10]. We used the wine dataset [3] to prove the soundness of our implementation by comparing lower bounds with Figure 1 of [15].

As an initialization for both MVB+ and MVB, we choose real random numbers for γ_{dgk} satisfying $\sum_k \gamma_{dgk} = 1$ and then initialize parameters, m_{gk}, l_{gk}, a_{gk}, and b_{gk}, based on randomly initialized γ_{dgk}. Hyperparameter values are set as follows: $\alpha = 1$, $\mu_0 = 0$, $\lambda_0 = 1$, $a_0 = 20$, and $b_0 = 20$. These values are also regarded as initial values for hyperparameter reestimation in MVB+. Observed real values x_{dg} in each microarray data are normalized in the same manner with [15].

The running time of MVB+ and MVB is proportional to the product of the following four numbers: the number of samples, the number of genes, the number of processes, and the number of iterations. For example, in case of dataset D1, MVB+ requires about 174 minutes for 500 iterations when the number of processes is 10. We found that MVB+ increases running time by at most 10% when compared with MVB.

4.4 Results

Figure 1 presents lower bounds obtained for LK (top left panel), D1 (top right), D2 (bottom left), and D3 (bottom right). The horizontal axis shows the number of iterations, and the vertical axis shows the lower bounds. We tested the integers from 2 to 10 as the number of processes. 500 iterations of updates are enough for most cases. However, lower bounds for 1,000 iterations are shown only for LK dataset, because convergence is slow. As Figure 1 shows, lower bounds obtained by MVB+ (solid lines) are larger than that by MVB (dashed lines). We can conclude that lower bound improvement is achieved.

Previous research [15] discusses that we can select an appropriate number of processes by comparing lower bounds obtained for different numbers of processes. Figure 2 shows the average and the standard deviation of 10 lower bounds obtained by MVB+ (solid line) and MVB (dashed line) staring from 10 different initializations for each number of processes ranging from 2 to 10. The horizontal axis shows the number of processes. Lower bounds are recorded when a change no more than 1.0^{-6} can be observed. It seems difficult to find a unique significant peak in all cases. We may need other methods, e.g. a nonparametric Bayesian approach [11], to choose an appropriate number of processes for large datasets.

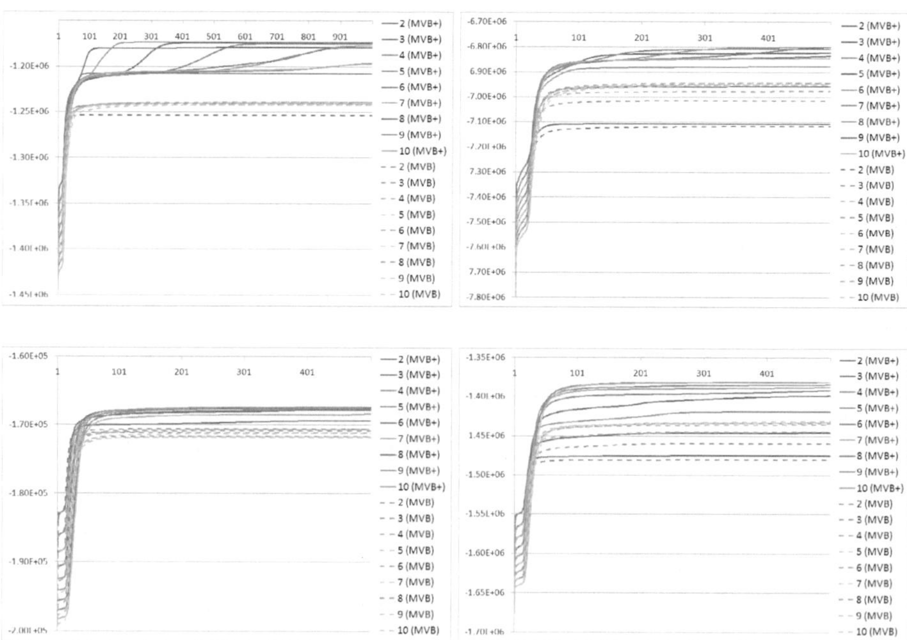

Fig. 1. Lower bounds obtained by MVB+ (solid lines) and MVB (dashed lines). Each graph gives the average of lower bounds obtained by inferences starting from 10 different random initializations. The number of processes ranges from 2 to 10. The datasets are LK (top left panel), D1 (top right), D2 (bottom left), and D3 (bottom left). MVB+ provides better results.

Table 2 provides precisions, recalls, and F-scores averaged over inferences starting from 100 different initializations for LK and D2 datasets. Actually, we discard 10 least frequently occurred clustering results among 100, because extraordinarily good or bad clusters are occasionally obtained for both MVB+ and MVB. We assume that an appropriate number of clusters is chosen beforehand in accordance with the true number of clusters. Namely, we set the number of processes to two and three for LK and D2 datasets, respectively. As Table 2 shows, MVB+ realizes better clustering than MVB. Since 100 clustering results given by MVB for LK dataset are the same, the standard deviation is zero.

Table 2. Precisions, recalls, and F-scores averaged over inferences starting from 100 different initializations. Standard deviations are also presented.

dataset	method	precision	recall	*F*-score
LK	MVB+	0.934±0.007	0.931±0.010	0.932±0.009
	MVB	0.930±0.000	0.924±0.000	0.927±0.000
D2	MVB+	0.837±0.038	0.822±0.032	0.829±0.033
	MVB	0.779±0.084	0.751±0.069	0.763±0.071

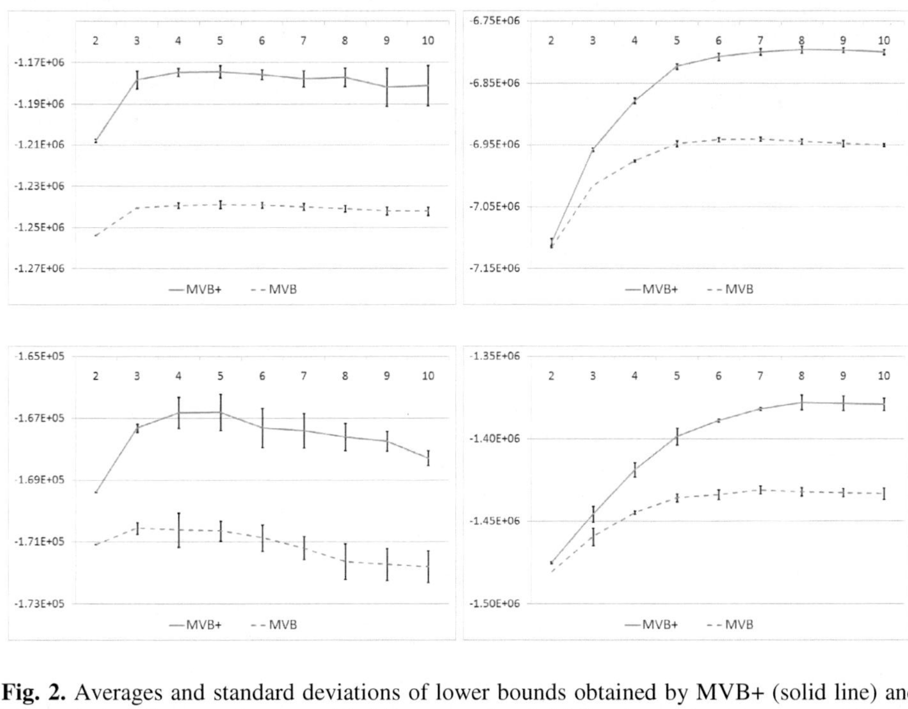

Fig. 2. Averages and standard deviations of lower bounds obtained by MVB+ (solid line) and MVB (dashed line) starting from 10 different random initializations. The datasets are LK (top left panel), D1 (top right), D2 (bottom left), and D3 (bottom left). MVB+ always provides better lower bounds than MVB for all numbers of processes. However, it is difficult to find a clear peak among different numbers of processes.

Our experiments also revealed an important difference between MVB+ and MVB. Figure 3 includes two images visualizing posterior parameters γ_{dgk} for LK dataset when the number of processes is two. These images are constructed as follows. We first select an inference result arbitrarily among a lot of results for each of MVB+ and MVB. Then, we choose the larger one among γ_{dg1} and γ_{dg2} for each pair of gene g and sample d. That is, we assign one among the two processes to each gene/sample pair. The assignments of different processes are shown by different colors. Rows and columns of images in Figure 3 correspond to genes and samples, respectively. Namely, each pixel in these images corresponds to a gene/sample pair.

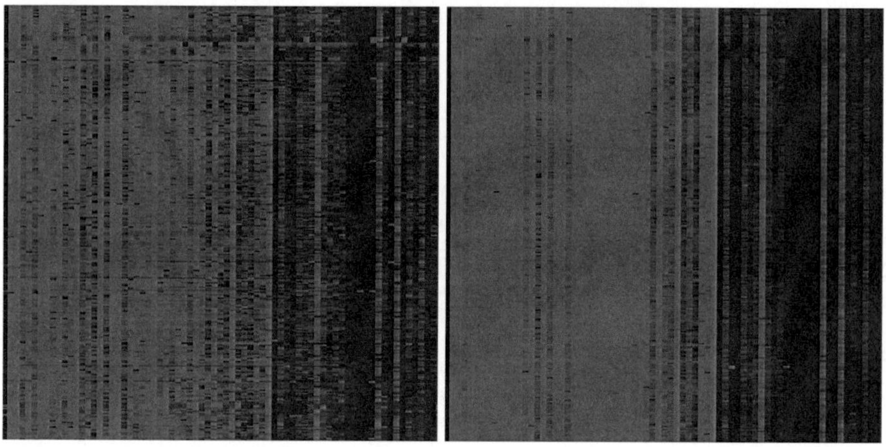

Fig. 3. Visualization of γ_{dgk} for the first 288 genes in LK dataset. The width of images is scaled to 400% for visibility. Each row corresponds to a gene, and each column to a sample. For each gene/sample pair, we choose one process based on γ_{dgk}. Different processes are shown by different colors. MVB+ (left panel) preserves diversity among genes better than MVB (right).

When we compare the image for MVB+ (left) with that for MVB (right), the former shows row-wise diversity in process assignments. Intuitively speaking, the visualization for MVB+ is more "noisy" than MVB. Therefore, it can be concluded that MVB+ preserves diversity among genes as diversity among process assignments. In contrast, MVB is likely to give almost the same process assignments for all genes. In fact, the rows in the visualization for MVB looks quite similar to each other. While we arbitrarily select an inference result for each of MVB+ and MVB, other results also lead to the same conclusion.

Since we can use the set of γ_{dgk} for a fixed g as a feature vector of gene g in clustering or classifying genes, this difference between MVB+ and MVB will show importance in such applications.

5 Conclusion

In this paper, we provide a hyperparamter reestimation technique *MVB+* for marginalized variational Bayesian inference of LPD. MVB+ achieves further maximization of lower bounds. Also for sample clustering, MVB+ gives better results. Further, MVB+ can preserve diversity among genes as diversity among process assignments.

Our experiments also show that it is difficult to guess an appropriate number of processes based on lower bounds for microarray data of realistically large size. It is a future work to devise a method for determining an appropriate number of processes. Further, with respect to computational efficiency, MVB+ consist of complicated numerical operations. Especially, evaluating digamma and trigamma functions from asymptotic series is time-consuming. Therefore, it is also an important future work to accelerate inferences, e.g. by using GPGPU [7].

References

1. Armstrong, S.A., Staunton, J.E., Silverman, L.B., Pieters, R., den Boer, M.L., Minden, M.D., Sallan, S.E., Lander, E.S., Golub, T.R., Korsemeyer, S.J.: MLL Translocations Specify a Distinct Gene Expression Profile that Distinguishes a Unique Leukemia. Nature Genetics 30, 41–47 (2002)
2. Arora, R., Ravindran, B.: Latent Dirichlet Allocation and Singular Value Decomposition Based Multi-document Summarization. In: 25th IEEE International Conference on Data Mining, pp. 713–718 (2008)
3. Blake, C.L., Newman, D.J., Hettich, S., Merz, C.J.: UCI Repository of Machine Learning Databases (1998), `http://archive.ics.uci.edu/ml/`
4. Blei, D.M., Ng, A.Y., Jordan, M.I.: Latent Dirichlet Allocation. Journal of Machine Learning Research 3, 993–1022 (2003)
5. Larlus, D., Jurie, F.: Latent Mixture Vocabularies for Object Categorization and Segmentation. Image and Vision Computing 27(5), 523–534 (2009)
6. Lienou, M., Maitre, H., Datcu, M.: Semantic Annotation of Large Satellite Images Using Latent Dirichlet Allocation. In: ESA-EUSC Workshop (2008)
7. Masada, T., Hamada, T., Shibata, Y., Oguri, K.: Accelerating Collapsed Variational Bayesian Inference for Latent Dirichlet Allocation with Nvidia CUDA Compatible Devices. In: International Conference on Industrial, Engineering & Other Applications of Applied Intelligent Systems (to appear, 2009)
8. Minka, T.: Estimating a Dirichlet Distribution (2000), `http://research.microsoft.com/en-us/um/people/minka/`
9. Rogers, S., Girolami, M., Campbell, C., Breitling, R.: The Latent Process Decomposition of cDNA Microarray Datasets. IEEE/ACM Trans. on Computational Biology and Bioinformatics 2, 143–156 (2005)
10. Smith, D.M.: Multiple-Precision Gamma Function and Related Functions. Transactions on Mathematical Software 27, 377–387 (2001)
11. Teh, Y.W., Jordan, M.I., Beal, M.J., Blei, D.M.: Hierarchical Dirichlet Processes. Journal of the American Statistical Association 101(476), 1566–1581 (2006)

12. Teh, Y.W., Newman, D., Welling, M.: A Collapsed Variational Bayesian Inference Algorithm for Latent Dirichlet Allocation. Advances in Neural Information Processing Systems 19, 1378–1385 (2006)
13. Wang, X.G., Grimson, E.: Spatial Latent Dirichlet Allocation. Advances in Neural Information Processing Systems 20, 1577–1584 (2008)
14. Xing, D.S., Girolami, M.: Employing Latent Dirichlet Allocation for Fraud Detection in Telecommunications. Pattern Recognition Letters 28, 1727–1734 (2007)
15. Ying, Y.M., Li, P., Campbell, C.: A Marginalized Variational Bayesian Approach to the Analysis of Array Data. BMC Proceedings 2(suppl. 4), S7 (2008)
16. Zinovyev, A.: http://www.ihes.fr/~zinovyev/princmanif2006/

Discovery of Correlated Sequential Subgraphs from a Sequence of Graphs

Tomonobu Ozaki[1] and Takenao Ohkawa[2]

[1] Organization of Advanced Science and Technology, Kobe University
[2] Graduate School of Engineering, Kobe University
1-1 Rokkodai, Nada, Kobe, 657-8501, Japan
{tozaki@cs.,ohkawa@}kobe-u.ac.jp

Abstract. *Dynamic graphs* or a sequence of graphs attract much attention recently. In this paper, as a first step towards finding significant patterns hidden in dynamic graphs, we consider the problem of mining successive sequence of subgraphs which appear frequently in a long sequence of graphs. In addition, to exclude insignificant patterns, we take into account the mutual dependency measured by θ-correlation coefficient among the components in patterns. An algorithm named CorSSS, which utilizes the generality ordering of patterns effectively, is developed for enumerating all frequent and correlated patterns. The effectiveness of CorSSS is confirmed through the experiments using real datasets.

Keywords: Graph mining, Correlation mining, Graph sequences.

1 Introduction

Graph-structured data is becoming increasingly abundant in many application domains such as bioinformatics, social network analysis, and so on. While several entity relations can be represented in graphs, all relations are not always static. For example, the following relations and structures vary over time: (1)daily communication networks by sending and receiving emails, (2)the link structures among web sites, and (3)biological networks such as metabolic pathways. The dynamic aspects in the above relations can be represented naturally in a sequence of graphs. Therefore, the discovery of some interesting dynamic aspect in a time series of graphs is recognized as one of important problems recently.

Motivated by this background, in this paper, we focus on a pattern mining problem of finding *frequently appearing successive sequences of subgraphs* in a long graph sequence. Because a sequence of subgraphs is very complex, it is easily imagined that huge number of patterns will be generated. To alleviate this problem, we take the concept of correlation mining into account. By employing θ-correlation coefficient[11] as a correlation measure, we accept only patterns whose components, *i.e.* subsequences of subgraphs, are correlated with each other. In addition, to relax and control the condition of this mutual dependency, a parameterized criterion on the correlation is proposed. To solve the problem of finding frequent and correlated sequential patterns efficiently, in this

R. Huang et al. (Eds.): ADMA 2009, LNAI 5678, pp. 265–276, 2009.

paper, we propose a novel algorithm named CorSSS. CorSSS reduces the search space effectively by using a tree-shaped data structure and by considering the generality ordering, *i.e.* a general-to-specific relationship, among patterns.

The rest of this paper is organized as follows. In section 2, some notations and definitions on sequences of graphs are introduced. The mutual dependency in a sequence of subgraphs is discussed in detail in section 3. Following the above preparation, our data mining problem is formally stated in section 4. An algorithm CorSSS for this problem will be proposed in this section. After mentioning related work in section 5, the experimental results are shown in section 6. Finally, in section 7, we conclude the paper and describe future work.

2 Sequences of Graphs

A *labeled graph* $g = (V_g, E_g, l_g)$ on a finite set of labels $\mathcal{L}$ consists of a vertex set V_g, an edge set $E_g \subseteq V_g \times V_g$, and a labeling function $l_g : V_g \cup E_g \rightarrow \mathcal{L}$ that assigns a label to each vertex and edge. A labeled graph $s = (V_s, E_s, l_s)$ is called a *subgraph* of other labeled graph $t = (V_t, E_t, l_t)$, denoted as $s \preceq t$, if there exists an injective function $f : V_s \rightarrow V_t$ such that $(1) \forall u \in V_s\ [l_s(u) = l_t(f(u))]$ and $(2) \forall (u, v) \in E_s\ [(f(u), f(v)) \in E_t \wedge l_s(u, v) = l_t(f(u), f(v))]$. If $s \preceq t$ holds, then t is said to be more specific than s. Hereafter, for the sake of simplicity, a labeled graph is referred to as a graph. A graph is also called a subgraph when it is treated as a part of pattern introduced below.

A *pattern* to be extracted is a successive sequence of n subgraphs $P = \langle p_1, p_2, \cdots, p_n \rangle$ where each p_i is a closed[16] and canonical subgraph[15,1]. While a closed subgraph is a maximally specific one among the set of subgraphs having the same support, a canonical subgraph is a representative among the set of subgraphs which are isomorphic to each other. The size of P, denoted as $|P|$, is defined as the number of subgraphs in P. The notation $P[i]$ is used to denote the i-th element in P, and $P[i, j] = \langle P[i], P[i+1], \cdots, P[j-1], P[j] \rangle$ $(1 \leq i < j \leq |P|)$ denotes a subsequence of P. Two subsequences $p_l(P) = P[1, |P|-1]$ and $p_r(P) = P[2, |P|]$ are called the *left and right parents* of P, respectively.

Given two patterns P and Q, if there exists an integer i $(1 \leq i \leq |Q| - |P| + 1)$ such that $\forall j = (1, \cdots, |P|)\ [\ P[j] \preceq Q[i+j-1]\ ]$, then Q is said to be more specific than P, and it is denoted as $P \sqsubseteq Q$. For any two patterns P and Q such that $P \sqsubseteq Q$, Q will be obtained from P by applying the following two kinds of specializations repeatedly.

1. Specialization by a new subgraph: Obtain a new pattern C' by adding a subgraph g to the head or tail of the current pattern $C = \langle c_1, c_2 \cdots, c_{|C|} \rangle$, *i.e.* $C' = \langle g, c_1, \cdots, c_{|C|} \rangle$ or $C' = \langle c_1, \cdots, c_{|C|}, g \rangle$.
2. Specialization by an existing subgraph: Obtain a new pattern $C' = \langle c_1, \cdots, c_{i-1}, c'_i, c_{i+1}, \cdots, c_{|C|} \rangle$ by replacing a subgraph c_i in the current pattern $C = \langle c_1, \cdots, c_i, \cdots, c_{|C|} \rangle$ with a more specific subgraph $c'_i (\succeq c_i)$.

Examples of patterns are shown in Fig. 1. In this figure, the size of P^3 is 4. The equations $P^1[2] = P^2[1] = P^3[2]$, $P^1 = P^3[1, 2]$ and $P^3 = p_l(P^4)$ hold. In

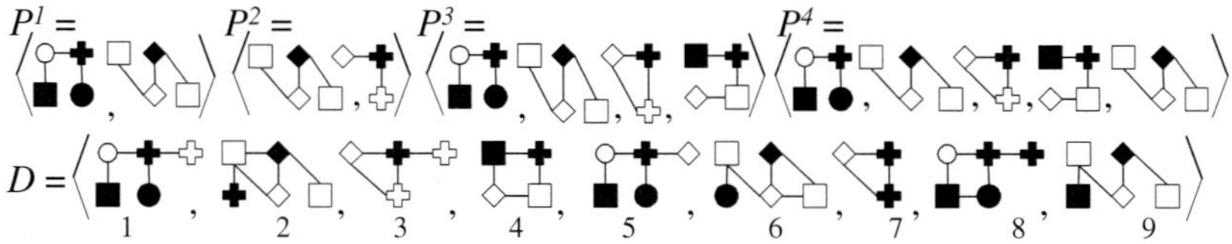

Fig. 1. Examples of Patterns in a Sequence of Graphs $\mathcal{D}$

addition, because of $P^2[1] \preceq P^3[2]$ and $P^2[2] \preceq P^3[3]$, $P^2 \sqsubseteq P^3$ holds. P^3 will be obtained from P^2 by (1)specializing $P^2[2]$ by adding an edge, (2)adding $P^3[1]$ to the head, and (3)adding $P^3[4]$ to the tail.

Given a long sequence of graphs $\mathcal{D}$, the *support* value $\sigma_\mathcal{D}(P)$ of a pattern $P = \langle p_1, \cdots, p_n \rangle$ in $\mathcal{D}$ is defined formally as follows:

$$\sigma_\mathcal{D}(P) = |O_\mathcal{D}(P)| \,/\, |\mathcal{D}| \quad \text{where}$$
$$O_\mathcal{D}(P) = \{j \mid 1 \leq j \leq |\mathcal{D}| - n + 1,\ P \sqsubseteq D[j, j+n-1]\}.$$

$O_\mathcal{D}(P)$ is a set of the leftmost positions of each occurrence of P in $\mathcal{D}$. For example in Fig. 1, $O_\mathcal{D}(P^1)$ is $\{1, 5, 8\}$ and thus $\sigma_\mathcal{D}(P^1)$ becomes 3/9.

3 Mutual Dependency of a Sequence of Subgraphs

To exclude uninteresting and trivial patterns, the relationship among the components, *i.e.* subsequences of subgraphs, in a pattern will be taken into account. We decide a pattern is interesting if its components are strongly correlated or tightly coupled with each other. On the contrary, patterns having redundant components from the aspect of correlation will be excluded. As a criterion to judge the mutual dependency among subsequences, a parameterized correlation criterion named (m, θ, k)-correlation is proposed where (1)$m \geq 1$ is a positive integer representing the size of subsequences, (2)$\theta (0 \leq \theta \leq 1)$ is the minimum threshold of θ-correlation coefficient[11] between two subsequences, and (3)$k \geq 1$ is a positive integer for the maximum number of *exceptions* per each subsequence.

First, we consider the components of a pattern. While it is natural to regard the subsequences as the components of a pattern, the appropriate size of the subsequences as a component depends on the application domains strongly and it is difficult to determine in advance. For example, a subgraph in a pattern, *i.e.* a subsequence whose size is 1, might be too short to be considered as a component. Thus, we parameterize the size of component to be considered. Given a positive integer $m \geq 1$, we regard that a pattern P $(|P| > m)$ is composed of a set $S(P, m) = \{P[i, i+m-1] \mid 1 \leq i \leq |P| - m + 1\}$ of all subsequences of length m in P. For example, $S(P^4, 2)$ in Fig. 1 becomes a set $\{\ P^4[1,2], P^4[2,3], P^4[3,4], P^4[4,5]\ \}$.

Next, the relationship between two components is considered. For two components $P_1 = P[i_1, j_1]$ and $P_2 = P[i_2, j_2]$ such that $i_2 - i_1 = d$, the *joint support* $\sigma_\mathcal{D}(P_1, P_2)$ and θ-correlation coefficient[11] $\phi_\mathcal{D}(P_1, P_2)$ are defined as follows:

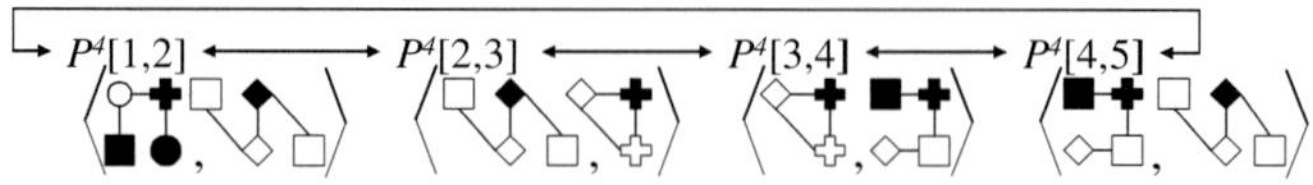

Fig. 2. A Dependency Graph and the (m, θ, k)-Correlation

$$\sigma_{\mathcal{D}}(P_1, P_2) = |O_{\mathcal{D}}(P_1, P_2)| \,/\, |\mathcal{D}| \quad \text{where}$$
$$O_{\mathcal{D}}(P_1, P_2) = \{x \in O_{\mathcal{D}}(P_2) \mid (x - d) \in O_{\mathcal{D}}(P_1)\}$$
$$\phi_{\mathcal{D}}(P_1, P_2) = \frac{\sigma_{\mathcal{D}}(P_1, P_2) - \sigma_{\mathcal{D}}(P_1)\sigma_{\mathcal{D}}(P_2)}{\sqrt{\sigma_{\mathcal{D}}(P_1)(1-\sigma_{\mathcal{D}}(P_1))\sigma_{\mathcal{D}}(P_2)(1-\sigma_{\mathcal{D}}(P_2))}}$$

A set $O_{\mathcal{D}}(P_1, P_2)$ corresponds to a set of the positions in $\mathcal{D}$ at which P_1 and P_2 appear together in considering the distance d between P_1 and P_2 in P.

The second parameter θ in (m, θ, k)-correlation is used to specify the minimum value of the correlation coefficient to judge there is a strong relationship between two components. Given a component $P_i \in S(P, m)$, a set of *uncorrelated* component in $S(P, m)$ with respect to the minimum correlation threshold $\theta (0 \leq \theta \leq 1)$ is defined as $UC(P_i, m, \theta) = \{P_j \in S(P, m) | P_i \neq P_j, \phi_{\mathcal{D}}(P_i, P_j) < \theta\}$. Each element in $UC(P_i, m, \theta)$ can be regarded as an *exception* in the mutual dependency. By the definition, the larger θ makes the larger $UC(P_i, m, \theta)$.

The strength of mutual dependency of a pattern will be assessed by using the third parameter. Given m, θ, and an positive integer $k \geq 1$, if the condition $\forall P_i \in S(P, m)[\ |UC(P_i, m, \theta)| < k\ ]$ holds, then P is said to be (m, θ, k)-correlated.

The (m, θ, k)-correlation can be interpreted by using the terminology in the graph theory (see Fig. 2). For a pattern P and three parameters m, θ and k, a *dependency graph* $G_P(m)$ is considered whose set of vertices is $S(P, m)$ and whose set of edges is $\{(P_i, P_j) \in S(P, m) \times S(P, m) \mid \phi_{\mathcal{D}}(P_i, P_j) \geq \theta\}$. That is to say, an edge is drawn from P_i to P_j if these two components are mutually correlated with each other. Then, a set $UC(P_i, m, \theta)$ corresponds to a set of vertices in $G_P(m)$ having no edge to P_i. Thus, the condition of (m, θ, k)-correlation, *i.e.* the size of all $UC(P_i, m, \theta)$ is less than k, can be judged whether $G_P(m)$ is a k-plex or not. If and only if $G_P(m)$ is a k-plex, then P must be (m, θ, k)-correlated. An extreme case of (m, θ, k)-correlation might be $(1, \theta, 1)$-correlation. A $(k = 1)$-plex is a clique, and a subsequence whose size is 1 means a subgraph. Therefore, in this case, all the pairs of subgraphs in a pattern have to be correlated. We show an example of dependency graph in Fig. 2. In this figure, $G_{P^4}(2)$ is a 2-plex and thus P^4 is $(2, \theta, 2)$-correlated.

4 Mining Correlated Sequential Subgraphs

In this section, after defining our data mining problem formally below, we propose the algorithms for solving the problem.

The ***FCSS*** **mining problem:** *Given a sequence of graphs* $\mathcal{D}$*, two numbers of the minimum support threshold* $\sigma (1/|\mathcal{D}| \leq \sigma \leq 1)$*) and the minimum correlation* *threshold* θ $(0 \leq \theta \leq 1)$*, and two positive integers of the size of components* $m \geq 1$ *and the maximum number of exceptions allowed* $k \geq 1$*, then the problem of "mining frequent correlated sequential subgraphs" (FCSS mining in short) is to find all the successive sequence of closed subgraphs* $P(|P| > m)$ *in* $\mathcal{D}$ *such that* P *is frequent* $(\sigma_{\mathcal{D}}(P) \geq \sigma)$ *and* (m, θ, k)*-correlated.*

4.1 A Naive Algorithm

Consider two patterns Q and $P = Q[i, i + |P| - 1]$ where Q is obtained by adding some subgraphs to the head and tail of P repeatedly. Since $S(P, m) \subseteq S(Q, m)$ holds, if P is not (m, θ, k)-correlated, then Q must not be (m, θ, k)-correlated. In addition, if P is not frequent, then Q is also not. By using these anti-monotone properties of the (m, θ, k)-correlation and the support value with respect to the "specialization by a new subgraph", a levelwise search algorithm shown in Fig. 3 can be constructed based on a sequential pattern mining algorithm GSP[10].

As an input, this algorithm takes a set $\mathcal{F}$ of frequent closed subgraphs. In each iteration of "levelwise_search", a candidate pattern P will be obtained by joining two frequent patterns s and t in F such that $p_r(s) = p_l(t)$ (line 4). In fact, P will be generated by adding the subgraph $t[|t|]$ to the tail of s. If a pattern P is not frequent, then it is ignored because no pattern containing P becomes frequent (line 5). Frequent patterns will be processed further according to their size. If the size of P is not enough, P is added to F' for the further expansion (line 6). Otherwise, the (m, θ, k)-correlation will be checked (line 7). Uncorrelated patterns will be ignored (line 8). This does not cause the incompleteness because of the anti-monotone property of (m, θ, k)-correlation with respect to the "specialization by a new subgraph".

In the check of (m, θ, k)-correlation, the support value of each component in a pattern is already calculated in the previous iterations in the levelwise search. In addition, other than for the combination of the first and last components, the joint support of components can be obtained from the left and right parents. These are strong advantages of employing the levelwise search.

4.2 Prefix Trees and Postfix Trees

While the *FCSS* mining problems can be solved by the naive levelwise algorithm shown in Fig. 3, it must be inefficient because the generality ordering *i.e.* a general-to-specific relationship, among patterns is not fully utilized. It is expected to get more efficient algorithm by introducing some operation on "specialization by an existing subgraph" to the naive algorithm. While the support value satisfies the anti-monotone property with respect to the "specialization by an existing subgraph", the (m, θ, k)-correlation does not unfortunately. Thus, we focus on the operation for utilizing the property of support value and propose to treat a set of patterns having the same parent as one unit by storing them into a tree-shaped data structure.

Given a set Pr, such that $\forall X, Y \in Pr[p_l(X) = p_l(Y)]$, of patterns having the same left parent in common, a tree-shaped data structure called a *prefix*

```
Algorithm Naive (𝓕, σ, θ, k, m)
levelwise_search(𝓕, σ, θ, k, m)
Procedure levelwise_search(F, σ, θ, k, m)
if F = ∅ then return
F' := ∅
for each s, t ∈ F such that p_r(s) = p_l(t)
    P := join(s, t) //create a new candidate
    if σ_D(P) < σ then continue
    if |P| ≤ m then F' := F' ∪ {P}
    else if P is (m, θ, k)-correlated then
        output P;  F' := F' ∪ {P}
levelwise_search(F', σ, θ, k, m)
```

Fig. 3. Pseudo Code of a Naive Algorithm

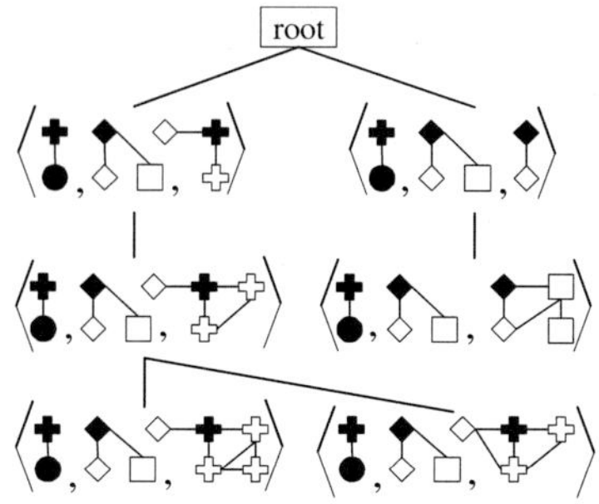

Fig. 4. An Example of Prefix Tree

tree is defined as $PrT(Pr) = (V_{Pr}, E_{Pr}, r)$ where V_{Pr} is a set of vertices, $E_{Pr} \subseteq V_{Pr} \times V_{Pr}$ is a set of parent-child relationships, and r is a root node. An example of prefix tree is depicted in Fig. 4. Each node $v \in V_{Pr} \setminus \{r\}$ contains a pattern. $gs(v)$ denotes the pattern in v. An edge $(u, v) \in E_{Pr}$ represents the generality relation between two patterns $U = gs(u)$ and $V = gs(v)$. It also corresponds to the generality relation between two subgraphs $U[\,|U|\,]$ and $V[\,|V|\,]$ because U and V have the same left parent in common. By definition, $U = gs(u) \sqsubseteq V = gs(v)$ holds if $(u, v) \in E_{Pr}$.

A *postfix tree* $PoT(Po) = (V_{Po}, E_{Po}, r)$ is defined in the similar manner for a set Po of patterns having the same right parent. The first subgraphs in each pattern are used to define a postfix tree. For any edge $(u, v) \in E_{Po}$ such that $U = gs(u)$ and $V = gs(v)$, $U[2, |U|] = V[2, |V|]$, $U[1] \preceq V[1]$ and $U \sqsubseteq V$ hold.

Note that, an edge (u, v) in prefix and postfix trees represents the "specialization by an existing subgraph". While the traversal from a parent to a child corresponds to the replacement of the *last* subgraph in a prefix tree, it means the specialization of the *first* subgraph in a postfix tree.

4.3 CorSSS : An Algorithm for *FCSS* Mining

In this subsection, we propose an algorithm CorSSS and explain it in detail.

The algorithm CorSSS for the problem of *FCSS* mining is shown in Fig. 5 and 6. This is an extension of the naive algorithm in Fig. 3 for considering the "specialization by an existing subgraph" by using prefix and postfix trees. CorSSS uses plural hash tables which store sets of prefix and postfix trees. The left and right parents of a pattern are used as the keys for hash tables. In the following discussion, for the sake of the simplicity, we do not distinguish a node u in a tree from a pattern $gs(u)$ in it.

First, in CorSSS, a prefix tree r generated from a set of frequent closed subgraphs $\mathcal{F}$ will be registered into an initial hash table L as a prefix tree. In addition, because the prefix and postfix trees are identical for a set of patterns

Algorithm CorSSS($\mathcal{F}$, σ, θ, k, m)
1: $r :=$ root of a prefix tree constructed from $\mathcal{F}$
2: $L :=$ a hash table which maps patterns to the roots of prefix trees
3: $R :=$ a hash table which maps patterns to the roots of postfix trees
4: put r in L with key $\langle\ \rangle$ (an empty pattern)
5: put r in R with key $\langle\ \rangle$ (an empty pattern)
6: levelwise_search(L, R, σ, θ, k, m)
Procedure levelwise_search(L, R, σ, θ, k, m)
1: **if** L or R is empty **then return**
2: $NL :=$ a hash table which maps patterns to the roots of prefix trees
3: $NR :=$ a hash table which maps patterns to the roots of postfix trees
4: **for each** KEY in a set of keys in R
5: $T_r :=$ the root of a postfix tree in R to which KEY is mapped
6: $T_l :=$ the root of a prefix tree in L to which KEY is mapped
7: loopPost(T_r, T_r, NL, NR, σ, θ, k, m)
8: levelwise_search(NL, NR, σ, θ, k, m)

Fig. 5. Pseudo Code of CorSSS

of size 1, r is also put into a hash table R as a postfix tree. Then, the procedure "levelwise_search" is invoked with two hash tables L and R.

In the procedure "levelwise_search", other procedure "loopPost" will be invoked for any combination of a postfix tree T_r in R and a prefix tree T_l in L having the same key KEY (line 4–7 in levelwise_search). Note that, every valid pattern P such that $p_r(P) = KEY$ has to be stored in T_r. As similar, T_l stores a complete set of valid patterns Q such that $p_l(Q) = KEY$. Therefore, the completeness of the algorithm is guaranteed by considering all combinations of two patterns t_r in T_r and t_l in T_l.

In the procedure "loopPost", all of patterns stored in a postfix tree T_r will be examined by the preorder traversal. During the traversal in T_r, by invoking the procedure "loopPre" for the preorder traversal of a prefix tree T_l (line 3 in loopPost), all the combination of t_r in T_r and t_l in T_l will be examine. In other words, all the combination will be considered by using the double traversals in the postfix and prefix trees (see Fig. 7). In "loopPre", a new candidate P will be generated by joining two patterns t_r and t_l (line 2). Hash tables NL and NR will be used in the next iteration of the levelwise search (line 8 in levelwise_search). If P is frequent, then it is added to a prefix tree in NL by using the key $p_l(P) = t_r$. As similar, P is also added to a postfix tree in NR by using the key $p_r(P) = t_l$. Adding P to NL and NR means that P will be specialized by "specialization by a new subgraph". In other words, not adding P to the hash tables means the pruning of P. Therefore, uncorrelated patterns of enough size need not be added to hash tables even if they are frequent. Remember that, the (m, θ, k)-correlation is anti-monotone with respect to the "specialization by a new subgraph".

In "loopPost", a new prefix tree NT_l is generated before invoking "loopPre" (line 2). Instead of T_l, NT_l is used for patterns t'_r in T_r such that $t_r \preceq t'_r$ (line 4). By copying each node in T_l into NT_l selectively (line 11 in loopPre), NT_l stores only patterns in T_l needed for a pattern t'_r. In "loopPre", if a pattern P obtained

Procedure loopPost(T_r, T_l, NL, NR, σ, θ, k, m) –Traversal in a postfix tree T_r–
1: **for each** $t_r \in T_r$'s children
2: NT_l := new root node of prefix tree
3: loopPre(t_r, T_l, NT_l, NL, NR, σ, θ, k, m)
4: loopPost(t_r, NT_l, NL, NR, σ, θ, k, m)

Procedure loopPre(t_r, T_l, NT_l, NL, NR, σ, θ, k, m)
–Traversal in a prefix tree T_l with t_r–
1: **for each** $t_l \in T_l$'s children
2: P := join($gs(t_r)$, $gs(t_l)$) //create a new candidate
3: **if** $\sigma_{\mathcal{D}}(P) < \sigma$ **then continue**
4: **if** $|P| \leq m$ **then**
5: add P to a prefix tree in NL with key $gs(t_r)$
6: add P to a postfix tree in NR with key $gs(t_l)$
7: **else if** P is (m, θ, k)-correlated **then**
8: output P
9: add P to a prefix tree in NL with key $gs(t_r)$
10: add P to a postfix tree in NR with key $gs(t_l)$
11: NT'_l := new node; $gs(NT'_l) := gs(t_l)$; add NT'_l to NT_l's children
12: loopPre(t_r, t_l, NT'_l, NL, NR, σ, θ, k, m)

Fig. 6. Pseudo Code of loopPost and loopPre in CorSSS

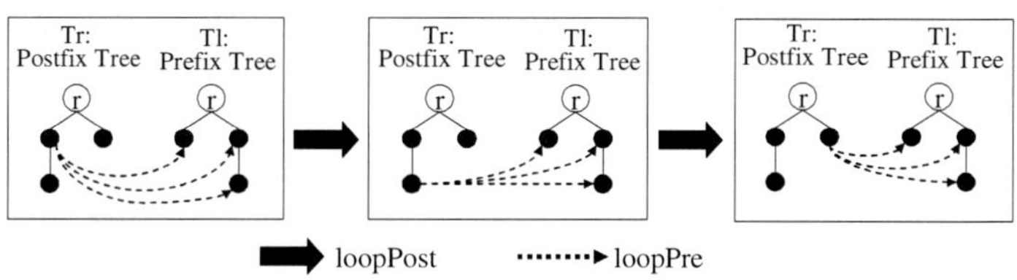

Fig. 7. Double traversals in the Prefix and Postfix Trees

by joining t_r and t_l is not frequent, then more specific patterns obtained from t'_r ($\succeq t_r$) and t'_l ($\succeq t_l$) need not be considered. By skipping the rest process in 'loopPre" (line 3 in loopPre), *i.e.* by not copying t_l into NT_l, the consideration of these patterns can be avoided. Furthermore, these patterns as well as P itself never be added to NL and RL. Thus, this skip operation can be regarded as a powerful pruning.

5 Related Work

Related to the frequent pattern mining from graph sequences, two algorithms Dynamic GREW[2] and GTRACE[4] have been proposed recently. These algorithms are different from CorSSS because the correlation is not considered.

Several data mining problems on the correlation mining in structured databases have been proposed. In the problem of *Correlated Graph Search*[5], all correlated subgraphs with a given query graph will be obtained by employing the

θ-correlation coefficient as a correlation measure. A set of mutually dependent subgraphs will be discovered in the problem of *HSG mining*[8]. In this problem, the h-confidence[13] is used to evaluate the degree of mutual dependency. In [14], a frequent hyperclique pattern miner named HFMG has been proposed to discover frequent and correlated subgraphs in a complex graph databases. In HFMG, the mutual dependency among substructures in a pattern is measured by the h-confidence, and only correlated patterns will be discovered. Although several algorithms for correlation mining in structured databases have been developed as mentioned above, to the best of the authors knowledge, there is no framework that focuses on graph sequences. In addition, while the correlation among components is considered, the discovered patterns have to satisfy the strict condition, *i.e.* all the combinations of components are correlated. On the contrary, the conditions on the correlation can be controlled precisely in CorSSS.

6 Experiments

CorSSS was implemented in Java and some experiments were conducted on a PC (CPU: Intel(R) Core2Quad 2.4GHz) with 4Gbytes of main memory. To obtain a set $\mathcal{F}$ of frequent closed subgraphs, we implemented a subgraph miner based on gSpan[15] and CloseGraph[16]. Two datasets were prepared for the experiments.

1. *MIT* : a dataset obtained from a part of the Reality Mining data[7] from MIT Media Lab. Each vertex is a mobile phone and its label is determined according to the total number of communications from and to it. Edges mean the communications between phones. The number of graphs is 1172. The average numbers of vertices and edges pre graph are 12.5 and 7.5, respectively.
2. *ENRON* : a dataset obtained from a part of ENRON email data[6] under a certain condition. Each vertex corresponds to a person and it is labeled by his/her position in the occupation. Each edge corresponds to the email communication between persons. The size of database is 795. The average numbers of vertices and edges per graph are 14.8 and 12.6, respectively.

The experimental results are shown in Table 1 where 'P' and '*Cand.*' denote the number of obtained patterns and the number of candidate patterns evaluated during the search process, respectively. Note that, *Cand.* includes patterns whose length is less than or equals to m. '*Time*' denotes the execution time in second after $\mathcal{F}$ is given. The number in parentheses denotes the ratio (in percentage) to the naive algorithm in Fig. 3. Two numbers in brackets are the number of frequent closed subgraphs and the execution time for obtaining them in second.

First, we consider the effects of parameters. If other parameters are the same, the increase of k and m seems to cause the increase of the number of obtained patterns and the execution time. The parameter θ gives a big impact on the result in some cases, such as *MIT* with $k = 1$, $m = 2$ and *ENRON* with $m = 2$.

In the experiments, it can be observed that the number of patterns increases with the increase of m. However, this relation does not always hold because

Table 1. Experimental Results

		MIT						*ENRON*					
		P	*Time*	*Cand.*	*P*	*Time*	*Cand.*	*P*	*Time*	*Cand.*	*P*	*Time*	*Cand.*
k	*m*	$\theta = 0.7$			$\theta = 0.5$			$\theta = 0.7$			$\theta = 0.5$		
		$\sigma = 0.05$, $[\vert\mathcal{F}\vert=28$, 0.4 sec.]						$\sigma = 0.1$, $[\vert\mathcal{F}\vert=55$, 0.3 sec.]					
1	1	0	0.2	0.7	0	0.1	0.7	0	0.2	1.8	0	0.2	1.8
			(77.0)	(87.2)		(70.9)	(87.2)		(73.6)	(60.1)		(89.6)	(60.1)
	2	173	0.4	4.5	1724	1.2	5.8	1160	0.8	17.5	173450	19.5	246.5
			(110.6)	(90.8)		(85.3)	(92.7)		(72.2)	(65.5)		(78.6)	(95.1)
2	1	270	0.5	4.5	270	0.5	4.5	875	1.2	17.5	875	1.2	17.5
			(66.8)	(89.0)		(67.0)	(89.0)		(65.1)	(62.7)		(65.7)	(62.7)
	2	1524	1.1	19.1	2963	2.1	20.3	7635	4.5	115.3	202564	26.2	376.9
			(76.5)	(91.8)		(89.6)	(92.2)		(50.8)	(71.5)		(76.7)	(88.3)
		$\sigma = 0.025$, $[\vert\mathcal{F}\vert=42$, 0.5 sec.]						$\sigma = 0.05$, $[\vert\mathcal{F}\vert=160$, 0.5 sec.]					
1	1	0	0.2	1.3	0	0.2	1.3	0	0.7	7.0	0	0.7	7.0
			(69.9)	(76.4)		(94.2)	(76.4)		(55.9)	(27.3)		(55.7)	(27.3)
	2	343	0.5	11.6	2348	1.5	13.1	5004	4.3	80.9	–	–	–
			(89.1)	(84.4)		(94.8)	(85.9)		(58.4)	(43.9)	–	–	–
2	1	553	0.9	11.6	553	0.8	11.6	3020	9.3	80.9	3020	9.3	80.9
			(82.8)	(82.0)		(74.5)	(82.0)		(56.7)	(39.9)		(56.6)	(39.9)
	2	4057	2.0	63.3	5724	3.0	64.6	35077	37.1	680.9	–	–	–
			(80.2)	(87.9)		(85.1)	(88.1)		(57.0)	(53.8)	–	–	–
		$\sigma = 0.01$, $[\vert\mathcal{F}\vert=100$, 0.6 sec.]						$\sigma = 0.025$, $[\vert\mathcal{F}\vert=416$, 1.3 sec.]					
1	1	0	0.3	4.6	0	0.4	4.6	0	2.1	26.3	0	2.1	26.3
			(90.3)	(46.4)		(105.0)	(46.4)		(49.5)	(15.2)		(49.5)	(15.2)
	2	1019	1.1	40.7	4082	2.9	42.3	27939	16.5	411.3	–	–	–
			(81.9)	(65.6)		(95.9)	(66.4)		(52.1)	(25.2)	–	–	–
2	1	1458	2.8	40.7	1458	2.8	40.7	12014	27.6	411.3	12014	27.6	411.3
			(77.0)	(60.4)		(77.2)	(60.4)		(51.0)	(23.1)		(51.1)	(23.1)
	2	13539	6.3	274.8	15328	7.8	276.2	188441	131.8	4209.0	–	–	–
			(78.6)	(75.7)		(80.8)	(75.8)		(48.1)	(31.4)	–	–	–

only patterns having more than m of size are accepted. In addition, in case of *ENRON* with $\theta = 0.7$, the numbers of obtained patterns in the condition of $k = 1$, $m = 2$ are more than those in $k = 2$, $m = 1$, even if the former condition requires each subsequence in a pattern to have larger number of correlated subsequences. These results show that the condition $m = 1$ is too restrictive in some cases. In other words, a subgraph in a pattern is too short to be considered as a component. It can be expected that a more flexible pattern mining can be realized in CorSSS by specifying the appropriate size of components.

Compared with the case of $k = 1$, a lot of patterns were obtained in the condition of $k = 2$. For example, in *MIT* with $m = 2$, $\theta = 0.7$, about 10 times of patterns were obtained on average. From these results, we believe that CorSSS succeeds in excluding uninteresting patterns by considering the mutual dependency. On the other hand, no pattern was obtained if we set $k = 1$ and $m = 1$. Thus, some appropriate combination of k and m has to be considered.

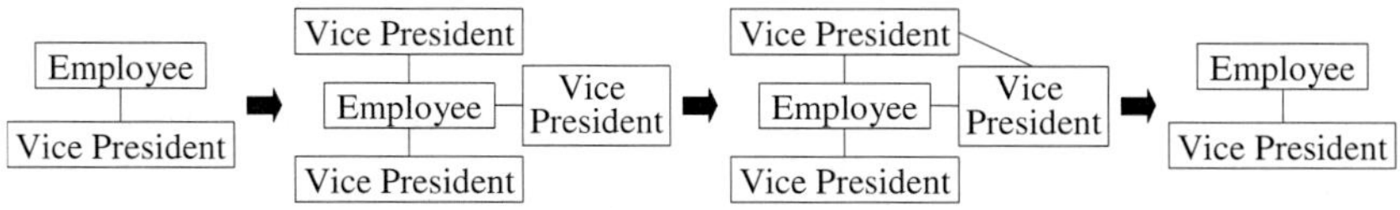

Fig. 8. An Example of Obtained Pattern

Then, we discuss the performance of CorSSS. While CorSSS failed to obtain the results in some parameter settings in *ENRON* due to the memory overflow, most problems were solved within a feasible computation time. Compared with the naive algorithm, the execution time of CorSSS in *ENRON* decreases to 60.7% on average and to 48.1% in the maximum. As similar, the number of candidates decreases to 46.6% on average and 15.2% in the maximum in *ENRON*. These results show the effectiveness of the consideration of generality ordering of patterns.

On the other hand, large reductions of execution time and the number of generated candidates were not observed in *MIT*. Furthermore, CorSSS needs more computation time than a naive algorithm in some cases. We consider the reasons for these results are: (1)the search space is too small to make the difference, and (2)some computational overhead is required to construct prefix and postfix trees. While the improvement of performance in CorSSS depends on the structure of prefix and postfix trees strongly, a naive way is employed for the construction of these trees in the current implementation. Thus, further improvement of the performance can be expected by developing some method for structuring more sophisticated trees.

An example of obtained pattern from $ENRON(\sigma = 0.025, \theta = 0.5, k = 2, m = 2)$ is shown in Fig. 8. Although the precise assessments were not made, the pattern might capture some situation in which some communications between an employee and a vice president cause the discussions by other vice presidents.

7 Conclusion

In this paper, we formulate a new data mining problem of finding frequent and correlated sequential subgraphs in a long graph sequence. To solve this problem, an algorithm CorSSS was developed which uses (1)the (m, θ, k)-correlation to measure the degree of mutual dependency among the components, and (2)a tree-shaped data structure for utilizing the generality ordering among patterns.

For future work, (1)further experiments with large-scale real world datasets, (2)the detailed assessment of the obtained patterns, and (3)the comparisons with the alternative approaches [2,4] are necessary. In addition, in order to exclude uninteresting patterns further and to obtain the more significant patterns, we also plan to incorporate some interestingness measures such as [17,9] and the concept of condensed representation[12,3] into the proposed framework.

References

1. Borgelt, C.: On canonical forms for frequent graph mining. In: Working Notes of the 3rd International ECML/PKDD Workshop on Mining Graphs, Trees and Sequences, pp. 1–12 (2005)
2. Borgwardt, K.M., Kriegel, H.-P., Wackersreuther, P.: Pattern mining in frequent dynamic subgraphs. In: Proc. of the 6th International Conference on Data Mining, pp. 818–822 (2006)
3. Gao, C., Wang, J., He, Y., Zhou, L.: Efficient mining of frequent sequence generators. In: Proc. of the 17th International Conference on World Wide Web, pp. 1051–1052 (2008)
4. Inokuchi, A., Washio, T.: A Fast Method to Mine Frequent Subsequences from Graph Sequence Data. In: Proc. of the 8th IEEE International Conference on Data Mining, pp. 303–312 (2008)
5. Ke, Y., Cheng, J., Ng, W.: Correlation search in graph databases. In: Proc. of the 13th ACM SIGKDD International Conference on Knowledge Discovery and Data Mining, pp. 390–399 (2007)
6. Klimt, B., Yang, Y.: The enron corpus: A new dataset for email classification research. In: Proc. of the 15th European Conference On Machine Learning, pp. 217–226 (2004)
7. MIT Media Lab. Reality Mining, `http://reality.media.mit.edu/`
8. Ozaki, T., Ohkawa, T.: Mining correlated subgraphs in graph databases. In: Proc. of the 12th Pacific-Asia Conference on Knowledge Discovery and Data Mining, pp. 272–283 (2008)
9. Sakurai, S., Kitahara, Y., Orihara, R.: A sequential pattern mining method based on sequential interestingness. International Journal of Computational Intelligence 4(4), 252–260 (2008)
10. Srikant, R., Agrawal, R.: Mining sequential patterns: Generalizations and performance improvements. In: Proc. of the 5th International Conference on Extending Database, pp. 3–17 (1996)
11. Tan, P.-N., Kumar, V., Srivastava, J.: Selecting the right interestingness measure for association patterns. In: Proc. of the 8th ACM SIGKDD International Conference on Knowledge Discovery and Data Mining, pp. 32–41 (2002)
12. Wang, J., Han, J.: BIDE: Efficient mining of frequent closed sequences. In: Proc. of the 20th International Conference on Data Engineering, pp. 79–90 (2004)
13. Xiong, H., Tan, P.-N., Kumar, V.: Hyperclique pattern discovery. Data Mining and Knowledge Discovery 13(2), 219–242 (2006)
14. Yamamoto, T., Ozaki, T., Ohkawa, T.: Discovery of internal and external hyperclique patterns in complex graph databases. In: Workshop Proceedings of the 8th IEEE International Conference on Data Mining, pp. 301–309 (2008)
15. Yan, X., Han, J.: gSpan: Graph-based substructure pattern mining. In: Proc. of the 2002 IEEE International Conference on Data Mining, pp. 721–724 (2002)
16. Yan, X., Han, J.: CloseGraph: mining closed frequent graph patterns. In: Proc. of the 9th ACM SIGKDD International Conference on Knowledge Discovery and Data Mining, pp. 286–295 (2003)
17. Yun, U.: WIS: Weighted interesting sequential pattern mining with a similar level of support and/or weight. ETRI Journal 29(3), 336–352 (2007)

Building a Text Classifier by a Keyword and Wikipedia Knowledge*

Qiang Qiu, Yang Zhang**, Junping Zhu, and Wei Qu

College of Information Engineering, Northwest A&F University
712100 Yangling, Shaanxi
{qiuqiang, zhangyang, zhujunping, lex}@nwsuaf.edu.cn

Abstract. Traditional approach for building text classifiers usually require a lot of labeled documents, which are expensive to obtain. In this paper, we propose a new text classification approach based on a keyword and Wikipedia knowledge, so as to avoid labeling documents manually. Firstly, we retrieve a set of related documents about the keyword from Wikipedia. And then, with the help of related Wikipedia pages, more positive documents are extracted from the unlabeled documents. Finally, we train a text classifier with these positive documents and unlabeled documents. The experiment result on 20Newsgroup dataset show that the proposed approach performs very competitively compared with NB-SVM, a PU learner, and NB, a supervised learner.

Keywords: text classification, keyword, unlabeled document, Wikipedia.

1 Introduction

With the rapid growth of the World Wide Web, text classification has become one of the key techniques for organizing online information. Traditional supervised learning techniques typically require a large number of labeled text documents for training a good text classifier. For real-life text classification applications, it is usually very expensive to label text documents manually. In order to reduce the amount of labeled documents for training, semi-supervised learning algorithms based on both labeled and unlabeled data have been used for text classification [1,2,3]. Another way is to develop classification algorithm based on a set of labeled positive examples and a large amount of unlabeled documents [4,5,6,7,8,9,10,11]. Recently, some researchers have proposed labeling keywords to build a classifier instead of labeling documents manually [12,13]. Some proposed methods launches text classification tasks with only unlabeled documents and the title word of each category for learning [14,15].

Recently, many researchers have proposed using Wikipedia knowledge to improve the performance of text classifiers [16,17]. Also, Wikipedia knowledge is used to extend domain-specific thesaurus [18,19]. Because of the plentiful knowledge inside Wikipedia, the approach proposed in this paper utilizes a keyword and Wikipedia to build a text

* This work is supported by Young Cadreman Supporting Program of Northwest A&F University (01140301).

** Corresponding author.

R. Huang et al. (Eds.): ADMA 2009, LNAI 5678, pp. 277–287, 2009.

classifier. This approach does not require any labeled documents. Firstly, some related Wikipedia articles about the keyword are extracted from the Wikipedia. And then, with the help of the related Wikipedia documents, more positive documents are mined from the unlabeled documents. Finally, we train a text classifier with these positive documents and unlabeled documents.

In order to measure the classification performance of the proposed approach, we make experiments on the 20Newsgroup dataset, with promising experiment result. The experiment result shows that our proposed approach could help to build excellent classifiers, and it could be more applicable to real-life text classification applications.

The rest of the paper is organized as following. Section 2 reviews the related work. Section 3 presents our approach to extract related Wikipedia articles by the keyword. Section 4 presents our approach to mine positive documents from unlabeled documents with the help of Wikipedia articles. Section 5 gives our classification algorithm. Our experiment result is shown in section 6, and finally, section 7 concludes this paper and gives our further work.

2 Related Work

With a small set of labeled documents, *Blum et al.* proposed co-training method to build a text classifier [1]. *Nigam et al.* used an algorithm for learning from labeled and unlabeled documents based on the combination of Expectation Maximization (EM) and a Naive Bayesian classifier [3].

In learning from large set of positive and unlabeled examples, a number of practical PU [11] learning algorithms were proposed [4,5,6]. These PU learning algorithms all conformed to the theoretical results presented in [4] by following a common two-step strategy: (1) identifying a set of reliable negative documents N from the unlabeled set U; and then (2) building a set of classifiers by iteratively applying EM or SVM, and then selecting a good classifier from the set. Furthermore, there are a few successful algorithms proposed by the research community to cope with the situation when there is only a small set of positive documents [10,11]. These two algorithms focus on enlarging positive examples.

With unlabeled training dataset, some approaches have been proposed to build a classifier by labeling keywords instead of labeling training documents [12,13]. The approach in [12] consists of three steps: (1) providing a small set of keywords for each class; (2) using these keywords to label raw training examples (P and N) from the unlabeled documents U; (3) building a classifier by NB-EM algorithm. However, it is too difficult for a user to provide sufficient keywords for accurate learning. So, the approach in [13] utilizes clustering and feature selection technique to label some words for each category. And then, the training examples are extracted based on the labeled words and NB-EM algorithm is used to build a classifier.

The works in [14,15] are more related to our research. The approaches proposed in [14,15] launch text classification tasks with only unlabeled documents and the title word of each category for learning. The approach in [14] contains two steps: (1) using a bootstrapping technique to label documents; (2) applying a feature projection technique to build a robust text classifier. The approach in [15] firstly integrates WordNet, information retrieval and DocMine algorithm [20] to label some positive examples. And then, with the help of the labeled positive documents, more positive documents are extracted from the unlabeled documents. Finally, a state-of-art PU learning algorithm is employed to build a good text classifier.

However, the approach in [14] depends on title words and the number of keywords. It is difficult for user to determine the number of the candidate words according to different kinds of data set [14]. The approach in [15] needs WordNet and a search engine, which is too difficult for non-expert users. Furthermore, WordNet has limited converge, since it is a manually constructed dictionary, and therefore laborious to maintain [16]. Recently, many researchers study using Wikipedia knowledge to improve text classification. The approach in [16] constructs a thesaurus of concepts from Wikipedia, and then expand the BOW representation with semantic relations. A mapping algorithm in [18] has been proposed to extend a domain-specific thesaurus with valuable information from Wikipedia.

Because of the plentiful knowledge inside Wikipedia, the approach proposed in this paper utilizes the keyword and knowledge from Wikipedia to build text classifiers. The proposed approach release users from labeling documents or labeling keywords manually. Also, applying this approach, users neither need to provide candidate words nor to build a search engine. This approach enables users to build an excellent text classifier easily.

3 Extracting Related Documents from Wikipedia by Keyword w

Wikipedia[1] is a multilingual, web-based, free content encyclopedia. Each Wikipedia article describes a single topic and belongs to at least one category of Wikipedia. Hyperlinks between articles keep some semantic relations. The approach proposed in this paper utilizes the relation among hyperlinks to extract related Wikipedia articles of the keyword w. Firstly, we extract the Wikipedia article $Page_w$ which topic is the keyword w. However, some keywords have ambiguous senses. For example, the keyword *bank* may refer to either a *financial institution* or a *shallow area in a body of water*. We use its common sense. For example, for the keyword *bank*, its common sense is *financial institution*. In Wikipedia disambiguation pages, the link of common sense is the first sense link, which corresponds to the target Wikipedia article $Page_w$ of the keyword w. And then, according to the hyperlinks of the article $Page_w$, more Wikipedia articles $Page_{link}$ are extracted. The hyperlinks of a Wikipedia article contains out-links and in-links. The out-links consist of the links that the Wikipedia article link to. And the in-links consist of the links which link to the Wikipedia article. We extract more related articles $Page_{link}$ from Wikipedia in the following way:

1) Extracting the out-links whose hypertext contains the keyword w. For example, *retail banking* is one out-link of article *bank*, and *retail banking* contains the keyword *bank*. We regard the link *retail banking* as one target link.

2) Extracting the links which belongs to both the in-links and the out-links of the article. For example, *financial* is one out-link of the article *bank*, and also, it is one

[1] http://www.wikipedia.org

in-link of the article *bank*. That is to say, article *financial* and article *bank* links to each other. We regard the link *financial* as one target link.

In Wikipedia, each link corresponds to an article. So, we regard the article $Page_w$ and the extracted articles $Page_{link}$ as related documents, $RD = Page_w \cup Page_{link}$, about the keyword w.

4 Extracting Positive Documents $TotalPos$ from Unlabeled Documents U by RD

In last section, a set of related Wikipedia articles, RD, could be extracted from Wikipedia. However, the set RD could not adequately represent the whole positive category. Moreover, the examples in RD and the positive examples in unlabeled documents U could not be considered to be drawn from the same distribution. In order to improve the performance of the text classifier, more positive documents should be extracted from U with the help of RD. Here, a new approach, which applies iteration method to extract the positive examples, is proposed. The approach consists of four steps:(1) Selecting a set of representative features RF from RD; (2) With the help of RF, splitting the documents U into three parts, say, strong positive examples SP, strong negative examples SN and unsure documents UD; (3) Building a classifier based on SP and SN and classifying the positive examples PE from UD; (4) With PE, extracting more positive examples iteratively until no more positives could be extracted. The overall set of positive examples, $TotalPos$, consist of SP and PE extracted in each iteration.

4.1 Selecting Representative Features RF from RD

We argue that the examples in RD and the hidden positive examples in U as well as in test documents share the same representative features. A set of representative features, RF, are selected from RD. The set RF consists of top *k* features with the highest

Algorithm 1. Feature Selection *(selectFeature)*

Input:

P: the set of positive documents;
U: the set of unlabeled documents;
k: the number of features;

Output:

RF: the set of represent features selected from P;

$RF \leftarrow \Phi, F \leftarrow \Phi$;
for each word feature $w_i \in P$ **do**
  computing $score(w_i)$ according to (1);
  $F \leftarrow F \cup \{w_i\}$;
end for
sorting the features descendingly according to their scores to a list L;
$RF = \{$ first k features in list $L\}$;
return RF;

scores. The $score()$ function, which gives high weight to features that occur frequently in the positive set RD and un-frequently in the whole document $U \cup RD$, is defined as:

$$score(w_i) = \frac{df(w_i, P)}{|P|} \log \frac{|P| + |U|}{df(w_i, P) + df(w_i, U)}. \quad (1)$$

Here, $df(w_i, P)$, and $df(w_i, U)$ are document frequency of w_i in P and U respectively. Algorithm 1 gives the details of our feature selection algorithm.

4.2 Splitting U

Once the set of representative keywords is determined, according to the predefined threshold t and the similarity between RF and each document in U, we split U into strong positive SP, strong negative SN, and unsure documents UD. The splitting rules could be found in algorithm 2. Here, $W_{RF,j}$ is the score of w_j in RF, and $W_{d_i,j}$ is the number of observation of term w_j in document d_i.

Algorithm 2. Splitting Unlabeled Documents *(splitDocuments)*

Input:

U: the set of unlabeled documents;
RF: the set of selected features;
t: the predefined threshold;

Output:

SP: the set of strong positives extracted from U;
SN: the set of strong negatives extracted from U;
UD: the set of unsure documents in U;

$SP \leftarrow \Phi, SN \leftarrow \Phi, UD \leftarrow \Phi$;
for each document $d_i \in U$ **do**
  $simi(RF, d_i) = \frac{\sum_j W_{RF,j} W_{d_i,j}}{\sqrt{\sum_j W_{RF,j}^2}\sqrt{\sum_j W_{d_i,j}^2}}$;
end for
$value \leftarrow \max(simi(RF, d_i)), d_i \in U$;
for each document $d_i \in U$ **do**
  if $simi(RF, d_i)/value \geq t$ **then**
    $SP \leftarrow SP \cup \{d_i\}$;
  else
    if $simi(RF, d_i)/value = 0$ **then**
      $SN \leftarrow SN \cup \{d_i\}$;
    else
      $UD \leftarrow UD \cup \{d_i\}$;
    end if
  end if
end for
return SP, SN, UD;

4.3 Extracting Positive Examples PE from UD

In last subsection, we split U into SP, SN and UD. However, the set UD may still contains some positive examples. In order to further improve the performance of classification, more positive examples PE are extracted from UD. Firstly, a Naive Bayesian (NB) [11] classifier is trained based on SP and SN. And then, we iteratively extract negative examples from UD until no more negatives could be extracted. The rest of UD is considered as the set PE of positive examples. Algorithm 3 gives the details.

Algorithm 3. Extracting Positive Examples from UD *(extractPos)*

Input:

SP: the set of strong positives;
SN: the set of strong negatives;
UD: the set of unsure documents;

Output:

PE: the set of positive examples extracted from UD;

```
1: ND ← Φ, PE ← Φ;
2: Build a NB classifier based on SP, SN;
3: for each document d_i ∈ UD do
4:     if NB.classify(d_i)==negative then
5:         ND ← ND ∪ {d_i}, UD ← UD − {d_i};
6:     end if
7: end for
8: if |ND| > 0 then
9:     SN ← SN ∪ ND;
10:    goto step 1;
11: else
12:    PE ← UD;
13: end if
14: return PE;
```

4.4 Extracting More Positive Examples Iteratively

After some positive examples PE are mined from U in the above way, the positive examples $SP \cup PE$ might be just a small part of positive documents in U. As the performance of text classifiers depends on the number of positive training examples, in order to further improve the performance of the classifier, we iteratively enlarge the set of positive documents extracted from U. Threshold t increases in the proportion of $\lambda(\geq 1)$ with the number of iterations increasing, so as to improve the precision of the extracted positive examples and reduce the number of iterations. Algorithm 4 describes how to extract overall set of positive examples, $TotalPos$, from the unlabeled documents U with the related Wikipedia articles RD. We regard the rest of U as negative examples $TotalNeg = U - TotalPos$.

5 Learning Algorithm

Given a set of training documents D, each document is considered as a set of words, drawn from the same vocabulary $V = \{w_1, w_2, ..., w_{|V|}\}$. In this paper, we only

Algorithm 4. Extracting Overall set of Positive Examples *(extractTotalPos)*

Input:

RD: the set of related Wikipedia articles;
U: the set of unlabeled documents;
k: the number of features;
t: the predefined threshold;
λ: the growth factor;

Output:

$TotalPos$: the overall set of positive examples;

1: $TotalPos \leftarrow \Phi$;
2: $U \leftarrow U - TotalPos$;
3: Obtaining the representative features RF by calling $selectFeature(RD, U, k)$; //algorithm 1
4: Obtaining strong positives SP, strong negatives SN and unsure documents UD by calling $splitDocuments(U, RF, t)$; //algorithm 2
5: Extracting positive examples PE from UD by calling $extractPos(SP, SN, UD)$; //algorithm 3
6: **if** $|PE| > 0$ **then**
7: $\quad TotalPos \leftarrow TotalPos \cup SP \cup PE$;
8: $\quad t \leftarrow \lambda * t, RD \leftarrow PE$;
9: $\quad$ goto step 2;
10: **else**
11: $\quad TotalPos \leftarrow TotalPos \cup SP$;
12: **end if**
13: **return** $TotalPos$;

consider binary classification. We computer the posterior probability, $Pr(c_j|d_i)$, for classification, where c_j is a class label and d_i is a document. Based on the multinomial model of Naive Bayesian algorithm [11,21], we have

$$Pr(c_j) = \frac{\sum_{i=1}^{|D|} Pr(c_j|d_i)}{|D|}. \tag{2}$$

and with Laplacian smoothing [11]

$$Pr(w_t|c_j) = \frac{1 + \sum_{i=1}^{|D|} N(w_t, d_i) Pr(c_j|d_i)}{|V| + \sum_{s=1}^{|V|} \sum_{i=1}^{|D|} N(w_s, d_i) Pr(c_j|d_i)}. \tag{3}$$

Assuming the probability of word event is independent of each other, we obtain the NB classifier [11]:

$$Pr(c_j|d_i) = \frac{Pr(c_j) \prod_{k=1}^{|d_i|} Pr(w_{d_i,k}|c_j)}{\sum_{r=1}^{|C|} \prod_{k=1}^{|d_i|} Pr(w_{d_i,k}|c_r)}. \tag{4}$$

When classifying a testing document d, the class with the highest $Pr(c_j|d)$ is taken as the class label of document d.

The Expectation-Maximization (EM) algorithm is a popular iterative algorithm for maximum likelihood estimation in problems with missing data [11]. The positive documents $TotalPos$ and negative documents $TotalNeg$ labeled by our proposed approach may contain some noise. Therefor, EM is employed to revise the class label of each document in $TotalPos$ and $TotalNeg$. It iterates over two steps, the Expectation step and the Maximization step. EM employs equations (1) and (2) for the Expectation step, and equation (3) for the Maximization step [11]. In EM algorithm, the parameters of the NB classifier will converge after a number of iterations. Based on [11], we give our learning algorithm, which is listed as following:

Algorithm 5. Learning Algorithm

Input:

$TotalPos$: the set of positive documents;
$TotalNeg$: the set of negative documents;

Output:

NBC: a Naive Bayesian classifier;

Building a NB Classifier NBC using $TotalPos$ and $TotalNeg$ based on equation (1) and (2);
while NBC parameters change **do**
  for each $d_i \in TotalPos \cup TotalNeg$ **do**
    Classifying d_i using NBC;
    if the class label of d_i changes **then**
      Updating $Pr(c_j)$ and $Pr(w_t|c_j)$;
    end if
  end for
end while
return NBC;

6 Evaluation Experiments

In this section, we evaluate our proposed approach, and compare it with NB-SVM [9], a PU learning algorithm, and NB [11], a supervised learning algorithm, as our approach, NB-SVM [9], and NB [11] are all based on Naive Bayesian algorithm.

6.1 Dataset

We used 20Newsgroup dataset in our experiment and employed its *bydate*[2] version. There are 20 categories in the 20Newsgroups dataset, and 10 categories are selected randomly to be used in our experiment. Each category has approximately 1000 articles. For each category, 60% of the documents are used as training dataset, and the remaining 40% of the documents as the test documents.

6.2 Experiment Settings

The preprocessing includes stop word removing and stemming. In algorithm 1, for selecting a set of representative features from positive documents, we set $k = 35$. When

[2] http://people.csail.mit.edu/jrennie/20Newsgroups/

extracting positive examples iteratively, we set threshold $t = 0.30$ and growth factor $\lambda = 1.5$. When experimenting with our approach, the training dataset is taken as unlabeled data.

For PU learning, 400 positive documents are selected randomly for each category from training dateset, and the rest of training examples are taken as unlabeled data. The NB-SVM [9] of LPU[3] system is employed to build a text classifier with these 400 positive examples and unlabeled examples. For supervised learning, NB is applied for text classification. We conduct five trails of experiments, and the averaged results are reported here.

6.3 Experimental Results

In this paper, F_1 is used to measure the performance of our text classifier, as it is widely used by the research community of text classification [4]. Table 1 gives the experimental results. In table 1, column 1 lists the category name, column 2 lists the corresponding keyword, column 3, 4, and 5 shows the classification performance of our proposed approach, NB-SVM, and NB respectively.

Table 1. Classification result ($F_1(\%)$)

Category	Keyword	Wikipedia	NB-SVM	NB
rec.sport.baseball	baseball	95.92	85.66	96.70
soc.religion.christian	christian	94.84	85.99	96.52
sci.crypt	encrypt	90.82	84.30	94.46
sci.electronics	electronics	41.61	62.55	85.67
comp.graphics	graphics	52.03	74.89	87.16
talk.politics.guns	guns	92.26	84.52	94.13
sci.med	medicine	83.35	71.70	91.50
talk.politics.mideast	mideast	92.43	89.25	94.96
rec.motorcycles	motorcycles	92.97	87.57	95.30
sci.space	space	86.60	78.77	91.54

It is obviously from table 1 that the proposed approach outperforms NB-SVM, a *PU* classifier based on 400 positive examples. Specifically, the performance is above 80% for most of the categories. However, for category *sci.electronics* and *comp.graphics*, the performance is not good.

Except category *sci.electronics* and *comp.graphics*, the performance of the proposed approach is very close to NB, a supervised classifier. The difference between the performance of our proposed approach and NB algorithm is less than 6%. Here, we train the NB algorithm based on 600 positive examples and 5400 negative examples. However, our proposed approach require only a keyword. Comparing our proposed approach with NB-SVM and NB, we believe that our proposed approach requires less user effort to launch text classification tasks.

[3] http://www.cs.uic.edu/ liub/LPU/LPU-download.html

7 Conclusion and Future Work

In many real-world text classification applications, it is often expensive to obtain enough labeled training examples for building a good text classifier. The approach proposed in this paper utilizes keyword and knowledge from Wikipedia to build text classifiers. Comparing with NB-SVM, a PU algorithm and NB, a supervised learning algorithm, for most of the categories in our experiment, the classification performance of the proposed approach outperforms NB-SVM, and its performance is close to NB. The proposed approach could be more applicable to real-life text classification applications, as it doesn't require any labeled training documents.

In this paper, for ambiguous terms, we chose the common sense. In our future work, we plan to study how to find the specific sense of the ambiguous terms with respective to a certain set of unlabeled documents, and extract more related Wikipedia articles.

References

1. Blum, A., Mitchell, T.: Combining labeled and unlabeled data with co-training. In: Proceedings of the eleventh annual conference on Computational learning theory (1998)
2. Ghani, R.: Combining labeled and unlabeled data for multiclass text categorization. In: International Conference on Machine Learning (2002)
3. Nigam, K., McCallum, A.K., Thrun, S., Mitchell, T.: Text classification from labeled and unlabeled documents using EM. Machine learning 39 (2000)
4. Liu, B., Lee, W., Yu, P., Li, X.: Partially Supervised Classification of Text Documents. In: International Conference on Machine Learning, pp. 387–394 (2002)
5. Li, X., Liu, B.: Learning to Classify Texts Using Positive and Unlabeled Data. In: International joint Conference on Artificial Intelligence, pp. 587–594 (2003)
6. Yu, H., Han, J., Chang, K.C.-C.: PEBL: Positive Example Based Learning for Web Page Classification Using SVM. In: Proceedings of the eighth ACM SIGKDD international conference on Knowledge discovery and data mining, pp. 239–248 (2002)
7. Fung, G.P.C., Yu, J.X., Lu, H., Yu, P.S.: Text Classification without Negative Examples. Proc. 21st Int'l Conf. Data Engineering (2005)
8. Yu, H., Han, J.: PEBL: Web Page Classification without Negative Examples. IEEE Trans. Knowledge and Data Engineering (2004)
9. Li, X., Liu, B.: Learning from Positive and Unlabeled Examples with Different Data Distributions. In: Gama, J., Camacho, R., Brazdil, P.B., Jorge, A.M., Torgo, L. (eds.) ECML 2005. LNCS, vol. 3720, pp. 218–229. Springer, Heidelberg (2005)
10. Fung, G.P.C., et al.: Text Classification without Negative Examples Revisit. IEEE Transactions on Knowledge and Data Engineering 18(1), 6–20 (2006)
11. Li, X., Liu, B., Ng, S.-K.: Learning to Classify Documents with Only a Small Positive Training Set. In: The European Conference on Machine Learning, pp. 201–213 (2007)
12. McCallum, A., Nigam, K.: Text classification by bootstrapping with keywords, EM and shrinkage. In: ACL Workshop on Unsupervised Learning in Natural Language Processing (1999)
13. Liu, B., Li, X., Lee, W.S., Yu, P.S.: Text Classification by Labeling Words. In: Proc. 19th National Conference on Artificial Intelligence (2004)
14. Ko, Y., Seo, J.: Text classification from unlabeled documents with bootstrapping and feature projection techniques. Information Processing and Management (2009)

15. Qiu, Q., Zhang, Y., Zhu, J.: Build a text classifier by a keyword and unlabeled documents. In: Pacific-Asia Conference on Knowledge Discovery and Data Mining (2009)
16. Wang., P., Hu, J., Zeng, H.J., Chen, Z.: Using Wikipedia knowledge to improve text classification. In: Knowledge information System (2008)
17. Wang., P., Hu, J., Zeng, H.J., Chen, L.: Improving Text Classification By Using Encyclopedia Knowledge. In: IEEE International Conference on Data Mining (2007)
18. Medelyan, O., Milne, D.: Augmenting domain-specific thesauri with knowledge from Wikipedia. In: Proceedings of the NZ Computer Science Research Student Conference, Christchurch, NZ (2008)
19. Milne, D., Medelyan, O., Witten, I.H.: Mining domain-specific thesauri from Wikipedia: A case study. In: Proceedings of the IEEE/WIC/ACM International Conference on Web Intelligence (2006)
20. Barbara, D., Domeniconi, C., Kang, N.: Mining Relevant Text from Unlabeled Documents. In: Proceedings of the Third IEEE International Conference on Data Mining (2003)
21. McCallum, A., Nigam, K.: A comparison of event models for naive bayes text classification. In: AAAI 1998 workshop on learning for text categorization (1998)

Discovery of Migration Habitats and Routes of Wild Bird Species by Clustering and Association Analysis

MingJie Tang[1,3], YuanChun Zhou[1], Peng Cui[2,3], Weihang Wang[1,3], Jinyan Li[4], Haiting Zhang[1,3], YuanSheng Hou[5], and BaoPing Yan[1]

[1] Computer Network Information Center, Chinese Academy of Sciences
[2] Institute of Zoology, Chinese Academy of Sciences
[3] Graduate University of Chinese Academy of Sciences
[4] School of Computer Engineering, Nanyang Technological University
[5] Bureau of Qinghai Lake National Nature Reserve
100190 Beijing
{tangrock,supercat0325}@gmail.com, yczhou@sdb.cnic.cn, cuipeng@ioz.ac.cn, jyli@ntu.edu.sg, zht@sdb.cnic.cn, houyuanseng@163.com, ybp@cnic.cn

Abstract. Knowledge about the wetland use of migratory bird species during the annual life circle is very interesting to biologists, as it is critically important for conservation site construction and avian influenza control. The raw data of the habitat areas and the migration routes can be determined by high-tech GPS satellite telemetry, that usually are large scale with high complexity. In this paper, we convert these biological problems into computational studies, and introduce efficient algorithms for the data analysis. Our key idea is the concept of hierarchical clustering for migration habitat localization, and the notion of association rules for the discovery of migration routes. One of our clustering results is the Spatial-Tree, an illusive map which depicts the home range of bar-headed geese. A related result to this observation is an association pattern that reveals a high possibility of bar-headed geese's potential migration routes. Both of them are of biological novelty and meaning.

Keywords: Clustering, Sequence mining, Bird Migration, Habitat, Route, Scientific data, Qinghai Lake.

1 Introduction

The Asian outbreak of highly pathogenic avian influenza H5N1disease in poultry in 2003 and 2004 was unprecedented in its geographical extent, and its transmission to human beings showed an ominous sign of life-threatening infection [1]. Research findings indicate that the domestic ducks in southern China played a central role in the reproduction and maintenance of this virus, and wild birds may have contributed to the wide spread of the virus. This assumption had led to another question: how to define and identify the habitat, migration distance and time. Indeed, understanding of the species' habitat is critical for us to find the roots of the answers, like answers to how the wild life and domestic poultry intersect together, what is the possibility of H5N1 spilling over from the poultry sector into some wild bird species [2].

R. Huang et al. (Eds.): ADMA 2009, LNAI 5678, pp. 288–301, 2009.

The spatial data analysis on the specie's transmission coordinates together with their layered maps can be conducted by GIS (Geographic Information System) including ESRI'S ARC\INFO 7.1.2 and ArcView 3.1 (Research Institute, Inc., Redlands, California, U.S.A.) [5]. However, there has been lack of a persuasive way to identify the stop area of the species and the wintering areas. The situation becomes further complicated when the scientist come to lineate the migration routes from the accumulated data points. Therefore, a bird migration data analysis system is desired, by which data can be systematically analyzed, and knowledge patterns are subsequently available for deep biological studies. In this work, we address the following three problems which are arisen from the bird migration data analysis.

Discovery of Bird Habitat. The habitat range of an animal is defined as the area explored by this individual during its normal activities (i.e., food gathering, mating and caring for young, Burt 1943, Powell 2000). Understanding the factors that determine the spatial coverage and distribution of animals is fundamental not only to theoretical science, but also to real-life applications such as conservation and wildlife management decision makings [5].

Analysis on the Site Connectedness between Habitats. Site connectedness is a measure relating to the accessibility, for the migrating storks, of the site to its neighboring stay sites [6]. The sites with lower connectedness are considered to those at higher risk of being isolated from the migration route network.

Identification of Migration Routes. To help conserve species that migrate long distances, it is essential to have a comprehensive conservation plan that includes identification of migration routes. This information is of an added importance for many rare wild bird species [4].

Our computational approaches to these problems are integrated into a data mining system. It consists of four major components: data preprocessing, clustering, habitat range estimation, and association rules analysis. The function of the clustering component is to cluster the data points and meanwhile identify the candidates of the habitats. Intuitively, a potential habitat is a region where wild bird species prefer to stay a long time, and it mathematically corresponds to a dense region of points over the entire area. For this purpose, we propose a new hierarchical clustering algorithm which can find the habitats with different levels of densities. The component of habitat range estimation is aimed to determine the precise home range and time duration of the birds on top of the clustering results. As bird's migration between the habitats can be considered as a sequence pattern, we apply an existing sequence mining technique to discover interesting associations between the habitats. This is the goal of the association rules analysis. Besides, a visualization technique is developed for an easy view of the distribution of the bird habitats and migration routes which is helpful to gain more insights into findings. With this visualization tool, all of our results can be embedded into the Google Map (One web GIS from Google).

We have conducted a pilot experiment on a real-world database to evaluate our system. Our computational results on the bird habitat, site connectedness and migration route are interesting and have been confirmed to have biological novelty. These results would be useful in future for the scientists to estimate the risk of virus infection of wild birds from poultry or the other way around.

The main contributions of our work are summarized as follows: (i) A new hierarchical clustering algorithm is proposed, and it is used to discover bird habitats, (ii) Association analysis is introduced to reveal the site connectedness between habitat areas, and (iii) Bird migration routes are rigorously studied by sequence mining algorithms.

Paper organization. Section 2 presents a short background introduction to satellite tracking technologies for monitoring migration routes of wild bird species, and gives a brief overview to clustering and association mining algorithms. In Section 3, the telemetry bird migration data is described. Section 4 presents the overall diagram of our data mining system and describes the computational techniques in each component. Our computational results and their evaluation are presented in Section 5. In Section 6, we summarize our major contribution, and point out our future work.

2 Background and Related Work

2.1 Satellite Tracking of Wild Bird Species

Recent advances in the technology of satellite tracking have allowed researchers to continuously track the movements of individual birds over a broad spatial scale without conducting extensive field observations after the birds have been equipped with satellite transmitters. The applications of satellite tracking to bird migration studies have enabled considerable progress to be made with regard to elucidating the migration routes and stay sites of various migratory bird species, with important implications, for example, for conservation [6]. Traditionally, most of biologist have to count those location plots in a certain area and then utilize kernel model to calculate the home range of bird species [3,5,6]. Until recently, Hiroto Shimazaki1 *et al.* [6] proposed a method to examine the location data points based on the idea of clustering. At the first step, their method groups the location points with similar characteristics in approaching speed and departure speed by using the ISODATA algorithm [9]. And then the extent of stay sites is determined by specifying the area attainable by a bird moving speed. At last, they evaluate the site connectedness between stopover sites. However they do not make full use of the bird tracking data—features such as latitude and longitude have not been used to get the habitat range. The identification of the migration routes has not been touched either. As shown in the previous studies that satellite tracking is a powerful to monitor birds' migration behavior, and the data is valuable to make significant contribution to biological research, yet, to the best of our knowledge, it has been long lack of a data mining system capable of conducting systematic migration data analysis.

2.2 Overview to Clustering Algorithms

Clustering is an extensively studied topic in the machine learning and data mining field. A clustering algorithm refers to a method that subgroups a set of data points according to a distance or density metrics. Clustering analysis can be used as a stand-along tool to get insight into the distribution of the data points in a data set, or can be used as a data preprocessing step for other types of data analysis. Various techniques have been explored for clustering spatial data sets. For instance, an improved k-medoid method, called CLARANS [12] was proposed recently. SNN [13] was also developed to cluster the earth science data. DBSCAN [10] and IncrDBSCAN [11] have been proposed to process the spatial data sets as well. Meanwhile, several hierarchical clustering approaches have been long investigated, including the agglomerative approach (eg. AGNES) and the divisive approach (eg. DIANA). Detailed description of AGNES and DIANA can be found at [14]. In this paper, our new idea is to combine DBSCAN with a hierarchical clustering approach to find the habitats with different levels of densities.

2.3 Association Analysis

Association rules mining and sequence mining are pioneer research topics in data mining, and they are still attracting lots of attentions. The classic association rule mining algorithms include Apriori[15] and FP-tree[16]. GSP [17] was the first approach to the discovery of frequent sequence patterns. Zaki then propose the SPADE algorithm [18] to find frequent sequence with a faster speed. The PSP (Prefix Tree For Sequential Patterns) approach [21] is much similar to the GSP algorithm, but it stores the database on a more concise prefix tree with the leaf nodes carrying the supports of the sequences. In this paper, we make use of these algorithms for bird migration routes analysis.

3 Bird Migration Data

Our studies are conducted at the Qinghai Lake National Nature Reserve, Qinghai province, China. Qinghai Lake, the largest salt lake in China with an area of 525 Km2, is located in the middle of Qinghai Province. The bird movement data are from 29 bar-headed geese (Anser indicus) from Qinghai Lake. Fourteen of them were captured on March 25-31, 2007, and the others were captured on March 28 - April 3, 2008. Each bird was weighed, measured and equipped with a 45g solar-powered portable transmitter terminal (PTT:9 North Star Science and Technology, LLC, Baltimore, Maryland USA) and 1 Microwave Telemetry (PTT-100, Columbia, Maryland USA). Transmitter signals were received by Argos data system (CLS America Inc., Maryland, USA) and transmitter locations were estimated. Argos classified the location accuracy into seven categories: 3, 2, 1, 0 and LA, B, Z with the approximation for class 3 < 150 m, class 2 = 150-350 m, class 1 = 350- 1000 m, class 0 > 1000 m. We also bind the GPS (Global Position System) location equipment on the PTTs. We call the location data as LG.

Table 1. Relational representation of our bird migration data

Obs	animal	ppt	date	time	Latitude	longitude	K94	Speed
85	BH07_67695	67695	2008-03-02	3:27:10	29.275	88.731	LZ	32
86	BH08_67688	67688	2008-03-02	4:27:10	30.275	89.25	KG	43

Our data sets received from Western Ecological Research Center are represented by the form shown in Table 1, which consists of 66796 and 22951 location data records for the 2007 survey and 2008 survey, respectively. About 90.1% of data records in the four categories of 0-3 are with high quality, which are used in our study; the remaining LA, B, Z categories were dismissed due to high noise. We note that PTT were deployed on 14 bar-headed geese from Qinghai Lake in March 2007. Three PTTs were still active as of 1 Nov 2008, and three PTTs were lost before the birds returned to their wintering place. In addition, among the PTTs deployed on the 15 bar-headed geese from Qinghai Lake in March 2008, nine out of them are still active as of Nov 1, 2008. Most of them have arrived at the winter area by Nov 1 2008.

We also note that for the satellite transmitters are expensive, it was impossible for us to use this equipment to track all the birds. Instead, only some key species were tracked. But many water bird species are highly faithful to the sites they use throughout their annual cycle (both within and between years) [6,7]. Such fidelity can be explained as a result of various selective pressures that flavor individuals which have an intimate knowledge of their environment. For most birds from the same population, they have the similar migration routes and habitat area [21]. Thus, although the number of our data samples is limited, the reliability and credibility of our survey are high.

4 Framework of Our Bird Migration Data Mining System

We propose a data mining system to discover the habitat area and migration route efficiently. A new hierarchical clustering algorithm is developed in the system to find sub-areas with a dense location points relative to the entire area. Then the Minimum Convex Polygon Home Range of bird species is calculated. Then, association analysis is used to discover the site connectedness and migration route between the discovered habitats. Figure 2 shows a diagram and component flow of our system, which consists of four phases: preprocessing, clustering, home range calculation and sequence mining. Each phase is described in detail in the subsequent subsections.

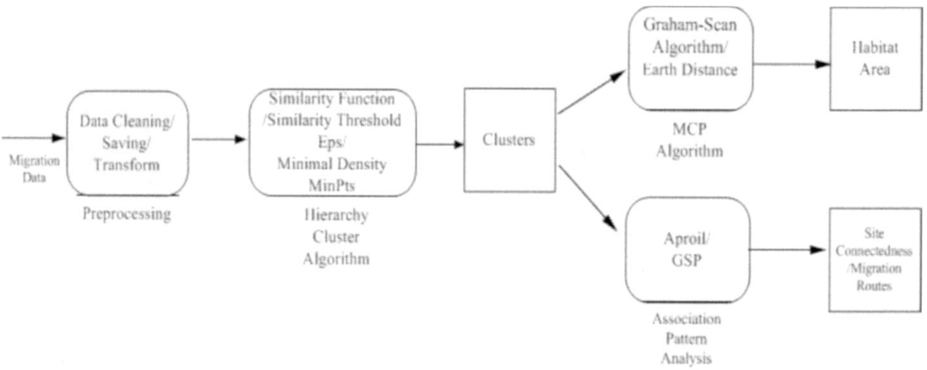

Fig. 1. System Framework of Our Bird Migration Data Mining System

4.1 Preprocessing

The raw data are downloaded from the USGS website. We focus on dynamic attributes such as latitude, longitude, time and speed. Outlier records are removed, and missing values are estimated and considered. The processed data are then stored at a relation database for further use.

4.2 Clustering Phase: Hierarchical Clustering and Spatial-Tree Building Based on DBSCAN

The objective of this phase is to mine interesting clusters from the preprocessed data set. As there are many choice of clustering algorithms, we require a clustering algorithm to satisfy the following criteria: (i) The algorithm should not require manual setting on the number of clusters. It is unreasonable to determine these parameters manually in advance. (ii) Since we only want to find important habitat area, the algorithm should filter out those with lower density. (iii) The location data are very large, the algorithm should be capable of handing a large data set within reasonable time and space constraints.

The DBSCAN [10] algorithm is a good choice as it meets all of these requirements. It does not need to input the number of clusters as a predefined parameter. According to the density-based definition, the density associated with a point is obtained by counting the number of points in a region of a specified radius, Eps, around the point. Points whose densities are above a specified threshold, MinPts, are classified as core points, while noise points are defined as non-core. Core points within the same radius of Eps to each other are merged together. Non-core and non-noise points, which are called border points, are assigned to the nearest core points. Those core points build the skeleton of a cluster. The algorithm makes use of the spatial index structure (R*-tree) to locate points within the Eps distance from the core points of the clusters. The time complexity of DBSCAN is O(N*logN). It is accepted in our application.

Biologists need to evaluate the "core areas", and then to identify the actual areas that are used within bird home ranges. The core area usually defined as areas concentrated by individual at each wetland. For example, the fostering place would be the core region, but the foraging area would be the out-of-core region. Motivated by this requirement, we introduce a Hierarchical DBSCAN (HDBSCAN) clustering approach, which can build up a Spatial-Tree encoding every cluster node like a Huffman tree code in a top-down manner.

The pseudo code of the HDBSCAN algorithm is shown in Fig 2. It adopts a Breath First Search-like strategy that clusters the data sets by using DBSCAN. Inputs are parameter Eps and Minpts for DBSCAN, together with bird migration data and predefined tree height. By one "first in, first out" queue "Q", spatial tree in Fig 2 is the output results. For each node in the same level of the tree, two pointers "front" and "last" point out their level (line3), and those nodes share the same DBSCAN parameter. At first, the DBSCAN are applied on those nodes (line 7). The clustering results are then put into "Q" (lines 11-15). If the depth of tree reaches the predefined tree height (lines 8-9), the hierarchical algorithm returns. Meanwhile, the id of a cluster is joined by its own cluster label and its father id as the tree grows. For instance in the

```
  Input: Location data: LD, Parameter: Eps and Minpts,
S-Tree: Height
  Output: LD with cluster lable and Spatial_Tree was
built
   DBSCAN_ OBJECT Root=Joint(LD,Eps,Minpts); // root
   node of Tree
   ENQUEUE(Q, Root) ;        // push DBSCAN object into
   Queue
   front:=0, last:=0, level=0;
   while(Queue<>empty and front<=last) DO
     DBSCAN_ OBJECT node= DEQueue(Q); // Pull data from
   Queue
      front++;                //
      Data_OBJECT Childern    =DBSCAN.getCluster(node);
   //Call DBSCAN
      if(level > Height)
         break;

      For i FROM 1 TO Childern.size  DO
         Data child=Childern.get(i);
         DBSCAN_ OBJECT Root=Joint(child,Eps,Minpts);
         ENQUEUE(Q,DBSCAN_ OBJECT) ;
      end For

      if(front>last) // members in one level have been
   searched
          last= Q.size()+front-1;
          level ++;
      end if
end while
```

Fig. 2. Pseudo code of the HDBSCAN algorithm

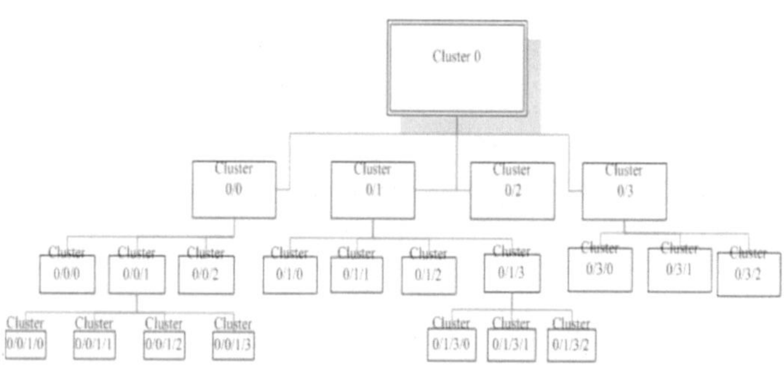

Fig. 3. An Spatial Tree Example: A Huffman coding-like Structure built by HDBSCAN

Fig 3, the left most leave node in the tree is encoded by his father id "0/0/1" and its own id "/0". Thus, its id is "0/0/1/0". By this Huffman encoding-like method, the cluster id is unique and the spatial tree is easy to manage.

4.3 Habitat Home Range Calculation Phase

In this phase, we use the idea of MCP (Minimum Convex Polygon) to circle the clusters and spherical geometry to obtain bird species' home range. There are two algorithms that compute the convex hull of a set of n points. Graham's scan runs in O(nlgn) time complexity, and the Jarvis's march runs in O(nh) time complexity, where h is the number of vertices of the convex hull. In our work, points with maximum or minimum latitude were found at first hand, and then we utilize Graham-Scan to compute the MCP. The run time is limited to O(n). A much more technical description of this approach can be referred to [23]. A closed geometric figure on the surface of a sphere is formed by the arcs of greater circles. The spherical polygon is a generalization of the spherical triangle [24]. If Φ is the sum of the radian angles of a spherical polygon on a sphere of earth radius R, then the area is:

$$s = [\Phi - (n - 2)\pi] * R^2 \tag{1}$$

4.4 Phase for Association Analysis

In this phase, association analysis is explored to discover site connectedness and bird migration routes. As illustrated in the Fig.4, points scattered around map are bird location sites, and their color stands for the discovered clusters labeled from the clustering phase. An arrow points out a bird migration route, which is considered as the pattern in the domain of data mining. Mining those spatial-temporal relationships between discovered habitats would be important for understanding how the different biological habitat elements interact with each other.

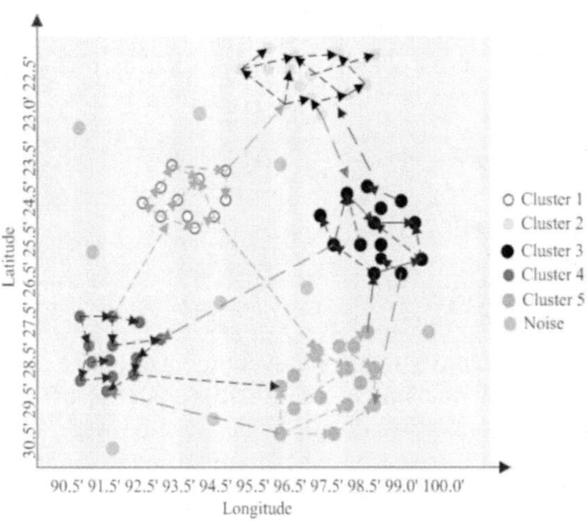

Bird ID	Visit Time	Visit Habitat
A	T1	Cluster 1
A	T2	Cluster 2
B	T1	Cluster 1
B	T2	Cluster 3
B	T4	Cluster {5,2}
C	T3	Cluster 2
C	T4	Cluster {5,3,2}
…	…	…
N	T_n	Cluster {1,2,5}

Fig. 4. Bird migration routes between clusters are converted into records in the right table

Biologists are interested in two types of spatial-temporal association patterns that involve sequences of events extracted from the clustered areas:

- **Non-sequential pattern**- relationships among the habitats for different birds, ignoring the temporal properties of the data. It can reveal the site connectedness.
- **Sequential pattern-** temporal relationships among the habitats for different birds, which are associated with migration routes.

One way to generate associative patterns from the migration data is to transform the spatial-temporal datasets in the Fig 4 into a set of transactions as in the Table 2. The main advantage of such approach is that we can use many of the existing algorithms to discover the association patterns that exist in the data. Different cluster areas that form the movement patterns can be recorded as the items for a bird transaction.

Table 2. Transforming the migration data into market-basket type transactions

Bird ID	T1	T2	T3	T4	T5	
A	Cluster 1	Cluster 2	θ	Cluster {2,1}	Cluster{2,3}	....
B	Cluster 1	Cluster{3,}	θ	Cluster{5,2}	Cluster {7,9}	....
C	θ	θ	Cluster{2}	Cluster{5,3,2}	Cluster {5,10}	....
....	....	....	....	....	....	....
N	Cluster{1,2,5}	Cluster {3,6}	θ	Cluster 10	Cluster{5,10}	....

Non-sequential associations among events only concern with the spatial cluster areas, irrespective to the timing information. The abstracted events can be transformed into a transaction format. Such representation allows us to apply the existing association rule mining algorithms. In this paper, we make use of the pioneering algorithm Apriori [15] to extract the association patterns. The following three interestingness measures are suggested to evaluate the association patterns such as one like: *cluster area A → cluster area B.*

$$\text{Support}=P(A,B) \tag{2}$$

$$\text{Confidence}=P(A,B)/P(A) \tag{3}$$

$$\text{Lift}=p(B|A)/P(B) \tag{4}$$

The support of a rule A → B is the probability that a transaction contains the code {A, B}. The confidence value of the rule denotes the conditional probability of {B} given {A}. Lift is computed to judge the correlation or the dependence between {A} and {B}. The association rule can be ranked based on an individual interestingness measure or their combinations.

If temporal information is incorporated, we can derive sequential associations among the events (cluster areas) using the existing sequential pattern discovery algorithms, such as GSP [17]. We choose to use the GSP algorithm, which was initially proposed by Agrawal et al. for finding frequent sequential patterns in the market-basket data. In the GSP approach, a sequence is represented as an ordered list of itemsets, s = <s1, s2, …, sn>. Each element sn of the sequence is subject to three timing constraints: window-size (i.e. maximum time interval among all items in the element), min-gap (i.e. minimum time difference among successive elements) and max-gap (maximum time difference among successive elements). In our paper, we have set the window-size to be 1 day, min-gap to be 0 and the max-gap to be 2 days. The above interestingness measures for non-sequential pattern need to be changed accordingly so as to measure sequential patterns. For instance, given one candidate sequence: *cluster area* ***A*** → *cluster area* ***B*** → *cluster area* ***C*** → *cluster area* ***D***. the confidence and lift are computed as follows:

$$\text{Confidence}=P(A\text{->}B\text{->}C\text{->}D)/P(A\text{->}B\text{->}C) \quad (5)$$

$$\text{Lift}= P(A\text{->}B\text{->}C\text{->}D)/ \{P(A\text{->}B\text{->}C)*P(D)\} \quad (6)$$

5 Experiment Results

We conducted many experiments to evaluate our system on the data sets of bar-headed goose (See details in the section 3). In this section, we first report an efficiency result of our HDBSCAN algorithm, we then give an interpretation on the results of the habitats discovered by using our HDBSCAN algorithm, a new hierarchical clustering approach. Then, we analyze the associative pattern to reveal the bar-headed goose migration site connectedness and routes. Finally, as a part of discussion, we present some implication advice for other research topics as well. Meanwhile, we combine our system with Goolge Map for a visualization of the distribution of the habitats and migration routes.

5.1 Spatial Distribution of Bar-Headed Goose

The Spatial-Tree of 2007 is built on the annual migration data from 2007-03 to 2008-03. In Fig. 6, the left panel is the Spatial-Tree and the right panel is the spatial distribution associated with certain nodes. The convex home range is depicted with polygons with different colors and the description is presented when the user clicks the marker with certain index. Due to page limitation, we only present the first node member associated with bar-headed goose over wintering, post-breeding, and stop over sites in the Fig 6 and details description in Table 3. From Fig 6, we can clearly find the breeding area Qinghai Lake with index 3, post breeding area Zhalin-Eling Lake with index 4. The maximum one is the wintering area in Tibet river valley with index 6 covering 9254 Km^2. It is interesting to note that one species (No. BH07_67693) moved to cluster with index 1 within Mongolia rather than stay in Qinghai Lake for breeding. The average range of the habitat area is 29045.38 Km^2.

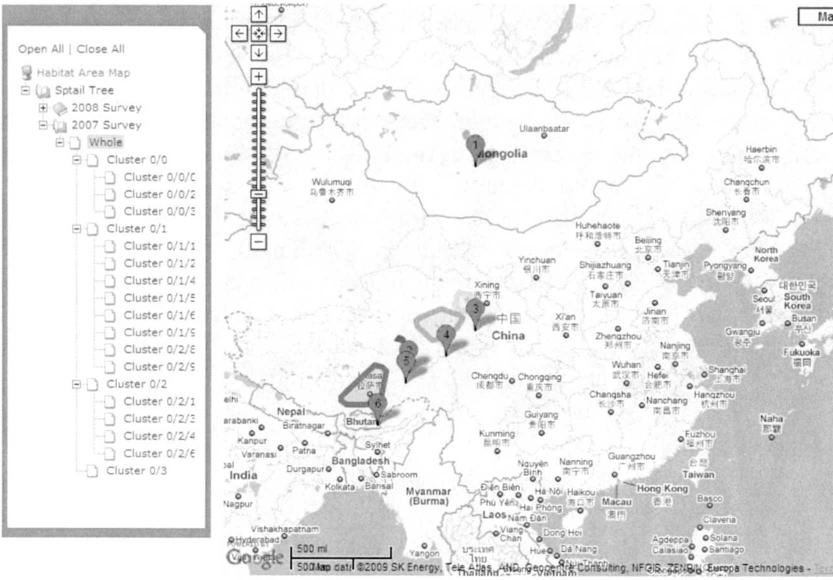

Fig. 6. Overview of 2007 bar-headed goose Spatial-Tree

Table 3. A part of results about discovered Migration habitat for bar-headed goose from Qinghai Lake, China

Cluster ID	Location Num	Home Range (Km^2)	Habitat Area Center		Bird Migration Time		Birds Num	Geography Description
			Longitude	Latitude	End	Begin		
0/0/	3368	16583.45	99.8082	36.9504	2007-10-24	2007-3-25	14	Qinghai Lake
0/1/	5394	62698.99	97.2455	35.0769	2007-12-14	2007-6-9	11	Zhalin-Eling Lake
0/3/	359	8206.649	93.4613	32.9937	2007-12-13	2007-10-11	7	YangZhiRiver YuanTou
0/2/	9069	85172.48	90.9785	29.6358	2008-2-25	2007-10-21	8	Tibet river valley
0/5/	56	361.828	99.865	47.9858	2007-6-6	2007-5-7	1	Mongolia
0/4/	34	1248.863	93.8064	33.77	2007-10-28	2007-10-13	3	TuoTuo River area

5.2 Site Connectedness of Bar- Headed Goose

As mentioned above, we transform the bird migration pattern into transaction database from in different levels of Spatial-Tree, separately. A part of association rules results from the Apriori shows some interesting patterns, where the CID means the cluster area id. Those association rules can effective evaluate the site connectedness. Those association rules can effective evaluate the site connectedness. For example, if we observe one associate rule {CID(0/0/1/) and CID(0/0/0/) and CID(0/0/3/) } -> { CID(0/1/4/) } with minimum support 21.4%. This can reveal the high site connectedness of stop area around the Qinghai Lake and Chaka salt Lake area. Thus, Qinghai lake Reserve is situated at the optimal location for storks preparing for the autumnal migration toward winter sites.

5.3 Migration Routes of Bar- Headed Goose

As described in the migration route mining in Section 4.3, bird migration between habitats could be regard as sequence. The discovered frequent sequence with higher confidence and lift would investigate a few interesting biological phenomena. A part of results are illustrated in Table 4. visualizing those sequences would help ornithologist to understand. From our observation, it is clearly that bar-headed goose departed the breeding place {ClD(0/0/)}in Qinghai Lake and then arrived at the post breeding area in Zhalin-Eling Lake area or Huangheyuan wetland {CID(0/1/)} and stayed there for about two months before heading to the south. Then they follow cluster 3 {CID(0/3/) } which is served as the stopover area, and finally arrive at the winter area {CID(0/2/)}. The movement of bar-headed goose depicted in this study conforms to the Central Asian Flyway [22]. There are eight birds migrated to Tibet river valley, and stay there from 2007-10-21 to 2008-02-25, with a total of 127 days over winter rather than fly to north-eastern India and Bangladesh. Mean fall migration duration was about 7 days. The migration distance is (1500 km (311km+382km+758km).

Table 4. Part of Sequential results in level 2 of 2007 Spatial-Tree, Minimum Support is 20% and Minimum Confidence is 30%

Rules ordered by support	1. [CID (0/0/1/)-> CID (0/0/0/)] (support=50%) 2. [CID (0/0/0/)-> CID (0/0/1)] (support=35.7%) 3. [CID ([0/1/1/)-> CID (0/2/1//)->CID (0/2/7/)-> CID (0/2/1/)] (support=21.4%)
Rules ordered by Confidence	1. [CID(0/1/1/)-> CID 0/2/1/-> CID 0/2/7/-> CID 0/2/1/] (confidence=100%) 2. [CID (0/0/0/)-> CID (0/0/0/)->CID (0/0/1/)-> CID (0/0/0/)-> CID (0/1/3/)] (confidence=100%) 3. [CID (0/0/1/)-> CID (0/1/6/) CID (0/1/3/) (confidence=75%) 4. [CID (0/0/1/)-> CID (0/0/0/)-> CID (0/1/4/)] (confidence=42%)
Rules ordered by Lift	1. CID (0/0/1/)-> CID (0/0/0/)->CID (0/0/1/)-> CID (0/3/0/) (lift=33.4%) 2. CID (0/1/1/)-> CID (0/2/1/)->CID (0/2/7/)-> CID (0/2/1/) (lift=25.4%)

5.4 Discussion and Implications for Habitat Conservation and Avian Influenza

Our cluster based approach for discovering the bar headed goose approximately depicts the geographical distribution of this species of wild water bird. Both of the cluster results in 2007 and in 2008 match greatly, which indicates that some certain habitats, such as the Qinghai Lake, DaLing Lake and the Tibet river valley, are of vital importance for some species. What is more, the discovered migration routes are critical for finding an adequate compromise between habitat protection and economic development in the regions along their migration routes. Wide areas of MCP prove that it is necessary to build a broad network to cover the different core region areas. The clustering results displayed in the GIS pave the way for human beings to construct a systematic nature reserve in future. In addition, scientists would like to do much more research, such as virus, plant and environment quality survey, to discover the way of highly pathogenic avian influenza disperse in the wild bird species' MCP.

6 Conclusion and Future Work

The satellite tracking has been used successfully to record the migration routes and stopover sites of a number of birds. Such information allows the development of a future plan for protecting the breeding and stopover sites. The proposal of our computational ideas and methods is motivated by the long-time lack of an efficient data analyzing approach which actually can help researchers to do this work systematically.

In this paper, we have suggested to explore the field by using the location data information as a supplement data mining process which can provide an alternative approach for traditional bird telemetry data analysis: visual observation from the location points. In order to discover the core range of the birds, a new clustering strategy has been introduced. This clustering strategy can effectively manage the different cluster areas and can discover the core areas in some larger habitat. Using association rule analysis, site connectedness of habitat and the autumnal migration routes for the bar-headed goose were investigated. Clustering and association rule mining do provide an effective assistance for biologists to discover new habitats and migration routes.

In the future, we plan to extend our current work to address several unresolved issues. Specifically, we intend to use Hidden Makov Model [25] to predict the bird movements. Also, we would like to compare the cluster area spatial distribution between different years, in an aim to discover the habitat changing trend of bird species. Finally, we intend to extend our analysis to other species in the Qinghai Lake to identify the cross habitat for different species.

Acknowledgments. This work was supported by Special Project of Informatization of Chinese Academy of Sciences in" the Eleventh Five-Year Plan", No.INFO-115-D02. The authors would like to thank the researchers in Western Ecological Research Center, Patuxent Wildlife Research Center and United States Geological Survey for sharing the data sets. Special Thanks to Professor Ying Liu at Graduate University of Chinese Academy of Sciences for her constructive suggestions in building our system.

References

1. Liu, J., et al.: Highly pathogenic H5N1 influenza virus infection in migratory birds. Science 309, 1206 (2005)
2. Li, Z.W.D., Mundkur, T.: Numbers and distribution of waterbirds and wetlands in the Asia-Pacific region: results of the Asian Waterbird Census 2002–2004. Wetlands International, Kuala Lumpur (2007)
3. Worton, B.J.: kernel methods for estimating the utilization distribution in home-range studies. Ecology 70, 164–168 (1989)
4. Kanai, Y., et al.: Discovery of breeding grounds of a Siberian Crane Grus leucogeranus flock that winter sin Iran, via satellite telemetry. Bird Conservation International 12, 327–333 (2002)
5. Mathevet, R., Tamisier, A.: Creation of a nature reserve, its effects on hunting management and waterfowl distribution in the Camargue (southern France). Biodiv. Conserv. 11, 509–519 (2002)

6. Shimazaki, H., et al.: Migration routes and important stopover sites of endangered oriental white storks (Ciconia boyciana) as revealed by satellite tracking
7. Ball, G.H., Hall, D.J.: ISODATA: a novel method of data analysis and pattern classification. Technical Report of Stanford Research Institute, Menlo Park, CA, Stanford Research Institute, 66 (1965)
8. Ester, M., et al.: A density-based algorithm for discovering clusters in large spatial databases with noise. In: Proc. 2nd Int. Conf. on Knowledge Discovery and Data Mining, Portland, OR, pp. 226–231 (1996)
9. Ester, M., Kriegel, H., Sander, J., Xu, X.: Incremental Clustering for Mining in a Data Warehousing Environment VLDB (1998)
10. Ng, R.T., Han, J.: Efficient and Effective Clustering Methods for Spatial Data Mining. In: Proc. 20th Int. Conf. on Very Large Data Bases, Santiago, Chile, pp. 144–155 (1994)
11. Ertöz, L., Steinbach, M., Kumar, V.: Finding topics in collections of documents: A shared nearest neighbor approach. In: Proceedings of Text Mine 2001, First SIAM International Conference on Data Mining, Chicago, IL,USA (2001)
12. Kaufman, L., Rousseeuw, P.J.: Finding Groups in Data: An Introduction to Cluster Analysis. John Wiley and Sons, Inc., New York (1990)
13. Agrawal, R., Srikant, R.: Fast Algorithms for Mining Association Rules. In: Proc. of the 20th VLDB Conference (1994)
14. Han, J., Pei, J., Yin, Y., Mao, R.: Mining frequent patterns without candidate generation: A frequent-pattern tree approach. International Journal of Data Mining and Knowledge Discovery 8(1), 53–87 (2004)
15. Agrawal, R., Srikant, R.: Mining sequential patterns. In: Proceedings of the 11th International Conference on Data Engineering (ICDE 1995) Taipei, Taipei, Taiwan, pp. 3–14 (1995)
16. Zaki, M.J.: Efficient Enumeration of Frequent Sequences. In: 7th International Conference on Information and Knowledge Management, Washington DC, November 1998, pp. 68–75 (1998)
17. Koperski, K., Han, J.: Discovery of Spatial Association Rules in Geographic Information Databases. In: Egenhofer, M.J., Herring, J.R. (eds.) SSD 1995. LNCS, vol. 951, pp. 47–66. Springer, Heidelberg (1995)
18. Pei, J., Han, J., Mortazavi-Asl, B., Pinto, H.: Prefixspan: Mining sequential patterns efficiently by prefix-projected pattern growth. In: Proceedings of the 2001 International Conference on Data Engineering (ICDE 2001), pp. 214–224 (2001)
19. Muzaffar, S.B., Johny, T.: Seasonal movements and migration of Pallas's Gulls Larus ichthyaetus from Qinghai Lake, China. Forktail 24, 100–107 (2008)
20. Miyabayashi, Y., Mundkur, T.: Atlas of key sites for Anatidae in the East Asian Flyway. Wetlands International—AsiaPacific, Tokyo: Japan, and Kuala Lumpur, Malaysia (1999), `http://www.jawgp.org/anet/aaa1999/aaaendx.htm` (accessed March 11, 2008)
21. Shan, G.H., JunYu, L., QiYing, L.: Introduction to ACM international Collegiate Programming Contest, 2nd edn., pp. 100–102 (in Chinese)
22. Weisstein, E.W.: Spherical Polygon. From MathWorld–A Wolfram Web Resource, `http://mathworld.wolfram.com/SphericalPolygon.html`
23. Daniel Sheldon, M.A.: Saleh Elmohamed, Dexter Kozen. In: Collective Inference on Markov Models for Model Appendix: Springer-Author Discount

GOD-CS: A New Grid-Oriented Dissection Clustering Scheme for Large Databases

Cheng-Fa Tsai and Chien-Sheng Chiu

Department of Management Information Systems
Pingtung University of Science and Technology, Pingtung, 91201, Taiwan
{cftsai,m9756001}@mail.npust.edu.tw

Abstract. Many recent clustering schemes have been applied in previous investigations to resolve the issues of high execution cost and low correction ratio in arbitrary shapes. Two conventional approaches that can solve one of these problems accurately are K-means and DBSCAN. However, DBSCAN is inefficient while K-means has poor accuracy. ANGEL and G-TREACLE have been proposed to improve current clustering tribulations, but require complicated procedures and numerous thresholds. This work presents a new clustering technique, called "GOD-CS", which employs grid-based clustering, neighborhood 8-square searching and tolerance rate to reduce these problems. Simulation results indicate that GOD-CS clusters large databases very quickly, while having almost identical or even better clustering performance in comparison to several existing well-known approaches with the original patterns in a simple procedure. Thus, GOD-CS performs well and is simple to implement.

Keywords: data clustering, data mining, grid-based clustering.

1 Introduction

Data clustering plays a significant role in numerous emerging business applications, since many applications need to classify clusters. Data mining is a common application of data clustering in the enterprise, and significantly raises business profitability. However, owing to the fast technological progress, the amount of data that can be stored in databases increases steeply, making data clustering a very useful process. Consequently, an efficient and effective clustering algorithm is strongly stipulated.

Clustering approaches can be divided into four types, namely partitioning, hierarchical, density-based and grid-based. Partitioning approaches run faster than other clustering approaches, but perform poorly in filtering noise and in shaping arbitrary patterns. They include K-means [6], which is a classical partitioning algorithm. Density-based approaches can recognize arbitrary shapes takes as its advantage, but have high computational cost. These methods include DBSCAN [5], in which time cost rises exponentially with the size of the data sets. Hierarchical approaches are not good at clustering in large database due to poor execution cost. Finally, grid-based approaches also cannot perform precise clustering, and therefore have poor accuracy.

R. Huang et al. (Eds.): ADMA 2009, LNAI 5678, pp. 302–313, 2009.

In summary, most existing clustering algorithms have high computational cost or poor pattern recognition. These issues need to be urgently addressed. This study presents a new improved grid-based clustering scheme, called "Grid-Oriented Dissection Clustering Scheme"(GOD-CS). GOD-CS adopts grid-based clustering, neighborhood 8-square searching technique and tolerance rate to reduce the threshold with only two parameters, and speed up the execution time and improve performance. Experimental results reveal that GOD-CS clusters groups effectively and efficiently in arbitrary patterns, and has an execution time of approximately one-fifth G-TREACLE [1] with a comparable correction rate.

2 Related Works

Clustering schemes can be categorized into several categories, namely partitioning, hierarchical, density-based and grid-based algorithm.

The K-means clustering method is fast, but has a low correction rate. K-means characterizes k cluster centers to find neighborhood points, and fine-tunes the center and neighborhood points of the resulting k clusters. K-means is easy to implement, and has fast execution with only one threshold, but easily classifies some points into the wrong cluster without detecting noise.

DBSCAN, which is an example of a density-based algorithm, recognizes arbitrary shapes almost perfectly with two thresholds (searching radius and minimum points that should be included), but has a very high time cost, making it unpopular for use in business applications.

Hierarchical algorithms agglomerate or divide groups into clusters according to similarities between groups. However, this process involves recursively compare the properties, resulting in a huge time complexity.

Finally, grid-based algorithm have rapid implementation times, since they assign each object to a grid to decrease processing cost, but have low precise rate and poor noise detection. Three new grid-based algorithms, namely GDH [3], ANGEL [2] and G-TREACL, have brought significant improvements in clustering technique over the past three years. The GDH algorithm adopts cubes, gradient decrease function and density attractors with three thresholds to perform clustering quickly and accurately. Although GDH has a low execution cost and a reasonable correction rate, both these factors could still be improved, and it requires too many thresholds. ANGEL is a hybrid density-based and grid-based method based on two means with 5 thresholds, and has a higher speed and a higher correction rate than GDH. G-TREACLE has recently been developed to raise clustering efficiency by employing a tree rather than DBSCAN to in shaping edges, but still stipulates five thresholds.

3 The Proposed GOD-CS Clustering Algorithm

This section describes the GOD-CS algorithm in detail in two parts. First, the process of GOD-CS is presented step by step with examples. Second, the refined programming code of GOD-CS is shown and explained in detail.

The GOD-CS algorithm has two parameters (*grid_size*, *tolerance_value*), and involves three simple steps: (1) initialization; (2) expansion with high density cubes, and (3) revising the high-density cubes, and defining the edges of clusters. In the first step, the initialization step formats a checker from the scale of original dataset, and then assigns each point to a grid. The cube with the highest density is identified, and is utilized to define populated cubes. In the second step, the high-density cubes, which have been defined in step 1, are combined with each linked neighborhood to form clusters. Step 3 revises the previously identified cubes to find those high-density cubes that were defined as non-high-density cubes. The implementation comprises three steps, which are described in detail as follows:

(1) Initialization: First, the checker is formed by entering the grid size. Fig. 1 depicts the checker.

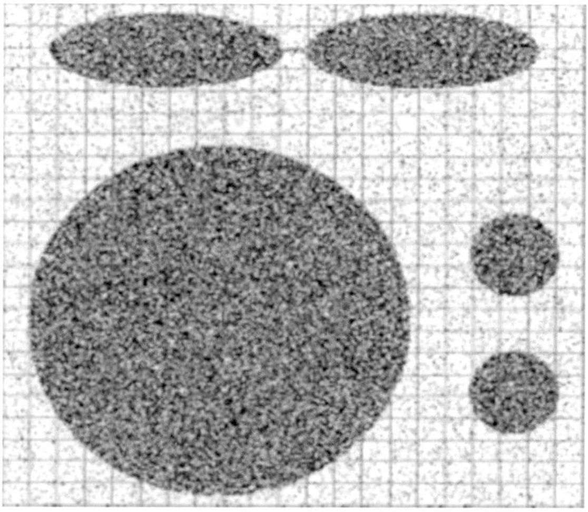

Fig. 1. Diagram of the pattern format after cutting as grids

The next stage is to find the highest density cube, i.e. the one with the most points, by comparing all cubes by density. High-density cubes are defined by multiplying the tolerance by with the number of points in the densest cube. An instance to explain this process in GOD-CS is given below.

Every number in Fig. 2 denotes the density of a cube, calculated from the points included in the square. Picture (a) illustrates the original format. The highlighted number in picture (b) with the red color (i.e. the number located in the second column and third row in picture (b)) represents the highest density cube, which is adopted to identify high-density cubes using the following formula:

$$SC = HDC \times TV \qquad (1)$$

where SC indicates the standard of checking high-density cubes; HDC denotes the highest density cube, and TV represents the tolerance value. In the case

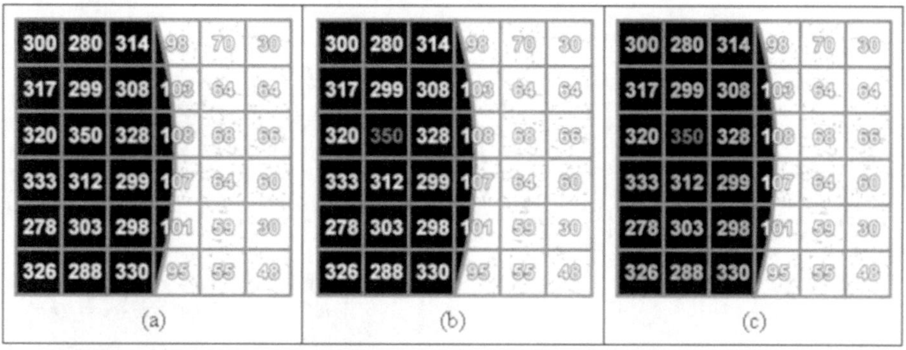

Fig. 2. Definition of high-density cubes

of Fig. 2, the bound for high-density cubes is found as 350*0.5=125. Fig. 2 (c) displays the high-density cubes, i.e. those with density of at least SC =125, highlighted in light blue (the left three columns).

(2) Expanding with high-density cubes: This process links high-density cubes mutually by using 8 neighborhood squares to speed up clustering time. A seed sheet is first generated to store every required expanding cube called as a seed such as the concept of DBSCAN. One previously unprocessed high-density cube is then selected as the starting cube for expansion to form a cluster. This starting cube is immediately added to the seed sheet. The seed sheet is then expanded policy by checking whether the 8 neighboring squares of each cube in the sheet should be added into seed sheet, and expanding every qualified cube in the sheet. Checked cubes are then deleted from the seed until the sheet is empty, thus forming a cluster. Finally, one cube is selected from the remaining high-density cubes as another starting cube, the identical expanding policy is run recursively until all high-density cubes have been processed. All the clusters are formed by following this process.

Fig. 3 shows the process of "Expanding with high-density cubes", which is composed of three steps. The right side of Fig. 3 depicts this process graphically. Steps 1, 2 and 3, which are in the left side of Fig.3, are run recursively until all the high-density cubes are processed. Significantly, Step 3 also runs itself recursively until the seed sheet is empty.

(3) Revising high-density cubes and defining cluster edges: To avoid poor tolerance values, which lead to false high-density cubes, the high-density cubes need to be revised, the edges of every cluster need to be defined correctly. Otherwise, some cubes in the cluster would be treated as noises. The proposed method redefines high-density cubes by discovering whether the number of high-density cubes among the 8 cubes neighboring a low-density cube is greater than 5. A cube whose 8 neighborhood cubes include at least 5 high-density cubes is also considered as a high-density cube. Fig. 4 presents two pictures simulating two distinct cases (the black background cubes denote high-density cubes, while

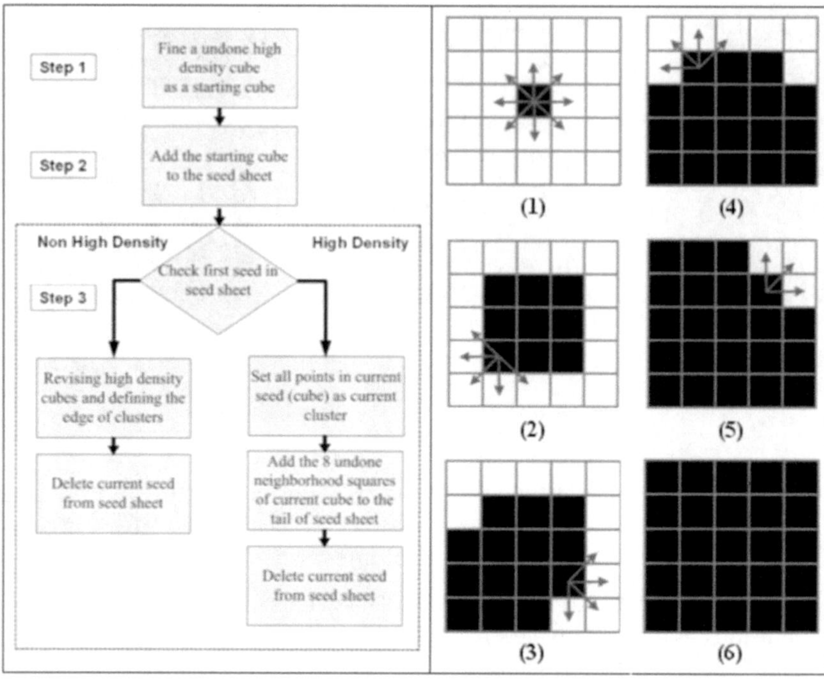

Fig. 3. The process and concept of expansion with high-density cubes

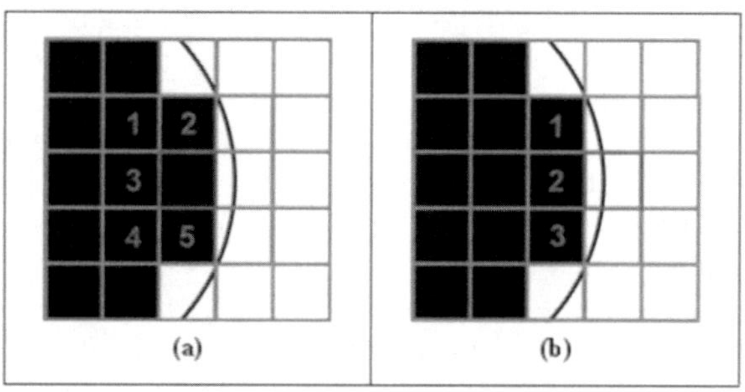

Fig. 4. High-density and low-density cubes, showing a cluster as including at least 5 high-density cubes

the white background cubes represent low-density cubes). Picture (a) shows a high-density cube on the edge of a cluster, and picture (b) illustrates the situation in a low-density cube. These two pictures in Fig. 4 indicate that every cube in a cluster has at least 5 neighboring high-density cubes, and every cube not in a cluster has no more than 3 neighboring high-density cubes.

Fig. 5 depicts the revising process with an example. The cluster has 5 high-density cubes, but is revised as a low-density with white background color. The

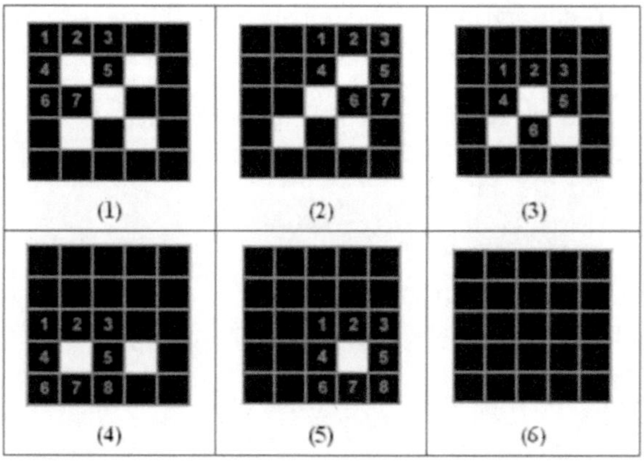

Fig. 5. Example to illustrate revision policy

number by each step in Fig. 5 denotes the number of neighboring high-density cubes near to the deviated low-density cube. The revision policy fully revises the deviated problem in the example of Fig. 5. Finally, the real edges of clusters can be defined during the expansion period after revising the high-density cubes.

The GOD-CS clustering algorithm can be described as follows:

```
GOD-CS(grid_size, tolerance_value)
    (1) Initialization(grid_size, tolerance_value);
    public cc=0;
    public seed_sheet=null;
    FOR each cube as cb DO
        IF cb.done==false && cb.high-density==true THEN
            seed_sheet.add(cb);
           (2) Expanding with high-density cubes();
        END IF
       cc++;
    END FOR
END GOD-CS
```

GOD-CS function is the majority method for executing our proposed clustering scheme. However, **(1) Initialization ()**, **(2) Expanding with high density cubes ()** and **(3) Revising high density cubes and defining the edge of clusters ()** are three sub-functions of GOD-CS, which are utilized each other. **(1) Initialization ()** and **(2) Expanding with high density cubes ()** are called by **GOD-CS ()**, afterward, **(3) Revising high density cubes and defining the edge of clusters ()** is employed by **(2) Expanding with high density cubes ()**.

grid_size is a parameter denoting the side of each square, *tolerance_value* is the second parameter of GOD-CS which means the range of being high-density cubes comparing with the greatest-density cube. Additionally, *cc* represents as a public variable to count the current cluster ID. Then, *seed_sheet* is the place which can store cubes that should be expanded.

```
(1) Initialization(grid_size, tolerance_value)
    CreateGridStructure(Find boundary of x&y_axis());
    FOR each point as cp DO
       PutPointsIntoRightGrid(cp);
    END FOR
    FOR each cube as cb DO
        max_grid_num = highest-cb.point_num;
    END FOR
    FOR each cube as cb DO
        IF cb.point_num >= max_grid_num * tolerance THEN
            Mark cb as high-density;
        END IF
    END FOR
END (1)
```

cp means the current point in a for-loop of each point, while *cb* stands for current cube in the for-loop of each cube.

```
(2) Expanding with high-density cubes()
    WHILE seed_sheet.size()!=0 DO
        seed_sheet.first.done=true;
        IF seed_sheet.first.high-density==true THEN
            FOR each point in seed_sheet.first as sp DO
                set sp as current_clusterid;
            END FOR
            FOR each 8 neighborhood cube as ncb DO
                IF ncb.done==false THEN
                    seed_sheet.add(ncb);
                END IF
            END FOR
        ELSE THEN
            (3) Revising high-density cubes
            and defining the edge of clusters();
        END IF
        seed_sheet.delete(first);
    END WHILE
END (2)
```

sp is a temporary location to place every point in first seed (cube). *ncb* means a neighborhood cube in the for-loop of each neighborhood cube.

```
(3)Revising high-density cubes and defining the edge of clusters()
   count=0;
   FOR each 8 neighborhood cube of seed_sheet.first as ncb DO
```

```
        IF ncb.high-density==true THEN
            count++;
        END IF
    END FOR
    IF count >=5 THEN
        set seed_sheet.first as high-density cube
        FOR each point in seed_sheet.first as cp DO
            cp=current_clusterid;
        END FOR
        FOR each 8 neighborhood cube of seed_sheet.first as ncb DO
            IF ncb.done==false THEN
                seed_sheet.add(ncb);
            END IF
        END FOR
    ELSE THEN
        set seed_sheet.first as boundary of a cluster;
    END IF
END (3)
```

count is initially given as zero which is used for counting the number of high-density neighborhood cubes.

4 Performance Studies

In this investigation, GOD-CS was implemented in a Java-based program, and the run on a desktop computer with 256MB RAM, an Intel 1.5GHz CPU on Microsoft

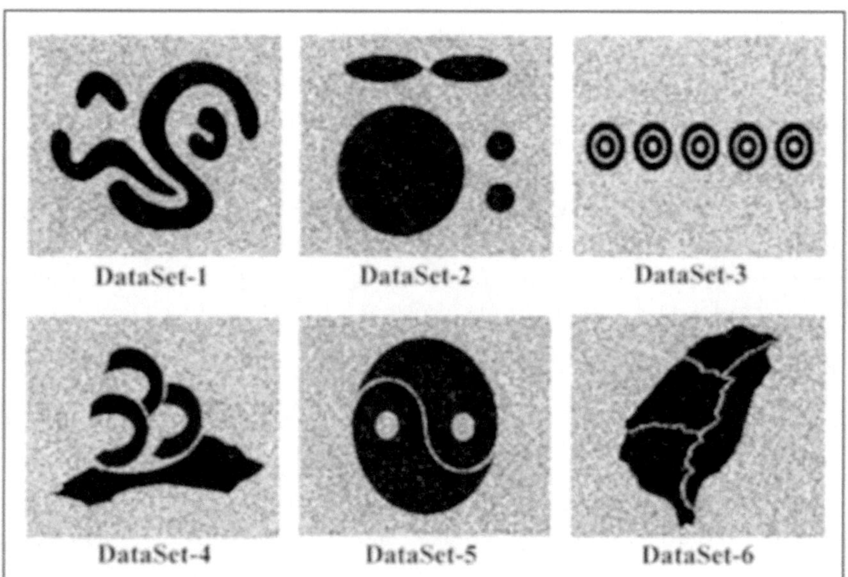

Fig. 6. The original datasets for experiment

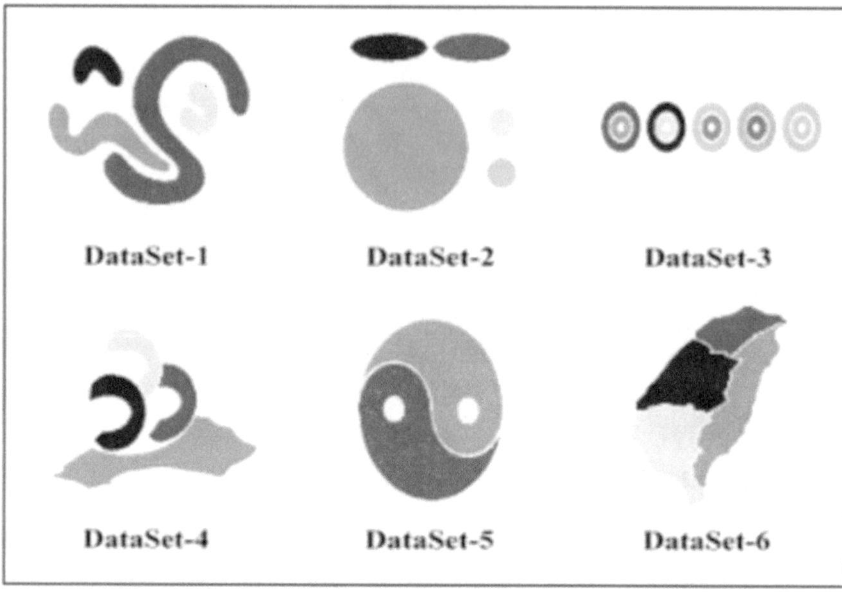

Fig. 7. The clustered datasets by GOD-CS

Windows XP professional Operation System. Six synthetic datasets were used to compare the performance with GOD-CS, K-means, DBSCAN, GDH, ANGEL and G-TREACLE. Fig. 6 illustrates the original six tested datasets. Experiments were performed on six synthetic datasets, which were of two types, consisting of 230,000 or 575,000 objects.

The noise filtering rate (NFR) and clustering correctness rate (CCR) were compared in every algorithm for the six datasets. In detail, NFR denotes the capability of successful detecting noises by the algorithms, and CCR represents the proportion of the algorithms in accurate clustering points. Fig. 6 and Fig. 7 illustrate six examined datasets, the former represents the original datasets, while the latter shows the clustered datasets by GOD-CS. Tables 1 and 2 list two comparison charts of the performance in the six synthetic datasets with time cost, CCR and NFR respectively. They depict K-means executed fast in low CCR and NFR, afterward, GDH, ANGEL and G-TREACLE perform excellent in arbitrary shapes. However, GOD-CS outperforms them in time cost in arbitrary patterns. Fig. 8 and Fig. 9 are two curve-charts revealing the information of time cost in GOD-CS, K-means, DBSCAN, GDH, ANGEL and G-TREACL using 230,000 objects datasets as well as 575,000 objects datasets respectively, which indicate GOD-CS experience the lowest time cost.

In summary, simulation results reveal that GOD-CS has lower execution cost than K-means, DBSCAN, GDH, ANGEL and G-TREACLE as well as comparable CCR and NFR. Thus, GOD-CS can cluster arbitrary patterns quickly.

Table 1. Comparisons with GOD-CS, K-means, DBSCAN, GDH, ANGEL and G-TREACLE using 230,000 objects datasets with 15% noise; item 1 denotes time cost in seconds; item 2 represents the CCR (%), while item 3 indicates NFR (%)

Algorithm	Item	DataSet-1	DataSet-2	DataSet-3	DataSet-4	DataSet-5	DataSet-6
K-means	1	8.406	13.782	9.718	20.829	2.75	7.344
	2	50.032%	56.241%	51.144%	58.108%	49.957%	59.056%
	3	0%	0%	0%	0%	0%	0%
DBSCAN	1	1,209.25	1,213.907	1,221.875	1,214.687	1,206.031	1,210.547
	2	99.999%	99.96%	99.999%	99.991%	99.96%	99.999%
	3	95.08%	96.513%	95.22%	95.38	96.41%	95.813%
GDH	1	3.453	5.875	8.985	25.969	16.672	19.172
	2	99.031%	99.712%	98.009%	98.642%	97.791%	94.431%
	3	96.036%	97.406%	98.766%	99.256%	99.283%	99.336%
ANGEL	1	3.14	3.782	6.734	6.859	9.672	11.359
	2	99.05%	99.051%	99.03%	99.271%	99.025%	98.412%
	3	96.683%	98.11%	98.656%	99.01%	99.08%	99.12%
G-TREACLE	1	2.112	2.022	4.458	4.679	6.223	7.256
	2	99.02%	99.245%	99.012%	99.376%	99.5%	98.995%
	3	98.779%	99.1%	99.032%	98.578%	98.862%	99.102%
GOD-CS	1	0.89	0.922	0.929	0.938	0.89	0.922
	2	98.3%	98.095%	98.272%	98.952%	99.16%	98.851%
	3	99.66%	99.754%	99.41%	99.659%	99.4%	99.544%

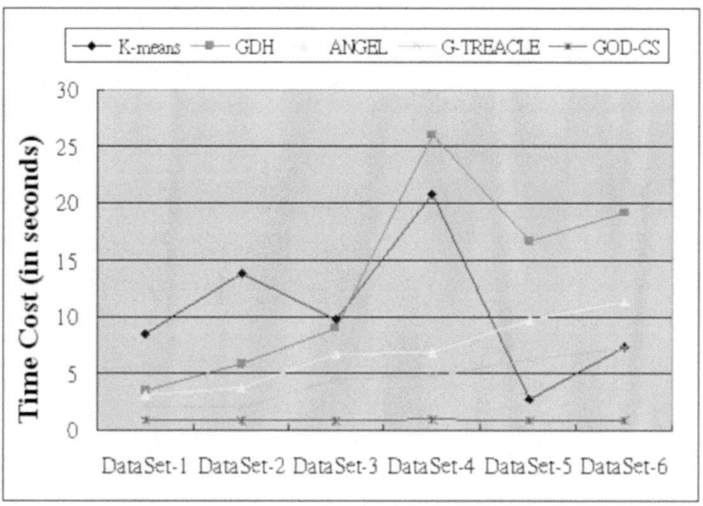

Fig. 8. Time cost comparison with K-means, GDH, ANGEL, G-TREACLE and GOD-CS using 230,000 objects datasets with 15% noise

Table 2. Comparisons with GOD-CS, K-means, DBSCAN, GDH, ANGEL and G-TREACLE using 575,000 objects datasets with 15% noise; item 1 denotes time cost in seconds; item 2 represents the CCR (%), while item 3 is NFR (%)

Algorithm	Item	DataSet-1	DataSet-2	DataSet-3	DataSet-4	DataSet-5	DataSet-6
	1	18.531	16.391	59.437	43.203	7.828	19.906
K-means	2	49.925%	51.149%	60.837%	57.612%	50.007%	54.49%
	3	0%	0%	0%	0%	0%	0%
	1	7,480.231	7,460.109	7,497.906	7,470.813	7,410.094	7,436.843
DBSCAN	2	99.998%	99.968%	99.997%	99.989%	99.954%	99.999%
	3	95.34%	96.53%	95.102%	95.093%	96.335%	95.906%
	1	8.188	9.516	13.359	31.75	26.297	51.469
GDH	2	99.213%	99.642%	98.229%	98.153%	96.456%	96.4%
	3	96.618%	97.477%	98.932%	99.408%	99.736%	99.71%
	1	7.922	8.212	10.876	12.554	15.437	19.663
ANGEL	2	99.45%	99.651%	99.432%	99.574%	99.482%	99.221%
	3	98.836%	99.11%	99.021%	99.22%	99.23%	99.032%
	1	6.156	5.594	7.776	8.469	10.64	15.75
G-TREACLE	2	99.392%	99.511%	99.375%	99.767%	99.754%	99.127%
	3	98.694%	99.051%	98.894%	98.377%	98.74%	98.949%
	1	1.359	1.375	1.609	1.625	1.360	1.313
GOD-CS	2	99.142%	99.344%	98.713%	99.212%	99.483%	99.467%
	3	99.513%	99.661%	99.392%	99.56%	99.591%	99.381%

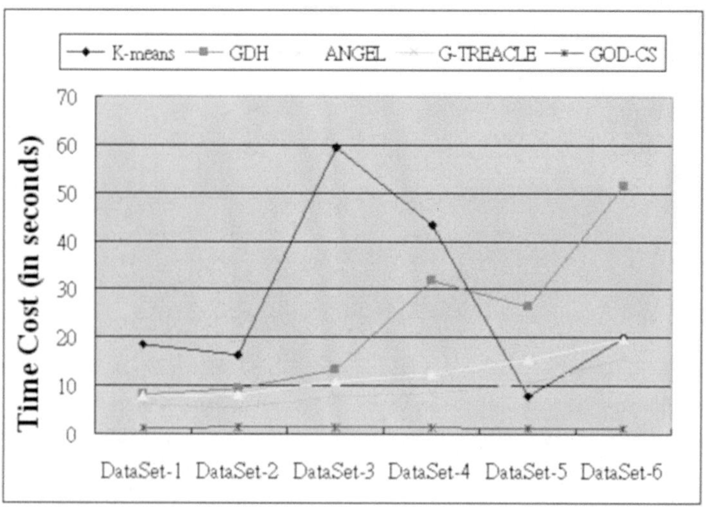

Fig. 9. Time cost comparison with K-means, GDH, ANGEL, G-TREACLE and GOD-CS using 575,000 objects datasets with 15% noise

5 Conclusion

Many clustering algorithms have high computational cost or problems with pattern recognition, which need to be urgently addressed. This work solve these problems using a new clustering algorithm called GOD-CS, based on grid-oriented clustering algorithm, the notion of tolerance value and searching the neighboring 8 squares. The proposed approach not only significantly reduces the execution cost, but also has excellent noise filtering capability. Although the grid-based clustering approach could lead to sub-optimal edge processing, in terms of dissecting into tiny cubes, the accuracy rate rises considerably with decreased run time when searching the neighboring 8 squares technique. Simulation results demonstrate that the execution cost of GOD-CS is approximately one-tenth to one-third of those of K-means, DBSCAN, GDH, ANGEL and G-TREACLE. However, the accuracy rate of GOD-CS is almost over 99% of those of the other algorithms in every case examined, revealing that the merits of GOD-CS far outweigh drawbacks: GOD-CS has a very high accuracy with a significantly lower execution cost than other existing algorithms.

Acknowledgement. The authors would like to thank the NSC, Taiwan for financially supporting this research under Contract No. NSC-96-2221-E-020-027.

References

1. Tsai, C.F., Yen, C.C.: G-TREACLE: A New Grid-Based and Tree-Alike Pattern Clustering Technique for Large Databases. In: Washio, T., Suzuki, E., Ting, K.M., Inokuchi, A. (eds.) PAKDD 2008. LNCS (LNAI), vol. 5012, pp. 739–748. Springer, Heidelberg (2008)
2. Tsai, C.F., Yen, C.C.: ANGEL: A New Effective and Efficient Hybrid Clustering Technique for Large Databases. In: Zhou, Z.-H., Li, H., Yang, Q. (eds.) PAKDD 2007. LNCS (LNAI), vol. 4426, pp. 817–824. Springer, Heidelberg (2007)
3. Wang, T.P., Tsai, C.F.: GDH: An Effective and Efficient Approach to Detect Arbitrary Patterns in Clusters with Noises in Very Large Databases. In: Degree of master at Pingtung University of Science and Technology, Taiwan (2006)
4. Borah, B., Bhattacharyya, D.K.: An Improved Sampling-Based DBSCAN for Large Spatial Databases. In: Proceedings of International Conference on Intelligent Sensing and Information, pp. 92–96 (2004)
5. Ester, M., Kriegel, H.P., Sander, J., Xu, X.: A Density-Based Algorithm for Discovering Clustering in Large Spatial Databases with Noise. In: Proceedings of the 2nd International Conference on Knowledge Discovery and Data Mining, pp. 226–231 (1996)
6. McQueen, J.B.: Some Methods of Classification and Analysis of Multivariate Observations. In: Proceedings of the 5th Berkeley Symposium on Mathematical Statistics and Probability, pp. 281–297 (1967)

Study on Ensemble Classification Methods towards Spam Filtering

Jinlong Wang[1], Ke Gao[1], Yang Jiao[1], and Gang Li[2]

[1] School of Computer Engineering, Qingdao Technological University, 266033 Qingdao, Shandong, China
[2] School of Information Technology, Deakin University 3125 Victoria, Australia
wangjinlong@gmail.com, qd_cap@yahoo.com.cn, jy_game_over@163.com, gang.li@deakin.edu.au

Abstract. Recently, many scholars make use of fusion of filters to enhance the performance of spam filtering. In the past several years, a lot of effort has been devoted to different ensemble methods to achieve better performance. In reality, how to select appropriate ensemble methods towards spam filtering is an unsolved problem. In this paper, we investigate this problem through designing a framework to compare the performances among various ensemble methods. It is helpful for researchers to fight spam email more effectively in applied systems. The experimental results indicate that online based methods perform well on accuracy, while the off-line batch methods are evidently influenced by the size of data set. When a large data set is involved, the performance of off-line batch methods is not at par with online methods, and in the framework of online methods, the performance of parallel ensemble is better when using complex filters only.

Keywords: spam email filtering, ensemble, classification.

1 Introduction

For the past decade, much progress has been made in designing accurate classifiers for text emails [1, 2, 3, 4, 5, 6]. Machine learning technologies, such as Naive Bayes [7, 8], Support Vector Machine (SVM) [9, 10], AdaBoost [11], PPM [12], *etc*, have achieved outperforming results in the anti-spam filtering task. However, even with the latest development of feature selection techniques and classification algorithms, it is still difficult to further improve the performance of single filters. Thus, many scholars attempt to combine filters to enhance the performance. The methods have been widely applied in machine learning as well as text categorization related fields.

Generally, the ensemble methods perform a more generalizing capability than single filters [13, 14]. Based on the combining style, the ensemble methods can be divided into two kinds: serial ensemble and parallel ensemble. The two kinds of ensemble methods have been widely used in real spam filtering application and development. On the serial side, there are many well-known commercial applications such as IronPort from Cisco[1],

[1] http://www.ironport.com/

R. Huang et al. (Eds.): ADMA 2009, LNAI 5678, pp. 314–325, 2009.

Barracuda Spam Firewall from Barracuda[2], *etc*. On the parallel side, IrnoMail from CipherTrust[3] shows the excellent performance. These tools are using a variety of techniques in spam filtering. Through detecting various features from an e-mail message[4], the weakness of single filters can be alleviated, accordingly the risk of misclassification can be reduced.

Summarized from the commercial applications on anti-spam task, we can find that people always focus on enhancing overall system performance by combining filters, and comparing whether the misclassification rate has been reduced between different ensemble methods. Many researchers have compared the performance between online ensemble classification [15] and also compared the performance between the same nature filters [16]. While all these approaches seem appealing, it is difficult to make a fair comparison of the differences between diverse ensemble methods, and no effective method has been proposed to evaluate these methods in different cases. This paper will attempt to fill the void by designing a reasonable framework to compare the performance between various ensemble methods. The experimental results on different datasets will lead to some conclusions which can help researchers to choose reasonable ensemble methods and make spam filter system more effective. The results will be much helpful for spam blog filtering, and other related applications in text categorization.

The remainder of this paper is organized as follows. Section 2 describes the research framework, including ensemble mode design and ensemble learning mode design. In Section 3, we present the experimental methods for comparison analysis. In Section 4, we describe the experimental results and analyze the result for obtaining some valuable conclusions. Finally, Section 5 concludes this paper.

2 Research Framework

2.1 Serial and Parallel Ensembles Design

Filter ensembles can be divided into two kinds, serial ensemble and parallel ensemble.

- Serial ensemble: Filters work in a serial way, processing messages in turn. If the spamminess score produced satisfies threshold, the ensemble system will evaluate this message straightforwardly. Otherwise the message will be turned into the next filter till the last filter with a final evaluation.
- Parallel ensemble: Each filter works separately and obtains its spamminess, then combine all the results in a certain way – linear or non-linear combination – as the final result.

According to the complexity of implementing filters, we arranged filters into two kinds: simple filters and complex filters. Generally, simple filters, which classify based on explicit rules, are generated using a particular part of an e-mail. Though with less

[2] http://www.barracudanetworks.com/

[3] http://www.securecomputing.com/index.cfm?skey=1612

[4] For the convenience of discussion, we use "message", "e-mail", or "e-mail message" to represent e-mails without difference.

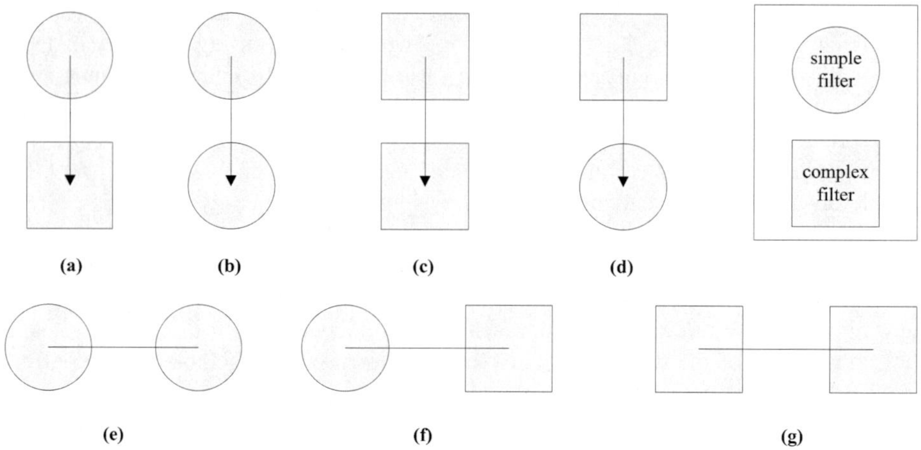

Fig. 1. Serial and parallel methods

accuracy, this kind of filters can gain lower cost in the process of learning and classifying. Complex filters, which classify based on implicit rules (training model), are generated according to textual content of an e-mail. This kind of filters need much more time in learning and classifying process in spite of higher accuracy.

Assuming in the extreme cases, we can initially parallel simple filters and complex filters respectively, and then answer spam or ham by hybriding them in a serial ensemble. This method is equivalent to using enhanced simple and complex filters in serial ensemble. Or, we can combine each filter in the following sequences, simple, complex, simple and then complex. Because obtaining the accurate classification of a simple filter is difficult, the ensemble style described above can be viewed as two complex filters hybrided in a serial ensemble. The ensemble methods are shown as Fig. 1:

1. Serial ensemble: first simple next complex(a), both simple(b), both complex(c), first complex next simple(d).
2. Parallel ensemble: simple and simple(e), simple and complex(f), complex and complex(g).

2.2 Designing Ensemble Learning Modes

It is ideal that the system can get the user's feedback immediately. However, this kind of nearly real-time processing would probably bring the system an extreme high load. In fact, users usually won't give feedback to the filter system immediately. Instead, they often send many feedbacks with some delay or even don't send any feedback at all. Therefore, we should trade off between real-time feedback and off-line batch learning to reduce the load in learning procedure while reserving the high accuracy. Our methods are shown as follows:

1. Chronological batch mode[5]
   Similar to online mode, it inputs into filter n messages in a chronological sequence, m_0 through m_{n-1}. Here, m_i does not mean a single e-mail message, but means a partition of messages. Then for each partition m_i, the classifier would produce a class based on messages prior to m_i. Assume 10000 messages are divided into 10 partitions according to the chronological order. We adopt the front partition as training model, and then classify the next ten percent messages using the obtained model. After that, we add this partition of messages to tune the model and process the next partition of messages. At last we evaluate performances of each phrase.
2. Online mode
   We start with an initial model produced by very limited instances or just use an empty model. When a message is classified it will be added to the training set immediately. This kind of method can generate a more real-time model than chronological batch method.

2.3 Designing Nature Based Ensembles

It is believe that for ensemble learning to be effective, it is important to produce individual learners with strong generalizing capability and much difference [17, 18]. According to [13, 19], we design two types of ensemble methods in this paper:

- Same nature filter ensemble: combining filters based on the same nature of model, such as discriminative model or generative model.
- Different nature filter ensemble: combining filters based on different learning methods.

3 Experimental Methods

3.1 Filters and Pre-processing

We compared seven filters according to our experiment methods. Table 1 shows the information of filters.

There are three simple filters compared in the experiment: a body full word filter based on term frequency statistics (BodyFullWordFilter); a subject full word filter based on term frequency statistics (SubjectFullWordFilter); and a re-filter based on user behavior (ReBasedFilter), by judging whether there contains character "re" in mail subject to give an answer. These three simple filters are proposed by [20].

There are four complex filters compared in the experiment, SVM (discriminative model) based filter, Bogofilter and SpamBayes are open-source filters based on Bayesian model, PPM (compression method) based filter. ROSVM is proposed by D. Sculley from Tufts [21][6]. Prediction by Partial Matching (PPM) [22] is a finite-context statistical modeling technique, it predicts the next symbol depending on many context models which has a fixed alphabetical order and number, calculates the count of each context

[5] In order to distinguish it with the traditional batch methods, we use the term "chronological batch" to indicate the off-line related experimental methods.

[6] http://www.eecs.tufts.edu/~dsculley/onlineSMO/

Table 1. The filtering list

Filter Name	Learning Method	Input Content
BodyFullWordFilter	word statistic	Body
SubjectFullWordFilter	word statistic	Subject
ReBasedFilter	user behavior	Subject
ROSVM	SVM	Subject/Body
PPM	Compression	Subject/Body/Original
Bogofilter	Naive Bayes	Original text
Spambayes	Naive Bayes	Original text

model predicts in input string, and produces a probability. The code of PPM can be downloaded from the website[7]. Bogofilter[8] and Spambayes[9] are two open-source filters based on Bayesian model, which performed well in TREC spam track. In our experiment, we adopt TREC version of the two filters.

3.2 Evaluation Measures

The TREC evaluation measures are mainly discussed in our experiment. Filters, due to the spamminess output, can be regarded as soft-classifier [19]. By setting different threshold value, filters can be characterized by the set of false positive rate and false negative rate, thus, we can obtain the ROC curve [23] to performance evaluating. (1-ROCA)%, which indicates the area above ROC curve, is a very effective evaluation measure. (1-ROCA)% has also a probabilistic interpretation [19]: it is the probability that the classifier will award a random spam message a lower value of spamminess than a random ham message.

Another evaluation measure is sm%@hm%=.1, which presents how much the spam misclassification rate is when ham misclassification rate is 0.1% [15][10]. Other than the overall performance of filters, considering the different cost for misclassification, we will evaluate the spam misclassification rate when the ham misclassification rate is rather low.

$$sm\% = \frac{|spam\ misclassified\ as\ ham|}{|all\ the\ spam|}$$

$$hm\% = \frac{|ham\ misclassified\ as\ spam|}{|all\ the\ ham|}$$

4 Experimental Results and Analysis

4.1 Datasets and Experimental Settings

Table 2 shows the data sets, which are appropriate for online setting experiments, retain the original format of e-mails. Considering that SVM based filter is adopted in

[7] http://ai.ijs.si/andrej/psmslib.html
[8] http://bogofilter.sourceforge.net/
[9] http://spambayes.sourceforge.net/
[10] http://plg1.cs.uwaterloo.ca/~trlynam/spamjig/spamfilterjig-full/

Table 2. The dataset characteristics

Corpus Name	Spam	Legitimate	Main language
SpamAssassin	1885	4149	en
trec06p	24192	12910	en
trec06c	42854	21766	cn
trec07p	50199	25220	en

Table 3. The range of each filter's spamminess

Filter Name	Spamminess Range
BodyFullWordFilter	-
SubjectFullWordFilter	t>0.6 || t<0.4
ReBasedFilter	t<0.5
ROSVM	-
PPM	t>0.55 || t<0.45
Bogofilter	t>0.6 || t<0.4
Spambayes	-

experimental settings, we choose trec06p and trec06c[11] to be a reference collection for establishing term-index table. The test corpus are SpamAssassin[12] and trec07p[13].

We investigate the pros and cons of various ensemble methods in experiments. According to analysis above, two experiments are arranged in the following: the first one is designed for validating the difference between online learning mode and chronological batch learning mode. After identifying which learning mode better, we discuss and analyze the performance of serial ensembles and parallel ensembles with the test results generated through the second experiment.

[15] discussed the ensemble methods based on results from filters in parallel style, they proposed four kinds of ensemble methods: Logistic Regression (LR) [24], SVM, log-odds and vote. In general, LR gives a superior result, but it is only suitable in batch mode. Log-odds performs moderately and suits for both online test as well as batch test. Considering that there are two modes in our experiments, we adopt the log-odds method [15] for result ensemble.

$$L_n = log\frac{|i < n|s_i \leq s_n \; and \; ith \; message \; is \; spam| + \varepsilon}{|i < n|s_i \geq s_n \; and \; ith \; message \; is \; ham| + \varepsilon}$$

The formula transforms spamminess to a score that history related only. Assume that a filter reports spamminess of the n-th message, after that, the formula produces L_n score according to user feedback and spamminess of all messages prior. It is more likely to be spam if the numerator of the formula is large, otherwise ham.

In order to facilitate the implementation of experimental settings, we adopt the result ensemble format which means each filter works separately and then we combine all the

[11] http://plg1.cs.uwaterloo.ca/cgi-bin/cgiwrap/gvcormac/foo06

[12] http://spamassassin.org/publiccorpus/, the order of e-mails were adjusted by received time

[13] http://plg.uwaterloo.ca/~gvcormac/

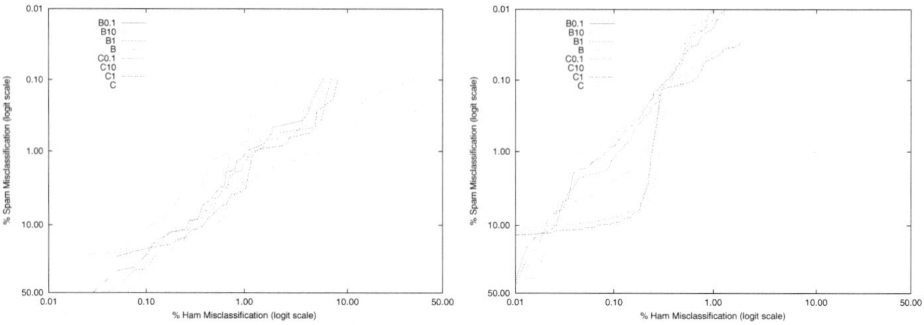

(a) SpamAssassin (b) trec07p

Fig. 2. Experiment 1 result on different datasets

results on demand. The spamminess of seven filters were tuned to the range [0,1]. For serial ensembles, the system will export judgment straightforwardly if the spamminess produced by present filter is high enough, otherwise the message will go to the next filter. The range of "high enough" are shown in Table 3.

4.2 Experiment 1

The experiment is designed to help us infer a more appropriate setting for ensemble system, by comparing the performance under different learning modes. We can find a trade-off between online mode and chronological batch mode which will not only reduce learning costs but also keep a high level accuracy.

We choose three complex filters – PPM, ROSVM, Bogofilter – as the base to combine. Under the chronological batch situation, due to the initial n messages are classified without any trained model, we have to cut off them. For corpus SpamAssassin and trec07p respectively, we cut off the initial 1000 and 10000 messages, retain the posterior 5034 and 65419 messages. For ROSVM under chronological batch situation, due to training a SVM model cost too much time, we adopt Fibonacci-like intervals [15] to determine the training set at 1000(10000), 2000(20000), and 5000(50000). Fig.2 shows the results, Table 4 and Table 5 show the statistical analysis, respectively on two corpus. Under serial and parallel ensemble methods, we implement two learning mode tests: online mode vs. chronological batch mode. The identifiers **B** and **C** in Fig.2, Table 4 and Table 5 represent the parallel and serial ensemble, respectively.

The result shows that online mode achieves better results than chronological batch mode no matter which ensemble method is used. From another view, the online mode can get a lower sm% while hm% is rather low. Comparing with the SpamAssassin test, we find some interesting phenomena on trec07p test. From Table 5, the performance between online mode and chronological batch mode are nearly the same, meanwhile, the performance between 10% and 0.1% of chronological batch mode are much close too. This indicates that the corpus size has certain influences on chronological batch mode tests. On the other hand, because of the cost of adapting model is very expensive, the online mode costs much more resources than chronological batch mode. According to results shown above, we can infer that, for the small corpora or personal mail system,

Table 4. Statistical results on SpamAssassin in experiment 1

Ensemble Method	Learning Mode	Batch Size	(1-ROCA)%	sm%@hm%=.1
serial ensemble	online	(C)	0.1474 (0.0746 - 0.2908)	19.76 (13.14 - 28.62)
serial ensemble	chronological batch	10%(C10)	1.5904 (1.2861 - 1.9653)	95.73 (88.28 - 98.53)
serial ensemble	chronological batch	1%(C1)	0.1886 (0.1095 - 0.3244)	20.78 (9.33 - 40.06)
serial ensemble	chronological batch	0.1%(C0.1)	0.1627 (0.0909 - 0.2910)	18.74 (13.25 - 25.83)
parallel ensemble	online	(B)	0.0601 (0.0293 - 0.1231)	14.56 (3.95 - 41.42)
parallel ensemble	chronological batch	10%(B10)	0.3340 (0.2191 - 0.5089)	62.80 (23.50 - 90.27)
parallel ensemble	chronological batch	1%(B1)	0.1826 (0.0989 - 0.3369)	32.10 (11.30 - 63.68)
parallel ensemble	chronological batch	0.1%(B0.1)	0.1580 (0.0801 - 0.3117)	30.43 (10.23 - 62.66)

Table 5. Statistical results on trec07p in experiment 1

Ensemble Method	Learning Mode	Batch Size	(1-ROCA)%	sm%@hm%=.1
serial ensemble	online	(C)	0.0335 (0.0242 - 0.0464)	7.79 (6.70 - 9.04)
serial ensemble	chronological batch	10%(C10)	0.0506 (0.0372 - 0.0688)	13.71 (12.70 - 14.79)
serial ensemble	chronological batch	1%(C1)	0.0381 (0.0271 - 0.0534)	9.26 (7.64 - 11.19)
serial ensemble	chronological batch	0.1%(C0.1)	0.0355 (0.0250 - 0.0505)	8.24 (7.15 - 9.49)
parallel ensemble	online	(B)	0.0159 (0.0071 - 0.0356)	0.60 (0.42 - 0.86)
parallel ensemble	chronological batch	10%(B10)	0.0177 (0.0088 - 0.0359)	2.53 (1.68 - 3.79)
parallel ensemble	chronological batch	1%(B1)	0.0163 (0.0076 - 0.0349)	1.25 (0.64 - 2.43)
parallel ensemble	chronological batch	0.1%(B0.1)	0.0153 (0.0069 - 0.0339)	0.84 (0.47 - 1.50)

it is more appropriate to adopt online mode ensemble method; for the large corpora or enterprise mail system, it is better to adopt chronological batch mode, in this case we can not only keep a high accuracy but reduce the system load effectively.

4.3 Experiment 2

From the result of experiment 1, we know that online mode performs better than chronological batch mode. So we adopt online mode for experiment 2. From this experiment, we discuss the performance of filters in serial as well as in parallel, and analyze performance based on accuracy obtained by corpus test.

According to the analysis above, the experiment involves seven ensemble methods of filters from "a" to "g". In order to find out the potential influences on performance by taking different nature filter's combination, we consider the experiment for both the same and different nature filters to combine. All the possible combinations are shown in Table 6. The test on corpus is shown in Fig. 3, Table 7 and Table 8.

Comparing from the view of serial or parallel ensemble:

- Serial ensemble of simple filters compare with parallel ensemble of simple filters, that is, b1 compare with e1, b2 compare with e2;
- Serial ensemble of simple-complex filters compare with parallel ensemble of simple-complex filters, that is, a, d compare with f;
- Serial ensemble of complex filters compare with parallel ensemble of complex filters, that is, c1 compare with g1, c2 compare with g2.

Table 6. All the ensemble methods

Ensemble Method	Filters	Nature
serial ensemble(a)	ReBasedFilter + PPM	different
serial ensemble(b1)	SubjectFullWordFilter + BodyFullWordFilter	same
serial ensemble(b2)	ReBasedFilter + BodyFullWordFilter	different
serial ensemble(c1)	Bogofilter + Spambayes	same
serial ensemble(c2)	Bogofilter + PPM + ROSVM	different
serial ensemble(d)	PPM + ReBasedFilter	different
parallel ensemble(e1)	SubjectFullWordFilter + BodyFullWordFilter	same
parallel ensemble(e2)	ReBasedFilter + BodyFullWordFilter	different
parallel ensemble(f)	ReBasedFilter + PPM	different
parallel ensemble(g1)	Bogofilter + Spambayes	same
parallel ensemble(g2)	ROSVM + PPM + Bogofilter	different

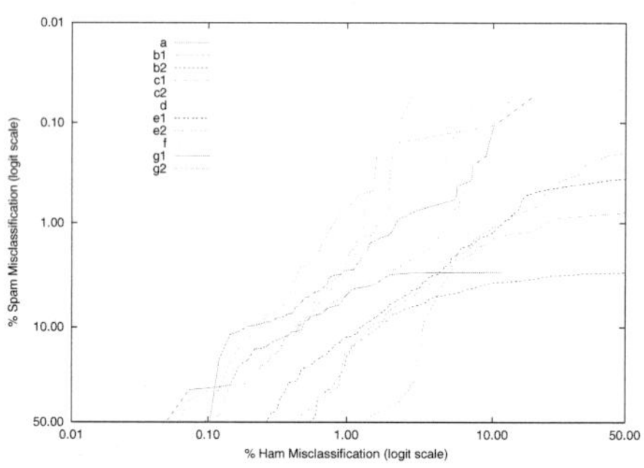

(a) SpamAssassin

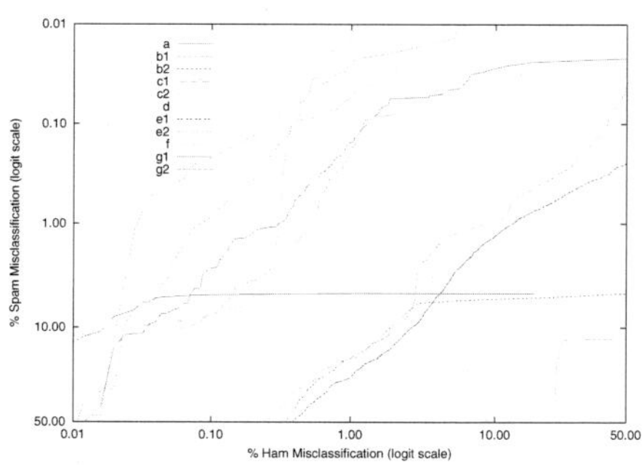

(b) trec07p

Fig. 3. Experiment 2 result on different datasets

Table 7. Statistical results on SpamAssassin(experiment 2)

Ensemble Method	(1-ROCA)%	sm%@hm%=.1
a	2.3028 (1.8129 - 2.9211)	31.46 (15.32 - 53.80)
b1	2.5894 (2.1863 - 3.0644)	99.05 (97.08 - 99.69)
b2	3.0230 (2.4944 - 3.6595)	85.84 (61.02 - 95.91)
c1	0.5251 (0.3528 - 0.7811)	54.32 (27.32 - 79.00)
c2	0.2080 (0.1359 - 0.3181)	38.73 (10.92 - 76.52)
d	5.9323 (5.2548 - 6.6911)	99.89 (nan - nan)
e1	0.9633 (0.7276 - 1.2745)	81.01 (55.01 - 93.70)
e2	1.0799 (0.8475 - 1.3752)	76.13 (48.36 - 91.57)
f	0.2117 (0.1335 - 0.3355)	49.12 (20.70 - 78.12)
g1	0.2383 (0.1514 - 0.3749)	63.82 (5.18 - 98.27)
g2	0.1243 (0.0670 - 0.2306)	33.00 (8.18 - 73.14)

Table 8. Statistical results on trec07p(experiment 2)

Ensemble Method	(1-ROCA)%	sm%@hm%=.1
a	3.6971 (3.5422 - 3.8584)	4.99 (4.79 - 5.20)
b1	27.5255 (27.0841 - 27.9714)	99.91 (99.86 - 99.95)
b2	4.3732 (4.2152 - 4.5368)	76.31 (69.66 - 81.88)
c1	0.0908 (0.0693 - 0.1189)	7.64 (5.09 - 11.32)
c2	0.0393 (0.0294 - 0.0525)	7.71 (6.58 - 9.02)
d	0.8319 (0.7416 - 0.9332)	99.97 (99.95 - 99.99)
e1	1.2416 (1.1692 - 1.3184)	73.31 (67.21 - 78.65)
e2	0.8089 (0.7465 - 0.8764)	74.43 (69.66 - 78.68)
f	0.0157 (0.0066 - 0.0377)	0.22 (0.16 - 0.31)
g1	0.0480 (0.0330 - 0.0698)	2.76 (1.44 - 5.24)
g2	0.0209 (0.0110 - 0.0400)	0.88 (0.54 - 1.46)

Under the (1-ROCA)% measure, all the results show that parallel based ensembles perform substantially better than serial based ensembles on both corpus.

However, it seems as if the sm%@hm%=.1 result cannot keep pace with (1-ROCA)% on SpamAssassin test. All the sm%@hm%=.1 are far from good because the values are too large, in addition, we cannot infer any tendency when observing them due to their distribution are not clear. On the other hand, the same thing on trec07 corpus are vastly different. The distribution of values are highly correspond with (1-ROCA)%. So, We modestly think the exception would be related to corpus size.

Comparing from the view of combining filters in the same or different nature:

- Same nature filters based on serial ensemble compare with different nature filters based on serial ensemble, that is, b1 compare with b2, c1 compare with c2;
- Same nature filters based on parallel ensemble compare with different nature filters based on parallel ensemble, that is, e1 compare with e2, g1 compare with g2.

In general, under both evaluation measures, almost all the results show that different nature combination performs better than the same nature combination when using complex filters only.

From the experiments, we can conclude that in the online mode, it is suitable for parallel ensemble. In addition, it is more appropriate to adopt the combination of different nature filters, if people use complex filters for combination.

5 Conclusion

Recently, the study on how to enhance the spam classification performance by combining a variety of machine learning algorithms becomes a hot topic. Summarized from the commercial applications on anti-spam task, we can find that people always focus on improving overall system performance by combining filters, and comparing whether the misclassification rate has been reduced between different ensemble methods. However, no one compared the differences between diverse ensemble methods yet.

In this paper, we test two main aspects on two benchmark corpora SpamAssassin and trec07p respectively through adjusting the frequency of feedback learning with combining various filters. The experimental results show that online mode performs better than chronological batch mode. Chronological batch model is obviously influenced by the size of data set. When processing a large data set, the performance of chronological batch is less superior to online method. As for online methods, combining filters in parallel mode can achieve better performance. Through combining different nature complex filters, we can obtain a better performance. According to the results based on the particular experimental methods, researchers can design much more reasonable ensemble systems in order to reduce the risk of system error or failure.

Time efficiency is a challenging problem for ensemble systems, however, because the experimental methods are based on results generated by each filter, we cannot measure the time cost of ensemble filtering. In future work, we will combine filters under programming level in order to evaluate the time costs.

Acknowledgement. This paper was partially supported by the National Natural Science Foundation of P.R.China (No.60802066), Science and Technology Planning Project (No.07-2-3-1-jch) of Qingdao China, the Excellent Young Scientist Foundation of Shandong Province of China under Grant (No.2008BS01009) and the Science and Technology Planning Project of Shandong Provincial Education Department (No.J08LJ22).

References

[1] Zhang, L., Zhu, J., Yao, T.: An evaluation of statistical spam filtering techniques. ACM Transactions on Asian Language Information Processing (TALIP) 3(4), 243–269 (2004)
[2] Goodman, J., Cormack, G.V., Heckerman, D.: Spam and the ongoing battle for the inbox. Communications of the ACM 50(2), 24–33 (2007)
[3] Cormack, G.V.: Email spam filtering: A systematic review. Found. Trends Inf. Retr. 1(4), 335–455 (2007)
[4] Yu, B., Xu, Z.b.: A comparative study for content-based dynamic spam classification using four machine learning algorithms. Knowledge-Based Systems 21(4), 355–362 (2008)
[5] Marsono, M.N., El-Kharashi, M.W., Gebali, F.: Targeting spam control on middleboxes: Spam detection based on layer-3 e-mail content classification. Computer Networks 53(6), 835–848 (2009)
[6] Guzella, T., Caminhas, W.: A review of machine learning approaches to spam filtering. Expert Systems with Applications 36(7), 10206–10222 (2009)
[7] Sahami, M., Dumais, S., Heckerman, D., Horvitz, E.: A bayesian approach to filtering junk e-mail. In: AAAI workshop on Learning for Text Categorization, pp. 55–62 (1998)
[8] Androutsopoulos, I., Koutsias, J., Chandrinos, K.V., Paliouras, G., Spyropoulos, C.D.: An evaluation of naive bayesian anti-spam filtering. In: Workshop on Machine Learning in the New Information Age, June 2000, pp. 9–17 (2000)
[9] Joachims, T.: Making large-scale support vector machine learning practical. In: Advances in kernel methods: support vector learning, pp. 169–184. MIT Press, Cambridge (1999)
[10] Drucker, H., Vapnik, V., Wu, D.: Support vector machines for spam categorization. IEEE Transactions on Neural Networks 10, 1048–1054 (1999)
[11] Carreras, X., Marquez, L.: Boosting trees for anti-spam email filtering. In: RANLP 2001: Proceedings of the 4th International Conference on Recent Advances in Natural Language Processing (2001)

[12] Bratko, A., Filipič, B., Cormack, G.V., Lynam, T.R., Zupan, B.: Spam filtering using statistical data compression models. The Journal of Machine Learning Research 7, 2673–2698 (2006)
[13] Zhou, Z.H., Wu, J., Tang, W.: Ensembling neural networks: Many could be better than all. Artificial Intelligence 137(1-2), 239–263 (2002)
[14] Zhou, Z.H.: Ensemble learning. In: Li, S.Z. (ed.) Encyclopedia of Biometrics, Springer, Berlin (2009)
[15] Lynam, T.R., Cormack, G.V., Cheriton, D.R.: On-line spam filter fusion. In: SIGIR 2006: Proceedings of the 29th annual international ACM SIGIR conference on Research and development in information retrieval, pp. 123–130. ACM, New York (2006)
[16] He, J., Thiesson, B.: Asymmetric gradient boosting with application to spam filtering. In: CEAS 2007: Proceedings of the Fourth Conference on Email and Anti-Spam (2007)
[17] Liu, W., Wang, T.: Ensemble learning and active learning for spam filtering. In: TREC 2007: Proceedings of the sixteenth Text Retrieval Conference Proceedings (2007)
[18] Dietterich, T.G.: Ensemble methods in machine learning. In: Kittler, J., Roli, F. (eds.) MCS 2000. LNCS, vol. 1857, pp. 1–15. Springer, Heidelberg (2000)
[19] Cormack, G.V., Bratko, A.: Batch and on-line spam filter comparison. In: CEAS 2006: Proceedings of the Third Conference on Email and Anti-Spam (2006)
[20] Liu, W., Wang, T.: An ensemble learning method of multi filter for spam filtering. In: NCIRCS 2008: Proceedings of the 3rd National Conference on Information Retrieval and Content Securit. (2008)
[21] Sculley, D., Wachman, G.M.: Relaxed online svms for spam filtering. In: SIGIR 2007: Proceedings of the 30th annual international ACM SIGIR conference on Research and development in information retrieval, pp. 415–422. ACM, New York (2007)
[22] Bratko, A., Filipič, B.: Spam filtering using compression models. Technical Report IJS-DP-9227, Department of Intelligent Systems, Jožef Stefan Institute, Ljubljana, Slovenia (2005)
[23] Fawcett, T.: Roc graphs: Notes and practical considerations for researchers. Technical Report HPL-2003-4, HP Laboratories (2004)
[24] Paul Komarek, A.M.: Fast robust logistic regression for large sparse datasets with binary outputs. In: Artificial Intelligence and Statistics (2003)

Crawling Deep Web Using a New Set Covering Algorithm

Yan Wang[1], Jianguo Lu[1,2], and Jessica Chen[1]

[1] School of Computer Science, University of Windsor
N9B 3P4 Windsor, Ont. Canada
{wang16c,jlu,xjchen}@uwindsor.ca
[2] Key Lab of Novel Software Technology, Nanjing, China

Abstract. Crawling the deep web often requires the selection of an appropriate set of queries so that they can cover most of the documents in the data source with low cost. This can be modeled as a *set covering* problem which has been extensively studied. The conventional set covering algorithms, however, do not work well when applied to deep web crawling due to various special features of this application domain. Typically, most set covering algorithms assume the uniform distribution of the elements being covered, while for deep web crawling, neither the sizes of documents nor the document frequencies of the queries is distributed uniformly. Instead, they follow the power law distribution. Hence, we have developed a new set covering algorithm that targets at web crawling. Compared to our previous deep web crawling method that uses a *straightforward* greedy set covering algorithm, it introduces *weights* into the greedy strategy. Our experiment carried out on a variety of corpora shows that this new method consistently outperforms its un-weighted version.

Keywords: deep web crawling, set covering problem, greedy algorithm.

1 Introduction

Deep web [1] is the web that is dynamically generated from data source such as databases or file system. Unlike surface web where data are available through URLs, data from a deep web are guarded by a search interface. The amount of data in deep web exceeds by far that of the surface web. This calls for deep web crawlers to excavate the data so that they can be reused, indexed, and searched upon in an integrated environment.

Crawling deep web [2,3,4,5,6] is the process of collecting hidden data by issuing queries through various search interfaces including HTML forms, web services and programmable web APIs. Crawling deep web data is important for several reasons, such as indexing deep web data sources, or backing up data.

Crawling the deep web often requires the selection of an appropriate set of queries so that they can cover most of the documents in the data source with low cost. Focusing on querying *textual data sources*, we provide a solution to this

R. Huang et al. (Eds.): ADMA 2009, LNAI 5678, pp. 326–337, 2009.

problem. Since it is not possible to select queries directly from the entire data source, we can make our selection from a *sample* of the database. It is shown that queries selected from a sample data source can perform on the total data source as well as on the sample one [7]. This leads to the following 4-step framework for crawling deep web [7]:

- Randomly selecting documents to build a sample database (*SampleDB*) from the original corpus (called *TotalDB*).
- Creating a set of queries called *query pool* (*QueryPool*) based on the *SampleDB*.
- Selecting a proper set of queries based on the *SampleDB* and the *QueryPool*.
- Mapping the selected queries into *TotalDB*.

An essential task in this framework is to select a proper set of queries based on the *SampleDB* and the *QueryPool* so that the cost of mapping the selected queries into *TotalDB* can be minimized. This can be modeled as a set covering problem which has been extensively studied. The conventional set covering algorithms, however, do not work well when applied to deep web crawling due to various special features of this application domain. Typically, most set covering algorithms assume the uniform distribution of the elements being covered, while for deep web crawling, neither the sizes of documents nor the document frequencies of the queries is distributed uniformly. Instead, they follow the power law distribution. In this regard, we have developed a new set covering algorithm that targets at web crawling. Compared to our previous deep web crawling method that uses a *straightforward* greedy set covering algorithm, it introduces *weights* into the greedy strategy. Our experiment carried out on a variety of corpora shows that this new method consistently outperforms its un-weighted version.

2 Problem Formalization

We have shown in [7] the criteria to select $SampleDB$ and $QueryPool$. Our task here is to select from $QueryPool$ an appropriate set of queries so that they can cover *all* the documents in $SampleDB$. In terms of efficiency, we would like to keep the *overlap* minimal, where the overlap refers to the number of documents covered by more than one queries.

This can be modeled as a Set Covering Problem (SCP) [8] as below:

Definition 1. *Let* $A = (a_{ij})$ *be a 0-1* $m \times n$ *matrix, and* $c = (c_j)$ *be an m-dimensional integer vector. Let* $M = \{1, ..., m\}$ *and* $N = \{1, ..., n\}$. *The value* c_j $(j \in N,\ c_j > 0)$ *represents the cost of column* j. *We say that a column* j $(j \in N)$ *covers a row* i $(i \in m$ *if* a_{ij} *=1. SCP calls for a minimum-cost subset* S $(S \subseteq N)$ *of columns, such that each row is covered by at least one column.*

Example 1. *Table1 gives a matrix A, where each column represents a query in* $QueryPool = \{q_1, q_2, \ldots, q_5\}$, *and each row represents a document of* $SampleDB$ $= \{d_1, \ldots, d_9\}$. $c_j = \sum a_{ij}$ *is the document frequency (df) of the term. One solution of the problem is* $Q = \{q_3, q_4, q_5\}$, *which can be obtained by the greedy algorithm [7].*

Table 1. Matrix A: the input matrix for set covering algorithms

doc number	q_1	q_2	q_3	q_4	q_5
d_1	0	0	1	0	0
d_2	0	0	1	1	0
d_3	1	0	1	0	1
d_4	0	0	1	0	1
d_5	1	0	0	0	1
d_6	1	1	0	1	0
d_7	0	0	0	1	0
d_8	1	1	0	0	1
d_9	0	0	1	1	1

3 Weighted Greedy Algorithm

Constructing the set of queries in a step-by-step manner, simple greedy algorithm of the set covering problem selects the most cost effective query in each step. Let Q be a set of queries already selected. According to simple greedy algorithms, we select the next query q to cover as many as possible new documents (i.e. documents not covered by any query in Q) per *unit cost*. Here, the cost is the document frequency df, and *unit cost* can be represented by $1/df$. In other words, we select q to maximize the value of new/df where new is the number of documents covered by q but not by any query in Q.

As an improvement of simple greedy algorithms, we introduce the weight of queries into the greedy strategy.

If a document can only be matched by one query, apparently that query must be included into Q. In general, when selecting a query, we should pay more attention to cover small documents since usually they can be matched by only very few queries. We assign a weight to each document, where small documents have larger weights. With this intuition, we introduce the weight of a document:

Definition 2. *Let $D = \{d_1, ..., d_m\}$ be the SampleDB and $QP = \{q_1, ..., q_n\}$ be the QueryPool. We consider each document as a set of terms and use the notation $q_j \in d_i$ to indicate that a term q_j occurs in the document d_i. The weight of a document with respect to QP and d_i $(1 \leq i \leq m)$, denoted by $dw_{d_i}^{QP}$ (or dw for short), is the inverse of the number of terms in QP that occurs in the document d_i, i.e.*

$$dw_{d_i}^{QP} = \frac{1}{|d_i \cap QP|}. \tag{1}$$

Definition 3. *The weight of a query q_j $(1 \leq j \leq n)$ in QP with respect to D, denoted by $qw_{q_j}^{QP}$ (or qw for short), is the sum of the document weights of all documents containing term q_j, i.e.,*

$$qw_{q_j}^{QP} = \sum_{q_j \in d_i, d_i \in D} dw_{d_i}^{QP}. \tag{2}$$

Table 2. Matrix B: the initial weight table of the example corresponding to Matrix A

doc number	q_1	q_2	q_3	q_4	q_5
d_1	0	0	1	0	0
d_2	0	0	0.5	0.5	0
d_3	0.33	0	0.33	0	0.33
d_4	0	0	0.5	0	0.5
d_5	0.5	0	0	0	0.5
d_6	0.33	0.33	0	0.33	0
d_7	0	0	0	1	0
d_8	0.33	0.33	0	0	0.33
d_9	0	0	0.33	0.33	0.33
df	4	2	5	4	5
qw	1.49	0.66	2.66	2.16	1.99
df/qw	2.66	3.00	1.87	1.84	2.50

Table 3. Matrix B: The second-round weight table of the example

doc number	q_1	q_2	q_3	q_5
d_1	0	0	1	0
d_3	0.33	0	0.33	0.33
d_4	0	0	0.5	0.5
d_5	0.5	0	0	0.5
d_8	0.33	0.33	0	0.33
df	4	2	5	5
qw	1.16	0.33	1.83	1.66
df/qw	3.42	6.00	2.72	3.00

As for greedy strategy, we prefer queries q_j with larger number of qw. However, a larger number of qw should be obtained by fewer number of dw. In other words, we prefer queries with smaller df/qw. Our weighted greedy algorithm is based on choosing the next query with the smallest df/qw.

Example 2. *Based on the matrix in Table 1, the weights of the documents are shown in the top part of Table 2. The document frequency, the weights of the queries, and their quotient are listed at the bottom of the table. For example, the weight of d_1 is one, the weight of d_2 is 1/2, and the weight of q_2 is the sum of the weights of the documents that is covered by q_2, i.e., 0.66.*

The more detailed weighted greedy algorithm is given in Algorithm 1.

Example 3. *Here we give an example to show how the weighted greedy method works. Table 1 is the matrix A of the example, Table 2 shows the initial values of weights and Table 4 shows the result from the weighted greedy method for the example.*

For the weighted greedy algorithm, at the beginning, we calculate the value of df/qw for each query from Table 2 and find that q_4 has the minimal df/qw value

Algorithm 1. Weighted Greedy Algorithm.

Input: *SampleDB*, *QueryPool* QP, $m \times n$ Matrix A, where m=|SampleDB| and n=|QP|

Output: A set of queries Q

1. $Q = \phi$;
2. Let $B = (b_{ij})$ be a $m \times n$ matrix and $b_{ij} = a_{ij} \times dw_{d_i}^{QP}$;
3. Based on the matrix B, we calculate the query weight for each term and move the q_j that minimizes $\frac{df_j}{qw_{q_j}^{QP}}$ into Q ;
4. Check if the queries in Q can cover all documents in *SampleDB*. If yes, the process ends;
5. Update matrix B by removing the selected query and the documents that are covered by the query. Go to Step 3.

Table 4. The result of the example by using the weighted greedy method

column	df	qw	df/qw	cost	unique rows
q_4	4	2.16	1.84	4	4
q_3	5	1.83	2.72	9	7
q_1	4	0.83	4.8	14	9

(1.84) hence q_4 is selected as the first query. Then the column of q_4 and the covered rows, i.e., d_2, d_6, d_7 and d_9 are removed from the matrix B and the resulting matrix is shown in Table3. In the second round, q_3 becomes the second query because it has the minimal value for df/qw (2.72) and the matrix B is updated again. Finally, there are only two rows d_5 and d_8 left in the matrix B. q_1 is selected for its minimal df/qw value (4.80). After the third round, the selected queries can cover all documents and the weighted greedy algorithm terminates. For convenience to compare the two algorithms, we also give one solution for the example from the greedy algorithm as shown in Table 5. Greedy algorithm can produce several solutions depending on which query is selected in the first step of the algorithm. Initially all the queries has the same value for df/new, hence an arbitrary query can be selected. In our example, we select the query that is the same as the one of the weighted greedy algorithm. From Table 2, we can see that only q_3 and q_4 can cover d_1 and d_7 respectively. So q_3 and q_4 are required and they should be selected as early as possible. The useful information can be used by the weighted greedy method because such required query usually has a smaller df/qw value and it could be selected earlier.

Table 5. The result of the example by using the greedy method

column	df	new rows	df/new	cost	unique rows
q_4	4	4	1	4	4
q_5	5	4	1.25	9	8
q_3	5	1	5	14	9

4 Experiment

We have run our experiments on the same data as that of [7] from four corpora: Reuters, Gov, Wikipedia and Newsgroup. These are standard test data used by many researchers in information retrieval. All *SampleDB* used have sample size 3000 and *relative size* [7] 20. We have used our search engine implemented in Lucene [9] to do all experiments in order to obtain the details of a data source such as its size. The details of the corpora are summarized in Table 6.

Table 6. Summary of test corpora

Name	Size in documents	Size in MB	Avg file size(KB)
Reuters	806,791	666	0.83
Wikipedia	1,369,701	1950	1.42
Gov	1,000,000	5420	5.42
Newsgroups	30,000	22	0.73

Since the algorithms are partly evaluated in terms of HR over OR, we give the definitions for HR and OR here.

Definition 4. *(Hit Rate, HR) Given a set of queries* $Q=\{q^1,q^2,...,q^k\}$ *and a database DB. The hit rate of Q on DB, denoted by HR(Q,DB), is defined as the ratio between the number of unique data items collected by sending the queries in Q to DB and the size of the data base DB, i.e., hit rate at the k-th step is:*

$$HR(Q,DB) = \frac{|\bigcup_{p=1}^{k} S(q^p, DB)|}{|DB|} \qquad (3)$$

Definition 5. *(Overlapping Rate, OR) Given a set of queries* $Q=\{q^1,...,q^k\}$*, the overlapping rate of Q on DB, denoted by OR(Q,DB), is defined as the ratio between the total number of collected data items and the number of unique data items retrieved by sending queries in Q to DB. i.e.,overlapping rate at the k-th step is:*

$$OR(Q,DB) = \frac{\sum_{p=1}^{k} |S(q^p, DB)|}{|\bigcup_{p=1}^{k} S(q^p, DB)|} \qquad (4)$$

Table 7 shows that our weighted greedy method is much better than the greedy method. As the result of the greedy method depends on which query is slected in the first step of the algorithm, we have given some statistic results here, such as the standard deviation, maximum and minimal values. Our weighted greedy method also has such problem: initially there can be more than one

Table 7. The results based on 100 times running in *SampleDB* (G:greedy method; WG: weighted greedy method; Diff: difference)

	Reuters			Wiki			Gov			Newsgroup		
	G	WG	Diff	G	WG	Diff	G	WG	Diff	G	WG	Diff
MaxCost	14677	11421	0.22	11939	9468	0.20	18284	13853	0.24	13016	11067	0.14
MinCost	13613	11421	0.16	10514	9466	0.09	15945	13853	0.13	11955	11063	0.07
AveCost	14151	11421	0.19	11252	9467	0.15	16992	13853	0.18	12567	11065	0.11
SD	255.6	0	0.02	262.5	0.98	0.02	428.5	0	0.02	213	2	0.02

queries having the same value of df/qw. However, seldom occurs in practice. From Table 7 we can see that (i) even the maximum cost of our weigted greedy method is better than the minimal cost of the greedy method; (ii) on average our method outperforms the normal greedy method by approximately 16%.

The difference between the results of the two methods can be explained as follows. The term with lower weight may cause less overlap if it is selected as a query because there will be fewer other terms contained by the same document which can be selected as queries in future. Based on this observation, we can see that *df/qw* of a term can somehow show an *overlapping possibility* of the term. At the same time, the value of *df/qw* of a term can also represent a *requirement degree*. For example, if the df of a term is 1 and the corresponding document also contains the term, the value of *df/qw* of the term is 1 (the minimal value of *df/qw*). In this case, of course, the term is *required* and should be selected as early as possible.

Figure 1 and Figure 2 show the relationship between the df and the number of terms and the relationship between the document size and the number of documents in the *SampleDB*. From these two figures, we can see that both follow

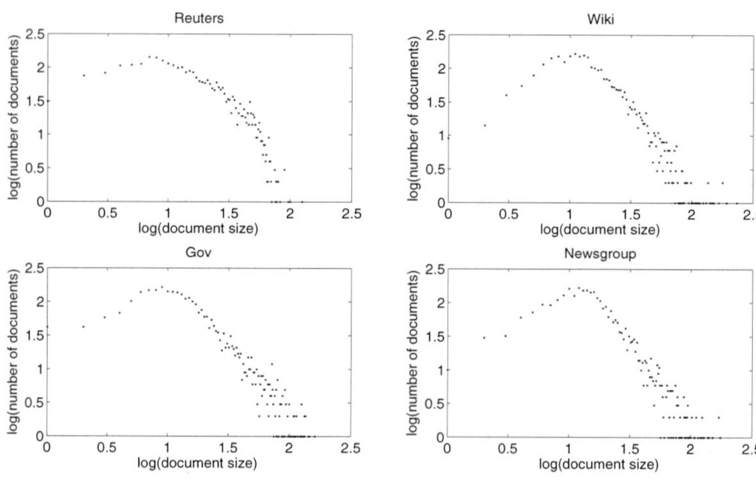

Fig. 1. The distribution of document size in *SampleDB*

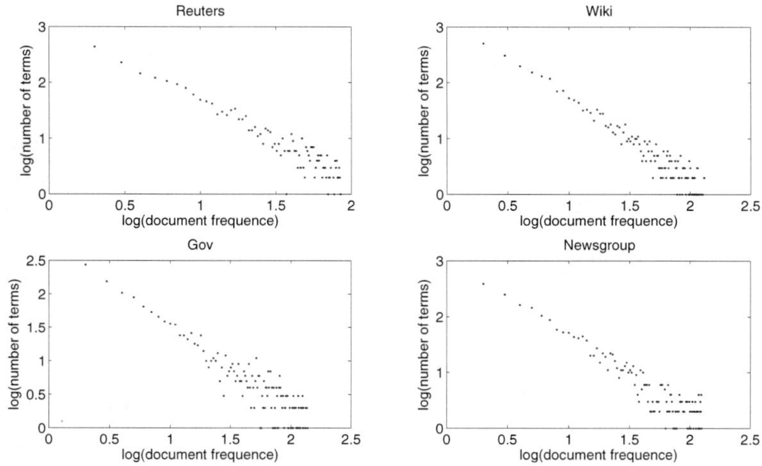

Fig. 2. The distribution of df in *SampleDB*

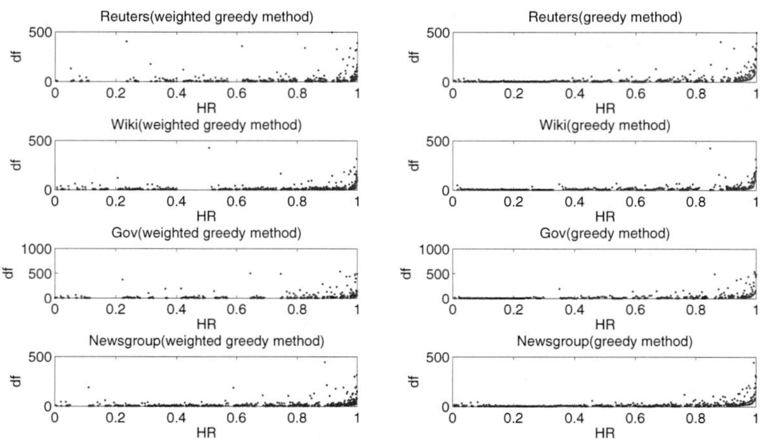

Fig. 3. The df changes in SampleDB

the power law distribution: most of the documents only contain few terms; most of the terms are contained by very few documents; but a few high df terms are in most of the documents. In such a situation, documents only containing one or two high df terms can have a smaller value of df/qw and thus are more desired. If such required high df terms can be selected earlier, the result should be better in *SampleDB*. Figure 3 shows that the weighted greedy method can select many high df terms earlier than the greedy method in *SampleDB*.

Figure 4 is derived from the experiments in *SampleDB*. In this figure, we can see that the values of df/tw of the terms selected by the weighted greedy method are much smaller than those selected by the greedy method especially when HR

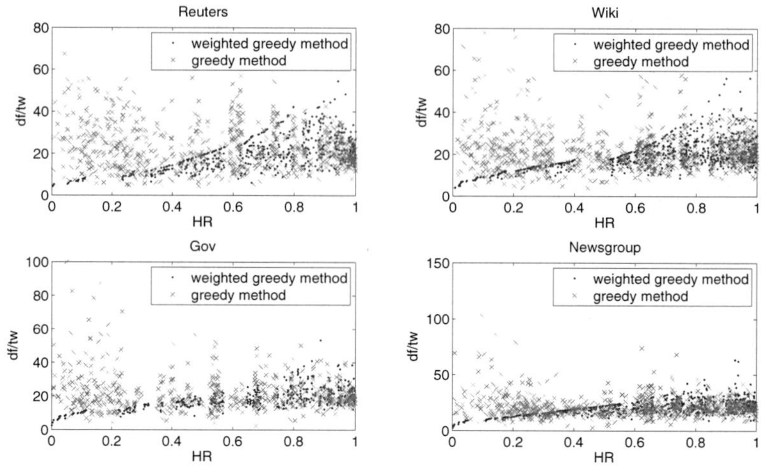

Fig. 4. Comparison on df/qw between the two methods

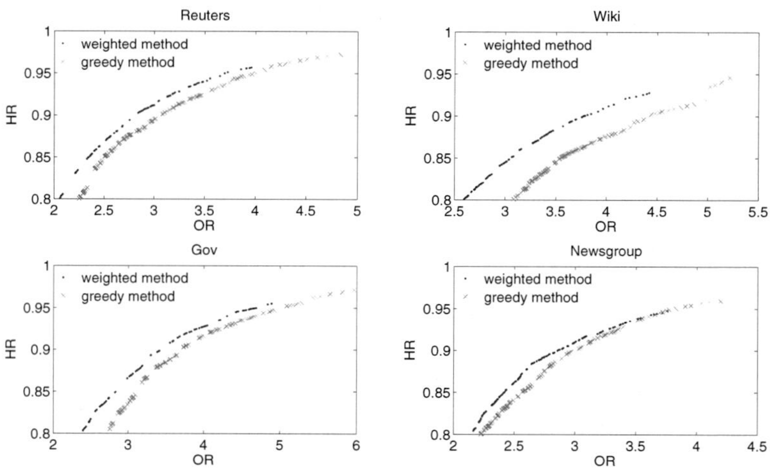

Fig. 5. Comparison on mapping results HR from 89%

ranges from 0% to 60% in *SampleDB*. The smaller number of *df/tw* implies that the terms selected by the weighted greedy method introduce less overlapping. Furthermore, as each *SampleDB* is a representative of its corresponding *TotalDB*, a term having a small overlapping possibility and a low requirement degree in *SampleDB* will have these properties preserved in *TotalDB*.

Now we discuss the performance of the two methods in *TotalDB*. Figure 5 shows that the weighted greedy method has a better performance than the greedy method and the mapping coverages of the two methods are good enough. For example, on Wikipedia corpus, when HR is 90%, the weighted greedy method is 15.1% better than the greedy method. The size of Wikipedia corpus is 1.36 million of documents. Thus, we can save 0.15*0.9*1.3 million of documents which is a significant saving.

5 Related Work

There are many research works on deep web crawling. Basically, two research challenges have been substantially addressed. One is learning and understanding the interface and the returning result so that query submission [10,11] and data extraction [12] can be automated. The other is how to harvest hidden documents as many as possible with a low cost [2,5,6,13]. Here we discuss several closely related works in the second area. The deep web is guarded by search interface, hence difficult to access. A sampling-based method could be a good choice for deep web crawling.

Generally speaking, the sampling-based methods have two directions. One is to use Information Retrieval Technologies to crawl the deep web [14,15,16]. In [15], the general idea is that firstly some carefully designed queries are issued to the database and then some samples as returns from the database are obtained. Secondly those samples are analyzed and further classified by some classification algorithms to obtain some typical terms that can accurately describe the database. Finally, those typical terms are used as queries to harvest the database. The other one is to use some selection methods to create a sample that can be representative of the database and the principle of human language to crawl the deep web [2,17,5]. For example, the Zip's law can be used to evaluate the frequency of a term in the database based on a sample [5]. In [17], the authors present a new technique to automatically create a description (a sample) for the database. They argue that accurate description can be learned by sampling a text database with simple keyword-based queries. Actually our framework for crawling the deep web is based on Callan and Connells research work in [17].

For the first kind of methods, usually they are based on existing domain knowledge. The authors suppose that the database is so heterogeneous and it is hard to obtain a high HR with queries selected randomly from a dictionary hence the emphasis is to have a high HR by issuing queries. For the second kind of methods, the authors try to minimize the number of queries with a high HR. On the contrary, we argue that the bottleneck to deep web crawling is the number of documents crawled, not the queries issued. Therefore our algorithm tries to minimize the documents retrieved, not the queries sent. Another difference from the Ntoulas et al's method [5] is that they estimate the returns and overlaps of the i-th query based on the documents downloaded by the previous $i-1$ queries. This approach requires the downloading of almost all the documents, hence it is not efficient. Our approach only requires a small portion of the data source, and learn the appropriate queries from the sample database.

6 Conclusions

In an earlier paper [7], we proposed a deep web crawling method based on sampling. In that paper, we showed that it is effective to learn appropriate queries from a sample data source, and empirically identified the appropriate sizes of the sample and the query pool. This paper presents a better algorithm, the weighted greedy algorithm, to select the queries from a sample data source. The weighted greedy method has a much better result than the greedy method in *SampleDB* from the four corpus so that the result from the weighted greedy method can beat the result from the greedy method in *TotalDB*. In the *SampleDB*, the weighted greedy method can select the term as a query which has lower overlapping possibility and higher requirement degree as earlier as possible and such properties can be successfully described by the query weight.

Acknowledgments. This work is supported by Key Laboratory of Novel Software Technology and the Natural Sciences and Engineering Research Council of Canada.

References

1. Bergman, M.K.: The deepweb: Surfacing hidden value. The Journal of Electronic Publishing 7(1) (2001)
2. Barbosa, L., Freire, J.: Siphoning hidden-web data through keyword-based interfaces. In: Proc. of SBBD (2004)
3. Chang, C.H., Kayed, M., Girgis, M.R., Shaalan, K.F.: A survey of web information extraction systems. IEEE Transactions on Knowledge and Data Engineering 18(10), 1411–1428 (2006)
4. Liddle, S.W., Embley, D.W., Scott, D.T., Yau, S.H.: Extracting data behind web forms. In: Olivé, À., Yoshikawa, M., Yu, E.S.K. (eds.) ER 2003. LNCS, vol. 2784, pp. 402–413. Springer, Heidelberg (2003)
5. Ntoulas, A., Zerfos, P., Cho, J.: Downloading textual hidden web content through keyword queries. In: Proc. of the Joint Conference on Digital Libraries (JCDL), pp. 100–109 (2005)
6. Wu, P., Wen, J.R., Liu, H., Ma, W.Y.: Query selection techniques for efficient crawling of structured web sources. In: Proc. of ICDE, pp. 47–56 (2006)
7. Lu, J., Wang, Y., Iiang, J., Chen, J., Liu, J.: An approach to deep web crawling by sampling. In: Proc. of Web Intelligence, pp. 718–724 (2008)
8. Caprara, A., Toth, P., Fishetti, M.: Algorithms for the set covering problem. Annals of Operations Research 98, 353–371 (2004)
9. Hatcher, E., Gospodnetic, O.: Lucene in Action. Manning Publications (2004)
10. Knoblock, C.A., Lerman, K., Minton, S., Muslea, I.: Accurately and reliably extracting data from the web: a machine learning approach. IEEE Data Engineering Bulletin 23(4), 33–41 (2000)
11. Nelson, M.L., Smith, J.A., Campo, I.G.D.: Efficient, automatic web resource harvesting. In: Proc. of RECOMB, pp. 43–50 (2006)
12. Alvarez, M., Pan, A., Raposo, J., Bellas, F., Cacheda, F.: Extracting lists of data records from semi-structured web pages. Data Knowl. Eng. 64(2), 491–509 (2008)

13. Ipeirotis, P.G., Jain, P., Gravano, L.: Towards a query optimizer for text-centric tasks. ACM Transactions on Database Systems 32 (2007)
14. Gravano, L., Ipeirotis, P.G., Sahami, M.: Query- vs. crawling-based classification of searchable web databases. Bulletin of the IEEE Computer Society Technical Committee on Data Engineering 25(1), 1–8 (2002)
15. Caverlee, J., Liu, L., Buttler, D.: Probe, cluster, and discover: focused extraction of qa-pagelets from the deep web. In: Proc. of the 28th international conference on Very Large Data Bases, pp. 103–114 (2004)
16. Ibrahim, A., Fahmi, S.A., Hashmi, S.I., Choi, H.: Addressing effective hidden web search using iterative deepening search and graph theory. In: Proc. of IEEE 8th International Conference on Computer and Information Technology Workshops, pp. 145–149 (2008)
17. Callan, J., Connell, M.: Query-based sampling of text databases. ACM Transactions on Information Systems, 97–130 (2001)

A Hybrid Statistical Data Pre-processing Approach for Language-Independent Text Classification

Yanbo J. Wang[1], Frans Coenen[2], and Robert Sanderson[2]

[1] Information Management Center, China Minsheng Banking Corp., Ltd.
Room 606, Building No. 8, 1 Zhongguancun Nandajie,
100873 Beijing, China
wangyanbo@cmbc.com.cn
[2] Department of Computer Science, University of Liverpool,
Ashton Building, Ashton Street, Liverpool, L69 3BX, UK
{Coenen,Azaroth}@liverpool.ac.uk

Abstract. Data pre-processing is an important topic in Text Classification (TC). It aims to convert the original textual data in a data-mining-ready structure, where the most significant text-features that serve to differentiate between text-categories are identified. Broadly speaking, textual data pre-processing techniques can be divided into three groups: (i) linguistic, (ii) statistical, and (iii) hybrid (i) & (ii). With regard to language-independent TC, our study relates to the statistical aspect only. The nature of textual data pre-processing includes: Document-base Representation (DR) and Feature Selection (FS). In this paper, we propose a hybrid statistical FS approach that integrates two existing (statistical FS) techniques, DIAAF (Darmstadt Indexing Approach Association Factor) and GSSC (Galavotti·Sebastiani·Simi Coefficient). Our proposed approach is presented under a statistical "bag of phrases" DR setting. The experimental results, based on the well-established associative text classification approach, demonstrate that our proposed technique outperforms existing mechanisms with respect to the accuracy of classification.

Keywords: Associative Classification, Data Pre-processing, Document-base Representation, Feature Selection, (Language-independent) Text Classification.

1 Introduction

Text mining is a promising topic of current research in data mining and knowledge discovery. It aims to extract various types of hidden, interesting, previously unknown and potentially useful knowledge from sets of collected textual data. In a natural language context, a given textual dataset is usually refined to produce a document-base, i.e. a set of electronic documents that typically consists of thousands of documents, where each document may contain hundreds of words. One important aspect of text mining is Text Classification (TC) – "*the task of assigning one or more predefined categories to natural language text documents, based on their contents*" [10]. Broadly speaking, TC studies can be separated into two divisions: *single-label* that assigns exactly one pre-defined category to each "unseen" document; and

R. Huang et al. (Eds.): ADMA 2009, LNAI 5678, pp. 338–349, 2009.

multi-label that assigns one or more pre-defined category to each "unseen" document. With regard to single-label TC, two distinct approaches can be identified: *Binary* TC which in particular assigns either a pre-defined category or the complement of this category to each "unseen" document; and *multi-class* TC which simultaneously deals with all given categories and assigns the most appropriate category to each "unseen" document. This paper is concerned with the single-label multi-class TC approach.

Text mining requires the given document-base to be first pre-processed, where the (unstructured) original textual data is converted in a (structured) data-mining-ready format, and the most significant text-features that serve to differentiate between text categories are identified. Thus the entire process of TC, in general, can be identified as textual data (document-base) *pre-processing* plus traditional *classification*. Broadly speaking, textual data pre-processing techniques can be divided into three groups: (i) linguistic, (ii) statistical, and (iii) hybrid (i) & (ii). Both the linguistic and the hybrid aspects pre-process document-bases depending on the rules and/or regularities in semantics, syntax and/or lexicology of languages. Such techniques are designed with particular languages and styles of language as the target, and involve deep linguistic analysis. For the purpose of building a language-independent text classifier that can be applied to cross-lingual, multi-lingual and/or unknown lingual textual data collections, this paper is only concerned with the statistical aspect of textual data pre-processing.

In [17] the nature of textual data pre-processing is characterized as: *Document-base Representation* (DR) which designs an application oriented data model to precisely interpret a given document-base in an explicit and structured manner; and *Feature Selection* (FS) which extracts the most significant information (text-features) from the given document-base. In DR the Vector Space Model (VSM) [20] is considered appropriate for many text mining applications, especially when dealing with TC problems. The VSM is usually presented in a binary format, where "*each coordinate of a document vector is zero (when the corresponding attribute is absent) or unity (when the corresponding attribute is present)*" [14]. In TC, two common DR approaches that are used to define VSM are the "bag of words" and the "bag of phrases". The motivation for the latter approach is that phrases seem to carry more contextual and/or syntactic information than single words. In [22] Scheffer and Wrobel argue that the "bag of words" representation does not distinguish between "*I have no objections, thanks*" and "*No thanks, I have objections*", where the "bag of phrases" approach seems to deal with this kind of situation better. Hence the experimental work in this paper is designed with respect to the "bag of phrases" DR setting.

In theory, the textual attributes of a document can include every text-feature (word or phrase) that might be expected to occur in a given document-base. However, this is computationally unrealistic, so it requires some FS mechanism (during the pre-processing phase) to identify the *key* text-features that will be useful for a particular text mining application, such as TC. In the past, a number of approaches have been proposed for TC, under the heading of statistical FS. Two major ones are the Darmstadt Indexing Approach Association Factor (DIAAF) and the Galavotti·Sebastiani·Simi Coefficient (GSSC). Other existing methods include: Relevancy Score (RS), Mutual Information (MI), etc.

Classification Rule Mining (CRM) is a well-established area in data mining and knowledge discovery for identifying hidden classification rules from a given class-database (i.e. usually a relational data table with a set of pre-defined class labels), the objective being to build a (rule based) classifier to categorize "unseen" data records. It should be noted that CRM refers to the rule based approach of the traditional classification problem. Approaches that are parallel to CRM include: probabilistic classification, support vector machine based classification, neural network based classification, etc. One CRM implementation mechanism is to employ association rule mining [1] methods to identify the desired classification rules, i.e. associative classification [2]. Coenen *et al.* [5] and Shidara *et al.* [24] indicate that results presented in the studies of [15, 16, 28] show that in many cases associative classification offers greater classification accuracy than other classification approaches, such as C4.5 [19] and RIPPER (Repeated Incremental Pruning to Produce Error Reduction) [7].

In the past decade, associative classification has been proposed for application in TC (e.g. [6, 29]). In [3] Antonie and Zaïane argue: an associative text classifier "*is fast during both training and categorization phases*", especially when handling large document-bases; and such classifiers "*can be read, understood and modified by humans*". In comparison, TC techniques other than the rule based, i.e. probabilistic based, support vector machine based, neural network based, etc., do not present the classifier in a human readable fashion, so that users do not see why the classification predictions have been made. Given the advantages offered by associative classification with respect to TC, this approach is adopted in our study to support the investigation of statistical (textual) data pre-processing for language-independent TC.

In this paper, we propose a statistical FS approach, which combines the ideas of DIAAF and GSSC mechanisms, namely Hybrid DIAAF/GSSC. The evaluation of Hybrid DIAAF/GSSC, under a statistical "bag of phrases" DR setting, is conducted using the TFPC (Total From Partial Classification) [5] associative classification algorithm; although any other associative classifier generator could equally well have been used. The experimental results demonstrate that Hybrid DIAAF/GSSC based textual data pre-processing approach outperforms alternative techniques with respect to the accuracy of classification. This in turn improves the performance of language-independent TC. The rest of this paper is organized as follows. In section 2, we describe the statistical "bag of phrases" DR approach. In section 3, we review the DIAAF and the GSSC statistical FS mechanisms. The Hybrid DIAAF/GSSC approach is proposed in section 4. Section 5 presents the experimental results. Finally our conclusions and open issues for further research are provided in section 6.

2 Statistical "Bag of Phrases" Document-Base Representation

In the "bag of phrases" DR approach, each element in a document vector represents a phrase describing an ordered combination of words appearing contiguously in sequence. Preliminarily, some definitions with regard to the statistical aspect are given as follows.

- **Words:** Words in a document-base are defined as continuous sequences of alphabetic characters delimited by non-alphabetic characters, e.g. punctuation marks, white space and numbers.
- **Noise Words (N):** Common and rare words are collectively defined to be *noise* words in a document-base. Note that noise words can be identified by their *support* value, i.e. the percentage of documents in the training dataset in which the word appears.
- **Upper Noise Words:** Common (upper noise) words are words with a support value above a user-supplied Upper Noise Threshold (UNT).
- **Lower Noise Words:** Rare (lower noise) words are words with a support value below a user-supplied Lower Noise Threshold (LNT).
- **Potential Significant Words:** A potential significant word, also referred to as a *key* word, is a non-noise whose *contribution* value exceeds some user-specified threshold *G*. The contribution value of a word is a measure of the extent to which the word serves to differentiate between classes and can be calculated in a number of ways (noted as various statistical FS mechanisms).
- **Significant Words (G):** The first *K* words (i.e. the first *k* words for each pre-defined class) that are selected from the ordered list of potential significant words (in a descending manner based on their contribution value) are defined to be significant words.
- **Ordinary Words (O):** Other non-noise words that have not been selected as significant words.
- **Stop Marks (S):** Not actual words but six key punctuation marks (, . : ; ! and ?). All other non-alphabetic characters are ignored.

In [6] the authors (based on the above definitions) propose a statistical "bag of phrases" (DR) approach for TC, namely DelSNcontGO: phrases are Delimited by stop marks (S) and/or noise words (N), and (as phrase contents) made up of sequences of one or more significant words (G) and ordinary words (O); sequences of ordinary words delimited by stop marks and/or noise words that do not include at least one significant word (in the contents) are ignored. The experimental results presented in [6] show that DelSNcontGO performs well with respect to the accuracy of classification. In this paper, this statistical "bag of phrases" DR approach will be further concerned in the section of experimental results.

3 Statistical Feature Selection

Statistical FS techniques automatically compute a weighting score for each text-feature in a document. A significant text-feature can be identified when its weighting score exceeds a user-defined weighting threshold. Methods under this heading do not involve linguistic analysis but focus on some document-base statistics. With regard to TC, the common intuitions of various methods here can be described as: (i) the more times a text-feature appears in a class the more relevant it is to this particular class; and (ii) the more times a text-feature appears across the document-base in documents of all classes the worse it is at discriminating between the classes.

A number of mechanisms have been proposed in statistical FS. Two major statistical models can be identified: Darmstadt Indexing Approach Association Factor (DIAAF) and Galavotti·Sebastiani·Simi Coefficient (GSSC).

- **DIAAF:** The Darmstadt Indexing Approach (DIA) [11] was originally "*developed for automatic indexing with a prescribed indexing vocabulary*" [12]. In a machine learning context, Sebastiani [23] argues that this approach "*considers properties (of terms, documents, categories, or pairwise relationships among these) as basic dimensions of the learning space*". Examples of the properties include the length of a document, the frequency of occurrence between a text-feature and a class, etc. One of the pair-wise relationships considered is the term-category relationship, noted as the DIA Association Factor (DIAAF) [23], which can be applied to select significant text-features for TC problems. The calculation of the DIAAF score, and reported in [10], can be specified in probabilistic form using:

  $$diaaf\_score(u_h, C_i) = \mathrm{P}(C_i \mid u_h) \,,$$

  where u_h represents a text-feature in a given document-base $Đ$ ($Đ = \{D_1, D_2, \ldots, D_{m-1}, D_m\}$), and C_i represents a set of documents (in $Đ$) labeled with a particular text-class. The DIAAF weighting score expresses the proportion of a feature's occurrence in the given class divided by a feature's document-base occurrence.

- **GSSC:** The GSS (Galavotti·Sebastiani·Simi) Coefficient defined in [13] represents the core calculation as well as a simplified variant of both the Chi-square Statistics (χ^2) and the Correlation Coefficient (CC) statistical FS mechanisms. In [27, 30], the authors state: (i) the well-established χ^2 statistic can be applied to measure the lack of independence between a term u_h and a pre-defined class C_i; and (ii) if the feature/term and the class are independent, the calculated χ^2 score has a natural value 0. In [18] Ng *et al.* introduce CC as a refined variant of χ^2 to generate a better set of key/significant features and improve the performance of the χ^2 approach. Ng *et al.* argue that "*words that come from the irrelevant texts or are highly indicative of non-membership in*" a class C_i are not as useful; and indicate that CC "*selects exactly those words that are highly indicative of membership in a category, whereas the χ^2 metric will not only pick out this set of words but also those words that are indicative of non-membership in the category*". In [13] Galavotti *et al.* provide an explanation of the rationale to further refine the CC approach, and demonstrate that this very simple approach (GSSC) can produce a comparable performance to the χ^2 metric. The GSSC is defined in probabilistic form using:

  $$gssc\_score(u_h, C_i) = \mathrm{P}(u_h, C_i) \times \mathrm{P}(\neg u_h, \neg C_i) - \mathrm{P}(u_h, \neg C_i) \times \mathrm{P}(\neg u_h, C_i) \,,$$

  where $\neg u_h$ represents a document that does not involve the feature u_h, and $\neg C_i$ ($Đ - C_i$) represents the set of documents labeled with the complement of the pre-defined class C_i.

Existing statistical FS techniques other than DIAAF and GSSC include: Mutual Information (MI), Relevancy Score (RS), etc. In this section, we further provide a brief review of MI and RS. Note that both MI and RS are referenced in the evaluation section of this paper (section 5).

- **MI:** Early work on Mutual Information (MI) can be found in [4, 9]. This statistical model is used to determine whether a genuine association exists between two text-features or not. In TC investigations, MI has been broadly employed in a variety of approaches to select the most significant text-features that serve to classify documents. The calculation of the MI score between a text-feature u_h and a pre-defined text-class C_i, as reported in [10], is achieved in probabilistic form using:

$$mi\_score(u_h, C_i) = log(P(u_h \mid C_i) / P(u_h)) .$$

  This score expresses the proportion (in a logarithmic terms) of the probability with which the feature occurs in documents of the given class divided by the probability with which the feature occurs in the document-base.

- **RS:** The initial concept of Relevancy Score (RS) was introduced by Salton and Buckley [21] as relevancy weight. It aims to measure how "unbalanced" a text-feature (term) u_h is across documents in a document-base Ɖ with and without a particular text-class C_i. Salton and Buckley define a term's relevancy weight as: "*the proportion of relevant documents in which a term occurs divided by the proportion of nonrelevant items in which the term occurs*". In [26] the idea of RS was proposed based on relevancy weight with the objective of selecting significant text-features in Ɖ for the TC application. A term's relevancy score can be defined as: the number of relevant (the target text-class associated) documents in which a term occurs divided by the number of non-relevant documents in which a term occurs. Fragoudis *et al.* [10] and Sebastiani [23] show that the RS score can be calculated in probabilistic form using:

$$relevancy\_score(u_h, C_i) = log((P(u_h \mid C_i) + d) / (P(u_h \mid \neg C_i) + d)) ,$$

  where d is a constant damping factor. In [26] the value of d was initialized as 1/6. For the simplicity, we choose 0 as the value of d in our study.

4 Proposed Statistical Feature Selection

With regard to language-independent TC, in this section, we introduce a new statistical FS technique. In the previous section, two statistical FS techniques DIAAF and GSSC were presented in detail. The newly proposed mechanism is a variant of the original GSSC approach that makes use of the DIAAF approach, namely Hybrid DIAAF/GSSC.

Recall that the probabilistic formula for calculating the DIAAF score is given by:

$$diaaf\_score(u_h, C_i) = P(C_i \mid u_h) .$$

Recall that the probabilistic formula for GSSC is:

$$gssc\_score(u_h, C_i) = P(u_h, C_i) \times P(\neg u_h, \neg C_i) - P(u_h, \neg C_i) \times P(\neg u_h, C_i) .$$

Substituting each of the four probabilistic components in GSSC by its DIAAF related function, a DIAAF based formula is derived in a GSSC fashion:

$$diaaf\text{-}gssc\_score(u_h, C_i) = P(C_i \mid u_h) \times P(\neg C_i \mid \neg u_h) - P(\neg C_i \mid u_h) \times P(C_i \mid \neg u_h) .$$

An example of Hybrid DIAAF/GSSC score calculation is provided in Table 1. Given a document-base *Đ* containing 100 documents equally divided into 4 classes (i.e. 25 per class), and assuming that text-feature (word) u_h appears in 30 of the documents, then the Hybrid DIAAF/GSSC score per class can be calculated as shown in the Table.

Table 1. Hybrid DIAAF/GSSC score calculation

Class	# docs per class	# docs with u_h per class	# docs without u_h per class	# docs with u_h in other classes	# docs without u_h in other classes	# docs with u_h in *Đ*	# docs without u_h in *Đ*	Hybrid DIAAF/ GSSC Score
1	25	15	10	15	60	30	70	0.357
2	25	10	15	20	55	30	70	0.119
3	25	5	20	25	50	30	70	-0.119
4	25	0	25	30	45	30	70	-0.357

The algorithm for identifying significant text-features (i.e. *key* words in the current context, with regard to sections 2 – Potential Significant Words) in *Đ*, based on Hybrid DIAAF/GSSC, is given as follows:

```
Algorithm: Key Word Identification - Hybrid DIAAF/GSSC
Input: (a) A document-base Đ (the training part, where
the noise
           words have been removed);
       (b) A user-defined significance threshold G;
Output: A set of identified key words S_KW;
Begin Algorithm:
(1) S_KW ← an empty set for holding the identified key
    words in Đ;
(2) C ← catch the set of pre-defined text-classes within
    Đ;
(3) W_GLO ← read Đ to create a global word set, where the
    word
      document-base support supp_GLO is associated with each
word u_h
      in W_GLO;
(4) for each C_i ∈ C do
(5)     W_LOC ← read documents that reference C_i to create a
    local
            word set, where the local support supp_LOC is
      associated
            with each word u_h in W_LOC;
(6)     for each word u_h ∈ W_LOC do
```

```
            contribution ← (u_h.supp_LOC / u_h.supp_GLO) ×
   (((|Đ| - |C_i|)
                    -(u_h.supp_GLO - u_h.supp_LOC)) / (|Đ| -
            u_h.supp_GLO)) -
                  ((u_h.supp_GLO - u_h.supp_LOC) / u_h.supp_GLO) ×
                  ((|C_i| - u_h.supp_LOC) / (|Đ| - u_h.supp_GLO));
            if (contribution ≥ G) then
                  add u_h into S_KW;
      end for
end for
return (S_KW);
End Algorithm
```

The intuition behind the Hybrid DIAAF/GSSC approach is:

1. The contribution of term u_h for class C_i tends to be high if the ratio of the class based term support to the document-base term support is high,
2. The contribution of term u_h for class C_i tends to be high if the ratio of the class-complement based term support of non-appearance to the document-base term support of non-appearance is high,
3. The contribution of term u_h for class C_i tends to be high if the ratio of the class-complement based term support to the document-base term support is low, and
4. The contribution of term u_h for class C_i tends to be high if the ratio of the class based term support of non-appearance to the document-base term support of non-appearance is low.

5 Experimental Results

This section presents an evaluation of the proposed statistical FS approach, using three well-known text collections (i.e. Usenet Articles, Reuters-21578 and MedLine-OHSUMED). The aim of this evaluation is to assess the approach with respect to the accuracy of classification in statistical "bag of phrases" DR setting. All evaluations described in this section were conducted using the TFPC[1] associative classification algorithm; although any other classifier generator could equally well have been used. All algorithms involved in the evaluation were implemented using the standard Java programming language. The experiments were run on a 1.87 GHz Intel(R) Core(TM)2 CPU with 2.00 GB of RAM running under the Windows Command Processor. For the experiments four individual document-bases (textual datasets) were used. Each was prepared/extracted (as a subset) from one of the above mentioned text collections. The preparation of Usenet Articles ("20 Newsgroups") based document-bases adopted the approach of Deng *et al.* [8], where the entire collection was randomly split into two document-bases covering 10 classes each: 20NG.D10000.C10 and 20NG.D9997.C10. The preparation of Reuters-21578 and the MedLine-OHSUMED document-bases recalled the idea of Wang *et al.* [25], where the Reuters.D6643.C8 and OHSUMED.D6855.C10 document-bases were generated.

---

[1] TFPC software may be obtained from
http://www.csc.liv.ac.uk/~frans/KDD/Software/Apriori-TFPC/aprioriTFPC.html

The experiments reported below were designed to evaluate the proposed Hybrid DIAAF/GSSC FS approach, in comparison with alternative mechanisms (i.e. DIAAF, GSSC, MI, and RS), with regard to the DelSNcontGO statistical "bag of phrases" DR approach. Accuracy figures, describing the proportion of correctly classified "unseen" documents, were obtained using the Ten-fold Cross Validation (TCV). A support threshold value of 0.1% and a Lower Noise Threshold (LNT) value of 0.2% were used, as suggested in [6]. A confidence threshold value of 50% was used (as proposed in the published evaluations of a number of associative classification studies [5, 15, 28]). The Upper Noise Threshold (UNT) value was set to 20%. The parameter *K* (maximum number of selected final significant words) was chosen to be 1,000. Note that the value of *K* was changed to be 900 instead of 1,000 for OHSUMED.D6855.C10. The reason to decrease the value of *K* here was that 1,000 selected final significant words generated more than 2^{15} significant phrases; and, for reasons of computational efficiency, the TFPC associative classifier limits the total number of identified attributes[2] (significant phrases) to 2^{15}. To ensure that there are enough candidate final significant (potential significant) words to be selected for each category, the *G* parameter was given a minimal value (*almost zero*) so that the *G* parameter could be ignored.

Table 2. Classification accuracy – comparison of the five statistical FS techniques in the statistical "bag of phrases" DR setting

	DIAAF	**GSSC**	**RS**	**MI**	**Hybrid DIAAF/GSSC**
20NG.D10000.C10	76.36	0	76.36	76.36	**76.43**
20NG.D9997.C10	81.45	0	81.45	81.45	**81.62**
Reuters.D6643.C8	87.57	0	87.79	87.79	**88.23**
OHSUMED.D6855.C10	78.83	0	79.64	79.53	**79.74**
Average Accuracy	81.05	0	81.31	81.28	**81.51**
# of Best Accuracies	0	0	0	0	**4**

The results presented in Table 2 are the classification accuracy values (obtained by different statistical FS mechanisms in the DelSNcontGO statistical "bag of phrases" DR setting), based on the 4 extracted/prepared document-bases. From Table 2 it can be seen that the proposed Hybrid DIAAF/GSSC mechanism outperforms other alternative approaches:

1. The number of instances of best classification accuracies obtained throughout the 4 document-bases can be ranked in order as: Hybrid DIAAF/GSSC (all cases), and DIAAF, GSSC, RS and MI (none of any case), which demonstrates the stability of Hybrid DIAAF/GSSC's good performance;

[2] The TFPC algorithm stores attributes as a signed short integer.

2. The average accuracy of classification throughout the 4 document-bases can be ranked in order as: Hybrid DIAAF/GSSC (81.51%), RS (81.31%), MI (81.28%), DIAAF (81.05%), and GSSC (0%), which shows the overall advantage of the proposed mechanism; and
3. The column of GSSC is shown with value '0' for all the records. The reason of this is that when applying the GSSC feature selection technique, with the TFPC associative text classifier, too many rules were generated thus causing computational difficulty and consequently no results were obtained.

6 Conclusions

This paper is concerned with an investigation of statistical feature selection for (single-label multi-class) language-independent text classification. A description of the statistical document-base representation in terms of "bag of phrases" was provided in section 2. Both the DIAAF and GSSC statistical FS approaches were reviewed in section 3. A new statistical FS technique (Hybrid DIAAF/GSSC) was consequently introduced in section 4, which integrates the ideas of DIAAF and GSSC. From the experimental results, it can be seen that the proposed Hybrid DIAAF/GSSC approach outperforms existing mechanisms regarding the DelSNcontGO (statistical) "bag of phrases" DR setting and the TFPC associative text classification. This in turn improves the performance of language-independent text classification.

The results presented in this paper corroborate that the traditional text classification problem can be solved, with good classification accuracy, in a language-independent manner. Further research is suggested to identify the improved statistical textual data pre-processing approach in terms of (statistical) document-base representation and (statistical) feature selection, and improve the performance of language-independent text classification.

Acknowledgment. The authors would like to thank Professor Paul Leng of the Department of Computer Science at the University of Liverpool, and Dr. Jiongyu Li and Fan Li of the Information Management Center in the China Minsheng Banking Corp., Ltd. for their support with respect to the work described here.

References

1. Agrawal, R., Imielinski, T., Swami, A.: Mining Association Rules between Sets of Items in Large Database. In: Proceedings of the 1993 ACM SIGMOD International Conference on Management of Data, Washington, DC, USA, May 1993, pp. 207–216. ACM Press, New York (1993)
2. Ali, K., Manganaris, S., Srikant, R.: Partial Classification using Association Rules. In: Proceedings of the Third International Conference on Knowledge Discovery and Data Mining, Newport Beach, CA, USA, August 1997, pp. 115–118. AAAI Press, Menlo Park (1997)
3. Antonie, M.-L., Zaïane, O.R.: Text Document Categorization by Term Association. In: Proceedings of the 2002 IEEE International Conference on Data Mining, Maebashi City, Japan, December 2002, pp. 19–26. IEEE Computer Society, Los Alamitos (2002)

4. Church, K.W., Hanks, P.: Word Association Norms, Mutual Information, and Lexicography. In: Proceedings of the 27th Annual Meeting on Association for Computational Linguistics, Vancouver, BC, Canada, pp. 76–83. Association for Computational Linguistics (1989)
5. Coenen, F., Leng, P., Zhang, L.: Threshold Tuning for Improved Classification Association Rule Mining. In: Ho, T.-B., Cheung, D., Liu, H. (eds.) PAKDD 2005. LNCS, vol. 3518, pp. 216–225. Springer, Heidelberg (2005)
6. Coenen, F., Leng, P., Sanderson, R., Wang, Y.J.: Statistical Identification of Key Phrases for Text Classification. In: Proceedings of the 5th International Conference on Machine Learning and Data Mining, Leipzig, Germany, July 2007, pp. 838–853. Springer, Heidelberg (2007)
7. Cohen, W.W.: Fast Effective Rule Induction. In: Proceedings of the 12th International Conference on Machine Learning, Tahoe City, CA, USA, July 1995, pp. 115–123. Morgan Kaufmann, San Francisco (1995)
8. Deng, Z.-H., Tang, S.-W., Yang, D.-Q., Zhang, M., Wu, X.-B., Yang, M.: Two Odds-radio-based Text Classification Algorithms. In: Proceedings of the Third International Conference on Web Information Systems Engineering Workshop, Singapore, December 2002, pp. 223–231. IEEE Computer Society, Los Alamitos (2002)
9. Fano, R.M.: Transmission of Information – A Statistical Theory of Communication. The MIT Press, Cambridge (1961)
10. Fragoudis, D., Meretaskis, D., Likothanassis, S.: Best Terms: An Efficient Feature-Selection Algorithm for Text Categorization. Knowledge and Information Systems 8(1), 16–33 (2005)
11. Fuhr, N.: Models for Retrieval with Probabilistic Indexing. Information Processing and Management 25(1), 55–72 (1989)
12. Fuhr, N., Buckley, C.: A Probabilistic Learning Approach for Document Indexing. ACM Transactions on Information System 9(3), 223–248 (1991)
13. Galavotti, L., Sebastiani, F., Simi, M.: Experiments on the Use of Feature Selection and Negative Evidence in Automated Text Categorization. In: Borbinha, J.L., Baker, T. (eds.) ECDL 2000. LNCS, vol. 1923, pp. 59–68. Springer, Heidelberg (2000)
14. Kobayashi, M., Aono, M.: Vector Space Models for Search and Cluster Mining. In: Berry, M.W. (ed.) Survey of Text Mining – Clustering, Classification, and Retrieval, pp. 103–122. Springer, New York (2004)
15. Li, W., Han, J., Pei, J.: CMAR: Accurate and Efficient Classification based on Multiple Class-association Rules. In: Proceedings of the 2001 IEEE International Conference on Data Mining, San Jose, CA, USA, November-December 2001, pp. 369–376. IEEE Computer Society Press, Los Alamitos (2001)
16. Liu, B., Hsu, W., Ma, Y.: Integrating Classification and Association Rule Mining. In: Proceedings of the Fourth International Conference on Knowledge Discovery and Data Mining, August 1998, pp. 80–86. AAAI Press, Menlo Park (1998)
17. Mladenic, D.: Text-learning and Related Intelligent Agents: A survey. IEEE Intelligent Systems 14(4), 44–54 (1999)
18. Ng, H.T., Goh, W.B., Low, K.L.: Feature Selection, Perceptron Learning, and a Usability Case Study for Text Categorization. In: Proceedings of the 20th Annual International ACM SIGIR Conference on Research and Development in Information Retrieval, Philadelphia, PA, USA, July 1997, pp. 67–73. ACM Press, New York (1997)
19. Quinlan, J.R.: C4.5: Programs for Machine Learning. Morgan Kaufmann Publishers, San Francisco (1993)
20. Salton, G., Wong, A., Yang, C.S.: A Vector Space Model for Automatic Indexing. Information Retrieval and Language Processing 18(11), 613–620 (1975)

21. Salton, G., Buckley, C.: Term-weighting Approaches in Automatic Text Retrieval. Information Processing & Management 24(5), 513–523 (1988)
22. Scheffer, T., Wrobel, S.: Text Classification beyond the Bag-of-words Representation. In: Proceedings of the Workshop on Text Learning, held at the Nineteenth International Conference on Machine Learning, Sydney, Australia (2002)
23. Sebastiani, F.: Machine Learning in Automated Text Categorization. ACM Computing Surveys 34(1), 1–47 (2002)
24. Shidara, Y., Nakamura, A., Kudo, M.: CCIC: Consistent Common Itemsets Classifier. In: Proceedings of the 5th International Conference on Machine Learning and Data Mining, Leipzig, Germany, July 2007, pp. 490–498. Springer, Heidelberg (2007)
25. Wang, Y.J., Sanderson, R., Coenen, F., Leng, P.H.: Document-Base Extraction for Single-Label Text Classification. In: Proceedings of the 10th International Conference on Data Warehousing and Knowledge Discovery, Turin, Italy, September 2008, pp. 357–367. Springer, Heidelberg (2008)
26. Wiener, E., Pedersen, J.O., Weigend, A.S.: A Neural Network Approach to Topic Spotting. In: Proceedings of the 4th Annual Symposium on Document Analysis and Information Retrieval, Las Vegas, NV, USA, April 1995, pp. 317–332 (1995)
27. Yang, Y., Pedersen, J.O.: A Comparative Study on Feature Selection in Text Categorization. In: Proceedings of the Fourteenth International Conference on Machine Learning, Nashville, TN, USA, July 1997, pp. 412–420. Morgan Kaufmann Publishers, San Francisco (1997)
28. Yin, X., Han, J.: CPAR: Classification based on Predictive Association Rules. In: Proceedings of the Third SIAM International Conference on Data Mining, San Francisco, CA, USA, May 2003, pp. 331–335. SIAM, Philadelphia (2003)
29. Yoon, Y., Lee, G.G.: Practical Application of Associative Classifier for Document Classification. In: Lee, G.G., Yamada, A., Meng, H., Myaeng, S.-H. (eds.) AIRS 2005. LNCS, vol. 3689, pp. 467–478. Springer, Heidelberg (2005)
30. Zheng, Z., Srihari, R.: Optimally Combining Positive and Negative Features for Text Categorization. In: Proceedings of the 2003 ICML Workshop on Learning from Imbalanced Data Sets II, Washington, DC, USA (2003)

A Potential-Based Node Selection Strategy for Influence Maximization in a Social Network

Yitong Wang[1] and Xiaojun Feng[2]

[1] Computer Science of Fudan University, Shanghai
ytwang@fudan.edu.cn
[2] Computer Science of Fudan University, Shanghai
072021144@fudan.edu.cn

Abstract. Social network often serves as a medium for the diffusion of ideas or innovations. The problem of influence maximization which was posed by Domingos and Richardson is stated as: if we can try to convince a subset of individuals to adopt a new product and the goal is to trigger a large cascade of further adoptions, which set of individuals should we target in order to achieve a maximized influence? In this work, we proposed a potential-based node selection strategy to solve this problem. Our work is based on the observation that local most-influential node-selection adopted in many works, which is very costly, does not always lead to better result. In particular, we investigate on how to set two parameters(θ_v and b_{uv}) appropriately. We conduct thorough experiments to evaluate effectiveness and efficiency of the proposed algorithm. Experimental results demonstrate that our approximation algorithm significantly outperforms local-optimal greedy strategy.

Keywords: social network, greedy algorithm, viral marketing, influence maximization, information diffusion.

1 Introduction

Social network is a social structure made of nodes that are tied by one or more specific types of relationship. In recent work, motivated by application to marketing, Domingos and Richardson posed a fundamental algorithmic problem for such systems [10], [20]. The premise of viral marketing is that by initially targeting a few "influential" members of the network, e.g. giving them free samples of the product and these "influential" members could trigger a cascading of influence since their friends will recommend the product to other friends and many individuals will try it as a result. So, now the problem is how should we choose k (the value of k is pre-defined) influential individuals as initial target set to maximize the process? Several node-selection strategies have been proposed based on node degree or characteristic of diffusion. Actually, we have to consider both factors during the process of node-selection: structure of network and diffusion model. There are two basic models: Linear Threshold Model(LTM) and

R. Huang et al. (Eds.): ADMA 2009, LNAI 5678, pp. 350–361, 2009.

Independent Cascade Model(ICM). We will introduce these models in Section 2. While a climbing-hill greedy algorithm that select the "most influential" node at each step has been proposed, we argue that local optimal (most-influential node which could provide the largest marginal increase at present) does not always lead to a global optimal and has an unacceptably high cost. Motivated by this observation, we propose a new algorithm by identifying some most potential nodes to maximize the spread of influence in a social network.

Section 2 introduces two most commonly used diffusion models and related work as well. In Section 3, we propose TW (Target Wise) greedy algorithm and present a new estimate b_{uv} of node influence. Randomized process is also described in Section 3. In Section 4, experiments are conducted on two datasets to evaluate the effectiveness of the proposed TW greedy algorithm. Finally, Section 5 draws some conclusions and discusses future works.

2 Background Knowledge

2.1 Two Fundamental Diffusion Models

We usually model the whole social network as a directed/undirected graph with each node representing an individual and edge between nodes representing some kind of relationship (friends, co-authorships etc.). Each node is marked active (an adoption of an idea or innovation) or inactive. When most adjacent nodes (neighbors) of one node are active, this node will also tend to be activated according to a pre-defined threshold.

Linear Threshold Model. LTM (linear Threshold Model) is at the core of many diffusion models based on node specific threshold [7], [8]. For a given social network that is modeled as an undirected graph $G(V, E)$, we define $N(v) = \{u|(u, v) \in E)\}$ as the neighbor set of node v and b_{uv} as influence of active node u on its inactive neighbor $v(\sum_{u \in N(v)} b_{uv} \leq 1)$. We also define $A(v)$ as set of active nodes in $N(v)(A(v) \subseteq N(v))$. Theoretically, for each node v, a node-specific threshold θ_v is defined. For a given node if $\sum_{u \in A(v)} b_{uv} \geq \theta_v$, node v becomes active. Intuitive meaning is that for an inactive node v, if total influence exerted by all its *active* neighbors exceeds a pre-defined threshold θ_v, node v becomes active. In turn, it will exert influence on its inactive neighbors and bring some inactive neighbors active again. This process will continue until no node can be activated. This diffusion model could be described briefly as follows: Given initial set A_0, b_{uv} and node-specific threshold θ_v, at step t of diffusion, activate nodes that meet the threshold based on A_{t-1}, and add these newly activated nodes to A_{t-1} to form A_t. The process continues until no more nodes can be activated. Obviously, θ_v is node-specific and highly dependent on each individual's personality. Moreover, each individual could exert different influence on his/her neighbors under various relation semantics (friends, co-authorship etc.).

Independent Cascading Model. Another diffusion model is Independent Cascade Model(ICM) based on interacting particle systems. It was first proposed in [11] and [12], and the diffusion process is as follows: if node v is activated at step t and it then tries to activate all its inactive neighbors with success probability p_{vu} for each inactive neighbor u. If it is successful, then u will be active in step $t+1$, else, v failed and will no longer have chance to activate u. It is worth noting that p_{vu} is a system variable and is independent to other attempts to activate u made previously but failed. Each active node v has only one chance to activate its inactive neighbor u. David Kempe and Jon Kleinberg [14] think, however, p_{vu} should decrease with the diffusion process unfolding. That is, if node u has been attempted to be activated many times but all failed, the influence of newly active node v exerted on u will be weakened. And yet another new model named Decreasing Cascade Model was proposed. Influence maximization under the three models discussed so far are all NP-complete, see [12], [13]. While some other models are proposed [1], [2], [3], [4], [5], [6], [7], [8] and [9], they are all variations of the two core models we have introduced.

2.2 Related Work

A climbing-up Greedy Algorithm (KK Greedy Algorithm). For a predefined k value, we have to choose one node to place it into target set at each step. A naive approximate solution to influence maximization is to select the "most influential" node at each step. We use S to represent target set and define:

1) $S_0 = \emptyset$

2) $I(S)$: Set of active nodes throuth diffusion starting from target set S

3) $m(u|S) = |I(S \cup \{u\}| - |I(S)|$: node influence of u given S

Node influence of u given S is defined as *extra* nodes activated if adding u to S. $I(S)$ represents set of active nodes produced through diffusion starting from a given target set S. It is evident that "most influential" node is the node with the maximum node influence at the moment with current target set S. It is a local optimal. The goal of influence maximization is to choose S_k wisely with the goal of maximizing $I(S_k)$. Kempe and Kleinberg in [13] and [14] proposed a climbing-up greedy algorithm (we name the algorithm by initials of authors) based on the naive idea: choose most influential node at each step: starting from S_0, at step i, we choose node u as ith member of target set due to local optimal strategy: $u = \arg\max_v m(v|S_{i-1})$ and form S_i as $S_i = S_{i-1} \cup \{u\}$. S_k is the target set which could produce approximate maximized final influence. We say "greedy" since the most "influential" node we chose at each step is a local optimal in term of "influence" based on status of each S_i. However, it is very clear that the most influential node is very costly to obtain since we have to *compute* $I(S_i \cup \{u\})$ *for every inactive node u currently in the graph.* Even for a given node u, $I(S_i \cup \{u\})$ is also very costly since we need to trace the whole diffusion process.

Set Cover Greedy Algorithm and Shapley Value-Based Node Selection. Set Cover Greedy algorithm [19] was proposed for influence maximization under ICM. It kept choosing node with highest "uncovered degrees", once a node is chosen, all its neighbors as well as itself are labeled as "covered". This procedure continues until k nodes are chosen. This algorithm is computationally fast and the underlying idea is that selected nodes cover the graph as much as possible. Obviously, "cover" does not imply "active". The paper [21] considers a special case of information maximization problem. For co-authorship network, given a value for k; we need to find a set of k researchers who have coauthored with maximum number of other researchers. The work in [21] does not involve diffusion process. The problem lies in the characteristic function, which is rather simple in this special case but computationally heavy in the general case. This makes it inapplicable to general cases. Heuristic approach is yet another node-selection strategy totally based on node degree.

3 Target Wise Greedy Algorithm

KK greedy algorithm is very costly. Computing of most influential node (local optimal) is extremely expensive since we have to consider all inactive nodes in G currently. Moreover, local optimal does not guarantee better results. We think it is more appropriate to choose some inactive nodes that might not be optimal at starting phase but could trigger more nodes in later stage of diffusion. Motivated by this understanding, we propose a novel algorithm named TW (Target wise) greedy algorithm under linear threshold diffusion model. We divide k steps of node selection in TW algorithm into two phases: "most potential" node selection (Phase 1) and "most influential" node selection (Phase 2). We introduce a parameter c ([0, 1]) to indicate the percentage of k steps in Phase 2. It is clear that TW greedy algorithm will be degenerated into KK greedy algorithm when c equals 1.

3.1 Potential-Based Node Selection Strategy

We define "potential" of node u as

$$p(u) = \sum_{v \in N(v),\ v \notin A(u)} b_{uv} \ . \tag{1}$$

$p(u)$ is the total possible influence that u could exert to all its inactive neighbors when u becomes active. The node with the maximum "potential value" is the most potential node. Obviously, the value $p(u)$ is determined by two factors: number of inactive neighbors of u at present as well as value of each b_{uv}. The more its inactive neighbors and the bigger each b_{uv}, node u has bigger "potential". As the title implies, we think the most potential node we selected in Phase 1 can accumulate some "influence" for future use since once it is chosen in target set, it could exert great influence on its inactive neighbors.

Target Wise Greedy Algorithm. We proposed target wise greedy algorithm based on potential-based node-selection strategy. The algorithm is depicted as Algorithm.1. The reason that we partition whole node-selection into two phases

Algorithm 1. Target Wise Greedy Algorithm

Input: Graph $G(V, E)$, threshold θ, influence b_{uv}, target set size k, parameter c

Initialize $S_0 = \emptyset$, $k_1 = k - \lceil ck \rceil$, $k_2 = k - k_1$
for $i = 1$ **to** k_1 **do**
  Choose node u, $u = \arg\max_v p(v)$
  $S_{i+1} = S_i \bigcup u$
  Update $p(v)$ for each v which has not been activated
end for
for $i = 1$ **to** k_2 **do**
  Choose node u, $u = \arg\max_v m(v|S_i)$
  make $S_{i+1} = S_i \bigcup u$
end for

is that: we want to select some nodes with great "potential" to bring much more nodes into active in later steps, while at present these "potential nodes" might not have largest influence. Exact running time of Target Wise greedy algorithm is difficult to obtain since it is related to how many nodes remained inactive at present as well as how to partition k steps into two phases in TW algorithm.

New Estimate of b_{uv}. Parameter b_{uv} is defined as the influence of active node u on inactive neighbor v and usually is estimated as 1/d(v) (d(v) is the degree of v), which means that for inactive node v, all its neighbors play the same roles and have the same influence on v. Obviously , this assumption does not hold in many real applications. We propose a new estimate of b_{uv} by taking account of not only how many neighbors of v but also how these neighbors connect to each other. We first review the concept of neighbor graph of a node: $NG(v)$ and then present the new estimate of b_{uv}. It is worth to be noting that degree of node in new estimate of b_{uv} is calculated based on neighbor graph.

$$NG(v) = G'(V', E'),\ V' = \{v\} \cup N(v),\ E' = \{(x, y)|x, y \in V',\ (x, y) \in E\}$$

$$b_{uv} = \frac{degree(u)}{\sum_{w \in N(v)} degree(w)} . \tag{2}$$

Fig. 1 gives neighbor graph of a node v. According to definition of b_{uv} given above, degree of u_1, u_2, u_3 in $NG(v)$ are 2, 2, 1, and accordingly, b_{u_1v}, b_{u_2v} and b_{u_3v} are 0.4, 0.4 and 0.2 respectively.

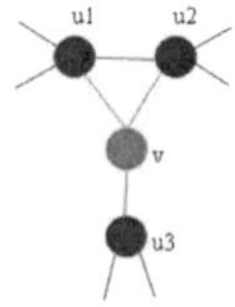

Fig. 1. An example of neighbor graph for node v

4 Experiments and Evaluations

4.1 Datasets

We conduct experiments on two datasets of social network and some statistics are shown in Table.1. We only consider an undirected graph in this paper. Dataset I is a medium-size dataset on yeast protein interaction, in which its social network effects are discussed in [15] and [16]. Dataset II [17] and [18] is a relatively large dataset. It can be seen from the table that dataset II is rather sparse while dataset I is much denser. We study influence maximization under linear threshold model. We first set b_{uv} as $1/d(v)$ and θ_v as $1/2$, which are commonly adopted in literatures. Parameter $c(0 \le c \le 1)$ indicates that Phase 2 has $\lceil ck \rceil$ steps and Phase 1 has $k - \lceil ck \rceil$ steps. We will evaluate experimental results in terms of final influence. As for running time, it is very clear that TW greedy algorithm is much faster than that of KK greedy algorithm since the cost of selecting "most potential" node is much lower than selecting "most influential" node as illustrated in their definition.

Table 1. Statistics of two data sets in Experiments

Data Set	node	edge	Average degree
Yeast protein interaction	2361(73 isolated point)	13292	11.6
Collaboration network in computational geometry	7343(1185 isolated point)	11898	3.9

4.2 Experimental Results

Joint Effects of *c* and *k*. We first investigate joint effect of parameter c and k based on final influence on two datasets and results are shown in Fig.2 and Fig.3. It is clear from Fig.2 that while KK greedy algorithm ($c = 1$) works better than the algorithm that are totally based on "potential" node selection($c = 0$), the proposed TW algorithm (including both most potential nodes and most

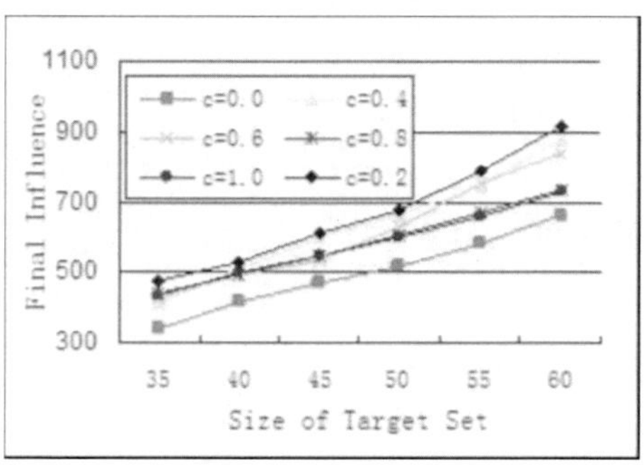

Fig. 2. Influence Curves with Various k and c values on dataset I

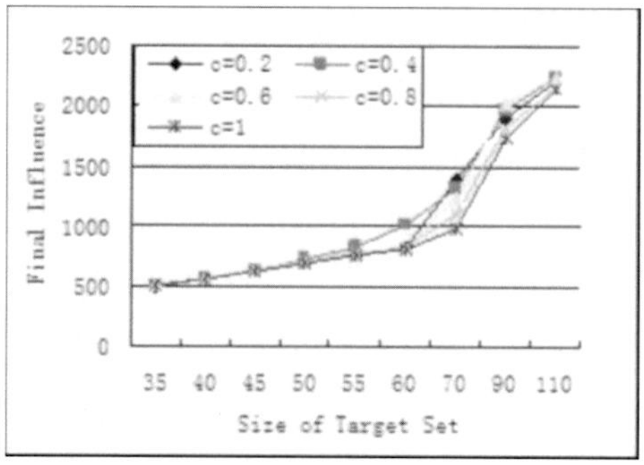

Fig. 3. Final Influence with difference c and k on dataset II

influential nodes) could beat KK greedy algorithm once c is positive and works much better when c is in the range 0.2 to 0.6.The proposed TW greedy algorithm achieves the best performance when c is around 0.2.

Fig.3 demonstrates almost the same phenomenon as that in Fig.2 except that the best performance is achieved when c is between 0.4 and 0.6. It indicates that when k is of appropriate medium-size, the proposed TW greedy algorithm outperforms KK greedy algorithm greatly even without the best c and the best performance could achieve around 15%-20% improvement in terms of final influence. Since dataset II is rather sparse, we could see from Fig.3 that when k is small, there is no big difference between TW and KK greedy algorithm and even for medium-size k, the improvement is also limited comparing with improvement in Fig.2.

Comparison between Different Algorithms. We compare TW greedy algorithms with other three related algorithms on dataset I. Our comparison is among four algorithms: TW greedy algorithm (c is 0.2), KK greedy algorithm ($c = 1$), heuristic and SCG algorithm (Set Cover Greedy) introduced in [19]. Heuristic

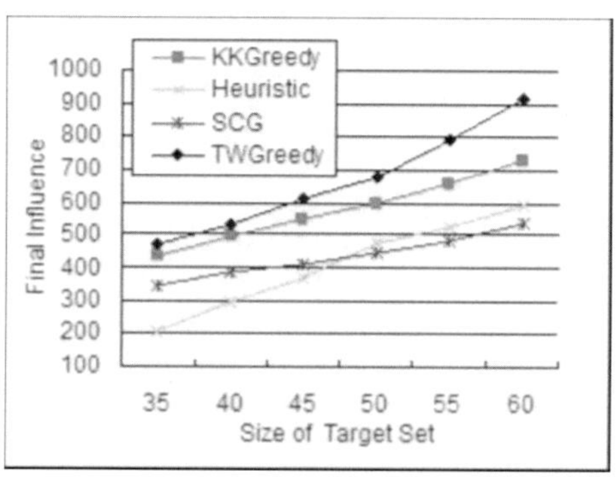

Fig. 4. Influence curves for different algorithms

algorithm is solely based on highest-degree node-selection. It was demonstrated in [14] that KK greedy algorithm is better than the heuristic algorithm since diffusion process should be also considered in addition to network feature. Our experimental results shown in Fig.4 also agree with this point.

We can see very clear from Fig.4 that TW greedy algorithm outperforms other three algorithms greatly. With increase of k (size of target set), advantages over other algorithms become even bigger and more evident.

Detailed Comparison between TW Greedy Algorithm and KK Greedy Algorithm. We investigate in detail about the difference between TW greedy algorithm and KK greedy algorithm to understand the merits of our approach more intuitively. The evaluation metric is average influence of set S(target set), which is defined as follow: $AI(S) = |I(S)|/|S|$, where AI represents "average influence". Average influence reveals how many active nodes could actually be triggered by each member of the target set on the average.

In TW greedy algorithm, we partition k steps of node-selection into two phases, $k - \lceil ck \rceil$ steps of most potential nodes selection and $\lceil ck \rceil$ steps of most influential nodes selection. We also partition k steps node-selection of KK greedy algorithm accordingly into $k - \lceil ck \rceil$ steps and $\lceil ck \rceil$ steps. We compare corresponding average influence of two target sets produced by two algorithms in terms of average influence at two phases. It is clear from Table.2 that given a fixed c (e.g. c is 0.1), average influence of TW greedy algorithm is a little bit lower than that of KK greedy algorithm at Phase 1(10.4 vs. 12.17), but at phases 2, average influence of TW greedy algorithm achieves much better (36.8 vs. 10.4). So in summary, TW greedy algorithm achieves much better performance since in Phase2, "influential" nodes could activate many more nodes due to "potential" accumulated in Phase1. We also found that average influence of TW greedy algorithm decreases with the increase of c. It is reasonable since the smaller c is; the bigger "potential" accumulated which will in turn influence more nodes.

Table 2. TW Greedy vs. KK Greedy based on average influence of two target sets (k is 50) on Dataset I

A(S)	TW Greedy Potential Accumulation	Potential Display	KK Greedy Corresponding Phase 1	Corresponding Phase2
c=0.1	10.4	36.8	12.17778	10.4
c=0.2	10.475	24.5	12.45	10.2
c=0.3	10.02941	20.9375	12.55882	10.8125
c=0.4	10.76667	17.15	12.8	10.8
c=0.5	10.48	13.16583	13.16	10.80769
c=0.6	10.84211	13.64516	13.26316	11.22581
c=0.7	11.8	12.28571	13.8	11.22857
c=0.8	12	12.175	14.6	11.35
c=0.9	13.6	11.82222	15.2	11.64444

Improvement Based on New b_{uv}. The effects of new estimate b_{uv} on two datasets are show in Fig.5 and Fig.6. We also present comparisons of TW greedy algorithm ($c = 0.5$) vs. KK greedy algorithm ($c = 1$) with old ($1/d(v)$) and new

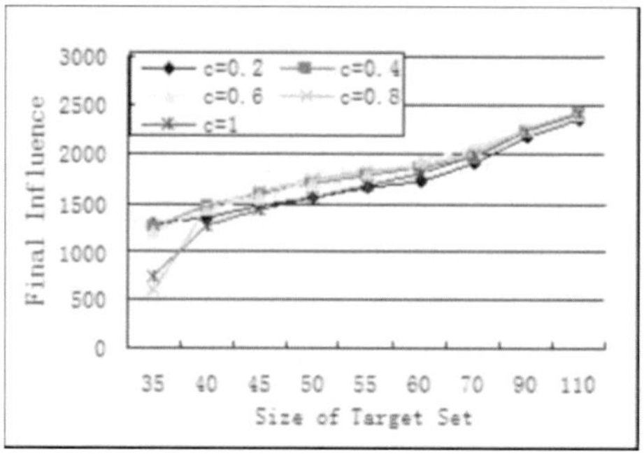

Fig. 5. TW greedy vs. KK greedy with new estimates of b_{uv} on dataset II

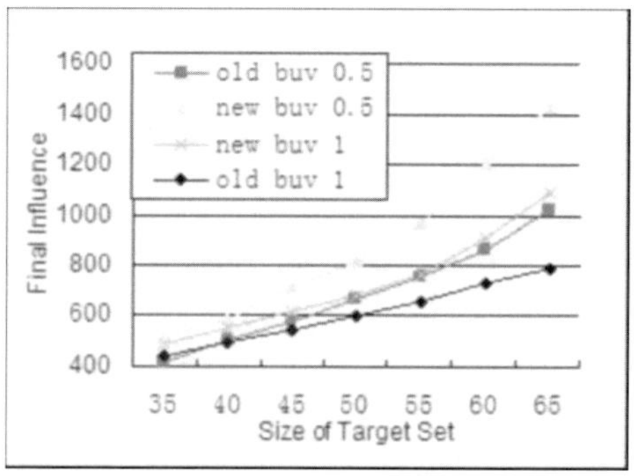

Fig. 6. TW greedy vs. KK greedy with two estimates of b_{uv} on dataset I

estimates (defined in section 3.2)of parameter b_{uv} on two datasets as shown in Fig.5 and Fig.6. It is very clear from the two figures as well as Fig.3 that the new estimate of b_{uv} is very effective on improvement of final influence. By combing Fig.3 and Fig.5, it is very clear that the proposed new b_{uv} is very effective on performance improvement. This is also true for Fig.6. Even for KK greedy algorithm, with the help of new estimate of b_{uv}, its performance also improves greatly. As we can see, the two datasets are very different in terms of network structure as well as data volume, however, the best performances are all achieved with c as 0.5 when new estimate b_{uv} is applied. Another message conveyed from Fig.5 and Fig.6 is that the improvement of TW over KK is more evident on dataset I than that of dataset II since structure of network plays a very important role during the diffusion process.

Randomize Node Specific Threshold θ_v. The randomized results shown in Fig.7 and Fig.8 are based on the average results of 5 attempts of randomizing. Combing Fig.3 and Fig.7, it is very clear that the effect of randomizing is very effective and double the final influence. Even for very small k, with randomized

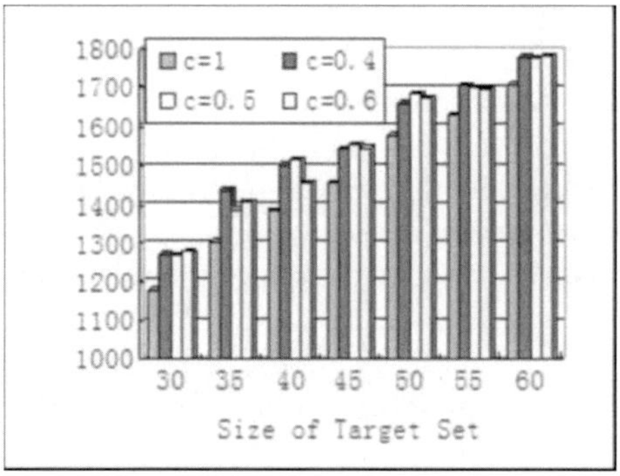

Fig. 7. TW greedy vs. KK greedy with randomized θ_v and old b_{uv} on dataset II

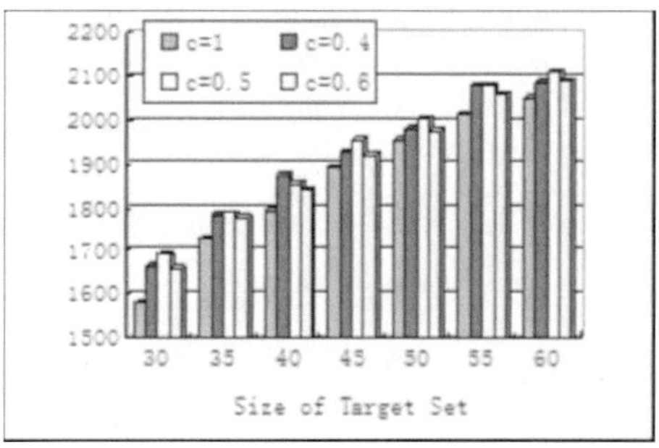

Fig. 8. TW greedy vs. KK greedy with randomized θ_v and new b_{uv} on dataset II

θ_v, the proposed TW achieves great results and outperforms KK greedy algorithm significantly.

It is evident from Fig.8 that while both randomized θ_v and new b_{uv} give better results independently, combing them could give much better results. Moreover, Fig.8 also conveys the message that the proposed TW algorithm is very stable and robust under new b_{uv} and randomized θ_v for different datasets.

5 Conclusion

In this paper, we proposed an approximate algorithm to influence maximization under a linear threshold model by identifying some "potential" nodes into the target set. While solutions to influence maximization under LTM and ICM models are NP-hard, we could only approximate the optimal result with various heuristics. KK greedy algorithm achieves 63% of the optimal result based on local optimal in terms of "influence", that is, at each step, it chooses the most "influential node" which could provide the largest marginal increase at present. However, by detailed check, we found that this local optimal is rather costly and not necessary. The proposed TW greedy algorithm is based on identifying some most "potential" nodes at the starting stage of node selection. Moreover, selecting the most "potential" node is much cheaper and easier than selecting the most "influential node" as their definitions clearly show. We conducted experiments on two datasets to evaluate joint effects of parameters introduced (c, k). Comparison of the proposed TW greedy algorithm with other related algorithms demonstrates that our approach outperforms them significantly both in running time and final influence. We also investigate how to set appropriate parameters b_{uv} and θ_v. Our underlying idea is that b_{uv} is related not only to number of neighbors of v but also how these neighbors connect to each other. θ_v is highly node specific and should be randomized. We think the new estimate of b_{uv} and randomized are more reasonable and experimental results demonstrate that they are very effective on performance improvement independently and combine them can give a even better performance. While the proposed TW greedy algorithm is very effective and efficient, there is still much work needed to further investigate, such as how to extend influence maximization to a virtual social web (social network on the Web), where relationships between nodes are uncertain and community information are included.

Acknowledgement. This work is supported by China NSF Project (No. 60603042).

References

1. Granovetter, M.: Threshold models of collective behavior. American Journal of Sociology 83(6), 1420–1443 (1978)
2. Schelling, T.: Micromotives and Macrobehavior. Norton (1978)
3. Berger, E.: Dynamic Monopolies of Constant Size. Journal of Combinatorial Theory Series B 83, 191–200 (2001)
4. Morris, S.: Contagion. Review of Economic Studies 67 (2000)
5. Peleg, D.: Local Majority Voting, Small Coalitions, and Controlling Monopolies in Graphs: A Review. In: 3rd Colloq. On Structural Information and Communication (1996)
6. Macy, M., Willer, R.: From Factors to Actors: Computational Sociology and Agent-Based Modeling. Ann. Rev. Soc (2002)
7. Valente, T.: Network Models of the Diffusion of Innovations. Hampton Press (1995)
8. Peyton Young, H.: The Diffusion of Innovations in Social Networks. Santa Fe Institute Working Paper 02-04-018 (2002)
9. Watts, D.: A Simple Model of Global Cascades in Random Networks. Proc. Natl. Acad. Sci. 99, 5766–5771 (2002)
10. Domingos, P., Richardson, M.: Mining the Network Value of Customers. In: ICDM (2001)
11. Goldenberg, J., Libai, B., Muller, E.: Talk of the Network: A Complex Systems Look at the Underlying Process of Word-of-Mouth. Marketing Letters 12(3), 211–223 (2001)
12. Goldenberg, J., Libai, B., Muller, E.: Using Complex Systems Analysis to Advance Marketing Theory Development. Academy of Marketing Science Review (2001)

13. Kempe, D., Kleinberg, J., Tardos, E.: Influential nodes in a diffusion model for social networks. In: Caires, L., Italiano, G.F., Monteiro, L., Palamidessi, C., Yung, M. (eds.) ICALP 2005, vol. 3580, pp. 1127–1138. Springer, Heidelberg (2005)
14. Kempe, D., Kleinberg, J., Tardos, E.: Maximizing the spread of influence in a social network. In: Proc. 9th Intl. Conf. on KDD, pp. 137–146 (2003)
15. Sun, S., Ling, L., Zhang, N., Li, G., Chen, R.: Topological structure analysis of the protein-protein interaction network in budding yeast. Nucleic Acids Research 31(9), 2443–2450 (2003)
16. Software package Protein Interaction Network PIN
17. Beebe, N.H.F.: Nelson H.F. Beebe's Bibliographies Page (2002)
18. Jones, B.: Computational Geometry Database (February 2002); FTP / HTTP
19. Estevez, P.A., Vera, P., Saito, K.: Selecting the Most Influential Nodes in Social Networks. In: Proceedings of International Joint Conference on Neural Networks, Orlando, Florida, USA, August 12-17 (2007)
20. Richardson, M., Domingos, P.: Mining knowledge-sharing sites for Viral Marketing. In: Eighth Intl. Conf. on knowledge Discovery and Data Mining (2002)
21. Rama Suri, N., Narahari, Y.: Determining the Top-k Nodes in Social Networks using the Shapley Value (Short Paper). In: Padgham, Parkes, Müller, Parsons (eds.) AAMAS 2008, Estoril, Portugal, May 12-16, pp. 1509–1512 (2008)

A Novel Component-Based Model and Ranking Strategy in Constrained Evolutionary Optimization

Yu Wu*, Yuanxiang Li, and Xing Xu

State Key Lab. of Software Engineering
Wuhan University,
430072 Wuhan, Hubei
{wy08_whu,yxli62,whuxx84}@yahoo.com.cn
http://www.sklse.org/

Abstract. This paper presents a component-based model with a novel ranking method (CMR) for constrained evolutionary optimization. In general, many constraint-handling techniques inevitably solve two important problems: (1) how to generate the feasible solutions, (2) how to direct the search to find the optimal feasible solution. For the first problem, this paper introduces a component-based model. The model is useful for exploiting valuable information from infeasible solutions and for transforming infeasible solutions into feasible ones. Furthermore, a new ranking strategy is designed for the second problem. The new algorithm is tested on several well-known benchmark functions, and the empirical results suggest that it continuously found the optimums in 30 runs and has better standard deviations for robustness and stability.

Keywords: component-based model, constraint handling, rank strategy, evolutionary algorithm.

1 Introduction

In the last decades, constrained optimization problems (*COPs*) have received much attention by most researchers and practitioners. The models for most real-world applications, are established in form of constraints imposed upon the optimization function. The general constrained optimization problems can be represented as follows:

$$\begin{array}{ll} Minimize & f(X), \quad X = [x_1, x_2, ..., x_D]^T \in \Re^D \\ \text{s.t.} & g_i(X) \leq 0, i = 1, 2, ..., m; \\ & h_j(X) = 0, j = m+1, ..., m+p; \\ & L_k \leq x_k \leq U_k, k = 1, 2, ..., D \end{array} \tag{1}$$

* This work was supported in part by the National Natural Science Foundation of China *(60773009)*, the National High Technology Research and Development Program of China *(863 Program)(2007AA01Z290)*.

R. Huang et al. (Eds.): ADMA 2009, LNAI 5678, pp. 362–373, 2009.

Where X is the vector of decision variable. The objective function f is defined on the search space $S \subseteq \Re^D$, and the feasible solutions exist in the feasible region $F \subseteq S$. Usually, S is an n-dimensional space defined by the lower and upper bounds, whereas F is restricted by a set of additional constraints including inequality and equality ones. m and p respectively are the number of inequality constraints and of equality constraints. It is common practice to transform equality constraints into form of inequalities: $h_j(X) - \varepsilon \leq 0$, where ε is the tolerance allowed (a very small value). This allows us to deal only with inequality constraints. In *COPs*, $G(X) = \sum_{i=1}^{m+p} max[0, g_i(X)]$ reflects the degree of constraint violation of the solution X.

Many evolutionary algorithms (EAs) have been broadly applied to solve *COPs*. The main challenge in these research is simultaneously handling the constraints as well as optimization of the objective function. According to a comprehensive survey provided by Michalewicz, Schoenauer and Coello [1], the constraint-handling methods usually can be classified into four categories:

Penalty function—is the most widely used technique in the conventional methods. The main difficulty of this method is tuning an appropriate penalty coefficient that adjusts the strength of the penalty. Various designs of the penalty coefficient have been proposed, such as static, dynamic [2], self-adaptive [3], death penalties. There are many other highly competitive techniques in recently including the use of adaptive penalty function with a "Segregational" selection operator, the use of multimembered evolution strategy with a stochastic ranking scheme[4], and so on.

Special representations and operators–are used to avoid generating and rejecting a large number of infeasible solutions. Michalewicz and Janikow presented a series of methods(GENOCOP I, II, III). Nevertheless, "homomorphous maps" [5] is the most competitive method in decoder technologies.

Separation of constraints and objectives—handle the value of objective function and the constraints separately. In this category, there are many different approaches including co-evolution, superiority of feasible points, behavioral memory and multiobjective optimization [6,7,8]. The last type of approach has been very popular in the recent years.

Hybrid methods— are coupled with another heuristic or mathematical programming approach, mainly related to Ant System, Simulated Annealing, Artificial Immune System, and Cultural Algorithms.

In the above survey on these methods, we found that two crucial concepts—feasible solution and infeasible solution, frequently appeared. Here constraints are only used to see whether a candidate solution is feasible or not. Moreover, most approaches usually take advantage of the valuable information of feasible solutions, but don't exploit adequately the information from infeasible ones.

In General, most of constraint-handling techniques previously discussed will inevitably solve two important problems: (1) how to generate the feasible solutions, (2) how to direct the search to find the optimal feasible solution.

For the first one, the short-cut is to exploit valuable information from infeasible solutions, and then direct transformation of infeasible solutions into feasible ones. A component-based model is presented for recognizing the characteristics of diverse information extracted from solutions. Supposing an infeasible solution which does not satisfied all constraints, the solution maybe consists of three parts: first part are several components related to non-satisfied constraints, another part are components related to satisfied constraints, the last one are components not related to all constraints. Therefore, maybe we only needed to change a few components of this solution for itself transformation into feasible one. The reason is that some non-satisfied constraints are merely influenced by the changed components not by others. In other words, the correlative components are changed so that the status of constraints vary from not-satisfied to satisfied. At last, in the component-based model, the interrelationship between each dimensional component of solution and constraints is revealed. A novel measurement of feasibility is defined. Different from traditional measurement, the definition in this paper is only related to a component of solution but not to a whole solution. The feasibility of components is measured so as to direct which component needs to be transformed at a smaller cost. According to feasible components on various dimensions, subpopulation classification and improved genetic operators are adopted for quickly transforming infeasible component into feasible one.

For the second problem, two situations in *COPs* are greatly difficult to solve. One situation is disjoint feasible regions in the whole search space, and the other is the optimum feasible lies on the boundary of the feasible space. In these cases, quality measure for the solutions is dominated by the objective function. Furthermore, reasonable exploration of infeasible regions may act as bridges connecting different feasible regions or as arrows pointing to the boundary of the feasible space. To address these concerns, we devise a novel ranking strategy with a balance between preserving feasible solutions and rejecting infeasible ones. A common rule, in which all infeasible solutions are regarded worse than feasible ones, is not followed by our strategy. Demarcated by the objective function value of the best feasible solution in the current population, infeasible solutions with superior objective function value are set at a higher rank so that they are given higher probabilities to survive for population diversity.

2 Generation of Feasible Solutions

How to generate the feasible solutions? We can take full advantage of valuable information of infeasible solutions. For the characters of diverse information extracted from solutions, a component-based model is presented. Two main mechanisms are applied in this model–extraction and subpopulation classification. The former takes each dimensional component as a unit to extract feasible(or infeasible) information. The information also can be expressed as the interrelationship between each dimensional component and constraints. The goal of extraction mechanism is to realize which component of a solution needs to be transformed at a smaller cost. The latter mechanism divides the population into a lot of subpopulation based on the former's work, and then classifies each individual into these subpopulation according as the feasibility of each dimensional component. The purpose of classified subpopulation is to direct the transformation of the infeasible component into the feasible one.

2.1 Extraction Technology

For a constrained optimization problem, each constraint is only correlative with certain components, namely, each component only impacts upon some specific constraints. Therefore, the interrelationship between each dimensional component of solution and constraints is revealed. Taking $g01$ function listed in literature [4] as an example.

The first dimensional component x_1 appears in some inequality constraints including g_1, g_2 and g_4. It means that the change to x_1 will impact on whether the constraints($g_1; g_2; g_4$) are satisfied or not. Therefore, if the component $x_1$of a solution is satisfied with the constraints($g_1; g_2; g_4$), some helpful information can be acquired from this component and can direct other x_1 of solutions which didn't meet the g_1, g_2, g_4 requirement. By analogy, the correlative constraints of each dimensional component in this test function can be deduced in Table 2. Supposing the correlative constraint set of the ith dimensional component is marked as CRG_i. The population size is represented as N_p and the dimensionality of constrained problem is D.

Table 1. Interrelationship between Each Dimensional Component and Constraints

x_i	x_1	x_2	x_3	x_4,x_5	x_6,x_7	x_8,x_9	x_{10}	x_{11}	x_{12}	x_{13}
CRG_i	$\{g_1, g_2, g_4\}$	$\{g_1, g_3, g_5\}$	$\{g_2, g_3, g_6\}$	$\{g_7\}$	$\{g_8\}$	$\{g_9\}$	$\{g_1, g_2, g_4, g_7\}$	$\{g_1, g_3, g_5, g_8\}$	$\{g_2, g_3, g_6, g_9\}$	$\emptyset$

At the early evolutionary stage, few feasible solutions are generated. Simultaneously, many infeasible solutions are usually consisted by two kinds of components—ones satisfied their all correlative constraints, and the other didn't satisfy a correlative constraint at least. Analyzing for generation of feasible solution at a small cost, the better way may be only change the last kind of components. After the above deduction, it is known that the ith dimensional component needn't to change when all correlative constraints are satisfied, or need to change when one correlative constraint isn't satisfied at least. However, if the correlative constraint set is empty, it means that the feasibility measure of solution isn't relevant to this component. In order to determine whether the component to change, some definitions are given.

definition 1 **(Feasible Component)**
For $x_i \in X, i = 1, ..., D$, $\forall g_j \in CRG_i$, if $max[0, g_j] = 0$, then x_i is a feasible component.
definition 2 **(Infeasible Component)**
For $x_i \in X, i = 1, ..., D$, $\exists g_j \in CRG_i$, if $max[0, g_j] \neq 0$, then x_i is an infeasible component.
definition 3 **(Feasible Solution)**
For $X_i, i = 1, ..., N_p$, $\forall x_j \in X_i$, if x_j is a feasible component, then X_i is a feasible solution.

definition 4 **(Infeasible Solution)**
For $X_i, i = 1, ..., N_p$, $\exists x_j \in X_i$, if x_j is an infeasible component, then X_i is an infeasible solution.

According to these definitions, a feasible solution is constituted by feasible components, and an infeasible solution maybe constituted by feasible and infeasible components. Therefore, the infeasible solution merely needs to be modified its infeasible components for transforming into feasible solution; the rest feasible components don't need change.

2.2 Subpopulation Classification

Subpopulation technique is widely applied in many evolutionary algorithms. But a similar approach in niche evolutionary algorithm has received considerable attentions by lots of researchers. In order to maintain the diversity of population, the niche algorithm based on crowding mechanism divides population into a number of niches by the use of the resemblance among solutions [11]. The divided niche seem an another form of subpopulation and is usually measured by Haiming distance or fitness distance. In this paper, the classified subpopulation is designed as a genetic breeding pool. A lot of characteristic information between feasible components and feasible solutions will be collected in the pool and be ready for generation of feasible components. In other words, many infeasible components combining with genetic operators can learn some feasible characters from the subpopulation and generate new feasible components. Different from niche algorithm, the number of subpopulations in this paper depends on the dimensionality of constrained problem.

Based on the various dimensional components, classified subpopulations are applied to direct the transformation of infeasible component into feasible one. Each dimensional subpopulation will include many solutions which contain the relevant dimensional feasible components. Namely, if the *ith* dimensional component of a solution is feasible, the *ith* subpopulation includes this solution. If the solution includes many various dimensional feasible components, it will be

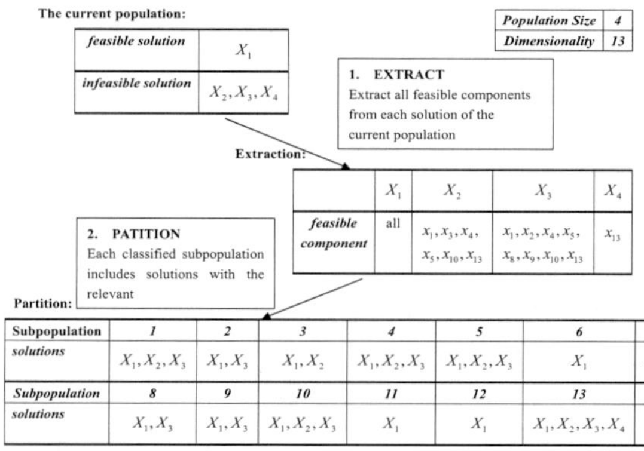

Fig. 1. An example on Subpopulation Classification

conserved in many related subpopulations. The example of the above proposed test function is displayed in Figure 1.

According to the number of solutions in each classified subpopulation and feasibility proportion of the current population, various kinds of multi-parent crossover operators are designed for solution feasibility or population diversity. An opposition-based mutation with a probability is embedded in CMR to accelerate its convergence speed. The high performance of this mutation has been verified in [9]. The generation for the new offspring is shown in Algorithm 1.

Let us define the meeting of the following terms:

$Numfc_j$ is the number of solutions in the jth subpopulation; P_m is mutation rate and P_fr is feasibility proportion of the current population; $[Min_j^t, Max_j^t]$ is the range of the jth component in the tth generation; rd_1, rd_2, rd_3 are different random numbers sampled in [-0.5,1.5], simultaneously, $rd_1 + rd_2 + rd_3 = 1$.

3 Search for Optimal Feasible Solution

How to direct the searching for the optimum feasible? The key technique is how to balance the dominance role of penalty and objective function. Many researchers have done some study on ranking methods for theirs simplicity and effectiveness. The ranking ways are usually on the basis of various lexicographic orders, such as stochastic ranking, pareto ranking. In general, feasible solutions are ranked highest and better than all infeasible solutions. However, infeasible solutions with superior objective function value are more efficient to guide the population toward the optimum feasible, especially when the feasible regions are disjoint or the optimum lies on the boundary of the feasible region. Therefore, we tend to remain the important feasible and infeasible solutions for ranking strategy.

A novel ranking strategy is designed to accomplish the above goal. The essential comparison rules between adjacent pairs can be summarized as the following

Algorithm 1. Offspring Generation with Genetic Operators

for $i = 0$ to N_p **do**
  for $j = 0$ to D **do**
    if $rand(0, 1) < P_{fr} \wedge Numfc_j \geq 3$ **then**
      select three parents $P_{r_1j}, P_{r_2j}, P_{r_3j}$ randomly from *the jth subpopulation*;
      $x_{ij}^* = rd_1 P_{r_1j} + rd_2 P_{r_2j} + rd_3 P_{r_3j}$;
    else
      select three parents $P_{r_1j}, P_{r_2j}, P_{r_3j}$ randomly from *the current population*;
      $x_{ij}^* = rd_1 P_{r_1j} + rd_2 P_{r_2j} + rd_3 P_{r_3j}$;
    end if
    if $rand(0, 1) \leq P_m$ **then**
      $x_{ij}^* = Min_j^t + Max_j^t - x_{ij}$;
    end if
  end for
end for

three points: 1) two feasible solutions are compared only based on their objective function values; 2) while at least one's objective function value of infeasible solution is less than the value of best feasible solution of the current population, two infeasible solutions are compared only based on their objective function values; 3) In the remaining situations, two solutions are compared based on the amount of their constraint violations.And the comparison formulation can be described as follow.

Supposing an adjacent pair are represented as X and Y. The objective function value of best feasible solution in the current population is f_{min}.
(i) X and Y are both feasible solutions:

$$better(X,Y) = \begin{cases} 1, & if \quad f(X) \leq f(Y); \\ 0, & otherwise. \end{cases}$$

(ii) X and Y are both infeasible solutions:

$$better(X,Y) = \begin{cases} 1, & if \quad f(X), f(Y) \leq f_{\min} \quad \wedge \quad f(X) \leq f(Y); \\ 0, & if \quad f(X), f(Y) \leq f_{\min} \quad \wedge \quad f(X) > f(Y); \\ 1, & if \quad f(X), f(Y) > f_{\min} \quad \wedge \quad \mathrm{G}(X) \leq G(Y); \\ 0, & if \quad f(X), f(Y) > f_{\min} \quad \wedge \quad \mathrm{G}(X) > G(Y); \\ 1, & if \quad f(X) \leq f_{\min} \quad \wedge \quad f(Y) > f_{\min}; \\ 0, & if \quad f(X) > f_{\min} \quad \wedge \quad f(Y) \leq f_{\min}. \end{cases}$$

(iii) X is a feasible solutions, and Y is an infeasible solution:

$$better(X,Y) = \begin{cases} 1, & if \quad f(Y) > f_{\min}; \\ 0, & if \quad f(Y) \leq f_{\min} \quad \wedge \quad f(X) > f(Y); \end{cases}$$

And vice versa when X and Y are respectively the infeasible solutions and feasible solution. After the above comparisons, the Ranking result can be concluded. Infeasible solutions with superior objective function value are ranked highest, followed by all feasible solutions, and other infeasible solutions with greater constraint violation value are ranked to the lowest level.

Considering few feasible solutions for the population at the early evolutionary stage, the ranking strategy should pay more attention to feasibility or constraint violation for a solution. So the whole ranking method is described in Algorithm 2. Where Pr is a proportion constant in $[0, 1]$. In this section, the main steps of CMR algorithm can be described as follows.

Algorithm 2. Ranking Method

if $P_{fr} \leq P_r$ **then**
  compare the adjacent pair according to the amount of their constraint violations, regardless of feasible or infeasible solutions;
else
  compare the adjacent pair according to the above formulations i),ii),iii);
end if

The Operational Process of CMR Algorithm

```
program procedure
  begin
    1) Create random initial population POP(0).
    2) Extract interrelationship between each dimensional component
       and constraints of COP.
    3) Calculate f(X),G(X) of each solution in POP(0)and check the
       feasibility of components or solutions.
    4) If stopping criterion has been met, stop; otherwise continue.
    5) Apply the proposed subpopulation classification.
    6) Apply the genetic operators including multiparent crossover
       and mutation to generate subpopulation SubPOP(t).
    7) Calculate f(X),G(X) of SubPOP(t) and check the feasibility of
       components or solutions.
    8) Apply the proposed ranking strategy in interim population
       incorporating POP_t with SubPOP(t), and select the former Np
       solutions as the members of POP(t+1).
    9) Go to step 4).
  end.
```

4 Experiment Verification

Twelve benchmark test functions from Runarsson and Yao [4] are applied in this paper, and the results of the CMR algorithm are compared against five state-of-the-art algorithms: the SR[4], the KM [5], the SMES[10], the VY[7] and the SAFF [3]. For each test case, 30 independent runs are performed. In the following experiments, the parameters for the CMR algorithm are as follows: the population size $N_p = 60$, the maximum generations is 5000, the mutation rate $P_m = 0.25$ and $P_r = 0.3, \varepsilon = 10^{-4}$. The experiments are performed on a computer with Intel Core-2 CPU 1.83GHz and 1GB of RAM, by using the visual C++ compiler.

4.1 Comparison between Six Algorithms

Table 3 summarizes the results from the conducted experiment. The statistical results include the known optimal solutions for each test function, the *best, mean, worst* objective function values, and the standard deviations. "—" means that solutions were not found or not available.

In the comparison, CMR can consistently find the optimal solutions in four test functions (g01, g03, g08, and g11) as other compared algorithms. All *best, mean* and *worst* objective function values of CMR were equivalent to the optimums for the above functions. Especially, CMR has better capability to deal with function g03 and has slightly better standard deviations than SR, SMES and others. For test functions g02, g07, g09, and g12, the near-optimal solutions

Table 2. Statistical Results for g01-g12 Functions out of 30 Independent Runs

Alg.	*Best*	*Mean*	*Worst*	*st.dev*	*Best*	*Mean*	*Worst*	*st.dev*
	g01(-15.0000)				**g02(-0.803619)**			
VY	**-15.000**	—	-12.000	8.5E-01	-0.803190	—	-0.672169	3.2E-02
KM	-14.786	-14.708	—	—	-0.799530	-0.796710	—	—
SAFF	**-15.000**	**-15.000**	**-15.000**	0.0E+00	-0.802970	-0.790100	-0.760430	1.2E-02
SMES	**-15.000**	**-15.000**	**-15.000**	0.0E+00	-0.803601	-0.785238	-0.751322	1.6E-02
SR	**-15.000**	**-15.000**	**-15.000**	0.0E+00	-0.803515	-0.781975	-0.726288	2.0E-02
CMR	**-15.000**	**-15.000**	**-15.000**	0.0E+00	-0.799209	-0.786728	-0.776587	1.7E-01
	g03(-1.0000)				**g04(-30665.539)**			
VY	**-1.0000**	—	-0.786	4.8E-02	-30665.531	—	-30651.960	3.3E+00
KM	**-1.0000**	**-1.0000**	—	—	-30664.500	-30655.300	—	—
SAFF	**-1.0000**	**-1.0000**	**-1.0000**	7.5E-05	-30665.500	-30665.200	-30663.300	4.8E-01
SMES	**-1.0000**	**-1.0000**	**-1.0000**	1.6E-02	**-30665.539**	**-30665.539**	**-30665.539**	0.0E+00
SR	**-1.0000**	**-1.0000**	**-1.0000**	1.9E-04	**-30665.539**	**-30665.539**	**-30665.539**	2.0E-05
CMR	**-1.0047**	**-1.0047**	**-1.0047**	3.7E-06	-30645.145	-30643.233	-30640.038	1.5E+00
	g05(5126.498)				**g06(-6961.814)**			
VY	5126.510	—	6112.223	3.4E+02	-6961.179	—	-6954.319	1.2E+00
KM	—	—	—	—	-6952.100	-6342.600	—	—
SAFF	5126.989	5432.080	6089.430	3.8E+03	-6961.800	-6961.800	-6961.800	0.0E+00
SMES	5126.599	5174.492	5304.167	5.0E+01	**-6961.814**	-6961.284	-6952.482	1.8E+00
SR	**5126.497**	5128.881	5142.472	3.5E+00	**-6961.814**	-6875.940	-6350.272	1.6E+02
CMR	5128.283	5129.274	5153.765	2.7E+01	-6955.983	-6955.335	-6955.237	2.8E-01
	g07 (24.306)				**g08(-0.095825)**			
VY	24.411	35.882	—	2.6E+00	**-0.095825**	—	**-0.095825**	0.0E+00
KM	24.620	26.826	—	—	**-0.095825**	-0.089157	—	—
SAFF	24.480	26.580	28.400	1.1e+00	**-0.095825**	**-0.095825**	**-0.095825**	0.0e+00
SMES	24.327	26.475	26.426	1.8E+00	**-0.095825**	**-0.095825**	**-0.095825**	0.0e+00
SR	24.307	24.374	24.642	6.6E-02	**-0.095825**	**-0.095825**	**-0.095825**	2.6e-17
CMR	24.404	24.408	24.770	7.3E-01	**-0.095825**	**-0.095825**	**-0.095825**	0.0E+00
	g09(680.630)				**g10(7049.248)**			
VY	680.762	—	684.131	7.4E-01	7060.553	—	12097.408	7.9E+02
KM	680.910	681.160	—	—	7147.900	8163.600	—	—
SAFF	680.640	680.720	680.870	5.9e-02	7061.340	7627.890	8288.790	3.7e+02
SMES	680.632	680.643	680.719	1.5e-02	7051.903	7253.047	7638.366	1.3e+02
SR	**680.630**	680.656	680.763	3.4e-02	7054.316	7559.192	8835.655	5.3e+02
CMR	680.780	680.785	680.792	7.2E-03	7056.234	7059.342	7061.658	2.1E+00
	g11(0.7500)				**g12(-1.0000)**			
VY	**0.7490**	—	0.8090	9.3E-03	—	—	—	—
KM	**0.7500**	**0.7500**	**0.7500**	—	-0.9999	—	—-	—
SAFF	**0.7500**	**0.7500**	**0.7500**	0.0E+00	—	—	—	—
SMES	**0.7500**	**0.7500**	**0.7500**	0.0E+00	**-1.0000**	**-1.0000**	**-1.0000**	0.0E+00
SR	**0.7500**	**0.7500**	**0.7500**	8.0E-05	**-1.0000**	**-1.0000**	**-1.0000**	0.0E+00
CMR	**0.7500**	**0.7500**	**0.7500**	0.0E+00	-0.9798	-0.9798	-0.9798	2.3E-05

of CMR were found in all 30 runs. The gap between the *best* solution and the optimum value is less than 0.1. For test functions g05, g06, and g10, only SR and SMES have produced the solutions extremely close to the optimum value known. However, CMR algorithm reached better *mean* and *worst* solutions in these functions, and the difference fluctuated in a small range. For the test function g04, the optimal solution was not found in CMR. Nevertheless, the probability of

Table 3. Experimental results on 12 benchmark functions with varying P_m

	P_m				
Func	*0.0*	*0.2*	*0.4*	*0.7*	*1.0*
g01	**-15.0000**	**-15.0000**	-14.9996	-10.1795	—
g02	-0.742069	-0.792659	-0.772180	-0.609778	-0.248819
g03	**-1.0050**	**-1.0050**	**-1.0050**	**-1.0049**	—
g04	-30491.794	-30643.148	-30617.990	-30655.365	-29621.527
g05	5148.744	5128.719	5181.846	5132.800	—
g06	-6956.674	-6932.642	-6884.239	-6953.286	—
g07	24.820	24.450	24.362	24.575	—
g08	**-0.095824**	**-0.095824**	-0.095737	-0.095785	-0.091875
g09	680.788	680.773	680.771	680.770	754.070
g10	7377.779	7410.935	7143.321	—	—
g11	0.8347	0.8727	**0.7489**	0.7618	**0.7499**
g12	-0.9799	-0.9799	-0.9799	-0.9737	-0.9794

reaching the optimal solution would increase when the maximum of generation number is set more than 8000.

In order to illustrate the effect of the mutation with exploitation capability in CMR, a set of experiments have been performed. Table 4 summarizes the mean of the objective function values in the case of the mutation probability P_m being set to 0.0, 0.2, 0.4, 0.7 and 1.0 over 30 runs.

From Table 4, we observed phenomenon when the parameter was fixed to different values. Whatever the mutation probability P_m was set, similar results were obtained for functions g03, g07, g08, g09, and g12. When this parameter gradually increased, the results were obtained better and better for functions g04 and g11, but degraded its performance for functions g01 and g02. More importantly, when this parameter is specified as $0 \leq P_m \leq 0.4$, this case provides better quality results for all test functions.

The experimental results illustrate the performance of CMR algorithm is similar to the compared algorithms in terms of the solutions quality. With slightly better standard deviations, CMR seems more robust and stable in obtaining consistent results.

4.2 Experiments of Feasibility Proportion

In order to detect the ability of CMR in balancing the objective function and the constraint violations, the feasibility proportion $P_f r$ in the population of each test function is memorized at each generation over 30 runs. Figure 2 shows the change curves of $P_f r$ on function g01, g04, g05, g10.

As shown in these curve figures, the feasible solutions can quickly be found at the early stage. Especially, for test functions g01, g04, g10, the feasibility proportion in the population attained a high value at the first 300 generations. The reason is that the subpopulation classification collects lots of characteristics between feasible components and feasible solutions. And the crossover operator,

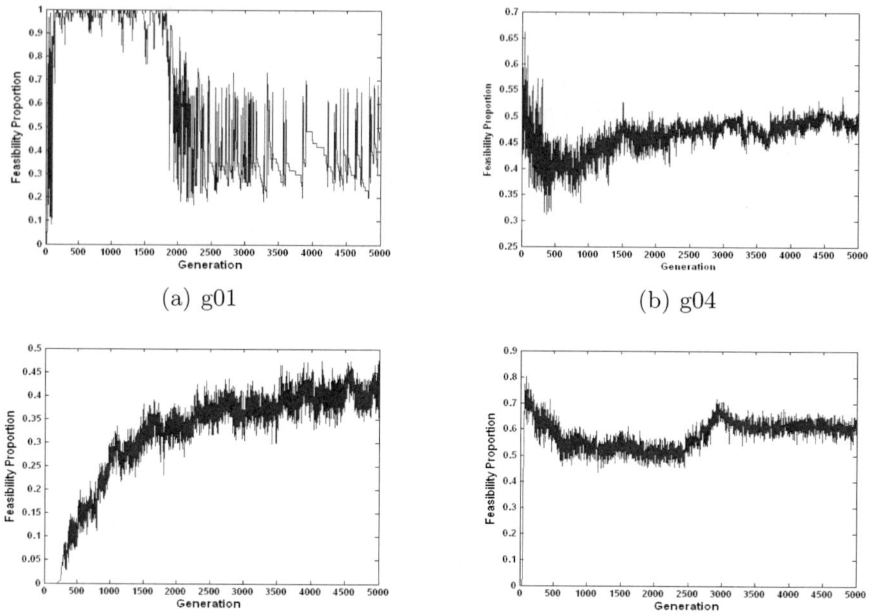

(a) g01 (b) g04

(c) g05 (d) g10

Fig. 2. Feasibility proportion versus generation for test functions out of 30 runs

based on the feasible components in various dimensional partition regions, is efficient to transform infeasible solutions into feasible ones. Also, note that a relatively stable feasibility proportion can be attained gradually as the iteration increases in four functions g04, g05, g10. Considering the diversity for the population, reasonable exploration of infeasible regions is allowed by the ranking strategy.

In all experiments, feasible solutions are continuously found for all the test functions in 30 runs. These results revealed that CMR has the substantial capability to deal with various kinds of *COPs*.

4.3 Conclusion

This paper has presented a novel algorithm for constrained evolutionary optimization, which is based on a component-based model and a new ranking method. Extraction and subpopulation classification introduced in this model, are two main technologies. The performance of this algorithm has been extensively investigated by experimental studies of 12 well-known benchmark test functions. The experimental results illustrate the CMR performance in terms of the quality of the resulting solutions, especially for robustness and stability in obtaining consistent results. It has much smaller magnitude of standard deviations.

References

1. Coello, C.A.C.: Theoretical and numerical constraint handling techniques used with evolutionary algorithms: A survey of the state of the art. J. Computer Methods in Applied Mech. 191(11-12), 1245–1287 (2002)
2. Kazarlis, S., Petridis, V.: Varying fitness functions in genetic algorithms: Studying the rate of increase of the dynamic penalty terms. J. Computer Science 1498, 211–220 (1998)
3. Farmani, R., Wright, J.A.: Self-adaptive fitness formulation for constrained optimization. J. IEEE Trans. Evolutionary Computation 7(5), 445–455 (2003)
4. Runarsson, T.P., Yao, X.: Stochastic ranking for constrained evolutionary optimization. J. IEEE Trans. Evolutionary Computation 4(3), 284–294 (2000)
5. Koziel, S., Michalewicz, Z.: Evolutionary algorithms, homomorphous mappings, and constrained parameter optimization. J. Evolutionary Computation 7, 19–44 (1999)
6. Zhou, Y., Li, Y., He, J., Kang, L.: Multiobjective and MGG evolutionary algorithm for constrained optimization. In: IEEE Conference on Evolutionary Computation 2003, pp. 1–5. IEEE Press, Los Alamitos (2003)
7. Venkatraman, S., Yen, G.G.: A generic framework for constrained optimization using genetic algorithms. J. IEEE Trans. Evolutionary Computation 9(4), 424–435 (2005)
8. Runarsson, T.P., Yao, X.: Search biases in constrained evolutionary optimization. J. IEEE Trans. Evolutionary Computation 35(2), 233–243 (2005)
9. Rahnamayan, S., Tizhoosh, H.R., Salama, M.M.A.: Opposition-based differential evolution algorithms. In: IEEE Conference on Evolutionary Computation 2006, pp. 6756–6763. IEEE Press, Canada (2006)
10. Mezura-Montes, E., Coello, C.A.C.: A simple multimembered evolution strategy to solve constrained optimization problems. J. IEEE Trans. Evolutionary Computation 9(1), 1–17 (2005)
11. Du, T., Fei, P., Shen, Y.: A Modified Niche Genetic Algorithm Based on Evolution Gradient and Its Simulation Analysis. In: Third International Conference on Natural Computation, China, vol. 4, pp. 35–39 (2007)

A Semi-supervised Topic-Driven Approach for Clustering Textual Answers to Survey Questions*

Hui Yang[1,**], Ajay Mysore[1], and Sharonda Wallace[2]

[1] Department of Computer Science, San Francisco State University, 94132, USA
[2] Human Nutrition & Food Science, California State Polytechnic University, 91768, USA
{huiyang,ajay0419}@sfsu.edu, spwallace@csupomona.edu

Abstract. We propose an algorithm to effectively cluster a specific type of text documents: textual responses gathered through a survey system. Due to the peculiar features exhibited in such responses (e.g., short in length, rich in outliers, and diverse in categories), traditional unsupervised and semi-supervised clustering[*] techniques are challenged to achieve satisfactory performance as demanded by a survey task. We address this issue by proposing a semi-supervised, topic-driven approach. It first employs an unsupervised algorithm to generate a preliminary clustering schema for all the answers to a question. A human expert then uses this schema to identify the major topics in these answers. Finally, a topic-driven clustering algorithm is adopted to obtain an improved clustering schema. We evaluated this approach using five questions in a survey we recently conducted in the U.S. The results demonstrate that this approach can lead to significant improvement in clustering quality.

1 Introduction

Facilitated by the World Wide Web and rapid advances in storage media, storing, sharing, and gathering information has become unprecedentedly convenient and even trivial. Examples include web blogging, social networks (e.g., facebook.com), and online surveys concerning commercial products or an emerging scientific discipline (e.g., nutritional genomics). Such information is routinely available in unstructured and textual format, and copious in volume. In this article we address a specific type of text documents: textual responses to the open-ended questions in an online survey questionnaire, where an open-ended question is one that solicits a short textual answer. An example of an open-ended question is "*What are your ethical concerns about the integration of nutritional genomics into the dietetics practice?*" Once the data gathering stage of a survey is completed, it is important to analyze these textual answers towards gaining an objective and comprehensive understanding of the re-

[*] This work is partially supported by the SFSU CCLS mini grant program.
[**] Corresponding author.

R. Huang et al. (Eds.): ADMA 2009, LNAI 5678, pp. 374–385, 2009.

sponses. Let us take the above question as one example. One will be interested in the following questions: (1) How many different concerns have been expressed? (2) What are these concerns? And (3) what are the dominant concerns? To answer such questions, text clustering can be employed. This technique categorizes a collection of text documents into a number of meaningful clusters through uncovering the underlying similarity amongst documents [Dhillon et al. 2001, Parsons et al. 2004, Steinbach et al. 2000, Zhong 2006].

These traditional clustering techniques, however, failed to deliver satisfactory clustering results. We applied three different clustering algorithms—modified k-means, single link based clustering, and co-occurring term pattern based clustering (Section 2.2)—to the datasets we have collected through an online survey. For each of these survey questions, these algorithms (1) failed to cluster a large fraction of its answers, (2) could identify up to nine times the number of the real clusters, and (3) resulted in a low clustering accuracy. (See Section 3 for a detailed discussion.) This poor performance is due to the following peculiar characteristics of the input text:

- In contrast to traditional text documents, answers to survey questions are often short and right to the point, with an average of two to three sentences. In addition, many answers only contain a short phrase. For instance, "not sure" or "insurance costs" are two answers to the above question.
- These answers tend to be similar to spoken English and involve a relatively small vocabulary (400~1,200 unique English words as against ~10,000 in traditional documents).
- Answers to the same question often cover an unusually wide range of variety. In addition, there exist a relatively large number of outliers that are unlike any of the other answers. In other words, each of these outliers corresponds to one unique cluster. We term such clusters as *singleton clusters* or *singletons*. Let us again use the above question to illustrate this. Its 396 answers spread over 122 clusters, among which 49 are singletons.

The issue here is how to improve the clustering quality. Clearly, one cannot just rely on the above (unsupervised) clustering techniques, as their performance counteracts the often complicated and time-consuming process of designing an effective survey questionnaire and then marketing it to the right population [Galloway 1997].

To address this issue, we have proposed a semi-supervised, topic-driven clustering approach. It is composed of three steps. The first step employs an unsupervised clustering algorithm such as k-means [Dhillon et al. 2001] to derive a preliminary clustering schema over all the answers to a question. Using this schema, a human specialist then identifies a list of major topics presented in these answers and an estimated number of answers associated with each topic. The final step finalizes the clustering schema by incorporating these major topics. The rationale behind this approach is twofold. First, the clustering schema derived in the first stage, though preliminary, should be able to sufficiently reduce a human's cognitive workload to a manageable level towards identifying the list of important topics. This was validated in our evaluation studies. Second, the topics obtained in the second stage are expected to capture the underlying structure hidden in the answers. Therefore, the integration of such knowledge significantly improves the clustering quality. This was also validated by our empirical results.

The proposed approach is closely related to two clustering techniques: semi-supervised clustering [Basu et al. 2002, Basu et al. 2004, Blum et al. 1998, Cohn et al. 2003, Wagstaff et al. 2001, Zeng et al. 2003, Zhong 2006] and topic-driven clustering [Zhao et al. 2005]. The proposed approach is different from the former in that it uses topic information as prior (or background) knowledge instead of labeled answers. Labeling answers can be a serious issue here since survey answers often spread over a large number of categories as discussed earlier. Topic identification on the other hand is much more feasible for a human. The proposed approach is also different from the topic-driven clustering technique in that it does not require the topics given a priori. Instead, the topics are derived by collaboration between computers and humans. Finally, this approach unifies the two clustering techniques. Also note that although this approach does not have constraints over the specific algorithms one can use in the first and third stages, it is important to adopt an algorithm that is suitable for the documents at hand. This will greatly reduce the workload from the human expert in the second step. Our evaluation has shown that a straightforward single-link based algorithm (Section 2.2) in general outperforms the commonly recognized top performer--the k-means algorithm [Steinbach et al. 2000].

We have evaluated this approach using the answers to five open-ended questions in a survey we have recently conducted in the United States. This survey was designed to assess the training needs of health professionals in the U.S. in an emerging discipline—Nutritional Genomics [Ngx-url, Wallace et al. 2007]. The evaluation results demonstrate that this proposed approach can lead to significant improvement in clustering quality when compared to that of unsupervised clustering algorithms.

2 Algorithm

Given a collection of answers to an open-ended survey question, where each answer corresponds to one short text document, the proposed algorithm employs three main steps to cluster these answers into meaningful groups. Denote this collection of answers as $\boldsymbol{A}$. These three steps are: Step I—generating a preliminary clustering solution of $\boldsymbol{A}$ using an unsupervised clustering algorithm; Step II—extracting the list of main topics in $\boldsymbol{A}$, which is manually carried out by a domain expert based on the previously derived clustering solution; and Step III—generating the final clustering solution by incorporating the identified topics. Next, we first formulate the problem and describe the data preprocessing tasks, and then detail the three main steps.

2.1 Problem Formulation and Data Preprocessing

Given an input dataset of N textual answers, denoted as $\boldsymbol{A}=\{a_1, a_2, \ldots, a_N\}$, the goal is to cluster these N answers into k groups such that answers in the same group are related to the same topic and answers in different groups are related to different topics. A number of preprocessing tasks are performed before we proceed to generate the preliminary clustering solution, including synonym replacement, stopword removal, and stemming. We use the Moby Thesaurus [Ward 2002] to identify the words and phrases that are semantically similar and replace them by a representative entry. For instance, phrases "lots of", "a great deal of" and "huge" bear a similar meaning and will be replaced by the word "big". This step is necessary because, as mentioned in

Section 1, this type of documents resembles spoken-English and usually contains short phrases such as "not sure", and "lots of", and "FYI". If one conducts stopword removal first, phrases such as "lots of" will be converted to "lots" and can lead to unnecessary ambiguity. The next task, stopword removal, removes the words that frequently occur in documents but do not bear significant semantic meanings, e.g., the word "the". Finally, stemming reduces the syntactic variants of a word to its root. For example, the stemming task will reduce "computer", "computing", and "computed" to their common root "compute". We use Martin Porter's stemming algorithm for this purpose [Porter 1980]. To represent each answer in the preprocessed input dataset ***A***, we adopt the vector space model [Salton 1989]. In this model, each answer ***a*** in ***A*** is represented as a vector $<tfidf_1, tfidf_2, \ldots, tfidf_m>$, where $tfidf_i$ is the term frequency-inverse document frequency (TFIDF) of the ith term in ***a*** and computed as $tf_i / \log_2 (N/df_i)$. Here, tf_i is the number of times that the ith term occurs in ***a***, N the number of answers in ***A*** (i.e., |***A***|), and df_i the number of documents that contain the ith term. Note that TFIDF measures the statistical significance of a term when being used to separate one document from the others in the same collection. This vector-based representation of ***A*** will be used as the input to the algorithms in both Step 1 and Step 3.

2.2 Step 1—Generating a Preliminary Clustering Solution

An unsupervised clustering algorithm is used to generate a preliminary clustering solution for ***A***. Specifically, we have realized three such algorithms: a modified k-means algorithm, a single-link based algorithm, and frequent co-occurring term pattern based algorithm.

The Modified *k*-means Algorithm: This algorithm bears similarity with the traditional *k*-means algorithm [Banerjee et al 2003, Dhillon et al. 2001] with two major modifications: (1) it does not require one to specify the value of k; and (2) the number of clusters generated in each iteration varies in the clustering process. Specifically, the algorithm proceeds as follows: It begins by randomly selecting a relative large number of seeds (i.e., initial cluster centers). Each document (answer) is then assigned to the cluster whose center is the closest to the document and their cosine similarity is greater than δ. Given two documents d_i and d_j in the vector space model, their **cosine similarity** is $cos(d_i, d_j)=d_i^t \cdot d_j$. This measure ranges between 0 and 1, where 0 indicates that the two documents have nothing in common and 1 that the two documents are identical. We set δ to 0.395 in our evaluation. Once every document has been put into its corresponding cluster, the algorithm identifies the document that is closest to the cluster centroid and takes it as the new cluster center. At this point, it examines the pair-wise similarity between clusters based on the cosine similarity between the cluster centers. If this similarity is ≥δ, it merges the two clusters and re-computes the center of the merged cluster. Once this merging process is done, a collection of new cluster centers will be identified. These new cluster centers are then used to re-cluster the documents in the next iteration in a similar fashion as described above. This process continues until there is no change in the clustering solution. Note that if the initial number of seeds is set to the number of documents in the answer collection, this algorithm also resembles an agglomerative, hierarchical clustering algorithm [Kaufman and Rousseeuw 1990].

The Single-link Based Clustering Algorithm: In this approach, we start with a randomly chosen answer *a*. We then identify all the answers whose cosine similarity with *a* is greater than δ (=0.395 in our evaluation). We say these answers are similar to *a*.

These answers together with a form an initial cluster. For each answer in this initial cluster, we then identify all the un-clustered answers that are similar to this answer and include them in the cluster until the cluster cannot be expanded any further. We then start to form another cluster by starting with another un-clustered answer in the same manner as with the answer a. We continue this process until there are no answers left. This approach assumes that if two answers are similar to the same answer, these two answers are also similar to each other. This assumption may not be true for traditional documents that are long and written in formal English; however, survey answers seem to be tolerant of this assumption.

Co-occurring Term Pattern Based Clustering: This approach clusters the answers based on the frequently co-occurring term patterns [Beil et al. 2002]. A frequent co-occurring term pattern is a set of terms that co-occur in at least s answers in the collection. Clearly, they are a special case of frequent association patterns [Agrawal et al. 1994]. We use the Apriori algorithm to identify these frequent co-occurring patterns, where s is the minimum support [Borgelt et al. 2002]. Once these patterns are generated, we next use them to cluster the input answers based on the observation that documents containing the same co-occurring patterns are likely to be similar to each other. Furthermore, the longer these shared patterns are, the more similar these documents are. To realize this, we rank the co-occurring patterns by length (number of terms involved) and frequency. We start from the longest frequent pattern. We then identify all the answers in which this pattern spans over no larger than a certain sliding window (e.g., 20 terms). These answers will form one cluster. We then proceed to the second ranked pattern, and use it to identify the next cluster. We continue this process until we have exhausted all the frequent patterns or the input documents. We maintain exclusive cluster membership for each answer by removing a clustered answer from further consideration. The rationale behind the sliding window is based on the observation that if the term-span of a co-occurring pattern varies significantly in two answers, e.g., one spans over 5 terms and the other 20 terms, these two answers are unlikely to be similar based on this co-occurring pattern.

2.3 Step 2—Extracting the Main Topics

In this stage the domain expert uses the clustering solution generated by one of the unsupervised clustering algorithms to identify a set of topics in $\boldsymbol{A}$. We discussed in Section 1 that it is not feasible for a domain expert to label even a small number of documents due to the unusual large number of categories embedded in $\boldsymbol{A}$. However, if provided with a preliminary clustering solution, the human expert finds it much manageable to identify a list of main topics in $\boldsymbol{A}$, and an estimated number of answers associated with each topic. Not surprisingly, the expert often chooses from the survey answers the representative terms and uses them to construct the topics. As a result, the resulting topics are usually short, e.g., "Patient Confidentiality/Documentation". For each topic, the expert relies heavily on the preliminary clustering solution to estimate the number of relevant answers. It is evident that better the quality of clusters (generated in Step 1) is, the easier and less time consuming this step will be. Among the three algorithms discussed in Section 2.2, the single link based algorithm is found to outperform the other two. This will be discussed further in Section 3. This list of identified topics is used in the third step to obtain an improved clustering solution.

2.4 Step 3—Topic-Driven Clustering

Given an input dataset of N textual answers $\boldsymbol{A}=\{a_1, a_2, \ldots, a_N\}$ in their vector space representation, let $\boldsymbol{T}=\{(t_1, x_1), (t_2, x_2), \ldots, (t_M, x_M)\}$ be the M main topics identified in the previous step, where t_i is the ith topic and x_i the estimated number of answers related to t_i. We first transform $\boldsymbol{A}$ and $\boldsymbol{T}$ into the TFIDF vector space as described in Section 2.1. The goal of this step is to cluster $\boldsymbol{A}$ into M clusters $\{\boldsymbol{C_1}, \boldsymbol{C_2}, \ldots, \boldsymbol{C_M}\}$, where the ith cluster $\boldsymbol{C_i}$ contains all the answers that are closely related to the ith topic t_i. We adopt three simple topic-driven clustering algorithms to obtain these clusters.

Topic to Answer Clustering (T2A): For each topic (t_i, x_i) in $\boldsymbol{T}$, this method identifies the top x_i answers in $\boldsymbol{A}$ that are the most similar to t_i and form a cluster. The cosine similarity between an answer and t_i is used. This will result in M clusters, each corresponding to one of the M topics. It is however possible that some answers can be put into more than one cluster. To address this issue, this method identifies all such answers and assigns each of them to the topic with which the answer has the highest cosine similarity. Note that this method can leave some answers unlabeled.

Answer to Topic Clustering (A2T): Unlike the T2A method, this approach strives to label an answer by the most relevant topic. Additionally, it does not use the x_i's in $\boldsymbol{T}$. It works as follows: for each answer in $\boldsymbol{A}$, it compares it with all the M topics in $\boldsymbol{T}$ and then labels this answer by the topic that gives the highest cosine similarity. This approach does not leave any answer unlabeled; however, a drawback of this method is that an answer can be labeled by a remotely relevant topic due to the singleton phenomenon discussed in Section 1.

T2A+A2T: For a given topic, it is straightforward that the above two methods—T2A and A2T—can associate this topic with two different sets of answers. Furthermore, T2A relies on the x_i's in $\boldsymbol{T}$ when constructing clusters; and A2T might label an answer by an unrelated topic. To combine the strength of T2A and A2T, and at the same time to overcome their drawbacks, we implement this third topic-based clustering method. Let $C_{i,T2A}$ and $C_{i,A2T}$ denote the clusters of the ith topic t_i, constructed by the T2A and A2T methods respectively. This method first identifies the union of of $C_{i,T2A}$ and $C_{i,A2T}$. For every answer in this union, it then computes its cosine similarity with t_i. The mean (μ) and standard deviation (σ) of these similarity measurements are also computed. Finally, the answers with cosine similarity $\geq(\mu- \sigma)$ are selected and clustered for the ith topic. The rationale here is that the unselected answers are likely not related to the given topic due to low similarity score. Eliminating them therefore is expected to improve the clustering quality.

3 Evaluation

In this section, we report the evaluation results of the proposed 3-step algorithm by applying it to analyze the textual answers gathered in a nationwide survey. The purpose of this survey is to assess the educational and training needs of health professionals in nutritional genomics [Ngx-url, Wallace et al. 2007]. This survey consists of 19 likert-scale and 5 open-ended questions. Approximately 2,500 invitees took the survey with a varying completion ratio.

Table 1 lists the five open-ended questions in the survey, which are paraphrased for brevity. For each question, we have manually identified and confirmed the true clustering

Table 1. Summary of the open-ended questions and the textual answers, where N is the total number of answers, |V| is the total number of unique words, K is the number of true clusters, N_c is the average number of answers in a cluster, and S is the total number of singleton clusters

	Description	N	|V|	K	N_C	S
Q1	What are the main sources where you read or heard about nutritional genomics or personalized nutrition in the past 2 years?	198	416	72	2.75	48
Q2	What was your response to the questions asked by your clients about nutritional genomics or personalized nutrition?	349	795	140	2.5	27
Q3	As a practicing dietitian, what are the reasons for you to enroll or not to enroll in a continuing education certificate program to enhance your knowledge in this area in nutritional genomics?	1196	1286	163	7.34	28
Q4	What are your personal comments on nutritional genomics that you'd like to share?	578	964	109	5.3	34
Q5	What are your ethical concerns about the integration of nutritional genomics into the dietetics practice?	396	951	122	3.25	49

solution for all the collected answers using the solutions produced by the 3-step algorithm described in Section 2. Information on these true clustering solutions is also given in Table 1. One can observe the characteristics discussed in Section 1 with regard to this type of text documents, such as small vocabulary and presence of outliers (singletons).

To evaluate the effectiveness of the proposed 3-step algorithm, we compare the clustering solutions generated in Step 1 (unsupervised clustering) and Step 3 (topic-driven clustering) with the true clustering solutions, respectively. Such comparisons will reveal the positive effect and necessity of integrating background knowledge, i.e., topics in this study. Specifically, we have employed the following measurements to examine different aspects of a clustering solution:

- **Normalized mutual information (NMI):** Given the set of survey answers ***A***, let ***TC***=*{TC*$_1$, *TC*$_2$, ..., *TC*$_M$*}* be the true clustering solution, and ***C***=$\{C_1, C_2, \ldots, C_N\}$ be the solution produced by a clustering algorithm (e.g., K-means and T2A), where M and N can be different. NMI is computed as follows [Strehl et al. 2002]:

$$NMI = \frac{\sum_{h,l} n_{h,l} \log\left(\frac{n \cdot n_{h,l}}{n_h n_l}\right)}{\sqrt{\left(\sum_h n_h \log \frac{n_h}{n}\right)\left(\sum_l n_l \log \frac{n_l}{n}\right)}}$$

  where n is the total number of textual answers collected for a given survey question; $h \in [1,M]$ and identifies a true cluster in ***TC***; $l \in [1,N]$ and identifies a cluster in ***C*** generated by the algorithm; n_h is the number of answers in the hth cluster in ***TC***; n_l is the number of answers in the lth cluster in ***C***; and $n_{h,l}$ is the number of answers shared by the hth cluster in ***TC*** and the lth in ***C***. The values of NMI range from *0* to *1*. It is *1* when ***TC*** and ***C*** are a perfect match with each other and near *0* when ***C*** corresponds to a random partitioning of the input answers. We will use the same notations here to explain the other measurements below.

- **Clustering accuracy:** For each cluster C_l in ***C***, we match it with a true cluster TC_h in ***TC*** that has the highest document overlap with C_l amongst all the M true clusters. We say that an answer in C_l is accurately labeled if it also belongs to TC_h. The clustering accuracy of a given algorithm is then defined as the percentage of all the n input answers that are accurately labeled.
- **Accuracy without singletons:** As observed in Table 1, the true clustering solution of a survey question contains a relatively large number of singleton clusters. These singletons, as discussed later, can distort the NMI and clustering accuracy measurements by boosting them to a higher value (Tables 2-5). We hence design this measurement. It is computed in the same manner as for the clustering accuracy except that all the singleton clusters in ***C*** will be ignored.
- **Document-based coverage:** $=(\Sigma_l|C_l|)/n$, where $l\in[1,N]$, $C_l\in$ ***C***, and $|C_l|$ is number of answers in the lth cluster. It computes the percentage of answers that are labeled by a clustering algorithm. A close-to-1 value is preferred.
- **Cluster-based coverage:** $=N/M$, i.e., the ratio between the number of clusters identified by a clustering algorithm and the number of true clusters.
- **Singleton coverage:** the ratio between the number of singleton clusters identified by a clustering algorithm and the number of true singleton clusters.

Shown in Table 2 are the NMI values of the three unsupervised algorithms (columns 2-4) and three topic-driven ones (columns 5-7). The algorithms rendering the highest three NMI values are in bold for each question. Note that for the co-occurring pattern based algorithm, the results reported in this section are based on a minimum support of 5% and a sliding window size of 20 terms. Table 2 presents a mixed picture, where it seems that the integration of topics does not help much except for the first question. This is counterintuitive. Upon a closer look at the NMI formulation above, one can notice that a higher NMI value will be resulted if the two clustering solutions (i.e., ***CT*** and ***C***) consist of many singletons. This is the case for question 1,

Table 2. The NMI values of all the algorithms for each question with the top three in bold. NMI can be misleading in the case of analyzing survey answers. See text for details.

	K-means	Link Based	Co-occurring	T2A	A2T	T2A+A2T
Q1	0.415900	0.733703	0.351774	**0.794156**	**0.798996**	**0.750222**
Q2	0.308552	**0.766750**	0.624530	**0.738135**	0.729898	**0.738335**
Q3	0.161796	**0.529060**	0.466802	**0.484268**	0.435182	**0.464216**
Q4	0.239076	**0.588691**	**0.609263**	0.546632	**0.587964**	0.573043
Q5	0.262372	**0.770540**	0.165190	**0.686937**	0.656214	**0.673072**

Table 3. "Accuracy without singletons" for all the algorithms, with the top 3 in bold

	K-means	Link Based	Co-occurring	T2A	A2T	T2A+A2T
Q1	50.70%	30.30%	51.78%	**58.04%**	**56.57%**	**54.29%**
Q2	**42.53%**	18.29%	30.10%	33.98%	**39.60%**	**40.12%**
Q3	21.41%	9.95%	35.06%	**38.64%**	**37.23%**	**41.16%**
Q4	22.84%	7.79%	27.23%	**39.06%**	**40.83%**	**39.54%**
Q5	11.87%	15.91%	22.94%	**49.30%**	**40.92%**	**44.82%**

which has the largest ratio of singletons (48/198) among all the questions in terms of its true clustering solution. Furthermore, the unsupervised algorithms tend to generate a large number of singletons as observed in Table 4. This means that NMI values can be misleading when measuring the quality of a clustering solution to a set of survey answers. To gain a more accurate understanding of the advantage of incorporating topics in the clustering process, we should take singletons out of the equation. We use the "accuracy without singleton" measure as described above.

Table 3 lists the "accuracy without singletons" measurement for the three unsupervised algorithms (columns 2-4) and three topic-driven algorithms (columns 5-7). For each question, the top three values are in bold. This table shows several interesting results. First, the three topic-driven algorithms outperform the unsupervised ones over all the questions except the Q2. This might be caused by that Q2 has an unusually large number of true clusters (140) as against its relatively small number of answers (349). Second, the single link based algorithm in general achieves the lowest accuracy among all the algorithms. This, however, should not be taken as the single evidence against this algorithm due to the singleton effect. As shown in Tables 4-5, when one takes the other measures into account, this algorithm outperforms the k-means and co-occurring term patterns algorithm on the whole. Third, compared with other questions, Q1 in general has a higher accuracy with regard to all the clustering algorithms. This can be attributed to the unique nature of Q1. In contrast to other questions whose answers are often in complete sentences, the answers to Q1 are mainly short phrases such as "classes or seminars" and "Journal of the American Dietetic Association". Finally, the three topic-driven algorithms are comparable when singletons are not included. Table 3 demonstrates the necessity of integrating domain knowledge.

Although the "accuracy without singletons" measurement is more informative on the clustering quality when compared to NMI, it only reflects one facet of a clustering algorithm. Especially when one considers the unique properties of survey answers (e.g., many singleton clusters), it is necessary to examine other aspects of a clustering

Table 4. Summary of results generated by the unsupervised clustering algorithms

	Q1	Q2	Q3	Q4	Q5
K-means					
1. Document-based coverage	35.86%	23.43%	21.41%	22.84%	11.87%
2. Cluster-based coverage	36.12%	56.33%	16.63%	30.00%	35.54%
3. Clustering accuracy	67.61%	67.07%	41.02%	50.76%	95.74%
4. Singleton-based coverage	25%	66.67%	35.71%	44.11%	46.94%
Link Based					
1. Document-based coverage	100%	100%	100%	100%	100%
2. Cluster-based coverage	140.3%	71.94%	358.9%	259.1%	261.2%
3. Clustering accuracy	70.85%	80.00%	58.95%	55.54%	88.89%
4. Singleton-based coverage	168.7%	800%	2092%	811.8%	589.8%
Term Co-Occurrence					
1. Document-based coverage	28.28%	82.57%	82.27%	75.61%	85.86%
2. Cluster-based coverage	16.7%	296.4%	427.6%	887.2%	663%
3. Clustering accuracy	53.57%	75.43%	56.30%	71.16%	74.12%
4. Singleton-based coverage	2.1%	485.1%	746.4%	564.7%	355.1%

solution. Tables 4-5 catalog the other four measures for all the algorithms, where Table 4 lists the results for the three unsupervised algorithms in Step 1 of the proposed algorithm and Table 5 for the three topic-driven algorithms in Step 3.

As shown in Tables 4-5, the k-means and term co-occurrences based algorithms fail to label a large portion of input answers (up to ~90%) as evidenced by the low document-based coverage, whereas the link-based algorithm has a full coverage. On the other hand, among the three topic-driven algorithms, A2T and T2A+T2A show a satisfying coverage of input documents. In terms of the cluster-based coverage, k-means normally identifies a small fraction; co-occurrence based algorithm varies significantly from question to question but in general significantly outnumbers the number of true clusters; the link-based algorithm is reasonably satisfying; finally all the topic-driven algorithms cover most of the true clusters. A similar observation can be made for all the six clustering algorithms with respect to the singleton-based coverage, except that the link-based algorithm tends to drastically outnumber that of the true singletons in one case. Finally, based on the clustering accuracy (with singletons

Table 5. Summary of results generated by semi-supervised clustering algorithms

	Q1	Q2	Q3	Q4	Q5
Topic to Answer (T2A)					
1. Document-based coverage	72.22%	73.14%	56.69%	62.46%	69.69%
2. Cluster-based coverage	94.37%	99.27%	96.32%	96.37%	90.09%
3. Clustering accuracy	93.00%	78.13%	57.82%	60.11%	73.55%
4. Singleton-based coverage	104.2%	418.5%	464.3%	223.5%	181.6%
Answer to Topic (A2T)					
1. Document-based coverage	100%	98.86%	97.91%	96.2%	98.74%
2. Cluster-based coverage	97.2%	98.5%	96.31%	97.27%	91.74%
3. Clustering accuracy	70.71%	57.51%	39.70%	44.25%	49.62%
4. Singleton-based coverage	58.3%	229.6%	103.6%	55.88%	69.39%
T2A+A2T					
1. Document-based coverage	88.38%	95.43%	94.65%	91.87%	90.15%
2. Cluster-based coverage	98.59%	97.1%	95.10%	98.18%	90.91%
3. Clustering accuracy	72.57%	61.98%	44.35%	43.87%	57.70%
4. Singleton-based coverage	66.67%	270.3%	128.6%	67.65%	93.88%

considered), the link-based algorithm generally outperforms the other two unsupervised algorithms (Table 4), and the 3 topic-driven algorithms outperform all the unsupervised algorithms. All in all, we can conclude that among the three unsupervised algorithms, the link-based algorithm tends to give the best clustering solution. Furthermore, by incorporating the background knowledge, i.e., topics, one can significantly improve the clustering quality with respect to a variety of measurements as discussed earlier.

4 Related Work

Semi-supervised clustering is a methodology that integrates prior (or background) knowledge in the clustering process towards improving the clustering quality. The background knowledge is commonly assumed in the form of a small set of labeled instances. Based on this knowledge, one can derive the deterministic relationship between two labeled instances, including the *must-link* relationship where two instances should belong to the same cluster and the *cannot-link* relationship where two instances should not belong to the same cluster. Previous studies in this area can be categorized into three main approaches: seeded, constraint-based, and feedback-based approaches. Seeded approaches utilize the labeled instances as seeds to construct an initial set of clusters [Basu et al. 2002]. Constraint-based approaches explicitly adjust the objective functions such that the prior knowledge (e.g., must-links and cannot-links) is kept invariant during the clustering process [Wagstaff et al. 2001]. Feedback-based approaches on the other hand first generate a clustering schema and then use the labeled instances to revise this schema [Cohn et al. 2003]. Recently, Zhao et al. proposed a semi-supervised clustering algorithm where the background knowledge was in the form of topics instead of labeled instances [Zhao et al. 2005]. They discussed three approaches that exploit both the similarity between topics and documents and the similarity between documents.

5 Conclusion

We have described a semi-supervised, topic-driven approach for clustering textual answers collected in a survey system. This approach first utilizes an unsupervised clustering algorithm to identify a clustering solution. This solution is then provided to a human specialist to identify a list of main topics. The final step incorporates these topics to arrive at an alternative clustering solution. We evaluated this approach using five datasets we have gathered through an online survey system. The results are promising and show that the proposed approach can significantly improve the clustering quality. We would like to note that instead of inviting a human expert to label the "more informative" instances in an unlabeled data, it seems more natural and straightforward for us humans to identify main topics. This is especially the case when dealing with survey answers. Furthermore, to reduce the cognitive load laid upon the human expert in this process, it is critical to employ an algorithm that is effective to the dataset of interest. Finally, we are interested in examining the possibility of using the topics identified in the second step to directly improve the clustering solutions obtained in the first step.

References

1. Agrawal, R., Srikant, R.: Fast Algorithms for Mining Association Rules. In: Proc. of the 20th VLDB Conf. (1994)
2. Banerjee, A., Dhillon, I., Sra, S., Ghosh, J.: Generative Model-Based Clustering of Directional Data. In: Proc. 9th ACM SIGKDD Int. Conf. Knowledge Discovery Data Mining, pp. 19–28 (2003)
3. Basu, S., Bilenko, M., Mooney, R.: A probabilistic framework for semi-supervised clustering. In: Proc.of the 10th ACM SIGKDD Int. Conf. Knowledge Discovery Data Mining, pp. 59–68 (2004)

4. Basu, S., Banerjee, A., Mooney, R.: Semi-supervised clustering by seeding. In: Proc. 19th Int. Conf. Machine Learning, pp. 19–26 (2002)
5. Beil, F., Ester, M., Xu, X.: Frequent Term-Based Text Clustering. In: SIGKDD 2002 (2002)
6. Blum, A., Mitchell, T.: Combining labeled and unlabeled data with co-training. In: The 11th Annual Conf. CLT, pp. 92–100 (1998)
7. Borgelt, C., Kruse, R.: Induction of Association Rules: Apriori Implementation. In: The 15th Conference on Computational Statistics (2002)
8. Choudhary, B., Bhattacharyya, P.: Text clustering using semantics. In: The Eleventh International WWW Conference (2002)
9. Cohn, D., Caruana, R., McCallum, A.: Semi-supervised clustering with user feedback (Tech. Rep. TR2003-1892). Cornell University (2003)
10. Dhillon, I., Modha, D.: Concept decompositions for large sparse text data using clustering. Machine Learning 42, 143–175 (2001)
11. Galloway, A.: A workbook on Questionnaire Design & Analysis, `http://www.tardis.ed.ac.uk/~kate/qmcweb/qcont.htm`
12. Ward, G.: The Moby Thesaurus List (English) (2002), `http://www.gutenberg.org/etext/3202`
13. Jian, W., Li, Z., Hu, X.: Ontology Based Clustering for Improving Genomic IR. In: 20th IEEE Int'l Sym. on Comp. Based Med. Sys. (2007)
14. Kaufman, L., Rousseeuw, P.J.: Finding Groups in Data: An Introduction to Cluster Analysis. John Wiley & Sons, Chichester (1990)
15. Genomics Survey, `http://dhcp-hensill4f-235-208.sfsu.edu/`
16. Parsons, L., Ehtesham, L., Haque, Liu, H.: Subspace Clustering for High Dimensional Data: A Review. SIGKDD Exploration 1(6), 90–105 (2004)
17. Porter, M.: An algorithm for suffix stripping. Program 14(3), 130–137
18. Salton, G.: Automatic Text Processing: The Transformation, Analysis, and Retrieval of Information by Computer. Addison-Wesley, Reading (1989)
19. Sholom, M., Indurkhya, N., Zhang, T., Damerau, F.: Text Mining Predictive Methods for Analyzing Unstructured Information. Springer, Heidelberg (2004)
20. Steinbach, M., Karypis, G., Kumar, V.: A Comparison of Document Clustering Techniques. In: Proc. TextMining Workshop, KDD (2000)
21. Strehl, A., Ghosh, J.: Cluster ensembles–a knowledge reuse framework for combining partitions. Journal of MLC 3, 583–617 (2002)
22. Wagstaff, K., Cardie, C., Rogers, S., Schroedl, S.: Constrained k-means clustering with background knowledge. In: ICML (2001)
23. Wallace, S., Wakimoto, P., Yang, H., Rodriguez, R.: Development of an on-line survey to assess training needs in nutritional genomics. In: Experimental Biology Annual Meeting, p. 53 (2007)
24. Zeng, H., Wang, X., Chen, Z., Lu, H., Ma, W.: Cbc: Clustering based text classification requiring minimal labeled data. In: ICDM (2003)
25. Zhao, Y., Karypis, G.: Topic-Driven Clustering for Document Datasets. In: SIAM International Conference on Data Mining, pp. 358–369 (2005)
26. Zhong, S.: Semi-supervised Model-based Document Clustering: A Comparative Study. Machine Learning 1(65) (2006)

An Information-Theoretic Approach for Multi-task Learning

Pei Yang[1], Qi Tan[1,2], Hao Xu[1], and Yehua Ding[1]

[1] School of Computer Science, South China University of Technology, 510640 Guangzhou, Guangdong
[2] School of Computer Science, South China Normal University, 510631 Guangzhou, Guangdong
{yangpei,xuhao}@scut.edu.cn, tanqi@scnu.edu.cn, dyh@hgsoft.com.cn

Abstract. Multi-task learning utilizes labeled data from other "similar" tasks and can achieve efficient knowledge-sharing between tasks. In this paper, a novel information-theoretic multi-task learning model, i.e. IBMTL, is proposed. The key idea of IBMTL is to minimize the loss mutual information during the classification, while constrain the Kullback Leibler divergence between multiple tasks to some maximal level. The basic trade-off is between maximize the relevant information while minimize the "dissimilarity" between multiple tasks. The IBMTL algorithm is compared with TrAdaBoost which extends AdaBoost for transfer learning. The experiments were conducted on two data sets for transfer learning, Email spam-filtering data set and sentiment classification data set. The experimental results demonstrate that IBMTL outperforms TrAdaBoost.

Keywords: Multi-task learning, transfer learning, information entropy.

1 Introduction

A real world learning task can often be viewed as consisting of multiple correlated subtasks. For example, in a study of the effectiveness of cardiac treatments, one may have multiple sets of data, each collected at a particular hospital; rather than designing individual classifiers for each of these classification tasks, it is desirable to share data across tasks to enhance overall generalization performance. The phenomenon we are observing here is a sort of "borrowing strength". The classifiers with a paucity of data borrow inferential strength from the similar classifiers with abundance of data. This represents a typical example of a general learning scenario called multi-task learning (MTL) (Caruana, 1997).

Hierarchical Bayesian models provide the flexibility to model both the individuality of tasks, and the correlations between tasks. The hierarchical model can achieve efficient information-sharing between tasks. The estimation of a learning task is affected by both its own training data and by data from the other tasks related through the common prior.

Often, the common prior in a hierarchical Bayesian model is specified in a parametric form with unknown hyper-parameters, for example, a Gaussian distribution with

R. Huang et al. (Eds.): ADMA 2009, LNAI 5678, pp. 386–396, 2009.

unknown mean and variance. Information is transferred between tasks by learning those hyper-parameters using data from all tasks. However, it is often difficult to know what the true distribution should be like, and an inappropriate prior could be misleading. Further, the model parameters of individual tasks may have high complexity, and therefore no appropriate parametric form can be found easily.

In this paper we propose a novel information-theoretic approach for multi-task learning. From the point of view of information theory, the precision of the classifier can be judged by the loss in mutual information during the classification. Let the random variable sets X, Y and T be the set of instances, features and classification labels respectively. Our model make an assumption that the relevant learning tasks should have the "similar" conditional distributions $p_i(y|t)$. The basic idea of our model can be formulated as to constrain the KL divergence of the conditional distribution between the multiple tasks to some maximal level, and then try to minimize the loss in mutual information, $I(X;Y|T)$. It doesn't need to know what the true distribution of the data should be like.

2 Related Work

Multi-task learning has been the focus of much interest in the machine learning community over the last decade. Multi-task learning is based on the assumption that multiple tasks share certain structures. Therefore, tasks can mutually benefit from these shared structures. Typical approaches to information transfer among tasks include: sharing hidden nodes in neural networks (Baxter, 2000; Caruana, 1997); placing a common prior in hierarchical Bayesian models (Bakker & Heskes, 2003; Yu et al., 2005; Zhang et al., 2006); sharing parameters of Gaussian processes (Lawrence & Platt, 2004); sharing a common structure on the predictor space (Ando & Zhang, 2005); and structured regularization in kernel methods (Evgeniou et al., 2005), among others.

Caruana trained a neural network on several tasks simultaneously as a way to induce efficient internal representations for the target task (Caruana, 1997). Modeling data from related scenarios is typical done via hierarchical Bayesian modeling. Baxter proposed hierarchical Bayesian inference as a model for studying multi-tasks learning (Baxter, 2000). Parameters that are shared between tasks are treated as hyper-parameters at a higher level than the task-specific model parameters. Heskes presented a Bayesian framework for learning to learn (Heskes, 2000). In their model all tasks are combined in a single back-propagation neural network. The hidden-to-output weights, being specific to each task, play the role of model parameters. The input-to-hidden weights, which are shared between all tasks, are treated as hyper-parameters. Yu et al. exploited the equivalence between parametric linear models and nonparametric Gaussian processed (Yu et al., 2005). Zhang et al. proposed an efficient Bayesian hierarchical model for recommendation systems (Zhang et al., 2007). Roy et al. provided an effective method for transfer learning using the Dirichlet Process mixture model as a generative model of data sets (Roy & Kaelbling, 2007). Yu et al. introduced a robust framework for Bayesian multi-task learning, t-processes (TP), for multi-task learning (Yu et al., 2007). TP allows the system to effectively distinguish good tasks from noisy or outlier tasks.

3 IBMTL

3.1 Basic Concepts

In classification task, let X, Y and T be the instance set, feature set and prediction labels respectively. The objective function I(X;Y)-I(T;Y) can be regarded as the loss in mutual information after categorization. A good categorization should keep the mutual information between data and features, and minimize the information loss. In other words, I(T; Y) should be close to I(X; Y). Therefore, in the information bottleneck classification setting, the quality of the categorization should be judged by the loss in mutual information between the original instances and categorized instances. This idea can be formulated in theorem 1.

Theorem 1. *If random variables Y, X, T form a Markov chain (denoted by* $Y \leftrightarrow X \leftrightarrow T$ *), then*

$$I(X;Y \mid T) = I(X;Y) - I(T;Y) \tag{1}$$

Here we omitted the proof as it is straightforward.

3.2 Model

Assume that there are n relevant learning tasks in the system. Let X_i be the instances set for the *i*th learning task. Let Y_i be the features set for the *i*th learning task. Let T_i be the prediction labels for the *i*th learning task. The conditional distributions of T_i given X_i is denoted by $p_i(t|x)$. The conditional distributions of Y_i given T_i is denoted by $p_i(y|t)$. We denote expectation by *E*.

Definition 1. *The expected value of the conditional distribution* $\overline{p}(y \mid t)$ *for all the tasks is defined by:*

$$\overline{p}(y \mid t) = \frac{1}{n}\sum_{i=1}^{n} p_i(y \mid t) \tag{2}$$

Figure 1 shows the graphical representation of the IBMTL model:

- For each learning task i, the edge from X_i to T_i denotes *X* were classified into *T*, and the edge from T_i to Y_i denotes T_i preserves the information contained in X_i about Y_i. The variational variables in this scheme are the conditional distribution $p_i(t|x)$ $(i=1,\dots,n)$.
- The model makes an assumption that all the conditional distributions $p_i(y|t)$ $(i=1,\dots,n)$ are "similar" in relevant learning tasks. Thus the conditional distribution $p_i(y|t)$ plays the role as the shared structures among multiple relevant tasks.

The Kullback Leibler (KL) divergence measure, also known as the relative entropy between $p_1(x)$ and $p_2(x)$, measures the "distance" between its two arguments. The KL arises in many fields as a natural divergence measure between two distributions. Here we use KL divergence, $D(p_i(y \mid t) \,\|\, \overline{p}(y \mid t))$, to measure the similarity between multiple tasks.

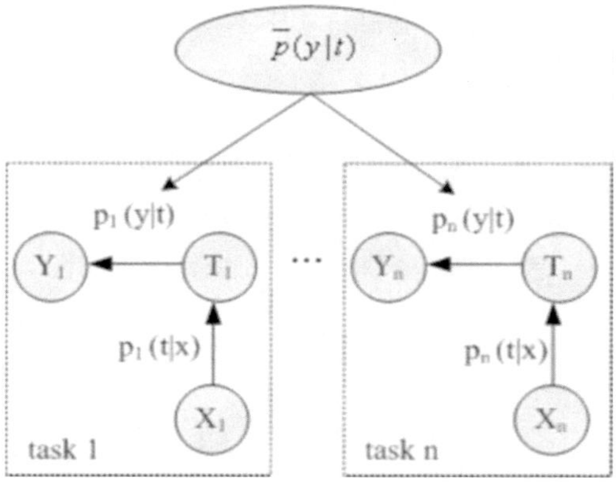

Fig. 1. Illustration of the IBMTL model

According to the IBMTL model, the objective function can be defined as:

$$\underset{\{p_1(t|x),\cdots,p_n(t|x)\}}{\arg\min} \quad Q \,. \tag{3}$$

Where

$$Q = \sum_{i=1}^{n} Q_i \,, \tag{4}$$

$$Q_i = I(X_i, Y_i \mid T_i) + \beta D(p_i(y \mid t) \,\|\, \bar{p}(y \mid t)) \,, \tag{5}$$

$$\text{s.t. } \forall i, \sum_t p_i(t \mid x) = 1 \,,$$

$$0 \le \beta \le 1 \,.$$

The variational variables in this scheme are the conditional distribution, $p_i(t \mid x)(i = 1, \cdots, n)$. β is a multiplier controlling the trade-off of the generalization ability between the single task and the whole tasks. The key idea of our model can be formulated as a variational principle of minimizing the loss in the mutual information, *I(X;Y|T)*, and constrain Kullback Leibler divergence of conditional distribution between the specific task and the whole tasks, $D(p_i(y \mid t) \,\|\, \bar{p}(y \mid t))$, to some maximal level.

Assume that every instance $x \in X$ belongs to precisely one class $t \in T$, then:

$$p(t \mid x) = \begin{cases} 1 & if \ x \in t \\ 0 & otherwise \end{cases} . \tag{6}$$

The objective function in Equation (5) is difficult to be optimized. Now we are to rewrite it into anther form.

Theorem 2. *Given the Markov chain condition* $Y \leftrightarrow X \leftrightarrow T$ *, then*

$$Q_i = \sum_t \sum_{x \in t} p_i(x)[(1-\beta) D(p_i(Y \mid x) \,\|\, p_i(Y \mid t)) + \beta D(p_i(Y \mid x) \,\|\, \bar{p}(Y \mid t))] \,. \tag{7}$$

Proof. Given the Markov chain condition $Y \leftrightarrow X \leftrightarrow T$, we have:

$$p(x, y \mid t) = p(x \mid t) p(y \mid x). \tag{8}$$

Then the first item at the right side of Eq. (5) can be rewritten as

$$\begin{aligned}
I(X_i;Y_i \mid T_i) &= \sum_{x,y,t} p_i(x, y, t) \log \frac{p_i(x, y \mid t)}{p_i(x \mid t) p_i(y \mid t)} \\
&= \sum_{x,y,t} p_i(x, y, t) \log \frac{p_i(y \mid x)}{p_i(y \mid t)} \\
&= \sum_{x,y,t} p_i(t) p_i(x \mid t) p_i(y \mid x) \log \frac{p_i(y \mid x)}{p_i(y \mid t)} \\
&= \sum_{x,t} p_i(x) p_i(t \mid x) \sum_{y} p_i(y \mid x) \log \frac{p_i(y \mid x)}{p_i(y \mid t)} \\
&= \sum_{t} \sum_{x \in t} p_i(x) D(p_i(Y \mid x) \,\|\, p_i(Y \mid t)).
\end{aligned} \tag{9}$$

The second item at the right side of Eq. (5) can be written as

$$\begin{aligned}
&\beta D(p_i(y \mid t) \,\|\, \overline{p}(y \mid t)) \\
&= \beta \sum_{t,y} p_i(t, y) \log \frac{p_i(y \mid t)}{\overline{p}(y \mid t)} \\
&= \beta \sum_{t,y} \sum_{x \in t} p_i(x, y) \log \frac{p_i(y \mid t)}{\overline{p}(y \mid t)} \\
&= \beta \sum_{t} \sum_{x \in t} p_i(x) \sum_{y} p_i(y \mid x) [\log \frac{p_i(y \mid x)}{\overline{p}(y \mid t)} - \log \frac{p_i(y \mid x)}{p_i(y \mid t)}] \\
&= \beta \sum_{t} \sum_{x \in t} p_i(x) [D(p_i(Y \mid x) \,\|\, \overline{p}(Y \mid t)) - D(p_i(Y \mid x) \,\|\, p_i(Y \mid t))].
\end{aligned} \tag{10}$$

According to Eq. (9) and (10), we have

$$\begin{aligned}
Q_i = \sum_{t} \sum_{x \in t} p_i(x) [&(1-\beta) D(p_i(Y \mid x) \,\|\, p_i(Y \mid t)) \\
&+ \beta D(p_i(Y \mid x) \,\|\, \overline{p}(Y \mid t))].
\end{aligned}$$

□

3.3 Algorithm

For simplicity, set:

$$D(x,t) = (1-\beta) D(p_i(Y \mid x) \,\|\, p_i(Y \mid t)) + \beta D(p_i(Y \mid x) \,\|\, \overline{p}(Y \mid t)). \tag{11}$$

Theorem 2 provides an alternative way to reduce the objective function value. From Equation (7), we know that minimizing $D(x,t)$ for a single instance x could reduce the objective function Q_i for the ith task. The following theorem 3 will show that the global objective function Q would also monotonically decrease.

Let X_i^r and X_i^s be the train set and the test set for task i respectively. Let $X_i = X_i^r \cup X_i^s$ be the data set for task i. Based on theorem 2, a formal description of the framework is given in Algorithm 1. In each iteration, the algorithm keeps the prediction labels for train set unchanged since their true labels are already known, while choosing the best category T for each data instance x in the test set to minimize the function $D(x,t)$. Thus we have:

$$h_i^{(k+1)}(x) = \underset{t}{\operatorname{argmin}} D_i^{(k)}(x,t) \quad \forall x \in X_i^s . \tag{12}$$

As we have discussed above, this process is able to decrease the global objective function.

Algorithm 1. The IDMTL Algorithm

Input: n learning tasks, $(X_i, Y_i, T_i)\ (i=1,\cdots,n)$; multiplier β .

Output: the final hypothesis $h_i^* : X_i^r \to T_i \quad (i=1,\cdots,n)$.

1. Set the count of iteration k=0. Initialize $p_i^{(k)}(y\,|\,t)$ based on the train set $X_i^r(i=1,\cdots,n)$.
2. Initialize the average conditional distribution $\overline{p}^{(k)}(y\,|\,t)$ based on Equation (2).
3. **do** { k++;
4.   **for** each learning task, $i=1,\cdots,n$ do
5.   **for** each instance $x \in X_i^s$ do
6. $h_i^{(k)}(x) = \arg\min_t D_i^{(k-1)}(x,t)$
7.   **end for**
8.   Update $p_i^{(k)}(y\,|\,t)$ based on the data set $X_i = X_i^r \cup X_i^s$.
9.   **end for**
10.   Update $\overline{p}^{(k)}(y\,|\,t)$ based on Equation (2).
11.   Set COUNT = the number of instances needed to adjust label.
12. } **while** (COUNT != 0)
13. Return the final hypothesis $h_i^{(k)} : X_i^r \to T_i \quad (i=1,\cdots,n)$.

3.4 Convergence

Since algorithm 1 is iterative, it is necessary to discuss its property of convergence. The following theorem shows that the global objective function in algorithm 1 monotonically decreases, which establishes that the algorithm converges eventually. Note that, although the algorithm is able to minimize the objective function value, it is only able to find a locally minimal one. Finding the global optimal solution is NP-hard.

Theorem 3. The global objective function monotonically decreases in each iteration of Algorithm 1.

$$Q^{(k)} \geq Q^{(k+1)}. \tag{13}$$

Proof. Since

$$h_i^{(k+1)}(x) = \underset{t}{\operatorname{argmin}} D_i^{(k)}(x,t) \quad \forall x \in X_i^s,$$

We have

$$D_i^{(k)}(x,t) \geq (1-\beta) D(p_i(Y|x) \| p_i^{(k)}(Y|h_i^{(k+1)}(x))) + \beta D(p_i(Y|x) \| \overline{p}^{(k)}(Y|h_i^{(k+1)}(x)))$$

The objective function for the ith task can be written as

$$Q_i^{(k)} = \sum_{t \in h_i^{(k)}} \sum_{x \in t} p_i(x) D_i^{(k)}(x,t)$$

$$\geq \sum_{t \in h_i^{(k)}} \sum_{x \in t} p_i(x)[(1-\beta) D(p_i(Y \mid x) \| p_i^{(k)}(Y \mid h_i^{(k+1)}(x)))$$

$$+\beta D(p_i(Y \mid x) \| \overline{p}^{(k)}(Y \mid h_i^{(k+1)}(x))]$$

$$= \sum_{t \in h_i^{(k+1)}} \sum_{x \in t} p_i(x) D_i^{(k)}(x,t).$$

Thus, we have

$$Q_i^{(k)} - Q_i^{(k+1)} \geq \sum_{t \in h_i^{(k+1)}} \sum_{x \in t} p_i(x) D_i^{(k)}(x,t) - \sum_{t \in h_i^{(k+1)}} \sum_{x \in t} p_i(x) D_i^{(k+1)}(x,t)$$

$$= \sum_t \sum_y p_i^{(k+1)}(t,y)[(1-\beta)\log \frac{p_i^{(k+1)}(y \mid t)}{p_i^{(k)}(y \mid t)} + \beta \log \frac{\overline{p}^{(k+1)}(y \mid t)}{\overline{p}^{(k)}(y \mid t)}]$$

$$= (1-\beta) D(p_i^{(k+1)}(y \mid t) \| p_i^{(k)}(y \mid t)) + \beta \sum_t \sum_y p_i^{(k+1)}(t,y) \log \frac{\overline{p}^{(k+1)}(y \mid t)}{\overline{p}^{(k)}(y \mid t)}$$

$$\geq \beta \sum_t \sum_y p_i^{(k+1)}(t,y) \log \frac{\overline{p}^{(k+1)}(y \mid t)}{\overline{p}^{(k)}(y \mid t)}.$$

Set $d = \arg\min_i p_i^{(k+1)}(t)$, we have

$$Q^{(k)} - Q^{(k+1)} = \sum_{i=1}^{n}(Q_i^{(k)} - Q_i^{(k+1)}) \geq \beta\sum_{i=1}^{n}\sum_t\sum_y p_i^{(k+1)}(t,y)\log\frac{\overline{p}^{(k+1)}(y|t)}{\overline{p}^{(k)}(y|t)}$$

$$= \beta\sum_t\sum_y[\sum_{i=1}^{n} p_i^{(k+1)}(t)p_i^{(k+1)}(y|t)]\log\frac{\overline{p}^{(k+1)}(y|t)}{\overline{p}^{(k)}(y|t)}$$

$$\geq \beta\sum_t\sum_y p_d^{(k+1)}(t)\sum_{i=1}^{n} p_i^{(k+1)}(y|t)\log\frac{\overline{p}^{(k+1)}(y|t)}{\overline{p}^{(k)}(y|t)}$$

$$= n\beta\sum_t\sum_y p_d^{(k+1)}(t)\overline{p}^{(k+1)}(y|t)\log\frac{\overline{p}^{(k+1)}(y|t)}{\overline{p}^{(k)}(y|t)}$$

$$= n\beta\sum_t p_d^{(k+1)}(t)D(\overline{p}^{(k+1)}(y|t) \,||\, \overline{p}^{(k)}(y|t)) \geq 0. \qquad \square$$

4 Experimental Evaluation

In order to know how well IBMTL works, the detailed experiments are performed. In our experiments, we compare the IBMTL algorithm with TrAdaBoost (Dai et al., 2007) which extends AdaBoost for transfer learning. TrAdaBoost allows users to utilize a small amount of newly labeled data to leverage the old data to construct a high-quality classification model for the new data. The basic idea of TrAdaBoost is to select the most useful diff-distribution instances as additional training data for predicting the labels of same-distribution techniques. To fit our multi-task learning scenario, in TrAdaBoost, the data set for the main task is regarded as the same-distribution data and all the train set for the other tasks is regarded as the diff-distribution data.

For implementation details, we use SVM classifier as the basic learners in TrAd aBoost. The maximum number of iterations is set to 50. In each iteration we sampled 30 percent of the instances to build the classifier. In IBMTL, we set β=0.2 manually. The performance in error rate was the average of 10 repeats by random.

4.1 Email Spam-Filtering Data Set

This dataset is provided by the 2006 ECML/PKDD discovery challenge. The task is to deal with personalized spam-filtering and generalization across related learning tasks. The challenge is that the distributions of emails for different users are different. There are 15 inboxes from different users. Each inboxes includes 400 emails.

The emails are in a bag-of-words vector space representation. Attributes are the term frequencies of the words. We removed words with less than four counts in the data set resulting in a dictionary size of about 150,000 words. The data set files are in the sparse data format used by SVMlight. Each line represents one email, the first token in each line is the class label (spam or non-spam). The tokens following the label information are pairs of word IDs and term frequencies in ascending order of the word IDs.

We included a standard feature selection mechanism, where for each dataset we selected the 2000 words with the highest contribution to the mutual information between the words and the documents. More formally stated, for each dataset, we sorted all words by,

$$I(y) \equiv p(y)\sum_{x\in X} p(x \mid y)\log\frac{p(x \mid y)}{p(x)},$$

and selected the top 2000.

The error rate was used to evaluate the classification performance of different algorithms. To observe the impact of the number of tasks on the performance, we randomly select n (=3, 6, 10) tasks from 15 tasks to conduct the experiments. The experimental results were shown in from figure 2 to figure 4. The x-axis represents the ratio of test data to data set. The y-axis represents the classification error rate. The ratio of test data to the data set is gradually increases from 0.5 to 0.98. The error rate was average over all the test data of the n tasks. All the three figures show that IBMTL significantly outperform TrAdaBoost. It demonstrates that the Kullback Leibler divergence, $D(p_i(y \mid t) \,\|\, \overline{p}(y \mid t))$, is an appropriate function to measure the similarity between the multiple task. By learning the classifiers in parallel under a unified representation, the transferability of expertise between tasks is exploited to the benefit of all. This expertise transfer is particularly important when we are provided with only a limited amount of training data for learning each classifier. By exploiting data from related tasks, the training data for each task is strengthened and the generalization of the resulting classifier-estimation algorithm is improved.

4.2 Sentiment Classification Data Set

This data set (John et al., 2007) contains product reviews downloaded from Amazon.com from 4 product types (domains): Kitchen, Books, DVDs, and Electronics. Each domain has several thousand reviews, but the exact number varies by domain. Reviews contain star ratings (1 to 5 stars). Reviews with rating > 3 were labeled positive, those with rating < 3 were labeled negative. After this conversion, each domain has 1000 positive and 1000 negative examples.

The task is to automatically classify the reviews on a product into positive and negative. The challenge is that the distributions of review-data among different types of products can be very different.

Our pre-processing included lowering the upper case characters and ignoring all words that contained digits or non alpha-numeric characters. We removed words with less than four counts in the data set. The feature selection process is the same as what we have done for the spam data set.

The experimental results were shown in from figure 5. The ratio of test data to the data set is gradually increases from 0.2 to 0.95. Figure 5 also shows that IBMTL gain the better generalization than that of TrAdaBoost.

Since our algorithm is an iterative algorithm, an important issue for IBMTL is the convergence property. Theorem 3 has already proven the convergence of IBMTL theoretically. The experimental results show that IBMTL always achieves almost convergence points within 20 iterations. This indicates that IBMTL converges very fast. We believe that 20 iterations is empirically enough for IBMTL.

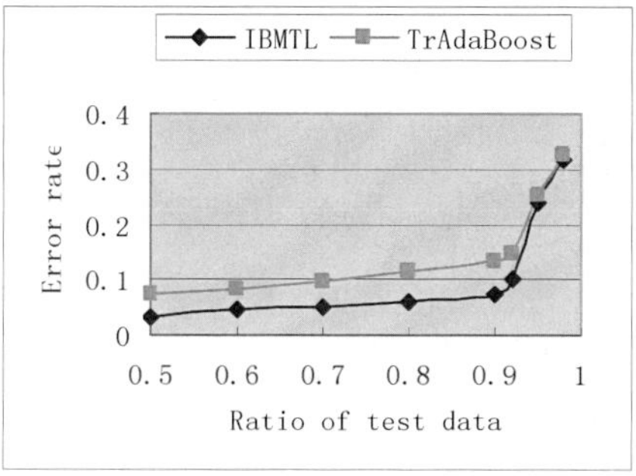

Fig. 2. Error rate curves on Spam (n=3)

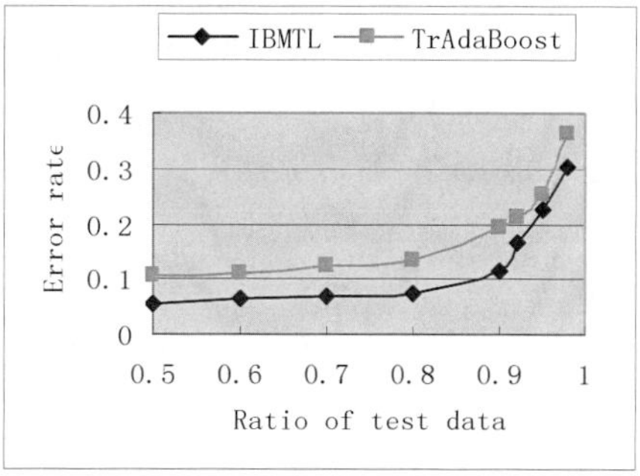

Fig. 3. Error rate curves on Spam (n=6)

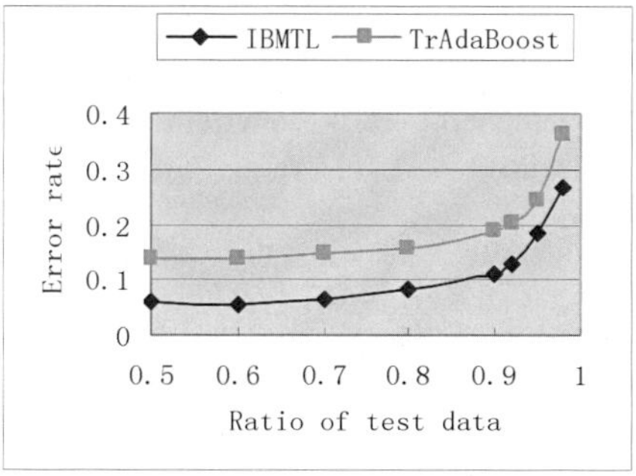

Fig. 4. Error rate curves on Spam (n=10)

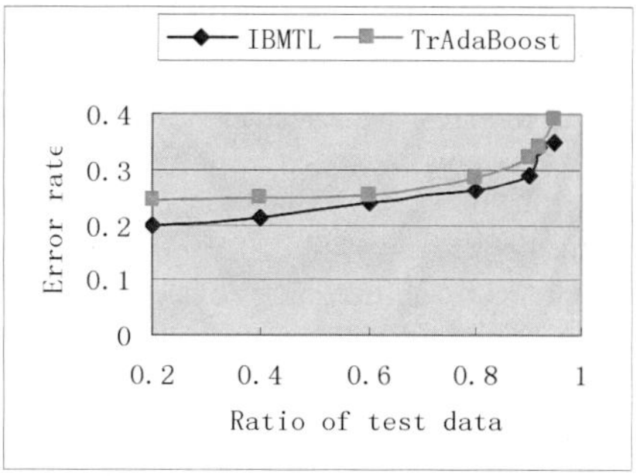

Fig. 5. Error rate curves on Sentiment

5 Conclusions and Future Work

An information-theoretic approach for multi-task learning is presented in this paper. Multi-task learning can mine the hidden relation between tasks. Based on the sharing structure multi-task learning can achieve efficient knowledge-sharing between tasks. The experiments also demonstrate that multi-task learning often help to improve performance especially when the data is limited and multiple relative tasks exists.

However, multi-task learning would hurt the performance when the tasks are too dissimilar. An idea multi-task learning algorithm should be able to measure the extent of relatedness between tasks and automatically identify similarities between tasks and only allow similar task to share information. The Kullback Leibler divergence is used to measure the similarity between the multiple tasks. Thus, it would help to select the good tasks. As part of ongoing work we are exploring the boundary between positive transfer and negative transfer and learning how to pick out good tasks from noisy tasks.

References

1. Ando, R.K., Zhang, T.: A framework for learning predictive structures from multiple tasks and unlabeled data. The Journal of Machine Learning Research 6(1), 1817–1853 (2005)
2. Caruana, R.: Multi-task learning. Machine Learning 28(1), 41–75 (1997)
3. Bakker, B., Heskes, T.: Task clustering and gating for Bayesian multitask learning. The Journal of Machine Learning Research 4(12), 83–89 (2003)
4. Baxter, J.: A model of inductive bias learning. Journal of Artificial Intelligence Research 12, 149–198 (2000)
5. Dai, W.Y., Yang, Q., Xue, G.R., et al.: Boosting for Transfer Learning. In: Proc of the 24th international conference on Machine learning, pp. 193–200. ACM Press, New York (2007)
6. Evgeniou, T., Micchelli, C.A., Pontil, M.: Learning multiple tasks with kernel methods. Journal of Machine Learning Research 6, 615–637 (2005)
7. Heskes, T.: Empirical bayes for learning to learn. In: Proceedings of the Seventeenth International Conference on Machine Learning, pp. 367–374. ACM Press, New York (2000)
8. Lawrence, N.D., Platt, J.C.: Learning to learn with the informative vector machine. In: Proceedings of the 21st International Conference on Machine Learning (2004)
9. Roy, D.M., Kaelbling, L.P.: Efficient Bayesian task-level transfer learning. In: Proc. of the 20th Joint Conference on Artificial Intelligence, pp. 2599–2604. ACM Press, New York (2007)
10. Yu, S.P., Tresp, V., Yu, K.: Robust multi-task learning with t-processes. In: Proc. of the 24th international conference on Machine learning, pp. 1103–1110. ACM Press, New York (2007)
11. Yu, K., Tresp, V., Schwaighofer, A.: Learning Gaussian processes from multiple tasks. In: Proceedings of the 22nd international conference on Machine learning, pp. 1012–1019. ACM Press, New York (2005)
12. Zhang, Y., Koren, J.: Efficient Bayesian hierarchical user modeling for recommendation systems. In: Proceedings of the 30th annual international ACM SIGIR conference on Research and development in information retrieval, pp. 47–54. ACM Press, New York (2007)
13. Blitzer, J., Dredze, M., Pereira, F.: Biographies, Bollywood, Boom-boxes and Blenders: Domain Adaptation for Sentiment Classification. In: Association of Computational Linguistics (ACL) (2007)

Online New Event Detection Based on IPLSA

Xiaoming Zhang and Zhoujun Li

School of Computer Science and Engineering, Beihang University,
100083 Beijing
yolixs@163.com

Abstract. New event detection (NED) involves monitoring one or multiple news streams to detect the stories that report on new events. With the overwhelming volume of news available today, NED has become a challenging task. In this paper, we proposed a new NED model based on incremental PLSA(IPLSA), and it can handle new document arriving in a stream and update parameters with less time complexity. Moreover, to avoid the limitation of TF-IDF method, a new approach of term reweighting is proposed. By dynamically exploiting importance of documents in discrimination of terms and documents' topic information, this approach is more accurate. Experimental results on Linguistic Data Consortium (LDC) datasets TDT4 show that the proposed model can improve both recall and precision of NED task significantly, compared to the baseline system and other existing systems.

Keywords: New event detection, PLSA, Term reweighting, TDT.

1 Introduction

The Topic Detection and Tracking (TDT) program, a DARPA funded initiative, aims to develop technologies that search, organize and structure multilingual news-oriented textual materials from a variety of sources. A topic is defined as a seminal event or activity, along with directly related events and activities [1]. An earthquake at a particular place could be an example of a topic. The first story on this topic is the story that first carries the report on the earthquakes' occurrence. The other stories that make up the topic are those discussing the death toll, the rescue efforts, and the commercial impact and so on. In this paper we define New Event Detection (NED), is the task of online identification of the earliest story for each topic as soon as that report arrives in the sequence of documents.

NED systems are very useful in situations where novel information needs to be ferreted out from a mass of rapidly growing data. Examples of real-life scenarios are financial markets, news analyses, intelligence gathering etc [2]. In the other side, NED is an open challenge in text mining. It has been recognized as the most difficult task in the research area of TDT. A performance upper-bound analysis by Allan et al. [3] provided a probabilistic justification for the observed performance degradation in NED compared to event tracking, and suggested that new approaches must be explored in order to significantly enhance the current performance level achieved in NED.

R. Huang et al. (Eds.): ADMA 2009, LNAI 5678, pp. 397–408, 2009.

Generally speaking, NED is difficult for several reasons: First, has to be done in an online fashion, which imposes constraints on both strategy and efficiency. Second, similar to other problems in text mining, we have to deal with a high-dimensional space with tens of thousands of features. And finally, the number of topics can be as large as thousands as in newswire data.

In this paper we reduce the dependence on TF-IDF weighting by exploiting new approaches, which exploits latent analysis of stories and reweighs terms based on other information of documents and terms. We also use the age character of story to detect new event. Experiments show that this approach can improve recall and precision of NED greatly, and it can also identify new stories or old stories as well.

2 Relate Works

In currently NED systems, there are mainly two methods to compare news story on hand with previous topic. First, each news story is compared to all the previous received stories. Papka et al. proposed Single-Pass clustering on NED [11]. The other method organizes previous stories into clusters which correspond to topics, and new story is compared to the previous clusters instead of stories. Lam et al build up previous query representations of story clusters, each of which corresponds to a topic [12]. In this manner comparisons happen between stories and clusters, and concept terms besides named entities of a story are derived from statistical context analysis on a separate concept database. Nevertheless, it has been proved that this manner is less accurate [5, 6].

Recent years, most work focus on proposing better methods on comparison of stories and document representation. Stokes et al. [14] utilized a combination of evidence from two distinct representations of a document's content. A marginal increase in effectiveness was achieved when the combined representation was used. In paper [15], a ONED framework is proposed. It combines indexing and compression methods to improve the document processing rate by orders of magnitude.

However, none of the systems have considered that terms of different types (e.g. Noun, Verb or Person name) have different effects for different classes of stories. Some efforts have been done on how to utilize named entities to improve NED[7,8,9,10]. Yang et al. gave location named entities four times weight than other terms and named entities [7]. DOREMI research group combined semantic similarities of person names, location names and time together with textual similarity [8,9]. UMass [10] research group split document representation into two parts: named entities and non-named entities. Paper [4] assumes that every news story is characterized by a set of named entities and a set of terms that discuss the topic of the story. It is assumed that a new story can at most share one of either the named entity terms or the topic terms with a single story. Further, these two stories must themselves be on different topics. Named entities are also used in other researches [17,18].

There are other methods that use indexing-tree or probabilistic model to improve NED. In paper [16] a new NED model is proposed to speed up the NED task by using news indexing-tree dynamically. Probabilistic models for online clustering of documents, with a mechanism for handling creation of new clusters have been developed. Each cluster was assumed to correspond to a topic. Experimental results did not show any improvement over baseline systems [19].

In practice, it is not possible to determine the true optimal threshold, because there is no knowledge about incoming and future news documents. Thus, to preserve the high performance for nonoptimal thresholds, an event analysis algorithm has to be resilient. In this paper we use IPLSA to exploit the latent analysis between stories, and it can reduce the dependence on optimal threshold because it can identify new story or old story more accurately. By using incremental approach, IPLSA can reduce the time of parameters reestimating greatly. We reduce the dependence of term weight on TF-IDF by introducing another term weighting algorithm, which exploit the discrimination of documents and topics in terms' weight.

3 PLSA Methods and Existing Incremental PLSA Methods

The Probabilistic Latent Semantic Analysis (PLSA) model incorporates higher level latent concepts or semantics to smooth the weights of terms in documents [20]. The latent semantic variables can be viewed as intermediate concepts or topics placed between documents and terms. Meanwhile, the associations between documents, concepts, and terms are represented as conditional probabilities and are estimated by the EM algorithm, an iterative technique that converges to a maximum likelihood estimator under incomplete data [21]. After the PLSA parameters have been estimated, the similarities between new documents (called query documents in [20]) and existing documents can be calculated by using the smoothed term vectors. The PLSA algorithm, which can be used in text classification and information retrieval applications [22], achieves better results than traditional VSM methods [20].

3.1 PLSA Model

PLSA is a statistical latent class model that has been found to provide better results than LSA for term matching in retrieval applications. In PLSA, the conditional probability between documents d and words w is modeled through a latent variable z, which can be loosely thought of as a class or topic. A PLSA model is parameterized by P(w|z) and P(z|d) , and the words may belong to more than one class and a document may discuss more than one "topic". It is assumed that the distribution of words given a class, P(w|z) is conditionally independent of the document, i.e., P(w|z, d) = P(w|z). Thus the joint probability of a document d and a word w is represented as:

$$P(w,d) = P(d)\sum_{z} P(w \mid z)P(z \mid d) \tag{1}$$

The parameters of a PLSA model, P(w|z)an d P(z|d), are estimated using the iterative Expectation-Maximization (EM)algorit hm to fit a training corpus D by maximizing the log-likelihood function *L*:

$$L = \sum_{d \in D} \sum_{w \in d} f(d,w) \log P(d,w) \tag{2}$$

Where f(d,w)is the frequency of word w in document d [11]. Starting from random initial values, the EM procedure iterates between 1)the E-step, where the probability that a word w in a particular document d is explained by the class corresponding to z is estimated as:

$$P(z \mid w,d) = \frac{P(w \mid z)P(z \mid d)}{\sum_{z'} P(w \mid z')P(z' \mid d)} \tag{3}$$

and 2)t he M-step, where parameters P(w|z)an d P(z|d)are re-estimated to maximize L:

$$P(w \mid z) = \frac{\sum_{d} f(d,w)P(z \mid w,d)}{\sum_{w'}\sum_{d} f(d,w')P(z \mid w',d)} \tag{4}$$

$$P(z \mid d) = \frac{\sum_{w} f(d,w)P(z \mid w,d)}{\sum_{z'}\sum_{w} f(d,w)P(z' \mid w,d)} \tag{5}$$

Although PLSA has been successfully developed, there are two main shortcomings. First, the PLSA model is estimated only for those documents appearing in the training set. PLSA was shown to be a special variant of LDA with a uniform Dirichlet prior in a maximum a posteriori model [23]. Secondly, PLSA lacks the incremental ability, i.e. it cannot handle new data arriving in a stream.

To handle streaming data, a naive approach is that it can re-train the model using both existing training data and new data. However, it is apparently not efficiently since it is very computationally expensive. What is more, for some practical applications, this is infeasible since the system needs real-time online update. Therefore, we need a fast incremental algorithm without compromising NED performance.

3.2 Existing PLSA Incremental Methods

There are some existing works on incremental learning of PLSA. Paper [24] provided a simple update scheme called Fold-In. The main idea is to update the P(z|d) part of the model while keeping P(w|z) fixed. However, P(w|z) can change significantly during EM iteration and affect P(z|d) as well. Thus, the result of Fold-In might be biased.

Tzu-Chuan Chou et al [25] proposed Incremental PLSA (IPLSA), a complete Bayesian solution aiming to address the problem of online event detection. For the time complexity, the algorithm needs $O(n_{iter}\ (n_{nd} + n_{od})\ (n_{nw} + n_{ow})\ K)$ operations to converge whenever there are new documents added, where n_{nd} is the number of new documents, and nod is the number of old documents, and n_{nw} is the number of new words and now is the number of old words, and K is the number of latent topics, and niter is the number of iterations. Note that the computational complexity is the same as that of the batched PLSA algorithm, although less EM iterations are needed. In the other paper, Chien and Wu proposed another PLSA incremental learning algorithm named MAP-PLSA [26], and the complexity is also great.

4 A Novel Incremental PLSA Model with Time Window

There are some issues should be considered for the incremental task in NED systems:

- Word-topic and document-topic probabilities for new documents.
- Efficiency of parameters updating.

In order to address the above problems a novel incremental PLSA learning algorithm and a new term weighting algorithm are proposed. When a new document arrives, the probability of a latent topic given the document P(z|d) is updated accordingly, and so does the probability of words given a topic, P(w|z) [27]. The formulae for incremental update are as follows:

- E-step

$$P(z|q,w)^{(n)} = \frac{P(z|q)^{(n)} P(w|z)^{(n)}}{\sum_{z'} P(z'|q)^{(n)} P(w|z')^{(n)}} \tag{6}$$

- M-step

$$P(z|q)^{(n)} = \frac{\sum_{w} n(q,w) \times P(z|q,w)^{(n)}}{\sum_{z'} \sum_{w'} n(q,w') \times P(z'|q,w')^{(n)}} \tag{7}$$

$$P(w|z)^{(n)} = \frac{\sum_{d} n(d,w) \times P(z|d,w)^{(n)} + \alpha \times P(w|z)^{(n-1)}}{\sum_{d} \sum_{w'} n(d,w') \times P(z|d,w')^{(n)} + \alpha \times \sum_{w''} P(w''|z)^{(n-1)}} \tag{8}$$

Where the superscript (n – 1) denotes the old model parameters and (n) for the new ones, $w' \in d$ and $w'' \in W$ are words in this document and all other words in the dictionary, respectively. The values of α are hyper-parameters that manually selected based on empirical results. The time complexity of this algorithm is $O(n_{iter} \cdot n_{nd} \cdot \|n_{nd}\| \cdot K)$, where niter is the number of iterations, n_{nd} is the number of new documents, $\|n_{nd}\|$ is the average number of words in these documents and K is the number of latent topic z.

When the PLSI algorithm is used in calculation of similarity between any two documents, a document d is represented by a smoothed version of the term vector $(P(w_1|d),P(w_2|d),P(w_3|d)\ldots.)$, where

$$P(w|d) = \sum_{z \in Z} P(w|z)P(z|d) \tag{9}$$

Then, after weighting by the IDF, the similarity between any two documents can be calculated by the following cosine function:

$$sim(\vec{d_1},\vec{d_2}) = \frac{\vec{d_1} \bullet \vec{d_2}}{|\vec{d_1}| \times |\vec{d_2}|}. \tag{10}$$

Where

$$\vec{d} = (P(w_1|d) \times we_1, P(w_2|d) \times we_2, \ldots.) \tag{11}$$

Where we_i is the weighting of w_i. In our approach the weightings of named entities are different to topic terms (terms in the document not identified as named entities). Because different types of entities have different effect to NED, for example, to distinguish stories about earthquakes in two different places, the named entities may play import roles.

As for IPLSA model, the number of documents to be processed affects the time complexity greatly. We can decrease the time complexity by reducing the number of documents to be processed. The rationale behind our approach is that some stories belonging to the same topic are very similar, and we can combine them to be one story. Then the weights of terms in the new story are updated accordingly. But how do we combine two stories to be one story. In the approach, if the similarity between two stories is greater than 2*θ(θ is the threshold value that determine whether a story is the first story of an event), then the two stories are combined.

In the online NED, events have "aging" nature, and an old inactive event less likely attracts new stories than recently active events. Therefore, temporal relations of stories can be exploited to improve NED. Using a lookup window is a popular way of limiting the time frame that an incoming story can relate to. An example of a window-based NED system is shown in Fig.1. In each advance of the window, which can be measured in time units or by a certain number of stories, the system discards old stories and fold in new ones.

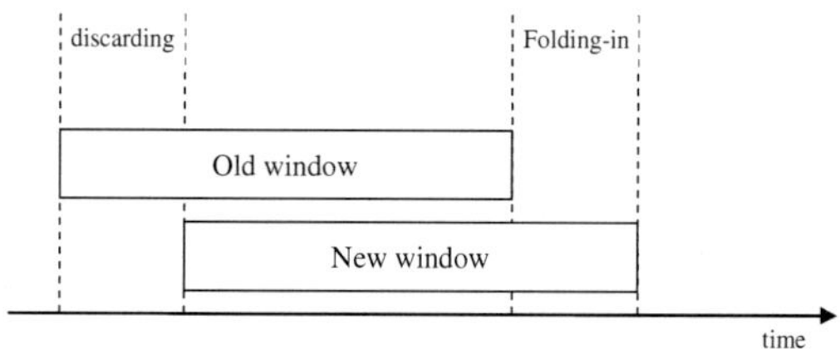

Fig. 1. The discarding and folding-in of a window

In our approach, when a story is labeled as a first story of an event, the widow advances and the EM algorithm reestimates all the parameters of the PLSA algorithm. Because we always hold stories in the window, so the number of story is not increase with time advance. As a result, the time complexity doesn't increase along with time advancing. The similarity between two documents in the same time window is modified by substituting (21) as follows, where T(d) is the time stamp of document d:

$$sim(\vec{d_1},\vec{d_2}) = \frac{1}{1+\left(\frac{|T(d_1)-T(d_2)|}{window \quad size}\right)^{\frac{1}{2}}} * \frac{\vec{d_1}\bullet\vec{d_2}}{|\vec{d_1}|\times|\vec{d_2}|} \tag{12}$$

The similarity between a new document and an event is defined as the maximum similarity between the new document and documents previously clustered into the event. A document is deemed the first story of a new event if its similarity with all the events in the current window is below a predetermined threshold θ; otherwise, it is assigned to the event that is the most similar.

5 Term Reweighting

We use χ2 statistic to compute correlations between terms and topics, and use it to select features form documents. For each document, we only keep the top-K terms with the largest χ 2 values rather than all the terms. Here K is a predetermined percent constant. Only the top-K terms are used to compute the similarity values of document pairs. Reducing the number of saved terms can reduce the time complexity of incremental PLSA model greatly.

TF-IDF has be the most prevalent terms weighting method in information retrieval systems. The basic idea of TF-IDF is that the fewer documents a term appears in, the more important the term is in discrimination of documents. However, the TF-IDF method can't weight terms of following classes properly:

- Terms that occurs frequently within a news category.
- Terms that occurs frequently in a topic, and infrequently in other topics.
- Terms with low document frequency, and appear in different topics.

Besides, this method rarely considers the importance of document weight. In fact, documents are also important in discrimination of terms. The main assumption behind document weighting is as following: the more information a document gives to terms the more effect it gives to latent variable, and the less information a document gives to terms the less effect it gives to latent variable.

To address above problems, we propose that term weight is constituted of following parts.

$$W(i,j) = LT(i,j) \times GT(i) \times GD(j) \times KD(i) \tag{13}$$

The notation used in the follow equations is defined as:

tf_{ij}: the frequency of term i in document j
df_i: the number of documents that contain term i.
df_{ci}: the number of documents containing term i within cluster c.
gf_i: the frequency of term i in document collection.
sgf: the sum frequency of all terms in document collection.
N_c: the number of documents in cluster c.
N_t: the total number of documents in collection.

We replace the TF in TF-IDF with following:

$$LT(i,j) = \frac{\log(tf_{ij}+1)}{\log dl_j + 1} \tag{14}$$

Entropy theory is used to set GT(i) and GD(j), and it replaces IDF with following:

$$GT(i) = \frac{H(d) - H(d \mid t_i)}{H(d)} = 1 - \frac{H(d \mid t_i)}{H(d)} \tag{15}$$

$$H(d \mid t_i) = -\sum p(j \mid i) \log p(j \mid i) \tag{16}$$

$$p(j \mid i) = \frac{tf_{ij}}{gf_i} \qquad H(d) = \log(N_t) \tag{17}$$

GD(j) is defined as following, and it mainly computer the importance of document to terms:

$$GD(j) = \frac{H(t) - H(t \mid d_j)}{H(t)} = 1 - \frac{H(t \mid d_j)}{H(t)} \tag{18}$$

$$H(t) = -\sum p(t) \times \log p(t) = -\sum_{i=1}^{n} \frac{gf_i}{sgf} \times \log \frac{gf_i}{sgf} \tag{19}$$

$$H(t \mid d_j) = -\sum p(i \mid j) \log p(i \mid j) \quad , \quad p(i \mid j) = \frac{tf_{ij}}{dl_j} \tag{20}$$

KD(i) is used to enhance weight of term that occurs frequently in a topic, and infrequently in other topics, and it is defined as following:

$$KD(i) = KL(P_{ci} \parallel P_{ti}) \tag{21}$$

$$P_{ci} = \frac{df_{ci}}{N_c}, P_{ti} = \frac{df_i}{N_t} \quad , \quad KL(P \parallel Q) = \sum p(x) \log \frac{p(x)}{q(x)} \tag{22}$$

6 Evaluation

In the evaluation, we used the standard corpora TDT-4 from the NIST TDT corpora. Only English documents tagged as definitely news topics (that is, tagged YES) were chosen for evaluation.

6.1 Performance Metrics

We follow the performance measurements defined in [19]. An event analysis system may generate any number of clusters, but only the clusters that best match the labeled topics are used for evaluation.

Table 1. A Cluster-Topic Contingency Table

	In topic	Not in topic
In cluster	a	b
Not in cluster	c	d

Table 1 illustrates a 2-2 contingency table for a cluster-topic pair, where a, b, c, and d represent the numbers of documents in the four cases. Four singleton evaluation measures, Recall, Precision, Miss, and False Alarm, and F1, are defined as follows:

- Recall =a/(a+c) if (a+c) > 0; otherwise, it is undefined.
- Precision=a/(a+b) if (a+b) > 0; otherwise, it is undefined.
- Miss =c/(a+c) if (a+c) > 0; otherwise, it is undefined.
- False Alarm=b/(b+d) if (b+d) > 0; otherwise, it is undefined.
- F1= 2 *Recall *Precision/(Recall+ Precision).

To test the approaches proposed in the model, we implemented and tested three systems:

System 1 (SS), in which the story on hand is compared to each story received previously, and use the highest similarity to determine whether current story is about a new event.

System 2 (SC), in which the story on hand is compared to all previous clusters each of which representing a topic, and the highest similarity is used for final decision for current story. If the highest similarity exceeds threshold θ, then it is an old story, and put it into the most similar cluster; otherwise it is a new story and creates a new cluster.

System 3 (IPLSA), implemented based on IPLSA proposed in section 4, and terms are reweighted according to section 5, and 40 latent variable z is used in this system.

6.2 Evaluation Base Performance and Story Indentation

In the first experiment we test the recall, precision, miss and F1 of these three systems. Figure 2 summarizes the results of Systems based on CS, SS and IPLSA. All these systems conducted multiple runs with different parameter settings; here we present the best result for each system with respect to the F1 measure. As shown, the results of the evaluations demonstrate that the proposed IPLSI algorithm outperforms the SS, CS model greatly.

Table 2. New event detection results

	CS	SS	IPLSA
Recall(%)	62	70	88
Precision(%)	57	68	85
Miss(%)	38	30	12
False Alarm(%)	0.162	0.160	0.157
F_1	0.59	0.68	0.86

In the second experiment, we mainly test the scores distribution of new stories and old stories. The main goal of our effort was to come up with a way to correctly identify new stories. In practice, it is not possible to determine the true optimal threshold, because there is no knowledge about incoming and future news documents. Thus, to preserve the high performance for nonoptimal thresholds, an event analysis algorithm has to be resilient. This is means that concentration of new story scores and old story scores in the low level and high level respectively can make a good identification of new event. To understand what we had actually achieved by using the proposed model, we studied the distribution of the confidence scores assigned to new and old stories for the three systems for the TDT4 collection (Figures 1 and 2 respectively).

In figure 3, we observe that the scores for a small fraction of new stories that were initially missed (between scores 0.8 and 1) are decreased by the system based on proposed model and a small fraction (between scores 0.1 and 0.3) is increased by a large amount. As to the old story scores, there is also a major impact of using IPLSA based system. In figure 4, we observe that the scores of a significant number of old stories (between scores 0.2 and 0.4) have been decreased. This had the effect of increasing the score difference between old and new stories, and hence improved new event identification by the minimum cost.

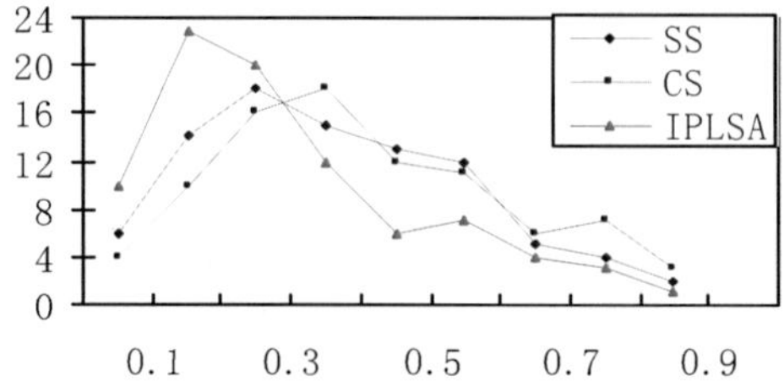

Fig. 3. Distribution of new story scores for the SS, CS IPLSA based model systems

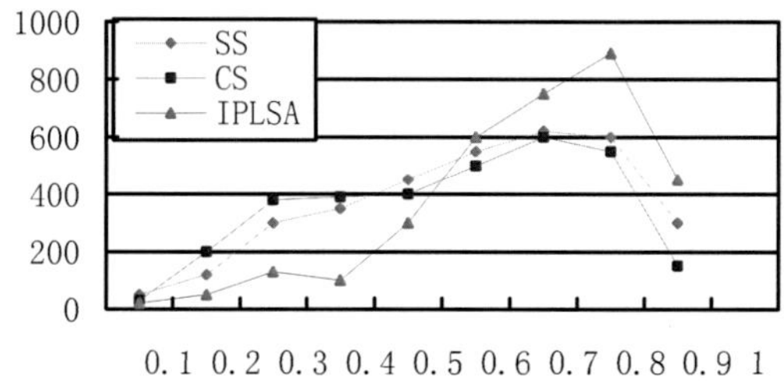

Fig. 4. Distribution of old story scores for the SS, CS IPLSA based model systems

7 Conclusion

We have shown the applicability of IPLSA techniques to solve the NED problem. Significant improvements were made over the 'SS' and 'SC' systems on the corpora tested on. We also have presented a new algorithm of term weighting. This algorithm adjusts term weights based on term distributions between the whole corpus and a cluster story set, and it also uses document information and entropy theory. Our experimental results on TDT4 datasets show that the algorithm contributes significantly to improvement in accuracy. NED requires not only detection and reporting of new events, but also suppression of stories that report old events. From the study of the distributions of scores assigned to stories by the 'SS', 'SC' and IPLSA model systems, we can also see that IPLSA model can do a better job of detecting old stories (reducing false alarms). Thus it can be believed that attacking the problem as "old story detection" might be a better and more fruitful approach.

References

1. Allan, J.: Topic Detection and Tracking: Event-Based Information Organization. Kluwer Academic Publishers, Dordrecht (2002)
2. Papka, R., Allan, J.: On-line New Event Detection Using Single Pass Clustering TITLE2: Technical Report UM-CS-1998-021 (1998)
3. Allan, J., Lavrenko, V., Jin, H.: First story detection in tdt is hard. Washiongton DC. In: Proceedings of the Ninth International Conference on Informaiton and Knowledge Management (2000)

4. Giridhar, K., Allan, J., Andrew, M.: Classification Models for New Event Detection. In: Proceeding of CIKM (2004)
5. Yang, Y., Pierce, T., Carbonell, J.: A Study on Retrospective and On-line Event Detection. In: Proceedings of SIGIR, Melbourne, Australia, pp. 28–36 (1998)
6. Allan, J., Lavrenko, V., Malin, D., Swan, R.: Detections, Bounds, and Timelines: Umass and tdt-3. In: Proceedings of Topic Detection and Tracking Workshop (TDT-3), Vienna, VA, pp. 167–174 (2000)
7. Yang, Y., Zhang, J., Carbonell, J., Jin, C.: Topic-conditioned Novelty Detection. In: Proceedings of the 8th ACM SIGKDD International Conference, pp. 688–693 (2002)
8. Juha, M., Helena, A.M., Marko, S.: Applying Semantic Classes in Event Detection and Tracking. In: Proceedings of International Conference on Natural Language Processing, pp. 175–183 (2002)
9. Juha, M., Helena, A.M., Marko, S.: Simple Semantics in Topic Detection and Tracking. Information Retrieval, 347–368 (2004)
10. Giridhar, K., Allan, J.: Text Classification and Named Entities for New Event Detection. In: Proceedings of the 27th Annual International ACM SIGIR Conference, New York, NY, USA, pp. 297–304 (2004)
11. Papka, R., Allan, J.: On-line New Event Detection Using Single Pass Clustering TITLE2: Technical Report UM-CS-1998-021 (1998)
12. Lam, W., Meng, H., Wong, K., Yen, J.: Using Contextual Analysis for News Event Detection. International Journal on Intelligent Systems, 525–546 (2001)
13. Thorsten, B., Francine, C., Ayman, F.: A System for New Event Detection. In: Proceedings of the 26th AnnualInternational ACM SIGIR Conference, pp. 330–337. ACM Press, New York (2003)
14. Nicol, S.a., Joe, C.: Combining Semantic and Syntactic Document Classifiers to Improve First Story Detection. In: Proceedings of the 24th Annual International ACM SIGIR Conference, pp. 424–425. ACM Press, New York (2001)
15. Luo, G., Tang, C., Yu, P.S.: Resource-Adaptive Real-Time New Event Detection. In: SIGMOD, pp. 497–508 (2007)
16. Kuo, Z., Zi, L.J., Gang, W.: New Event Detection Based on Indexing-tree and Named Entity. In: Proceedings of SIGIR, pp. 215–222 (2007)
17. Makkonen, J., Ahonen-Myka, H., Salmenkivi, M.: Applying semantic classes in event detection and tracking. In: Proceedings of International Conference on Natural Language Processing, pp. 175–183 (2002)
18. Makkonen, J., Ahonen-Myka, H., Salmenkivi, M.: Simple semantics in topic detection and tracking. In: Information Retrieval, pp. 347–368 (2004)
19. Zhang, J., Ghahramani, Z., Yang, Y.: A probabilistic model for online document clustering with application to novelty detection. In: Saul, L.K., Weiss, Y., Bottou, L. (eds.) Advances in Neural Information Processing Systems, vol. 17, pp. 1617–1624. MIT Press, Cambridge (2005)
20. Hofmann, T.: Probabilistic Latent Semantic Indexing. In: Proc. ACMSIGIR 1999 (1999)
21. Dempster, A.P., Laird, N.M., Rubin, D.B.: Maximum Likelihood from Incomplete Data via the EM Algorithm. J. Royal Statistical Soc. B 39, 1–38 (1977)
22. Brants, T., Chen, F., Tsochantaridis, I.: Topic-Based Document Segmentation with Probabilistic Latent Semantic Analysis. In: Proc. 11th ACM Int'l Conf. Information and Knowledge Management (2002)
23. Girolami, M., Kaban, A.: On an Equivalence Between PLSI and LDA. In: Proc. of SIGIR, pp. 433–434 (2003)

24. Thomas, H.: Unsupervised Learning by Probabilistic Latent Semantic Analysis. Maching Learning Journal 42(1-2), 177–196 (2001)
25. Chou, T.C., Chen, M.C.: Using Incremental PLSA for Threshold Resilient Online Event Anlysis. IEEE Transaction on Knowledge and Data Engineering 20(3), 289–299 (2008)
26. Chien, J.T., Wu, M.S.: Adaptive Bayesian Latent Semantic Analysis. IEEE Transactions on Audio, Speech, and Language Processing 16(1), 198–207 (2008)
27. Wu, H., Yongji, W., Xiang, C.: Incremental probabilistic latent semantic analysis for automatic question recommendation. In: Proceedings of ACM conference on Recommender systems, Lausanne, Switzerland, October 23-25 (2008)
28. Yang, Y., Pedersen, J.: A Comparative Study on Feature Selection in Text Categorization. In: Fisher, J.D.H. (ed.) The Fourteenth International Conference on MachineLearning, pp. 412–420. Morgan Kaufmann, San Francisco (1997)

Discovering Knowledge from Multi-relational Data Based on Information Retrieval Theory

Rayner Alfred*

Universiti Malaysia Sabah,
Center for Artificial Intelligence, Locked Bag 2073,
88999 Kota Kinabalu, Sabah, Malaysia
ralfred@ums.edu.my

Abstract. Although the *TF-IDF* weighted frequency matrix (vector space model) has been widely studied and used in document clustering or document categorisation, there has been no attempt to extend this application to relational data that contain one-to-many associations between records. This paper explains the rationale for using *TF-IDF* (term frequency inverse document frequency), a technique for weighting data attributes, borrowed from Information Retrieval theory, to summarise datasets stored in a multi-relational setting with one-to-many relationships. A novel data summarisation algorithm based on *TF-IDF* is introduced, which is referred to as *Dynamic Aggregation of Relational Attributes* (*DARA*). The *DARA* algorithm applies clustering techniques in order to summarise these datasets. The experimental results show that using the *DARA* algorithm finds solutions with much greater accuracy.

Keywords: Vector Space Model, Information Retrieval, Clustering, Data Summarization, Knowledge Discovery.

1 Introduction

Structured data such as data stored in a relational database that have one-to-many relationships between records stored in the target and non-target tables can be transformed into a single table. This process is known as propositionalisation [10]. Another model that can be used to represent individual object is vector space model [14]. Vector space model (or term vector model) is an algebraic model for representing text documents (and any objects, in general) as vectors of identifiers. Although the vector space model has been widely studied and used in document clustering or document categorisation, there has been no attempt to extend this application to relational data that contain one-to-many associations between records.

In this paper, we propose a data summarisation approach to summarise data stored in non-target tables by clustering them into groups. In order to assist the clustering process, a technique borrowed from information retrieval theory

* Head of Machine Learning Research Group, Center for Artificial Intelligence, Universiti Malaysia Sabah, Kota Kinabalu, Sabah, Malaysia.

R. Huang et al. (Eds.): ADMA 2009, LNAI 5678, pp. 409–416, 2009.

has been used to construct vectors of patterns that represent records stored in non-target tables. The approach is based on the idea that, for each unique record stored in non-target tables, there is a vector of patterns that represents this record. With this representation of data, it is possible to compare records in order to cluster them. Thus, clustering can then be applied to summarise data stored in a relational database following the principles of information retrieval theory [16].

In a relational database, records stored in a target table are associated with one or more records stored in a non-target table. The one-to-many relationship that exists between these records makes the vector space model suitable to represent data stored in multiple tables. For instance, an analogy of the representation of data for text documents and the representation of data for relational data with one-to-many relations in the form of a vector space model, for the purpose of summarisation, can be illustrated through the properties outlined below. With the assumption that the target table T in relational databases has one-to-many relationships with the non-target tables NT, the following analogies can be made:

1. *word* is to the representation of data for text documents as *instance of features* is to the representation of data for datasets stored in a non-target table with one-to-many relations.
2. *document* is to the representation of data for text documents as *individual record* is to the representation of data for datasets stored in a non-target table with one-to-many relations.
3. *collection of documents* is to the representation of data for text documents as the *target table* is to the representation of data for datasets stored in a relational database.

Since the representation of data for text documents can be represented as vectors of terms, in our implementation, the *Dynamic Aggregation of Relational Attributes* (*DARA*) system adopts the vector space model to represent data stored in multiple tables with one-to-many relations.

Section II introduces the framework of our data summarisation approach, *DARA*. The data summarisation method employs the *TF-IDF* weighted frequency matrix [14] to represent the relational data model, where the representation of data stored in multiple tables will be analysed and it will be transformed into data representation in a vector space model. Then, section III describes the *DARA* algorithm, which summarises structured data stored in a multi-relational setting, in more detail. Section IV describes the experimental design to demonstrate that the data summarisation method using the proposed *DARA* algorithm can improve the classification task for data stored in multiple tables with one-to-many relations. The performance accuracy of the $J48$ classifier for the classification tasks using these summarised data will be presented and finally, this paper is concluded in section V.

2 The Data Summarization Algorithm

This section defines the framework of the data summarisation algorithm. This algorithm, as shown in Table 1, is based on the concepts of clustering and aggregation. The recursive function $ALGORITHM1 : SUMMARISE$ accepts 4 input parameters $\langle S, RDB, T_T, L_{T_T} \rangle$, where S is the schema of the database, RDB is the relational database, T_T is the target table and L_{T_T} is the list of tables visited. The schema of the database provides information on tables that can be linked to the target table through primary and foreign keys. As a result of the application of the algorithm, the target table will receive the additional fields that represent the result of data summarisation, S_{d_i} from the non-target tables, T_{NT_i} (Line 10 in Table 1). The list of visited tables, L_{T_T}, is used to handle a situation when we have a serial relationship of non-target tables with cycles.

The data summarisation algorithm first propagates the unique identifier label of each individual, stored in the target table T_T, to the other, non-target, tables T_{NT} in a relational database (Line 04 in Table 1). By doing this, all tables can be virtually joined to relate records stored in the target table with records stored in the non-target tables. For each non-target table T_{NT} that can be associated with the target table T_T, it then checks if the non-target table T_{NT} has a one-to-many relationship with another non-target table. If this is the case, recursive function $SUMMARISE$ is called again and this non-target table becomes the target table inserted, as one of the parameters, into the recursive $SUMMARISE$ function (Line 07 in Table 1). The calling of the recursive $SUMMARISE$ function stops when there is no other unvisited non-target tables that can be linked to the current non-target table. It will then perform the data summarisation using the operation called $DARA$ to summarise each non-target relation based on the unique identifier label of each individual, I. The summarised data, S_{d_i}, obtained from each non-target table T_{NT_i} is gathered and a new field is appended to T_T that corresponds to the new summarised data S_{d_i} ($UT_T \longleftarrow T_T$ join S_{d_i}) to represent the result of the newly summarised data.

The $ALGORITHM2 : DARA$, shown in Table 1, accepts two input parameters, a non-target table T_{NT} and a foreign key F_{Key} of T_{NT} that is associated with the primary key of the target table T_T. The $DARA$ algorithm transforms the representation of the data stored in the non-target table from a relational model to a vector space model. The $DARA$ algorithm performs two main tasks to prepare the data for the transformation process. The two tasks in question are data discretisation and feature construction. Next, the $DARA$ algorithm transforms the representation of each individual record stored in the non-target table T_{NT} into a vector of patterns, $I_{feature-vector} \longleftarrow I_{one-to-many}$. Each transformed record will be used to form the $(n \times p)$ TF-IDF weighted frequency matrix [13] to represent n records with p patterns. The notation $V_f \longleftarrow V_f \cup I_{feature-vector}$ in the algorithm performs this task. After the representation of the data has been changed to a vector space format, these records are clustered, $S_d \longleftarrow$ cluster(V_f), as a means of summarising the records stored in the non-target table T_{NT}. The summarised data S_d is then returned to the $SUMMARISE$ function. The DARA algorithm that transforms the data representation of the

Table 1. Dynamic Aggregation of Relational Attributes (DARA) Algorithm

ALGORITHM 1: SUMMARISE(S, RDB, T_T, L_{T_T})
INPUT: S: Database Schema, RDB: Database, T_T: Target Table, L_{T_T}: List of T_T visited
OUTPUT: UT_T: Updated Target Table
$L_{T_T} \leftarrow L_{T_T} \cup T_T$
Find all n non-target tables T_{NT_i} related to T_T
FOR each T_{NT_i} in RDB related to T_T, $T_{NT_i} \neq T_T$ and $i \leq n$
    Propagate identifier labels for all target records I in T_T to T_{NT_i}
FOR each T_{NT_i} in RDB related to T_T, $T_{NT_i} \neq T_T$ and $i \leq n$
  IF T_{NT_i} has further relationship with T_{NT_k}, $i \neq k$ and $T_{NT_k} \notin L_{T_T}$
    T_{NT_i} = SUMMARISE(S, RDB, T_{NT_i}, L_{T_T})
  Find F_{Key_i}; F_{Key} is the foreign key in T_{NT_i}
  $S_{d_i} \longleftarrow DARA(T_{NT_i}, F_{Key_i})$
  $UT_T \longleftarrow T_T$ join S_{d_i}
RETURN UT_T
ALGORITHM 2: DARA(T_{NT}, F_{Key})
INPUT: T_{NT}: Non-target Table ($T_{NT} \neq T_T$), F_{Key}. **OUTPUT**: Sd: Summarised data
$V_f \longleftarrow \emptyset$; Initialise feature vector matrix, V_f
$S_d \longleftarrow \emptyset$; Initialise summarised data, S_d
prepare-data(T_{NT}) ; Discretisation and feature construction
FOR a set of records that uses F_{Key} representing I in T_{NT}
    $I_{feature-vector} \longleftarrow I_{one-to-many}$;
    $V_f \longleftarrow V_f \cup I_{feature-vector}$
RETURN $S_d \longleftarrow$ cluster(V_f)

data stored in the non-target table T_{NT_i} is explained in greater detail in the next subsection.

3 Dynamic Aggregation of Relational Attributes (DARA)

There are three main stages in the proposed data summarisation approach: *data preparation*, *data transformation* and *data clustering*.

3.1 Data Preparation Stage

The *DARA* algorithm performs two types of data preparation process, that include features discretisation and features construction [8,7] processes. Discretisation is a process of transforming continuous-valued features to nominal. Alfred and Kazakov have discussed the discretisation methods for continuous attributes in a multi-relational setting in greater detail [3]. Alfred has also discussed how feature construction affects the performance of the proposed *DARA* method in summarising data stored in a multi-relational setting with one-to-many relations [1].

3.2 Data Transformation Stage

At this stage, the representation of data stored in a relational database is changed to a vector space data representation.

Data Transformation Process

In a relational database, a single record, R_i, stored in the target table can be associated with other records stored in the non-target table. Let R denote a set of m records stored in the target table and let S denote a set of n records $(T_1, T_2, T_3, ..., T_n)$, stored in the non-target table. Let S_i be a subset of S, $S_i \subseteq S$, associated through a foreign key with a single record R_a stored in the target table, where $R_a \in R$. Thus, the association of these records can be described as $R_a \longleftarrow S_i$. In this case, we have a single record stored in the target table that is associated with multiple records stored in the non-target table. The records stored in the non-target table that correspond to a particular record stored in the target table can be represented as vectors of patterns. As a result, based on the vector space model [13], a unique record stored in non-target table can be represented as a vector of patterns. In other words, a particular record stored in the target table that is related to several records stored in the non-target table can be represented as a *bag of patterns*, i.e., by the patterns it contains and their frequency, regardless of their order. The *bag of patterns* is defined as follows:

Definition 3.1. In a *bag of patterns* representation, each target record stored in the non-target table is represented by the set of its pattern and the pattern frequencies.

The process of encoding these patterns into binary numbers depends on the number of attributes that exist in the non-target table. For example, there are two different cases when encoding patterns for the data stored in the non-target table. In the first case (case I), a non-target table may have a single attribute. In this case, the $DARA$ algorithm transforms the representation of the data stored in a relational database without constructing any new feature to build the $(n \times p)$ TF-IDF weighted frequency matrix, as only one attribute exists in the non-target table. In the other case (case II), it is assumed that there is more than one attribute that describes the contents of the non-target table associated with the target table. The $DARA$ algorithm may construct new features [2], which results in more riched representation of each target record in the non-target table. All continuous values of the attributes are discretised and the number of bins is taken as the cardinality of the attribute domain. After encoding the patterns as binary numbers, the algorithm determines a subset of the attributes to be used to construct a new feature.

Data Representation in a Vector Space Model

In this subsection, we describe the representation of data for records stored in multiple tables with one-to-many relations. Let DB be a database consisting of n records. Let $P := P_1,...,P_m$ be the set of different patterns existing for the ith target record O_i in DB and let each target record O_i has zero or more occurrences of each pattern P_i, such that $|P_i| \geq 0$, where $i = 1, ..., m$. Each target record $O_i \in DB$, where $i = 1, ..., n$ can be described by maximally m different patterns where each pattern having its frequency,

$$O_i := P_1(O_i) : |P_1(O_i)| : |O_b(P_1)|, ..., P_m(O_i) : |P_m(O_i)| : |O_b(P_m)| \quad (1)$$

where $P_j(O_i)$ represents the jth pattern of the ith target record and $|P_j(O_i)|$ represents the frequency of the jth pattern of the ith target record, and finally $|O_b(P_j)|$ represents the number of target records exist in the data that have the jth pattern. If all different patterns exist for O_i, then the total different patterns for O_i is $|O_i| = m$ else $|O_i| < m$. In the $DARA$ approach, the vector-space model [13] is applied to represent each target record. In this model, each target record O_i is considered as a vector in the pattern-space. In particular, the TF-IDF weighted frequency matrix borrowed from [13] is employed, in which each target record O_i, $i = 1, ..., n$ can be represented as

$$\left(rf_1 \cdot \log_2 \left(\frac{n}{of_1} \right), ..., rf_m \cdot \log_2 \left(\frac{n}{of_m} \right) \right) \quad (2)$$

where rf_j is the frequency of the jth pattern in the target record, of_j is the number of target records that contain the jth pattern and n is the number of target records.

3.3 Data Clustering Stage

After transforming the dataset, the newly transformed data (in the form of vector space model) is taken as an input to a clustering algorithm. The idea of this approach is to transform the data representation for all target records in a multi-relational environment into a vector space model and find the similarity distance measures for all target records in order to cluster them. These target records are then grouped based on the similarity of their characteristics, taking into account all possible representations and the frequency of each representation for all target records.

4 Experiments with Data Summarisation

After summarising the data stored in the non-target tables, the actual data mining takes place in the modeling phase, based on identified goals and the assessment of the available data, an appropriate mining algorithm is chosen and run on the prepared data. In this work, the summarised data are coupled with the $C4.5$ classifier (from $WEKA$) [17], as the induction algorithms that are run on the $DARA$'s transformed data representation. Then the effectiveness of the data transformation with respect to $C4.5$ is evaluated. The $C4.5$ learning algorithm [12] is a top-down method for inducing decision trees. These experiments are designed to investigate two main factors:

1. The feasibility of using data summarisation to support the data mining task (e.g. classification) in a multiple tables environment.
2. The performance of the *DARA* algorithm compared to other relational data mining approaches including Progol [15], Tilde [4], Foil [6], RDBC [9], RElaggs [11].

These experiments use datasets from the Mutagenesis database (*B*1, *B*2, *B*3) [15], Financial database (Discovery Challenge PKDD 1999) and Hepatitis database (PKDD 2005). Table 2 shows the comparison between the results obtained in these experiments and the other previously published results on Mutagenesis and Financial datasets, such as *Progol* [15], *Foil* [6], *Tilde* [4], *RDBC* [9] and *RElaggs* [11]. Referring to table 2, the DARA algorithm produces better results compared to the other approaches on relational data mining.

Table 2. Results Previously Published of Mutagenesis (*B*1, *B*2, *B*3) and Financial (PKDD 1999) Datasets

		Mutagenesis		
Setting	Financial	B1	B2	B3
DARA	**93.2**	**88.8**	**88.3**	**88.7**
Progol	-	76.0	81.0	83.0
Foil	74.0	83.0	75.0	83.0
Tilde	81.3	75.0	75.0	85.0
RDBC	-	83.0	84.0	82.0
RElaggs	99.9	86.7	87.8	86.7

In short, some of the findings that can be concluded from these experiments are outlined as follows: (1) Data summarisation for multiple tables with a high number of one-to-many relationship is feasible in order to get higher accuracy estimations. (2) Without considering the class information, the *DARA* algorithm produced higher accuracy estimation results compared to the other relational data mining approaches.

5 Conclusions

This paper introduced the concept of data summarisation that adopts the *TF-IDF* weighted frequency matrix concept borrowed from the information retrieval theory [13] to summarise data stored in relational databases with a high number of one-to-many relationships among entities, through the use of a clustering technique. Clustering algorithms can be used to generate summaries based on the information contained in the datasets that are stored in a multi-relational environment. This paper also outlined an algorithm called *Dynamic Aggregation of Relational Attributes* (*DARA*) that transforms the representation of data stored in relational databases into a vector space format data representation that is suitable in clustering operations. By clustering these multi-association occurrences of an individual record in the multi-relational database, the characteristics of records stored in non-target tables are summarised by putting them into groups that share similar characteristics. In this paper, experimental results show that using the *DARA* algorithm finds solutions with much greater accuracy. The experiments show that data summarisation improves the performance accuracy of the prediction task. By clustering these records based on the multi-instances that are related to them, the records can be summarised by putting them into groups that share similar characteristics.

References

1. Alfred, R.: A genetic-based feature construction method for data summarisation. In: Tang, C., Ling, C.X., Zhou, X., Cercone, N.J., Li, X. (eds.) ADMA 2008. LNCS, vol. 5139, pp. 39–50. Springer, Heidelberg (2008)
2. Alfred, R., Kazakov, D.: Clustering Approach to Generalised Pattern Identification Based on Multi-Instanced Objects with DARA. In: Ioannidis, Y., Novikov, B., Rachev, B. (eds.) ADBIS 2007. LNCS, vol. 4690. Springer, Heidelberg (2007)
3. Alfred, R., Kazakov, D.: Discretisation Numbers for Multiple-Instances Problem in Relational Database. In: Ioannidis, Y., Novikov, B., Rachev, B. (eds.) ADBIS 2007. LNCS, vol. 4690, pp. 55–65. Springer, Heidelberg (2007)
4. Blockeel, H., Raedt, L.D.: Top-Down Induction of First-Order Logical Decision Trees. Artif. Intell. 101(1-2), 285–297 (1998)
5. Blockeel, H., Sebag, M.: Scalability and Efficiency in Multi-Relational Data Mining. In: SIGKDD Explorations, vol. 5(1), pp. 17–30 (2003)
6. Finn, P.W., Muggleton, S., Page, D., Srinivasan, A.: Pharmacophore Discovery Using the Inductive Logic Programming System PROGOL. Machine Learning 30(2-3), 241–270 (1998)
7. Guyon, I., Elisseeff, A.: An Introduction to Variable and Feature Selection. Journal of Machine Learning Research 3, 1157–1182 (2003)
8. Jain, A.K., Duin, R.P.W., Mao, J.: Statistical Pattern Recognition: A Review. IEEE Transactions on Pattern Analysis and Machine Intelligence 22(1), 4–37 (2000)
9. Kirsten, M., Wrobel, S.: Relational Distance-Based Clustering. In: 8th International Conference on Inductive Logic Programming, pp. 261–270 (1998)
10. Kramer, S., Lavrac, N., Flach, P.: Propositionalisation Approaches to Relational Data Mining. In: Deroski, S., Lavrac, N. (eds.) Relational Data mining. Springer, Heidelberg (2001)
11. Krogel, M.A., Wrobel, S.: Transformation-Based Learning Using Multirelational Aggregation. In: Rouveirol, C., Sebag, M. (eds.) ILP 2001. LNCS, vol. 2157, pp. 142–155. Springer, Heidelberg (2001)
12. Quinlan, R.J.: C4.5: Programs for Machine Learning. Morgan Kaufmann Series in Machine Learning (1993)
13. Salton, G., McGill, M.: Introduction to Modern Information Retrieval. McGraw-Hill Book Company, New York (1984)
14. Salton, G., Wong, A., Yang, C.S.: A Vector Space Model for Automatic Indexing. Commun. ACM 18(11), 613–620 (1975)
15. Srinivasan, A., Muggleton, S., Sternberg, M.J.E., King, R.D.: Theories for Mutagenicity: A Study in First-Order and Feature-Based Induction. Artif. Intell. 85(1-2), 277–299 (1996)
16. van Rijsbergen, C.J.: Information Retrieval. Butterworth (1979)
17. Witten, I.H., Frank, E.: Data Mining: PracticalMachine Learning Tools and Techniques with Java Implementations. Morgan Kaufmann, San Francisco (1999)

A Secure Protocol to Maintain Data Privacy in Data Mining

Fodé Camara[1], Samba Ndiaye[1] and Yahya Slimani[2]

[1] Cheikh Anta Diop University, Department of Mathematics and Computer Science, Dakar, Senegal
fode.camara@ucad.edu.sn, ndiayesa@ucad.sn
[2] Department of Computer Science, Faculty of Sciences of Tunis, 1060 Tunis, Tunisia
yahya.slimani@fst.rnu.tn

Abstract. Recently, privacy issues have becomes important in data mining, especially when data is horizontally or vertically partitioned. For the vertically partitioned case, many data mining problems can be reduced to securely computing the scalar product. Among these problems, we can mention association rule mining over vertically partitioned data. Efficiency of a secure scalar product can be measured by the overhead of communication needed to ensure this security. Several solutions have been proposed for privacy preserving association rule mining in vertically partitioned data. But the main drawback of these solutions is the excessive overhead communication needed for ensuring data privacy. In this paper we propose a new secure scalar product with the aim to reduce the overhead communication.

1 Introduction

Motivated by the multiple requirements of data sharing, privacy preserving and knowledge discovery, Privacy Preserving Data Mining (PPDM) has been studied extensively in data mining community. In many cases, multiple parties may wish to share aggregate private data, without leaking any sensitive information at their end. For example, different superstores with sensitive sales data may wish to coordinate among themselves in knowing aggregate trends without leaking the trends of their individual stores. This requires secure and cryptographic protocols for sharing the information across the different parties. The data may be distributed in two ways across different sites: horizontal partitioning where the different sites may have different sets of records containing the same attributes; and vertical partitioning where the different sites may have different attributes of the same sets of records. For the vertically partitioned case, many primitive operations such as computing the scalar product can be useful in computing the results of data mining algorithms. For example, [1] uses the secure scalar product over the vertical bit representation of itemset inclusion in transactions, in order to compute the frequency of the corresponding itemsets. This key step is applied repeatedly within the framework of a roll up procedure of itemset counting. The efficiency of secure scalar products is important because they can

R. Huang et al. (Eds.): ADMA 2009, LNAI 5678, pp. 417–426, 2009.

generate excessive overhead communication. Several works have been carried out on privacy preserving association rule mining for vertically partitioned data. Most of them suffer from communication overhead. We challenge this drawback by proposing a new secure scalar product protocol that reduce drastically the communication cost between sites. This feature distinguishes our proposal from the existing ones.

The remainder of this paper is organized as follows. Section 2 provides the state-of-the-art in privacy preserving data mining. In Section 3, we present our proposed approach and we give examples to illustrate the practicability of our protocol. Evaluation of computation and communication costs of our protocol is presented in Section 4. Section 5 discusses the security aspect of our proposal. Section 6 evaluates this work by comparing it with related topics in the privacy preserving association rule mining community. Finally, Section 7 concludes with a discussion of the contributions of our proposal and our current research plans.

2 Privacy and Data Mining

The problem of privacy preserving data mining has become more important in recent years because of the multiple requirements of data sharing, privacy preserving and knowledge discovery. Two main problems are addressed in privacy preserving data mining: the first one is the protection of sensitive raw data; the second one is the protection of sensitive knowledge contained in the data, which is called knowledge hiding in database. We can classify the techniques of knowledge hiding in database in two groups: approaches based on data modification and the approaches based on data reconstruction. The basic idea of data reconstruction is to modify directly the original database D, to obtain a new database D'. According to the way of modifying the original database we can still push classification by distinguishing two families of techniques: techniques based on the distortion and the techniques based on the blocking of the data. The distortion changes a value of attribute by a new value [2] (i.e. change of value 1 in 0), while blocking [3], is the replacement of an existing value of attribute by a special value noted by "?". However approaches based on data modification cannot control the side effects. For example in association rule, hiding a non sensitive rule R_i witch had confidence bigger than threshold $minconf$ (i.e. $conf(R_i) > minconf$), can give a new value such that $conf(R_i) < minconf$. They also make many operations of I/O especially when the original database is too large. Another solution concerns the approaches based on data reconstruction [4]. The basic idea of these approaches is to extract first the knowledge K from the original database D. The new database D' is then reconstructed from K. The basic idea of data reconstruction is inspired by the recent problem of *Inverse Frequent Set Mining*. Opposite approaches use techniques based on data reconstruction that control directly the side effects. The main proposal to solve the problem of the protection of sensitive raw data is the secure multi-party computation. A Secure Multi-party Computation (SMC) problem deals with computing any function on any input, in a distributed network where each participant holds one of the inputs, while ensuring that no more information is revealed to a participant in the computation than can be inferred from that participants input and output. Secure two party computation was first investigated by Yao [5,6] and was later generalized to multi-party computation [7,8]. For example, in a 2-party setting, Alice and Bob may have two inputs x and y, and may wish to both compute the function $f(x, y)$ without revealing x or y to each other. This problem can also be generalized across k parties by designing the k argument function $h(x_1, ..., x_k)$. Several data mining algorithms may be viewed in the context of repetitive computations of many such primitive functions like the *scalar product*, *secure sum*, and so on. In order to compute the function $f(x, y)$ or $h(x_1, ..., x_k)$ a protocol will have to designed for exchanging information in such a way that the function is computed without compromising privacy. The problem of distributed privacy preserving data mining overlaps closely with a field in cryptography for determining secure multi-party computations. A broad overview of the intersection between the fields of cryptography and privacy-preserving data mining may be found in [9]. Clifton et al. [10] give a survey of multi-party computation methods.

3 Proposed Approach

3.1 Problem Definition

Scalar product is a powerful component technique. Several data mining problems can essentially be reduced to securely computing the scalar product. To give an idea of how a secure scalar protocol can be used, let us look at association rule mining over vertically distributed data. The association rule mining problem can be formally stated as follows [11]: Let $I = \{i_1, ..., i_n\}$ be a set of items. Let D be a set of transactions, where each transaction T has an unique identifier TID and contains a set of items, such that $T \subseteq I$. We say that a transaction T contains X, a set of items in I, if $X \subseteq T$. An association rule is an implication of the form $X \Rightarrow Y$, where $X \subseteq I$, $Y \subset I$, and $X \cap Y = \emptyset$. The rule $X \Rightarrow Y$ holds in a database D with confidence c, if $c\%$ of transactions in D that contain X tend to also contain Y. The association rule $X \Rightarrow Y$ has support s in D, if $s\%$ of transactions in D contain $X \cup Y$. Within this framework, we consider mining boolean association rules. The absence or presence of an attribute is represented by a value taking from $\{0, 1\}$. Transactions are represented under the form of strings of 0 and 1, while the database can be represented as a matrix of $\{0, 1\}$. The association rule mining algorithm over vertically distributed databases is based on the classic Apriori algorithm of Agrawal and Srikant [12]. The key issue of this algorithm is the problem of finding the frequent itemsets in a database. To determine if a given itemset is frequent or not, we count the number of transactions, in D, where the values for all the attributes in this itemset are 1. This problem can be transformed into a simple mathematical problem, using the following definitions: Let $l + m$ be the total number of attributes, where Alice has l attributes, $\{A_1, ..., A_l\}$, and Bob has the remaining m attributes, $\{B_1, ...B_m\}$. Transactions are a sequence of $l + m$ $1's$ or $0's$. Let $minsupp$ be the minimal support, and n the total number of transaction in database D. Let $\overrightarrow{X}$ and $\overrightarrow{Y}$ be the columns in D, i.e., $x_i = 1$ if row i has value 1 for item or attribute X. The scalar product of two vectors $\overrightarrow{X}$ and $\overrightarrow{Y}$ of cardinality n is defined as follows: $\overrightarrow{X} \bullet \overrightarrow{Y} = \sum_{i=1}^{n} x_i \times y_i$. Determining if the 2-itemset $< XY >$ is frequent can be reduced to test if $\overrightarrow{X} \bullet \overrightarrow{Y} \geq minsupp$. The generalization of this process to a w-itemset is straightforward. Assume Alice has p attributes $a_1, ..., a_p$ and Bob has q attributes $b_1, ..., b_q$. We want to compute the frequency of the w-itemset $< a_1...a_p, b_1...b_q >$, where $w = p + q$. Each item in $\overrightarrow{X}$ (respectively in $\overrightarrow{Y}$) is composed of the product of the corresponding individual elements, i.e., $x_i = \prod_{j=1}^{p} a_j$ (respectively $y_i = \prod_{j=1}^{q} b_j$).

Now, we can formalize our problem as follows: Assume that 2 parties, for example Alice and Bob, such that each has a binary vector of cardinality n, e.g. $\overrightarrow{X} = (x_1, ..., x_n)$ and $\overrightarrow{Y} = (y_1, ..., y_n)$. The problem is to securely compute the scalar product of these two vectors, e.g. $\overrightarrow{X} \bullet \overrightarrow{Y} = \sum_{j=1}^{n} x_i \times y_i$.

3.2 Security Tools

To define our secure scalar product protocol, we have to use a semantically secure additive homomorphic public-key cryptosystem. Indeed to ensure security in data transmission, we chose a public-key cryptosystem. The security of a public-key cryptosystem is determined by a security parameter k. For a fixed k, it should take more than polynomial in k operations to break the cryptosystem [13](Section 2). The public-key cryptosystem that have chosen is homomorphic. This choice is justified by the fact that, if given $Enc(x)$ and $Enc(y)$, one can obtain $Enc(x \perp y)$ without decrypting x, y for some operation $\perp$. Furthermore, the homomorphic public-key cryptosystem is additive; that means that a party can add encrypted plaintexts by doing simple computations with ciphertexts, without having the secret key. Informally, this means that for probabilistic polynomial-time adversary, the analysis of a set of ciphertexts does not give more information about the cleartexts than what would be available without knowledge of the ciphertexts. This property is very important in our binary context because an attacker could always encrypt 0 and 1 by using the public key, and then compare the resulting ciphertexts with the received message to decide the value of a bit. One of the most efficient currently known semantically secure homomorphic cryptosystems is Paillier cryptosystem [14]. This cryptosystem has all the four properties described above. In Paillier's case the plaintext space is defined by $P(sk) = Z_N$, with $N \geq 2^{1024}$, e.g. N is a hard-to-factor modulus.

3.3 Algorithm

Before to give our proposed algorithm, we will suppose the following. First, Alice generates a pair of key and a random value r and it computes $Enc_{pk}(x_i, r)$ which she sends to Bob. This message is computationally indistinguishable from the received message since the semantic security of the encryption system guarantees that no extra information is revealed. She sends also the public key to Bob. Bob

Algorithm 1. Private Scalar Product Protocol

Require: N=2 (number of sites; Alice and Bob), Alice vector: $\overrightarrow{X} = (x_1, ..., x_n)$, Bob vector: $\overrightarrow{Y} = (y_1, ..., y_n)$

1: **for** Alice **do**
2: Generates a pair of key (sk, pk);
3: Generates $(Enc_{pk}(x_1), ..., Enc_{pk}(x_n))$ using the public key pk;
4: Sends $(Enc_{pk}(x_1), ..., Enc_{pk}(x_n))$ to Bob;
5: **end for**
6: **for** Bob **do**
7: Computing $\prod_{i=1}^{n} p_i$, where $p_i = Enc_{pk}(x_i)$ if $y_i = 1$ and $p_i = 1$ if $y_i = 0$ then uses the additive property of homomorphic encryption for computing $\prod_{i=1}^{n} p_i = Enc_{pk}(\overrightarrow{X} \bullet \overrightarrow{Y})$;
8: Sends $Enc_{pk}(\overrightarrow{X} \bullet \overrightarrow{Y})$ to Alice;
9: **end for**
10: **for** Alice **do**
11: Computes $Dec_{sk}(Enc_{pk}(\overrightarrow{X} \bullet \overrightarrow{Y}))$;
12: Sends the final result to Bob;
13: **end for**

uses the additive property of homomorphic encryption and computes $\prod_{i=1}^{n} p_i$, with $p_i = Enc_{pk}(x_i)$ if $y_i = 1$ otherwise $p_i = 1$. A public-key cryptosystem is additive homomorphic when $Enc_{pk}(x_1, r_1) \times Enc_{pk}(x_2, r_2) \times ... \times Enc_{pk}(x_n, r_n) = Enc_{pk}(x_1 + x_2 + ... + x_n, r_1 \times r_2 \times ... \times r_n)$, where $+$ is a group operation and $\times$ is a groupoid operation. For the sake of simplicity of notations, we will not explicitly include, in the rest of the paper, the randomness as an input of the encryption functions. Then, in step 5, Bob sends $Enc_{pk}(\overrightarrow{X} \bullet \overrightarrow{Y})$ to Alice. Having the secret key, Alice deciphers the final result what she sends to Bob. Now we describe our proposed algorithm.

To illustrate the behavior of our algorithm, we consider the following first scenario: we suppose that Alice has the vector $\overrightarrow{X} = (1, 0, 0, 1)^T$ and Bob has $\overrightarrow{Y} = (1, 0, 0, 1)^T$. We want to compute securely the scalar product of $\overrightarrow{X} \bullet \overrightarrow{Y}$. We obtain the following:

1. Alice view's: Alice generates a pair of key (sk, pk); to do that it uses the public key pk and generates $(Enc_{pk}(x_1), Enc_{pk}(x_2), Enc_{pk}(x_3), Enc_{pk}(x_4))$, what she sends to Bob.
2. Bob view's: Bob computes $\prod_{i=1}^{n} p_i$, where $p_1 = Enc_{pk}(x_1)$ (because $y_1 = 1$), $p_2 = 1$, $p_3 = 1$ and $p_4 = Enc_{pk}(x_4)$. He uses the additive property of homomorphic encryption to compute $Enc_{pk}(x_1) \times Enc_{pk}(x_4) = Enc_{pk}(x_1 + x_4) = Enc_{pk}(\overrightarrow{X} \bullet \overrightarrow{Y})$. Finally Bob sends $Enc_{pk}(\overrightarrow{X} \bullet \overrightarrow{Y})$ to Alice.
3. Alice view's: Using the secret key, Alice computes $Dec_{sk}(Enc_{pk}(\overrightarrow{X} \bullet \overrightarrow{Y}))$ and sends the result to Bob.

As second scenario, we suppose that Alice has the vector $\overrightarrow{X} = (1, 1, 1, 1)^T$ and Bob has $\overrightarrow{Y} = (1, 1, 0, 1)^T$. As for the first scenario, We want to compute securely the scalar product of $\overrightarrow{X} \bullet \overrightarrow{Y}$. This computation gives:

1. Alice view's: Using the public key pk, Alice generates a pair of key (sk, pk); then it generates $(Enc_{pk}(x_1), Enc_{pk}(x_2), Enc_{pk}(x_3), Enc_{pk}(x_4))$, what she sends to Bob.
2. Bob view's: Bob computes $\prod_{i=1}^{n} p_i$, where $p_1 = Enc_{pk}(x_1)$ (because $y_1 = 1$), $p_2 = Enc_{pk}(x_2)$, $p_3 = 1$ and $p_4 = Enc_{pk}(x_4)$. Using the additive property of homomorphic encryption, it computes $Enc_{pk}(x_1) \times Enc_{pk}(x_2) \times Enc_{pk}(x_4) = Enc_{pk}(x_1 + x_2 + x_4) = Enc_{pk}(\overrightarrow{X} \bullet \overrightarrow{Y})$. Finally, Bob sends $Enc_{pk}(\overrightarrow{X} \bullet \overrightarrow{Y})$ to Alice.
3. Alice view's: Using the secret key, Alice computes $Dec_{sk}(Enc_{pk}(\overrightarrow{X} \bullet \overrightarrow{Y}))$ and sends the result to Bob.

In this second scenario, although Alice has a constant vector, Bob cannot deduce in no manner this one. Semantic security guarantees that it is not possible for Bob to distinguish between the encryption of a 0 or a 1 value, when r is randomly chosen.

4 Computation and Communication Evaluation

From the communication view point, our protocol needs the following messages: (i) for each entry of the vector, our protocol requires one message; (ii) one message to send the public key; (iii) Bob needs to send $Enc_{pk}(\overrightarrow{X} \bullet \overrightarrow{Y})$ to Alice; (iv) finally, Alice must send the result of the scalar product to Bob. Hence, the number of messages is $n+3$, where n is the dimension of the vector. In this case we obtain a total communication cost $O(n)$. From the computation view point, Alice performs, in our protocol, n encryptions and 1 decryption. Bob performs less $n-1$ additions. Therefore the computational complexity of our protocol is linear, e.g. $O(n)$.

5 Security Analysis

In this section, we will verify the security of our protocol. This security depends on the semantic secure public-key cryptosystem used. A public-key cryptosystem is semantically secure (IND-CPA secure) when a probabilistic polynomial-time adversary cannot distinguish between random encryptions of two elements, chosen by herself. Paillier [14] recalls the standard notion of security for public key encryption schemes in terms of indistinguishability, or semantic security as follows: We consider chosen-plaintext attacks (CPA), because homomorphic schemes can never achieve security against chosen-ciphertext attacks. To define security, we use the following game that an attacker A plays against a challenger:

$$
\begin{aligned}
(pk, sk) &\leftarrow KG(.) \\
(St, m_0, m_1) &\leftarrow A(find, pk) \\
b \leftarrow \{0,1\} \ at\ random; c^* &\leftarrow Enc_{pk}(m_b) \\
b' &\leftarrow A(guess, c^*, St).
\end{aligned}
$$

The advantage of such an adversary A is defined as follows:

$$Adv(A) = \mid Pr[b' = b] - \tfrac{1}{2} \mid .$$

A public key encryption scheme is said to be $\varepsilon - indistinguishable$ under CPA attacks if $Adv(A) < \varepsilon$ for any attacker A which runs in polynomial time.

From this definition, it is quite obvious that the role of the randomness r is crucial to ensure (semantic) security of a public key encryption scheme. In effect, a deterministic scheme can never be semantically secure, because an attacker A could always encrypt m_0 and m_1 by using pk, and then compare the resulting ciphertexts with the challenge one c^*, to decide the value of the bit b.

At this step, we will now give a proof of security for the entire protocol.

Proof. To analyze security let us examine the information propagated by each site taking part in the protocol. All propagated informations in our protocol can be summarized as follows:

1. Alice's view: In steps 1, 2 and 3, for each $x_i, i \in \{1..n\}$, Alice generates a random number r and computes $Enc_{pk}(r, x_i)$; the result of this computation is sent to Bob.
2. Bob's view: Bob receives $Enc_{pk}(x_i, r)$. This message is computationally indistinguishable from the received message since the semantic security of the public key encryption system guarantees that no extra information is revealed [14].
3. In step 4, Bob computes $Enc_{pk}(\overrightarrow{X} \bullet \overrightarrow{Y}) = \prod Enc_{pk}(x_i)$ if $y_i = 1$. The homomorphic property of encryption system guarantees that this computation does not reveal the values of $x_{i \in \{1..n\}}$.
4. Then, in step 5, Bob sends $Enc_{pk}(\overrightarrow{X} \bullet \overrightarrow{Y})$ to Alice. The security of the remaining stages is not important because the result of the scalar product is not private.

6 Related Works

Let us reconsider the most popular secure scalar product in order to compare them with our protocol. The protocol in [1] is an algebraic solution which uses a matrix of decision of dimension $n \times n/2$, where n is the dimension of the vectors. Each part encrypts its vector using the matrix of decision. In first step, the primary site namely Alice, generates $n/2$ random values and performs n^2 multiply and addition operations. The responder site, namely Bob, performs $n^2 + 2$ multiply and addition operations. Finally Alice performs $n/2$ multiply and addition operations to obtain the final result. The protocol in [1] thus has a computational complexity of $O(n^2)$. The communication cost is $O(n)$, i.e. $3n/2+2$ messages if we consider that each entry of the vector requires a message. Therefore, the communication overhead is $n/2+1$. In [15], the authors proposed a secure scalar product protocol based on relation $2\sum_{j=1}^{n} x_i y_i = \sum_{j=1}^{n} x_i^2 + \sum_{j=1}^{n} y_i^2 - \sum_{j=1}^{n} (x_i - y_i)^2$. In this protocol each site needs n messages towards the other for send his vector. Bob needs to send $\sum_{j=1}^{n} y_i^2$, whereas Alice sends the result to Bob. In this case the communication cost is $2n+2$ and the communication overhead is $n+1$ messages. In order to determine the computational overhead, we performed the following analysis. Alice will compute $\sum_{j=1}^{n} x_i^2$, and Bob will compute $\sum_{j=1}^{n} y_i^2$ locally (i.e. $2n-2$ additions and $2n$ multiply operations). Then, we execute the protocol *Add Vectors Protocol* which requires n permutations of values and n encryptions. Finally, the computational complexity is $O(n)$. If Alice has a constant vector (i.e made up only of 1 values), Bob could deduce the values from Alice. To solve this problem the authors in [15] propose that Alice generates a random vector $\overrightarrow{r}$ that is added to her random vector $\overrightarrow{X}$. Thus, even if Alice has a constant vector $\overrightarrow{X}$, the fact that $\overrightarrow{r}$ was randomly generated will ensure that $(\overrightarrow{X}+\overrightarrow{r})$ will be always a vector with non-equal values. Now, they can use the scalar product protocol twice for computing $(\overrightarrow{X}+\overrightarrow{r}) \bullet \overrightarrow{Y}$ and $\overrightarrow{r} \bullet \overrightarrow{Y}$. Then, Alice can obtain $\overrightarrow{X} \bullet \overrightarrow{Y} = (\overrightarrow{X}+\overrightarrow{r}) \bullet \overrightarrow{Y} - \overrightarrow{r} \bullet \overrightarrow{Y}$ and send it to Bob. This solution obviously will double the cost of the scalar product protocol, but according to the authors it still will be efficient than all existing ones. In [8], Du and Atallah propose a protocol called *Permutation Protocol.* This last uses an additive homomorphic encryption as in [15] and our protocol. In first stage Alice generates a pair of key (sk, pk) and crypt its vector with the public key pk. Then it sends its n encrypted values and the public key to Bob; this phase needs $n+1$ messages. Using the public key it crypt its vector, then uses the property of additive homomorphism to compute $Enc_{pk}(\overrightarrow{X}) \times Enc_{pk}(\overrightarrow{Y}) = Enc_{pk}(\overrightarrow{X}+\overrightarrow{Y})$. It permutes the entries and sends $\sigma(Enc_{pk}(\overrightarrow{X}+\overrightarrow{Y}))$ to Alice (σ represents the permutation operation). Site Alice performs $Dec_{sk}(\sigma(Enc_{pk}(\overrightarrow{X}+\overrightarrow{Y})))$ to obtain the result, then sends it to Bob. The total number of messages exchanged in this protocol is thus $2n+2$ messages. If we note by *encrypted_msg_time* the time to encrypt a message, *decrypted_msg_time* the time to decrypt a message and *permutation_time* the time to permute the entries, we have $2n \times encrypted\_msg\_time + decrypted\_msg\_time + permutation\_time = O(n)$. Now let us compare our protocol with the protocols described above. For better analyzing the efficiency of our protocol, we also compare it with $DNSP$ that is the scalar product computation without privacy constraint. In the $DNSP$ model, Bob would send his vector to Alice to compute the scalar product and Alice will need to send the result to Bob. This process requires $n+1$ messages. $DNSP$ requires n multiply and $n-1$ addition operations. Hence, the computational complexity is $O(n)$. Table 1 summarizes comparisons between our protocol and four other ones.

The table 1 highlights the differences between our protocol and four other ones under three metrics: computational complexity, communication cost and the communication overhead. If the computational complexity does not constitute

	DNSP	[8]	[1]	[15]	Our protocol
Communication cost	$n+1$	$2n+2$	$3n/2+2$	$2n+2$	$n+3$
Communication overhead	0	$n+1$	$n/2+1$	$n+1$	2
Computational complexity	$O(n)$	$O(n)$	$O(n^2)$	$O(n)$	$O(n)$

Fig. 1. Communication overhead and computational complexity

really a problem because there are several parallel architectures and data mining algorithms, communications will create a bottleneck that can decrease drastically the overall performance of a data mining process. In table 1 we observe that our protocol has a communication overhead of 2. Our protocol needs $n+3$, while $DNSP$ model needs $n+1$ without privacy. Table 1 shows clearly that our protocol outperforms of all other protocols.

7 Conclusion and Future Works

The secure scalar product is a very important issue in privacy preserving data mining. Recently, several protocols have been proposed to solve this problem. The evaluation of these protocols shows that they generate an important communication overhead. In this paper, we proposed a private scalar product protocol based on standard cryptographic techniques. This proposal has two features: (i) it is secure; (ii) it reduces the amount of communication between sites. The current version of this protocol uses only binary contexts. In the future, we plan to extend this protocol for numerical data and different kinds of databases (dense and sparse databases).

References

1. Vaidya, J.S., Clifton, C.: Privacy preserving association rule mining in vertically partitioned data. In: The Eighth ACM SIGKDD International Conference on Knowledge Discovery and Data Mining, Edmonton, Alberta, Canada, July 23-26, pp. 639–644 (2002)
2. Oliveira, S.R.M., Zaiane, O., Saygin, Y.: Secure Association Rule Sharing. In: Dai, H., Srikant, R., Zhang, C. (eds.) PAKDD 2004. LNCS, vol. 3056, pp. 74–85. Springer, Heidelberg (2004)
3. Saygin, Y., Verykios, V., Clifton, C.: Using Unknowns to prevent discovery of Association Rules. ACM SIGMOD Record 30(4) (2001)
4. Guo, Y.H., Tong, Y.H., Tang, S.W., Yang, D.Q.: Knowledge hiding in database. Journal of Software 18(11), 2782–2799 (2007)
5. Yao Andrew, C.C.: Protocols for secure computations. In: Proc. of the 23rd Annual IEEE Symposium on Foundations of Computer Science, Chicago, Illinois, November 1982, pp. 160–164 (1982)
6. Yao, A.C.C.: How to generate and exchange secrets. In: Proc. of the 27th Symposium on Foundations of Computer Science (FOCS), Toronto, Canada, October 1986, pp. 162–167 (1986)
7. Goldreich, O.: Secure multi-party computation - working draft (2000), `http://citeseer.ist.psu.edu/goldreich98secure.html`
8. Du, W., Atallah, M.J.: Secure multi-party computation problems and their applications: a review and open problems. In: New Security Paradigms Workshop, Cloudcroft, New Mexico, September 2001, pp. 11–20 (2001)
9. Pinkas, B.: Cryptographic Techniques for Privacy-Preserving Data Mining. ACM SIGKDD Explorations 4(2) (2002)
10. Clifton, C., Kantarcioglu, M.: Tools for privacy preserving distributed data mining. SIGKDD Explorations 4(2), 28–34 (2003)

11. Agrawal, R., Imielinski, T., Swami, A.N.: Mining association rules between sets of items in large databases. In: Buneman, P., Jajodia, S. (eds.) Proceedings of the ACM SIGMOD International Conference on Management of Data, Washington, D.C., May 1993, pp. 207–216 (1993)
12. Agrawal, R., Srikant, R.: Fast algorithms for mining association rules. In: Proceedings of the 20th International Conference on Very Large Data Bases, Santiago, Chile, September 1994, pp. 487–499 (1994)
13. Goethals, B., Laur, S., Lipmaa, H., Mielikäinen, T.: On private scalar product computation for privacy-preserving data mining. In: Park, C.-s., Chee, S. (eds.) ICISC 2004. LNCS, vol. 3506, pp. 104–120. Springer, Heidelberg (2005)
14. Paillier, P.: Public-key cryptosystems based on composite degree residuosity classes. In: Stern, J. (ed.) EUROCRYPT 1999. LNCS, vol. 1592, pp. 223–238. Springer, Heidelberg (1999)
15. Amirbekyan, A., Estivill-Castro, V.: A New Efficient Privacy-Preserving Scalar Product Protocol. In: Proc. Sixth Australasian Data Mining Conference (AusDM 2007), Gold Coast, Australia (2007)

Transfer Learning with Data Edit*

Yong Cheng[1] and Qingyong Li[2]

[1] Department of Computer Sciences,
Beijing University of Chemical Technology, Beijing
chengyong@mail.buct.edu.cn
[2] School of Computer and Information Technology,
Beijing Jiaotong University, Beijing
liqy@bjtu.edu.cn

Abstract. We often face the situation where very limited labeled data are available to learn an effective classifier in target domain while there exist large amounts of labeled data with similar feature or distribution in certain relevant domains. Transfer learning aims at improving the performance of a learner in target domain given labeled data in one or more source domains. In this paper, we present an algorithm to learn effective classifier without or a few labeled data on target domain, given some labeled data with same features and similar distribution in source domain. Our algorithm uses the data edit technique to approach distribution from the source domain to the target domain by removing "mismatched" examples in source domain and adding "matched" examples in target domain. Experimental results on email classification problem have confirmed the effectiveness of the proposed algorithm.

Keywords: Transfer Learning, Data Edit, Semi-supervised Learning.

1 Introduction

Most machine learning approaches in theory and practice work with the basic assumption that the learning tasks belong to the same domain and share some common features. If the distribution keeps unchangeable, these approaches can work very well and achieve their best performance. However, there exist several applications that challenge these traditional machine learning approaches. Firstly, there may be a lack of labeled training examples due to huge costs of manually labeling the training example; secondly, in certain situation where the feature spaces in training set and test set are not the same; thirdly, the distribution is not stable and can be changed easily. Thus, there is a need for developing new machine learning techniques to cope with such situations.

Recently, the *Transfer Learning* is proposed to face these challenges [2]. Simply speaking, transfer learning uses the experiences gathered from previous learning tasks in order to improve the performance of a new learning task. In machine learning community, the *transfer learning* emphasizes the transfer of knowledge

* This work is supported by the young teacher project in BUCT under contract (Number QN0824).

R. Huang et al. (Eds.): ADMA 2009, LNAI 5678, pp. 427–434, 2009.

across domains, tasks and distributions that are similar but not the same [2][4]. It provides a possible solution to the problems described in the above. It is worthy of noting that the transfer learning relies on the assumption that all tasks are mutually related or the domains are similar and relevant.

There exist a large body of literatures on the study of transfer learning. The work that is closely related to ours is *self-taught learning*, where only a few labeled examples in target domain are available. In [1], Dai et al. proposed a boosting algorithm named *TrAdaBoost* to deal with the inductive transfer learning problems. Like classical AdaBoost algorithm, *TrAdaBoost* trains the base classifier on the weighted source and target data in an iterative way. In [3], Liao et al. proposed an active learning algorithm called "Migratory-Logit" or "M-Logit" to select unlabeled data in the target domain to be labeled with the help of the source domain data. Wu and Dietterich also use source domain data to improve the performance of SVM classifier [5]. S. J. Pan et al provides a good overview on transfer learning and also proposes a classification framework based on the number of labeled data in source and target domain [2].

Differing from the above work, the main contribution of this paper is a new algorithm proposed to learn robust classifier on the target data with the help of "valuable" labeled examples selected from the source domain. In detail, we use the data edit technique to evaluate the most "valuable" labeled examples and change the distribution of training set from source domain to target domain as smoothly as possible. Finally, we derive the classifier based on the training set where the distribution is very close to that of the target domain. The experimental results on email classification problem validate our algorithm.

The remainder of the paper is organized as follows. In Section 2 and Section 3, we formulate the transfer learning problem and present a new transfer leraning algorithm with the data edit technique. Next, the experimental results on email classification problem are presented in Section 4. Finally, Section 5 concludes the paper.

2 Problem Description

In this section, we introduce some notations that are used in the remainder of the paper. At first, we give the definition of "domain" and "task".

Definition 2.1 (Domain). A domain $\mathcal{D}$ consists of two components: $\mathcal{D} =< \mathcal{X}, \mathcal{P}(X) >$, $\mathcal{X}$ is the feature space, $\mathcal{P}(X)$ is the marginal probabilistic distribution where $X = \{x_1, x_2, \cdots, x_n\} \in \mathcal{X}$.

Definition 2.2 (Task). Given a domain $\mathcal{D} =< \mathcal{X}, \mathcal{P}(X) >$, a "task" is defined as $\mathcal{T} =< \mathcal{Y}, f >$, of which, $\mathcal{Y}$ is the label space, f is a function, which is learnt from training set $\{x_i, y_i\}$, $x_i \in X$, $y_i \in \mathcal{Y}$, can predict the label of a new example, say x.

The problem can be described as follows, given a source domain $\mathcal{D}_S$ and a target domain $\mathcal{D}_T$. The two domains $\mathcal{D}_S$ and $\mathcal{D}_T$ are mutually related in the sense that they share the same representation and similar distribution. All data in the source domain are labeled, while no or very few examples in target domain are labeled. Let $\mathcal{P}(X_S)$ and $\mathcal{P}(X_T)$ be the marginal probabilistic distributions of the source domain $\mathcal{D}_S$ and the target domain $\mathcal{D}_T$ respectively. Our goal is to predict the labels of data in target domain with the help of labeled data in source domain.

3 New Approach for Transfer Learning

3.1 Motivation

In general, a classifier trained on the data from source domain will perform badly on the data from target because of the distribution gap of two different domains. Our algorithm tries to learn a classifier that can transfer the knowledge at instance level from source domain to target domain. In other words, the classifier trained iteratively on a pool of training instances that consists of labeled examples in source domain and examples with "predicted" labels in target domain. The distribution of the pool will gradually approach that of target domain by adding "good" examples with "predicted" labels from target domain and removing "bad" examples from source domain. In our algorithm, the "good" and "bad" examples are evaluated using the data edit technique. After the training process is finished, we believe that the distribution of the pool is a good approximation of that of target domain. Thus, the classifier can work on the target domain very well.

3.2 Data Edit Technique

We use data edit technique to examine the separability degree in feature space. Data edit, or separability index is a non-parametric statistics method that characterize class separability from a given learning samples [6]. The technique can be briefly introduced here. At first, a neighborhood structure, for example, neighborhood graph or relative neighborhood graph is built to describe the local relationship of data in the domain. Then, some edges are removed to get the clusters of examples and the number of edges that connect examples with same or different labels is calculated. Finally, the law of the edge proportion that must be removed under the null hypothesis is established for the analysis of whether the classes are separable or not. Without loss of generality, we use relative neighborhood graph to express the proximity of examples in the paper. More details about data edit are omitted due to the limit of space. Reader of interest can refer [6] to learn more details.

3.3 Algorithm Description

Initially, the training set consists of all labeled data, i.e. all data in source domain and sometimes partial data in target domain. Based on the initial training set, the initial classifier is built to classify the unlabeled data in target data. It is obvious that the classification performance is not very well because the initial classifier mainly reflects the data distribution in source domain. Even though, there still exist some unlabeled examples that are assigned to correct labels. We assume that the feature spaces in source and target domain are smooth. Thus, from the perspective of local learning, there exist some labeled examples in source domain and the unlabeled examples in the target domain are highly consistent with their neighborhood regarding the labels.

To improve the performance of classifier on data from target domain, the pool of training need to be modified to weaken the influences from source distribution and increase the influences from target distribution. Thus, "good" examples with "predicted" labels will be kept in the pool while "bad" examples from source domain are removed in an iterative way. Thus, how to pick these "good" and "bad" examples in the sense of matching the training pool is key to our approach. As introduced above, The pickup process of adding and removing examples is performed using the data edit techniques.

Now, let's consider binary classification, the data editing technique is required to perform hypothesis verification process. We know the mean and variance of statistics J is given by formulation (1) and (2). The explanation of the parameter in the equation is omitted due to the limit of space, you can refer [6] for more details.

$$\mu = S_0\pi_1\pi_2 \tag{1}$$

$$\sigma^2 = S_1\pi_1^2\pi_2^2 + S_2\pi_1\pi_2(\frac{1}{4} - \pi_1\pi_2) \tag{2}$$

The critical value for $J_{1,2}$ at the significance level α_0 is calculated below:

$$J_{1,2;\alpha_0/2} = S_0\pi_1\pi_2 - u_{1-\alpha_0/2}\sqrt{S_1\pi_1^2\pi_2^2 + S_2\pi_1\pi_2(\frac{1}{4} - \pi_1\pi_2)} \tag{3}$$

$$J_{1,2;1-\alpha_0/2} = S_0\pi_1\pi_2 + u_{1-\alpha_0/2}\sqrt{S_1\pi_1^2\pi_2^2 + S_2\pi_1\pi_2(\frac{1}{4} - \pi_1\pi_2)} \tag{4}$$

Now, let's consider two typical examples x_s and x_t in the training set, of which x_s comes from source domain, while x_t is from the target domain. If the observation value of J that associates with x_s locates in the left rejection region, then there are significantly less cut edges than expected under hypothesis H_0, hence, we consider it as example that perfectly matches the distribution in the source domain. It is obvious that this example x_s needs to be removed in order to change smoothly the distribution of pool set to the target domain. If the observation value of J locates in the right acceptance region, we will keep the example in the training set for next iteration. As for the example x_t, we will be very careful to perform the pickup process, because the label given by the classifier is not very reliable and confident. A commonly used method is to rank the observation values of J and pick up the unlabeled examples and their predicted labels with most confidence and add to the training set. Our algorithm works in an iterative way, with the iteration process going, the "mismatched" data from source domain have been removed and unlabeled example with "reliable" labels are added to the training set, the distribution of training set will approach gradually to the distribution of target domain. Thus, it is more likely to build a classifier with better performance. The algorithm is described in detail below.

Algorithm 1. Transfer Learning Algorithm with Data Edit

Input: Labeled data in source domain $\mathcal{X}_{L_S}$, Labeled data in target domain $\mathcal{X}_{L_T}$, Unlabeled data in target domain $\mathcal{X}_{U_T}$, Base classifier *learner*(), Maximum iteration times K, Source confidence level $\alpha_s[K]$, Target confidence level $\alpha_t[K]$;

Output: The learned hypothesis h;

t = 1;
$L_S \leftarrow X_{L_S}$, $L_T \leftarrow X_{L_T}$, $S_R = T_A = \emptyset$;
repeat
    $\alpha_s = \alpha_s[t]$, $\alpha_t = \alpha_t[t]$;
    $TS \leftarrow L_S \cup L_T$;
    Train a base classifier h on TS: $h \leftarrow learner(TS)$;
    Build a RNG or KNN Graph G for $D(\mathcal{X}_L \cup \mathcal{X}_U)$;
    Classify examples $x_t \in (\mathcal{X}_U - L_T)$ with base classifier h;
    for *each example* $x_s \in L_S$ **do**
        Find the neighborhood $\mathcal{N}_s$ of x_s in G;
        Compute the distribution function of J_s under H_0;
        Compute the observation value o_s of J_s;
        if o_s *locates the left rejection region with confidence level* α_s **then**
            $S_R \leftarrow S_R \cup x_s$;
        end
    end
    for *each example* $x_t \in (\mathcal{X}_U - L_T)$ **do**
        Find the neighborhood $\mathcal{N}_t$ of x_t in G;
        Compute the distribution function of J_t under H_0;
        Compute the observation value o_t of J_t;
        if o_t *locates the left rejection region with confidence level* α_t **then**
            $T_A \leftarrow T_A \cup x_t$;
        end
    end
    $L_S \leftarrow L_S - S_R$, $L_T \leftarrow L_T \cup T_A$;
    $t \leftarrow t + 1$;
until $t > K$ *or Convergence* ;
return hypothesis h;

It is worthy of noting that the confidence level in data editing is not fixed. In fact, after several iterations, it will become more and more difficult to find the labeled examples with confidence level α_0 under hypothesis H_0. In our algorithm, we adopt a strategy to regulate the confidence level when the unlabeled examples are added to the training set. Thus, the initial confidence level α_0 for the removing and adding data is required to set at the beginning of the algorithm.

4 Experimental Results

We use email classification problem to verify the algorithm proposed. Email classification problem is the task of automatically deciding whether a received email is spam or not. The task is an ideal problem to conduct our experiments.

In general, the criterion of different people to decide a received email as spam is different. Obviously, the spam distributions of different people are various. There exist certain situations where the labeled emails are rare, for instance, when a user installs a new kind email client or anti-virus software, at the beginning, there is no labeled emails available, however, it is very easy to get some spam considered by other people. Our algorithm can provide an effective way to train classifier with impressive performance by transferring the knowledge from a user to another user.

Our experiments are tested on the Enron email data set which is a standard email benchmark available in the machine learning community[1]. The raw Enron contains 619, 446 messages belonging to 158 users. We cleaned the email corpus by removing most folders from each user, but left two folders, namely "inbox" and "deleted-items" before further analysis. The spam distribution of each user is different and independent. In order to train effective classifier, we also ignore some users have no large number of emails. The selected users are summarized in table 1.

Table 1. Selected Users in Enron Dataset

User Name	Num of Inbox	Num of Deleted-Items	Distribution	Number of Threads
shackleton-s	1058	655	61.76% 38.24%	unknown
giron-d	377	415	47.60% 52.24%	632
lavorato-j	277	715	27.92% 72.08%	416
taylor-m	406	632	39.11% 60.89%	unknown
dasovich-j	1388	1564	47.01% 52.98%	unknown
lay-k	1373	1126	54.94% 45.06%	910
beck-s	752	309	70.88% 29.12%	2630
farmer-d	85	380	18.28% 81.72%	3387
kaminski-v	560	1792	23.81% 76.19%	5550
kitchen-l	141	202	41.10% 58.89%	unknown
lokay-m	56	352	13.72% 86.27%	913
sanders-r	81	530	13.25% 86.74%	unknown

In each experiment, two users with similar email spam distribution are selected, one user as the source domain, and another user as the target domain. All emails in source domain are labeled as spam and norm ones, while only partial emails in target domain are labeled. At the beginning, all labeled data in source and target domain form the the initial training set, then the initial classifier is trained in an iterative way. In the data set from target domain, 1%, 2%, 3% and 5% examples are labeled in order to evaluate the performance of proposed algorithm. The evaluation criteria is classification error rate. In all experiment, the Naïve Bayes and SVM classifier are adopted as the base learner for comparison. When removing examples from source domain and adding example from target domain, the initial confidence in data editing technique is set to $\alpha = 0.01$. In each experiment, the algorithms are performed for 20 times then the mean classification error is calculated.

[1] http://www.cs.cmu.edu/~enron/

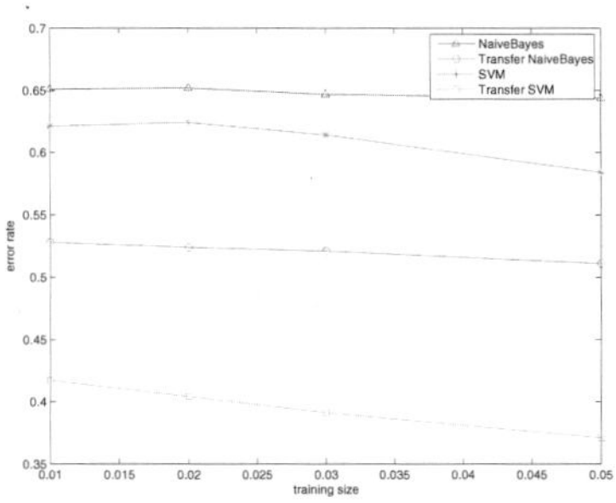

Fig. 1. Shackleton-s vs beck-s

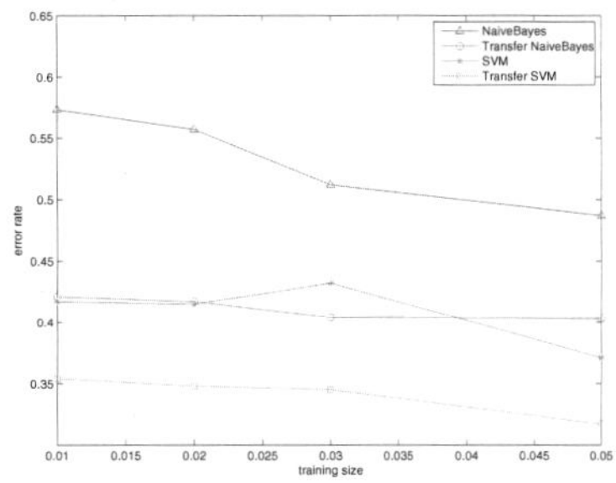

Fig. 2. Giron-d vs farmer-d

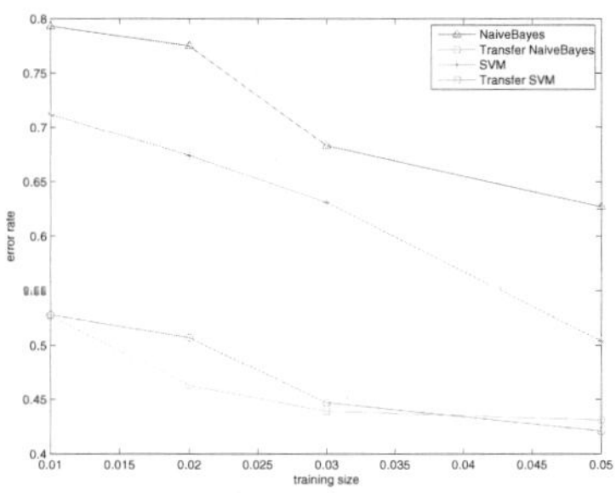

Fig. 3. Lavorato-j vs kaminski-v

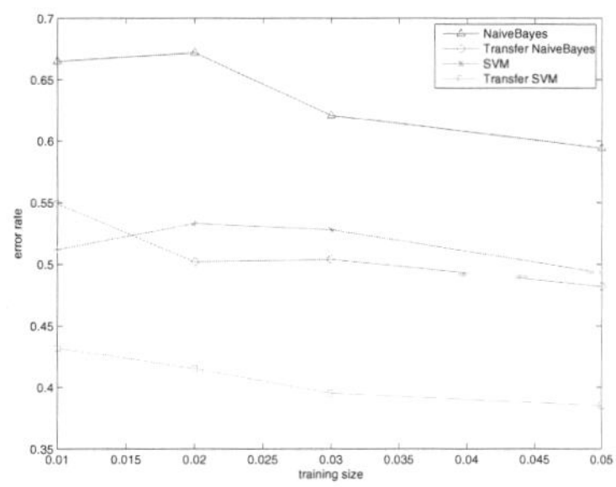

Fig. 4. Taylor-m vs kitchen-l

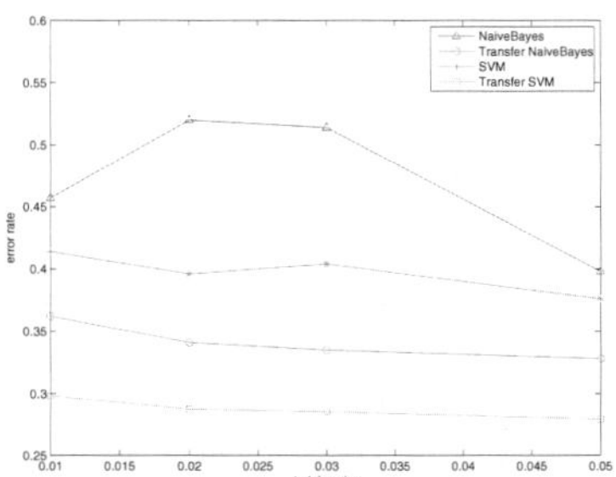

Fig. 5. Dasovich-j vs lokay-m

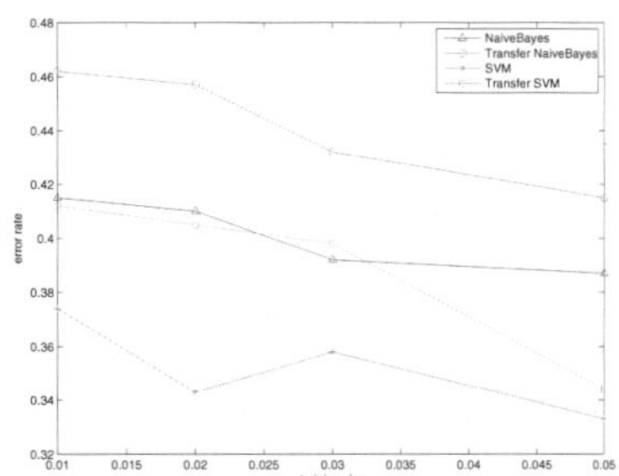

Fig. 6. Lay-k vs sanders-r

In the email data set, 12 users are selected to form 6 transfer learning tasks. Each task includes one email user as the source domain and another user as the target. The experimental results are listed from figure (1) to figure (6). In all figures, the labeled emails of the first user form the the source domain and emails of the second user form the target domain. It is obvious that the transfer learning version of classifier outperforms its standard counterpart in most cases except figure (6). In our experiments, the SVM classifier has better classification accuracy than the Naïve bayes classifier in both the standard and the transfer version. However, when there is no plenty of data for training, the performance will certainly become worser as shown in figure (3). The results shown in the figures confirms the effectiveness of transfer learning with data edit techniques, our approach is very useful when there is little labeled data in the target domain while plenty of labeled data in the similar but not same domain is available.

5 Conclusions

In this paper, a new transfer learning algorithm is proposed to learn effective classifier from a few labeled data set in target domain with plenty of labeled data in source domain. The algorithm uses the data editing technique to ensure the smooth shift of distribution from source domain to target domain. The data editing technique has been confirmed to remove labeled examples that perfectly match the source distribution and add new unlabeled examples that match the target distribution. The experimental results on email classification problem have confirmed the effectiveness of our approach.

References

1. Dai, W., Yang, Q., Xue, G.-R., Yu, Y.: Boosting for transfer learning. In: Ghahramani, Z. (ed.) Proceedings of the 24th International Conference on Machine Learning (ICML 2007), June 20-24, pp. 193–200. ACM, New York (2007)
2. Jialin, S., Yang, Q.: A survey on transfer learning. Technical Report HKUST-CS08-08, Hong Kong University of Science and Technology (2008)
3. Liao, X., Xue, Y., Carin, L.: Logistic regression with an auxiliary data source. In: Raedt, L.D., Wrobel, S. (eds.) Proceedings of the 22nd International Conference on Machine Learning (ICML 2005), August 7-11, pp. 505–512. ACM, New York (2005)
4. Ling, X., Dai, W., Xue, G.-R., Yang, Q., Yu, Y.: Spectral domain-transfer learning. In: Proceedings of the 14th ACM SIGKDD International Conference on Knowledge Discovery and Data Mining, Las Vegas, Nevada, August 2008, pp. 488–496 (2008)
5. Wu, P., Dietterich, T.G.: Improving svm accuracy by training on auxiliary data sources. In: Brodley, C.E. (ed.) Proceedings of the 21st International Conference on Machine Learning (ICML 2004), July 4-8. ACM, New York (2004)
6. Zighed, D.A., Lallich, S., Muhlenbach, F.: Separability index in supervised learning. In: Elomaa, T., Mannila, H., Toivonen, H. (eds.) PKDD 2002. LNCS, vol. 2431, pp. 475–487. Springer, Heidelberg (2002)

Exploiting Temporal Authors Interests via Temporal-Author-Topic Modeling

Ali Daud[1], Juanzi Li[1], Lizhu Zhou[1], and Faqir Muhammad[2],

[1] Department of Computer Science & Technology, 1-308, FIT Building, Tsinghua University, 100084 Beijing, China
[2] Department of Mathematics & Statistics, Allama Iqbal Open University, 44000 Sector H-8, Islamabad, Pakistan
ali_msdb@hotmail.com, ljz@keg.cs.tsinghua.edu.cn, dcszlz@tsinghua.edu.cn, aioufsd@yahoo.com

Abstract. This paper addresses the problem of discovering temporal authors interests. Traditionally some approaches used stylistic features or graph connectivity and ignored semantics-based intrinsic structure of words present between documents, while previous topic modeling approaches considered semantics without time factor, which is against the spirits of writing. We present Temporal-Author-Topic (TAT) approach which can simultaneously model authors interests and time of documents. In TAT mixture distribution over topics is influenced by both co-occurrences of words and timestamps of the documents. Consequently, topics occurrence and correlations change over time, while the meaning of particular topic almost remains unchanged. By using proposed approach we can discover topically related authors for different time periods and show how authors interests and relationships change over time. Experimental results on research papers dataset show the effectiveness of proposed approach and dominance over Author-Topic (AT) model, due to not changing the meaning of particular topic overtime.

Keywords: Temporal Authors Interests, Topic Modeling, Unsupervised Learning.

1 Introduction

Enormous information on the web has provided us with many challenging knowledge discovery problems. For example, in Social Networks authors interests' mining is an important issue discussed these years from reviewer and program committee member recommendation point of view. Unfortunately, most of the existing topic modeling approaches for authors interests discovery ignored the time factor. However, Web is highly dynamic, so ignorance of time factor is not a realistic assumption to be made by now. Most of the datasets such as research papers, blogs and news do not have static co-occurrence patterns; they are instead dynamic. The data are collected over time and data patterns keep on changing, by showing rising or falling trends overtime.

R. Huang et al. (Eds.): ADMA 2009, LNAI 5678, pp. 435–443, 2009.

Previously, two major frameworks used to identify the authors interests and relationships are 1) stylistic features (such as sentence length), author attribution and forensic linguistics to identify what author wrote a given piece of text [7,9] and 2) Graph connectivity based approaches as a basis for representation and analysis for relationships among authors [8,13,18]. Above mentioned frameworks based on writing styles and network connectivity ignored the semantics-based intrinsic structure of words present between the documents. While, semantics-based topic modeling approaches AT and TOT models [14,17] did not consider authors interests with time information.

Here it is necessary to mention that exploitation of author interests (who is writing on what topic without any discrimination between renowned and not-renowned publication venues) and expert finding [5] (who is most skilled on what topic with the discrimination between renowned and not-renowned publication venues) are notably two different knowledge discovery problems.

In this paper, we investigate the problem of authors interests discovery with the time factor. We combined time factor with the static authors interests modeling idea of AT model to propose Temporal-Author-Topic (TAT) approach, whose generative model is based on similar framework with ACT1 model [16]. Proposed approach models the temporal authors interests and relationships with respect to time and can provide promising answers to the questions about temporal authors interests. Experimental results and discussions elaborate the importance of problem and usefulness of our approach over AT model.

The novelty of work described in this paper lies in the; formalization of the temporal authors interests discovery problem, proposal of TAT approach, and experimental verification of the effectiveness of our approach on the real world dataset. To the best of our knowledge, we are the first to deal with the temporal authors interests discovery problem by proposing a topic modeling approach, which can implicitly capture word-to-word, word-to-author and author-to-author relationships by taking time factor into account.

The rest of the paper is organized as follows. In Section 2, we introduce generative probabilistic modeling towards temporal authors interests discovery and illustrate our proposed approach with its parameters estimation and inference making details. In Section 3, dataset, parameter settings, baseline approach with empirical studies and discussions about the results are given. Section 4 brings this paper to the conclusions.

2 Probabilistic Generative Topic Modeling

In this section, before describing our Temporal-Author-Topic (TAT) approach, we will first describe how documents and authors interests are modeled in the past.

2.1 Modeling Documents and Authors Interests with Topics

Latent Dirichlet Allocation (LDA) [4] is a state of the art topic model used for modeling documents by using a latent topic layer between them. It is a Bayesian network that generates a document using a mixture of topics. For each document d, a topic mixture multinomial distribution θ_d is sampled from Dirichlet α, and then a latent topic z is chosen and a word w is generated from topic-specific multinomial distribution Φ_z over words of a document for that topic.

The Author model [12] was proposed to model documents text and its authors interests. For each document d, a set of authors' $\mathbf{a}_d$ is observed. To generate each word an author a, is uniformly sampled from the set of authors, and then a word w is generated from an author-specific multinomial distribution Φ_a over words of a document for that topic. Later, words and authors of documents are modeled by considering latent topics to discover the research interests of authors [15]. In AT model, each author (from set of A authors) of a document d is associated with a multinomial distribution θ_a over topics is sampled from Dirichlet α and each topic is associated with a multinomial distribution Φ_z sampled from Dirichlet β over words of a document for that topic. In this model time information was not considered.

2.2 Modeling Temporal Authors Interests with Topics

In TAT approach for modeling temporal interests of authors, we viewed a document as a composition of words with each word having the publishing year of document as time stamp along with its authors. Symbolically, a collection of $\mathbf{D}$ documents can be written as: $\mathbf{D} = \{(\mathbf{w}_1,\mathbf{a}_1,y_1),(\mathbf{w}_2,\mathbf{a}_2,y_2),\ldots,(\mathbf{w}_d,\mathbf{a}_d,y_d)\}$, where $\mathbf{w}_d$ is word vector chosen from a vocabulary of size V, $\mathbf{a}_d$ is author vector and y_d is the time stamp of document d.

TAT considers that an author is responsible for generating some latent topics of the documents on the basis of semantics-based intrinsic structures of words with time factor. In the proposed model, each author (from set of A authors) of a document d is associated with a multinomial distribution θ_a over topics and each topic is associated with a multinomial distribution Φ_z over words and multinomial distribution Ψ_z with a time stamp for each word of a document for that topic. So, θ_a, Φ_z and Ψ_z have a symmetric Dirichlet prior with hyper parameters α, β and γ, respectively. The generating probability of the word w with year y for author a of document d is given as:

$$P(w,y|a,d,\emptyset,\Psi,\theta) = \sum_{z=1}^{T} P(w|z,\emptyset_z)P(y|z,\Psi_z)P(z|a,\theta_a) \quad (1)$$

The generative process of TAT is as follows:
For each author $a = 1,\ldots, K$ of document d
Choose θ_a from Dirichlet (α)
For each topic $z = 1,\ldots, T$
Choose Φ_z from Dirichlet (β)
Choose Ψ_z from Dirichlet (γ)
For each word $w = 1,\ldots, N_d$ of document d
Choose an author a uniformly from all authors $\mathbf{a}_d$
Choose a topic z from multinomial (θ_a) conditioned on a
Choose a word w from multinomial (Φ_z) conditioned on z
Choose a year y associated with word w from multinomial (Ψ_z) conditioned on z

Gibbs sampling is utilized [1,10] for parameter estimation and inference making in our approach, which has two latent variables z and a; the conditional posterior distribution for z and a is given by:

$$P(z_i = j,\ a_i = k|w_i = m, y_i = n, \quad \mathbf{z}_{-i}, \boldsymbol{a}_{-i}, \mathbf{a}_d) \infty \frac{n_{-i,j}^{(wi)} + \beta}{n_{-i,j}^{(\cdot)} + W\beta} \frac{n_{-i,j}^{(yi)} + \gamma}{n_{-i,j}^{(\cdot)} + Y\gamma} \frac{n_{-i,j}^{(ai)} + \alpha}{n_{-i,\cdot}^{(ai)} + A\alpha} \quad (2)$$

where $z_i = j$ and $a_i = k$ represent the assignments of the ith word in a document to a topic j and author k respectively, $w_i = m$ represents the observation that i^{th} word is the mth word in the lexicon, $y_i = n$ represents i^{th} year of paper publishing, attached with the nth word in the lexicon and z_{-i} and a_{-i} represents all topic and author assignments not including the ith word. Furthermore, $n_{-i,j}^{(wi)}$ is the total number of words associated with topic j, excluding the current instance, $n_{-i,j}^{(yi)}$ is the total number of years associated with topic j, excluding the current instance and $n_{-i,j}^{(ai)}$ is the number of times author k is assigned to topic j, excluding the current instance, V is the size of the lexicon, Y is the number of years and A is the number of authors. "." Indicates summing over the column where it occurs and $n_{-i,j}^{(.)}$ stands for number of all words and years that are assigned to topic z respectively, excluding the current instance.

During parameter estimation, the algorithm needs to keep track of W x T (word by topic), Y x T (year by topic) and T x A (topic by author) count matrices. From these count matrices, topic-word distribution Φ, topic-year distribution Ψ and author-topic distribution θ can be calculated as:

$$P(w|z) = \emptyset_{zw} = \frac{n_{-i,j}^{(wi)} + \beta}{n_{-i,j}^{(.)} + V\beta} \quad (3)$$

$$P(y|z) = \Psi_{zy} = \frac{n_{-i,j}^{(yi)} + \gamma}{n_{-i,j}^{(.)} + Y\gamma} \quad (4)$$

$$P(z|a) = \theta_{az} = \frac{n_{-i,j}^{(ai)} + \alpha}{n_{-i,.}^{(ai)} + A\alpha} \quad (5)$$

where, $\emptyset_{zw}$ is the probability of word w in topic z, Ψ_{zy} is the probability of year y for topic z and θ_{az} is the probability of topic z for author a. These values correspond to the predictive distributions over new words w, new years' y and new topics z conditioned on w, y and z.

Now finally by deriving Bayes' Theorem, we can obtain the probability of an author a given topic z and year y as:

$$P(a|z,y) = \frac{P(z,y|a).P(a)}{P(z,y)}, \text{ where} \\ P(z,y|a) = P(z|a).P(y|a) \text{ and } P(y|a) = \textstyle\sum_z P(y|z).P(z|a) \quad (6)$$

Here, for calculating *P(a)* we simply used the number of publications of one author in a year. For more simplicity some works assume it uniform [3] and Propagation approach can also be used to calculate it in a more complex way [11].

3 Experiments

3.1 Dataset, Parameter Settings and Baseline Approach

Our collection of DBLP [6] dataset contains D = 90,124 papers with K= 112,317 authors for Y = 2003-2007 years, to use almost smooth yearly data distribution for getting equalized impact of text and authors information. We processed the dataset by a) removing stop-words, punctuations and numbers b) down-casing the obtained words and c) removing words and authors that appears less than three times in the dataset. This led to a vocabulary size of 10,788, a total of 469,821 words and 26,073 authors.

In our experiments, for 150 topics T the hyper-parameters α, β and γ were set at $50/T$, .01, and 0.1 (our approachs application is not sensitive to hyper parameters unlike some other topic model applications so no need of parameter optimization). Topics are set at 150 on the basis of human judgment of meaningful topic plus measured perplexity [4], a standard measure for estimating the performance of probabilistic models with the lower the best, for the estimated topic models.

We attempted to qualitatively compare TAT approach with AT model and used same number of topics for evaluation. Dataset was portioned by year and for each year all the words and authors were assigned to their most likely topics using AT model. The number of Gibbs sampler iterations used for AT is 1000 and parameter values same as the values used in [15].

3.2 Results and Discussions

Topically related Authors for Different Years. We discovered and probabilistically ranked authors related to a specific area of research on the basis of latent topics for different years. Table 1 illustrates 2 different topics out of 150, discovered from the 1000th iteration of a particular Gibbs sampler run. The words associated with a topic are quite intuitive and precise in the sense of conveying a semantic summary of a specific area of research. The authors associated with each topic for different years are quite representative. Here it is necessary to mention that top 10 authors associated with the topics for different years are not the experts of their fields, instead are the authors who produced most words for that topic in a specific year. For example, Baowen Xu is known for software engineering, by anlaysing DBLP data we have found that he has published papers having high probability words related to Data Mining (DM) topic during the years he is related to that topic, while he is not related to DM topic later because of not producing high probability words for that topic.

Table 1. An illustration of 2 discovered topics from 150-topics. Each topic is shown with the top 10 words (first column) and authors that have highest probability conditioned on that topic for each year (second to sixth column). Titles are our interpretation of the topics.

Data Mining (DM)											
Topic 142		**Year 2003**		**Year 2004**		**Year 2005**		**Year 2006**		**Year 2007**	
Words	**Prob.**	**Authors**	**Prob.**	**Authors**	**Prob.**	**Authors**	**Prob.**	**Authors**	**Prob.**	**Authors**	**Prob.**
mining	0.20871	Jiawei Han	0.2	Jiawei Han	0.2	Jiawei Han	0.2	Jiawei Han	0.2	Jiawei Han	0.2
patterns	0.07798	Francesco Bonchi	0.0845	Hui Xiong	0.1023	Francesco Bonchi	0.0646	Hui Xiong	0.0561	Hui Xiong	0.0866
rules	0.05193	Baowen Xu	0.0732	George Karypis	0.0683	Srinivasan Parth.	0.0634	Francesco Bonchi	0.0439	Jian Pei	0.0311
frequent	0.04291	Hui Xiong	0.0539	Reda Alhajj	0.0573	Baowen Xu	0.0509	Srinivasan Parth.	0.0431	Christopher Kruegel	0.0264
pattern	0.04155	George Karypis	0.0525	Francesco Bonchi	0.0515	Jian Pei	0.0458	Reda Alhajj	0.0315	Reda Alhajj	0.0243
association	0.04121	Jian Pei	0.0507	Srinivasan Parth.	0.0385	Reda Alhajj	0.0424	Olfa Nasraoui	0.0266	Martin Ester	0.0237
discovery	0.023	Srinivasan Parth.	0.0474	Shiwei Tang	0.0385	Shiwei Tang	0.0414	Jian Pei	0.0264	Francesco Bonchi	0.0203
databases	0.02283	Takeaki Uno	0.0375	Baowen Xu	0.0324	George Karypis	0.0321	Martin Ester	0.0256	Olfa Nasraoui	0.0197
rule	0.01908	Jeffrey Xu Yu	0.0341	Jianyong Wang	0.0297	Hui Xiong	0.0275	Shiwei Tang	0.0246	Won Suk Lee	0.0197
discovering	0.01619	Won Suk Lee	0.0327	Jian Pei	0.0262	Ke Wang	0.0266	Jianyong Wang	0.019	Takeaki Uno	0.0181
XML Databases (XMLDB)											
Topic 18		Year 2003		Year 2004		Year 2005		Year 2006		Year 2007	
Words	Prob.	Authors	Prob.	Authors	Prob.	Authors	Prob.	Authors	Prob.	Authors	Prob.
XML	0.16606	Surajit Chaudhuri	0.2	Surajit Chaudhuri	0.2	Surajit Chaudhuri	0.2	Jianhua Feng	0.2	Surajit Chaudhuri	0.2
query	0.08406	Divesh Srivastava	0.1038	Jayant R. Haritsa	0.1161	Philip A. Bernstein	0.106	Surajit Chaudhuri	0.1239	Jianhua Feng	0.157
database	0.05458	Raymond K. Wong	0.0825	Carlos A. Heuser	0.0727	Kevin Chen-Chuan	0.0961	Raymond K. Wong	0.0804	Kevin Chen-Chuan	0.0961
databases	0.05113	Jayant R. Haritsa	0.0774	Tok Wang Ling	0.0719	Dan Suciu	0.0912	Dimitri Theodoratos	0.0586	Christoph Koch	0.0959
relational	0.04016	Dan Suciu	0.0676	Raymond K. Wong	0.0675	Tok Wang Ling	0.0847	Raghu Ramakrishnan	0.0529	Dan Suciu	0.0946
queries	0.03467	Christoph Koch	0.06	Kevin Chen-Chuan	0.0437	Donald Kossmann	0.0811	Sourav S. Bhowmick	0.0518	Donald Kossmann	0.0721
schema	0.02981	Carlos A. Heuser	0.0533	Raghu Ramakrishnan	0.0427	Dimitri Theodoratos	0.0788	Divesh Srivastava	0.0487	Carlos A. Heuser	0.0711
querying	0.02636	Hongjun Lu	0.0486	Sourav S. Bhowmick	0.0418	Jianhua Feng	0.0757	Christian S. Jensen	0.046	Divesh Srivastava	0.0601
documents	0.02401	Elke A. Rund.	0.0414	Christoph Koch	0.0392	Jayant R. Haritsa	0.0726	Erik Wilde	0.0441	Dimitri Theodoratos	0.0526
views	0.02338	Jixue Liu	0.0402	Divesh Srivastava	0.0358	Divesh Srivastava	0.0615	Katsumi Tanaka	0.0436	Erik Wilde	0.0494

In addition, by doing analysis of authors home pages and DBLP data, we found that all highly ranked authors for different years have published papers on their assigned topics; *"no matter where they are publishing and they are old or new researchers"*. For example, Jianhua Feng (a new researcher) for XMLDB topic started writing on this topic after 2004 and then has published many papers in the following years especially in 2006 (he is ranked first) and in 2007 (he is ranked second). He published most papers in WAIM (country level conference), DASFAA (continent level conference) and some in WWW (world level conference) and here we considered world level conference among the best (world class conferences) in that area of research. While, other top ranked authors for this topic Surajit Chaudhuri (an old researcher) in 2006 (he is ranked second) and in 2007 (he is ranked first) published many papers in SIGMOD (world level conference), ICDE (world level conference) and VLDB (world level conference) and produced many words for XMLDB topic. He is continuously publishing over the years for this topic. Here, Jianhua Feng and Surajit Chaudhuri produced most words for this topic and ranked higher without the discrimination of where they published and from when they are publishing. This matches well with the statement stated above and provides qualitative supporting evidence for the effectiveness of the proposed approach.

A direct comparison with the previous approaches is not fair in terms of perplexity [4], as previous topic modeling approaches were unable to discover temporal authors interests with considering time factor. To measure the performance in terms of precision and recall [2] is also out of question due to unavailability of standard dataset and use of human judgments cannot provide appropriate (unbiased) answers for evaluating temporal authors interests. So, we compared TAT approach with AT model [14], TAT approach can have same meaning for particular topic overtime, but by ignoring time factor AT model changed the meaning of particular topic overtime (inability to discover similar topics for different years). It concludes that approaches which did not consider time factor are unable to discover approximately similar topics for different years. We can say that the time-based solution provided by us is well justified and produced quite promising and functional results.

Table 2. Top ten associated authors with Jiawei Han for different years

Jiawei Han									
2003		2004		2005		2006		2007	
Jian Pei	0.3180	Jian Pei	0.3279	Jian Pei	0.3575	Jian Pei	0.3435	Jian Pei	0.3183
Jeffrey Xu Yu	0.3271	Srinivasan Parthasarathy	0.3916	Francesco Bonchi	0.4648	Srinivasan Parthasarathy	0.4015	Won Suk Lee	0.3598
Francesco Bonchi	0.3781	Francesco Bonchi	0.4057	Srinivasan Parthasarathy	0.4725	Olfa Nasraoui	0.4305	Olfa Nasraoui	0.3824
Srinivasan Parthasarathy	0.4168	Olfa Nasraoui	0.4273	Olfa Nasraoui	0.5465	Hui Xiong	0.4329	Hui Xiong	0.4561
Hui Xiong	0.4567	Anthony K. H. Tung	0.4422	Haixun Wang	0.5632	Francesco Bonchi	0.4549	Martin Ester	0.4654
Haixun Wang	0.4786	Gao Cong	0.4608	Gang Chen	0.5753	Martin Ester	0.4661	Gang Chen	0.5005
Won Suk Lee	0.4807	Joshua Zhexue Huang	0.4616	Philip S. Yu	0.5866	Gang Chen	0.4993	Agma J. M	0.5137
Kuniaki Uehara	0.4844	Martin Ester	0.4666	Hui Xiong	0.6060	Agma J. M. Traina	0.5119	Zhoujun Li	0.5354
Olfa Nasraoui	0.4956	Show-Jane Yen	0.4697	S. Muthukrishnan	0.6148	Mete Celik	0.5184	Haixun Wang	0.5480
Takeaki Uno	0.5170	Wynne Hsu	0.4730	Jeffrey Xu Yu	0.6215	Efim B. Kinber	0.5281	Takeaki Uno	0.5531

Temporal Social Network of Jiawei Han. TAT approach can also be used for dynamic correlation discovery between authors for different years, as compared to only discovering static authors correlations [14]. To illustrate how it can be used in this respect, distance between authors i and j is calculated by using eq. 7 for author-topic distribution for different years.

$$sKL(i,j) = \sum_{z=1}^{T} \left[\theta_{iz} log \frac{\theta_{iz}}{\theta_{jz}} + \theta_{jz} log \frac{\theta_{jz}}{\theta_{iz}} \right] \tag{7}$$

We calculated the dissimilarity between authors; smaller dissimilarity value means higher correlation between the authors. Table 2 shows topically related authors with Jiawei Han for different years. Here, it is obligatory to mention that top 10 authors related to Jiawei Han are not the authors who have co-authored with him mostly, but rather are the authors that tend to produce most words for the same topics with him. Again the results are quite promising and realistic as most of the authors related to Jiawei Han for different years are also related to DM topic. Especially Jian Pei who is ranked first for Jiawei Hans associations had co-authored with him in 2003 (1), 2004 (6), 2005 (2), 2006 (3), 2007 (1) papers. Comparatively, AT model [14] is unable to discover temporal social network of authors, as AT has not considered time information, so unable to fine similar topics for different years.

Temporal Interests. Now by using TAT we will show topic-wise and author-wise temporal interests. In figure 1 (left side), for DM topic Jian Pei has a stable publishing interest shows his consistency to retain his position, while Franscesco Bonchi and George Karypis either started writing less related to this topic or some other auhtors have influenced (pushed back) their interests ranking by writing more on the same topic.

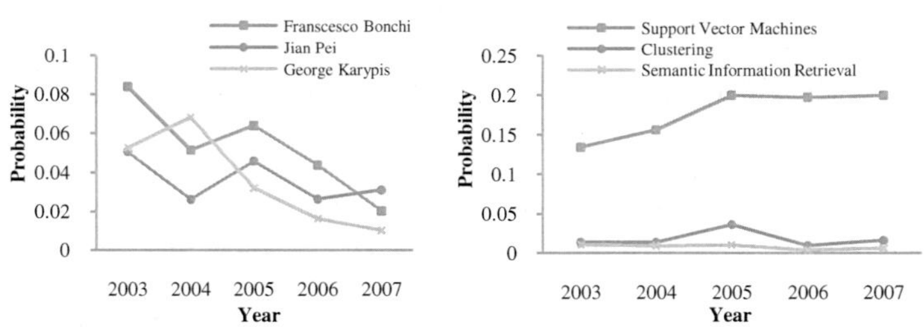

Fig. 1. Topic-wise Authors Interests for Data Mining (Left) and Author-wise Interests Thorsten Joachims (Right)

Thorsten Joachims is pioneer of Support Vector Machines (SVM) and still strongly publishing related to that topic (figure 1 (right side) shows clearly the importance of temporal authors interests discovery problem and effectiveness of proposed approach by matching well with the real world situation). For second and third related interests topics clustering and semantic information retrieval he has published very little, by analyzing his publications in DBLP data we found that he used SVM as a tool to perform clustering and information retrieval tasks. Comparatively, AT model [14] is unable to discover topic wise temporal authors interests, as AT has not considered time information.

4 Conclusions

This study deals with the problem of temporal authors interests discovery through modeling documents, authors and time simultaneously, which is significant. Initially we discussed the motivation for temporal authors interests modeling. We then introduced TAT approach which can discover and probabilistically rank authors related to specific knowledge domains for different time periods. We demonstrated the effectiveness of proposed approach by discovering temporal interests with respect to topics, authors and by presenting authors temporal social network (associations). TAT approach can handle the problem of AT model of change in the meaning of topic overtime. Empirical results and discussions prove the effectiveness of proposed approach.

Acknowledgements. The work is supported by the National Natural Science Foundation of China under Grant (90604025, 60703059), Chinese National Key Foundation Research and Development Plan under Grant (2007CB310803) and Higher Education Commission (HEC), Pakistan. We are thankful to Jie Tang and Jing Zhang for sharing their codes, valuable discussions and suggestions.

References

1. Andrieu, C., Freitas, N.D., Doucet, A., Jordan, M.I.: An Introduction to MCMC for Machine Learning. Journal of Machine Learning 50, 5–43 (2003)
2. Azzopardi, L., Girolami, M., Risjbergen, K.V.: Investigating the Relationship between Language Model Perplexity and IR Precision-Recall Measures. In: Proc. of the 26th ACM SIGIR, Toronto, Canada, July 28-August 1 (2003)
3. Balog, K., Bogers, T., Azzopardi, L., Rijke, M.D., Bosch, A.V.D.: Broad Expertise Retrieval in Sparse Data Environments. In: Proc. of the SIGIR, pp. 551–558 (2007)
4. Blei, D.M., Ng, A.Y., Jordan, M.I.: Latent Dirichlet Allocation. Journal of Machine Learning Research 3, 993–1022 (2003)
5. Daud, A., Li, J., Zhou, L., Muhammad, F.: A Generalized Topic Modeling Approach for Maven Search. In: Proc. of the APWeb-WAIM, Suzhou, China, April 2-4 (2009)
6. DBLP Bibliography Database, http://www.informatik.uni-trier.de/~ley/db/
7. Diederich, J., Kindermann, J., Leopold, E., Paass, G.: Authorship Attribution with Support Vector Machines. Applied Intelligence 19(1) (2003)
8. Erten, C., Harding, P.J., Kobourov, S.G., Wampler, K., Yee, G.: Exploring the Computing Literature using Temporal Graph Visualization. Technical Report, Department of Computer Science, University of Arizona (2003)
9. Gray, A., Sallis, P., MacDonell, S.: Softwareforensics: Extending Authorship Analysis Techniques to Computer Programs. In: Proc. of the 3rd IAFL, Durham NC (1997)
10. Griffiths, T.L., Steyvers, M.: Finding Scientific Topics. In: Proc. of the National Academy of Sciences, USA, pp. 5228–5235 (2004)
11. Zhang, J., Tang, J., Li, J.: Expert Finding in a Social Network. In: Kotagiri, R., Radha Krishna, P., Mohania, M., Nantajeewarawat, E. (eds.) DASFAA 2007. LNCS, vol. 4443, pp. 1066–1069. Springer, Heidelberg (2007)
12. McCallum, A.: Multi-label Text Classification with A Mixture Model Trained by EM. In: Proc. of AAAI 1999 Workshop on Text Learning (1999)

13. Mutschke, P.: Mining Networks and Central Entities in Digital Libraries: A Graph Theoretic Approach applied to Co-author Networks. Intelligent Data Analysis, 155–166 (2003)
14. Rosen-Zvi, M., Griffiths, T., Steyvers, M.: Smyth. P.: The Author-Topic Model for Authors and Documents. In: Proc. of the 20th conference on UAI, Canada (2004)
15. Steyvers, M., Smyth, P., Griffiths, T.: Probabilistic Author-Topic Models for Information Discovery. In: Proc. of the 10th ACM SIGKDD, Seattle, Washington (2004)
16. Tang, J., Zhang, J., Yao, L., Li, J., Zhang, L., Su, Z.: ArnetMiner: Extraction and Mining of Academic Social Networks. In: Proc. of the 14th ACM SIGKDD (2008)
17. Wang, X., McCallum, A.: Topics over Time: A Non-Markov Continuous-Time Model of Topical Trends. In: Proc. of the 12th ACM SIGKDD, pp. 424–433 (2006)
18. White, S., Smyth, P.: Algorithms for Estimating Relative Importance in Networks. In: Proc. of the 9th ACM SIGKDD, pp. 266–275 (2003)

Mining User Position Log for Construction of Personalized Activity Map

Hui Fang[1], Wen-Jing Hsu[1], and Larry Rudolph[2]

[1] Singapore-MIT Alliance, Nanyang Technological University
N4-02b-40, 65 Nanyang Drive, Singapore 637460
fang0025@ntu.edu.sg, hsu@pmail.ntu.edu.sg
[2] VMware Inc. 5 Cambridge Center, Cambridge, MA, USA
rudolph@vmware.com

Abstract. Consider a scenario in which a smart phone automatically saves the user's positional records for personalized location-based applications. The smart phone will infer patterns of user activities from the historical records and predict user's future movements. In this paper, we present algorithms for mining the evolving positional logs in order to identify places of significance to user and representative paths connecting these places, based on which a personalized activity map is constructed. In addition, the map is designed to contain information of speed and transition probabilities, which are used in predicting the user's future movements. Our experiments show that the user activity map well matches the actual traces and works effectively in predicting user's movements.

1 Introduction

Personal positioning refers to the inference of a mobile user's current position based on user's historical positional records and occasional measurements. It has inspired many location-aware applications for personal use with the development of wearable computers that are equipped with position sensors. See, e.g., [4]. The raw data collected from the position sensors, or the user's log, provides an important source for mining useful patterns for personal positioning.

The existing data mining techniques for personal positioning can be classified into two categories: (i) location prediction based on location-to-location transition probabilities [3], and (ii) position tracking based on Bayesian filters [2]. Techniques in the first category aim at inferring user's presence at certain locations (called the *significant places*) and the transitions among these places; but these techniques are generally not designed for the purpose of inferring accurate positions. Techniques in the second category involve online estimation of the user's exact positions, and as such, they usually require well-defined dynamics model of the target. A well-designed model of user activities thus constitute the basis for accurate prediction of user's nondeterministic movements.

The challenges for personal positioning lies in the construction and maintenance of this personalized map, and the modeling of the user movements. In

R. Huang et al. (Eds.): ADMA 2009, LNAI 5678, pp. 444–452, 2009.

a mobile and pervasive context, the device may be allowed to access some positioning sensors, but the large-scale map resources such as central geographic database may be unavailable. A smart device can assist the user's movement only with geometrically accurate map of his environment. In other words, a smart device needs to figure out the map of user's activities on its own in many cases. Unfortunately, the mobile user's positional records can increase quickly with regular uses, which poses difficulty in both storage and information retrieval. Therefore, the key problem here is to build an internal representation of user's environment from user's own movements. This map should be different from generic maps in two aspects: (i) it evolves from user's movement log; (ii) it highlights the geographical features (locations and paths) of significance specific to the user.

These challenges motivate the paper. We propose a new data mining approach to generate a map of user activities for use in personal positioning. Our construction of the user activity map includes three parts: identifying significant places, identifying representative paths, and building the speed and transition model. The experiments show that the position tracking based on user activity map provides sufficient accuracy.

The paper is organized as follows. Section 2 presents the problem statement and our approach. Section 2.1 shows the construction of significant places. Section 2.2 shows the construction of representative paths. Section 2.3 shows the construction of user activity map. Section 3 shows the experimental results. Finally Section 4 concludes the paper.

2 Patterns in User Activities

As "history often repeats itself", with more and more positional samples of regular use over a period of time, a repeated path will be accumulated into the user's trace log. The user's trace log is a sequence of position measurements

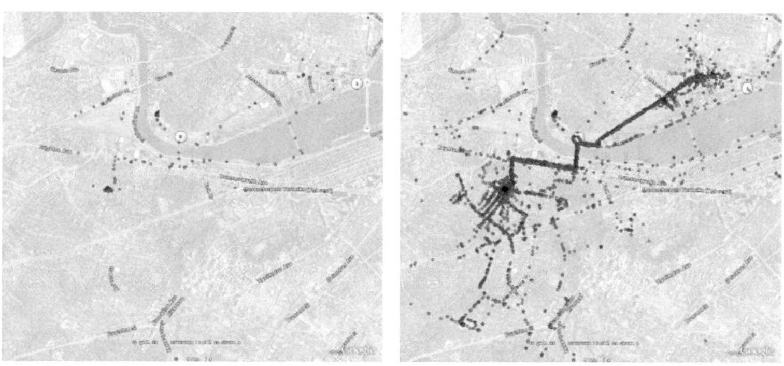

(a) one week. (b) three months.

Fig. 1. GPS samples of a user collected in Boston-Cambridge (MA) area. Each GPS sample is marked by a star. The background is shown by Google Earth.

ordered by time. Fig. 1(a) shows the GPS records gathered from a user over a week. Fig. 1(b) shows the GPS data of the same user over three months. The figures suggest that some repetitive patterns exist in the user's trace log.

The required map of user activities should identify the locations and paths that are significant to the user. In addition, people establish relationship between locations to learn the environment. For example, one can tell a direction by saying "five minutes walk from the shop, you will get to the post-office". The shop, the post-office, and the path between them, thus form the significant components of a map.

2.1 Identifying Significant Places

To extract patterns, one crucial step is to identify "significant places". A mobile user might stay at one location for some time, or he may visit this location frequently (while the device may keep recording). Thus, a location is identified to be significant if user's dwelling time at this location is sufficiently long, or if the density of the location is sufficiently high. A location is defined to be a tuple $(x, y; r)$ where (x, y) represents the coordinates of a position and r denotes the radius of the associated circle centered at (x, y). The *density* of a location is the number of sample points within the associated circular region. The *dwelling time* at each location is approximated by averaging the time staying at this position for each trace.

We choose two thresholds T_{dw}, λ_{min} for dwelling time and location density respectively to identify significant places. Figure 2(a) draws the number of significant places against the dwelling time threshold using the 3-month history of a voluntary user. For example, a point at $(60, 53)$ means that there are 53 significant places whose dwelling time is no less than 60 seconds. The relatively stable number of SPs implies that the user stays long enough only in places that are of significance to him. Figure 2(b) draws the number of significant places

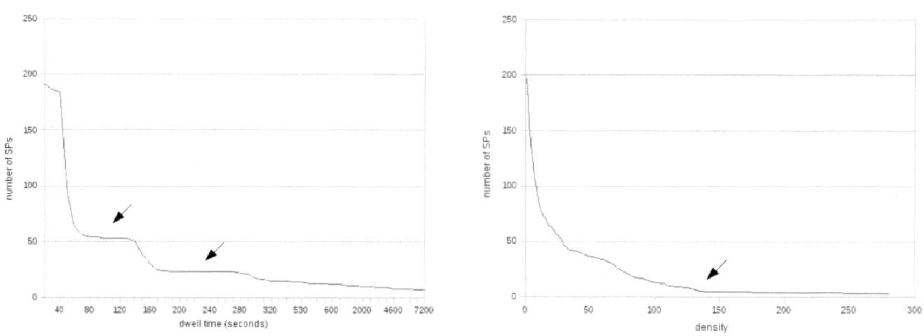

(a) The number of significant places remains invariant when the dwelling time is inside certain interval.

(b) The number of significant places obtained by setting density threshold.

Fig. 2. The number of significant places against T_{dw}, λ_{min}, resp

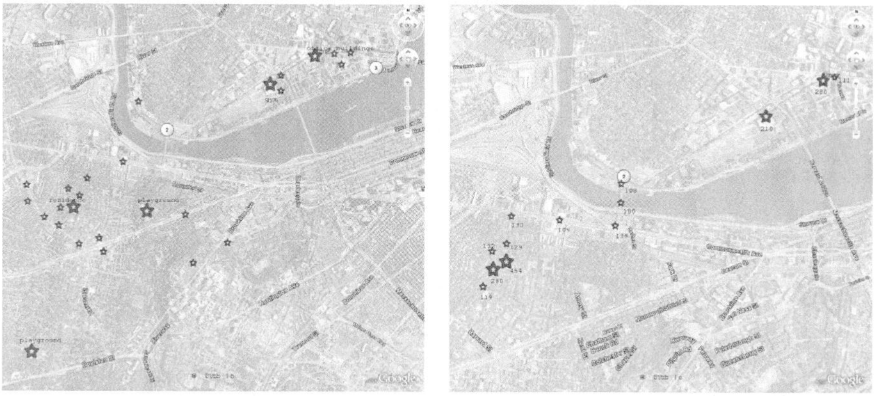

(a) Twenty-six locations are identified with dwelling time over 5 minutes, which are office, gym, playground, residence, and the path points nearing traffic lights.

(b) Thirteen locations are identified with density >= 100, which are residences, a gym, an office bldg., and the path points. Each is marked by a density number.

Fig. 3. Significant places are identified in Google Earth

against the density threshold using the 3-month history. The arrow shows the point from which the number of SPs changes slowly with the increase of density.

Figure 3(a) shows some locations whose dwelling time is more than 5 minutes are inferred from user's historical data. These places, identified as office, home, playground etc. on the map, are mostly important to the user's activities. Figure 3(b) also shows that the set of the significant places obtained by setting density threshold (=100) is rather similar to the set as seen in Figure 3(a). That is, important places to user, such as office and residence, are included by both schemes for identifying significant places.

2.2 Identifying Representative Paths

Multiple traces between two significant places do not necessarily correspond to the same path. Two nearby traces may actually correspond to two disjoint paths that only join at the two ends, or they may share some portion of a path but split at a branching point. In addition, noises are inherent in the trace samples. Consequently, identifying representative paths implies two closely related problems: (i) The first problem is to classify traces into different paths according to certain metric of trace similarity. (ii) The second problem is to form a representative path for those similar traces.

Our path reconstruction is carried out in two steps. Firstly, the user's log, or a sequence of position measurements ordered in time, is segmented into multiple disjoint traces according to their similarity under both temporal and spatial metrics. Secondly, similar traces will be consolidated, and some disjoint but geographically close traces can be transformed into connected paths. The resulting representative paths will be further used in map construction.

A path $\mathbb{P}$ is treated as a curve $x(t)$ in the plane. A trace of $\mathbb{P}$ is a sequence P of sample points $\{z_k\}_k$ from this path. Denote $z_k = x_k + n(t_k)$, where $x_k = x(t_k)$ is the path position at time point t_k, and $n(t_k)$ is called noise. Let $P^{(i)} = \{z_k^{(i)} : k = 1, 2, ..., K^{(i)}\}$ denote a trace. The *path reconstruction* problem is to estimate $x(t)$ from a set of N_s traces $\{P^{(i)}\}, 1 \leq i \leq N_s$ for the same path. The distance between two traces P, Q is defined by their Hausdorff distance $d_H(P, Q)$ [1].

A path reconstruction algorithm named PCM (pairwise curve-merging) is presented below. After path reconstruction, similar traces will be consolidated, and some disjoint but geographically close traces can be transformed into connected paths. The resulting representative paths will be further used in map construction.

PCM Algorithm: Reducing Multiple Traces to Repr. Paths: The basic idea of PCM algorithm is to iteratively compare every two traces and merge similar segments of the two traces. The algorithm is applied pairwise on the traces and a predefined distance threshold value, ϵ, is given. When the iterative process terminates, the remaining traces will all be distinct based on the distance metric.

The PCM algorithm (in Algorithm 1) compares every two traces P, Q for a given tolerance ϵ: each point p of P is compared with its nearest neighbor q in Q. If p is within the tolerance distance of q, and if p is closer to q than to any adjacent point of p in P, then p is replaced by q. Intuitively, the PCM algorithm enforces two conditions when carrying out the merging operation. Firstly, the algorithm repeatedly merges two curves whenever nearby (but not the same) points between the curves are found. Secondly, one polygonal curve partitions the plane into many zones by bisecting the line segments of the curve. Each point of the curve dominates one zone. For one point $p \in P$, only the points of Q that lie inside p's zone can replace p. The second condition ensures the order of the curve points.

Lemma 1 below shows that the process of the PCM can terminate. Furthermore, it shows that two nearby curves in Hausdorff distance will still remain close by after merging. We refer readers to [6] for the proof.

Lemma 1. *PCM can finish in finite number of steps. Moreover, let $\bar{P}, \bar{Q}$ denote the new curves obtained from P and Q, respectively, after executing subroutine $Merge(P, Q, \epsilon)$. If $d_H(P, Q) \leq \epsilon$, then*

$$\max\{d_H(\bar{P}, P), d_H(\bar{P}, Q), d_H(\bar{Q}, P), d_H(\bar{Q}, Q)\} \leq \epsilon.$$

The computational complexity of the PCM algorithm is estimated as follows. Let n be the number of curves, and m be the maximum number of sample points on a trace. The PCM algorithm at most runs $n(n-1)$ iterations of subroutine $Merge(P, Q, \epsilon)$. For $Merge(P, Q, \epsilon)$, we can firstly construct a Voronoi diagram for the points of Q to speed up the nearest-neighbor searching. The construction costs $O(m \log m)$ time, and each nearest-neighbor searching costs $O(\log m)$ time [1]. Since $Merge(P, Q, \epsilon)$ traverses all the points of P, the total running time is $O(m \log m)$. As a result, the running time of PCM algorithm is $O(n^2 m \log m)$.

Input: ϵ and a set of traces, M
Output: Updated M

changed = True;
while *changed* **do**
    changed = False;
    for $P, Q \in M$ *and* $P \neq Q$ **do**
        for *each point* $p \in P$ **do**
            q = the point in Q that is nearest to p;
            p_{left}, p_{right} = p's left and right point in P, resp.;
            if $p \neq q$ *and* $d(p, q) \leq \min\{\epsilon, d(p_{left}, q), d(p_{right}, q)\}$ **then**
                $p = q$;
            end
        end
        if $\bar{P} \neq P$ **then**
            changed = True;
        end
    end
end

Algorithm 1. Merging a set M of traces, by tolerance ϵ

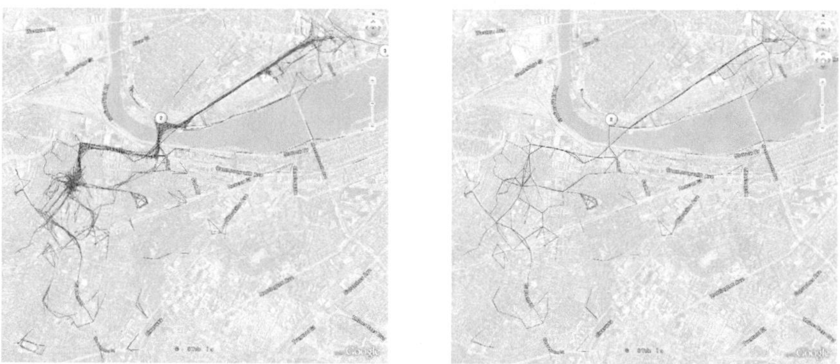

(a) A user's raw data segmented into 748 traces, with thresholds $T_{th} = 1200$ seconds, D_{th} =0.2 km, and a total of 6003 GPS sample points. The disjointed points in the raw data are pruned off.

(b) Paths are reconstructed from the user's traces. The parameter ϵ = 0.050km. The resulted 197 significant places are connected by 289 edges.

Fig. 4. Paths are reconstructed from the user's traces by the PCM algorithm, and shown in Google Earth

Figure 4(a) shows the user's traces given as the input of the pairwise curve merging algorithm. Figure 4(b) shows the corresponding output of the pairwise curve merging algorithm. To compare the merged results with the actual roads and streets, the output is also shown on the maps of Google Earth. It can be seen that the redundancy in the traces is dramatically reduced, and the reconstructed paths reflects the actual roads rather well.

2.3 Extracting Transition Probabilities in the Map

A map of user activities can be constructed after identifying significant places and representative paths from historical log. The map is a 2-dimensional directed graph $G =< V, E >$ without self-cycles, where V is the set of significant places and E is the set of edges. Each vertex $v \in V$ has associated information of the x-y coordinates of the center and the radius of the circle representing the place, dwelling time, and density. The density of a location is defined to be the number of samples within the associated circular region. Each edge $e =< v_1, v_2 >\in E$ records the connectivity from v_1 to v_2. Two vertices v_1, v_2 have an edge when there is any path crossing from v_1 to v_2. Moreover, each edge also records the edge width, the speed and the number of traversals, which are calculated based on the historical records.

A sample map is given in Figure 5(a). The user speed on the edge e is approximated by the average speed of the historical traces that constitute e. For each edge $e =< v_1, v_2 >$, its speed is calculated by the distance between v_1 and v_2, divided by the time difference.

The number of traversals over the edges provides an approximation of the transition probabilities among the vertices. Figure 5(b) shows a subset of the map which contains three vertices and their transitions. Specifically, the transition probability density function (*pdf*) from vertex *src* is approximated by

$$Prob(z|src) \sim normalization\{\lambda(src, z) :< src, z >\in E\},$$

where $\lambda(src, z)$ is the number of traversals on edge $< src, z >$. Let $\pi(v)$ denote the transition probability density function for vertex v. For example, $\pi(A) = [\frac{20}{87}, \frac{29}{87}, \frac{2}{87}, \frac{2}{87}, \frac{16}{87}, \frac{11}{87}, \frac{7}{87}]$ for vertex A in Figure 5(b).

Supported by the map with transition probabilities, certain inferences concerning the path likely to be used in between the significant places can be carried out. For example, a "shortest" path from the source vertex to the destination vertex can be obtained by applying a searching algorithm. We use a classical heuristic searching algorithm, $A*$ *search* [5], in order to minimize the total estimated cost. Just like using distance $d(v_1, v_2)$ as a cost function in the shortest-path problem, we interpret the transition probabilities as a part of the cost on the edges. The cost function is chosen to be

$$J(v_1, v_2) = \frac{d}{\log(1 + \lambda)},$$

where λ is the number of traversals on the edge $< v_1, v_2 >$, and d is the length of the edge. The rationale of this cost function is: when the user visits a path more often than the other paths from the same location, he/she is more likely to reuse this path than the other paths in future. Each use of a path is thus like shortening the distance between two locations by a $\frac{1}{\log(1+\lambda)}$ percent.

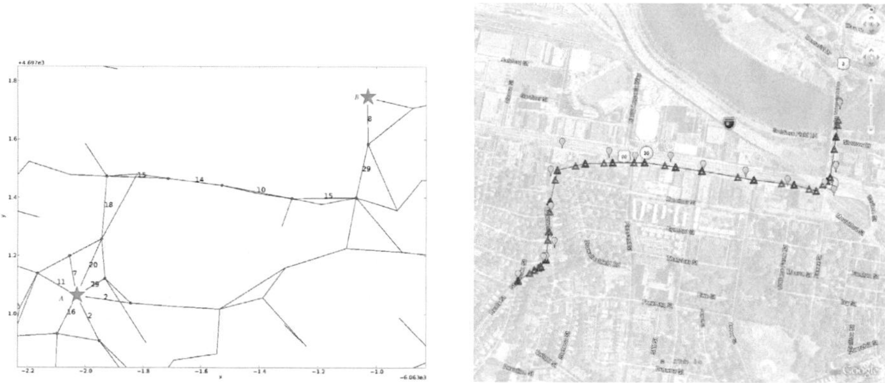

(a) A "shortest" path from source A to destination B is marked between two stars on the map. The numbers on the edge indicate the numbers of traversals. The numbers of traversals from vertex A to its neighbors are also marked.

(b) Estimating a trajectory using the map. Each prediction is indicated by a triangle, while each measurement is drawn as a balloon. The predicted trajectory is shown in solid lines.

Fig. 5. A snapshot of user activity map

Figure 5(b) shows a "shortest" path in the map for the given source and destination. The average speed for this path is evaluated from history to be 2.8 meters per second (by bicycle).

3 Experiments and Performance Evaluation

To verify the effectiveness of the user activity map, we set up experiments for a position tracking application as follows. Assume that a map of user activities is given (e.g., Figure 5(a)), which has been extracted from the user's historical positional data.

We test the tracking method for a given trace with the known path from source to destination. The position inference is activated periodically. The method takes into consideration user's speed model derived from the map. Specifically, the average velocity with its standard deviation can be statistically approximated from the historical traces. In this experiment, the estimated trajectory of the user is then compared with the actual trajectory at each time step.

Figure 5(b) draws all observed positions of a user's trajectory and a predicted trajectory. It shows that the predicted trajectory well matches the user's actual movements, which is even better than merely relying on GPS measurements. This is possible because GPS signals can be momentarily disturbed by external sources. The *Root Mean Squared Error* for the estimated trajectory and the actual trajectory is about 4.3 meters, which should work fairly well for location-based applications.

4 Discussions and Conclusion

This paper presents the method for mining the pattern of user activity from historical positional data. The method includes the algorithm for constructing significant places and representative paths. It also derives the information about user's speed and transition probabilities. The map of user activity is then applied to position tracking applications. The experiments show that the algorithm can be applied to personal position tracking, and it deals with user's non-linear movement behaviors fairly well. Thus we believe that our algorithm for mining the user activity map can form a basis for many promising personal location-aware applications.

References

1. Aurenhammer, F., Klein, R.: Voronoi diagrams. In: Handbook of Computational Geometry, pp. 201–290. Elsevier, Amsterdam (2000)
2. Doucet, A., Godsill, S., Andrieu, C.: On sequential monte carlo sampling methods for bayesian filtering. Stat. and Comput. 10(3), 197–208 (2000)
3. D'Roza, T., Bilchev, G.: An overview of location-based services. BT Technology Journal 21(1), 20–27 (2003)
4. Hoffmann-Wellenhof, B., Lichtenegger, H., Collins, J.: GPS: Theory and Practice, 3rd edn. Springer, New York (1994)
5. Russell, S., Norvig, P.: Artificial Intelligence: A Modern Approach, 2nd edn. Prentice Hall, Englewood Cliffs (2003)
6. Fang, H.: Mining user position log to generate activity map for personal location-aware applications. Technical Report NTUTR-09-FH02, Nanyang Technological Univ (February 2009)

A Multi-Strategy Approach to KNN and LARM on Small and Incrementally Induced Prediction Knowledge

JuiHsi Fu* and SingLing Lee

Chung Cheng University
168 University Road, Minhsiung Township,
62162 Chiayi, Taiwan
{fjh95p,singling}@cs.ccu.edu.tw

Abstract. Most classification problems assume that there are sufficient training sets to induce the prediction knowledge. Few studies are focused on the label prediction according to the small knowledge. Hence, a classification algorithm in which the prediction knowledge is induced by only few training instances at the initial stage and is incrementally expanded by following verified instances is presented. We have shown how to integrate kNN and LARM methods to design a multi-strategy classification algorithm. In the experiments on *edoc* collection, we show that the proposed method improves 4% in accuracy of low-confidence results of kNN prediction and 8% in accuracy of results of the dominant class bias of LARM prediction. We also show experimentally that the proposed method obtains enhanced classification accuracy and achieves acceptable performance efficiency.

Keywords: k Nearest Neighbors (kNN), Lazy Associative Rule Mining (LARM), Multi-Strategy Learning, Lazy Learning.

1 Introduction

Machine Learning approaches attempt to induce knowledge from solved cases, named as training sets (instances), and then use it to predict labels of unknown cases, named as testing sets (instances). In general, most approaches assume that there are sufficient training sets for building the prediction knowledge. However, obtaining enough training sets for learning problems is usually time-consuming since training instances are labeled and verified by human experts. Few studies are focused on the label prediction only depending on the small induced knowledge. In this case, the classification algorithm should be tailored to the unlabeled instance since the complete prediction knowledge could not be induced by the insufficient training sets. Also, desirable prediction results may not be achieved, so the training sets need to be expanded for updating the prediction knowledge. Thus, classification methods are required to perform efficiently when the procedure of updating

* This work is supported by NSC, Taiwan, under grant no. NSC 97-2221-E-194-029-MY2.

R. Huang et al. (Eds.): ADMA 2009, LNAI 5678, pp. 453–460, 2009.

prediction knowledge is frequently executed. In this paper, we are motivated to design a classification algorithm in which small prediction knowledge is induced at the initial step and incrementally updated by following training instances.

Generally speaking, eager and lazy are two major types of classification algorithms. Eager algorithms predict the label of a testing instance after building the prediction knowledge. Lazy algorithms generate the prediction knowledge until the testing instance arrives. In other words, eager classification depends on the whole training set, and lazy classification utilizes the training instances that are related to the testing instance. Especially, when only few training instances are available for inducing prediction knowledge, the prediction result of lazy classification could be more tailored to the testing instance than that of eager classification. Moreover, lazy classifiers apparently require no extra operation for inducing prediction knowledge when the training set is incrementally updated, but eager classifiers need to rebuild that additionally. Therefore, the lazy algorithm is a proper solution for the document classification problem in which small prediction knowledge induced at the initial stage is incrementally updated.

In recent years, lazy classification was already employed in decision trees [1] and Bayesian rules [2]. The instance-based algorithm, like k-nearest neighbors (kNN), is inherently a lazy algorithm. But, kNN prediction results might be not quite precise when those most relative neighbors have low similarity with the testing instance. With the advances in data mining, Veloso et al [3,4] proposed a lazy associative classifier that integrated associative rules and lazy classification for label prediction, named as Lazy Associative Rule Mining (LARM). It usually suffers from the dominant (most trained) class bias in prediction because of the biased instance proportion. Although each learning algorithm contains explicit or implicit biases and weaknesses in label prediction, it would be wise to combine different kinds of these appropriately[5].

Some algorithms that take advantages of biases of rule-based and metric-based methods have been designed in multi-strategy learning[6]. However, [7,8,9] are partial and entire eager classification. [10] proposes a prediction method that is based on rules mined from the k nearest instances of the testing instance. But, the prediction results strongly depend on the selection of k nearest neighbors. In this paper, a multi-strategy lazy learning algorithm based on kNN and LARM is carefully designed. Our main contribution is to improve low-confidence prediction in kNN and dominant class bias in LARM simultaneously.

This paper is organized as follows: in Sect. 2, we introduce kNN and LARM methods. Then, we propose a multi-strategy lazy algorithm in Sect. 3, and demonstrate the numerical analysis of the proposed method in Sect. 4. Finally, we give conclusions in Sect. 5.

2 Related Works

2.1 Metric-Based Classification Methods

Metric-based classification methods are instance-based learning algorithms [11]. They simply store the training instances and retrieve relative ones when a testing instance arrives. Apparently, metric-based methods have properties of lazy classification. For example, label prediction of kNN, a traditional metric-based method, depends on the most relative neighbors that are decided by a similarity metric. However, low similarity neighbors identified by kNN might have less confidence to decide the label of a testing instance. Hence, low confidence prediction is an interesting issue in kNN classification.

2.2 Rule-Based Classification Methods

Lazy Associative Rule Mining (LARM) is an integration method of Associative Ruling Ming [12] and lazy classification. LARM mines associative rules from the subset of the training set that is related to the testing instance. Originally, [13] denoted associative rules as class associative rules (CARs) of the form I $\longrightarrow$ y, where the antecedent (I) is composed of features (items) and the consequent (y) is a class label. Given a training set D and a testing instance x_t, D_{x_t} that is the set of training instances of which features co-occur in x_t is generated, and then associative rules are mined from D_{x_t} for predicting x_t's label. However, if class y_d is a dominant class, rules of the form I $\longrightarrow y_d$ would be induced more than those of the form I $\longrightarrow y_i$, where $y_i != y_d$. It tends to lead that low-confidence-value rules effect the prediction results more than high-confidence-value rules when the number of low-confidence-value rules are large enough. The reason is that the confidence value summed by low-confidence-value rules might be larger than that summed by high-confidence-value rules. Hence, LARM tends to suffer from the dominant class bias when the document proportion is unbalanced.

In the next section, we propose a multi-strategy lazy algorithm in order to improve classification results made by low-similarity references identified by kNN and to ease prediction results of the dominant class bias in LARM generated by the biased instance proportion.

3 The Proposed Method

Motivated by the weaknesses for each of LARM and kNN classification methods, we proposed a multi-strategy solution to improve the prediction ability of these two based methods. Compared to the specific references identified by kNN, LARM extracts more detailed information among training instances because all possible combinations of features are considered for label prediction. Clearly, our proposed multi-strategy learning is designed as that kNN prediction is performed at the first phase. When the references identified by kNN are low-confidence for classification, LARM takes responsibility, in turn, for predicting labels of testing instances. Therefore, kNN prediction at the first phase is able to ease the dominant class bias by high-similarity neighbors, and LARM prediction at the second phase might improve low-confidence kNN prediction by detailed rule mining.

At the first phase, given a training set D, a testing instance x_t, an integer k, Cosine Similarity metric, and a similarity threshold T_s, when testing instance x_t arrives, kNN calculates the similarity value, s_{ti}, of each pair (x_t, x_i), in which x_i is an instance in the training set D, and then extracts k most similar instances,

input : training set D, testing instance x_t, an integer k, minimum confidence $mconf$, similarity threshold T_s, and pruning threshold T_p
output: x_t's predicted label, y_t
begin
  if *the average similarity value of all k most similar neighbors* $\geqq T_s$ **then**
    $y_t \longleftarrow kNN(D, x_t, k, T_s)$;
  else
    $y_t \longleftarrow LARM(D, x_t, mconf, T_p)$;
  end
  Return y_t ;
end

Algorithm 1. The proposed multi-strategy classification algorithm

D_{x_t}. After that, if the average similarity value of instances in D_{x_t} exceeds T_s, x_t's predicted label is decide by the majority voting of all instances in D_{x_t}. Otherwise, going to the second phase.

At the second phase, only related training instances of which features co-occur in the testing instance are collected in D_{x_t}. Then, associative rules are mind by *apriori* algorithm [14] from D_{x_t}. It is observed that when most of itemsets of which sizes are j imply the same class label, they could dominate the induced result among all j-itemsets. In this case, the execution time of mining rules of $(j+1)$-itemsets can be saved. Hence, we define a stop criteria: when the size of certain j-itemsets that imply the same class is $Tp\%$ of the size of all j-itemsets, *apriori* algorithm is not necessary to mine $(j+1)$-rules.

The proposed multi-strategy method is presented in Alg. 1. Given instances (training set D and a testing instance x_t) and thresholds (similarity threshold T_s, min confidence $mconf$, pruning threshold T_p), kNN decides x_t's predicted label firstly, and LARM also makes the prediction if the average similarity value of all k most similar references is not qualified.

4 Experiments

In this section we present experimental results for the evaluation of the proposed classifiers in terms of classification accuracy and computational performance that is measured on the desktop (Windows 2003/Intel(R) Xeon(R) CPU 2.00GHz/4.00 GB RAM). Our evaluation is based on the comparison against based lazy classifiers LARM and kNN. The minimum confidence $mconf$ is 0.3 and the threshold Tp in the proposed method is assigned as 0.9. We use two datasets, *edoc* and *ModApte*, in our simulation. The first dataset, *edoc*, is collected by Chinese official documents in Chung Cheng University and composed of traditional Chinese characters and short contents. It is utilized for simulation of inducing small prediction knowledge since short document contents mostly contain insufficient information. Totally, there are 5,332 documents and 81 classes, and each document belongs to only one class. The document proportion is biased

Table 1. Classification accuracy of kNN methods in terms of different k and metrics on *edoc*, *ModApte1287*, and *ModApte2627*

	k			
	1	7	13	17
edoc	**66.04%**	63.97%	61.40%	60.07%
ModApte1287	48.87%	**52.84%**	51.67%	51.36%
ModApte2627	50.32%	53.18%	**54.17%**	53.45%

since 1,663 documents belong to the two specific classes. Each document content is segmented by Chinese sentence segmentation tool [15], and then 1-length Chinese terms are removed. Also, remaining terms are weighted by tfidf [16], $(tf * \frac{N}{DF})$, and 80% potentially significant ones are kept. The second dataset, *ModApte*, is the split of Reuters-21578 collection [17], and contains 13,307 documents and 115 categories. Especially, only 10%, and 20% of documents in each category are randomly picked to simulate the environment of induced the small prediction knowledge. So, instead of the whole *ModApte* collection, 1287, and 2627 documents in the two subsets, *ModApte1287* and *ModApte2627*, are simulated in our experiments. Also, only one category is labeled to each document, and the document proportion is biased since 631 and 1264 documents belong to the two specific categories (acq and earn). All terms in datasets are weighted by tfidf, $(tf * log_{10}\frac{N}{DF})$, and 80% potentially significant ones are kept. At the initial stage of the experiment, no labeled instance is put into the training set, and then the training set is continually expanded when labels of testing instances are verified. These two conditions are used for simulating the environment of incrementally updating the small prediction knowledge.

On each data collection, classification accuracy of kNN methods with different values of k is presented in Tab. 1 and the highest one is bold-faced. It is observed that, on *edoc*, the highest accuracy is achieved by $k = 1$. So, documents in *edoc* are difficultly classified since few instances that are similar to each other locate closely. For other datasets, the proper values of k are 7 and 13 on ModApte1287 and ModApte2627, respectively.

The similarity threshold, Ts, in the proposed method should be decided, such that the improvement made by the proposed method in low-confidence prediction and dominant class bias could be appropriately demonstrated. The classification accuracy and running time of our methods in different Ts are shown by Tab. 2. The running time of a method is the time spent by predicting labels of all testing instances and re-inducing classification knowledge after the label of each testing instance is verified. In terms of accuracy and efficiency of the label prediction, 0.4 is assigned to Ts on *edoc*, so the testing documents, of which the average similarity value of the references is less than 0.4, are required to be classified by LARM at the second phase. On *ModApte1287*, the proper Ts=0.4 obtains acceptable efficiency and high prediction accuracy. However, on *ModApte2627*, the highest classification accuracy, 54.25%, is achieved by Ts=0.1. Only little improvement is made since accuracy of the based classifier, kNN when k=13, is

Table 2. The prediction accuracy and efficiency of the proposed method by different values of Ts

	k=1		k=7		k=13	
	edoc		*ModApte1287*		*ModApte2627*	
T_s	accuracy	time	accuracy	time	accuracy	time
0.1	65.81%	0:2:47	52.76%	0:0:48	**54.25%**	0:3:29
0.4	**67.40%**	0:11:42	**53.38%**	0:2:8	53.98%	0:9:39
0.6	66.49%	0:22:34	51.83%	0:30:40	50.97%	2:51:58
0.9	64.80%	0:32:54	51.20%	0:41:4	50.55%	3:24:58

Table 3. Classification accuracy of LARM and the proposed method when the prediction biases to the dominant class C_a and C_b

	LARM	k=1	LARM	k=7	LARM	k=13
	edoc		*ModApte1287*		*ModApte2627*	
bias to C_a	56.11%	**64.89%**	44.49%	**48.28%**	54.83%	**55.27%**
bias to C_b	55,73%	**64.72%**	70.93%	**81.80%**	60.55%	**77.91%**
bias to C_a and C_b	55.96%	**64.83%**	58.28%	**64.80%**	58.52%	**67.75%**

54.17%. Hence, our method doesn't work well when the prediction knowledge is induced by larger than 2,627 documents in *ModApte* collection.

Prediction of the dominant class bias in LARM and the proposed method is shown in Tab. 3. The two dominant classes on *edoc*, *ModApte1287*, and *ModApte2627* are denoted by C_a and C_b. It is observed that the proposed method improves the prediction accuracy when the label of a testing instance is decided as C_a or C_b. This analysis tells that LARM prediction tends to suffer from the dominant class bias in the unbalanced document proportion. And, the proposed algorithm is able to avoid the bias since kNN prediction at the first phase is based on the specific references of the testing instance.

Low-confidence prediction is defined by that, in the metric-based prediction, the average similarity value of the selected references is less than Ts which is already decided by previous experiment results. Low-confidence prediction results performed by kNN and the proposed method, except on *ModApte2627*, are shown in Tab. 4. Clearly, it is found that in terms of classification accuracy, the improvement of low-confidence prediction is made by our method since LARM prediction at the second phase extracts sufficiently detailed information from training instances. Hence, this experiments demonstrate that the results of low-confidence prediction can be corrected at the second phase.

The last evaluation in Tab. 5 presents the comparison against the based classifiers LARM and kNN that are also designed by the single strategy. Compared with LARM and kNN, the proposed method, a multi-strategy algorithm, achieves better prediction results. Because only small prediction knowledge is induced for

Table 4. The accuracy of low-confidence prediction performed by kNN and the proposed method

kNN	our method	kNN	our method
Ts = 0.4, k=1, on *edoc*		Ts = 0.4, k=7, on *ModApte1287*	
35.44%	**39.70%**	26.57%	**32.19%**

Table 5. The prediction accuracy and efficiency performed by the based methods and the proposed classifier

	LARM		kNN		the proposed method	
	accuracy	time	accuracy	time	accuracy	time
edoc	64.10%	0:33:14	66.04%	0:2:57	**67.40%**	0:11:42
ModApte1287	49.81%	0:37:20	52.84%	0:0:48	**53.38%**	0:2:8
ModApte2627	48.99%	3:8:28	54.17%	0:3:31	54.25%	0:3:29

classification, the performance of all classifiers is not impressively accurate. However, it is noted that their classification performance is very efficient. Therefore, the experiments show that our method has the ability of obtaining acceptable accuracy and achieving reasonable efficiency in the environment where the small prediction knowledge induced at the initial stage is incrementally updated.

5 Conclusion

When the prediction knowledge is incrementally updated, lazy algorithms are suitable for classification since (1) lazy classification is tailored to the testing instance. (2) lazy approaches require no extra operation for re-inducing prediction knowledge. Notably, each kind of algorithms would encounter special biases and weaknesses. Therefore, we propose a multi-strategy lazy algorithm to avoid the dominant class bias in LARM and improve low-confidence classification results in kNN. In the simulation results, at first, we use numerical analysis to present the problems of low-confidence prediction in kNN and the dominant class bias in LARM. Then, the proposed multi-strategy method is demonstrated that it improves these two kinds of prediction results. Summarily, it is shown that the proposed multi-strategy lazy algorithm obtains reasonable accuracy and achieve reasonable efficiency in the environment of incrementally updating the induced small prediction knowledge.

References

1. Friedman, J.H., Kohavi, R., Yun, Y.: Lazy Decision Tree. In: 13th National Conference on Artificial Intelligence, pp. 717–724. AAAI Press and MIT, Boston (1996)
2. Zheng, Z., Webb, G.: Lazy learning of bayesian rules. Machine Learning 1, 53–84 (2000)

3. Veloso, A., Meira Jr., W., Zaki, M.J.: Lazy Associative Classification. In: 6th International Conference on Data Mining, pp. 645–654. IEEE Press, HongKong (2006)
4. Veloso, A., Meira Jr., W.: Lazy Associative Classification for Content-based Spam Detection. In: 4th Latin American Web Congress, pp. 154–161. IEEE Press, Cholula (2006)
5. Michell, T.M.: The need for biases in learning generalizations. Technique Report CBM-TR-117, Computer Science Department, Rutgers University, New Jersey (1980)
6. Michalski, R.S., Tecuci, G.: Machine learning: A multistrategy approach. Morgan Kaufmann, SanMateo (1994)
7. Domingos, P.: Unifying instance-based and rule-based induction. Machine Learning 24, 141–168 (1996)
8. Li, J., Ramamohanarao, K., Dong, G.: Combining the strength of pattern frequency and distance for classification. In: Cheung, D., Williams, G.J., Li, Q. (eds.) PAKDD 2001. LNCS, vol. 2035, pp. 455–466. Springer, Heidelberg (2001)
9. Golding, A.R., Rosenbloom, P.S.: Improving accuracy by combining rule-based and case-based reasoning. Artificial Intelligence 87, 215–254 (1996)
10. Wojan, A.: Combination of Metric-Based and Rule-Based Classification. In: Slezak, D., Wang, G., Szczuka, M., Duentsch, I., Yao, Y. (eds.) RSFDGrC 2005. LNCS, vol. 3641, pp. 501–511. Springer, Heidelberg (2005)
11. Michell, T.M.: Machine Learning. McGraw-Hill, New York (1997)
12. Kantardzic, M.: Data Mining: Concepts, Models, Methods, and Algorithms. Wiley Interscience, USA (2003)
13. Liu, B., Hsu, W., Ma, Y.: Integrating classification and association rule mining. In: ACM International Conference on Knowledge Discovery and Data Mining, pp. 80–86. ACM Press, New York (1998)
14. Agrawal, R., Imielinski, T., Swami, A.: Mining association rules between sets of items in large database. In: ACM-SIGMOD International Conference on Management of Data, pp. 207–216. ACM Press, Washington (1993)
15. Chinese Knowledge and Information Processing (CKIP) of Academia Sinica of Taiwan, A Chinese word segmentation system, `http://ckipsvr.iis.sinica.edu.tw`
16. Salton, G., McGill, M.J.: An introduction to modern information retrieval. McGraw-Hill, New York (1983)
17. Reuters-21578 Text Categorization Test Collection, `http://www.daviddlewis.com/resources/testcollections/reuters21578`

Predicting Click Rates by Consistent Bipartite Spectral Graph Model

Wensheng Guo[1,2] and Guohe Li[1]

[1] Department of Computer Science and Technology, China University of Petroleum-Beijing, 102249 Beijing
[2] Key Laboratory of Earth Prospecting and Information Technology Beijing, 102249 Beijing
{gwsice,ligh}@cup.edu.cn

Abstract. Search advertising click-through rate (CTR) is one of the major contributions to search ads' revenues. Predicting the CTR for new ads put a direct impact on the ads' quality. Traditional predicting methods limited to Vector Space Model fail to sufficiently consider the search ads' characteristics of heterogeneous data, and therefore have limited effect. This paper presents consistent bipartite graph model to describe ads, adopting spectral co-clustering method in data mining. In order to solve the balance partition of the map in clustering, divide-and-merge algorithm is introduced into consistent bipartite graph's co-partition, a more effective heuristic algorithm is established. Experiments on real ads dataset shows that our approach worked effectively and efficiently.

Keywords: Bipartite Graph, Spectral Clustering, CTR.

1 Introduction

Search engine advertising has become a significant element of major search engines today. To maximize revenue and user satisfaction, search advertising systems must predict the expected user behavior for each displayed advertisement and must maximize the expectation that a user will act (click) on it.

The search advertising system can make expected user behavior predictions based on historical click-through performance of the ad (CTR). However newly created ads, for lack of historical information, it is often difficult to make CTR predictions effectively, thus affecting the probability that a user will see and click on each ad. As a result the need for establish an effective prediction model through historical data analysis is urgent. In recent years, driven by actual demand, the study of CTR predicting made some achievements, including: Moira Regelson found that using historical data aggregated by cluster leads to more accurate estimates on average of term-level CTR for terms with little or no historical data[1]. Matthew Richardson used features of ads, terms, and advertisers to learn a model that accurately predicts the click-though rate for new ads[2]. In [3], the authors addressed the problem of computing

R. Huang et al. (Eds.): ADMA 2009, LNAI 5678, pp. 461–468, 2009.

CTR for existing ads using maximum likelihood estimation and also presented an algorithm learning an ensemble of decision rules that can be used for predicting the CTR for unseen ads.

Comparing with popular field of data mining such as social network [4], WWW [5] and telephone call graph, it is still lack of research in search advertisement field. As is a high-order dataset, it became a new data mining research field that using graph model and spectral co-clustering method to describe the complex relationship between heterogeneous data.

We summarized our key works on implementing bipartite graph spectral clustering in the search advertisement as well.

1. Propose a consistent bipartite graph model for search ads, then build a basic CTR predicting framework using co-clustering.
2. Introduce the divide-and-merge method to spectral clustering process to compute the ratio cut of the graph. The method can effectively determine whether the size of each clustering result is balanced.
3. Propose a heuristic to redistribute the tiny cluster in divide-and-merge process. It can effectively eliminate the diseased clustering result and get a better graph partition.

2 Predicting Framework

Precise matching with the query and search advertising is a complex heterogeneous object clustering problem. It includes the query words, search ads and feature terms of the three types of different objects. Weight of the edge between query and ads reflects the CTR of the ads in such query words, and the weight of edge between ads and its feature terms reflects the frequency of terms happened in the ads. It is clearly that the problems can be resolved with a star structure consistent bipartite undigraph [6]. Note that the query words collection $Q=\{q_1,q_2,\cdots,q_l\}$, ads collection $A=\{a_1,a_2,\cdots,a_m\}$ and terms collection $W=\{w_1,w_2,\cdots,w_n\}$ is the three groups of vertexes in the graph. $E=\{\{q_i,a_j\}:q_i\in Q,a_j\in A\}$ and $F=\{\{a_i,w_j\}:a_i\in W,w_j\in W\}$ are the edge collections in the graph. If the times which user clicks one ad a_j with query word q_i exceed the threshold value, there is an edge $\{q_i,a_j\}$ in the graph. If term w_j happens in ad a_i, there is an edge $\{a_i,w_j\}$ in the graph. There is no edge between both query words, both ads, both terms in the graph. In order to mine the rules among the query words, ads and terms and predict the CTR, it is necessary to co-cluster those triple heterogeneous objects. The optimal clustering result is corresponding to such a graph partition. That is, weight of the cross edge between different partition is minimum and the clustering result is as far as possible "balanced."

Clustering result with consistent bipartite graph reveals matching rules between query and advertisement, which can be directly used for advertising click-through rate on the new forecast. The steps of predicting by search ads log mining are as follow:

I. Training Phase

1. Establishing the consistent bipartite graph model of search ads by its log;
2. Using the method of semi-definite programming to cluster in the consistent bipartite graph, then obtaining the co-clustering results including query words, ads and terms;
3. After using the divide-and-merge algorithm to determine uneven cluster in the bipartite graph clustering and redistributing the tiny cluster by heuristic, final clustering result will be obtained.

II. Testing Phase

1. Pre-processing: feature extracting from the new ads and standardizing those feature terms as a sequence;
2. Comparing with cluster center: computing how the new ads is similar to each cluster center by the distance of new ads term sequence and terms in each cluster. The distances we used can be seen in Table 1;
3. CTR predicting: According to the category of new ads, obtaining the corresponding historical information of CTR from graph model. Then predict the CTR of new ads as well.

3 Spectral Co-clustering of Consistent Bipartite Graph

3.1 Algorithm Overview

As is shown in Section 2, consistent bipartite graph can not be simply co-clustered with the two bipartite graph (Q, A, E) and (A, W, F). This is because the two sub-graph clustering is often difficult to guarantee the optimal solution in the central object with entirely consistent. Bin Gao et al [6] purposed a semi definite planning (SDP) method in theory to solve the partitioning problem. There is a consistent bipartite graph co-partitioning algorithm for search ads as followed, which E is the adjacent matrix of the vertex set Q and A, F is the adjacent matrix of A and W:

1. Set parameters λ, θ_1 and θ_2, which $0 < \lambda < 1$ is the weight parameter for balance of the two sub-issues, while θ_1 and θ_2 both are planning constraints;
2. On the basis of adjacent matrix E and F, we denoted $D^{(1)}$, $D^{(2)}$, $L^{(1)}$, $L^{(2)}$ as the diagonal matrices and Laplacian matrices. As $\Pi_1 = \begin{bmatrix} D^{(1)} & 0 \\ 0 & 0 \end{bmatrix}_{s\times s}$, $\Pi_2 = \begin{bmatrix} 0 & 0 \\ 0 & D^{(2)} \end{bmatrix}_{s\times s}$, $\Gamma_1 = \begin{bmatrix} L^{(1)} & 0 \\ 0 & 0 \end{bmatrix}_{s\times s}$, $\Gamma_2 = \begin{bmatrix} 0 & 0 \\ 0 & L^{(2)} \end{bmatrix}_{s\times s}$ ($s = l + m + n$ in

section 2), we extended $D^{(1)}$, $D^{(2)}$, $L^{(1)}$, $L^{(2)}$ to $\Pi_1, \Pi_2, \Gamma_1, \Gamma_2$, computed Γ using $\Gamma = \frac{\lambda}{e^T \Pi_1 e}\Gamma_1 + \frac{1-\lambda}{e^T \Pi_2 e}\Gamma_2$;

3. We obtained $\min_H \begin{bmatrix} 0 & 0 \\ 0 & \Gamma \end{bmatrix} \bullet H$ using SDPA iterative algorithm, which subject to:

$$\begin{cases} \begin{bmatrix} -e^T\Pi_1 e & 0 \\ 0 & \Pi_1 \end{bmatrix} \bullet H = 0, \begin{bmatrix} -e^T\Pi_2 e & 0 \\ 0 & \Pi_2 \end{bmatrix} \bullet H = 0, \begin{bmatrix} 0 & e^T\Pi_1/2 \\ \Pi_1 e/2 & 0 \end{bmatrix} \bullet H = 0, \\ \begin{bmatrix} 0 & e^T\Pi_2/2 \\ \Pi_2 e/2 & 0 \end{bmatrix} \bullet H = 0, \begin{bmatrix} 1 & 0 \\ 0 & 0 \end{bmatrix} \bullet H = 0, \begin{bmatrix} 0 & e \\ e & 0 \end{bmatrix} \bullet H = \theta_1, \\ \begin{bmatrix} 0 & 0 \\ 0 & e \end{bmatrix} \bullet H = \theta_2, H \geq 0 \end{cases} \tag{1}$$

4. Extract vector h as a classification sign from matrix H;
5. Run clustering algorithm on h such as K-means to obtain the result of partition.

In the clustering process with bipartite graph of search ads, there is often a large subset with dense connections in graph or the remaining subgraph. In this case, the spectral clustering algorithm usually collects branches which is loosely connected with clustering center into several clusters respectively, and collects the subgraph with dense connections into one cluster. If all of these disconnected subgraphs are independent cluster, the size of cluster result that we obtained becomes huge difference.

To solve this problem, spectral algorithm distinguishes the uneven clustering result at first. In this paper, a Divide-and-Merge algorithm is used, see Section 3.2. On this basis, the algorithm uses a heuristic function to control and correct the cluster size.

3.2 Divide-and-Merge for Uneven Cluster

The divide-and-merge method for spectral clustering is also a two-phase algorithm [7]. In the first phase, the algorithm bisects the graph recursively: through sorting the eigenvalues of the Laplacian matrix, we obtained the optimal bisection by computing the second eigenvector[8]. The algorithm in the literature [7] applied to document clustering. In order to cluster the consistent bipartite graph better, we improved the algorithm as follow: In divide phase, we introduced a heuristic to redistribute the tiny cluster to eliminate the uneven result. In merge phase, we gave a recursive algorithm in which was introduced the ratio cut.

Algorithm 1. Divide-and-Merge: Divide Phase

```
while(clusters amount < cnum_0){
  Let A is the largest cluster's adjacency matrix;
  Compute the second largest eigenvector v' of D^{-1/2} A D^{-1/2};
  Let v = [D_1^{-1/2} v'_2 ; D_2^{-1/2} v'_1] and sort v; i = ratio_init;
  while(C is not discarded){
    Find 1/i ≤ t ≤ 1-1/i, the cut
    (S,T)=({1,...,t·n},{t·n+1,...,n})makes the cluster ratio is
    minimum;
    (C_1,...,C_l)=Redistribute(S,T);
    if(l>1) Split C into (C_1,...,C_l);
    else if(i=3) Discard the cluster C
    else i--;}}
```

It is very easy to lead to small clusters in the divide phase as we found. In fact, partition of graph through the second eigenvector always trends many disconnected subgraph. Therefore, we introduced a heuristic algorithm Redistribute (S,T) for processing those tiny clusters.

We also used a parameter $ratio\_init$ to control the graph bisection process. Terminal condition of the loop in algorithm is the denominator $i=3$. At this time the bisection will be continued no longer, and the divide phase is end.

Algorithm 2. Divide-and-Merge: Merge Phase

```
for each(C from leaf up to root){
  if(C is leaf) {OPT_n(C,1)=0; OPT_d(C,1)=|C|}
  else{
    let (C_1,...,C_l) be the children of C;
      for(i=1;i<=total below C;i++){
        for(all i_1+...+i_l=i){
          N(i_1,...,i_l) = Σ_{j≠j'} d(C_j,C_{j'}) + Σ_{j=1}^{l} OPT_n(C_j,i_j);
          D(i_1,...,i_l) = Σ_{j≠j'} |C_j|·|C_{j'}| + Σ_{j=1}^{l} OPT_d(C_j,i_j);
          if( OPT_n(C,i)/OPT_d(C,i) > N(i_1,...,i_l)/D(i_1,...,i_l) ){
            OPT_n(C,i)=N(i_1,...,i_l); OPT_d(C,i)=D(i_1,...,i_l); }}}}
```

Merge phase is designed to optimize the ratio cut of the final clustering, that is, the intercluster similarity. Therefore, we adopted the computing process from the leaf to the root. For obtaining the optimal ratio cut for a given cluster C which is divided into i subgraphs, it is needed to compute all of the $i_1 + \cdots + i_l = i$ partitions one by one. According to the recursive algorithm, we had a hypotheses that a good partition i_j can be only established by its subgraph C_j. So that the ratio cut of partition can be drawn as formula 2:

$$ratio\_cut = \frac{\sum_{j \neq j'} d(C_j, C_{j'}) + \sum_{j=1}^{l} OPT_n(C_j, i_j)}{\sum_{j \neq j'} \left|C_j\right| \cdot \left|C_{j'}\right| + \sum_{j=1}^{l} OPT_d(C_j, i_j)}. \tag{2}$$

4 Experiments

For the experiment we had in total 3284 ads in five categories. They were sampled from the finance channel and games channel of Yahoo by our spider and were labeled by artificial. Then we randomly selected 20 ads for each category to build training set. Here we carried out feature selection for characteristic so that only 332 terms were reserved. We also picked 50 query words for these ads and recorded the click rates with every query and advertisement. Each category has a category name (C_1 is Stock, C_2 is Currency, C_3 is Industry, C_4 is Puzzle, C_5 is Video) and the expected clustering results would be $\{C_1, C_2, C_3\}$ (which master categories are all Finance) and $\{C_4, C_5\}$ (which are Games). Ads should be clustered according to the categories they belong to.

4.1 Results of Spectral Clustering

Firstly we ran the spectral clustering algorithm as described in section 3. As there are three dimensional data such as query words, ads and terms in the results, we plotted the embeddings of query words and terms in a separate sub-figure to show the results distinctly, although the embedding values of the three heterogeneous objects were co-clustered together.

From Fig. 1 we found that ads from finance channel (the dark-colored points) were clustered together with their query words. The others (games channel, the white-colored points) were merged into one cluster. To summarize, this experiment showed that spectral clustering algorithm can avoid the case of ill partitioning to many extents, and these results confirmed that the consistent bipartite graph co-partitioning can mine the search advertisement data successfully.

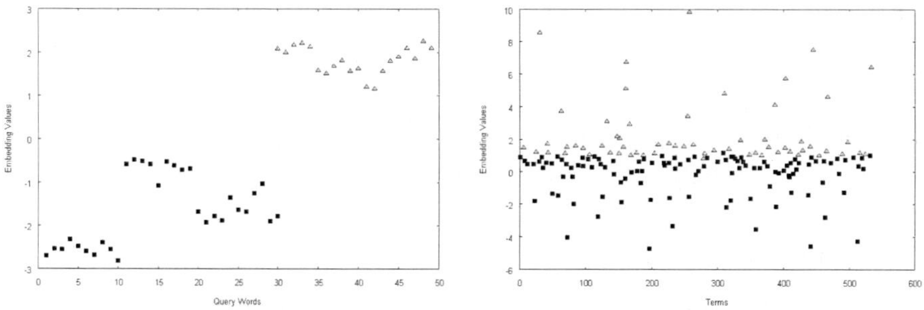

(a) Query Words (b) Terms

Fig. 1. Distribution of Co-Clustering Queries, ads and terms with Consistent Bipartite Graph

4.2 Predicting Click Rates

We compared predictions of CTR using several different methods across twenty-four one-week periods from March 2008 to September 2008. At this level of resolution the majority of terms won't display periodicity, so the periodicity considerations could be disregarded. For each period and each term, we made predictions of the current period's CTR based on the following methods:

1. clustering for terms using vector space model
2. hierarchical clustering for terms using vector space model
3. basic bipartite graph model
4. consistent bipartite graph using vector distance
5. consistent bipartite graph using edit distance
6. consistent bipartite graph using largest matching subgraph

All of these methods are clustering algorithms which are mining the search ads dataset. Each prediction method makes use of the historical data from the previous period in predicting the current period's CTR. Method 1 to 3 are using only terms in the ads to cluster, and method 4 to 6 are co-clustering algorithm which use all of the query words, ads and terms.

For each week, we computed the predicted rate using the methods described above. The predicted number of clicks is then equal to the predicted rate times the observed number of impressions in the current period. The results are in Table 1. In addition to studying how the impressions can impact on the predicted results, we divided the

Table 1. Mean error across terms and weeks

	Prediction Method	Mean error	Mean error (B1)	Mean error (B2)	Mean error (B3)
1	Vector space model	0.732	0.462	0.617	1.653
2	Hierarchical	0.811	0.458	0.704	2.220
3	Basic bipartite graph	0.775	0.419	0.684	1.771
4	CBG Vector distance	0.525	0.388	0.476	1.121
5	CBG Edit distance	0.430	0.263	0.421	0.874
6	CBG Subgraph match	0,434	0.247	0.535	0.825

terms into buckets based on the number of impressions in the reference week: 10 or fewer (B1), 11-50 (B2), and more than 50 (B3).

To summarize, our experiments on consistent bipartite graph of search ads are less error than traditional vector space model. The prediction results with the triples(query, ads, terms) is better than the one only used terms. Algorithm presented in this paper can well handle the high-order heterogeneous data of search advertisement.

5 Conclusion

In this paper, we proposed a consistent bipartite graph model to represent the search advertisement referring to query words, ads and terms. We used a divide-and-merge algorithm in consistent bipartite graph co-partitioning to balance the size of every cluster. The predicting framework in this paper is given a better method to model the heterogeneous data of search advertisement, and is also can be used in search ads system for online predicting as well. Experiments on real ads dataset showed that our approach worked effectively and efficiently.

References

1. Regelson, M., Fain, D.C.: Predicting click-through rate using keyword clusters. In: Richardson, M., Dominowska, E., Ragno, R. (eds.) Proceedings of the Second Workshop on Sponsored Search Auctions (2006)
2. Richardson, M., Dominowska, E., Ragno, R.: Predicting clicks: estimating the click-through rate for new ads. In: Proceedings of the 16th International Conference on World Wide Web (WWW 2007), pp. 521–530. ACM, New York (2007)
3. Dembczynski, K., Kotlowski, W., Weiss, D.: Predicting Ads' Click-Through Rate with Decision Rules. In: Proceedings of the 17th International Conference on World Wide Web (WWW 2008), Beijing, China (2008)
4. Kumar, R., Novak, J., Raghavan, P., Tomkins, A.: Structure and evolution of blogspace. Communications of the ACM 47(12), 35–39 (2004)
5. Ding, C., He, X., Zha, H.: A spectral method to separate disconnected and nearly-disconnected web graph components. In: Proceedings of the seventh ACM SIGKDD international conference on Knowledge discovery and data mining (KDD 2001), pp. 275–280. ACM Press, New York (2001)
6. Gao, B., Liu, T.Y., Zheng, X., Cheng, Q.S., Ma, W.Y.: Consistent Bipartite Graph Co-Partitioning for Star-Structured High-Order Heterogeneous Data Co-Clustering. In: Proceedings of the eleventh ACM SIGKDD international conference on Knowledge discovery in data mining, New York, NY, USA, pp. 41–50
7. Cheng, D., Vempala, S., Kannan, R., Wang, G.: A divide-and-merge methodology for clustering. In: Proceedings of the twenty-fourth ACM SIGMOD-SIGACT-SIGART symposium on Principles of database systems (PODS 20905), USA, pp. 196–205. ACM Press, New York (2005)
8. Fiedler, M.: Algebraic connectivity of graphs. Czech Math. J. 1973(23), 298–305 (1973)

Automating Gene Expression Annotation for Mouse Embryo

Liangxiu Han[1], Jano van Hemert[1], Richard Baldock[2], and Malcolm Atkinson[1]

[1] National eScience Centre, School of Informatics, University of Edinburgh, UK
{liangxiu.han,j.vanhemert}@ed.ac.uk, mpa@nesc.ac.uk
[2] MRC Human Genetics Unit, Institute of Genetic and Molecular Medicine, Edinburgh, UK
Richard.Baldock@hgu.mrc.ac.uk

Abstract. It is of high biomedical interest to identify gene interactions and networks that are associated with developmental and physiological functions in the mouse embryo. There are now large datasets with both spatial and ontological annotation of the spatio-temporal patterns of gene-expression that provide a powerful resource to discover potential mechanisms of embryo organisation. Ontological annotation of gene expression consists of labelling images with terms from the anatomy ontology for mouse development. Current annotation is made manually by domain experts. It is both time consuming and costly. In this paper, we present a new data mining framework to automatically annotate gene expression patterns in images with anatomic terms. This framework integrates the images stored in file systems with ontology terms stored in databases, and combines pattern recognition with image processing techniques to identify the anatomical components that exhibit gene expression patterns in images. The experimental result shows the framework works well.

Keywords: Gene Expression, Mouse Embryo, Pattern Recognition, and Wavelet Transform.

1 Introduction

Understanding the role of the expression of a given gene and interactions between genes in a mouse embryo requires monitoring the gene expression levels and spatial distributions on a large scale. The availability of high throughput instruments such as RNA in situ hybridization (ISH) method provides the possibility to construct a transcriptome-wide atlas of mouse embryos that can provide spatial gene pattern information for comprehensive analysis of the gene interactions and developmental mechanisms of the mouse embryo. The ISH employs probes to detect and visualise spatio-temporal gene patterns in tissues. The outputs of the ISH on tissues are images stained to reveal the presence of gene expression patterns. To understand gene functions and interactions of genes in depth, we need to transform the raw image data into knowledge. Annotating

R. Huang et al. (Eds.): ADMA 2009, LNAI 5678, pp. 469–478, 2009.

the raw images of the ISH provides a powerful way to address this issue. The process of annotating gene expression pattern is to label images with terms from the ontology for mouse anatomy development. If an image is tagged with a term, it means that the anatomical component is expressing as a gene.

Much effort has been invested into the curation of gene expression patterns in developmental biology, for example, the EUREXPress-II project [1] has built a transcriptome-wide atlas database for the developing mouse embryo established by ISH, which has collected more than 18,000 genes at one development stage of the mouse embryo and curated 4 Terabytes of images. The research work in [2] has produced 3375 genes for Genome-wide analysis on Drosophila. Many other gene expression pattern images generated via ISH such as flybase [3] and mouse atlas[4] also provide rich information for the genetic analysis on tissues. The current annotations of gene expressions are made manually by domain experts. With massive amount of curated images available for analysis, it is a huge task for domain experts. Therefore, developing efficiently automatic annotation technique is important. Some existing work [5][6][7] [8] has made attempts on the automating annotation of the gene expression patterns on fruit fly and mouse brain [9]and has provided potential opportunities for further genetic analysis. However, to date, no related work has been done on the automatic annotation of gene expressions for mouse embryos. Comparing with a fly embryo, a mouse embryonic structure [15][16] is more complicated and has more anatomic components, for example, the EURExpress data have 1,500 anatomical features used for the annotations of the mouse embryo.

In this paper, we have used image data from the EURExpress-II project [1] and proposed a new data mining framework for automatic annotation of gene expression patterns in images from developmental mouse embryos. The initial result from the pilot is promising and encouraging. The main contribution of our work consists of following aspects: (1) The combination of statistical pattern recognition and image processing methods can reduce the cost for processing large amount of data and improve the efficiency. We employ the image processing method to standardise and denoise images. The wavelet transform is used to generate and project features from spatial domain to wavelet domain. Considering the high-dimensional features, we use Fisher Ratio analysis to extract the significant features and build up the classifiers based on Linear Discriminant Analysis(LDA). Our classifiers have been evaluated with multi-objective gene expression patterns coexisting in images and the initial results have shown our proposed framework functioned well. (2) Due to multi-anatomical components coexisting in images, this is a typical multi-class classification problem. In this framework, we have formulated this multi-class classification into a two-class problem. We have trained one classifier for each anatomical component. As a result, multi-classifiers for multi-components have been constructed. Each classifier in our framework is a binary classifier, which will give an answer either 'yes' or 'no' when an un-annotated image is coming through. The main advantage is a strong extensibility of the framework. If a new anatomical component to be annotated appears, we can create a new classifier and directly plug it in and no need to train previous existed classifiers. The classification performance will not affected due to introducing a new class under the same observation dataset. Meanwhile, this design can also improve the scalability and parallel process capability. Classifiers can be arbitarily assembled and deployed based on requirements.

The rest of this paper is organised as follows: the problem domain analysis is described in Section 2; Section 3 presents the methodology used in this proposed framework; Section 4 describes the evaluation result; Section 5 presents the conclusion and future work.

2 Problem Domain Analysis

Currently, in the EURExpress database, 80% of images (4 Terabytes in total) have been manually annotated by human domain experts. For cost-effectiveness, our goal is to automatically perform annotation by classifying the remaining 20% into the correct terms of anatomical components (this would be still 85,824 images to be annotated with a vocabulary of 1,500 anatomical terms). In addition if this is successful we can also validate existing annotations to find errors and inconsistencies. This is a significant challenge.

- Firstly the images generated via ISH include variations arising from natural variation in the source embryos and experimental processing variation and distortion. The same anatomic components therefore may have variable shape, location and orientation.
- Secondly, each image for a given gene will in general be annotated with multiple anatomic terms. This means features for multiple anatomy components coexist in the image, which increases the difficulty of discrimination.
- Thirdly, the number of images associated with a given anatomy terms is uneven. Some of terms may be associated with many images others with only a small number.
- Finally, the dimensionality of each image is high and represented as pixels $m * n$. and in the EurExpress case typically 3Kx4K pixels.

To address these challenges, we propose a new extensible data mining framework that integrates both the images in the file systems and annotation databases and combines image processing with statistical pattern recognition techniques to automatically identify gene expressions in images, as shown in Fig. 1.

To automatically annotate the remained 20% images, we need to learn these annotations by machines first and then automate the classification process by the deployment of classifiers. This would require a training stage to train these annotated data and build up classifiers, a test and evaluation stage for evaluating the performance of classifiers and then finally a deployment stage for deploying the classifiers to perform the classification of un-annotated images.

The processes in the training stage include image integration, image processing, feature generation, feature selection and extraction, and classifier design.

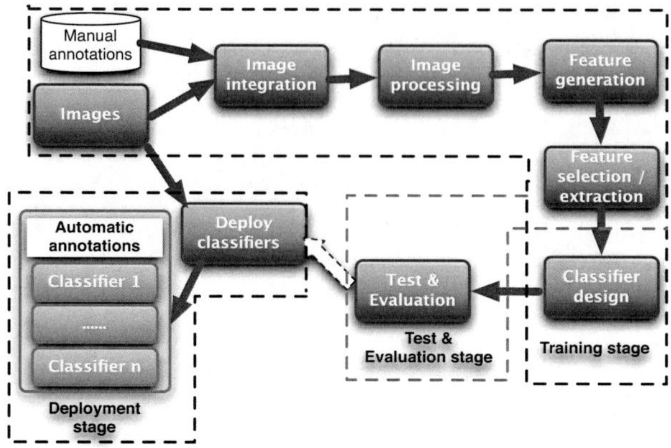

Fig. 1. The data mining framework of automating annotation of gene expressions

- **Image integration:** Before starting the data mining, we need to integrate data from different sources: the manual annotations have been stored in the database and the images are located in the file system. The outputs of this process are images with annotations.
- **Image processing:** The size of the images is variable. We apply median filtering and image rescaling to reduce image noise and rescale the images to a standard size. The outputs of this process are standardised and denoised images, which can be represented as two-dimensional arrays $(m * n)$.
- **Feature generation:** After image pre-processing, we generate those features that represent different gene expression patterns in images. We use wavelet transform to obtain features. The resulting features of wavelet transform are 2 dimensional arrays $(m * n)$.
- **Feature selection and extraction:** Due to the large number of features, the features need to be reduced and selected for building a classifier. This can be done by either feature selection or feature extraction or both. Feature selection selects a subset of the most significant features for constructing classifiers. Feature extraction performs the transformation on the original features for the dimensionality reduction to obtain a representative feature vectors for building up classifiers.
- **Classifier design:** The main task in this case is to classify images into the right gene terminologies. The classifier needs to take an image's features as an input and for each of anatomical features outputs a rating as 'not detected', 'possible', 'weak', 'moderate' or 'strong' (In the current experimental stage, we use two types 'detected as a gene' and 'not detected as a gene'). We have built separate classifiers for each of anatomical components and considered them independently.

The test and evaluation stage will use the result from the training stage to test images. During this stage, k-fold cross validation is used for evaluating the classification performance. With k-fold validation, the sample dataset is randomly split into k disjoint subsets. For each subset, we train a classifier using the data in the other k-1 subsets and then evaluate the classifier's performance on the data in that subset. Thus, each record of the data set is used once to evaluate the performance of a classifier. If 10-fold validation is used, we can build 10 classifiers each trained on 90% of the data and each evaluated on a different 10% of the data.

The deployment stage will deal with the configuration on how to deploy classifiers onto the system, apply classifiers to automatically perform annotation on un-annotated images, and deliver results to the users.

In the following sections, we will mainly focus on the major methods used in the training stage and evaluation stage because of their importance.

3 The Methodology

3.1 Feature Generation Using Wavelet Transform

We first obtain samples by integrating both images and manual annotations using a database SQL query to specify which images should be processed. These sample images are filtered and standardised in a uniform size suitable for the feature generation process.

To characterise multi-gene expression patterns in an embryo image, in this paper, we use wavelet transform to represent and generate features. Wavelet transform has been well-recognised as a powerful tool for applications in signal and image processing [10][11][12]. There are two major reasons for using the wavelet transform in our case: (1) Wavelet transform provides a mathematical tool for the hierarchical decomposition of functions, which can decompose images into space and frequency domains, obtain a projective decomposition of the data into different scales and therefore provide local information of images, unlike Fourier transform that only provides global information of images in frequency domain. (2) By using wavelet transform, the image can be decomposed into different subimages at subbands (different resolution levels). The resolutions of the subimages are reduced. On the other hand, the computational complexity will be reduced by operating on a lower resolution image.

In mathematics, wavelet transform refers to the representation of a signal in terms of a finite length or fast decaying oscillating waveform (known as the mother wavelet). This waveform is scaled and translated to match the input signal. In formal terms, this representation is a wavelet series, which is the coordinate representation of a square integrable function with respect to a complete, orthonormal set of basis functions for the Hilbert space of square integrable functions. The wavelet transform includes continuous wavelet transform and discrete wavelet transform. In this case, 2D discrete wavelet transform has been used to generate features from images.

In fact, wavelet transform of a signal can be represented as an input passing through a series filters with down sampling and deriving output signals based on scales (resolution levels). This can be done by iteration process. Fig. 2(a) shows the filter representation of wavelet transform on a 2D array input. LL is a low-low pass filter that is a coarser transform of the original 2D input and a

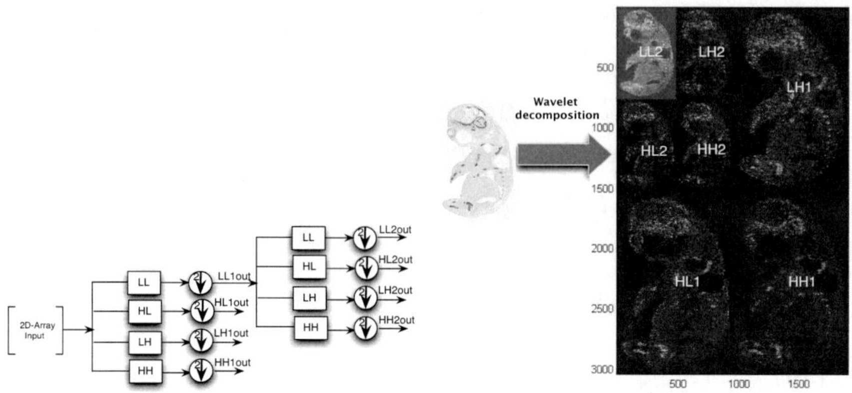

(a) Wavelet decomposition on 2D-array (b) Wavelet decomposition on an image

Fig. 2. Wavelet decomposition

circle with an arrow means down sampling by 2; HL is a high-low pass filter that transforms the input along the vertical direction; LH is a low-high pass filter that transforms the input along the horizontal direction ; and HH is a high-high pass filter that transforms the input along the diagonal direction. At the first iteration of applying these filters into the input (called wavelet decomposition), the result of wavelet transform will be $LL1out$, $HL1out$, $LH1out$, $HH1out$. At the second iteration, we can continue performing wavelet transformation on $LL1out$ and the output will be $LL2out$, $HL2out$,$HL2out$, $HH2out$. These steps can be continuously and the initial input signal therefore is decomposed into different subbands.

Mathematically, for a signal $f(x, y)$ with 2D array($M * N$), the wavelet transform results of applying filters at different resolution levels (e.g., $LL1out$, $HL1out$, $LH1out$, $HH1out$, $LL2out$, $HL2out$, ...) can be calculated as follows:

$$W_\phi(j_0, m, n) = \frac{1}{\sqrt{MN}} \sum_{x=0}^{M-1} \sum_{y=0}^{N-1} f(x, y)\phi_{j_0,m,n}(x, y). \tag{1}$$

$$W_\psi^i(j, m, n) = \frac{1}{\sqrt{MN}} \sum_{x=0}^{M-1} \sum_{y=0}^{N-1} f(x, y)\psi_{j,m,n}^i(x, y), i = (H, V, D). \tag{2}$$

where $W_\phi(j_0, m, n)$ is $LLout1$ and $W_\psi^i(j, m, n)$ respectively represents $HL1out$, $LH1out$ and $HH1out$ when the wavelet decomposition is performed along the vertical, horizontal and diagonal direction. j_0 is a scale as start point. $\phi_(j_0, m, n)$ and $\psi_{j,m,n}$ are wavelet basis functions. In this case, we use Daubecheis wavelet basis functions(db3) [13].

An example of wavelet transform on an embryo image at the second resolution level is shown in Fig. 2(b). The image is decomposed into four subbands (sub-images). The subbands $LH1$, $HL1$ and $HH1$ are the changes of the image along horizontal, vertical directions and diagonal directions with the higher frequency component of the image, respectively. After applying filters, the wavelet transform of $LL1$ is further carried out for the second level resolution as $LL2$, $LH2$, $HL2$ and $HH2$. If the resolution of the image is 3040x1900, the sizes of subimages downsampling by 2 at the second resolution level are respectively $LL2$(760x475) , $LH2$(760x475), $HL2$(760x475), HH(760x475), $LH1$(1520x950), $HL1$(1520x950) and $HH1$(1520x950). The total wavelet transform coefficients (features) for the image are 3040*1900=5,776,000.

3.2 Feature Selection and Extraction Using Fisher Ratio Analysis

Due to the resulting of high-dimentional features generated it is necessary to select the most discriminating features. We use Fisher ratio analysis [14] for feature selection and extraction. The Fisher ratio finds a separation space for discriminating features of two classes by maximizing the difference between classes and minimising within the class.

Assuming two classes, $C_1\{x_1, ..., x_i, ...x_n\}$ and $C_2\{y_1, ..., y_i, ...y_n\}$, the Fisher ratio is defined as the ratio of class-to-class variance to the variance of within classes. The Fisher Ratio can be represented as follows:

$$FisherRatio = \frac{(m_{1,i} - m_{2,i})^2}{(v_{1,i}^2 + v_{2,i}^2)}. \tag{3}$$

where $m_{1,i}$ represents the mean of samples at the ith feature in C_1, $m_{2,i}$ represents the mean of samples at the ith feature in C_2. $v_{1,i}$ represents the variance of samples at the ith feature in C_1. Similarly, $v_{2,i}$ represents the variance of samples at the ith feature in C_2.

3.3 Classifier Building Using LDA

We train each classifier for each anatomical component, and formulate our multi-class problem as a two-class problem. Namely, we treat and divide our sample dataset into two classes during each training: one class contains all of samples with a certain gene expression to be annotated and the other contains all of samples without that gene expression. In this case, we use Linear Discriminant Analysis(LDA) [14] for solving our classification problem. For a given two-class problem ($C_1\{x_1, ..., x_i, ...x_n\}$ and $C_2\{y_1, ..., y_i, ...y_n\}$), the linear discriminant function can be formulated as follows:

$$f(X) = W^t X + w_0. \tag{4}$$

The goal is to find W (weight vector) and w_0 (threshold) so that if $f(X) > 0$, then X is C_1 and if $f(X) < 0$ then X is C_2. The idea is to find a hyperplane that can separate these two classes. To achieve the goal, we need to maximise the target function denoted as follows:

$$T(W) = \frac{|W^t S_B W|}{|W^t S_W W|}. \quad (5)$$

where S_W is called the within-class scatter matrix and S_B is the between-class scatter matrix. They are defined respectively as follows:

$$S_B = (m_1 - m_2)(m_1 - m_2)^t. \quad (6)$$

where,
m_1 = mean of $x_i \in C_1$ and m_2 = mean of $y_i \in C_2$.

$$S_W = S_1 + S_2. \quad (7)$$

where,
$S_1 = \sum_{x \in C_1}(X - m_1)(X - m_1)^t$ and $S_2 = \sum_{y \in C_2}(Y - m_2)(Y - m_2)^t$.

4 Evaluation

We have implemented and deployed our data mining framework into our testbed (a distributed environment). Two databases were created: one for annotations of anatomical components and the other one for feature parameters that is used to store parameters and results from the processes of feature generation and extraction and classifier building. All of image files are located in a file system. Because the features generated are big, we store the features into files hosted in a file system, with references in annotation and parameter databases. Considering the large-scale data mining application in this case, 4 Terabytes data we have curated, we have modularized functional blocks shown in Fig. 1 in order to parallelize these processes in further experiments in near future.

Currently, we have built up 9 classifiers for 9 gene expressions of anatomical components(Humerus, Handplate, Fibula, Tibia, Femur, Ribs, Petrous part, Scapula and Head mesenchyme) and have evaluated our classifiers with multi-gene expression patterns in 809 images. We use the cross validation with 10 folds.

Table 1. The preliminary result of classification performance using 10-fold validation

Gene expression \ Classification Performance	Sensitivity	Specificity
Humerus	0.7525	0.7921
Handplate	0.7105	0.7231
Fibula	0.7273	0.718
Tibia	0.7467	0.7451
Femur	0.7241	0.7345
Ribs	0.5614	0.7538
Petrous part	0.7903	0.7538
Scapula	0.7882	0.7099
Head mesenchyme	0.7857	0.5507

Note: Sensitivity: true positive rate. Specificity: true negative rate.

The dataset (809 image samples) is divided into 10 subsets. 9 subsets are formed as a training set and one is viewed as a test set. The classification performance is computed based on the average correct or error rate across all 10 tries. The advantage of this method is every sample will be in a test set only once and 9 times in a training set.

The preliminary result of the 10-fold cross validation in our case is shown in table 1. The result shows the correct rate for identifying images with Humerus can achieve 75.25% and the correct rate for identifying images without Humerus gene expression can achieve 79.21%. Similarly, the correct rates for identifying with and without gene expressions on Handplate as 71.05% and 72.31% ; on Fibula as 72.73% and 71.8%; on Tibia as 74.67% and 71.8%; on Femur as 72.41% and 73.45%; on Ribs as 56.14% and 75.38%; on Petrous part as 79.03% and 75.38%; on Scapula as 78.82% and 55.07%. Except the ribs, all other gene expression can be identified well. The various morphologies and the number of ribs in images cause the lower identification rate.

5 Conclusion and Future Work

In this paper, we have developed a new data mining framework to facilitate the automatic annotation of gene expression patterns of mouse embryos. There are several important features of our framework: (1) the combination of statistical pattern recognition with image processing techniques can help to reduce the cost for processing large amount of data and improve the efficiency. We have adopted the image processing method to standardise and denoise images. Wavelet transform and Fisher Ratio techniques have been chosen for feature generation and feature extraction. The classifiers are constructed using LDA. (2) For enhancing the extensibility of our framework, we formulate our multi-class problem into a two-class problem and design our classifiers with a binary status:'yes' or 'no'. One classifier only identifies one anatomical component. Classifiers for each gene expression are independent on each other. If new anatomical component need be annotated, we do not have to train previous classifiers again. The classifiers can be assembled and deployed into the system based on user requirements. (3) We have evaluated our proposed framework by using images with multi-gene expression patterns and the preliminary result shows our framework works well for the automatic annotation of gene expression patterns of mouse embryos.

The future work will focus on the improvement of the classification performance and parallelise each functional block proposed in this framework in order to enhance the scalability for processing large-scale data of this case in further experiments later on.

Acknowledgments. This work is supported by the ADMIRE project, which is funded by EU Framework Programme 7 FP7-ICT-215024.

The authors acknowledge the support Lalir Kumar and Mei Sze Lam of the EurExpress team (FP6 funding) at the MRC Human Genetics Unit. The authors would also like to thank Jianguo Rao, Jeff Christiansen and Duncan Davidson at the MRC Human Genetics Unit, IGMM, for useful discussions and suggestions.

References

1. EURexpress II project, http://www.eurexpress.org/ee_new/project/index.html (retrieved March 08, 2009)
2. Lecuyer, E., Yoshida, H., Parthasarathy, N., Alm, C., Babak, T., Cerovina, T., Hughes, T.R., Tomancak, P., Krause, H.M.: Global Analysis of mRNA Localization Reveals a Prominent Role in Organizing Cellular Architecture and Function. Cell 131, 174–187 (2007)
3. Drysdale, R.: FlyBase: a database for the Drosophila research community. Methods Molec. Biol. 420, 45–59 (2008)
4. Lein, E.S., Hawrylycz, M.J., et al.: Genome-wide atlas of gene expression in the adult mouse brain. Nature 445, 168–176 (2006)
5. Grumbling, G., Strelets, V., Consortium, T.F.: FlyBase: anatomical data, images and queries. Nucleic Acids Research 34, D485–D488 (2006)
6. Harmon, C.L., Ahammad, P., Hammonds, A., Weiszmann, R., Celniker, S.E., Sastry, S.S., Rubin, G.M.: Comparative Analysis of Spatial Patterns of Gene Expression in Drosophila melanogaster Imaginal Discs. In: Speed, T., Huang, H. (eds.) RECOMB 2007. LNCS (LNBI), vol. 4453, pp. 533–547. Springer, Heidelberg (2007)
7. Pan, J.Y., Balan, A.G.R., Xing, E.P., Traina, A.J.M., Faloutsos, C.: Automatic Mining of Fruit Fly Embryo Images. In: Proceedings of the 12th ACM SIGKDD international conference on Knowledge discovery and data mining, pp. 693–698. ACM, New York (2006)
8. Zhou, J., Peng, H.: Automatic recognition and annotation of gene expression patterns of y embryos. Bioinformatics 23(5), 589–596 (2007)
9. Carson, J.P., Ju, T., Lu, H.-C., Thaller, C., Pallas, S.L., Crair, M.C., Warren, J., Chiu, W., Eichele, G.: A Digital Atlas to Characterize the Mouse Brain Transcriptome. PLoS Comput. Biology 1, 290–296 (2005)
10. Jawerth, B., Sweldens, W.: An Overview of Wavelet Based Multiresolution Analyses. SIAM Review 36(2), 377–412
11. Stollnitz, E., DeRose, T., Salesin, D.: Wavelets for Computer Graphics. Morgan Kaufmann Publishers, Inc., San Francisco (1996)
12. Mallat, S.G.: A Wavelet Tour of Signal Processing. Academic Press, London (1999)
13. Daubechies, I.: Ten Lectures on Wavelets. SIAM, Philadelphia (1992)
14. Duda, R.O., Hart, P.E.: Pattern Classification and Scene Analysis. Wiley, Chichester (1973)
15. Baldock, R., Bard, J., Burger, A., et al.: EMAP and EMAGE: A Framework for Understanding Spatially Organised Data. Neuroinformatics 1(4), 309–325 (2003)
16. Christiansen, J.H., Yang, Y., Venkataraman, S., et al.: EMAGE: A Spatial Database of Gene Expression Patterns during Mouse Embryo Development. Nucleic Acids Research 34, D637 (2006)

Structure Correlation in Mobile Call Networks

Deyong Hu, Bin Wu, Qi Ye, and Bai Wang

Beijing Key Laboratory of Intelligent Telecommunications Software and Multimedia
Beijing University of Posts and Telecommunications, 100876 Beijing
deron.hu@yahoo.com.cn, wubin@bupt.edu.cn, jack_hill@263.net,
wangbai@bupt.edu.cn

Abstract. So far researchers have done a large amount of work on the structure of social network, but neglected the correlation between two connected nodes from structure perspective. In this paper, our primary goal is to acquire structural properties and detect anomalies in mobile call networks. In order to investigate the structural properties, we define some metrics which are based on the clique size vectors of mobile call users. To evaluate our metrics, we explore several real-world mobile call networks. We find that people tend to communicate with the one who has similar structure. Moreover, we find that the connected people have similar structural changes on the whole. Utilizing these results, we detect anomalies and use a visualization toolkit to give a view of the anomalous scenarios in static and dynamic networks.

Keywords: Social Network, Correlation, Structural Property, Clique Size Vector, Anomaly, Visualization.

1 Introduction

The study of social network analysis on personal behavior, interactions between a person's social group and dynamics has been a long standing work for social network analysts. Recent research [1, 2, 3, 4] has primarily focused on the attributes of people, without taking their structures into account. Given that two people have interactions, are their structures similar to each other? Are their temporal structural changes similar to each other? We are not aware of any of the two research issues that were done before. In this paper, we solve these two issues in mobile call networks to see that whether people who have interactions have similar structures and similar structural changes and to detect anomalies in the networks.

We have three sources of data: (1) the CDR (Call Detail Records) of one network segment notated by sa in one city notated by A from Oct 2005 to Mar 2006. (2) the CDR of another network segment notated by sb in the same city A from Oct 2005 to Mar 2006. (3) the CDR in another city notated by B from Jul 2008 to Sep 2008. Our goal is to find the patterns in mobile call networks and detect anomalies using the patterns.

We analyze the structural properties through correlation analysis between two connected users in static and dynamic mobile call networks.

Characterizing the correlation in static mobile call networks could provide us the insight into the global structural properties. M. E. J. Newman in [5] analyzes the correlation

R. Huang et al. (Eds.): ADMA 2009, LNAI 5678, pp. 479–486, 2009.

in network from vertex degree perspective, however, we analyze the correlation in mobile call networks from vertex structure perspective. For the vertex structure, we consider the clique size vector of it. Not like one dimensional vertex degree, our problem falls into the domain of multidimensional clique size vector. In order to analyze the static structural properties, we adopt the correlation coefficient matrix of multidimensional random variable in probability theory.

Characterizing the correlation in dynamic mobile call networks could provide us the insight into structural changes properties. Amit A. Nanavati et al. [3] analyze the structure and evolution of massive telecom graphs, without analyzing the correlation from structural changes of mobile call users perspective. We track the structure of a mobile call user temporally and examine the structural changes between two consecutive snapshots. Moreover, we explore the correlation of the structural changes of two mobile call users who have interactions.

In this paper, we will show that there is indeed a very strong correlation between two connected users in mobile call networks from structure perspective. Users tend to interact with the one who has similar structure. The connected users have similar structural changes on the whole. Making use of the results, we detect some anomalous user pairs and explore the egocentric networks [6] of the two connected users through using the visualization toolkit JSNVA [7].

In short, our contributions are as follows: 1. We propose a new method for correlation analysis of graph structure, which extends traditional correlation analysis of degree [5]. To the best of our knowledge, the method to solve the particular problem that we tackle in this paper has not been studied before; 2. We find the static and dynamic structural properties of connected users through our correlation analysis in mobile call networks; 3. Utilizing the results, we detect some anomalies in the networks.

The outline of the paper is as follows: Section 2 surveys the related work. Section 3, we will provide the metrics for the problem. Section 4, datasets are described in detail and some experiments are set up. Section 5 concludes the paper.

2 Related Work

Aris Anagnostopoulos et al. [8] define fairly general models that replicate the sources of social correlation. Parag Singla et al. [1] attempt to address that if two people are connected, how the correlation of them will be from person's age, location, duration and interest perspective.

Anomaly detection refers to the problem of finding patterns in data that do not conform to expected behavior [9]. Alec Pawling et al. [10] discuss the use of data stream clustering algorithms for anomaly detection.

While initial work neglects the structure correlation between people who have interactions, however, we analyze the correlation from structure perspective. We attempt to address whether two connected people have similar structures. Moreover, for dynamics of network, we use similarity to define the temporal graph changes and attempt to address whether two connected people have similar structural changes. Furthermore, we attempt to address the anomalies in the networks using the results we get.

3 Metrics

A call graph G obtained from CDR is represented by a pair $\langle V, E\rangle$, where V is a set of vertices representing mobile call users and E is a set of edges representing mobile calls. If user a calls user b, then the vertex-pair $\langle a, b\rangle$ is called edge, and b is a neighbor of a. In particular, we will often consider all the neighbors of a, notated by $N_G(a)$. Let $G_i = \langle V_i, E_i\rangle$ be the subgraph at time-stamp i, where V_i is the user set at time-stamp i and E_i is the call set at time-stamp i. In this paper, edges are undirected.

For a given graph G, we can find all maximal cliques using the algorithm mentioned in [11]. The vertex structure can be depicted by its egocentric network and formalized by its clique size vector. In order to reflect the vertex structure more accurately, we consider the smallest clique size be not 3 but 2, as there are too many vertices whose degrees are 1. And we neglect the isolated vertices because our research issues are based on the edge (connected users). Let max be the largest clique size, then the size of vertex clique size vector is $max - 1$. The vertex clique size vector is formalized as follows: $X = (x_2, x_3, ..., x_{max})$ where x_i represents the number of maximal cliques containing the vertex with clique size i.

3.1 Correlation in Static Mobile Call Networks

In order to examine the correlation in static mobile call networks, we bring in the concept of the covariance matrix of multidimensional random variable in probability theory. Suppose $X = (X_2, X_3, ..., X_{max})$ is a $max - 1$ dimensional random variable, where X_i is a random variable to denote the number of maximal cliques with clique size i. We define matrix $C = (c_{ij})$ as covariance matrix for X, where c_{ij} is defined as follows:

$$c_{ij} = cov(X_i, X_j) = E\{[X_i - E(X_i)][X_j - E(X_j)]\} , \tag{1}$$

where $i, j = 2, 3, ..., max$. Obviously we can get $c_{ij} = c_{ji}$ and $D(X_i) = c_{ii}, i = 2, 3, ..., max$. Then we can compute correlation coefficient ρ_{ij} using the following equation:

$$\rho_{ij} = \frac{cov(X_i, X_j)}{\sqrt{D(X_i)}\sqrt{D(X_j)}} = \frac{c_{ij}}{\sqrt{c_{ii}}\sqrt{c_{jj}}} , \tag{2}$$

where $i, j = 2, 3, ..., max$ and $c_{ii} \neq 0, c_{jj} \neq 0$. Also, we can easily get $\rho_{ij} = \rho_{ji}$. For some i, if $c_{ii} = 0$ in covariance matrix C, which means that all the vertices have the same number of maximal cliques with clique size i, then the clique size i has no correlation with other clique size. So we define $\rho_{ij} = 0$ and $\rho_{ji} = 0$ for all $j = 2, 3, ..., max$ if $c_{ii} = 0$. Then we get the correlation coefficient matrix $P = (\rho_{ij})$ for $i, j = 2, 3, ..., max$ where $\rho_{ij} \geq -1$ and $\rho_{ij} \leq 1$.

3.2 Correlation in Dynamic Mobile Call Networks

We address vertex structural changes through the calculation of structure similarities between two consecutive snapshots in dynamic call graphs. In order to get a better reflection to the structural changes, we consider them in two aspects. One is the extent of the changes, and the other is the direction of changes.

We use Euclidean distance of the two vertex clique size vectors between two consecutive timestamps to depict the extent of the changes and the difference of vertex neighbors between two consecutive timestamps to depict the direction of changes. The extent of vertex u's structural changes $S_c(u,t)$ between time-stamp t and $t+1$ is defined as follows:

$$S_c(u,t) = [d(X_t, X_{t+1})]^2 = \sum_{i=2}^{max} (x_{t+1,i} - x_{t,i})^2 \ , \tag{3}$$

where $i = 2, 3, ..., max$, X_t denotes the vertex clique size vector at time-stamp t and $x_{t,i}$ denotes the number of maximal cliques containing u with clique size i at time-stamp t. The direction of vertex u's structural changes $S_d(u,t)$ between time-stamp t and $t+1$ is defined as follows:

$$S_d(u,t) = \frac{|N_{G_{t+1}}(u) \cap N_G(u)|}{|N_{G_{t+1}}(u) \cup N_G(u)|} - \frac{|N_{G_t}(u) \cap N_G(u)|}{|N_{G_t}(u) \cup N_G(u)|} = \frac{|N_{G_{t+1}}(u)| - |N_{G_t}(u)|}{|N_G(u)|} \ . \tag{4}$$

To have a better reflection to the changes, we combine $S_c(u,t)$ and $S_d(u,t)$ into $S(u,t)$, where $S(u,t)$ is defined as follows:

$$S(u,t) = \begin{cases} \frac{S_c(u,t)S_d(u,t)}{|S_d(u,t)|} & S_d(u,t) \neq 0 \ , \\ S_c(u,t) & S_d(u,t) = 0 \ . \end{cases} \tag{5}$$

Then from time-stamp 1 to T_{max}, the vertex u's structure similarity vector is defined as $S(u) = (S(u,1), S(u,2), ..., S(u, T_{max} - 1))$.

From time-stamp 1 to T_{max}, for two users u and v, correlation coefficient between $S(u)$ and $S(v)$ is used to depict their structural changes correlation. Let $cov(S(u), S(v))$ represent the covariance between $S(u)$ and $S(v)$, and sd_S represent the standard deviation of S. Then the structural changes correlation coefficient $\rho_S(u,v)$ is defined as follows:

$$\rho_S(u,v) = \begin{cases} \frac{cov(S(u),S(v))}{sd_{S(u)} sd_{S(v)}} & sd_{S(u)} \neq 0 \ and \ sd_{S(v)} \neq 0 \ , \\ 0 & sd_{S(u)} = 0 \ or \ sd_{S(v)} = 0 \ . \end{cases} \tag{6}$$

We define the average vertex structural changes correlation coefficient $\bar{\rho_S}$ in the temporal call graphs as follows:

$$\bar{\rho_S} = \frac{\sum_{\langle u,v \rangle \in E} \rho_S(u,v)}{|E|} \ . \tag{7}$$

4 Experiments

4.1 Datasets

Our analysis is based on three sources of datasets provided by telecom service providers in two cities in China. We acquire the mobile call pairs from the CDR (Call Detail

Records) of the datasets. And also we get rid of the call pairs with duration less than 5 seconds, which can not be considered as a valid call. Our study is not for all outer calls which contain long distance and international calls and so on, but for all local calls (intra-region calls). As sa segment dataset and sb segment dataset are from the same city A and at the same time period, they can be merged into another dataset called $merge$ dataset. In order to analyze the structural changes, we organize the call pairs from Oct 2005 to Mar 2006 monthly in A city and the call pairs form Jul 2008 to Sep 2008 half-monthly in B city. The call graphs are shown in Table 1 and 2.

Table 1. Call graphs in A city

Nodes(sa,sb,$merge$,B)	Edges(sa,sb,$merge$)	Period	Avg.Deg(sa,sb,$merge$)
30578,151136,198706	62688,300563,525082	Oct 2005	4.10,3.98,5.29
30075,153323,200418	62171,301195,522294	Nov 2005	4.13,3.93,5.21
29786,156153,202685	62330,316437,542335	Dec 2005	4.19,4.05,5.35
30073,154752,201881	61915,305771,528726	Jan 2006	4.12,3.95,5.24
29358,159733,207783	55191,290916,492510	Feb 2006	3.76,3.64,4.74
28085,160825,207728	51128,284983,471043	Mar 2006	3.64,3.54,4.54
44971,238718,298399	167424,976262,1659948	Oct 2005 to Mar 2006	7.45,8.18,11.13

Table 2. Call graphs in B city

Nodes	Edges	Period	Avg.Deg
930934	2797077	1st half of Jul 2008	6.01
921327	2671521	2nd half of Jul 2008	5.78
928975	2603032	1st half of Aug 2008	5.60
925983	2530894	2nd half of Aug 2008	5.47
934535	2831904	1st half of Sep 2008	6.06
935920	2826932	2nd half of Sep 2008	6.04
1129521	6032414	Jul 2008 to Sep 2008	10.68

4.2 Structure Correlation in Static Mobile Call Networks

After we compute the clique size vectors of all users, the correlation coefficient matrix is easily got using equations 1 and 2. Then we can have a good view of the structure similarity by drawing a scaled image of correlation coefficient matrix.

Fig. 1 (a), (b) and (c) depict the vertex structure correlation coefficient of sa, sb and $merge$ segments in A city respectively. Fig. 1 (d) depicts the vertex structure correlation coefficient in B city. The space of the value above 0.5 is about seventy-five percent of the whole space in average, that is to say two connected users have very similar clique size vectors, or the value above 0.5 is impossible to hold so much space. This means that two connected users have similar structures. However, we find in the first half part of Fig. 1 (b) that the value decreases fast with the growth of the distance away from the diagonal and the value above 0.5 is about fifty-five, which means that their structures don't have very strong correlation.

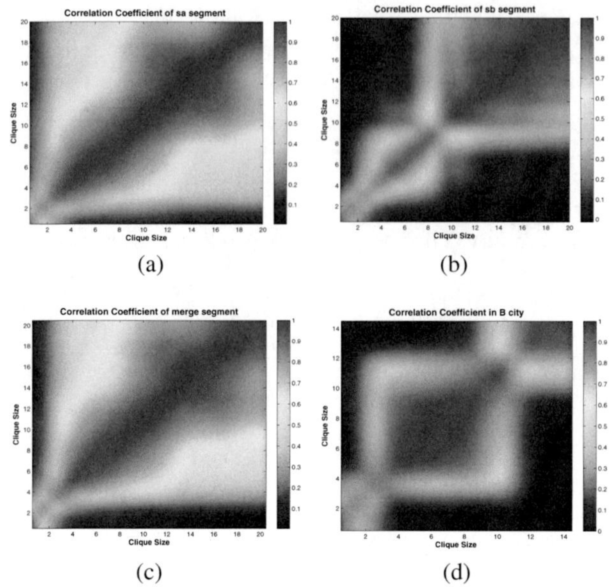

Fig. 1. Vertex Structure Correlation Coefficient

Anomaly Detection. The importance of anomaly detection is due to the fact that anomalies in data translate to significant (and often critical) actionable information in a wide variety of application domains [9]. As we find the pattern that the two connected users have similar structures, and we compute and sort the correlation coefficient between all two connected users, we detect some anomalies which don't conform to the pattern. Fig. 2 (a) shows the correlation coefficient distribution and Fig. 2 (b) shows the anomaly. From Fig. 2 (a), we can get that most two connected users have similar structures. From Fig. 2 (b) we can get that one user's structure is very sparse and the other user's structure is very dense. The correlation coefficient ρ of their structures is -0.61.

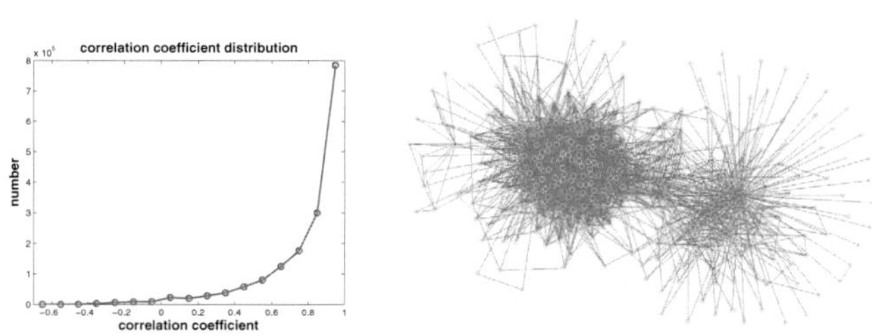

(a) distribution (b) anomaly

Fig. 2. The anomaly in static mobile call network

4.3 Structural Changes Correlation in Dynamic Mobile Call Networks

After we get the structure similarity vector S of all users using equations 3, 4 and 5, the structural changes correlation coefficient between two connected users can be easily got using equation 6. Then we can have $\bar{\rho}$ using equation 7.

In A city, the average correlation coefficient $\bar{\rho}$ is 0.1297 of sa segment, 0.1301 of sb segment and 0.1042 of $merge$ segment. In B city, the average correlation coefficient $\bar{\rho}$ is 0.1458. As the value of $\bar{\rho}$ is all more than zero and around 0.13, a conclusion that the two connected persons have similar structural changes on the whole could be made.

Anomaly Detection. As we find the pattern that the two connected users have similar structural changes, and we sort the structural changes correlation coefficient between all two connected users, we detect some anomalies which don't conform to the pattern. Fig. 3 shows the anomaly. In Fig. 3, for the two connected users u and v, the user u's structure similarity vector is $S(u) = (29, 29, 43, 79, -9)$ and the user v's structure similarity vector is $S(v) = (52, 67, -22, -73, 195)$, and the structural changes correlation coefficient ρ of them is -0.97. We can make a conclusion that they have different structural changes processes.

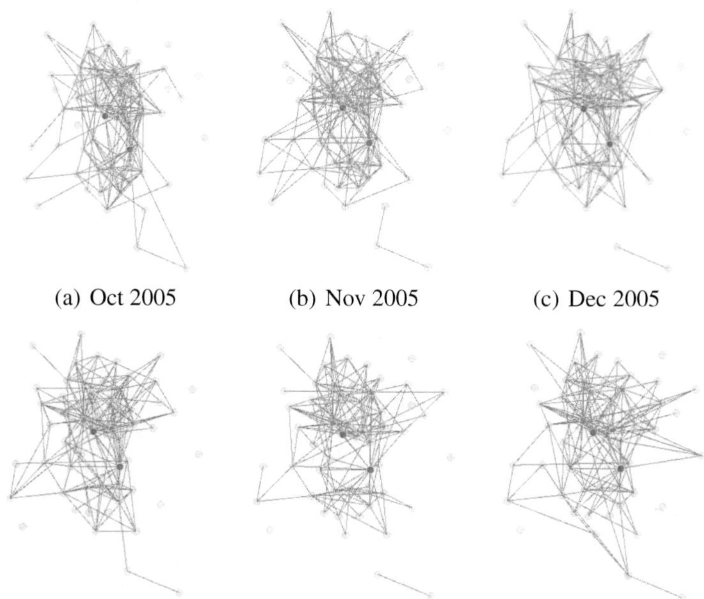

(a) Oct 2005 (b) Nov 2005 (c) Dec 2005

(d) Jan 2006 (e) Feb 2006 (f) Mar 2006

Fig. 3. The anomaly in dynamic mobile call network

5 Conclusion

In this paper, our analysis is based on the clique size vector. By defining some metrics for static and dynamic call graphs and exploring several real-world mobile call networks, we find that users who have interactions have similar structures and similar structural changes on the whole. With the results we find, some anomalies are detected in the static and dynamic networks. By visualizing the anomalous scenario, we can get details of the structures of the two connected users. To the best of our knowledge, this is the first study on our two research issues mentioned in this paper from structure perspective. We believe that these issues pave the way for further study of structural properties in mobile call networks.

Acknowledgement. This work is supported by the National Natural Science Foundation of China (No.60402011), the National Key Technology R&D Program of China (No.2006BAH03B05) and the Specialized Research Fund between Beijing University of Posts and Communications and IBM China Research Laboratory (No.JTP200806014-3).

References

1. Singla, P., Richardson, M.: Yes, there is a correlation:from social networks to personal behavior on the web. In: Proceeding of the 17th international conference on World Wide Web, pp. 655–664. ACM, New York (2008)
2. Leskovec, J., Horvitz, E.: Planetary-scale views on a large instant-messaging network. In: Proceeding of the 17th international conference on World Wide Web, pp. 915–924. ACM, New York (2008)
3. Nanavati, A.A., Singh, R., Chakraborty, D., Dasgupta, K., Mukherjea, S., Das, G., Gurumurthy, S., Joshi, A.: Analyzing the structure and evolution of massive telecom graphs. IEEE Transactions on Knowledge and Data Engineering 20(5), 703–718 (2008)
4. Gonzalez, M.C., Hidalgo, C.A., Barabási, A.L.: Understanding individual human mobility patterns. Nature 453(7196), 779–782 (2008)
5. Newman, M.E.J.: Assortative mixing in networks. Physical Review Letters 89(20), 208701 (2002)
6. Fisher, D.: Using egocentric networks to understand communication. IEEE Internet Computing 9(5), 20–28 (2005)
7. Ye, Q., Wu, B., Wang, B.: Jsnva: A java straight-line drawing framework for network visual analysis. In: Tang, C., Ling, C.X., Zhou, X., Cercone, N.J., Li, X. (eds.) ADMA 2008. LNCS, vol. 5139, pp. 667–674. Springer, Heidelberg (2008)
8. Anagnostopoulos, A., Kumar, R., Mahdian, M.: Influence and correlation in social networks. In: Proceeding of the 14th ACM SIGKDD international conference on Knowledge discovery and data mining, pp. 7–15. ACM, New York (2008)
9. Chandola, V., Banerjee, A., Kumar, V.: Anomaly detection: A survey. ACM Computing Surveys (2009)
10. Pawling, A., Chawla, N.V., Madey, G.: Anomaly detection in a mobile communication network. Computational & Mathematical Organization Theory 13(4), 407–422 (2007)
11. Tomita, E., Tanaka, A., Takahashi, H.: The worst-case time complexity for generating all maximal cliques and computational experiments. Theoretical Computer Science 363(1), 28–42 (2006)

Initialization of the Neighborhood EM Algorithm for Spatial Clustering

Tianming Hu[1], Ji Ouyang[1], Chao Qu[1], and Chuanren Liu[2]

[1] Dongguan University of Technology,
Dongguan, Guangdong
tmhu@ieee.org
[2] Beihang University,
Beijing

Abstract. Like other iterative refinement clustering algorithms, the Neighborhood Expectation-Maximization (NEM) algorithm is sensitive to the initial state of cluster separation. Therefore, the study of the initialization methods is of great importance for the success of finding a better sub-optimal solution in practice. However, existing initialization methods for mixture model based clustering using EM-style algorithms do not account for the unique properties of spatial data, such as spatial autocorrelation. To that end, this paper incorporates spatial information into the initialization and compares three representative initialization methods. Our experimental results on both synthetic and real-world datasets show that the NEM algorithm usually leads to a better clustering solution if it starts with initial states returned by the spatial augmented initialization method based on K-Means.

Keywords: Initialization, Neighborhood Expectation-Maximization, Spatial Clustering.

1 Introduction

Compared to conventional data, the attributes under consideration for spatial data include not only non-spatial normal attributes, but also spatial attributes that describe the object's spatial information such as location and shape. The assumption of independent and identical distribution is no longer valid for spatial data. In practice, almost every site is related to its neighbors. Traditional clustering algorithms, such as the Expectation-Maximization (EM) algorithm [1], do not consider spatial information and only optimize criteria like likelihood over non-spatial attributes. To that end, [2] proposed the Neighborhood Expectation-Maximization (NEM) algorithm, which extended EM by adding a spatial penalty term into the objective function. Such a penalty favors those solutions where neighboring sites are assigned to the same class.

Like other iterative refinement clustering algorithms such as K-Means and EM, NEM can obtain a sup-optimal solution in practice and its solution is dependent largely on the initial state of cluster separation. Since it is usually impossible to achieve the global optimization, which has been shown to be NP-hard,

R. Huang et al. (Eds.): ADMA 2009, LNAI 5678, pp. 487–495, 2009.

the study of the initialization methods is of great value, which aims to provide a better initial state for clustering algorithms to yield a sub-optimal solution in practice. However, conventional initialization methods do not account for spatial information. Along this line, we consider incorporating spatial information early into the initialization methods and compare three representative methods. Our experimental results show that such a practice generally provides a better initial state, from which NEM is able to obtain a better final cluster separation in the output solution. We choose NEM in our experiments mainly for two reasons: (1) NEM for spatial clustering is a representative of the model based clustering using EM-like algorithms. (2) The principle of incorporating spatial information early into initialization methods can be applied to other constrained or regularized model based clustering algorithms, such as in consensus clustering [3].

Overview. The rest of the paper is organized as follows. Section 2 introduces the problem background and related work. In Section 3, we present the three initialization methods and then describe the spatial augmented versions. Experimental evaluation is reported in Section 4. Finally, we draw conclusions in Section 5.

2 Background and Related Work

2.1 NEM for Spatial Clustering

In a spatial framework of n sites $S = \{s_i\}_{i=1}^n$, each site is described not only by a feature vector of normal attributes $\mathbf{x}_i \equiv \mathbf{x}(s_i)$, but also by its spatial information, e.g., (latitude, longitude). Often the neighborhood information can be represented by a contiguity matrix W with $W_{ij} = 1$ if s_i and s_j are neighbors and $W_{ij} = 0$ otherwise.

NEM belongs to the class of models that mainly come from image analysis where Markov random fields and EM style algorithms are intensively used. In detail, we assume the data $X = \{\mathbf{x}\}_{i=1}^n$ come from a mixture model of K components $f(\mathbf{x}|\Phi) = \sum_{k=1}^K \pi_k f_k(\mathbf{x}|\theta_k)$, where π_k is k-th component's prior probability, missing data(cluster label) $y \in \{1, ..., K\}$ indicates which component $\mathbf{x}$ comes from, i.e., $p(\mathbf{x}|y = k) = f_k(\mathbf{x}|\theta_k)$, and Φ denotes the set of all parameters. Let $\overline{P}$ denote a set of distributions $\{\overline{P}_{ik} \equiv \overline{P}(y_i = k)\}$ governing $\{y_i\}$. As highlighted in [4], the penalized likelihood U in NEM can be written as $U(\overline{P}, \Phi) = F(\overline{P}, \Phi) + \beta G(\overline{P})$, where $F(\overline{P}, \Phi) = E_{\overline{P}}[\ln(P(\{\mathbf{x}, y\}|\Phi))] + H(\overline{P})$, and $G(\overline{P}) = \frac{1}{2} \sum_{i=1}^n \sum_{j=1}^n W_{ij} \sum_{k=1}^K \overline{P}_{ik}\overline{P}_{jk}$. $\beta > 0$ is a fixed spatial coefficient that weighs the spatial penalty G and determines the degree of smoothness in the solution clustering. U can be maximized by alternately estimating its two parameters $\overline{P}$ in E-step and Φ in M-step.

2.2 Related Work

This paper focuses on the cluster initialization for NEM and we consider the subsequent cluster refinement in NEM a sub-task of the whole spatial clustering.

Input: a set of n instances $\{x_i\}_{i=1}^n$.
Output: K centers $\{m_k\}_{k=1}^K$.
Steps:
1. **For** $k = 1 : K$ **Do**
2. set $m_k = x_i$, where $i \sim \text{unif}(1, n)$.
3. **End of for**

Fig. 1. The random sampling method

In particular, we only investigate those inexpensive initialization methods that provide an initial set of centers. The data are then assigned to the closest center and thus an initial clustering is obtained for NEM to refine. Roughly speaking, the cluster initialization methods fall into three major families: random sampling, distance optimization and density estimation. They will be introduced in more detail later. There is another class of work that focuses on refining the initial state of clustering methods, such as Randomized Local Search [5]. Guided with a specific evaluation function, they perform a heuristic search for the sub-optimal setting of the initialization method. Finally, to avoid strong dependence on initial states and to escape local traps, stochastic versions of EM have been proposed. For instance, between E-step and M-step, the Stochastic EM algorithm [6] assigns data to a component stochastically according to current soft $\overline{P}(y)$ and hence obtains a hard version of $\overline{P}(y)$. In the Classification Annealing EM algorithm [6], the variances of random assignments are decreasing to zero as the number of iterations increases, which is achieved with a sequence of decreasing temperatures.

3 Cluster Initialization Methods

In this section, we outline three representative methods for clustering initialization. Then we present the spatial augmented versions of them.

3.1 Random Sampling

Based on the statistical assumption that a randomly sampled subset would reflect the distribution of the original set, the random sampling method returns K seed centers by uniformly selecting K input instances, as shown in Fig. 1. Despite its simplicity, it is very efficient and has shown to produce satisfactory results in prior studies, especially when K is relatively large compared to n, the dataset size. However, the assumption may fail when K is too small or much smaller than the "natural" number of clusters. Although it is a common practice to train the mixture model with random initialization several times and finally output the best solution, the computation cost is multiplied considerably.

3.2 Within-Cluster Scatter Minimization

Probably being the most widely used clustering algorithm, the K-Means algorithm can be regarded as a simplified version of EM on Gaussian mixture in that all components share the same prior probability and a fixed covariance matrix. As shown in Fig. 2, it essentially estimates its two parameters, K centers $\{m_k\}$ and a partition C, to minimize the within-cluster scatter $S_w(\{m_k\}, C) = \sum_{k=1}^{K} \sum_{C(x)=k} \text{dis}(x, m_k)$. Starting from an initial $\{m_k\}$, it derives a partition C in the E-step. Then it re-estimates the new centers using C in the M-step.

K-Means has shown satisfactory clustering results in various studies in the literature, especially when the data are convex in shape. It is often used first in mixture model based clustering with EM to provide an initial clustering and to speed up convergence subsequently [6]. Note that although K-means is used as the initialization method in EM, it needs initialization too.

Input:
$\{x_i\}_{i=1}^{n}$: a set of n instances.
$\{m_k\}_{k=1}^{K}$: the initial K centers.
Output: K centers $\{m_k\}_{k=1}^{K}$.
Variable: a clustering $C : \{x_i\}_{i=1}^{n} \rightarrow \{1, ..., K\}$
Steps:

While (criterion not met) **Do**
  For $i = 1 : n$ **Do**
    $C(x_i) = \text{argmin}_k \text{dis}(x_i, m_k)$
  End of for
  For $k = 1 : K$ **Do**
    $m_k = \text{mean}\{x_i \mid C(x_i) = k\}$
  End of for

Fig. 2. The K-Means method

3.3 Between-Cluster Scatter Maximization

Because many clustering methods essentially minimize the within-cluster scatter during the iterative refinement process, it may be beneficial to provide them with an initial complementary clustering that optimizes the between-cluster scatter. To avoid costly computation, the between-cluster scatter is often reduced to the between-center scatter, that is, we need to select K maximally separated instances $\{m_k = x_{k(i)}\}_{k=1}^{K}$ to maximize $\sum_{k=1}^{K} \sum_{j=1}^{K} \text{dis}(m_k, m_j)$. To minimize such a scatter, a greedy search method was proposed in [7], which utilizes the sorted pairwise distances for initialization. Termed KKZ in this paper, it is described in Fig. 3. For consistency, we augment K centers with $m_0 = 0$, that is, m_1 is set to the instance x with the largest $\text{dis}(x, 0)$. The computation could be expensive if K is large, for we need to compute the minimum distance to existing centers. However, since generation of new centers does not affect the existing

Input: a set of n instances $\{x_i\}_{i=1}^{n}$.
Output: K centers $\{m_k\}_{k=1}^{K}$.
Variables: minimum distance to existing centers $\{d(x_i)\}$.
Steps:

    For $k = 0 : K$ **Do**
        For $i = 1 : n$ **Do**
            $d(x_i) = \min_{l \in [1:k]} \mathrm{dis}(x_i, m_l)$
        End of for
        $m_{k+1} = \mathrm{argmax}_{x_i} \{d(x_i)\}$
    End of for

Fig. 3. The KKZ method

ones, the computational complexity of KKZ can be comparable to one K-Means iteration using a buffering mechanism that stores the distances between each instance and the existing centers.

3.4 Spatial Augmented Initialization

We can see that all initialization methods above only consider normal attributes without accounting for spatial information. If the positive autocorrelation is the major trend within data, then most sites would be surrounded by neighbors from the same class. Based on this observation, we propose to augment feature vector $\mathbf{x}_i$ of site s_i with $\mathbf{x}_{Ni}$, the average of its neighbors. That is, the augmented vector becomes $\mathbf{x}_i' = [\mathbf{x}_i, \alpha\mathbf{x}_{Ni}]$, where $\alpha > 0$ is a coefficient to weigh the impact of the neighbors, and $\mathbf{x}_{Ni} = \frac{\sum_{j=1}^{n} W_{ij}\mathbf{x}_i}{\sum_{j=1}^{n} W_{ij}}$. Then the initialization methods can be run on the augmented $\{\mathbf{x}_i'\}$.

As for the computational cost, the random sampling method remains unchanged. The cost is generally increased for the distance optimization methods, depending on the particular distance function used. In the case of squared Euclidean distance used in this paper, the cost is doubled, since the new distance computation is on vectors of double length.

4 Experimental Evaluation

In this section, we first introduce the clustering validation measures used in our experiments. Then we report comparative results on both synthetic and real datasets.

4.1 Performance Criteria

Let $C, Y \in \{1, ..., K\}$ denote the true class label and the derived cluster label, respectively. Then clustering quality can be measured with conditional entropy

$H(C|Y)$ and error rate $E(C|Y)$. Both reach a minimum value of 0 in case of a perfect match. When the target variable C is continuous, we calculate the weighted standard deviation defined as $S(C|Y) = \sum_{k=1}^{K} P_Y(k)\,\mathrm{std}(C|Y = k)$.

4.2 Comparison Methodology

We evaluate random sampling, K-Means and KKZ on both original and augmented data. For fair comparison, in all our experiments, we first compute the augmented version of vectors and then randomly draw K vectors. They are treated as the initial centers returned by the random sampling method on the augmented data. The first half of these vectors are treated as those on original data. These vectors also play the role of initial centers for K-Means that runs 10 iterations. Euclidean distance is used in K-Means and KKZ. For random sampling and K-Means, we report average results of 20 runs.

NEM's detailed settings are as follows. Gaussian mixture is employed as the model. β is tuned empirically for each dataset, though the optimal β is usually the same for all methods. The number of internal iterations of E-step is set to 10. The outer iteration in NEM is stopped when $|(U^t - U^{t-1})/U^t| < 0.0001$. In the augmented vector $\mathbf{x}'_i = [\mathbf{x}_i, \alpha\mathbf{x}_{Ni}]$, often $\alpha = 1$ led to the best results, so we only report results with $\alpha = 1$.

4.3 Experimental Results

In the following, we report comparative results on five datasets. Some data characteristics are listed in Table 1. The last row gives the smoothness of the target variable measured with contiguity ratio.

Table 1. Some data characteristics

Data	Im1	Im2	Satimage	House	Election
size	400	400	4416	506	3107
#attribute	1	1	4	12	3
#class	4	4	6	n/a	n/a
ratio	0.78	0.84	0.96	0.58	0.61

Synthetic Datasets. We first compare various initialization methods for NEM on a series of simulated images. The images are synthesized in the following way: First, a partition in four classes was simulated from a Potts Markov random field model with four neighbors contexts on a 20×20 rectangular grid. Then, the observations are simulated from this partition based on four Gaussian densities. Fig. 4 shows two sample partitions Im1 and Im2 of different smoothness, together with their observations. The observations for both partitions are drawn from four Gaussian densities: $N(0, 0.5^2), N(1, 0.5^2), N(2, 0.8^2), N(4, 0.8^2)$. Their clustering results are given in Table 2, where the best results are in bold numbers. "A-X" means initialization methods "X" on the augmented data. For comparison, we also list the results under supervised mode where each component's parameters

Table 2. Comparison with discrete target variables

Data	Method	Entropy	Error
Im1	Sup	0.8252	0.3400
$r = 0.78$	Rand	0.9333 ± 0.1112	0.4610 ± 0.0682
	A-Rand	0.9027 ± 0.0527	0.4450 ± 0.0336
	KMeans	0.8306 ± 0.0069	0.3890 ± 0.0080
	A-KMeans	$\mathbf{0.8145} \pm 0.0234$	$\mathbf{0.3747} \pm 0.0297$
	KKZ	0.9578	0.4425
	A-KKZ	0.8947	0.4025
Im2	Sup	0.8045	0.3275
$r = 0.84$	Rand	0.9900 ± 0.1899	0.4888 ± 0.1026
	A-Rand	0.9558 ± 0.1749	0.4755 ± 0.1091
	KMeans	$\mathbf{0.7405} \pm 0.0028$	0.3483 ± 0.0016
	A-KMeans	0.7540 ± 0.0175	$\mathbf{0.3442} \pm 0.0092$
	KKZ	0.9362	0.4100
	A-KKZ	0.7834	0.3550
Satimage	Sup	0.6278	0.2058
$r = 0.96$	Rand	0.5970 ± 0.0906	0.2691 ± 0.0591
	A-Rand	0.6243 ± 0.1031	0.2711 ± 0.0593
	KMeans	0.5170 ± 0.0689	$\mathbf{0.2016} \pm 0.0393$
	A-KMeans	$\mathbf{0.5142} \pm 0.0557$	0.2022 ± 0.0345
	KKZ	0.8468	0.3653
	A-KKZ	0.8630	0.4337

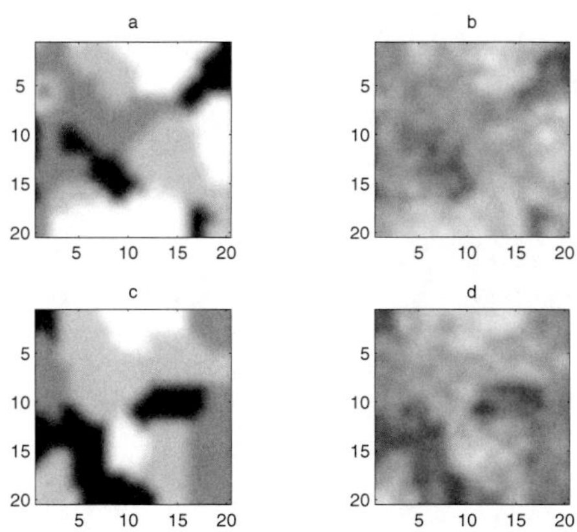

Fig. 4. (a) and (b) show Im1's true partition and observations, respectively. The counterpart of Im2 is shown in (c) and (d).

are estimated with all data from a single true class. One can see that among three methods, the initializations based on K-Means lead to the best results, though the improvement of using augmented data is more obvious with random sampling and KKZ.

Satimage Dataset. We then evaluate them on a real landcover dataset, Satimage, which is available at the UCI repository. It consists of the four multi-spectral values of pixels in a satellite image together with the class label from a six soil type set. Because the dataset is given in random order, we synthesize their spatial coordinates and allocate them in a 64×69 grid to yield a high contiguity ratio of 0.96 with four neighbor context. From Table 2, one can see that although

K-Means still gives best results, using augmented data fails to bring about improvement. The reasons may be that Satimage's contiguity ratio is so high that almost every site is surrounded by sites from the same class with very similar observations. Thus using augmented data does not make much a difference to the initialization results.

House Dataset. The initialization methods are also evaluated on a real house dataset with a continuous target variable. The 12 explanatory variables, such as nitric oxides concentration and crime rate, are originally used to predict the median value of owner-occupied homes, which is expected to has a small spread in each cluster of a reasonable partition. After normalizing the data to zero mean and unit variance, we try two Gaussian mixtures, one with two components, the other with four components. The results are shown in Table 3. One can see the random sampling method on the augmented data achieves best results, though its variance is also the highest.

Table 3. Comparison with continuous target variables

Method	House(K=2)	House(K=4)	Election(K=2)	Election(K=4)
Rand	8.0918 ± 0.0460	7.7830 ± 0.2855	0.1088 ± 0.0046	0.0992 ± 0.0037
A-Rand	**8.0011** ± 0.0565	**7.7768** ± 0.5366	0.1011 ± 0.0048	0.0953 ± 0.0045
KMeans	8.0633 ± 0.0003	7.8401 ± 0.1612	0.0968 ± 0.0009	0.0927 ± 0.0009
A-KMeans	8.0624 ± 0.0001	7.8170 ± 0.0992	**0.0965** ± 0.0002	**0.0919** ± 0.0008
KKZ	8.0632	7.8145	0.1077	0.1007
A-KKZ	8.0642	7.8145	0.1077	0.0962

Election Dataset. Finally we evaluate them on a real election dataset for 1980 US presidential election results covering 3107 counties. Originally the three attributes, fraction of population with college degree, fraction of population with homeownership and income, are used to predict voting rate. Here voting rate is used to evaluate clustering performance. Again, we normalize the data and test two Gaussian mixtures with two and four components respectively. Table 3 shows that the initializations with K-Means on the augmented data achieve the best results.

5 Conclusions

In this paper, we studied the initialization issue of the NEM algorithm for spatial clustering. To provide a good initial state, we propose to push spatial information early into the initialization methods. Along this line, we also evaluated three representative initialization methods including random sampling, K-Means and KKZ. As demonstrated by our experimental results on various data sets, starting from initial states from spatial augmented initialization methods, NEM generally leads to a better clustering result, especially on those data with low contiguity ratio. Among these three initialization methods, the experiments also highlighted that the initialization method based on K-Means usually provided the best initial state for NEM.

Acknowledgments. This work was partially supported by SRF for ROCS, Sci. Tech. Plan Foundation of Guangdong(No. 20070328005), and Sci. Tech. Plan Foundation of Dongguan(No. 2007108101022).

References

1. Dempster, A.P., Laird, N.M., Rubin, D.B.: Maximum likelihood from incomplete data via the EM algorithm. Journal of Royal Statistical Society B(39), 1–38 (1977)
2. Ambroise, C., Govaert, G.: Convergence of an EM-type algorithm for spatial clustering. Pattern Recognition Letters 19(10), 919–927 (1998)
3. Topchy, A., Jain, A.K., Punch, W.: Clustering ensembles: Models of consensus and weak partitions. IEEE Transactions on Pattern Aanalysis and Machine Intelligence 27(12), 1866–1881 (2005)
4. Neal, R., Hinton, G.: A view of the EM algorithm that justifies incremental, sparse, and other variants. In: Jordan, M. (ed.) Learning in Graphical Models, pp. 355–368. Kluwer Academic Publishers, Dordrecht (1998)
5. Franti, P., Kivijarvi, J.: Randomised local search algorithm for the clustering problem. Pattern Analysis & Applications 3(4), 358–369 (2000)
6. Celeux, G., Govaert, G.: A classification EM algorithm for clustering and two stochastic versions. Computational Statistics and Data Analysis 14(3), 315–332 (1992)
7. Katsavounidis, I., Kuo, C., Zhang, Z.: A new initialization technique for generalized lloyd iteration. IEEE Signal Processing Letters 1(10), 144–146 (1994)

Classification Techniques for Talent Forecasting in Human Resource Management

Hamidah Jantan[1,2], Abdul Razak Hamdan[2], and Zulaiha Ali Othman[2]

[1] Faculty of Computer Science and Mathematics,
Universiti Teknologi MARA (UiTM) Terengganu,
23000 Dungun, Terengganu, Malaysia
[2] Faculty of Information Science and Technology
Universiti Kebangsaan Malaysia (UKM)
43600 Bangi, Selangor, Malaysia
hamidahjtn@tganu.uitm.edu.my, {arh,zao}@ftsm.ukm.my

Abstract. Managing an organization's talents, especially in assigning the right person to the right job at the right time is among the top challenge for Human Resource (HR) professionals. This article presents an overview of talent management problems that can be solved by using classification and prediction method in Data mining. In this study, talent's performance can be predicted by using past experience knowledge in HR databases. For experiment purposes, we used the possible classification and prediction techniques in order to find out the suitable techniques for HR data. An example demonstrates the feasibility of the suggested classification techniques using selected employee's performance data. Finally, the initial experiment results show the potential classification techniques for talent forecasting in Human Resource Management (HRM).

Keywords: Classification, Prediction, Talent Forecasting, Human Resource Management.

1 Introduction

Nowadays, there are many areas adapting Data mining (DM) approach as their problem solver tool such as in finance, medical, marketing, stock, telecommunication, manufacturing, health care, customer relationship and many others. However, the application of DM have not attracted much attention in Human Resource Management (HRM) field[1,2]. HR data actually provide a rich resource for knowledge discovery and for decision support tools development. HRM is a comprehensive set of managerial activities and tasks concerned with developing and maintaining a workforce that is the human resource[3]. Recently, among the challenge of HR professionals is managing talent, especially to ensure the right person for the right job at the right time. This task involves a lot of managerial decisions, which are sometimes very uncertain and difficult. The current HR decision practices are depends on various factors such as human experience, knowledge, preference and judgment. These factors can cause inconsistent, inaccurate, inequality and unexpected decisions. As a result, especially in promoting individual growth and development, this situation can often

R. Huang et al. (Eds.): ADMA 2009, LNAI 5678, pp. 496–503, 2009.

make people feel injustice. Besides that, to identify the existing talent task in an organization is among top talent management challenges[4]. In this case, DM approach can be employed as a tool to predict the potential talent by using the past employee performance data in databases. For that reasons, in this study we attempts to use classification techniques in DM to handle issue on talent forecasting. The purpose of this paper is to suggest the possible classification techniques for talent forecasting throughout some experiments using selected classification algorithms.

This paper is organized as follows. The second section describes the related work on DM for talent management, talent forecasting and the possible techniques for classification. The third section discusses the experiment setup. Section 4 shows some experiment results and discussion. Finally, the paper ends with Section 5 where the concluding remarks and future research directions are identified.

2 Related Work

Recently, data mining has given a great deal of concerns and attentions in the information industry and in society as a whole. This is due to the wide accessibility of enormous amounts of data and the important need for turning such data into useful information and knowledge [5, 6].

2.1 Data Mining for Talent Management

HR application such as Decision Support System (DSS) interfaces with DM tool can help executives to make more informative and objective decisions. Besides that, it can help managers to retrieve, summarize and analyze information in the related data to make wiser and more informative decision. Over the years, data mining has evolved various techniques to perform the tasks including database oriented techniques, statistic, machine learning, pattern recognition, neural network, rough set and etc. DM has been applied in many fields as mentioned before, but its application in HRM is very rare[2]. In recent times, there are some research showing interest on solving HRM problems using DM approach[1]. In that case, DM techniques used are Decision tree, Rough set theory, Fuzzy DM and etc. Besides that, some of them use hybrid DM techniques to enhance the capability of the techniques[7]. In HRM, DM techniques are usually used in personnel selection, to choose the right candidates for a specific job. However, prediction applications in HRM are infrequent, there are some examples such as to predict the length of service, sales premiums, persistence indices of insurance agents and analyze mis-operation behaviors of operators[2]. In fact, until now there have been little discussions on talent forecasting, project assignment and recruitment of talent using DM approach.

In any organization, talent management is becoming an increasingly crucial approach in HR functions. Talent is considered as the capability of any individual to make a significant difference to the current and future performance of the organization[8]. In fact, managing talent involves human resource planning that regards processes for managing people in organization. Besides that, talent management can be defined as an outcome to ensure the right person is in the right job; process to ensure leadership continuity in key positions and encourage individual advancement; and decision to manage supply, demand and flow of talent through human capital engine[9]. Talent management is very important and need some attentions from HR professionals. TP Track Research Report finds that among the top current and future talent management challenges are developing existing talent; forecasting talent needs; identifying existing talent and etc[4]. In this study, we focus on talent management challenge which is to identify the existing talent regarding to the key talent in an organization by predicting their performance. In this case, we use the past data from the employee databases to implement the prediction process by using classification techniques. The talent management process consists of recognizing the key talent areas in the organization, identifying the people in the organization who constitute its key talent, and conducting development activities for the talent pool to retain and engage them and have them ready to move into more significant roles[9] as illustrated in Fig. 1. The processes can be incorporated with DM tool in order to solve some talent management tasks.

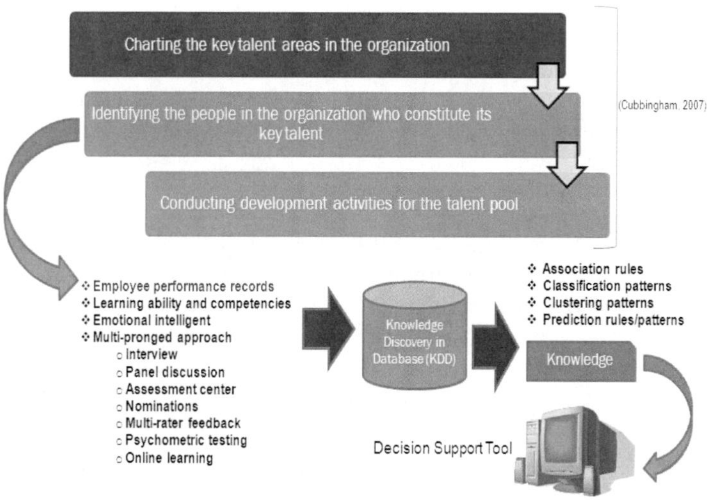

Fig. 1. Data Mining for Talent Management

Talent management processes involve HR activities that need to be integrated into an effective system[10]. In recent year, with the new demands and the increased visibility, HRM seeks a strategic role by turning to DM approach[1]. DM analysis tool can be used to identify generated patterns from the existing data in HR databases as useful knowledge. In this study, we focus on identifying the patterns that relate to the organization talent. The patterns can be generated by using some of the major data mining techniques such as clustering to list the employees with similar characteristics, to group the performances and etc. From the association technique, patterns that are discovered can be used to associate the employee's profile for the most appropriate program/job, associate employee attitude with performance and etc. For prediction and classification, the pattern can be used to predict the percentage accuracy in employee performance, predict employee's behavior and attitudes, predict the performance progress throughout the performance period, identify the best profile for different employee and etc.[7]. The matching of data mining problems and talent management needs are very crucial. For some reason, it is very important to determine the suitable data mining techniques with the problem to be solved.

2.2 Classification Techniques for Talent Forecasting

Prediction and classification abilities are among the methods that can produce intelligent decision. Besides that, these abilities are two forms of data analysis that can be used to extract models describing important data classes or to predict future data trends[6]. The classification process has two phases; the first phase is learning process where training data are analyzed by classification algorithm. The learned model or classifier is represented in the form of classification rules. The second phase is classification process, where the test data are used to estimate the accuracy of classification rules. If the accuracy is considered acceptable, the rules can be applied to the classification of new data. Some of the techniques that are being used for data classification are decision tree, Bayesian methods, Bayesian networks, rule-based algorithms, neural network, support vector machine, association rule mining, k-nearest-neighbor, case-based reasoning, genetic algorithms, rough sets, fuzzy logic.

In this study, we focus our discussion on three classification techniques for talent forecasting i.e. decision tree, neural network and k-nearest-neighbor. Decision tree and neural network are found useful in developing predictive models in many fields[5]. The advantage of decision tree technique is that it does not require any domain knowledge or parameter setting, and is appropriate for exploratory knowledge discovery. The second technique is neural-network which has high tolerance of noisy data as well as the ability to classify pattern on which they have not been trained. It can be used when we have little knowledge of the relationship between attributes and classes. The K-nearest-neighbor technique is an instance-based learning using distance metric to measure the similarity of instances. All these three classification techniques have their own advantages and for that reasons, we attempt to explore these classification techniques for HR data.

3 Experiment Setup

In this study, we have two phases of experiment; the first phase is to identify the possible techniques for classification. In this experiment, we focus our study on the accuracy of the classification technique in order to identify the suitable classifier algorithm for the selected HR data. The second phase of experiment is to compare the accuracy of the techniques with attribute reduction. The employee data contains 53 related attributes from five performance factors that are shown in Table 1. In this case, the performance factors are depends on the nature of the particular job; different employee has different of evaluation criteria. Due to the confidential and security of data, for the exploratory study purposes, we simulate one hundred employee performance data that are based on the talent performance factors. In this case study, the simulated data is for professional and management employees in higher education institution that known as academic staff.

Table 1. Attributes for Each Factor

Factor	Attributes	Variable Name	Meaning
Background	7	D1,D2,D3,D5,D6,D7,D8	Age ,Race, Gender, Year of service, Year of Promotion 1, Year of Promotion 2, Year of Promotion 3
Previous performance evaluation	15	DP1,DP2,DP3,DP4,DP5,DP6, DP7,DP8,PP9,DP10, DP11,DP12,DP13,DP14,DP15	Performance evaluation marks for 15 years
Knowledge and skill	20	PQA,PQC1,PQC2,PQC3,PQD1, PQD2,PQD3,PQE1,PQE2,PQE, PQE4,PQE5,PQF1,PQF2,PQG1, PQG2,PQH1,PQH2,PQH3, PQH4	Professional qualification (Teaching, supervising, research, publication and conferences)
Management skill	6	PQB,AC1,AC2,AC3,AC4, AC5	Student obligation and administrative tasks
Individual Quality	5	T1,T2,SO,AA1,AA2	Training, award and appreciation
Total	**53**		

The process of classification includes the input variables i.e. talent factors for academic staff; and the outcome of the classification process i.e. talent patterns for different positions in academic domain such as for senior lecturer, associate professor and professor. The most important performance factors are extracted from the previous performance data, knowledge and expertise records. Besides the performance factors, the background and management skill are also important in order to identify the possible talent for that position. In addition, to identify the talent patterns in the existing HR databases, several classification techniques are used for the simulated data. The selected classification techniques are based on the common techniques used for classification and prediction especially in DM. As we mentioned before, the classification and prediction techniques chosen are neural network which is quite popular in data mining community as pattern classification technique[11], decision tree as 'divide-and–conquer' approach from a set of independent instances for classification and nearest neighbor for classification that based on distance metric.

4 Result and Discussion

The experiment on each classification technique is using 10 fold cross validation for the training and test dataset. The first experiment uses all attributes in the dataset. The accuracy for each of the classification technique using full attributes is shown in Table 2. It shows that the accuracy of C4.5 is 95.1% and K-Star is 92.1% which can be considered as indicators for the suitable data mining techniques for HR data.

Table 2. The Accuracy for Each Classifier for Full Attributes

Classifier Algorithm	*Accuracy*
C4.5	95.14%
Random forest	74.91%
Multi Layer Perceptron (MLP)	87.16%
Radial Basis Function Network	91.45%
K-Star	92.06%

The second experiment is considered as relevant analysis in order to find out the accuracy of the selected classification technique using the dataset with attribute reduction. The purpose of attribute reduction is to choose the most relevant attributes only in the dataset. The reduction process is implemented using Boolean reasoning technique by selecting the shortest length of attributes. Table 3 shows the relevant analysis results for attribute reduction, 20 attributes are selected from background and previous performance evaluation input.

Table 3. Attributes for Each Factor with Reduction

Factor	**Attributes**	**Variable Name**
Background	5	D1,D5,D6,D7,D8
Previous performance evaluation	15	DP1,DP2,DP3,DP4,DP5,DP6, DP7,DP8,PP9,DP10, DP11,DP12,DP13,DP14,DP15
Total	**20**	

In this case, with attribute reduction, we can decrease the preprocessing and processing time and space. By using these attributes reduction input, the second stage of experiments is implemented. The purpose of this experiment is to find out the accuracy of the classification techniques with attribute reduction. Table 4 shows the accuracy of the classification technique using dataset with attribute reduction are lower than the accuracy using dataset with full attributes.

For this initial experiment results, the C4.5 for decision tree and K-star for Nearest Neighbor have the highest percentage of accuracy (Fig.2.) i.e. for both full attributes and attributes reduction. Besides that, the low accuracy in Table 4 shows that most of the attributes used are important in these experiments. This result also shows us the suitability of the classification techniques or classifier for the selected HR data.

Table 4. The Accuracy for Each Classifier with Attribute Reduction

Classifier Algorithm	*Accuracy*
C4.5	61.06%
Random forest	58.85%
Multi Layer Perceptron (MLP)	55.32%
Radial Basis Function Network	59.52%
K-Star	60.22%

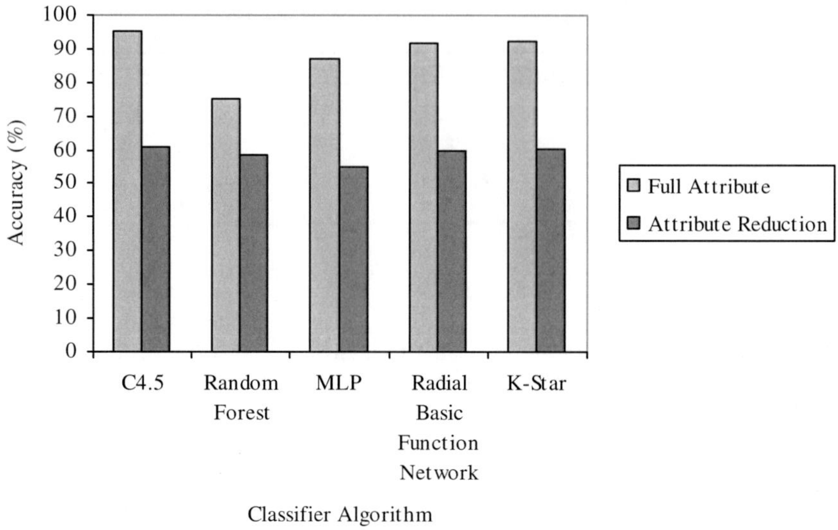

Fig. 2. The Accuracy of Classifier Algorithm for Full Attributes and Attribute Reduction

5 Conclusion

This paper has described the significance of the study and the literature study on DM in human resource, especially for talent forecasting. However, there should be more DM classification and prediction techniques to be applied to the different problem domains in HRM field of research; in order to broaden our horizon of academic and practice work on DM for HRM. Besides that, other DM classification techniques such as Support Vector Machine (SVM), Artificial Immune System (AIS), Fuzzy Clustering and etc, should be considered as alternative classifier for HR data. For that reason, the actual HR data can be tested using those techniques to find out the highest classification accuracy. In addition, the relevancy of the attributes should also be considered as a factor to the accuracy of the classifier. For future work, the attribute reduction process should be implemented using other reduction techniques and the number of training data should be more. Finally, the ability to continuously change and obtain new understanding of the classification and prediction techniques in HRM has become the major contribution to DM in HR.

References

1. Ranjan, J.: Data Mining Techniques for better decisions in Human Resource Management Systems. International Journal of Business Information Systems 3, 464–481 (2008)
2. Chien, C.F., Chen, L.F.: Data mining to improve personnel selection and enhance human capital: A case study in high-technology industry. Expert Systems and Applications 34, 280–290 (2008)

3. DeNisi, A.S., Griffin, R.W.: Human Resource Management. Houghton Mifflin Company, New York (2005)
4. A TP Track Research Report Talent Management: A State of the Art. Tower Perrin HR Services (2005)
5. Tso, G.K.F., Yau, K.K.W.: Predicting electricity energy consumption: A comparison of regression analysis, decision tree and neural networks. Energy 32, 1761–1768 (2007)
6. Han, J., Kamber, M.: Data Mining: Concepts and Techniques. Morgan Kaufmann Publisher, San Francisco (2006)
7. Hamidah, J., Razak, H.A., Zulaiha, A.O.: Data Mining Techniques for Performance Prediction in Human Resource Application. In: 1st Seminar on Data Mining and Optimization, pp. 41–49. FTSM UKM, Bangi (2008)
8. Lynne, M.: Talent Management Value Imperatives: Strategies for Execution. In: The Conference Board (2005)
9. Cubbingham, I.: Talent Management: Making it real. Development and Learning in Organizations 21, 4–6 (2007)
10. CHINA UPDATE: HR News for Your Organization: The Tower Perrin Asia Talent Management Study (2007)
11. Witten, I.H., Frank, E.: Data Mining: Practical Machine Learning Tools and Techniques. Morgan Kaufmann Publishers, San Francisco (2005)

A Combination Classification Algorithm Based on Outlier Detection and C4.5

ShengYi Jiang and Wen Yu

School of Informatics, Guangdong University of Foreign Studies,
510006 Guangzhou, Guangdong
{jiangshengyi,yuyu_588}@163.com

Abstract. The performance of traditional classifier skews towards the majority class for imbalanced data, resulting in high misclassification rate for minority samples. To solve this problem, a combination classification algorithm based on outlier detection and C4.5 is presented. The basic idea of the algorithm is to make the data distribution balance by grouping the whole data into rare clusters and major clusters through the outlier factor. Then C4.5 algorithm is implemented to build the decision trees on both the rare clusters and the major clusters respectively. When classifying a new object, the decision tree for evaluation will be chosen according to the type of the cluster which the new object is nearest. We use the datasets from the UCI Machine Learning Repository to perform the experiments and compare the effects with other classification algorithms; the experiments demonstrate that our algorithm performs much better for the extremely imbalanced data sets.

1 Introduction

A dataset is imbalanced if the classes are not equally represented and the number of examples in one class (majority class) greatly outnumbers the other class (minority class). Imbalanced data sets exist in many real-world domains, such as gene profiling, text classifying, credit card fraud detecting, and medical diagnosing. In these domains, the ratio of the minority to the majority classes can be drastic such as 1 to 100, or 1 to 1000.But what we are really interested in is the minority class rather than the majority class. Thus, we need a fairly high prediction for the minority class. However, the traditional data mining algorithm behaves undesirable in the instance of imbalanced data sets, since it tends to classify almost all instances as negative and maximize the overall prediction accuracy. A number of algorithms solving imbalance classification problem have been developed so far[1], the representatives are resampling methods[2],[3], boosting-based algorithm[4],[5], methods based on SVM such as kernel methods[6], and Knowledge acquisition via information granulation[7]. However, the technique applied in the previous work is to correct the skewness of the class distribution in each sampled subset by using over-sampling or under-sampling. Over-sampling may make the decision regions of the learner smaller and more specific, thus cause the learner to over-fit. There is an inherent loss of valuable information in the process of under-sampling[3].

Many classifiers such as C4.5 tree classifier perform well for balanced datasets but poorly for imbalanced datasets. To remedy drawbacks of the random sampling, we

R. Huang et al. (Eds.): ADMA 2009, LNAI 5678, pp. 504–511, 2009.

introduce a combination classification algorithm based on outlier detection and C4.5 which balance datasets by clustering them instead of simply eliminating or duplicating samples. The basic idea of our algorithm is to combine the one-pass clustering [8] and C4.5 to build decision trees on the two relatively balanced subsets respectively to achieve better classification accuracy. The experimental results demonstrate that the presented algorithm performs much better than C4.5 for the extremely imbalanced data sets.

2 A Clustering-Based Method for Unsupervised Outlier Detection

Some concepts about the clustering-based method for unsupervised outlier detection [8] are described briefly as follows:

Definition 1. For a cluster C and an attribute value $a \in D_i$, the frequency of a in C with respect to D_i is defined as: $Freq_{C|D_i}(a) = \left|\{object | object \in C, object.D_i = a\}\right|$.

Definition 2. For a cluster C, the cluster summary information (CSI) is defined as: $CSI = \{kind, n, ClusterID, Summary\}$,where $kind$ is the type of the cluster C with the value of 'Major' or 'Rare'), n is the size of the cluster $C(\, n = | C |\,)$, $ClusterID$ is the set of class label of the objects in cluster C, and $Summary$ is given as the frequency information for categorical attribute values and the centroid for numerical attributes: $Summary = \{< Stat_i, Cen > | Stat_i = \{(a, Freq_{C|D_i}(a)) | a \in D_i\}, 1 \le i \le m_C, Cen = (c_{m_C+1}, c_{m_C+2}, \cdots, c_{m_C+m_N})\}$.

Definition 3. The distance between clusters C_1 and C_2 is defined as $d(C_1, C_2) = \sqrt{\dfrac{\sum_{i=1}^{m} dif(C_i^{(1)}, C_i^{(2)})^2}{m}}$, where $dif(C_i^{(1)}, C_i^{(2)})$ is the difference between C_1 and C_2 on attribute D_i , For categorical attributes, $dif(C_i^{(1)}, C_i^{(2)}) = 1 - \dfrac{1}{|C_1| \cdot |C_2|} \sum_{p_i \in C_1} Freq_{C_1|D_i}(p_i) \cdot Freq_{C_2|D_i}(p_i) = 1 - \dfrac{1}{|C_1| \cdot |C_2|} \sum_{q_i \in C_2} Freq_{C_1|D_i}(q_i) \cdot Freq_{C_2|D_i}(q_i)$, while for numerical attribute, $dif(C_i^{(1)}, C_i^{(2)}) = \left|c_i^{(1)} - c_i^{(2)}\right|$.

From the definition 3, when the cluster only include one object, the definition of the distance between two objects and the distance between the object and the cluster can also be generated.

Definition 4. Let $C = \{C_1, C_2, \cdots, C_k\}$ be the results of clustering on training data D, $D = \bigcup_{i=1}^{k} C_i \ (C_i \bigcap C_j = \Phi, i \neq j)$. The outlier factor of cluster C_i, $OF(C_i)$ is defined as power means of distances between cluster C_i and the rest of clusters: $OF(C_i) = \sqrt{\dfrac{\sum_{j \neq i} d(C_i, C_j)^2}{k-1}}$.

The one-pass clustering algorithm employs the least distance principle to divide dataset. The clustering algorithm is described as follows:

(1) Initialize the set of clusters, S, as the empty set, and read a new object p.
(2) Create a cluster with the object p.
(3) If no objects are left in the database, go to (6), otherwise read a new object p, and find the cluster C_1 in S that is closest to the object p. Namely, find a cluster C_1 in S, such that $d(p,C_1) \le d(p,\overline{C})$ for all $\overline{C}$ in S.
(4) If $d(p,C_1) > r$, go to (2).
(5) Merge object p into cluster C_1 and modify the *CSI* of cluster C_1, go to (3).
(6) Stop.

3 Combination Classification Algorithm Based on Outlier Detection and C4.5

3.1 A Description of Algorithm

The classification algorithm is composed of model building and model evaluation, the details about model building are described as follows:

Step 1 Clustering: Cluster on training set *D* and produce clusters $C = \{C_1, C_2, \cdots, C_k\}$;

Step 2 Labeling clusters: Sort clusters $C = \{C_1, C_2, \cdots, C_k\}$ and make them satisfy: $OF(C_1) \ge OF(C_2) \ge \cdots \ge OF(C_k)$. Search the smallest b, which satisfies $\frac{\sum_{i=1}^{b}|C_i|}{|D|} \ge \varepsilon (0 < \varepsilon < 1)$, and label clusters $C_1, C_2, \cdots, C_b$ with 'Rare', while $C_{b+1}, C_{b+2}, \cdots, C_k$ with 'Major'.

Step 3 Building classification model: Build the decision tree named RareTree on the clusters labeled 'Rare' and the decision tree named MajorTree on the clusters labeled 'Major' respectively by C4.5 algorithm. Both the RareTree and MajorTree will be considered as the classification model of the whole data. If the clusters labeled 'Rare' or 'Major' are composed of only one class, the decision tree can not be built, thus, the only one class will be labeled as the default class of the cluster.

The details about model evaluation are described as follows:

When classify a new object p, compute the distance between object p and each cluster ($C = \{C_1, C_2, \cdots, C_k\}$) respectively. Find out the nearest cluster C_j, if C_j is labeled 'Rare', the decision tree RareTree will classify the object p; otherwise the decision tree MajorTree will classify the object p. Besides, if the decision tree can not be built on the corresponding clusters, label the object p the default class of the corresponding cluster.

3.2 Factors for the Classification Performance

①Selecting neighbor radius r

We use sampling technique to determine neighbor radius r. the details are described as follows:

(1) Choose randomly N0 pairs of objects in the data set D.
(2) Compute the distances between each pair of objects.
(3) Compute the average mathematical expectation EX and the average variance DX of distances from (2).
(4) Select r in the range of [EX-0.5*DX, EX].

The experiments demonstrate that as r locates in the range of [EX-0.5*DX, EX], the results are stable. We set r as EX-0.5*DX in the following experiments.

② Selecting Parameter ε

ε can be specified twice as much as the ratio of minority samples based on the prior knowledge about the data distribution. If there is no prior knowledge or the data set is moderately imbalanced, ε can be set as 30%.

4 Experimental Results

Here we use Recall, Precision, F-measure and Accuracy to evaluate our algorithm. If the number of minority class is more than one, the values of Recall, Precision and F-measure are set as the average weighted results of the minority classes respectively.11 extremely imbalanced data sets and 5 moderately imbalanced data sets from UCI machine learning repository[9] are chosen to run experiments respectively. The data set KDDCUP99 contains around 4900000 simulated intrusion records. There are a total of 22 attack types and 41 attributes (34 continuous and 7 categorical). The whole dataset is too large. We random sample a subset with 249 attack records (neptune, smurf) and 19542 normal records. A summary of 16 data sets is provided in table 1.

Table 1. Summary of Datasets

Data sets	Instance Size	Number of Features		Data sets	Instance Size	Number of Features
Anneal	798	38		Musk_clean2	6598	166
Breast	699	9		Mushroom	8124	22
Car	1354	6		Page-block	5473	10
Cup99	19791	40		Pendigits	3498	16
German	1000	20		Pima	768	8
Glass	214	9		Satimage	2990	36
Haberman	306	3		Sick	3772	29
Hypothyroid	3163	25		Ticdata2000	5822	85

We ran our algorithm, C4.5 and Ripper by conducting a 10-fold cross validation on each data set respectively in order to make the comparison. In our experiment, our algorithm is implemented by C++ program; C4.5 and Ripper are implemented as J48 and JRip respectively in WEKA.

4.1 Performance Comparison on Extremely Imbalanced Data Sets

The performance of the 11 extremely imbalanced data sets is reported in table 2.

Table 2. The classification evaluation of minority classes on the extremely imbalanced data

Data sets	Classification algorithm	Recall	Precision	F-measure	Accuracy
Anneal	C4.5	85.26	89.41	86.65	92.61
	Ripper	94.12	96.66	95.26	94.11
	Our algorithm	90.20	93.25	90.87	93.36
Car	C4.5	59.25	62.87	60.63	90.77
	Ripper	50.48	49.94	50.21	84.27
	Our algorithm	67.96	67.34	67.65	90.92
Cup99	C45	99.6	98.79	99.18	99.98
	Ripper	98.36	99.19	98.77	99.97
	Our algorithm	100	100	100	100
Glass	C4.5	41.03	52.31	45.7	67.29
	Ripper	43.6	53.58	46.86	66.82
	Our algorithm	58.97	59.04	58.65	66.36
Hypothyroid	C4.5	91.4	92.6	92	99.24
	Ripper	92.1	90.8	91.4	99.18
	Our algorithm	91.39	92.62	92	99.24
Musk_clean2	C4.5	89.4	90.3	89.8	96.88
	Ripper	84.9	90.6	87.6	96.3
	Our algorithm	92.53	93.63	93.08	97.88
Page-block	C4.5	81.46	84.54	82.84	96.88
	Ripper	84.31	84.47	84.35	97
	Our algorithm	84.11	84.49	84.19	97.06
Satimage	C4.5	49.1	50.9	50	86.12
	Ripper	52	62.8	56.9	86.15
	Our algorithm	49.46	52.67	51.02	85.08
Sick	C4.5	88.3	91.5	89.9	98.78
	Ripper	87.9	84.6	86.2	98.28
	Our algorithm	87.01	95.26	90.95	98.94
Ticdata2000	C4.5	0	0	0	93.97
	Ripper	1.1	23.5	2.2	93.87
	Our algorithm	1.33	25	2.53	93.38
Zoo	C4.5	64.71	75	69.19	92.08
	Ripper	64.71	76.99	62.12	89.11
	Our algorithm	76.47	74.51	74.78	92.08
Average	C4.5	68.14	71.66	69.63	92.24
	Ripper	68.51	73.92	69.26	91.37
	Our algorithm	72.68	76.16	73.25	92.21

4.2 Performance Comparison on Moderately Imbalanced Data Sets

The performance of the 5 moderately imbalanced data sets is reported in table 3.

Table 3. The classification evaluation on the moderately imbalanced data

Data sets	Classification algorithm	Recall	Precision	F-measure	Accuracy
Breast	C4.5	94.4	94.42	94.39	94.28
	Ripper	95.71	95.69	95.7	95.57
	Our algorithm	95.21	95.20	95.20	95.1
German	C4.5	70.36	68.55	69.01	70.3
	Ripper	70.84	68.45	68.91	70.8
	Our algorithm	72.63	71.17	71.57	72.53
Mushroom	C4.5	100	100	100	100
	Ripper	100	100	100	100
	Our algorithm	99.99	99.99	99.99	99.98
Pendigits	C4.5	94.2	94.22	94.2	94.17
	Ripper	93.76	93.8	93.77	93.74
	Our algorithm	94.88	94.91	94.89	94.86
Pima	C4.5	72.1	71.45	71.62	72.01
	Ripper	71.7	70.86	71.03	71.61
	Our algorithm	73.03	72.34	72.5	72.95
Average	C4.5	86.21	85.73	85.84	86.15
	Ripper	86.40	85.76	85.88	86.34
	Our algorithm	87.15	86.72	86.83	87.08

The experiments imply that the presented algorithm performs much better than C4.5 in terms of F-measure when the dataset is extremely imbalanced, while showing slight improvement or comparable results on moderately imbalanced data, and does not sacrifice one class for the other but attempts to improve accuracies of majority as well as minority class. Above all, for any degree of imbalance in dataset, our algorithm performs better or at least comparable to C4.5.

4.3 Performance Comparison with the Classifiers for Imbalanced Datasets

In order to make the comparison with the Classifiers for imbalanced datasets, we make some other experiments whose evaluation is the same as the kernel-based two-class classifier[6]. The performance comparison of each data set is reported in table 4, table 5 and table 6 respectively.

The results from the table 4 to table 6 imply that the algorithm presented in this paper performs better than the other Classifiers for imbalanced datasets such as k-nearest neighbor (1-NN and 3-NN), a kernel classifier and LOO-AUC+OFS in terms of Accuracy and is comparable in terms of Precision, F-measure and G-mean.

Table 4. Eight-fold cross-validation classification performance for Pima data set

Algorithm	Precision	F-measure	G-mean	Accuracy
KRLS with all data as centres	0.68±0.07	0.61±0.04	0.69±0.03	0.71±0.02
1-NN	0.58±0.06	0.56±0.04	0.65±0.02	0.66±0.02
3-NN	0.65±0.07	0.61±0.04	0.69±0.04	0.70±0.03
LOO-AUC+OFS(ρ=1)	0.70±0.09	0.63±0.05	0.71±0.03	0.72±0.03
LOO-AUC+OFS(ρ=2)	0.62±0.07	0.67±0.05	0.74±0.04	0.74±0.04
Our algorithm	0.62	0.62	0.7	0.74

Table 5. Two-fold cross-validation classification performance for Haberman data set

Algorithm	Precision	F-measure	G-mean	Accuracy
KRLS with all data as centres	0.63±0.07	0.41±0.05	0.54±0.04	0.61±0.03
1-NN	0.36±0.01	0.38±0.02	0.50±0.02	0.56±0.01
3-NN	0.30±0.07	0.22±0.04	0.38±0.04	0.51±0.03
LOO-AUC+OFS(ρ=1)	0.61±0.05	0.31±0.03	0.45±0.02	0.58±0.01
LOO-AUC+OFS(ρ=2)	0.51±0.02	0.44±0.06	0.57±0.05	0.63±0.03
Our algorithm	0.44	0.52	0.68	0.71

Table 6. Ten-fold cross-validation classification performance for Satimage data set

Algorithm	Precision	F-measure	G-mean	Accuracy
LOO-AUC+OFS(ρ=1)	0.6866	0.5497	0.6689	0.7187
LOO-AUC+OFS(ρ=2)	0.5279	0.5754	0.7708	0.7861
Our algorithm	0.59	0.58	0.72	0.87

5 Conclusion and Future Work

In this paper, we present a combination classification algorithm based on outlier detection and C4.5 which groups the whole data into two relatively balanced subsets by outlier factor. As a result, the skewness of the datasets of the both major clusters and rare clusters has been reduced so that the bias towards the majority class of the both decision trees built on the clusters respectively will be alleviated. The experiments on datasets from UCI imply that the performance of our algorithm is superior to C4.5, especially in case of extremely imbalanced data sets. In the further research, other classification algorithm will replace C4.5 to combine with data division by clustering to solve the imbalance classification problem.

Acknowledgments. This work is supported by the National Natural Science Foundation of China (No.60673191), the Natural Science Research Programs of Guangdong Province's Institutes of Higher Education (No.06Z012).

References

1. Weiss, G.M.: Mining with Rarity: A Uinfying Framework. Sigkdd Explorations 6(1), 7–19 (2004)
2. Marcus, A.: Learning when data set s are imbalanced and when costs are unequal and unknown. In: Proc. of t he Workshop on Learning from Imbalanced Data Sets II, ICML, Washington DC (2003)

3. Liu, X.-Y., Wu, J., Zhou, Z.-H.: Exploratory Undersampling for Class-Imbalance Learning. IEEE Transactions on Systems, Man, and Cybernetics, Part B: Cybernetics 39(2), 539–550 (2009)
4. Han, H., Wang, W.-Y., Mao, B.-H.: Borderline-SMOTE: A New Over-Sampling Method in Imbalanced Data Sets Learning. LNCS, pp. 878–887. Springer, Heidelberg (2005)
5. Guo, H., Viktor, H.L.: Learning from Imbalanced Data Set s with Boosting and Data Generation: The DataBoost-IM Approach. Sigkdd Explorations 6, 30–39 (2003)
6. Hong, X., Chen, S., Harris, C.J.: A Kernel-Based Two-Class Classifier for Imbalanced Data Sets. IEEE Transactions on Neural Networks 17(6), 786–795 (2007)
7. Su, C.-T., Chen, L.-S., Yih, Y.: Knowledge acquisition through information granulation for imbalanced data. Expert Systems with applications 31, 531–541 (2006)
8. Jiang, S., Song, X.: A clustering-based method for unsupervised intrusion detections. Pattern Recognition Letters 5, 802–810 (2006)
9. Asuncion, A., Newman, D.J.: UCI Machine Learning Repository (2007), `http://www.ics.uci.edu/~mlearn/MLRepository.html`

A Local Density Approach for Unsupervised Feature Discretization

ShengYi Jiang and Wen Yu

School of Informatics, Guangdong University of Foreign Studies,
510006 Guangzhou, Guangdong
{jiangshengyi,yuyu_588}@163.com

Abstract. Discretization is an important preprocess in data mining tasks. Considering the density distribution of attributes, this paper proposes a novel discretization approach. The time complexity is $O(m*n*\log n)$ as EW and PKID, so it can scale to large datasets. We use the datasets from the UCI repository to perform the experiments and compare the effects with some current discretization methods; the experimental results demonstrate that our method is effective and practicable.

1 Introduction

Various discretization methods have been developed so far. The representatives are equal width discretization(EW),equal frequency discretization(EF),statistic test method[1,2],entropy-based discretization[3,4],clustering based discretization[5] and so on. The quality of discretization method can only be executed through the learning process on discretized data afterwards. The existing methods of discretization can be classified mainly into two categories: unsupervised and supervised. Many experimental studies[6] show that supervised methods which use the class membership of examples perform much better than their unsupervised counterparts. But if no class information is available, unsupervised discretization is the sole choice. Histograms are one of the most useful tools in exploratory data analysis. The frequently used histograms are either equal width where the range of observed values is divided into k intervals of equal length, or equal frequency where the range of observed values is divided into k bins so that the counts in all bins are equal. But different data distributions require different ideal interval frequencies, so the unsupervised approach(EW,EF) is very sensitive to outliers. Data density distribution is closely related to the class label. As a result, both EW and EF potentially suffer much attribute information loss since k is determined without reference to the density distribution information. PKID (proportional k-interval discretization)[7] adjusts discretization bias and variance by tuning the interval size and number, and further adjusts the NaiveBayes probability estimation bias and variance to achieve lower classification error. But PKID also ignore the data distribution information as EW and EF. To remedy the above-mentioned drawbacks, we present a new approach called a local density approach for unsupervised feature discretization which has two merits: (1) It makes the best use of the density distribution of each numeric attribute to find the natural interval. (2) Our approach has the flexibility of having different number of partitions in different numeric attributes.

R. Huang et al. (Eds.): ADMA 2009, LNAI 5678, pp. 512–519, 2009.

2 The Local Density Approach for Unsupervised Feature Discretization

The basic idea of the presented discretization approach is to take account of the density distribution of all numeric attributes to find the natural interval. The density of the transition zone between any two intervals must be the minimum density since the density of the internal interval is much greater than boundary.

2.1 Description of the Discretization Approach

Our approach is programmed in C++ language. The discretization process can be described as follows: n is the instance number; m is the numeric attribute number. We assume that the numeric attribute A can be split into k intervals: $(b_i, b_{i+1}]\,(i = 0,1,\cdots,k-1)$, $b_0 = -\infty$, $b_k = +\infty$, freq[i] denotes the frequency of records contained in the interval $(b_i, b_{i+1}]\,(i = 0,1,2,\cdots,k-1)$, min[i] denotes the array element subscript of the minimum value in the interval $(b_i, b_{i+1}]\,(i = 0,1,2,\cdots,k-1)$, and max[i] denotes the array element subscript of the maximum value in the interval $(b_i, b_{i+1}]\,(i = 0,1,2,\cdots,k-1)$. MaxF and MinF are the parameters which describe the maximum interval frequency and the minimum interval frequency respectively.

An array of structures is defined to store the summary information of attributes.

```
struct density
{
    float x;  // attribute value
    int num;  // the frequency of attribute value x
    int sum;  // the number of instances contained within the neighbor of x
}stru,
```

We assume that the array a of structures named stru is used to store the summary information of attribute A including n records as a[0],a[1],…,a[n-1].

The details about our approach are as follows:

Step 1. Compute the neighbor radius r.

Step 2. Sort data in ascending order and remove the repetition: sort the members x of array a in ascending order, remove the same members x of original array a[0]~a[n-1] so that the members x of array a[0]~a[n-1] do not have repetition, the frequency of x are stored in the member num of array a[0]~a[n0-1], at the same time, set member sum of array a[0]~a[n0-1] as 0. Then a new array of structures a[0]~a[n0-1](n0 ≤ n) has been made.(n0 is the number of attribute A without repetition,)

Step 3. Compute the local density of each attribute value: compute the number of instances which are contained within the neighbor radius r of each point a[i].x and store it in a[i].sum,

Step 4. Compute the initial cutpoints: compute the local minimums $\{b_0, b_1, \cdots, b_{k-1}\}$ of the sequence a[i].sum (0<i<n0-1).

Step 5. Compute the frequency and the range of position in each initial interval: compute the frequency of each interval $(b_i, b_{i+1}]$ $(i = 0,1,\cdots,k-1)$ stored in freq[i] and the array element subscript of the minimum value in the interval $(b_i, b_{i+1}]$ stored in min[i], simultaneously, the array element subscript of the maximum value in the interval $(b_i, b_{i+1}]$ stored in max[i].

Step 6. Merge intervals of low frequency: calculate each interval's frequency, if the frequency of the interval $(b_i, b_{i+1}]$ as freq[i]/n is lower than MinF, merge the interval to the nearest adjacent interval, and update its interval frequency and the range of position; This step may be repeated several times, because it is possible that the interval frequency is still lower than MinF after merging.

Step 7. Split the interval of high frequency: recalculate each interval's frequency, if the frequency of the interval $(b_i, b_{i+1}]$ as freq[i]/n ≥ MaxF, the interval will be split into $\lceil \log(freq[i]) - \log(\log(freq[i])) \rceil$ subintervals which are the same interval frequency(like equal-frequency method).

Step 8. A sequence of values in the numeric attribute is replaced by the nominal attribute corresponding to the interval it belongs to.

For simplicity, we just describe this approach of only one numeric attribute's discretization processing; a multiplicity of numeric attributes can be discretized simultaneously.

2.2 Factors for the Discretization Performance

(1) The method of computing the neighbor radius r

We use sampling technique to determine neighbor radius r. the details are described as follows:

(a) Randomly choose N0 pairs of objects in the array a.
(b) Compute the distances between each pair of objects.
(c) Compute the average EX of distances from (b).
(d) Select r in the range of [0.5EX, EX].

The experiments demonstrate that as r locates in the range of [0.5EX, EX], the accuracy is stable which implies that our approach is robust. We set r as 0.7ex in the following experiments.

(2)Parameters in step 6 and step 7

In the following experiments, we set MaxF=0.4,MinF=0.03.The number of cut points in the high frequency interval is appropriate in the range from $\lceil \log(freq[i]) \rceil - 1$ to $\lceil \log(freq[i]) - \log(\log(freq[i])) \rceil - 1$, in the following experiments, we set the number of cut points as $\lceil \log(freq[i]) - \log(\log(freq[i])) \rceil - 1$.

3 Experimental Results

Here we use the results of classification tasks on discretized data to test the performance of our algorithm, we run experiments on 29 datasets from UCI machine learning repository[8] and a salary dataset which is from the real-life dataset. Table 1 describes each dataset. In our experiment, C4.5, Ripper, Naïve-Bayes, EW, PKID and MDL[6] are provided in WEKA.

Table 1. Summary of Experimental Datasets

Dataset	Nominal/Continuous attributes	Instance size	Number of Class
Adult	8/6	20000	2
Aneal	32/6	898	6
Austra	8/6	690	2
Breast	0/9	699	2
Credit	9/6	690	2
Dermatology	33/1	366	6
Diabetes	0/8	768	2
Ecoli	1/8	336	8
Flag	10/18	194	6
German	7/13	1000	2
Glass	0/9	214	6
Haberman	0/3	306	2
Heart	7/6	270	2
Hepatitis	6/13	155	2
Horse-colic	7/19	368	2
Hypothyroid	7/18	3163	2
Ionosphere	0/34	351	2
Iris	0/4	150	3
Labor	0/8	57	2
Letter-recognition	0/16	20000	26
Liver	0/6	345	2
Musk_clean2	0/188	6598	2
Pendigits	0/16	10992	10
Pima	0/8	768	2
Satimage	0/36	6435	6
Segment-test	0/19	2310	7
Sonar	0/60	208	2
Vehicle	0/18	846	4
Vowel-context	1/10	990	11
Wine	0/13	178	3
Salary	0/1	80	4

Table 2. The distribution of the salary data set

Professional title	Number of the people	Salary range
Professor	8	5266.08-4858.24
Associate professor	22	4272.34-3644.98
Instructor	40	3438.55-2885.22
Teaching assistant	10	2702.28-2420.35

Table 3. The cutpoints of the salary data set by different discretization methods

Discretization method	cutpoints
Our approach	2793.75, 3161.81, 3541.77,4565.29
MDL	2793.75, 3541.77, 4565.29
PKID	2793.75, 2913.11, 2974.46,3099.82,3541.77,3990.89,4155.65

Salary dataset contains salary data of 80 employees in a department of a university, it contains 2 attributes: professional title and salary. Table 2 gives the distribution of the salary data set and Table 3 gives the cutpoints of the salary data set by different discretization methods, our approach finds an optimal result that each interval only corresponds to one title and outperforms PKID.

Table 4. Accuracy of C4.5 with different discretization (10-fold cross validation)

Data sets	without discretization	Our approach	EWD	PKID	MDL
Adult	86.24%	84.66%	84.25%	85.95%	86.55%
Aneal	91.53%	90.01%	88.75%	92.36%	91.62%
Austra	84.93%	87.00%	84.93%	85.14%	85.22%
Breast	94.56%	95.77%	94.99%	94.41%	95.71%
Credit	85.22%	87.33%	84.64%	85.10%	87.10%
Dermatology	93.99%	93.96%	93.99%	93.85%	93.99%
Diabetes	72.40%	74.77%	74.61%	74.04%	78.13%
Ecoli	82.65%	77.29%	72.35%	65.77%	83.10%
Flag	59.33%	60.98%	61.65%	62.11%	62.63%
Haberman	71.57%	72.65%	71.34%	73.53%	71.34%
Hepatitis	64.19%	63.94%	65.24%	65.48%	65.24%
Hypothyroid	99.28%	98.49%	97.51%	95.23%	99.27%
German	70.50%	72.61%	71.70%	70.80%	72.10%
Glass	72.90%	61.87%	51.40%	51.21%	74.77%
Heart	76.67%	80.07%	73.33%	77.78%	81.48%
Horse-colic	67.93%	68.67%	67.12%	67.69%	66.30%
Iris	94.00%	96.00%	92.00%	91.53%	93.33%
Ionosphere	91.45%	87.15%	87.75%	89.12%	89.17%
Labor	73.68%	76.84%	64.91%	69.82%	80.70%
Letter Recognition	88.20%	81.33%	77.60%	77.92%	78.76%
Liver	66.67%	63.91%	56.81%	57.45%	63.19%
Musk_clean2	96.89%	94.66%	96.17%	93.86%	96.83%
Pendigits	94.92%	88.78%	85.10%	60.62%	89.43%
Pima	73.96%	73.91%	74.35%	73.93%	77.73%
Satimage	85.94%	84.29%	84.10%	79.42%	83.23%
Sonar	74.18%	71.59%	69.09%	68.89%	80.05%
Vehicle	72.72%	68.20%	70.60%	62.25%	70.60%
Vowel-context	80.11%	60.09%	75.05%	49.33%	79.23%
Wine	93.26%	82.98%	78.65%	79.72%	94.38%
average	81.37%	79.30%	77.59%	75.67%	81.76%

3.1 Performance Comparison

The experimental datasets were discretized separately by the proposed approach and other methods such as EWD, PKID and MDL. Each dataset is shuffled randomly for 10 times during each discretization to make sure that the class distribution in the training and test data are not biased or clustered in any form. We implement the classification algorithm on the discreted data and the numeric data respectively in order to make the comparison with the performance of different methods. The accuracy values of three classifications are reported in table 4, table 5 and table 6.

Table 5. Accuracy of RIPPER with different discretization(10-fold cross validation)

Data sets	without discretization	Our approach	EWD	PKID	MDL
Aneal	93.58%	91.49%	92.19%	94.50%	94.55%
Austra	84.93%	85.19%	85.25%	84.99%	86.01%
Breast	94.56%	95.67%	93.98%	93.83%	95.26%
Credit	85.22%	85.28%	84.83%	84.99%	86.64%
Dermatology	88.58%	89.54%	89.51%	89.73%	88.33%
Diabetes	72.40%	72.99%	73.11%	74.74%	77.51%
Ecoli	82.23%	80.51%	74.85%	78.69%	82.26%
Flag	59.90%	60.88%	61.65%	62.58%	61.70%
German	70.50%	70.00%	69.86%	70.15%	71.91%
Glass	72.90%	60.56%	50.19%	60.42%	70.65%
Haberman	72.81%	71.47%	73.33%	72.25%	72.12%
Heart	76.67%	77.33%	78.26%	78.89&	82.33%
Hepatitis	62.97%	67.29%	69.10%	69.74%	68.07%
Horse-colic	67.93%	72.80%	84.62%	72.20%	86.09%
Hypothyroid	99.16%	98.40%	97.19%	97.72%	99.14%
Ionosphere	91.45%	88.77%	87.44%	88.69%	91.74%
Iris	94.00%	95.33%	92.53%	92.60%	94.87%
Labor	73.68%	89.30%	83.51%	80.70%	87.02%
Liver	66.67%	58.58%	62.09%	54.84%	63.19%
Pendigits	94.27%	91.11%	89.44%	82.46%	90.46%
Pima	73.96%	73.46%	74.01%	74.05%	77.30%
Satimage	85.79%	82.83%	83.40%	76.30%	83.56%
Sonar	74.71%	70.87%	66.88%	57.50%	79.18%
Vehicle	68.91%	65.74%	63.42%	58.39%	67.74%
Vowel-context	70.64%	56.73%	66.65%	43.41%	73.62%
Wine	93.26%	88.03%	83.99%	90.00%	94.66%
Average	79.68%	78.47%	78.13%	76.22%	81.77%

According to the experimental results from table 4 to table 6, the local density approach for unsupervised feature discretization outperforms EW and PKID. Meanwhile the average accuracy of the three classifiers on nominal data discretized by our approach is better or at least comparable to that on numeric data without discretization.

Table 6. Accuracy of Naive-Bayes with different discretization(3-fold cross validation)

Data sets	without discretization	Our approach	EWD	PKID	MDL
Adult	83.38%	82.57%	81.86%	83.75%	83.88%
Aneal	79.91%	93.46%	92.17%	95.81%	95.79%
Austra	77.19%	86.17%	85.29%	86.42%	85.65%
Breast	96.01%	97.00%	97.25%	97.25%	97.22%
Credit	77.74%	86.30%	84.81%	85.67%	86.23%
Dermatology	97.46%	97.92%	97.68%	97.46%	97.89%
Diabetes	75.42%	74.11%	75.36%	73.54%	77.92%
Ecoli	84.88%	79.58%	80.51%	80.71%	85.77%
Flag	43.35%	59.69%	59.38%	59.95%	59.23%
German	74.52%	74.82%	75.43%	74.78%	74.67%
Glass	46.54%	65.47%	57.01%	65.00%	72.34%
Haberman	74.97%	74.28%	75.26%	72.06%	72.55%
Heart	84.33%	83.48%	83.30%	81.81%	83.04%
Hepatitis	69.74%	68.45%	68.28%	67.35%	71.10%
Horse-colic	67.28%	68.97%	68.97%	69.84%	66.44%
Hypothyroid	97.51%	97.10%	96.87%	97.62%	98.60%
Ionosphere	82.85%	89.63%	90.51%	88.15%	90.94%
Iris	95.40%	93.27%	94.13%	92.60%	94.47%
Labor	91.93%	95.44%	92.81%	91.93%	93.51%
Letter recognition	64.14%	71.26%	69.78%	73.21%	73.72%
Liver	55.77%	66.61%	64.12%	62.26%	63.19%
Musk_clean2	83.76%	84.01%	79.81%	90.96%	91.56%
Pendigits	80.37%	83.34%	85.63%	85.45%	87.18%
Pima	73.45%	74.22%	74.26%	72.04%	77.28%
Satimage	79.55%	81.45%	80.59%	82.18%	82.23%
Sonar	68.17%	75.91%	75.91%	72.55%	85.10%
Vehicle	44.53%	59.15%	60.07%	61.02%	62.07%
Vowel-context	61.89%	57.92%	62.37%	55.72%	64.19%
Wine	96.85%	96.97%	96.35%	95.06%	98.88%
Average	76.17%	79.95%	79.51%	79.73%	81.82%

4 Conclusion

In this paper, we propose the local density approach for unsupervised feature discretization, the time complexity is $O(m*n*\log n)$, and thus it scales up well in large datasets. Meanwhile, it adopts a more flexible strategy to handle interval size so as to efficiently update discretized intervals upon the data density distribution. The experiments demonstrate that our approach could find out the natural discretization intervals so that it could discretize the numeric attributes efficiently. The comparison of the experimental results on datasets from UCI with the counterparts shows that our approach outperforms the existing unsupervised methods such as EW and PKID in terms of classification accuracy.

Acknowledgments

This work is supported by the National Natural Science Foundation of China (No.60673191), the Natural Science Research Programs of Guangdong Province's Institutes of Higher Education (No.06Z012). The authors would like to thank Professor Joshua Zhexue Huang who is assistant director in the E-Business Technology Institute in the University of Hong Kong for his valuable comments.

References

1. Liu, H., Setiono, R.: Feature selection via discretization. IEEE Transactions on Know ledge and Data Engineering 9, 642–645 (1997)
2. Tay, E.H., Shen, L.: A modified Chi2 algorithm for discretization. IEEE Transactions on Knowledge and Data Engineering 14, 666–670 (2002)
3. Fayyad, U.M., Irani, K.B.: Multi-interval discretization of continuous valued attributes for classification learning. In: Proc. of the 13th International Joint Conference on Artificial Intelligence, pp. 1022–1029 (1993)
4. Clarke, E.J., Braton, B.A.: Entropy and MDL discretization of continuous variables for Bayesian belief networks. International Journal of Intelligence Systems 15, 61–92 (2000)
5. Höppner, F.: Objective Function-based Discretization, pp. 438–445. Springer, Heidelberg (2006)
6. Dougherty, J.R., Kohavi, S.M.: Supervised and Unsupervised Discretization of Continuous Features. Machine Learning. In: Proc of 12th International Conference, pp. 194–202. Morgan Kaufmann, San Francisco (1995)
7. Yang, Y., Webb, G.I.: A Comparative Study of Discretization Methods for Naive-Bayes Classifiers. In: Pacific Rim Knowledge Acquisition Workshop (PKAW 2002), Tokyo, pp. 159–173 (2002)
8. Asuncion, A., Newman, D.J.: UCI Machine Learning Repository (2007), `http://www.ics.uci.edu/~mlearn/MLRepository.html`

A Hybrid Method of Multidimensional Scaling and Clustering for Determining Genetic Influence on Phenotypes

Qiao Li[1], Wenjia Wang[1], Alexander J. MacGregor[2], and George Smith[1]

[1] School of Computing Sciences, University of East Anglia, NR4 7TJ Norwich, UK
[2] School of Medicine, University of East Anglia, NR4 7TJ Norwich, UK

Abstract. As a branch of modern biomedical study, genetic epidemiology research on complex diseases usually aims to identify the genetic influences on phenotypes. This paper presents a hybrid method named MDS-C, which combines the multidimensional scaling method and a clustering algorithm, to unveil the genetic relationships among phenotypes by using phenotypic information only. In MDS-C, the cross-twin cross-trait correlation between any two phenotypes is designed to measure the genetic similarity. MDS-C is verified by a series of simulation studies. Then it is applied to a real bone mineral density (BMD) dataset collected by the St Thomas' UK Adult Twin Registry. Its results suggest that the genetic influence on BMD is site-specific.

Keywords: Biomedical study, MDS-C, twins, BMD.

1 Introduction

Phenotypes in complex diseases usually are determined by a combination of multiple genetic and environmental factors [1]. One of important objectives of genetic epidemiology research on complex diseases is to separate and identify the genetic influence on the phenotypes. However, in some cases, the genetic information is not available, so the genetic influence on the phenotypes cannot be measured directly. In this situation, the knowledge of the genetic characteristics of twins can be used to separate the effects of genes and environment and to estimate the genetic influence on the phenotypes. The principle is based on the fact that there are two types of twins: monozygotic (MZ) twins sharing 100% of their genes, while dizygotic (DZ) twins only having 50% of their genes in common.

In this paper, we present a hybrid method named MDS-C, which combines the multidimensional scaling (MDS) [2, 3] with a clustering algorithm, to discover genetic similarity among phenotypes by using phenotypic information only. In the process of the MDS-C, we applied a modified MDS method using the cross-twin cross-trait correlation to calculate the genetic similarities among multiple phenotypes, and visually represents these phenotypes in a two dimensional Euclidian space. Then, a clustering algorithm, the average-linkage clustering [4], is applied onto the outputs of

R. Huang et al. (Eds.): ADMA 2009, LNAI 5678, pp. 520–527, 2009.

the MDS to discover the clusters of phenotypic variables. These phenotypic clusters suggest that the phenotypes in one cluster should be influenced by a same genetic factor, while the phenotypes in different clusters should be influenced by different genetic factors.

This paper is organized as follows. In Section 2, we introduce the methods involved in the analysis, namely multidimensional scaling and clustering, as well as the hybrid MDS-C method. The cross-twin cross-trait correlation used in MDS is presented in Section 2.1. The method of MDS-C is described in Section 2.2. In Section 3, we show the simulation studies of verifying MDS-C. Finally, in Section 4, the MDS-C is applied to a real twin dataset collected by St Thomas' Adult Twin Registry to discover the genetic influence in the appearance of the bone mineral density (BMD) on different human body sites.

2 Methods

2.1 Cross-Trait Cross-Twin Correlation

As we aim to find out the underlying genetic relationships among the phenotypes, the similarity measure between phenotypes must reflect the genetic influence. In this paper, we apply a metric, namely the cross-twin cross-trait correlation.

MZ Twin Pair No.	Phenotype A		Phenotype B		...
	Twin 1	Twin 2	Twin 1	Twin 2	...
1	A_{1T1}	A_{1T2}	B_{1T1}	B_{1T2}	
2	A_{2T1}	A_{2T2}	B_{2T1}	B_{2T2}	
.	.	.	.	.	
.	.	.	.	.	
n	A_{nT1}	A_{nT2}	B_{nT1}	B_{nT2}	

(a) MZ twins dataset

DZ Twin Pair No.	Phenotype A		Phenotype B		...
	Twin 1	Twin 2	Twin 1	Twin 2	...
1	A_{1T1}	A_{1T2}	B_{1T1}	B_{1T2}	
2	A_{2T1}	A_{2T2}	B_{2T1}	B_{2T2}	
.	.	.	.	.	
.	.	.	.	.	
m	A_{mT1}	A_{mT2}	B_{mT1}	B_{mT2}	

(b) DZ twins dataset

Fig. 1. A sample of phenotypes (A and B) in MZ (sample size: n pairs of twins) and DZ (sample size: m pairs) twin datasets

The twin data containing the selected phenotypes can be divided into two sub-datasets according to the twin types: one for MZ twins and the other for DZ twins (see Fig. 1). The genetic similarity between two phenotypes (e.g., bone mineral density on two different body sites) can be indicated by comparing the cross-twin cross-trait correlations in MZ and DZ twins. For example, we use A and B to represent any two phenotypes (Fig. 1), two sets of cross-twin cross-trait correlations are generated from MZ and DZ datasets. For MZ twins: $r_{MZ}(A_{T1}, B_{T2})$ and $r_{MZ}(A_{T2}, B_{T1})$, DZ twins: $r_{DZ}(A_{T1}, B_{T2})$ and $r_{DZ}(A_{T2}, B_{T1})$; where $r_{MZ}(A_{T1}, B_{T2})$ and $r_{MZ}(A_{T2}, B_{T1})$ denote the Pearson correlations between A_{T1} and B_{T2}, and between A_{T2} and B_{T1} in MZ twins, respectively. $r_{DZ}(A_{T1}, B_{T2})$ and $r_{DZ}(A_{T2}, B_{T1})$ are the Pearson correlations between A_{T2} and B_{T1}, and between A_{T2} and B_{T1} in DZ twins, respectively.

So, for phenotypes A and B, the mean values of cross-trait cross-twin correlations in MZ and DZ twins are:

$$\bar{r}_{MZ}(A,B) = \frac{r_{MZ}(A_{T1},B_{T2}) + r_{MZ}(A_{T2},B_{T1})}{2};\ \bar{r}_{DZ}(A,B) = \frac{r_{DZ}(A_{T1},B_{T2}) + r_{DZ}(A_{T2},B_{T1})}{2}. \tag{1}$$

Thus, genetic similarity (the general cross-twin cross-trait correlation) between any two phenotypes, e.g., A and B, can then be measured by:

$$r(A,B) = 2\times[\bar{r}_{MZ}(A,B) - \bar{r}_{DZ}(A,B)]. \tag{2}$$

Equation 2 applies the fact that MZ twins share 100% genes while DZ twins share 50% of them. This application of cross-twin cross-trait correlation is used as the metric of genetic dissimilarity between any two phenotypes in this paper.

2.2 MDS-C Method

MDS-C is a hybrid method designed to unveil the genetic relationships among phenotypes by using phenotypic information only. It includes two phases (i.e. MDS and clustering) to identify the genetically determined clusters existing in the phenotypes. Firstly, a modified MDS method using the cross-twin cross-trait correlation calculates the genetic similarities among multiple phenotypes, and represents these phenotypes in a two dimensional Euclidian space. Then, the average-linkage clustering is applied onto the outputs of the MDS to discover the clusters of phenotype variables. Phenotypes in one cluster should be influenced by a same genetic factor, while the phenotypes in different clusters are influenced by different genetic factors.

Multidimensional Scaling (MDS) is a method of visualizing the similarity of a set of objects into a low dimensional Euclidean space. The objects often represent the samples, and each of them is characterized by a collection of variables. In addition, when the relationships among variables are concerned, MDS can also be applied to capture the distance between variables, rather than record samples [5]. In the present paper, MDS is used to represent the genetic similarities among the phenotypic variables by Euclidean distances, and the genetic similarity between any pair of phenotypes is attained by the application of the cross-twin cross-trait correlation.

Three steps are considered in the procedure of the MDS [6]: Firstly, obtaining the comparative distances between all pairs of variables. In the present paper, cross-twin cross-trait correlations between all pairs of phenotypic variables are applied to represent the comparative distances; Secondly, estimating the constant as the difference between the comparative distances and the converted absolute distances that allows the variables to be fitted by a Euclidean space of the smallest possible dimensionality. The last step in MDS is to determine the Euclidian space in which the variables lie and obtain the projections of the variables on axes based on the absolute distances.

MDS is capable of producing the plot of approximating Euclidian configuration to represent the distance between variables. However, in this paper, though MDS convert the genetic similarity to the Euclidian distance, the relationships among the phenotypic variables are still vague. Clustering is then applied to the output of the MDS to determine clusters among the variables.

Clustering analysis is a process of automatically grouping objects into a number of clusters using a measure of association, so that the objects in one group are similar and objects belonging to different groups have less similarity [7]. Hierarchical clustering is a general approach to cluster analysis. The traditional representation of hierarchical clustering is a tree (called a dendrogram), with individual objects at the leaves and a single cluster containing all elements at the root. Cutting the tree at a given height will give a clustering result at a selected precision. Average-linkage clustering [4], which is used in this study, is a method of hierarchical clustering that specifies the distance between two clusters as the average distance between objects from one cluster and objects from another cluster. For example, the distance between clusters X and Y is:

$$D(X,Y) = \frac{1}{N_X \times N_Y} \sum_{i=1}^{N_X} \sum_{j=1}^{N_Y} d(x_i, y_j)\ ; \quad x_i \in X, y_j \in Y \cdot \tag{2}$$

Where $d(x_i, y_j)$ is the Euclidean distance between objects x_i and y_j; N_X and N_Y are the numbers of objects in clusters X and Y respectively.

As shown above, MDS-C is designed to identify the clusters of phenotypes that reveal the genetic relationships among these phenotypes. In this paper, we use cross-twin cross-trait correlation in the MDS-C to analyze twin datasets.

3 Simulation Designs

3.1 Simulation Scenarios

In order to evaluate the performance of the MDS-C, three different simulation scenarios are designed (see Fig. 2). Fig. 2 represents 10 phenotypes with three different relationships in three scenarios respectively. Scenario 1 attempts to represent the situation where 10 phenotypes ($P_1 \ldots P_{10}$) are grouped into 3 clusters determined by 3 shared genetic factors (Ac_1, Ac_2 and Ac_3) respectively. Residual genetic factors ($A_1 \ldots A_{10}$) account for the remaining heritability of each of the variables. In addition, individual environmental factors ($E_1 \ldots E_{10}$) influence each of the 10 phenotypic variables, respectively. The aim is to see if the MDS-C could distinguish the existence of the 3 genetically determined clusters of variables.

Scenario 2 represents the situation that 10 phenotypes are grouped into two clusters determined by two shared genetic factors respectively; the important feature is that, these two genetic factors are correlated to some degree. The cluster controlled by A_{c1} correlates to the cluster determined by A_{c2} at certain degree (in the present paper, the correlation is 0.2).

Scenario 3 is designed based on Scenario 1. This simulation is considered to represent the situation that underlying genetically determined clusters and environmentally determined clusters are both working on the phenotypes. Comparing to Scenario 1, the difference is that the group P1 … P5 and the group P6 … P10 are influenced by environmental factors (Ec_1 and Ec_2) respectively. The aim of this simulation study is to reveal the clusters controlled by the shared genetic factors only. The shared environmental factors are used to test whether they would confound the MDS-C.

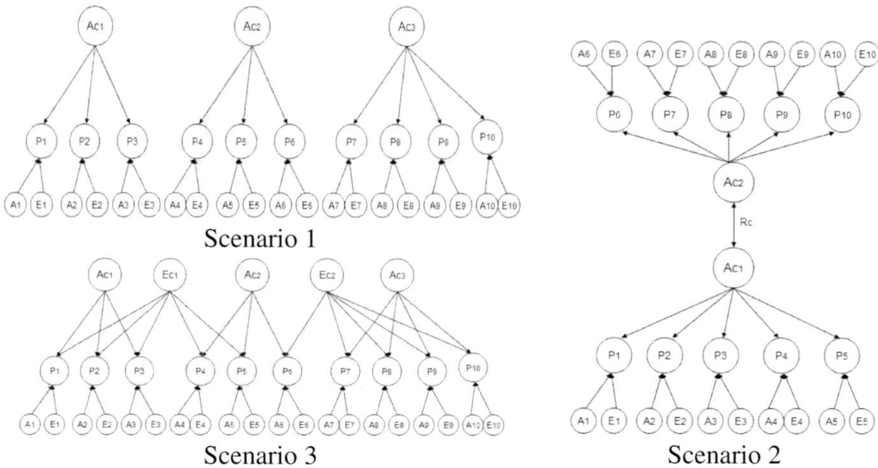

Fig. 2. P_i (i=1 …, 10) represents the disease phenotypic variables; Ac_n (n=1, 2, 3) represents the shared genetic factors; A_i and E_i represent individual genetic and environmental factors which influence each phenotype (P_i). Rc is the correlation coefficient between Ac; Ec_m (m=1, 2) the shared environmental factors.

3.2 Procedure of Simulations

All three simulated scenarios are generated and processed in the following steps:

Step 1. Data simulation: generate simulated twin data based on the models designed in each scenario;
Step 2. Genetic similarity calculation: calculate the cross-twin cross-trait correlations;
Step 3. MDS-C application;
Step 4. Assessment: As the underlying genetic models are known, we evaluate the performance of the MDS-C by comparing the clustering results with the simulated models. The validation coefficient $ARI_{(HA)}$ [8] is applied.

In the procedure, the values of genetic factors (Ac) shared within the clusters are stepped up from 0.1 to 0.5, simultaneously; The sample size of the simulated data is stepped up from 150 pairs (50 MZ and 100 DZ) to 1500 pairs (500 MZ and 1000 DZ) with a step size of 150 (50 for MZ and 100 DZ, respectively); at each point, the experimental procedure is implemented 1000 times to get average assessment measures.

3.3 Results

MDS-C is applied to detect the genetically determined clusters existing in the phenotypes. Fig. 3 uses the accuracy coefficients, $ARI_{(HA)}$ [8], to represent the performance of the MDS-C in three scenarios with different sample size and shared genetic factors.

In these three scenarios, the accuracy coefficients for different Ac values improve with the increase of the twin data sample size. The outputs of Scenario 1 and Scenario 3 are similar: when Ac_1, Ac_2 and Ac_3 are larger than 0.3 and the sample size exceeds 900 pairs (300 MZ and 600 DZ), the genetic determined clusters can be detected with more than 90% accuracy. In Scenario 2, the existence of the genetic correlation (R_c)

influences the performance of the MDS-C, as R_c increases the probability that the two clusters overlap with each other. However, even so, the accuracy in Scenario 2 can still reach a high level (larger than 0.8) when Ac $\geq$ 0.3. The results of these simulation experiments show that the MDS-C method can detect reliably the underlying genetic relationships among the phenotypes. Accordingly, we move our analysis to a real data application.

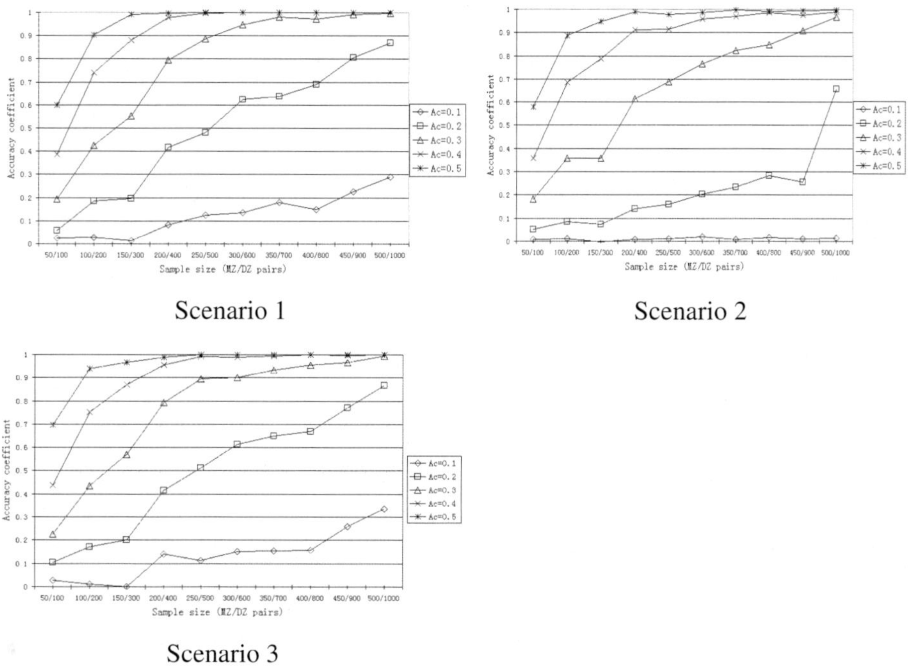

Fig. 3. Accuracy in the three scenarios

4 Bone Mineral Density (BMD) Data Analysis

The real bone mineral density (BMD) data comes from St Thomas' UK Adult Twin Registry. It contains 905 female twin pairs who were identified and invited to participate in the study. These twin pairs were 19-80 years old and were recorded by an extensive range of clinical phenotypes related to cardiovascular disease, obesity, diabetes and osteoporosis. Here we only use the data relevant to the study of bone mineral density (BMD).

This BMD data set contains the information of bone mineral density on different body sites (spine, hip, and forearm) of twins. The underlying genetic relationships of these bones' phenotypes are expected to be found. Table 1 shows the phenotypes existing in the BMD data. We apply MDS-C to this BMD dataset. As Fig. 4 shows, MDS-C reveals 3 clusters which are based on the genetic similarity among these phenotypes.

Table 1. BMD phenotypes on three body sites (the explanation for each phenotype is in the brackets)

Spine	Hip	Forearm
S1 (spine 1)	HN (hip neck)	RT (radius third)
S2 (spine 2)	HT (hip trochanter)	RM (radius mid)
S3 (spine 3)	HW(hip wards triangle)	RU (radius ultradistal)
S4 (spine 4)	HI (hip inter trochanter)	UT (ulna third)
		UM (ulna mid)
		UU (ulna ultradistal)

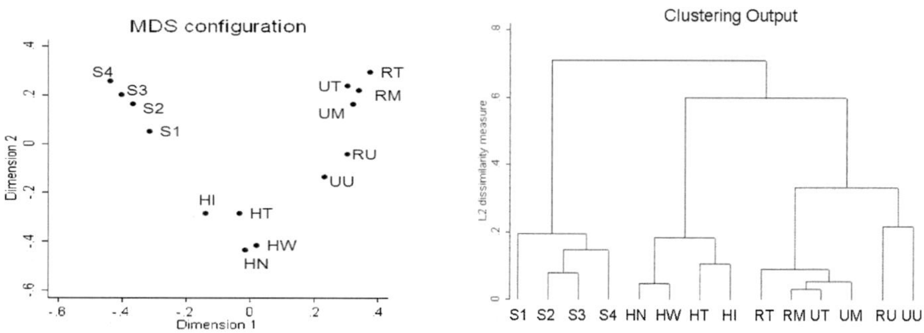

(a) (b)

Fig. 4. MDS-C output: (a) MDS output of BMD data analysis (b) Clustering dendrogram

According to the outputs, the BMD phenotypes from three sites of body (spine, hip and forearm) share or partly share three different genetic factors respectively. Furthermore, the genetic factors existing in hip and forearm correlate with each other to a minimal degree. The ultradistal region on both of radius and ulna may have less correlation to the genetic factor comparing with other phenotypes on the forearm.

5 Conclusion

In this paper, a hybrid method MDS-C is developed to discover genetic influence on phenotypes using phenotypic information only from the twin dataset. Another useful feature of this approach is its ability of reducing the dimensionality of the problem whilst preserving the patterns of genetic influence in phenotypes, and therefore it contributes to solve the challenge of the numerous phenotypes with limited sample size in current biomedical studies.

The MDS-C is verified by three simulated models. The results show that, the MDS-C is capable of unveiling the genetically determined clusters among phenotypes from the twin dataset by using cross-trait cross-twin correlation as the distance metric. The MDS-C is then applied to a real bone mineral density (BMD) data collected by St Thomas' UK Adult Twin Registry. The results suggest that the BMD phenotypes on 3 body sites (spine, hip and forearm) share 3 different genetic factors respectively, and imply that the genetic effects on BMD are site-specific. Further medical research can be carried on using this result as a guideline.

Acknowledgments

This work was supported by grant from the Arthritis Research Campaign, UK. The authors are grateful for the contribution of Vic Rayward-Smith.

References

1. Thomas, D.C.: Statistical Methods in Genetic Epidemiology. Oxford University Press, Oxford (2004)
2. Kruskal, J.B., Wish, M.: Multidimensional Scaling. Sage University Paper series on Quantitative Application in the Social Sciences, 07–011 (1978)
3. Borg, I., Groenen, P.: Modern Multidimensional Scaling: Theory and Applications. Springer, Heidelberg (2005)
4. Jain, A.K., Murty, M.N., Flynn, P.J.: Data Clustering: A Survey. ACM Computing Surveys 31 (1999)
5. Cox, T.F., Cox, M.A.A.: Multidimensional Scaling. CRC Press, Boca Raton (2001)
6. Torgerson, W.S.: Multidimensional scaling: I. Theory and method. Psychometrika 17, 401–419 (1952)
7. Kantardzic, M.: Data Minig: Concepts, Models, Methods, and Algorithms. IEEE Press, Los Alamitos (2003)
8. Hubert, L., Arabie, P.: Comparing partitions. J. Classif. 2, 193–218 (1985)

Mining Frequent Patterns from Network Data Flow*

Xin Li[1,2], Zhi-Hong Deng[1], Hao Ma[3], Shi-Wei Tang[1], and Bei Zhang[3]

[1] Key Laboratory of Machine Perception (Ministry of Education),
School of Electronics Engineering and Computer Science,
Peking University, 100871, Beijing
[2] Dalian Commodity Exchange,
116023, Dalian, Liaoning
[3] Computer Center, Peking University,
100871, Beijing
zhdeng@cis.pku.edu.cn

Abstract. The main objective of network monitoring is to discover the event patterns that happen frequently. In this paper, we have intensively studied the techniques used to mine frequent patterns from network data flow. We devel-oped a powerful class of algorithms to deal with a series of problems when min-ing frequent patterns from network data flow. We experimentally evaluate our algorithms on real datasets collected from the campus network of Peking Uni-versity. The experimental results show these algorithms are efficient.

1 Introduction

Network monitoring is critically important for network administrators to make sure the whole network running regularly and working availably. The main objectives in network monitoring can be summarized as two aspects: understanding traffic features which appear especially most frequently, and detecting outburst network anomalies [1-3]. No matter understanding traffic load or anomaly detection, it is necessary to find frequently happened features in data packets. Therefore, it is a primary task in network monitoring to mine frequent patterns from network data flow.

However, the varying and dynamic characteristics of network traffic, which are fast transfer, huge volume, shot-lived, inestimable and infinite, make it very difficult to implement efficient on-line mining algorithms. So we choose a sliding window as data processing model to make sure the mining result novel and integrated. Then, we develop a powerful class of algorithms, which are vertical re-mining algorithm, multi-pattern re-mining algorithm, fast frequent multi-pattern capturing algorithm, fast frequent multi-pattern capturing supplement algorithm to deal with problems when mining frequent patterns from network data flow. Finally, we evaluate our algorithms on real datasets collected from the campus network of Peking University. The experimental results show that these algorithms are efficient.

* Supported by the National High-Tech Research and Development Plan of China under Grant No.2009AA01Z136

R. Huang et al. (Eds.): ADMA 2009, LNAI 5678, pp. 528–535, 2009.

The remainder of the paper is organized as follows. The data processing model is given in Section 2. Four algorithms for mining frequent patterns from network data flow are developed in Section 3. Experimental results are presented in Section 4. Section 5 summarizes our study and points out some future research issues.

2 Data Processing Model

In this section, we introduce the sliding window model to handle network data flow. Sliding window [4] is a typical data model for flow control, especially for network data transfers. So we build a sliding window in fixed length, which means the size of this window is decided by a fixed length period of time, such as ten minutes or five minutes. The sliding window should contain all the fresh data in network; so once a new piece of data appears, the sliding window should move forward to capture it. Because there are hundreds even thousands of new data generated in one second in network data flow, to avoid updating sliding window too frequently, we use basic windows with equal length to divide a sliding window into several continuous partitions. So only when a new basic window comes, the sliding window will moves forward to capture it and drop the oldest one at the same time.

3 Algorithms

3.1 Vertical Re-mining Algorithms

Vertical method [5] [6] is one classical frequent pattern mining methods. A simple method is to run vertical algorithm for each sliding window. We call it **vertical re-mining algorithm**. By this algorithm, we can directly get complete set of frequent patterns. The problem is that, whenever the sliding window moves forward to a new position, we have to call this algorithm to scan the whole dataset in current sliding window and mine them from beginning to update results. However, one move only changes a small part of the sliding window, most of the data within the window unchanged. It also means that one move of the sliding window only brings a small change of the frequent patterns. So vertical re-mining algorithm will do a lot of repeated scan and repeated mining work. Obviously, it is inefficient.

3.2 Fast-Update Mining Algorithm

Since most part of two neighbouring sliding windows is the same, we develop a fast-update mining algorithm to avoid repeated scanning and mining between them. The fast-update algorithm works based on an independent coding technique within each basic window.

When we use vertical mining algorithm to mine frequent patterns from a sliding window, remember that each transaction is numbered with a unique ID at first in that sliding window, denoted by ID_{SW}. And each frequent pattern keeps a Tid-list, which contains all the ID_{SW} in which this pattern occurs. In our fast-update algorithm, we first number transactions based on basic window instead of sliding window, denoted by ID_{BW}. See example in Table 1. Suppose that there are thirty transactions in a sliding window and the sliding window is equably divided by five basic windows. So these transactions are numbered from 1 to 30 by ID_{SW}, but numbered from 1 to 6 by ID_{BW} within the corresponding basic window. So in our fast-update algorithm, each frequent pattern keeps *n* Tid_{BW}-lists, where *n* is the number of basic windows in a sliding window. The advantage of using ID_{BW} is that when a new basic window comes, all overlapped basic windows in the current sliding window just keep the original Tid_{BW}-lists without repeated computation. It only needs to collect the Tid_{BW}-list from the new incoming basic window, and updates the support by totalizing the length of all overlapped Tid_{BW}-lists and this new Tid_{BW}-list. So this independent coding greatly reduces the mining time.

Figure 1 shows the mining structure based on the data in Table 1. When we calls the **fast-update mining algorithm** for the first time, it runs the same processes with vertical mining in Table 1 except using the independent Tid_{BW}-lists to calculate the support instead of the uniform Tid_{SW}-list. Figure 1(a) shows the complete mining process during the first mining. Every pattern keeps five Tid_{BW}-lists there, and only the Tid_{BW}-lists belongs to the same basic window intersect together to obtain a new Tid_{BW}-list. When all the frequent patterns are mined out, we delete all the Tid_{BW}-lists but record their lengths in the pattern tree, showed in Figure 1(b). The support of each frequent pattern can be easy calculated by totalizing the record of each basic window together. From then on, whenever the sliding window moves to a new position, our fast-update mining algorithm only need to scan and calculate the Tid_{BW}-list for each frequent pattern in that new basic window, add the length of the new Tid_{BW}-list into pattern tree for every frequent pattern, and delete the oldest record at the same time. Then output the frequent patterns with the updated support in pattern tree and all the infrequent ones can be directly cut down from the pattern tree.

The fast-update mining algorithm gets the frequent patterns with their support in the current sliding window by just dealing with a small part of data within the newest basic window, so it improves the performance quite a lot compared with vertical re-mining algorithm. However, it only updates all frequent patterns that already exist in pattern tree. Besides these patterns, there are many other patterns, which were not frequent before but may become frequent later. To use the fast-update mining algorithm, we must find means to capture these new emerging frequent patterns. The situation is

Table 1. Transctions in A sliding window with five basic windows

ID_{SW}	ID_{BW}	Transaction	ID_{SW}	ID_{BW}	Transaction
1	1	A B	16	4	A C
2	2	A	17	5	C
3	3	A B	18	6	B C
4	4	B	19	1	C
5	5	B C	20	2	C
6	6	B C	21	3	A C
7	1	A B C	22	4	A
8	2	A	23	5	A B C
9	3	A C	24	6	A B C
10	4	A	25	1	A B
11	5	A	26	2	A
12	6	A B C	27	3	A B
13	1	A B C	28	4	A B
14	2	A	29	5	A C
15	3	B C	30	6	B C

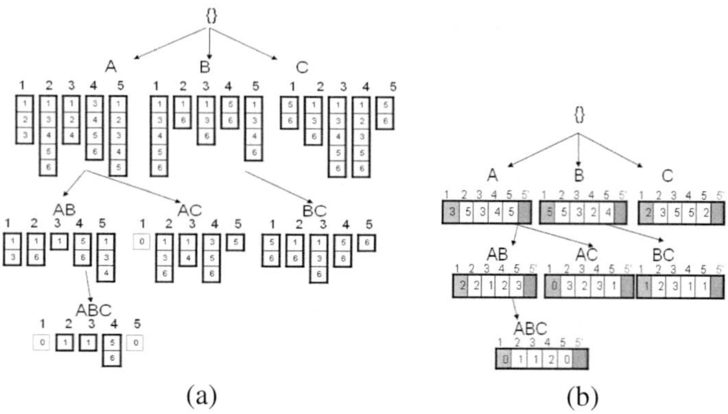

Fig. 1. The first running of fast update algorithm and the pattern tree

complex, so we divide this problem into two sub-problems: capturing new frequent 1-patterns and capturing new frequent multi-patterns, and settle them separately in the following two sections.

3.3 Capturing New Frequent 1-Patterns by Candidate Queue

The core advantage of fast-update mining algorithm is that it avoids the repeated scanning and mining of the data in all overlapped basic windows. However, fast-update mining algorithm fails to get the support of a frequent 1-pattern that is not in the pattern tree as not frequent in the last sliding window, but appears frequent in the new incoming basic window.

To avoid repeated scanning still, here we develop a method called candidate queue to capture all the new emerging frequent 1-patterns. Once there is one such 1-pattern appearing at the first time, we put it in a queue as a candidate pattern, and record its corresponding Tid_{BW}-list collected from that new incoming basic window. The next time another basic window comes, we check if it still keeps frequent locally in these two basic windows together. If not, we delete it from the candidate queue, else, record its new Tid_{BW}-list collected from the second basic window. The same step is performed until when there are n record of Tid_{BW}-lists for that 1-pattern in the queue (n is the number of basic windows in a sliding window). At this time, we can make sure it is a new frequent 1-pattern. So we allocate a new node in pattern tree to denote it and delete it from candidate queue. Since the pattern is placed in the pattern tree, we can use the fast-update mining algorithm to calculate its support quickly from then on.

Our candidate queue is worked based on the actual distribution of network flow. Usually, there are two types of 1-patterns transferred in network data flow: (1) common practice, appearing with a steadily frequency; (2) sudden events, appearing very frequently at some point, and maybe continue from then on or disappear at once. For the first type, if it is frequent, which means that network administrators may get interested with it, our pattern tree has already captured it. For the second type, if it is a meaningful sudden events, which means it suddenly becomes frequent at some point and continues for a while, our candidate queue will successfully capture it finally. In addition, once a 1-pattern appears frequent at some point, even re-scanning of the original data is almost useless and time wasting, because it hardly appears before that point.

The problem of capturing new frequent 1-patterns has been solved by candidate queue without repeated scanning. Here we introduce a **multi-pattern re-mining algorithm** to evade the problem of capturing new frequent multi-patterns which fast-update mining algorithm failed to mine. Simply speaking, the **multi-pattern re-mining algorithm** contains four steps:

(1) Scan the new incoming basic window BW to collect 1-patterns that exist in pattern tree, or exist in candidate queue, or appear frequent in BW with their Tid_{BW}-list;
(2) Use fast-update mining algorithm to calculate only frequent 1-patterns in pattern tree;
(3) Update candidate queue: a) add new 1-patterns into it, b) delete unsatisfied ones, c) for the ones that are frequent globally, delete them as well and allocate new nodes in pattern tree to denote them;
(4) Calculate the support of frequent (k+1)-patterns using the Tid-lists of frequent k-pattern.

3.4 Fast Frequent Multi-pattern Capturing Algorithm

The multi-pattern re-mining algorithm is not effective enough from the algorithm point of view. Because it can not avoid the repeated mining of multi-patterns, where fast-update mining algorithm has already deal with it. So to use the fast-update mining algorithm furthest, we still need to solve the problem of capturing new emerging multi-patterns that may be potentially frequent.

We first discuss the situations that may produce new frequent multi-patterns. Theoretically speaking, if a multi-pattern becomes frequent from infrequent, it must belong to one of the following situations.

(1) Combined by all new emerging frequent 1-patterns;
(2) Combined by new and old frequent 1-patterns which still keep frequent currently;
(3) Combined by all old frequent 1-patterns already in pattern tree and keeping frequent currently.

For the first situation, it is easy to calculate the support, because we have saved all the Tid_{BW}-lists of every new frequent 1-pattern in candidate queue, we can directly intersect them to obtain the Tid_{BW}-lists of this new multi-pattern as well as its support. However, for the last two situations, we lost the detailed information of old frequent patterns to calculate the new one. Because we can not store all patterns' Tid_{BW}-lists in the pattern tree, especially for the multi-pattern' Tid_{BW}-lists in the pattern tree which is space-consuming and may use up all the memory. So it is very difficult to capture this part of new multi-patterns, especially for the third situation that combined by all old frequent 1-patterns. Even we store the complete Tid_{BW}-lists for every old frequent 1-pattern in the pattern tree, we have to do the intersection from the first level of the pattern tree to the last level again to estimate if a new leaf node may generate which denotes a new frequent multi-pattern. So it does a lot of repeated calculations for all the multi-patterns already known frequent in the pattern tree.

Since the steps used to capture all new frequent multi-patterns are so complicated and may destroy the performance, we turn to observe the distribution and characteristics of network data flow again, and find that maybe we can ignore a part of new multi-patterns produced under some situations discussed before. We find that usually a server that appears very active in the network almost always is combined within a fixed pattern. That is, if the server is active, the corresponding pattern is active, vice versa. So we abstract this observation and give the following **supposition**:

If the sub-pattern of an infrequent multi-pattern is frequent all the while, there is hardly any chance for this multi-pattern to become frequent.

Based on the above supposition, we develop two algorithms: **fast frequent multi-pattern capturing algorithm** and **fast frequent multi-pattern capturing supplement algorithm**. The former one only consider new multi-patterns combined by new frequent 1-patterns besides using fast-update mining algorithm and candidate queue method; the latter does all the same process but makes a supplement by considering the new multi-patterns combined by new and old frequent 1-patterns. **The fast frequent multi-pattern capturing algorithm** is shown as follows.

```
Input: dataset in the new incoming basic window DB_b, the
minimum support ξ, the pattern tree P_tree, the candidate
queue Q_can.
Output: the complete set of frequent patterns
Scan DB_b once to find L[1]= {1-patterns in P_tree or
Q_can, or new local frequent ones in DB_b};
Scan DB_b again to collect TL[1] = {the Tid_BW-list of
each 1-pattern in L[1]};
Call fast-update mining algorithm on P_tree to update it;
Update Q_can
  Obtain new local frequent ones in L[1] ;
  Delete unsatisfied ones;
  Collect new frequent 1-patterns N[1], and delete them
  from Q_can.
Make nodes in P_tree for N[1]={ new frequent 1-patterns};
for ( k = 2; N[k-1]≠∅; k++) do begin
  for all p∈N[k-1] and q∈N[k-1], where p[1]= q[1],…,
  p[k-2] = q[k-2], p[k-1]< q[k-1] do begin
    c = p[1], p[2],…,p[k-1],q[k-1]; //Candidate k-
    pattern
    c.Tid_BW-lists= intersection(p.Tid_BW-lists, q.Tid_BW-
    lists);
    If (c.support ≥ |DB|×ξ) then
      N[k]= N[k] ∪{c};
      NL[k]= NL[k] ∪{c.Tid-list};
      Construct a new node for c in P_tree as the children
      of p;
    end if
  end for
  Delete NL[k-1]; // No use of NL[k-1] any more
end for
Answer = all nodes in P_tree.
```

The fast frequent multi-pattern capturing supplement algorithm is the same as the fast frequent multi-pattern capturing algorithm except increasing a sub-procedure before returning the final results. The increased sub-procedure is used to mine frequent multi-patterns composed by both old and new added frequent 1-patterns in pattern tree.

4 Performance Evaluation

To evaluate our algorithms, we gather network flow data from the campus network of Peking University. In our experiment, we set the size of the sliding window 5 minutes and the size of basic window 1 minute. Table 2 shows the dataset used in our experiment. By setting the minimum support to 0.1%, we can mine about 2,000 frequent patterns in every five minutes. Our experiments are performed on a Pentium(R) 4 CPU 3.00GHz computer with 512 memory, running Microsoft Windows XP. All algorithms are coded in C++ and compiled by Microsoft visual studio.NET 2003.

Table 2. The dataset used in our experiments

The end moment of SW	Number of transactions	The end moment of SW	Number of transactions	The end moment of SW	Number of transactions
16:01	3,640,511	16:06	3,711,287	16:11	3,661,745
16:02	3,661,418	16:07	3,698,435	16:12	3,663,615
16:03	3,682,325	16:08	3,685,583	16:13	3,665,485
16:04	3,703,232	16:09	3,651,824	16:14	3,667,355
16:05	3,724,139	16:10	3,659,875	16:15	3,669,229

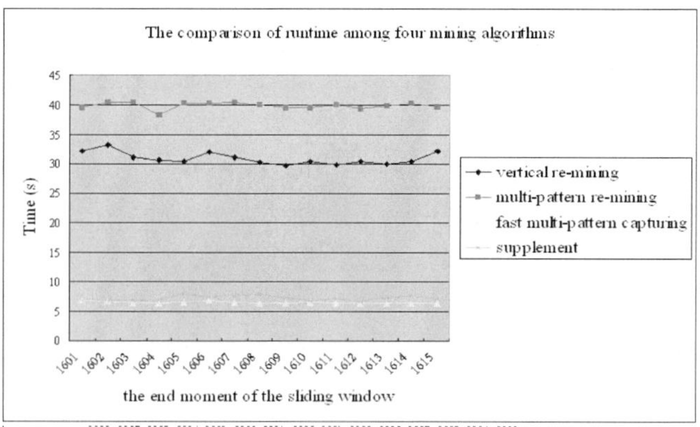

Fig. 2. Comparison of runtime of four mining algorithms

Figure 2 and Figure 3 shows the total comparison of our proposed four mining algorithms. We have conclusions as follows: (1) The vertical re-mining algorithm can find the complete set of frequent patterns, where time-consuming is on a middle level; (2) The multi-pattern re-mining algorithm can also find the complete set of frequent patterns but with a delay, because of the candidate queue's temporal cache operation. However, the efficiency is worst; (3) These two fast multi-pattern capturing algorithms speed up the total mining process while they cover more than 85% of the complete set of frequent patterns.

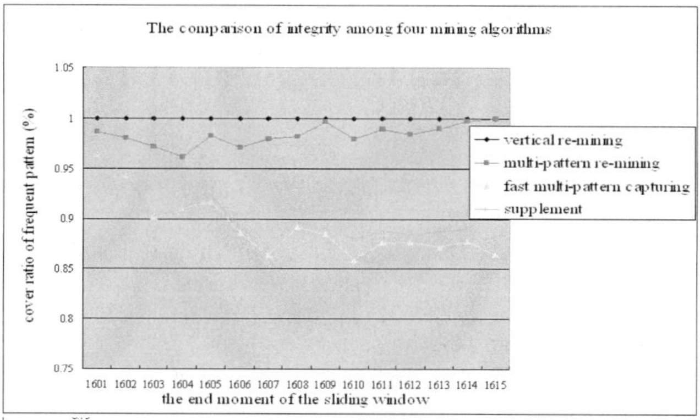

Fig. 3. Comparison of mining results of four mining algorithms

5 Conclusions

In this paper, we propose the problem of applying frequent pattern mining in the field of network monitoring. We proposed four algorithms to mine frequent patterns from Network flow data. The experimental results on real datasets show that our algorithms are effective. In future, we plan to adopt the methods of mining closed pattern [7] to improve this work.

References

1. Ghosh, A.K., Schwartzbard, A.: A study in using neural networks for anomaly and misuse detection. In: 8th USENIX Security Symposium, pp. 141–151. USENIX Association, Washington (1999)
2. Lippmann, R.P., Cunningham, R.K.: Improving intrusion detection performance using keyword selection and neural networks. Computer Networks 34, 597–603 (2000)
3. Crandall, J.R., Su, Z., Wu, S.F., Chong, F.T.: On Deriving Unknown Vulnerabilities from Zero-Day Polymorphic and Metamorphic Worm Exploits. In: 12th ACM Conference on Computer and Communications Security, pp. 235–248. ACM Press, New York (2005)
4. Chang, J., Lee, W.S.: Finding recent frequent itemsets adaptively over online data streams. In: 9th ACM SIGKDD International Conference on Knowledge Discovery and Data Mining, pp. 487–492. ACM Press, New York (2003)
5. Zaki, M.J.: Scalable algorithms for association mining. IEEE Transactions on Knowledge and Data Engineering 12(3), 372–390 (2000)
6. Zaki, M.J., Gouda, K.: Fast vertical mining using diffsets. In: 9th ACM SIGKDD International Conference on Knowledge Discovery and Data Mining, pp. 326–335. ACM Press, New York (2003)
7. Wang, J., Han, J., Pei, J.: CLOSET+: searching for the best strategies for mining frequent closed itemsets. In: 9th ACM SIGKDD international conference on Knowledge discovery and data mining, pp. 236–245. ACM Press, New York (2003)

Several SVM Ensemble Methods Integrated with Under-Sampling for Imbalanced Data Learning

ZhiYong Lin[1,2], ZhiFeng Hao[3], XiaoWei Yang[4], and XiaoLan Liu[4]

[1] School of Computer Science and Engineering, South China University of Technology, 510640 Guangzhou, Guangdong

[2] Department of Computer Science, Guangdong Polytechnic Normal University, 510665 Guangzhou, Guangdong

[3] Guangdong University of Technology, 510006 Guangzhou, Guangdong

[4] College of Science, South China University of Technology, 510640 Guangzhou, Guangdong

sophyca@yahoo.cn

Abstract. Imbalanced data learning (IDL) is one of the most active and important fields in machine learning research. This paper focuses on exploring the efficiencies of four different SVM ensemble methods integrated with under-sampling in IDL. The experimental results on 20 UCI imbalanced datasets show that two new ensemble algorithms proposed in this paper, i.e., *CABagE* (which is bagging-style) and *MABstE* (which is boosting-style), can output the SVM ensemble classifiers with better minority-class-recognition abilities than the existing ensemble methods. Further analysis on the experimental results indicates that *MABstE* has the best overall classification performance, and we believe that this should be attributed to its more robust example-weighting mechanism.

Keywords: Imbalanced data learning, Under-sampling, SVM, Ensemble.

1 Introduction

The main task of Imbalanced Data Learning (IDL) is to tackle the so-called "Class Imbalance Problem" (CIP), in which some classes have more learning examples than others [1]. In IDL, the overall performance of traditional learning algorithms will degenerate significantly, since they readily output classifiers which are biased towards majority class severely [1]. Therefore, how to deal with CIP efficiently has become one of the top challenges in machine learning research [2].

One of the most important approaches to address CIP is data preprocessing [1], which can be further subdivided into two categories: over-sampling [3] and under-sampling [4]. No evidences or theoretical justifications show that over-sampling prevails over under-sampling and vise versa. So, both of them are widely used in IDL [5]. Compared with over-sampling, one merit of under-sampling is that it can build a smaller balanced training sample and further reduce the subsequent training time. But, after being under-sampled, the majority class will lost some useful information for

R. Huang et al. (Eds.): ADMA 2009, LNAI 5678, pp. 536–544, 2009.

training, and this may degrade the classifier's performance to some extent. To overcome this drawback, we may consider combining under-sampling with ensemble learning [6-8]. More specifically, we can train multiple different classifiers on multiple different training subsets by means of under-sampling the majority class several times, and then combine them into an ensemble classifier. Making the best use of the information of the training data, such ensemble classifier usually achieves more satisfying overall classification performance.

Motivated by the recent encouraging results on the support vector machine (SVM) ensemble for pattern classification [6, 9-10], we focus on exploring the efficiencies of SVM ensemble integrated with under-sampling in IDL. Four different ensemble algorithms are studied, and two of them are newly proposed in this paper, which are Cluster Based Asymmetric Bagging Ensemble (*CABagE*) and Modified Asymmetric Boosting Ensemble (*MABstE*). The experimental results on 20 benchmark datasets from UCI [19] show that both *CABagE* and *MABstE* can output the SVM ensemble classifiers of better minority-class-recognition abilities, and *MABstE* has the best overall classification performance among of four ensemble algorithms.

2 SVM Ensemble Methods Integrated with Under-Sampling

Consider a binary CIP, whose training set is $\mathcal{TS}=\{(x_i,y_i)|x_i\in\mathcal{R}^d, y_i=\pm 1, i=1,\cdots,n\}$, we let $\mathcal{P}$ and $\mathcal{N}$ denote the positive (majority) and negative (minority) class training subset respectively, i.e., $\mathcal{TS}=\mathcal{P}\cup\mathcal{N}$. Assume $|\mathcal{P}|=n_+$ and $|\mathcal{N}|=n_-$, then the imbalance ratio *IR* is defined as n_-/n_+, which is far greater than 1 in a typical CIP.

For a binary CIP, SVM can output the classifier $sign(f(x))$ with the following form:

$$f(x)=\sum_{i=1}^{n}\alpha_i y_i K(x,x_i)+b \tag{1}$$

where K is a predefined kernel function (say, the most frequently used RBF kernel $K(\bar{x},\tilde{x})=\exp\{-\sigma\|\bar{x}-\tilde{x}\|^2\}$ with $\sigma>0$ as the kernel parameter). The coefficients α_is and the bias term b in (1) can be obtained by solving the following problem's Lagrangian dual [16]:

$$\begin{aligned}\min_{w,b,\xi}\ & \frac{1}{2}\|w\|^2+C_+\sum_{i:y_i=+1}\xi_i+C_-\sum_{i:y_i=-1}\xi_i\\ s.t\quad & y_i(<w,\phi(x_i)>+b)\ge 1-\xi_i\\ & \xi_i\ge 0,\ i=1,2,\cdots,n\end{aligned} \tag{2}$$

where $\phi: x\to\phi(x)$ is a mapping determined by the kernel function implicitly such that $K(\bar{x},\tilde{x})=<\phi(\bar{x}),\phi(\tilde{x})>$, and C_+ (C_-) is the penalty factor trading off between the classifier's margin and its empirical error for positive (negative) class.

2.1 Bagging-Style Ensemble Algorithms

Bagging is one of the most popular ensemble methods [11-12]. In standard bagging, multiple classifiers are trained on the bootstrap subsets of the whole training set in parallel, and then they are aggregated by the majority-voting mechanism. However, for imbalanced data, here we only consider under-sampling the negative class, i.e., all the positive examples are kept so as to form a relatively balanced training subset. This kind of bagging is also called as asymmetric bagging in [14]. Now, we present the first SVM ensemble approach based on the normal asymmetric bagging as follows, where *round*() denotes the (nearest integer) rounding function.

Normal **A**symmetric **Bag**ging **E**nsemble (**NABagE**)

1) Set T (the number of base SVM classifiers) and r (under-sampling ratio, $r \in [0,1]$). $t \Leftarrow 1$.
2) Draw a subset $\bar{\mathcal{N}}$ from $\mathcal{N}$ with size $|\bar{\mathcal{N}}| = round(rn_- + (1-r)n_+)$ by using random replacement sampling method (bootstrapping). Let $\mathcal{S} = \bar{\mathcal{N}} \cup \mathcal{P}$.
3) Train SVM on $\mathcal{S}$ and get the base classifier $h_t(x) = sign(f_t(x))$.
4) If $t<T$ Then $t \Leftarrow t+1$, goto 2); Else output the following SVM ensemble:

$$h(x) = sign(\sum_{t=1}^{T} h_t(x)) .$$

For drawing as many representative examples as possible from $\mathcal{N}$ (especially in the limitation of sample capacity), it would be better to partition $\mathcal{N}$ into several subsets by using clustering technique firstly, and then draw examples from these subsets proportionally. Base on this consideration, we proposed the following cluster based asymmetric bagging ensemble method:

Cluster **B**ased **A**symmetric **Bag**ging **E**nsemble (**CBABagE**)

1) Set T (the number of base SVM classifiers), r (under-sampling ratio, $r \in [0,1]$) and l (expected cluster size, $1 \le l \le n_-$). $t \Leftarrow 1$.
2) Use k-means clustering technique to partition $\mathcal{N}$ into k subsets $\mathcal{N}_1, \cdots, \mathcal{N}_k$, where $k = round(n_- / l)$.
3) Allocate the sub-sample size $m_i = round(m|\mathcal{N}_i|/n_-)$ for each $\mathcal{N}_i$ ($i = 1, \cdots, k$), where $m = round(rn_- + (1-r)n_+)$ is the size of total negative sample.
4) Draw a subset $\bar{\mathcal{N}}_i$ from $\mathcal{N}_i$ with size $|\bar{\mathcal{N}}_i| = m_i$ by bootstrapping for $i = 1, \cdots, k$. Let $\bar{\mathcal{N}} = \bar{\mathcal{N}}_1 \cup \cdots \cup \bar{\mathcal{N}}_k$ and $\mathcal{S} = \bar{\mathcal{N}} \cup \mathcal{P}$.
5) Train SVM on $\mathcal{S}$ and get the base classifier $h_t(x) = sign(f_t(x))$.
6) If $t<T$ Then $t \Leftarrow t+1$, goto 4); Else output the following SVM ensemble:

$$h(x) = sign(\sum_{t=1}^{T} h_t(x)) .$$

2.2 Boosting-Style Ensemble Algorithms

Boosting is another popular ensemble method, and AdaBoost is its outstanding representative [12-13]. Unlike Bagging, in AdaBoost, all the base classifiers are trained sequentially, not parallelly. In addition, an adaptive example-weighting mechanism is adopted in AdaBoost to improve the ensemble classifier's recognition ability on the hard classified examples. Similar to Normal Asymmetric Bagging Ensemble, we can easily get Normal Asymmetric Boosting Ensemble as follows:

Normal Asymmetric Boosting Ensemble (NABstE)

1) Set T (the number of base SVM classifiers) and r (under-sampling ratio, $r \in [0,1]$).
2) Initialize the weight value $w_i=1/n$ for every example ($1 \leq i \leq n$). $t \Leftarrow 1$.
3) Recompute the weight value of each negative example as follows:

$$\overline{w}_i = w_i / \sum_{i=1}^{n_-} w_i \quad (1 \leq i \leq n\text{-}).$$

4) Draw a subset $\overline{\mathcal{N}}$ from $\mathcal{N}$ with size $|\overline{\mathcal{N}}|=round(rn_- + (1-r)n_+)$ by replacement sampling according to the current weight distribution $\overline{w}$ over $\mathcal{N}$. Let $\mathcal{S}=\overline{\mathcal{N}} \cup \mathcal{P}$.
5) Train SVM on $\mathcal{S}$ and get the base classifier $h_t(x) = sign(f_t(x))$.
6) $\varepsilon_t \Leftarrow \sum_{i \in I} w_i$, $I = \{i \mid y_i \neq h_t(x_i)\}$; $\beta_t \Leftarrow 0.5\ln(1/\varepsilon_t - 1)$.
7) Adjust the weight values as follows:

$$w_i \Leftarrow w_i \exp\{-\beta_t y_i h_t(x_i)\} / \sum_{j=1}^{n} w_j \exp\{-\beta_t y_j h_t(x_j)\}, \ (1 \leq i \leq n).$$

8) If $t<T$ Then $t \Leftarrow t+1$, goto 3); Else output the following SVM ensemble:

$$h(x) = sign(\sum_{t=1}^{T} \beta_t h_t(x)).$$

Obviously, in *NABstE*, more weight a negative example gets, more frequently it will be sampled into the next training subset. However, such weighting mechanism may have some potential risk of giving too large weight to the outliers and thereby result in the overfitted ensemble classifier. To overcome this deficiency, we propose a kind of simple but more robust weighting mechanism as follows.

Firstly, we define a covering set $Cov(x)$ for every negative example x as $Cov(x) = \{x_i \mid D(x, x_i) \leq d_x, (x_i, y_i) \in \mathcal{N}\}$, where $D(\overline{x}, \tilde{x}) = \| \overline{x} - \tilde{x} \|_2$ is the Euclidean distance between $\overline{x}$ and $\tilde{x}$, and $d_x = \min\{D(x, x_j) \mid (x_j, y_j) \in \mathcal{P}\}$, i.e. the minimal value of the distances between x and all positive examples. Then, based on $Cov(x)$, we can reweight the negative examples in a more robust way as depicted at the step 3 in the following Modified Asymmetric Boosting Ensemble.

Modified Asymmetric Boosting Ensemble (MABstE)

1)~2) are the same as the corresponding steps in *NABstE*.

3) Recompute the weight value of each negative example as follows:

$$\overline{w}_i = \tilde{w}_i / \tilde{w}, \text{ where } \tilde{w} = \sum_{i=1}^{n_-} \tilde{w}_i \text{ and } \tilde{w}_i = \sum_{j \in J_i} w_j, J_i = \{ j \mid x_j \in Cov(x_i)\} \ (1 \leq i \leq n\text{-}).$$

4)~8) are the same as the corresponding steps in *NABstE*.

It is worthy to point out that $Cov(x)$ can be found in the feature space, since we can calculate the distance by using the "kernel trick" as follows:

$$\begin{aligned} D(\phi(\overline{x}), \phi(\tilde{x})) &= \| \phi(\overline{x}) - \phi(\tilde{x}) \| = \sqrt{<\phi(\overline{x}), \phi(\overline{x})> - 2 <\phi(\overline{x}), \phi(\tilde{x})> + <\phi(\tilde{x}), \phi(\tilde{x})>} \\ &= \sqrt{K(\overline{x}, \overline{x}) - 2K(\overline{x}, \tilde{x}) + K(\tilde{x}, \tilde{x})} \end{aligned} \quad (3)$$

3 Numerical Experiments

3.1 Settings

Twenty UCI imbalanced datasets summarized in Table 1 are used to evaluate the different SVM ensemble algorithms. For each dataset, its nominal attributes (if exist) are converted into numerical ones, and then all attributes are rescaled to [−1, 1]. We randomly divide each dataset into two subsets by stratified sampling. One containing 80% of the whole data is used for training, and the other containing the remaining 20% for testing. Then, we run the ensemble algorithms on the training set and evaluate them on the testing set. Such process is repeated 10 times for each dataset.

Table 1. Summary of datasets used in this paper

n: the size of each dataset; *Pos*: positive class label; *Neg*: negative class label.

No.	Dataset	*n*	*Pos, Neg*	*IR*	Source Dataset
1	Wine	178	3, remainder	2.71	Wine
2	GlassBWFP	214	building_windows_float _processed, remainder	2.06	Glass Identification
3	GlassVWFP	214	vehicle_windows_float _processed, remainder	11.59	Glass Identification
4	Newthyroid	215	hypo, remainder	5.14	Thyroid Disease
5	Haberman	306	die, survive	2.78	Haberman's Survival
6	EColiIMU	336	iMu, remainder	8.60	E. Coli Genes
7	EColiOM	336	om, remainder	15.67	E. Coli Genes
8	Ionosphere	351	bad, good	1.79	Ionosphere
9	wdbc	569	m, b	1.68	Breast Cancer Wisconsin (Diagnostic)
10	Wisconsin	683	4, 2	1.86	Breast Cancer Wisconsin(Original)
11	Abalone18-9	731	18, 9	16.39	Abalone
12	Abalone16-9	756	16, 9	10.29	Abalone
13	Pima	768	1, 0	1.88	Pima Indians Diabetes
14	VehicleVAN	846	van, remainder	3.25	Vehicle Silhouettes
15	German	1000	bad, good	2.33	German Credit Data
16	cmc	1473	no_use, remainder	1.34	Contraceptive Method Choice
17	yeastME2	1484	ME2, remainder	28.07	yeast
18	yeastVAC	1484	VAC, remainder	48.50	yeast
19	CarGood	1728	good, remainder	24.06	Car Evaluation
20	CarUnacc	1728	remainder, unacc	2.34	Car Evaluation

LIBSVM [18] is adopted as the trainer to produce the base SVM classifiers, where the RBF kernel with parameter σ =0.1 is used. The penalty factor for the negative class *C*- is fixed as 100, but for positive class C_+ is tuned such that $C_+ / C_- = |\mathcal{N}| / |\mathcal{P}|$, and it depends on the specific training set. For *T*, we consider four different values: 10, 50, 100 and 150. For *r*, we also consider four different values: 0, 0.2, 0.5 and 0.8. As for l, which is used in *CABagE* as the expected cluster size, we just fix it as 20. Thus, we get $4 \times 4 = 16$ different settings for every ensemble algorithm. For every setting, we run the ensemble algorithm 10 times on each dataset. So, there are total $16 \times 10 = 160$ testing results on each dataset.

3.2 Results

In IDL, we should adopt some reasonable metrics to measure the quality of classification. *True Positive Rate* (*TPR*) and *True Negative Rate* (*TNR*) are two important metrics. Both of them are expected to be high for a good classifier, but they are often conflict. *BAC* (Balance Accuracy), which is defined as 0.5(*TPR*+*TNR*), and *GMean*,

Table 2. Average *TPR* (%)

No.	*NABagE*	*CABagE*	*NABstE*	*MABstE*
1	99.38	***99.72***	99.58	99.58
2	71.15	***95.87***	68.51	84.33
3	21.67	***22.29***	13.96	16.88
4	89.00	91.50	89.50	***94.75***
5	22.85	***56.48***	21.52	36.99
6	48.23	63.02	50.63	***76.46***
7	77.92	81.67	78.54	***84.38***
8	90.00	***91.58***	91.35	90.98
9	96.56	***97.19***	96.65	96.98
10	96.49	***98.16***	94.77	97.65
11	42.60	50.29	42.16	***55.24***
12	29.77	34.38	26.48	***39.30***
13	60.41	***88.80***	59.08	78.40
14	98.16	***99.01***	97.16	97.93
15	63.10	***72.97***	64.84	66.23
16	56.98	***96.96***	56.37	64.58
17	20.75	20.88	17.44	***44.56***
18	18.50	***20.75***	6.63	5.88
19	60.58	76.35	62.88	***85.19***
20	98.73	***99.74***	97.99	98.70

Table 3. Average *TNR* (%)

No.	*NABagE*	*CABagE*	*NABstE*	*MABstE*
1	*97.03*	96.03	71.83	96.00
2	79.60	57.25	***80.27***	71.45
3	90.72	89.09	***95.88***	93.45
4	***99.48***	99.17	97.64	97.00
5	***91.53***	63.74	91.41	86.16
6	93.84	91.94	***94.49***	89.60
7	98.24	97.89	***99.04***	97.75
8	***95.30***	90.37	92.22	93.57
9	***97.64***	92.05	96.79	95.83
10	***96.26***	95.44	96.24	95.68
11	96.45	95.65	***97.78***	94.73
12	96.12	95.77	***98.53***	95.57
13	***83.47***	55.52	83.11	67.50
14	97.47	96.58	97.82	97.59
15	***66.97***	53.71	62.88	62.96
16	***77.65***	6.51	75.96	66.27
17	95.80	95.71	***97.63***	92.67
18	89.08	86.78	***96.18***	93.66
19	98.01	98.14	***99.28***	96.76
20	98.71	97.39	***98.88***	97.94

Table 4. Average *BAC* (%)

No.	*NABagE*	*CABagE*	*NABstE*	*MABstE*
1	***98.07***	98.00	76.75	97.98
2	74.30	76.74	74.35	***77.28***
3	***55.73***	55.22	55.02	54.23
4	93.45	***94.65***	93.31	93.95
5	56.70	***59.72***	56.32	58.74
6	69.38	76.85	71.88	***84.04***
7	87.32	89.03	87.93	***91.29***
8	***92.54***	91.06	90.24	92.20
9	***97.22***	95.08	97.08	96.56
10	95.65	96.35	95.50	***96.71***
11	68.91	71.91	68.52	***72.69***
12	62.22	64.08	61.76	***65.19***
13	71.11	72.12	71.12	***72.42***
14	97.59	***97.74***	97.42	97.34
15	***64.44***	63.61	63.92	64.20
16	***67.09***	51.53	66.84	66.78
17	58.15	58.20	57.31	***66.21***
18	***53.66***	53.25	50.91	49.72
19	77.18	84.95	77.90	***89.37***
20	***98.71***	98.63	98.50	97.59

Table 5. Average *GMean* (%)

No.	*NABagE*	*CABagE*	*NABstE*	*MABstE*
1	***98.17***	97.83	73.16	97.75
2	73.11	73.38	73.02	***76.86***
3	18.23	17.59	16.01	***20.09***
4	93.69	***94.99***	92.19	94.73
5	33.54	49.75	35.88	***54.40***
6	52.47	69.64	60.41	***81.63***
7	86.23	88.57	86.90	***90.01***
8	***92.54***	90.88	91.39	92.19
9	***97.08***	94.55	96.70	96.38
10	96.35	***96.78***	95.48	96.64
11	58.56	65.97	60.15	***70.71***
12	39.02	45.47	40.04	***53.80***
13	70.34	69.80	69.50	***72.39***
14	***97.80***	97.78	97.48	97.75
15	***64.69***	62.23	63.67	64.35
16	***66.27***	24.46	65.21	65.25
17	20.72	21.08	21.02	***58.01***
18	***15.83***	15.43	9.22	7.43
19	68.71	84.90	73.92	***90.18***
20	***98.72***	98.56	98.43	98.32

which is defined as $(TPR \times TNR)^{0.5}$, are two reasonable tradeoffs between *TPR* and *TNR* [15]. So, we decide to report our algorithms' performance on 20 datasets under *TPR*, *TNR*, *BAC* and *GMean*. All the metric values are reported as their averages over the 160 testing results, and they are listed in Table 2~5 respectively. In each table, the first column denotes the dataset number and we highlight the maximum of each row in boldfaced and italic type.

As mentioned before, the classifiers trained by using traditional learning algorithms are easily to bias towards the majority class, i.e., they often have high *TNR* but low *TPR*. According to Table 2 and 3, it seems that *NABagE* and *NABstE* still can not escape from such dilemma. Contrastively, Table 2 indicates that both *CABagE* and *MABstE* have the abilities to improve *TPR*. In other words, they may output the 'bias-rectified' classifiers. But, if we consider combining *TPR* and *TNR* into *BAC* as the measure metric, we can find that both *NABagE* and *MABstE* are competitive and have better performance according to Table 4. Similar conclusion can be drawn from Table 5 for *GMean* metric.

To compare these algorithms further, we calculate their average ranking values under different metrics by using Friedman's method [17]. Take *GMean* as an example, we sort the *GMean* values of four algorithms on the testing dataset in descending order for each run. Then, we assign rank 1 to the algorithm of highest *GMean* value, and rank 2 to the algorithm of second highest *GMean* value, … , until rank 4 to the algorithm of lowest *GMean* value. If some algorithms have the same *GMean* values, then we reassign each of them with their average ranking values. Finally, we average all the ranking values for each algorithm. The corresponding results are reported in Table 6. Notice that the smaller the average ranking values is, the better performance has for the corresponding algorithm. Under both *BAC* and *GMean* metrics, *MABstE* has the smallest average ranking values. Therefore, from a comprehensive point of view, we believe *MABstE* is more suitable for tackling CIP than the other three ensemble algorithms.

Table 6. Average ranking values

Metric	*NABagE*	*CABagE*	*NABstE*	*MABstE*
TPR	2.75	***1.69***	2.47	2.01
TNR	1.62	2.69	2.00	3.00
BAC	2.20	2.26	2.44	***2.18***
GMean	2.23	2.25	2.49	***2.12***

4 Conclusion

By adopting SVM as the base classifier trainer, we proposed and studied four ensemble algorithms integrated with under-sampling in this paper. We have carried out numerical experiments on 20 UCI imbalanced datasets to compare these ensemble algorithms.

First of all, the experimental results show that two new ensemble algorithms, including *CABagE* and *MABstE*, have higher *TPR* than the others. This implies that they can produce SVM ensemble classifiers which are less biased to the majority class. However, by using *BAC* and *GMean* as metrics, we found that both *MABstE* and *NABagE* are competitive and have better performance than the others. Further comparison between these four algorithms by way of Friedman's ranking show that *MABstE* overall outperforms the others. We believe that it should be attributed to our proposed more robust example-weighting mechanism, which may efficiently prevent the classifiers from overfitting the training data.

Acknowledgments. This work has been supported by Program for Excellent and Creative Young Talents in Universities of Guangdong Province (LYM08074).

References

1. Chawla, N.V., Japkowicz, N., Kotcz, A.: Editorial: Special Issue on Learning from Imbalanced Data Sets. ACM SIGKDD Explorations Newsletter 6, 1–6 (2004)
2. Yang, Q., Wu, X.: 10 Challenging Problems in Data Mining Research. International Journal of Information Technology & Decision Making 5, 597–604 (2006)
3. Chawla, N.V., Bowyer, K.W., Hall, L.O., Kegelmeyer, W.P.: Smote: Synthetic Minority Over-Sampling Technique. Journal of Artificial Intelligence Research 16, 341–378 (2002)
4. Kubat, M., Matwin, S.: Addressing the Curse of Imbalanced Training Sets: One-Sided Selection. In: Proceedings of the 14th International Conference on Machine Learning, pp. 179–186. Morgan Kaufmann, San Francisco (1997)
5. Batista, G.E.A.P.A., Prati, R.C., Monard, M.C.: A Study of the Behavior of Several Methods for Balancing Machine Learning Training Data. ACM SIGKDD Explorations Newsletter 6, 20–29 (2004)
6. Liu, Y., An, A., Huang, X.J.: Boosting Prediction Accuracy on Imbalanced Datasets with SVM Ensembles. In: Ng, W.-K., Kitsuregawa, M., Li, J., Chang, K. (eds.) PAKDD 2006. LNCS, vol. 3918, pp. 107–118. Springer, Heidelberg (2006)
7. Liu, X.Y., Wu, J.X., Zhou, Z.H.: Exploratory under-Sampling for Class-Imbalance Learning. IEEE Transactions on Systems, Man, and Cybernetics - Part B: Cybernetics 39, 539–550 (2009)
8. Dietterich, T.: Ensemble Learning. In: Arbib, M.A. (ed.) The Handbook of Brain Theory and Neural Networks, 2nd edn., pp. 110–125. The MIT Press, Cambridge (2002)
9. Wang, S.-J., Mathew, A., Chen, Y., Xi, L.-F., Ma, L., Lee, J.: Empirical Analysis of Support Vector Machine Ensemble Classifiers. Expert Systems with Applications 36, 6466–6476 (2008)
10. Kim, H.-C., Pang, S., Je, H.-M., Kim, D., Bang, S.Y.: Constructing Support Vector Machine Ensemble. Pattern Recognition 36, 2757–2767 (2003)
11. Breiman, L.: Bagging Predictors Machine Learning 24, 123–140 (1996)
12. Bauer, E., Kohavi, R.: An Empirical Comparison of Voting Classification Algorithms: Bagging, Boosting, and Variants. Machine Learning 36, 105–139 (1999)
13. Freund, Y., Schapire, R.E., Abe, N.: A Short Introduction to Boosting. Journal of Japanese Society for Artificial Intelligence 14, 771–780 (1999)
14. Tao, D., Tang, X., Li, X., Wu, X.: Asymmetric Bagging and Random Subspace for Support Vector Machines-Based Relevance Feedback in Image Retrieval. IEEE Transactions on Pattern Analysis and Machine Intelligence 28, 1088–1099 (2006)

15. Caruana, R., Niculescu-Mizil, A.: Data Mining in Metric Space: An Empirical Analysis of Supervised Learning Performance Criteria. In: Proceedings of the 10th ACM SIGKDD International Conference on Knowledge Discovery and Data Mining, pp. 69–78. ACM, New York (2004)
16. Cristianini, N., Shawe-Taylor, J.: An Introduction to Support Vector Machines. Cambridge University Press, Cambridge (2000)
17. Conover, W.J.: Practical Nonparametric Statistics, 3rd edn. Wiley, Chichester (1999)
18. Chang, C.-C., Lin, C.-J.: Libsvm: A Library for Support Vector Machines (2001), http://www.Csie.Ntu.Edu.Tw/~Cjlin/Libsvm
19. Asuncion, A., Newman, D.J.: UCI Machine Learning Repository. University of California, School of Information and Computer Science, Irvine (2007), http://www.ics.uci.edu/~mlearn/MLRepository.html

Crawling and Extracting Process Data from the Web

Yaling Liu and Arvin Agah

The University of Kansas, Department of Electrical Engineering & Computer Science,
1520 West 15th Street,
Lawrence, KS 66045-7621, USA
yliu6@eecs.ku.edu

Abstract. In this paper, we address the design and implementation of a supporting system for process-based searches. This supporting system can efficiently crawl the Web and extract processes from obtained data. The retrieved processes can then be used in a Process-Based Search Engine (PBSE). In this work, a process is defined as a sequence of activities for achieving a goal. A PBSE uses the extracted processes to transform an original query into multiple sub-queries, and then performs keyword search for each transformed sub-query. To facilitate effective process-based searches, a large number of high quality processes are required. This paper focuses on how to efficiently and effectively build a database of processes by exploring the Web.

Keywords: Process-based search, information extraction, supervised learning, wrapper generation.

1 Introduction

In earlier work, we have created a novel Web search system, referred to as "Process-Based Search Engine (PBSE)". PBSE requires a large number of high quality processes in order to successfully perform process-based searches. In this paper, we address the approaches to building this supporting system (a database of processes, referred to as Process Base).

Traditional general-purpose Web search engines (for example, Google) were designed based on the assumption that one query is answered by one document. In the PBSE, this traditional model of Web search is extended by assuming that one query may be answered by the combination of multiple documents. One way to implement this extended model is illustrated as follows.

1. Decide the multiple sub-concepts implied by the initial query, if they exist.
2. Transform the initial query into a sequence of sub-queries, each of which represents a sub-concept.
3. Use a traditional general-purpose Web search engine to search for each sub-query.
4. Integrate and format the results obtained from searching these sub-queries and display them to the user.

In the current implementation of the PBSE, processes are used to discover the sub-concepts implied in an initial query. Following is a simple example to illustrate this

R. Huang et al. (Eds.): ADMA 2009, LNAI 5678, pp. 545–552, 2009.

approach. When an initial query: "How to drive a car in Lawrence, KS," is fired to a search engine, a relevant process is first searched out from the process base. The title of the process is "How to drive a car." This process contains five steps: 1) Go to a driving school; 2) Take the driver's license test; 3) Buy a car; 4) Buy auto insurance; 5) Arrange parking spaces. These steps are considered as the sub-concepts of the initial query. Each step in the process is used to form one query sent to a traditional search engine. Obtained results from all sub-queries are integrated and presented to the user. We name this category of search as process-based search because of the adoption of processes in transforming the initial query. A system which implemented such functionalities is called Process-Based Search Engine (PBSE).

PBSE must have access to a large number of high quality processes before it can possibly perform effective process-based searches. This paper introduces the approach to semi-automatically building such a database of processes. There are two sources as follows which can be explored to achieve this.

1. Retrieve the processes explicitly published on the Web. For example, the how-to articles published on eHow (http://www.ehow.com).
2. Retrieve implicit processes. These processes are not explicitly published on any Web documents, but recorded as usage track logs on Web servers or Web search engines.

In this paper, we focus on the first source. As for the second one, it is possible to discover processes from the huge usage track logs kept by most Web search engines [1]. However this is not the focus of this writing.

2 Related Work

There has not been any existing work on accumulating processes from the Web because of the lack of need. However, process mining, also called workflow mining, or process discovery has been investigated for more than one decade. Process mining rediscovers explicit control flow models given their workflow event logs [2, 3, 4, 5, 6]. In a similar situation, the Web search engines keep query tracks [1]. However, due to the fact that the number of possible activities is too big to be managed, compared to the problem in process mining, it is difficult to extract processes by simply using the same techniques as process mining. In the area of software engineering, the process mining problem has also been studied [7, 8]. The big success of open source development in recent years has attracted interests from commercial software vendors. The same approaches of log analysis in process mining have been used to discover software processes in open source communities [9, 10].

3 Process Base

Web pages which possibly contain processes are very sparsely distributed on the Web. In order to efficiently extract more processes while downloading relatively fewer pages, a simple topic crawler has been developed, based on the following two observations: 1) The title of the homepage of a Web site generally describes the content stored on the site; 2) Inside a Web site, a seemly topic-unrelated link may be a navigation page which links to many relevant documents. We uses Google SOAP Search API to filter process-related Web sites, by search process-related keywords (for instance, "process", "howto" or "how to") in the homepage titles of Web sites. Then the filtered candidate sites are crawled in order of breadth-first.

The documents collected by the process crawler are sent to the process extractor. The process extractor uses supervised learning and wrapper induction approaches to extract processes from the crawled semi-structured documents. Because the PBSE needs the title of the processes (for instance, "How to drive a car", denoted by *process_title*) and the title of the steps (for instance, "Go to a driving school", denoted by *step_title*) to perform process-based searches, the process extractor must obtain both of these elements in order to claim a successful extraction.

The process extractor is developed based on the wrapper generation techniques [11], which have been used in crawling the deep Web [12, 13, 14]. Wrapper generation creates rules (for instance, the symbols or HTML tags which always occur before or after a retrieving target) to retrieve structured data from the response pages. The procedure of extracting processes contains three steps: 1) Learn rules from a couple of manually created samples; 2) Apply generated rules on documents collected by the process crawler to extract processes; 3) Validate and store obtained data into the process base. This procedure is based on the assumption that the documents collected from an identical source share common structures. They are dynamically rendered by identical programs or templates. Thus, a supervised learning program can be created to discover the regularities by observing a few positive samples, and then induce a wrapper which can extract the desired target from unseen documents.

```
function pt_wrapper_generator
input: Positive sample documents d_1, d_2, …, d_n and
processes p_1, p_2, …, p_n, identified from the sample
documents by a human user.
output: A disjunctive rule as shown in (1).
rule = NULL
for (each pair of processes p_i and p_j from p_1, p_2, …, p_n)
  t_i = title of p_i
  t_j = title of p_j
  pt_start = common consecutive tokens occurring
             before t_i in d_i and before t_j in d_j
  pt_end = common consecutive tokens occurring after
        pt_start + t_i in d_i and after pt_start + t_j in d_j
  if (pt_start is not empty and pt_end is not empty)
   then
    rule = rule OR <pt_start, pt_end>
return rule
```

Fig. 1. Process title wrapper generator

n ($1 < n < 10$) positive documents, denoted by $d_1, d_2, \ldots, d_n$, containing processes are randomly selected from an identical Web site as training samples. A human user identifies a process from each page. Identified processes, denoted by $p_1, p_2, \ldots, p_n$, contain at least *process_title* and *step_titles*. Let t_i denote the title of process p_i, $i = 1, 2, \ldots, n$.

The algorithm of generating *process_title* wrappers (illustrated in Figure 1) finds common consecutive tokens occurring before and after t_i in d_i and t_j in d_j respectively, $1 \leq i, j \leq n$. These two common strings are referred to as the "beginning landmark" and "ending landmark" of *process_title*. To minimize the error rate of the later extraction, landmarks with maximum length are identified and are output by the wrapper generator. For each pair of documents in the sample collection, a *process_title* wrapper may be generated, denoted by *<pt_start, pt_end>*. If more than one wrapper is generated, then a disjunctive rule combining all induced wrappers is created. Let *p* represent the number of identical wrappers generated. The disjunctive rule for extracting *process_title* is denoted as follows.

$$<pt\_start_1, pt\_end_1>\ OR\ <pt\_start_2, pt\_end_2>\ OR\ \ldots\ <pt\_start_p, pt\_end_p> \qquad (1)$$

```
function st_wrapper_generator
input: Positive sample documents d_1, d_2, …, d_n and
processes p_1, p_2, …, p_n, identified from the sample
documents by a human user.
output: A disjunctive rule as illustrated in (2).
rule = NULL
for (each process p_i from p_1, p_2, …, p_n)
  t_1 = the first step title of p_i
  t_2 = the second step title of p_i
  st_start = common consecutive tokens occurring
              before t_1 in d_i and before t_2 in d_i
  st_end = common consecutive tokens occurring
            after st_start + t_1 in d_i and
            after st_start +  t_2 in d_i
  if (the token occurring before st_start + t_1 in d_i
      is a number) then
      st_start_before_number = common consecutive token
            occurring before the first non-number token
            before st_start + t_1 in d_i and
            before st_start + t_2 in d_i
  if (st_start_before_number is not empty and st_start
      is not empty and st_end is not empty) then
    rule = rule OR <st_start_before_numbe,
                    st_start, st_end>
  else if (st_start is not empty and st_end is not
            empty) then
    rule = rule OR <st_start, st_end>
return rule
```

Fig. 2. Step title wrapper generator

Given a process *p* and its source document *d*, a *step_title* wrapper can be induced. Let s_i denote the title of step *i* in process *p*, $i = 1, 2, \ldots, m$. Similar to the algorithm of generating *process_title* wrappers, we look for the common strings occurring before and after s_i in *d*, $i = 1, 2, \ldots, m$. The difference is that, for *step_title*, a wrapper can be generated by observing the steps in one document, while for *process_title*, a wrapper is induced by observing process titles in at least two documents. These two discovered common strings are referred to as the "beginning landmark" and "ending landmark" of *step_title*. Again, to minimize the error rate, the longest landmarks are identified and are output by this component.

For each document in the positive sample collection, a *step_title* wrapper may be generated, denoted by <*st_start, st_end*>. If more than one wrapper is generated, a disjunctive rule combining all induced wrappers is created. Let q represent the number of identical wrappers generated. The disjunctive rule for extracting *step_title* is denoted as follows.

$$<st\_start_1,\ st\_end_1>\ OR\ <st\_start_2,\ st\_end_2>\ OR \ldots <st\_start_q,\ st\_end_q> \quad (2)$$

```
function process_extractor
input: disjunctive rules of process_title and
step_title wrappers, and crawled documents from
identical Web site.
output: extracted process candidates.
for (each crawled document d)
  process_title = NULL
  for (each process_title wrapper in the disjunctive
      rule)
     process_title = apply the process_title wrapper
                     on d
     if (process_title is not empty) then break
 step_list = NULL
 for (each step_title wrapper in the disjunctive rule)
    step_list = apply the step_title wrapper on d
                repeatedly from the beginning of d
                until the ending of d
   if (step_list contains more than one step)
       then break
 output the extracted process candidate
```

Fig. 3. Process extractor

Many *step_titles* are numbered in their beginning landmarks. Therefore, if numbers are found at the same position before all steps, then these numbers are skipped to assure a common string can be properly identified in front of the *step_title*. In this case, the representation of a *step_title* wrapper in (2) is replaced by (3). This type of *step_title* wrapper is referred to as "numbered wrapper" in contrast to the "non-numbered wrapper" in (2). Numbered wrappers and non-numbered wrappers can be mixed to form a disjunctive rule for extracting *step_title*. The algorithm for generating this rule is illustrated in Figure 2.

$$<st\_start\_before\_number,\ st\_start\_after\_number,\ st\_end> \quad (3)$$

After the wrappers for *process_title* and *step_title* for a specific Web site are generated, these wrappers are applied to each collected document from the identical Web site to extract process candidates, as illustrated in Figure 3.

4 Experiments

In the experiments, the keywords "how to" were used to search the Web. The process crawler and extractor were applied to the filtered top three Web sites to extract data. When crawling each Web site, the level of bread-first traverse was set to two. Table 1 shows the number of documents crawled from each site and the number of crawled documents that contain a process.

Table 1. Number of the crawled documents and processes

	www.eHow.com	www.wikiHow.com	www.howtodothings.com
Crawled documents	16,651	3,591	2,419
Crawled documents containing process	887	1,547	1,228

After crawling, a human user manually created six positive samples for each one of the three collected Web sites. These samples were used to generate a set of wrappers for each Web site. The created wrappers were then applied to the document collection crawled from the identical Web site to extract processes.

In the experiments, the performance of the process extractor was measured by precision and recall. The precision and recall measures have been used to evaluate various information retrieval algorithms. In the specific problem of this writing, precision is defined as the number of correctly extracted processes over the number of extracted processes. Recall is defined as the number of correctly extracted processes over the number of processes.

Table 2. Precision and recall of process_title extractions

	eHow	wikiHow	howtodothings
Processes	887	1,547	1,228
Extracted *process_title*	870	1,510	1,201
Correct extraction	870	1,510	1,201
Precision	100%	100%	100%
Recall	98%	98%	98%

Table 2 shows the precision and recall of the trained *process_title* extractor. As *process_title* is an element which is well regulated in HTML pages regardless of the differences of Web sites, the precisions for all three Web sites were perfect and the recalls were also close to 100%.

Table 3. Precision and recall of step_title extractions

	eHow	wikiHow	howtodothings
Processes	887	1,547	1,228
Extracted *step_title*	660	475	756
Correct extraction	640	247	696
Precision	97%	52%	92%
Recall	72%	16%	57%

Table 4. Precision and recall of process extractions

	eHow	wikiHow	howtodothings
Processes	887	1,547	1,228
Extracted *process*	660	475	737
Correct extraction	640	247	678
Precision	97%	52%	92%
Recall	72%	16%	55%

Table 3 shows the precision and recall of the trained *step_title* extractor. As the degrees of structure and regularity of *step_title* vary in different Web sites, the precisions and recalls differ dramatically. The *step_title* extractor for "eHow" exhibited high precision and relatively high recall, because the crawled Web pages from "eHow" are relatively well structured and follow the identical regularities, which make it easier for the program to generate effective wrappers. The precision of "www.howtodothings.com" is as high as 92% and therefore also acceptable, though only half of the targets have been found. The Web pages from "wikiHow" exhibit a relatively lower degree of regularities because of the characteristic of open editing, which caused the relatively low precision and recall.

Table 4 shows the precision and recall of the process extractor after validation. A process is defined as being extracted if its *process_title* and at least two *step_titles* are extracted. The extractions were successful for two Web sites out of three as indicated by the precisions and recalls. Overall, the proposed approach and developed program proved to be efficient and effective. The extracted processes in the experiments met the requirements of PBSE.

5 Conclusion

In this paper, a supporting system for process-based searches has been developed to efficiently and effectively retrieve processes from the Web. This supporting system consists of two major components: a process crawler and a process extractor. The process crawler collected process-related documents from the Web. Then the collected documents were sent to the process extractor. The process extractor was trained with a few manually created positive samples. A set of extraction rules was generated during the training. These rules were then applied to the crawled documents to extract processes. In the experiment, a total of 22,661 documents were downloaded, in which 3,662 documents contain process content. From the 3,662 documents, 1,565 processes have been successfully extracted using the extractor trained by only 18 manually created samples. Therefore the approach we proposed can provide valuable data to the PBSE and facilitate effective process-based searches.

References

1. Pass, G., Chowdhury, A., Torgeson, C.: A picture of search. In: 1st international Conference on Scalable information Systems, InfoScale 2006, vol. 152. ACM, New York (2006)
2. Agrawal, R., Gunopulos, D., Leymann, F.: Mining Process Models from Workflow Logs. In: Schek, H.-J., Saltor, F., Ramos, I., Alonso, G. (eds.) EDBT 1998. LNCS, vol. 1377, pp. 469–483. Springer, Heidelberg (1998)

3. van der Aalst, W.M., van Dongen, B.F., Herbst, J., Maruster, L., Schimm, G., Weijters, A.J.: Workflow mining: a survey of issues and approaches. Data Knowl. Eng. 47(2), 237–267 (2003)
4. Rembert, A.J.: Comprehensive workflow mining. In: 44th Annual Southeast Regional Conference. ACM-SE 44, pp. 222–227. ACM, New York (2006)
5. Alves de Medeiros, A.K., Weijters, A.J.M.M., van der Aalst, W.M.P.: Genetic process mining: an experimental Evaluation. Journal of Data Mining and Knowledge Discovery 14(2), 245–304 (2007)
6. Turner, C.J., Tiwari, A., Mehnen, J.: A genetic programming approach to business process mining. In: 10th Annual Conference on Genetic and Evolutionary Computation. GECCO 2008, pp. 1307–1314. ACM, New York (2008)
7. Cook, J.E., Wolf, A.L.: Discovering models of software processes from event-based data. ACM Trans. Softw. Eng. Methodol. 7(3), 215–249 (1998)
8. Cook, J.E., Wolf, A.L.: Software process validation: quantitatively measuring the correspondence of a process to a model. ACM Trans. Softw. Eng. Methodol. 8(2), 147–176 (1999)
9. Jensen, C., Scacchi, W.: Applying a Reference Framework to Open Source Software Process Discovery. In: 1st Workshop on Open Source in an Industrial Context, Anaheim, CA (2003)
10. Jensen, C., Scacchi, W.: Data Mining for Software Process Discovery in Open Source Software Development Communities. In: Workshop on Mining Software Repositories, Edinburgh, Scotland, pp. 96–100 (2004)
11. Muslea, I., Minton, S., Knoblock, C.: A hierarchical approach to wrapper induction. In: Etzioni, O., Müller, J.P., Bradshaw, J.M. (eds.) 3rd Annual Conference on Autonomous Agents. AGENTS 1999, pp. 190–197. ACM, New York (1999)
12. Raghavan, S., Garcia-Molina, H.: Crawling the Hidden Web. In: Apers, P.M., Atzeni, P., Ceri, S., Paraboschi, S., Ramamohanarao, K., Snodgrass, R.T. (eds.) 27th international Conference on Very Large Data Bases, pp. 129–138. Morgan Kaufmann Publishers, San Francisco (2001)
13. Zhao, H., Meng, W., Wu, Z., Raghavan, V., Yu, C.: Fully automatic wrapper generation for search engines. In: 14th international Conference on World Wide Web, pp. 66–75. ACM, New York (2005)
14. Mundluru, D., Xia, X.: Experiences in crawling deep web in the context of local search. In: 2nd international Workshop on Geographic information Retrieval. GIR 2008, pp. 35–42. ACM, New York (2008)

Asymmetric Feature Selection for BGP Abnormal Events Detection

Yuhai Liu[1], Lintao Ma[2], Ning Yang[2], and Ying He[3]

[1] Plexus SQA Team ,Qingdao R&D Center,
Alcatel-Lucent Technologies,
266101 Qingdao, Shandong
[2] Informaton Engineering Center,
Ocean University of China,
266071 Qingdao, Shandong
[3] Electronic Information and Science Department,
Qingdao University of China,
266071 Qingdao, Shandong
yuhailiu@alcatel-lucent.com, mlt_zq@yahoo.com.cn,
69808571@163.com

Abstract. Border Gateway Protocol (BGP), which is the defacto standard inter-domain routing protocol in the Internet today, has severe problems, such as worm viruses, denial of service (DoS) attacks, etc. To ensure the stability and security of the inter-domain routing system in the autonomy system, it is critical to accurately and quickly detect abnormal BGP events. In this paper, a novel feature selection algorithm based on the asymmetric entropy named FSAMI is proposed to evaluate the characteristics of describing the BGP abnormal events, which is independent on the machine learning methods. Meanwhile the under-sampling, neural network (NN) and feature selection are introduced to predict BGP abnormal activities to treat the imbalance problem. Numerical experimental results on RIPE archive data set show that the FSAMI method improves the g_means values of abnormal events detection and helps to improve the prediction ability.

Keywords: Feature Selection, Asymmetric Mutual Information, Neural Networks, BGP.

1 Introduction

With the rapid development of the information technologies, Internet has become an important part of society infrastructure. However, Border Gateway Protocol (BGP), which is the defacto standard inter-domain routing protocol in the Internet today, has severe problems, such as worm viruses, DoS attacks and lack of security mechanisms, etc. To ensure the stability and security of the inter-domain routing system in autonomy systems, it is critical to detect abnormal BGP routing dynamics.

In the past few years, several well-known BGP events have been reported. For example, in April 2001, a misconfiguration caused AS 3561 to propagate more than

R. Huang et al. (Eds.): ADMA 2009, LNAI 5678, pp. 553–560, 2009.

5,000 invalid route announcements from one of its customers, causing connectivity problems throughout the entire Internet [1]. In January 2003, the Slammer worm caused a surge of BGP updates [2]. In August 2003, the East Coast electricity blackout affected 3,175 networks and many BGP routers were shut down [3].

There has been a flurry of recent work focusing on the detection of routing anomalies using BGP update message data. A new approach proposed and implemented in [4] describes anomalies as deviations from normal behavior. When facing evolving network topologies and traffic conditions, this method may perform poorly and may not scale gracefully. In [5] the authors devise a mechanism to detect anomalies in BGP routing. In [6] the researchers use a change point detection algorithm to detect several types of network anomalies.

In order to detect abnormal routing dynamics of BGP, we introduce under-sampling, neural networks and feature selection to classify the BGP routing dynamics. Our BGP data was the RIPE archive which is a huge archive of BGP updates and routing tables continuously collected by RIPE monitors around the world. Each instance of BGP data contains 35 characteristics [7]. For building robust models, good features should be selected while still describing the data with sufficient accuracy.

Unlike the feature selection for imbalanced problems addressed in [8,9] by balancing the ration of training data set, here the mutual information entropy is extended to asymmetric mutual information entropy to rank the features and the distribution of mutual information is applied to feature selection. Since the BGP classes are imbalanced, under-sampling [10] is used in training neural networks that was used to classify BGP routing dynamics and detect the abnormal ones. Then, we use feature selection method to find optimum features. The effectiveness of our algorithm is analyzed theoretically and empirically.

The rest of this paper is organized as follows. Section 2 introduces BGP on the Internet. Section 3 presents the learning methods studied in this paper. Section 4 reports on the empirical study. Conclusion is drawn in section 5.

2 Learning Methods

2.1 Under-Sampling

Like over-sampling, under-sampling also changes the training data distribution such that the costs of the examples are explicitly conveyed by the appearances of examples. However, the working style of under-sampling opposites that of over-sampling in the way that the former tries to decrease the numbers of inexpensive examples while the latter tries to increase the number of expensive examples.

Concretely, the $k-th$ class will have N_k^* training examples after re-sampling, which is computed according to equation (1). Here the λ class has the smallest number of training examples to be eliminated, which is identified in equation (2).

$$N_k^* = [\frac{Cost[k]}{Cost[\lambda]} N_\lambda] \tag{1}$$

$$\lambda = \arg\min_{j} \frac{\frac{Cost[j]}{\min_{c} Cost[c]} N \arg\min_{c} Cost[c]}{N_j} \quad (2)$$

If $N_k^* < N_k$, $(N_k - N_k^*)$ training examples of the $k-th$ class should be eliminated.

2.2 Feature Selection Based on Asymmetric Mutual Information

For any pattern classification problem, the selection of useful features constitutes a significant part of the solution. Although feature selection for imbalanced data sets was studied in [8, 9], only the traditional feature selection method was used after they applied the over-sampling, under-sampling and asymmetric bagging to training data. In our scheme, the features were ranked by asymmetric mutual information (for short FSAMI), which is considered unequal probability distribution of training data. In addition, the FSAMI can be used for high dimension data set not like [8] that will take much time to evaluate every feature by machine learning methods, and is independent on the machine learning methods.

Only the definition of asymmetric mutual information is given here as it is one of the key points of these ideas and because of the confine of space for research proposal. Let *p* design the probability to be “abnormal”, 1-*p* being for the probability of “normal”. We need the maximal uncertainty to be reached for a given probability of “abnormal”, noted $p=w$. This criterion should verify the classical properties of the entropy measures shown bellow. More formally, we seek a non negative function *h* of *p*. In real application, *p* is estimated at each leaf by the frequency. This function *h* should respect the entropy properties, except that the maximum should be reached for $p=w$. If there are only two classes and the data distribution is balanced, p=0.5. But our BGP data is imbalanced. So the requested properties are [12]:

A rational function could be expressed as follow:

$$\lambda = \arg\min_{j} \frac{\frac{Cost[j]}{\min_{c} Cost[c]} N \arg\min_{c} Cost[c]}{N_j} \quad (3)$$

where the reference probability w is given by the user. By construction, H reaches is maximum at $w = (w_1, ..., w_k)$.

We adopt the above idea from asymmetric entropy in [19] to define asymmetric mutual information $I(X,Y)$ to be the difference between entropy on X, entropy on Y and entropy on (X,Y) that is $I(X,Y) = H(X) + H(Y) - H(X,Y)$. According to this measure, a feature X is regarded more correlated to Y than feature Z, if $I(X,Y) > I(Z,Y)$.

The feature selection by asymmetric mutual information procedures as follow:

1) Let the size of n training data set: X =($X_1, X_2 ..... X_n$). The training sample X_i=($x_i^1, x_i^2, ..., x_i^T$) has the T features with the target Y_i. Num_fs is the number of user preferring features

2) Rank the feature t by calculating asymmetric mutual entropy $I_t(X,Y) = \sum_{i=1:n} I(x_i^t, Y_i)$;

3) Order the features by the $I_t(X,Y)$;

4) Choose the top Num_fs features as the optimal features.

3 Empirical Results

3.1 Configuration

Back propagation (BP) neural network is used in the empirical study. Each network has one hidden layer containing ten units, and is trained to 200 epochs. Note that since the relative instead of absolute performance of the investigated methods are concerned, the architecture and training process of the neural networks have not been finely tuned. After under-sampling the training data, neural networks were achieved. Here we set 90% of the data as the training set, the left 10% as the testing set. Each trial was repeated 10 times. At last, we get the average results.

3.2 Data Set

The BGP data archive that we used to prepare database tables was the RIPE archive. We used the BGP update data from six randomly selected peers. (We did not use the Oregon Route Views archive, as it does not contain BGP updates for the Code Red and Nimda worm periods that we want to study).

BGP data is cleaned and processed according to [7]. To obtain worm events that can affect BGP, we collected BGP data as follows. Using the data from an eight-hour period immediately after each worm started to propagate as abnormal data, we also prepared data from ten randomly chosen "normal" days (dispersed within a two-year period from July 2001 to August 2003), in which no major events were known to have happened.

To test the rules obtained from the training, we further prepared data from the half day when the Slammer worm was active (January 25, 2003). To provide a basis of comparison when testing the rules, data from another set of four randomly chosen "normal" days were also collected. We choose 2001.12.10, 2001.12.12, 2002.8.10 as our BGP normal data set. These normal data set and abnormal data construct the BGP data set, marked as AS 513, AS 4777 and AS 13129.

To eliminate the effect of boundary, we delete the first row and the last row of data set. In total, the three BGP data used contains 5733 rows of normal data and 504 rows of (CodeRed and Nimda) worm data. Each worm and normal period is further divided into 1-minute bins, with each bin represented by exactly one data row. As a result, for each bin, a new row is added to a corresponding database table used for training or testing. When used for training, a new row will also be labeled as either "abnormal" or "normal". The 35 parameters of BGP data is illustrated in table 1.

Table 1. Parameter list

ID	Parameter	Definition
1	Announce	# of BGP announcements
2	Withdrawal	# of BGP withdrawals
3	Update	# of BGP updates (= Announce + Withdrawal)
4	AnnouPrefix	# of announced prefixes
5	WithdwPrefix	# of withdrawn prefixes
6	UpdatedPrefix	# of updated prefixes (= AnnouPrefix + WithdwPrefix)
7	WWDup	# of duplicate withdrawals
8	AADupType1	# of duplicate announcements (all fields are the same)
9	AADupType2	# of duplicate announcements (only AS-PATH and NEXT-HOP fields are the same)
10	AADiff	AADiff # of new-path announcements (thus implicit withdrawals)
11	WADupType1	# of re-announcements after withdrawing the same path (all fields are the same)
12	WADupType2	# of re-announcements after withdrawing the same path (only AS-PATH and NEXT-HOP fields are the same)
13	WADup	WADupType1 + WADupType2
14	WADiff	# of new paths announced after withdrawing an old path
15	AW	# of withdrawals after announcing the same path
16-35		the mean and the standard deviation of ten different types of inter-arrival time

3.3 Algorithm Effectiveness

In this paper, we adopt Geometric Mean, which considers both the accuracies of the minority class and the majority class equally. A+, the accuracy of the minority class, is computed by $\frac{TP}{(TP+FN)}$. A-, the accuracy of the majority class, is computed by $\frac{TN}{(FP+TN)}$. Then, geometric mean (g_means) is computed by $\sqrt{(A+)\times(A-)}$.

3.4 Experiment Results in BGP Data Set

In order to demonstrate the effect of imbalanced learning methods and FSAMI, we have performed the following series experiments by using neural networks approach (NN) as base classifiers.

The experiment is as follows:

1) un_NN is a baseline method, which applies the under-sampling method to training data sets with all features before using NN .
2) un_FS_NN is adopt the under-sampling before applying FSMI, which uses NN as base learners
3) un_aFS_NN is adopt the under-sampling before applying FSAMI, which uses NN as base learners

Experiments are performed to investigate if under-sampling and feature selection by asymmetric mutual information help to improve performance of abnormal BGP events detection. We test the learning methods on AS 513, AS 4777 and AS 13129 respectively.

The comparison of un_NN , un_FS_NN and un_aFS_NN algorithm which select 7 features, *w*=0.2 in 3 BGP data sets is shown in figure 1(a). Number 1, 2, 3 in *x*-axis respectively means 3 BGP data sets: AS 513, AS 4777 and AS 13129. The y-axis illustrates the average geometric mean of different algorithms. In figure 1(b), all algorithms use 9 features and *w*=0.2.

An analysis of the results from BGP data set in test data sets shows that, on average, the average geometric mean was approximately 32% with online algorithm, about 90% with un_FS_NN and 93% with un_aFS_NN.

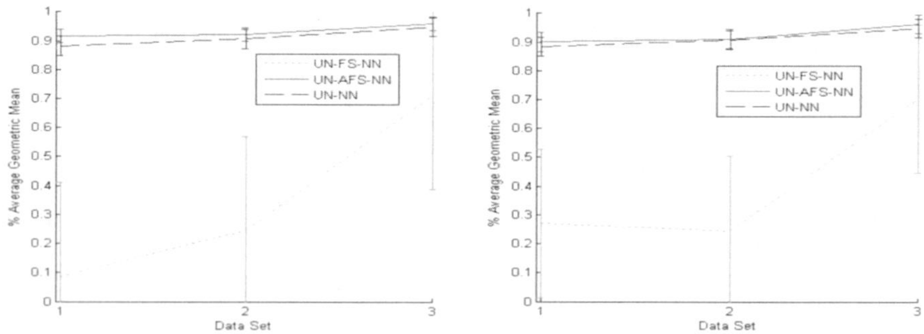

(a) 7 Features (b) 9Features

Fig. 1. The comparisons of un_NN, un_FS_NN and un_aFS_NN algorithm with different features and fixed reference probability w=0.2 in 3 BGP datasets

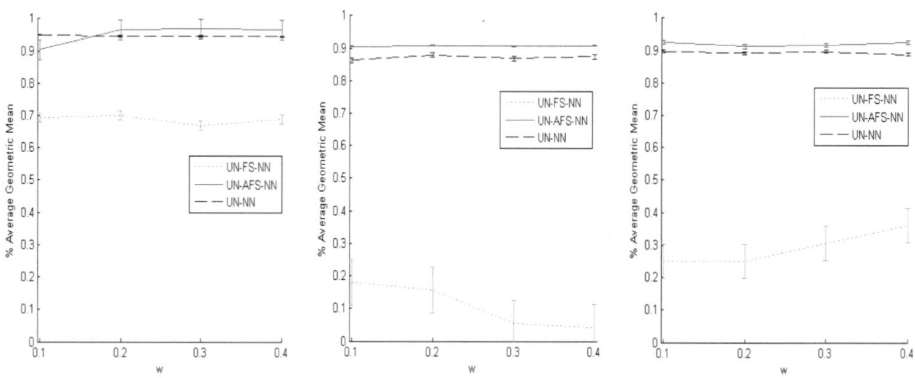

(a) Dataset: AS 513 (b) Dataset: AS 4777 (c) Dataset: AS 13129

Fig. 2. The comparisons of un_NN, un_FS_NN and un_aFS_NN algorithm with different reference probability w and fixed features (8) in 3 BGP datasets

Figure 2 show the results of 3 BGP data sets with different values of w, 8 features. One can observe that, except for un_FS_NN, different values of the reference probability do not highly affect the performance of the other methods. The x-axis uses different values of w, w respectively selects 0.1, 0.2, 0.3, 0.4. The y-axis illustrates the average geometric mean of different algorithms.

From figures 1-2, we can obviously find that:

1) applying asymmetric feature selection for BGP Abnormal Events Detection is affective, and aFs_NN does improve performance of Fs_NN and has a higher average geometric mean.
2) un_FS_NN does fail in every cases, it does not improve performance of un_NN as in [9] for protein sequence problem.
3) un_aFS_NN slightly improve results of un_NN with several features. We also observed that there was little difference results between un_aFS_NN.
4) The results obtained by un_FS_NN are great different.

The above results show that FSAMI and under-sampling perform better than the other several methods of un_NN, un_FS_NN. Here we give some insights on these results:

1) Since this is a abnormal events detection problem, we pay more attention to abnormal events. un_aFS_NN improves the g_means values of ordinary un_NN with several features. Simultaneously, the result shows un_aFS_NN is proper to solve the imbalanced abnormal BGP events detection problem. Under-sample wins in two aspects, one is that it makes the individual data subsets balanced, the second is that it pay more attention to the abnormal data by leaving the abnormal data always in the data set.
2) FSAMI achieves slightly better results than un_NN does. FSAMI using asymmetric mutual information as criteria also makes un_aFS_NN win in two aspects, one is that independent the machine learning, it will be feasible to high dimension data, the second is that different features selected for different individual data subsets, which improves their whole performance. The results improved by un_aFS_NN than un_NN are slightly, we consider the reason is that un_NN already get high values that are difficult to improve with several features.

4 Conclusion

To address the imbalance problem of BGP abnormal events detection, we propose to apply feature selection by asymmetric mutual information and under-sampling to the modeling of detection of BGP abnormal events. A novel algorithm FSAMI are compared with under-sampling of NN on a large BGP data set, experiments show that feature selection by asymmetric mutual information and under-sampling can improve the prediction ability of NN in terms of g_means.

This work proposed the feature selection by asymmetric mutual information to prediction of BGP abnormal events and furthermore extends FSAMI to under-sampling. Although this work does experiment by the under-sampling, asymmetric bagging and over-sampling are also good optional methods to verify.

References

1. Misel, S.: Wow: AS7007, http://www.merit.edu/mail.archives/nanog/1997-04/msg00340.html
2. Lad, M., Zhao, X., Zhang, B., Massey, D., Zhang, L.: An analysis of BGP update surge during Slammer attack. In: Proceedings of the International Workshop on Distributed Computing (IWDC), pp. 66–79. Springer, Heidelberg (2003)
3. Cowie, J., Ogielski, A., Premore, B., Smith, E., Underwood, T.: Impact of the 2003 blackouts on Internet communications. Technical report, Renesys (November 2003)
4. Feather, F., Maxion, R.: Fault detection in an Ethernet network using anomaly signature matching. In: Proceedings of ACM SIGCOMM, San Francisco, CA, September 1993, pp. 279–288 (1993)
5. Deshpande, S., Thottan, M., Ho, T., Sikdar, B.: A Statistical Approach to Anomaly Detection in Interdomain Routing. In: Proceedings of IEEE BROADNETS, San Hose, CA, October 2006, pp. 1–10 (2006)
6. Thottan, M., Ji, C.: Anomaly detection in IP networks. IEEE Trans. on Signal Processing 51(8), 2191–2203 (2003)
7. Li, J., Dou, D., Wu, Z., Kim, S., Agarwal, V.: An Internet Routing Forensics Framework for Discovering Rules of Abnormal BGP Events. ACM SIGCOMM Computer Communication Review 35(5), 55–66 (2005)
8. Li, G.-Z., Meng, H.-H., Lu, W.-C., Yang, J.Y., Yang, M.Q.: Asymmetric bagging and feature selection for activities prediction of drug molecules. BMC Bioinformatics, 108–114 (2008)
9. Al-Shahib, A., Breitling, R., Gilbert, D.: Feature Selection and the Class Imbalance Problem in Predicting Protein Function from Sequence. Applied Bioinformatics 4(3), 195–203 (2004); PubMed. (2005)
10. Zhou, Z.-H., Liu, X.-Y.: Training Cost-Sensitive Neural Networks with Methods Addressing the Class Imbalance Problem. IEEE Transactions On Knowledge And Data Engineering 18(1), 63–77 (2006)
11. Rekhter, Y., Li, T.: A border gateway protocol 4, BGP-4 (1995)
12. Marcellin, S., Zigued, D.A., Ritschard, G.: An asymmetric entropy measure for decision trees. In: IPMU 2006, July 2006, pp. 975–982. Paris, France (2006)

Analysis and Experimentation of Grid-Based Data Mining with Dynamic Load Balancing

Yong Beom Ma, Tae Young Kim, Seung Hyeon Song, and Jong Sik Lee

School of Information Engineering Inha University #253, YongHyun-Dong, Nam-Ku
402-751 Incheon, Republic of Korea
myb112@hanmail.net, silverwild@gmail.com,
songseunghyun@lycos.co.kr, jslee@inha.ac.kr

Abstract. Algorithms and methods for analyzing large amounts of data are studied and developed. This paper presents a Data Mining (DM) method operated in grid computing environment. Because DM technology uses large amounts of data and requires costs to compute, utilizing and sharing computing data and resources are key issues in DM. Therefore, a Dynamic Load Balancing (DLB) algorithm and a decision range readjustment algorithm are proposed and applied to the Grid-based Data Mining (GDM) method. And we analyzed the average waiting time for learning and computing time. For a performance evaluation, the system execution time, computing time, and average waiting time for learning are measured. Experimental results show that GDM with the DLB method provides many advantages in terms of processing time and cost.

Keywords: Data Mining, Dynamic Load Balancing, Grid Computing.

1 Introduction

Data Mining (DM) [6], [7] is the process of finding a correlation, pattern, or rule from large amounts of data. DM has been noticed because of the increasing demand for information acquisition and utilization in the past decade. Recently most of organizations restore excessive amounts of data by computerizing the most part of their operations. Thus, the need for data analysis methodologies that can discover useful knowledge from large amounts of data is increased. DM is widely used in a field of information technology recently because improved principles and methods can be applied to many research areas including machine learning, pattern recognition, statistics, and so on. However, DM technology use large amounts of data and requires costs to compute. It is necessary to process a DM operation efficiently. Therefore, a Grid-based Data Mining (GDM) and a Dynamic Load Balancing (DLB) are performed for effective task execution.

Grid computing [3], [4] is a new computing paradigm that has the potential to change the way computation and data access are performed. Grid computing can solve large-scale computing problems by utilizing geographically distributed resources owned by

R. Huang et al. (Eds.): ADMA 2009, LNAI 5678, pp. 561–568, 2009.

different users or organizations. The grid can also reuse data. A grid computing system requires a real-time linkage among multiple and geographically distributed systems. Thus, complex large-scale execution and the collaborative sharing of geographically dispersed data sets and computing resources are important in grid computing.

In this paper, we present a GDM with DLB. A decision range readjustment algorithm [12] is applied to the GDM method. This paper focuses on an analysis of the GDM with DLB for reducing the processing time for DM and cost required by data analysis and processing. To evaluate the performance, the system execution time, computing time, and average waiting time for learning are measured. For experiments in grid computing environment, an inter-federation communication system with High-Level Architecture (HLA) middleware for supporting communications among distributed components is employed. Finally, GDM with DLB is compared with other GDM methods.

2 Related Works

2.1 Overview of Load Balancing

Communication mechanisms and interconnection networks play a fundamental role in increasing overhead in a distributed computing environment such as a grid computing system. The overhead is caused by the exchange of load estimates, synchronization messaging, migration of tasks, and intercommunications. As such, there are two major research motivations to apply a load balancing paradigm [1], [8] to such an environment: (i) to improve the resource utilization of the system by redistributing the load among nodes, and (ii) to optimize the average response time of the system.

In recent years, massive amounts of data must be processed in biological applications. Therefore, DLB technology, which provides low processing and communication cost, is widely used. A notable example is random polling [11], a simple yet effective and commonly used DLB scheme. In the random polling scheme, when a processor becomes idle, the processor polls randomly determined processors until it finds a busy one. After the processor selects a donor processor, the processor receives one divided part from the selected processor. Zaki et al [10] showed that different load balancing schemes are optimal for different applications under varying program and system parameters. They selected the best scheme that automatically generates a parallel code by calling to a run-time library for load balancing. Therefore, a customized DLB can be realized and good performance is provided. Flexible Agent System for Heterogeneous Cluster (FLASH) [13] supports load balancing in the system with respect to heterogeneous clusters. FLASH offers an agent-based framework for the creation of distributed and load balanced applications.

This paper uses DLB for a large-scale work distribution. Equal amounts of workload are allocated to each middle group. A customized DLB that allocates workloads according to the status of subgroups is employed. The DLB method makes the execution time in each processor equivalent and provides benefits in processing time and communication cost.

3 Grid-Based Data Mining with Dynamic Load Balancing

In GDM, the workload of large-scale data processing is extremely high. The workload greatly influences the processing time for DM and the cost to evenly distribute such a workload. Therefore, in large-scale data processing for DM, it is necessary to disperse workloads through DLB. In this paper, we use an Artificial Neural Network (ANN) [2], [5] for DM. We define processing for DM as a DM job or a job and each job divided by a resource broker as a task. A DM job can be tested when a DM result integrated from the results of resource providers is obtained. Therefore, it is important to allocate the job to resource providers according to the status of the resource providers. We propose a customized DLB algorithm in order to disperse workloads of a large-scale DM job. We only consider the computing resources of the resource providers, because this paper focuses on efficient processing of a large-scale DM job. Other issues will be addressed in future works. A DM job is divided into several tasks that are uniformly assigned to resource brokers in the middleware layer. As a result, all resource brokers can receive equal amount of tasks. The allocated tasks are dispersedly reassigned to resource providers with inferior computing resources. Each resource broker monitors the computing abilities and updates the status information of the resource providers. This procedure is illustrated in Fig. 1.

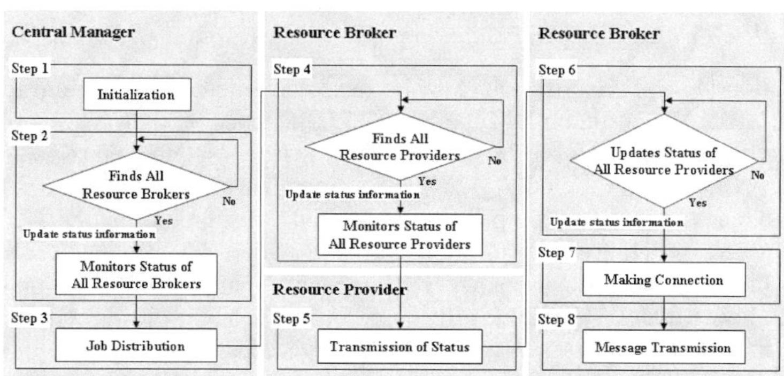

Fig. 1. Procedure of DLB

GDM with DLB monitors the computing abilities of resource providers, establishes a connection between the resource broker and the resource provider dynamically, and allocates a large-scale DM job to resource providers dynamically. In GDM with DLB, resource utilization can be very high and little waiting time is incurred, since DLB is only executed based on the computing ability of the resource providers. The processing time for DM can be reduced, since GDM with DLB can dynamically respond to requests to use resources. Meanwhile, some time is needed to monitor and update the status of the resource providers.

GDM with DLB flexibly divides a job into N number of tasks considering each node's ability and conditions whenever the divided job arrives. This method divides a large-scale DM job into several tasks considering the computing ability of each resource provider. In addition, GDM with DLB distributes tasks to suitable resource providers, thus distinguishing it from SLB.

4 Performance Analysis

We analyze the average waiting time for learning and the computing time in order to evaluate the performance of ANN optimum DM [12], GDM, GDM with SLB, and GDM with DLB. The average waiting time for learning represents the average waiting time from the aspect of resource providers. Waiting time for learning is generated when a job allocated to a resource provider exceeds its computing ability. Computing time represents the processing time of components and is the sum of the computing time based on each task carried out by the central manager, resource brokers, and resource providers.

Table 1 presents the results of an analysis of the average waiting time for learning. *WTL* is divided into four types according to the job allocation method; WTL_S, WTL_D, WTL_{SLB}, and WTL_{DLB}. N_{adj} varies for every experiment, because it is directly affected by the decision range readjustment algorithm [12]. Therefore, the average waiting time for learning is determined by the waiting time for learning (*WTL*) generated in each resource provider. *WTL* is obtained by equation (1), where *N* is the number of resource providers and *J* is the job allocated to a resource provider.

$$WTL = \sum_{i=1}^{N} \frac{J_i - CA_i}{CA_i}, \qquad 0 \leq WTL \tag{1}$$

N; number of decision range adjustments, WTL; waiting time for learning, J; allocated job, CA; computing ability of a resource provider

In equation (1), *WTL* is affected by *J*, because each method is different in the allocated job *J*. In ANN optimum DM method, WTL_S depends on the J_S, the job allocated to a resource provider. WTL_D, WTL_{SLB}, and WTL_{DLB} depend on the job allocated to each resource provider; J_D, J_{SLB}, and J_{DLB}. *WTL* is generated when a job allocated to a resource provider exceeds the computing ability of the resource provider. If *WTL* does not exceed the computing ability of the resource provider, *WTL* has a lower value than 0. This value is ignored because a *WTL* lower than 0 means that the waiting time for learning is lower than 0. In GDM, a resource provider having low computing ability generates more *WTL* because jobs are sequentially allocated to resource providers. In GDM with SLB, *WTL* varies according to the computing ability of the resource provider, because jobs are equally allocated to resource providers. However, a resource provider is not allocated a job that is too large, even if the computing ability is low, because the allocated job is almost equal to the average. Therefore, the average waiting time for learning of GDM with SLB is less than that of ANN optimum DM and more than that of GDM with DLB. In GDM with DLB, little *WTL* is generated, because *WTL* allocates jobs in consideration of the computing ability of the resource provider.

The difference in *WTL* was as follows: $WTL_S \doteqdot WTL_D \geq WTL_{SLB} > WTL_{DLB}$. Thus, the waiting time for learning of GDM with DLB is mostly affected by communication time and is certainly shorter than that of the other two methods. The rate of reduction of the average waiting time for learning is presented in Table 1 and WTL_{DLBi} / WTL_{Si} is close to 0. Therefore, the rate of reduction of the average waiting time for learning of GDM with DLB is the smallest among all of the methods. It appears that GDM with DLB provides a greater reduction of 1-*WTL/N* of the average waiting time for learning than the other methods.

Table 1. Analysis of average waiting time for learning (N; Number of resource providers, N_{adj}; Number of decision range adjustments, CA; Computing ability of a resource provider)

	Average waiting time for learning	Reduction rate of average waiting time for learning when compared with ANN optimum
ANN Optimum	$\left(\sum_{j=1}^{N_{adj}}\left((\sum_{i=1}^{N} WTL_{Si})/N\right)\right)/N_{adj}$	100%
GDM	$\left(\sum_{j=1}^{N_{adj}}\left((\sum_{i=1}^{N} WTL_{Di})/N\right)\right)/N_{adj}$	$(\sum_{i=1}^{N}\frac{WTL_{Di}}{WTL_{Si}})\times 100\%$
GDM with SLB	$\left(\sum_{j=1}^{N_{adj}}\left((\sum_{i=1}^{N} WTL_{SLBi})/N\right)\right)/N_{adj}$	$(\sum_{i=1}^{N}\frac{WTL_{SLBi}}{WTL_{Si}})\times 100\%$
GDM with DLB	$\left(\sum_{j=1}^{N_{adj}}\left((\sum_{i=1}^{N} WTL_{DLBi})/N\right)\right)/N_{adj}$	$(\sum_{i=1}^{N}\frac{WTL_{DLBi}}{WTL_{Si}})\times 100\%$

Table 2. Analysis of computing time (N_{adj}; Number of decision range adjustments, T_{RB}; Computing time of resource brokers), $CM_D \fallingdotseq CM_{SLB} \fallingdotseq CM_{DLB}$, $\frac{CM_D}{CM_S} \fallingdotseq \frac{CM_{SLB}}{CM_S} \fallingdotseq \frac{CM_{DLB}}{CM_S}$

	Computing time	Reduction rate when compared with ANN optimum
ANN Optimum	$N_{adj} \times (T_S + T_{RB} + CM_S)$	100%
GDM	$N_{adj} \times (T_G + T_{RB} + CM_D)$	$(\frac{T_D}{T_S} + \frac{CM_D}{CM_S})\times 100\%$
GDM with SLB	$N_{adj} \times (T_{SLB} + T_{RB} + CM_{SLB})$	$(\frac{T_{SLB}}{T_S} + \frac{CM_{SLB}}{CM_S})\times 100\%$
GDM with DLB	$N_{adj} \times (T_{DLB} + T_{RB} + CM_{DLB})$	$(\frac{T_{DLB}}{T_S} + \frac{CM_{DLB}}{CM_S})\times 100\%$

Table 2 presents the results of an analysis of computing time. N_{adj} is the number of decision range readjustments and T_{RB} is the computing time of the resource brokers. T_S, T_D, T_{SLB}, and T_{DLB} are the computing time of resource providers in each method. CM, the computing time of the central manager, is divided into four types according to the job allocation methods: CM_S, CM_D, CM_{SLB}, and CM_{DLB}. Computing time of resource providers includes learning and the testing time of resource providers, including the waiting time for learning. As mentioned before, waiting time for learning is the shortest in GDM with DLB. The computing time of resource providers in each method varies according to the waiting time for learning. *CM* is the time for resource allocation and job scheduling. We assume that *CM* is almost uniform in the four methods. Therefore, *CM* depends on *T*, the computing time of resource providers, and N_{adj}. Since the computing time of resource providers depends on the waiting time for learning, the difference in *T* is as follows: $T_S \fallingdotseq T_D \geq T_{SLB} > T_{DLB}$. Thus, the computing time is shortest in GDM with DLB. We assume that the computing time of the central manager is very similar in each method; the real reduction rate of each method is shown as Table 2. If we disregard the computing time of central mangers, the reduction rate of GDM with DLB is the lowest, because T_{DLB} is the lowest. There is likely a slight difference according to N_{adj}; we anticipate that GDM with DLB can provide a greater reduction of $1\text{-}T_{DLB}/T_S$, $1\text{-}T_{DLB}/T_D$, and $1\text{-}T_{DLB}/T_{SLB}$ for computing time than can ANN optimum DM, GDM, or GDM with SLB.

5 Experiments and Results

In order to evaluate the system performance, we experiment and compare four Grid-based Data Mining (GDM) methods, ANN optimum DM (ANN Optimum), GDM, GDM with SLB, and GDM with DLB. To simulate GDM, we design and develop a grid testbed using HLA middleware specifications and RTI (Run-Time Infrastructure) [9] implementation. The GDM system with DLB consists of some federates. Inter-federate communication works on the GDM system. The RTI message passing for data management among federates depends on the inter-federate communication inside the federation. For the platform setting, we develop a grid system using a RTI implementation that operates on Windows operating systems. A total of 16 federates are allocated to machines, respectively, and they are connected via a 10 Base T Ethernet network.

For experiments, we need to implement DM in grid computing environment. We assume that the reliable recognition rate is 90% for the experiment on measuring performances and do not consider environmental aspects such as a failure in the network connection. We also assume that the large-scale data has already been preprocessed prior to classification.

5.1 Comparison of Computing Time

This experiment compares the ANN optimum DM method operated in a single machine with GDM methods. In this experiment, we compare the computing time of ANN optimum DM (ANN Optimum), the GDM, GDM with SLB, and GDM with DLB. The computing time includes the processing time and the training and testing time of the learning machine only, and does not include communication time, waiting time, network interference, and so on. We calculate the computing time by subtracting the communication time including network interference and waiting time for learning from the system execution time. The computing time of ANN optimum DM, GDM, and GDM with SLB increases irregularly. However, the computing time of GDM with DLB is comparatively regular. In this experiment, GDM with DLB shows a 63.88% greater reduction than GDM and a 46.40% greater reduction than GDM with SLB. In addition, GDM with DLB shows a 67.65% greater reduction than ANN optimum DM. In terms of reduction of the computing time, GDM with DLB is a more effective method than ANN optimum DM, GDM, and GDM with SLB, respectively. Fig. 2 shows the computing time in each method.

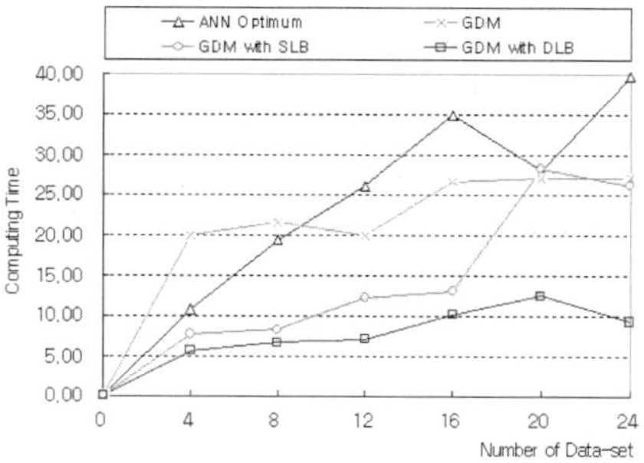

Fig. 2. The computing time

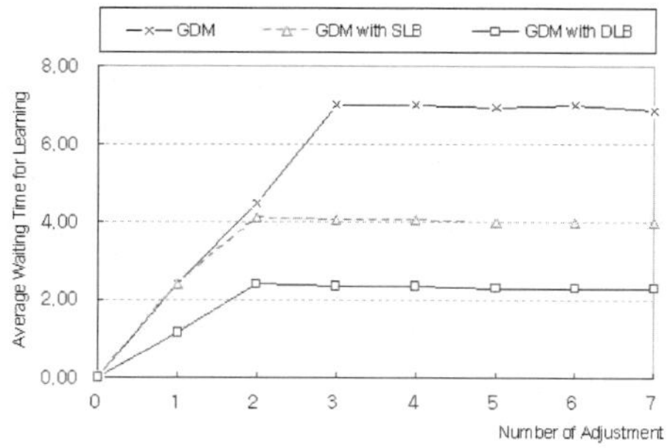

Fig. 3. The average waiting time for learning

5.2 Comparison of Average Waiting Time for Learning

The waiting time for learning is the sum of waiting time for learning of all learning machines. Each learning machine has a different learning time, because resources in a grid-based system have different computing abilities. Waiting time for learning is generated in each learning machine, because a learning machine that has completed its learning waits for the completion of others. Even if other learning machines already complete their computation, they wait for the completion of learning machine which has the longest computing time. Thus, task completion time of all learning machines is uniform. The average waiting time for learning is obtained by equation (2), where *WTL* is the waiting time for learning and *n* is the number of learning machines. Fig. 3 represents the average waiting time for learning of each GDM method. In section 4, we projected that the average waiting time for learning of GDM with DLB would be the shortest. In this experiment, GDM with DLB shows a 63.71% greater reduction than GDM and 43.03% more than GDM with SLB.

$$Average\ Waiting\ Time\ for\ Learning = \left(\sum_{i=1}^{N_{adj}} (\sum_{j=1}^{n} WTL_{ij}) / n \right) / N_{adj} \tag{2}$$

N_{adj}; number of decision range adjustments,
n; number of learning machines, WTL; waiting time for learning

6 Conclusion

This paper presents a GDM with DLB in order to provide improved performance. Generally, DM requires high performance computing ability for large-scale data processing.

We solve the computing problem of large-scale DM by a distributed environment such as grid computing environment, since the grid computing environment can effectively and rapidly process information by using distributed computing resources. Some advantages in terms of processing time for DM and cost are obtained by applying a decision range readjustment algorithm to DM. A DLB algorithm that can disperse the workload of large-scale DM job to the GDM method is implemented. The GDM with DLB method proposed in this paper divides a large-scale DM job into several tasks considering the computing ability of each learning machine. Tasks are then distributed to suitable learning machines.

For a performance evaluation, ANN GDM experiments were conducted on a grid testbed using HLA. The performance of GDM using DLB was compared with GDM with non-load balancing and GDM using SLB. Experimental results demonstrate that GDM with DLB outperformed GDM with non-load balancing and GDM with SLB, respectively.

References

1. Kumar, V., Grama, A., Rao, V.N.: Scalable Load Balancing Techniques for Parallel Computers. Journal of Distributed Computing 7 (1994)
2. Braspenning, P.J., Thuijsman, F., Weijters, A.J.M.M.: Artificial Neural Networks: An Introduction to ANN Theory and Practice. In: Neural Network School 1999. LNCS, vol. 931, pp. 1–66. Springer, Heidelberg (1995)
3. Berman, F., Fox, G., Hey, T.: Grid Computing: Making the Global Infrastructure a Reality. J. Wiley, Chichester (2003)
4. Foster, I., Kesselman, C.: The Grid: Blueprint for a New Computing Infrastructure. Morgan Kaufmann, San Francisco (1998)
5. Zurada, J.M.: Introduction to Artificial Neural Systems. Jaico Publishing House (1992)
6. Han, J., Kamber, M.: Data Mining: Concepts and Techniques, pp. 279–310. Morgan Kaufmann, San Francisco (2000)
7. Nadler, M., Smith, E.P.: Pattern Recognition Engineering, pp. 75–80. John Wiley & Sons Inc., Chichester (1992)
8. Kumar, V., Grama, A., Gupta, A., Karypis, G.: Introduction to Parallel Computing: Design and Analysis of Algorithms. The Benjamin/Cummings Publishing Company (1994)
9. Kapolka, A.: The Extensible Run-Time Infrastructure (XRTI): An Experimental Implementation of Proposed Improvements to the High Level Architecture, Master's Thesis, Naval Postgraduate School (2003)
10. Zaki, M.J., Li, W., Parthasarathy, S.: Customized Dynamic Load Balancing for a Network of Workstations. In: 5th IEEE International Symposium on High Performance Distributed Computing (1996)
11. Sanders, P.: A Detailed Analysis of Random Polling Dynamic Load Balancing. In: International Symposium on Parallel Architectures, Algorithms, and Networks (1994)
12. Ma, Y.B., Cho, K.C., Jang, S.H., Lee, J.S.: Grid-based ANN Data Mining for Bioinformatics Applications. In: International Conference on Hybrid Information Technology, Jeju (2006)
13. http://www.iti.uni-luebeck.de/iti/index.php?id=flash

Incremental Document Clustering Based on Graph Model

Tu-Anh Nguyen-Hoang[1], Kiem Hoang[2], Danh Bui-Thi[1], and Anh-Thy Nguyen[1]

[1] Faculty of Information Technology, University of Science, 70000 VNU-HCM, Vietnam
[2] Faculty of Computer Science, University of Information Technology, 70000 VNU-HCM, Vietnam
nhtanh@fit.hcmuns.edu.vn, tuanaivnn@yahoo.com, popstarsongngu@yahoo.com, kiemhv@uit.edu.vn

Abstract. In this paper, we propose a new approach based on graph model and enhanced IncrementalDBSCAN to solve incremental document clustering problem. Instead of traditional vector-based model, a graph-based is used for document representation. By using graph model, we can easily update graph structure when a new document is added to database. Meanwhile, IncrementalDBSCAN is an effective incremental clustering algorithm suitable for mining in dynamically changing databases. Similarity between two documents is measured by hybrid similarity of their adapting feature vectors and shared-phrase information. Our experimental results demonstrate the effectiveness of the proposed method.

Keywords: Incremental document clustering, graph model, shared phrases, enhanced IncrementalDBSCAN.

1 Introduction

With the tremendous growth of digital content in Internet, databases, and archives, document clustering has attracted a great deal of attention in information retrieval and text mining community. Document clustering (also called as Text clustering) is a more specific technique for unsupervised document organization, automatic topic detection, document content summarization, and efficient information retrieval or filtering. The goal of document clustering method is automatically group a set of documents into subsets or clusters, so that documents within a cluster have high similarity, but are very dissimilar to documents in other clusters.

Most of existing clustering methods process statically on the whole collection of documents (i.e., all documents are collected before clustering). However, in many applications the document collection is dynamic in the sense that new documents are continuously added to the database and need to be processed. Hence, an incremental clustering solution is needed to process documents as soon as they arrive. Various approaches have been proposed, including Suffix Tree Clustering [14], Single-Pass clustering [10], DC-tree clustering [13], SHC [5], and others.

R. Huang et al. (Eds.): ADMA 2009, LNAI 5678, pp. 569–576, 2009.

Furthermore document representation model is one of the important factors involved in text clustering. The widely used model is to represent documents by keyword vectors according to standard vector space model with TF-IDF term weighting [7]. However, this traditional vector model suffers from the fact that it loses important structural information in original text, such as the order in which terms appear or the locations of terms within the text. In order to overcome the limitations of vector space model, graph-based document representation model was introduced [4], [8], and [9]. The main benefit of graph-based techniques is that they allow us to keep inherent structural information of original document. Schenker et al. [8] used graph based k-means algorithm for clustering web documents. Yoo and Hu [12] represent a set of documents as bipartite graphs using domain knowledge in ontology. Then documents are clustered based on average-link clustering algorithm. However, these approaches relying on graph based model are not intended to be used in an incrementally growing set of documents.

This paper presents an incremental document clustering approach. Our method consists of three important parts. The first part concentrates on document representation model based on graph because graphs can model additional information which is often not present in commonly used data representations, such as vectors. To save clustering time, the method incrementally extracts shared phrases from graphs when building graphs representing documents. The second part proposed hybrid similarity measure between documents based on shared phrases and adapting document vectors. The last part presents the enhanced IncrementalDBSCAN algorithm for clustering document sets. It assigns document to their respective clusters incrementally. Though the performance of IncrementalDBSCAN algorithm [3] is excellent, it tends to merge document clusters which less connected into the large one. Then the enhanced IncrementalDBSCAN algorithm proposed by this study is used to cluster the document set.

The rest of paper is organized as follows. Section 2 describes graph based document representation model. Section 3 introduces our measure for calculating similarities between documents and incremental feature selection method. Section 4 presents the enhanced IncrementalDBSCAN for grouping documents. An extensive experimental evaluation on news articles is conducted and the results are reported in section 5. Finally, we conclude the paper and present our future work in section 6.

2 Graph Based Document Representation Model

In graphic model, documents are transformed into a graph or set of graphs. There are numerous methods for creating graphs from documents. The authors of [9] described six major algorithms: standard, simple, n-distance, n-simple distance, absolute frequency, and relative frequency. All these methods are based on adjacency of terms. In our case we used simple document representation and represent a set of documents as a graph to indicate proximity and occurrence relationship between terms.

The simple model is a labelled nodes directed graph $G = (V, E)$, where V is a set of *nodes*, E: is a collection of *edges*. Under this model, each node v represents a unique term in the whole document set. Each node is labelled with the term it represents. Each edge e is an ordered pair of nodes (v_i, v_j) representing a directed connection from v_i to v_j. If the term v_j immediately precedes the term v_i somewhere in a document, then there is a directed edge from v_i to v_j. The above definition of graph suggests that the number of nodes in a graph is the number of unique terms in the document set.

In order to deal with incrementally growing set of documents, we have used DIG (Document Index Graph) algorithm [4] to index the documents while maintaining the sentence structure in the original documents. This algorithm incrementally builds graph by processing one document at a time. When a new document appears, new terms are added to the graph as necessary and connected with other nodes to reflect the adjacency of terms. The DIG algorithm can be used for incremental finding overlapping subgraphs or shared phrases from previous documents when building the graph at the same time. These shared phrases are input of the process in next section.

3 Document Similarity Measure and Incremental Feature Selection

Similarity measure is one of decisive factors for the success of clustering method in general and document clustering method in particular. Based on the observation that the use of a combination of single terms and phrases together might improve clustering results, we used DIG algorithm [4] to incremental find shared phrases between documents. These shared phrases can provide both text and document structure information. Then, we cluster documents based on integration of two similarity measures (call as hybrid): their similarity on the shared phrases (sim_{sp}) and cosine similarity of document vectors (sim_{df}). Given two documents (d_1 and d_2), the hybrid similarity is defined as (λ=0.2 in the experiments)

$$sim(d_1,d_2)=\lambda \bullet sim_{df}(d_1,d_2)+(1-\lambda)\bullet sim_{sp}(d_1,d_2). \tag{1}$$

where $\lambda \in [0, 1]$ as weights.

The hybrid similarity measure takes into account the structural information of a document by considering shared phrases, thus capturing the subtopic structure of the document. Our hybrid similarity is a superior measure of term order similarity.

The similarity between documents on shared phrases (sim_{sp}) is derived from phrase-based similarity measure of [4] with some modifications.

$$sim_{sp}(d_1,d_2)=\frac{\sqrt{\sum_{i=1}^{P}\left(\frac{l_i}{avg(|s_i|)}\right).(f_{1i}+f_{2i})^2}}{\sum_j|s_{1j}|+\sum_k|s_{2k}|}. \tag{2}$$

where: P: the number of shared phrases, f_{1i}, f_{2i}: the frequencies of shared phrases i in the documents d_1 and d_2 respectively, l_i: the length of the shared phrase, $|s_{ij}|$: the length of the sentence j in the document d_i, avg($|s_i|$) : the average length of sentences containing shared phrases i.

The sim_{df} is calculated based on cosine correlation similarity measure [7] which is defined by the cosine of the angle between the two document vectors of single term weights. The weight of each term is usually calculated using TF-IDF weighting scheme. However, the TF-IDF weighting scheme is not suitable for an incremental document clustering algorithm [13] so we used proposed TF-IG (Term Frequency – Information Gain) function to identify the goodness of terms not only in a document but also in all clusters. This term weight scheme can reflect importance of a specific term in a document and easily involve in incremental clustering process. Besides that this TF-IG function can be used in incremental feature selection process. The term weights of document vector are calculated as

$$w_{ij} = \frac{IG(j)}{MinIG} \times tf_{ij} . \quad (3)$$

where w_{ij} is the weight of a single term t_j in a document d_i, the term frequency tf_{ij} denotes the frequency of term t_j in a document d_i, *MinIG* is the minimum information gain value of terms, and *IG(j)* is the Information Gain [11] of term t_j in all the collection.

In order to further improve clustering efficiency for our method, feature selection (or dimensionality reduction) is used to reduce the size of the feature space without sacrificing clustering quality. Based on the idea that clustering performance and feature selection can be reinforced by each other [6], we propose a novel incremental feature selection technique, which utilizes Information Gain (IG) - supervised feature selection method for text clustering. Our feature selection technique calculates the relevant score for each term using IG method whenever a new document is clustered and there are changes in cluster structure (i.e., a new cluster is formed or two clusters are merged). Then some irrelevant terms are removed based on the calculated score of each term. Finally, selected terms form new feature space and document vectors are adapted to cluster new coming document.

In our experiment, we select 10% single terms to form feature space and create adapting document vectors. Because the number of shared phrases between documents is relatively small, we do not carry out feature selection on them.

4 Clustering Algorithm

IncrementalDBSCAN algorithm [3] is an efficient incremental clustering algorithm suitable for mining in a data warehousing environment. This algorithm is based on the DBSCAN algorithm [2] which is a density based clustering algorithm. Due to its density-based qualities, in IncrementalDBSCAN the effects of inserting and deleting objects are limited only to the neighborhood of these objects. IncrementalDBSCAN requires only a distance function for pairs of objects and is applicable to any data set from a metric space. In addition, IncrementalDBSCAN algorithm is not easily affected by noise (or outliers) while noise is one of the well-known characters of document set. Since IncrementalDBSCAN algorithm has excellent performance on clustering, therefore, this study modifies original IncrementalDBSCAN algorithm to prevent it from merging less connected clusters into large one.

We list the basic idea of IncrementalDBSCAN algorithm and some our modifications for document set.

Let *D* be the set of objects and *p* be an object to be inserted. Then, $UpdSeed_{Ins} = \{q \mid q$ is the core object in $D \cup \{p\}$, $\exists q'$: q' is the core object in $D \cup \{p\}$ but not in *D* and $q \in N_{Eps}(q')\}$.

When inserting a new object *p* into the data set, new density connections can be established so clustering process just restricts on $UpdSeed_{Ins}$ set. When inserting an object *p* into the database *D*, there are some cases:

(1) ***Noise***: $UpdSeed_{Ins}$ is empty, i.e. there are no "new" core objects after insertion of *p*. Then, *p* is a noise object and nothing else is changed.
(2) ***Creation***: $UpdSeed_{Ins}$ just contains core objects which did not belong to any cluster before the insertion of *p*, i.e. they were noise objects or equal to *p*. In this case, a new cluster including these objects and *p* will be created.
(3) ***Absorption***: $UpdSeed_{Ins}$ contains core objects which were members of exactly one cluster *C* before insertion. The object *p* and some noise objects can be absorbed into the cluster *C*.
(4) ***Merge***: $UpdSeed_{Ins}$ contains the core objects which were members of several clusters before the insertion. All these clusters and the object *p* will be merged into one cluster.

When processing with document set IncrementalDBSCAN algorithm tends to merge document clusters which less connected into one. To overcome this weakness, we make some modifications for the case *(4)*. When $UpdSeed_{Ins}$ contains core objects which were members of several clusters before insertion, we first check the document density of these clusters. If the number of documents in a cluster containing core objects belonging to $UpdSeed_{Ins}$ is smaller than a specific value α, we do not merge the cluster into new one. Otherwise, these clusters having amount of documents not less than α and the new document *p* are merged into one cluster.

Based on hybrid similarity, the enhanced IncrementalDBSCAN algorithm assigns document to their respective clusters incrementally and effectively.

5 Experimental Evaluation

We conducted a series of experiments using our proposed graph model, hybrid similarity measure, and incremental clustering method. We first tested the effectiveness of the graph–based model document representation, and hybrid similarity between documents based on shared phrases versus single terms only. The second set of experiments was to evaluate the accuracy of the incremental document clustering method, based on enhanced IncrementalDBSCAN. The web pages of VnExpress[1], TuoiTre[2] are used as the targets for verification. Among news categories provided in these newspapers, documents in 7 different domains: economy, tourism, music, sports, fashion, informatics, and society are extracted as a data set of documents to be used in these experiments. After retrieving the data sets, we generate various document collection whose numbers of classes are 3 to 7 using the document sets. Clustering methods have been evaluated in many ways. Among them we adopted the widely used Entropy measure and F-measure [1] to score the cluster quality.

We conduct some experiences on the corpus sets using different document representation models. Table 1 presents the F-measure and Entropy results for the following data models: graph–based document representation model with the hybrid similarity measure based on shared phrases and adapting document vectors; vector space model, the cosine similarity measure with TF-IDF single term weights; and vector space model, the cosine similarity measure with our proposed TF-IG single term weights. We

[1] http://vnexpress.net
[2] http://tuoitre.com.vn

used incremental feature selection method for all models. IncrementalDBSCAN algorithm is used to clustering documents. As shown in Table 1, our proposed method consistently produces better clustering results for various corpus set. In addition, our TF-IG function gives improvement on the clustering quality compared to traditional TF-IDF term weights in vector model. It is obvious that the graph based representation model with hybrid similarity plays an important role in accurately judging the relation between documents.

Table 1. Performance evaluation of different document representation models

Corpus ID	No of Docs	Vector space model TF-IDF term weights		Vector space model TF-IG term weights		Proposed method : Graph model Hybrid similarity	
		F-measure	Entropy	F-measure	Entropy	F-measure	Entropy
DS31	1066	0.976	0.089	0.986	0.088	0.997	0.035
DS32	3023	0.847	0.058	0.858	0.052	0.995	0.047
DS51	1798	0.758	0.574	0.776	0.563	0.915	0.175
DS52	4082	0.815	0.396	0.834	0.336	0.969	0.141
DS71	5319	0.791	0.485	0.812	0.454	0.966	0.214

Table 2. Comparison of F-measure, Entropy, and number of cluster found for SHC and proposed method with enhanced IncrementalDBSCAN

Corpus ID	No of reference clusters	SHC algorithm			Proposed method : Enhanced IncrementalDBSCAN		
		No of clusters found	F-measure	Entropy	No of clusters found	F-measure	Entropy
DS31	3	7	0.956	0.019	3	0.997	0.035
DS32	3	12	0.958	0.021	3	0.995	0.047
DS51	5	21	0.843	0.115	6	0.915	0.175
DS52	5	45	0.899	0.122	6	0.969	0.141
DS71	7	83	0.856	0.196	8	0.966	0.214

In order to evaluate our proposed method, we compare it with SHC algorithm [5]– an incremental document clustering algorithm, based on the cluster cohesiveness measure using similarity histograms. SHC was reported to provide better clustering quality [5] than other methods such as Hierarchical Agglomerative Clustering (HAC), Single-Pass or *k*-Nearest Neighbor clustering. We used the same graph model and hybrid similarity between documents for SHC algorithm and our approach with enhanced IncrementalDBSCAN. Table 2 compares the F-measure as well as the Entropy for our method and SHC for all corpus sets. The parameters chosen for different algorithms were the ones that produced best results. The average performances of F-measure have increased 7%. The number of founded clusters is also under control. Entropy values are not good as SHC. Because Entropy favors small clusters and SHC algorithm tends to splitting data set to smaller clusters. Hence there is a large mismatch between the number of reference clusters and those found by the SHC algorithm.

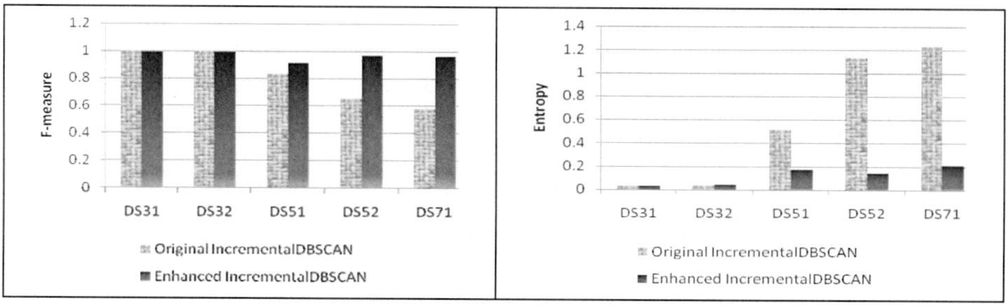

Fig. 1. Performance of enhanced IncrementalDBSCAN and original IncrementalDBSCAN

We compare the enhanced IncrementalDBSCAN algorithm with the original IncrementalDBSCAN algorithm in terms of Entropy and F-measure. Figure 1 shows enhanced IncrementalDBSCAN achieves improvement over original one, 15% averagely in terms of F-measure and 18% in terms of Entropy. We note that the enhanced IncrementalDBSCAN performs better than the original one on same data representation model, especially when the number of documents and clusters increase.

6 Conclusions

Due to massive volumes of unstructured data generated in the globally networked environment, the importance of document clustering will continue to grow. In this paper, we present our new approach for incrementally clustering documents. The proposed scheme consists of three main steps: (1) graph representation for documents, (2) incrementally extract features from graph sets and calculate pair wise document similarity based on shared phrases and gradually adapting feature space, (3) use enhanced Incremental DBSCAN algorithm to cluster documents based on these similarities. The results show that our approach is competitive to existing schemes in term of performance. Our method can provide a meaningful explanation of each document cluster by identifying its frequent shared phrases.

As for future research, some issues are still open:

- We need do more experiments with larger data to affirm the effectiveness of our method.
- Some method should be developed to automatically adapting the input parameters for Incremental DBSCAN over time.

References

1. Baeza-Yates, R., Ribeiro-Neto, B.: Modern Information Retrieval. Addison Wesley, Reading (1999)
2. Ester, M., Kriegel, H.-P., Sander, J., Xu, X.: A density based algorithm for discovering clusters in large spatial databases with noise. In: Proceedings of the 2nd Int. Conf. on Knowledge Discovery and Data Mining, pp. 226–231 (1996)

3. Ester, M., Kriegel, H.-P., Sander, J., Wimmer, M., Xu, X.: Incremental Clustering for Mining in a Data Warehousing Environment. In: Proceedings of VLDB, pp. 1–11 (1998)
4. Hammouda, M., Kamel, M.: Phrase-based Document Similarity Based on Index Graph Model. In: Proceedings of the 2002 IEEE international conference on Data mining (ICDM), pp. 203–210 (2002)
5. Hammouda, M., Kamel, M.: Incremental Document Clustering using Cluster Similariry Histograms. In: Proceedings of the IEEE/WIC international conf. on Web Intelligence, pp. 597–601 (2003)
6. Liu, T., Liu, S., Chen, Z., Ma, W.-Y.: An Evaluation on Feature Selection for Text Clustering. In: Proceedings of the 12th ICML, pp. 488–495 (2003)
7. Salton, G., Wong, A., Yang, C.S.: A vector space model for automatic indexing. Communication of ACM 18(11), 613–620 (1975)
8. Schenker, A., Last, M., Bunke, H., Kandel, A.: A Comparison of Two Novel Algorithms for Clustering Web Documents. In: Proceedings of 2nd International Conference on Web Document Analysis (IWWDA), pp. 71–74 (2003)
9. Schenker, A., Last, M., Bunke, H., Kandel, A.: Classification of Web Documents Using Graph Matching. International Journal of Pattern Recognition and Artificial Intelligence, Special Issue on Graph Matching in Computer Vision and Pattern Recognition 18(3), 475–479 (2004)
10. Yang, Y., Carbonell, J.G., Brown, R.G., Pierce, T., Archibald, B.T., Liu, X.: Learning Approaches for Detecting and Tracking News Event. IEEE Intelligent Systems 14(4) (1999)
11. Yang, Y., Pedersen, J.O.: A comparative study on feature selection in text categorization. In: Proceedings of ICML 1997, pp. 412–420 (1997)
12. Yoo, I., Hu, X.: Clustering Large Collection of Biomedical Literature Based on Ontology-Enriched Bipartite Graph Representation and Mutual Refinement Strategy. In: Ng, W.-K., Kitsuregawa, M., Li, J., Chang, K. (eds.) PAKDD 2006. LNCS, vol. 3918, pp. 303–312. Springer, Heidelberg (2006)
13. Wong., W., Fu, A.: Incremental Document Clustering for Webpage Classification. In: Proceedings of 2000 Int'l Conf. Information Soc. in the 21st Century: Emerging Technologies and New Challenges (2000)
14. Zamir, O., Etzioni, O., Madanim, O., Karp, R.M.: Fast and Intuitive Clustering of Web Documents. In: Proceeding of 3rd Int'l Conf. Knowledge Discovery and Data Mining, pp. 287–290 (1997)

Evaluating the Impact of Missing Data Imputation

Adam Pantanowitz and Tshilidzi Marwala*

School of Electrical & Information Engineering
University of the Witwatersrand, Johannesburg
Private Bag 3, Wits, 2050, South Africa
adam.pantanowitz@wits.ac.za, tmarwala@uj.ac.za

Abstract. This paper presents an impact assessment for the imputation of missing data. The assessment is performed by measuring the impacts of missing data on the statistical nature of the data, on a classifier, and on a logistic regression system. The data set used is HIV seroprevalence data from an antenatal clinic study survey performed in 2001. Data imputation is performed through the use of Random Forests, selected based on best imputation performance above five other techniques. Test sets are developed which consist of the original data and of imputed data with varying numbers of specifically selected missing variables imputed. Results indicate that, for this data set, the evaluated properties and tested paradigms are fairly immune to missing data imputation. The impact is not highly significant, with, for example, linear correlations of 96 % between HIV status probability prediction with a full set and with a set of two imputed variables using the logistic regression analysis.

Keywords: Impact, imputation, missing data, random forest, sensitivity.

1 Introduction

Missing data are a common difficulty encountered in many real-world situations and studies resulting in difficulty with data analysis, study and visualisation [1,2]. The missing information also reduces insight into the data, and the underlying cause for the fact that data are missing may make the data of particular interest. Decision making system often cannot make use of decision policies to make a decision without all the information at hand and such systems are thus reliant on data imputation techniques to determine the values of missing data. The *impact* of the imputation of missing data should be considered such that insight is gained into the validity of decisions made by such systems.

This paper evaluates the concept of missing data and presents a brief background of the methodology and main paradigm used. The data set is examined,

* Tshilidzi Marwala has since become Executive Dean of the Faculty of Engineering and the Built Environment at the University of Johannesburg. P.O. Box 524, Auckland Park, 2006, Johannesburg, South Africa.

R. Huang et al. (Eds.): ADMA 2009, LNAI 5678, pp. 577–586, 2009.

and thereafter feature selection on the data is described for the purpose of the impact assessment. The impact and sensitivity analysis is described and results are presented. Finally, conclusions are drawn.

2 Background

2.1 Missing Data

Missing data are an inherent problem common to data mining and collection for real world data sets, especially large ones. Missing data impacts on decision making systems and statistical methods fall short when data are unknown [1]. Studies have highlighted the need to research decision support systems when key information is missing or inaccessible [3]. The effect of missing data on such decision support systems is marked, and it is shown that results are degraded by simply assigning arbitrary values to the missing data elements. In the context of surveys, missing data may result for a number of reasons such as incomplete variable collection from subjects, non-response from subjects, poorly defined surveys, and data being removed for reasons such as confidentiality [1,2]. A discussion on the categorisation of missing data and the missing data mechanism is available in [4]. This discussion also evaluates a number of existing methods for dealing with missing data.

2.2 Random Forests

"Random Forest" (RF) is an algorithm first introduced in 2000 by Breiman [5] which generalises ensembles of decision trees through bagging (bootstrap aggregation), thus combining multiple random predictors in order to aggregate predictions [6,7]. RFs allow for complexity without over-generalising the training data [8]. RF can be used for both regression and classification, and has been used with success in the context of missing data [9]. "Out-of-bag" (oob) data, which are unused training data, are used in predicting variable importance.

RFs have been an area of research in the last few years for their advantageous features and success [5]. RFs are said to work fast, have excellent accuracy offering improvements over single classification and regression trees, be impervious to over-fitting the data, have no dimensionality problems, give an unbiased self-assessment and variable importance assessment, and have effective methods for missing data estimation and for outlier location [5,6]. These properties make the RF algorithm a logical candidate for this missing data study.

3 Methodology

This study is performed by making use of data imputed using the RF algorithm. For details of the techniques used for the imputation of the missing data, and for comparisons between RFs and other investigated imputation paradigms for the data set used, the reader is referred to [4]. The paradigms compared include

RFs, auto-associative neural networks with genetic algorithms, auto-associative neuro-fuzzy configurations, and two RF and neural network based hybrids. RFs are superior in imputing the missing data in terms of computation time and in terms of accuracy of prediction for the given data set [4]. RFs are thus used in their regression capacity for imputing the missing data for this impact study and for forming the sets discussed in section 3. RFs are further used in their classification capacity in classifying HIV status in the impact study in order to measure comparatively the impact of missing data on a classifier.

Once missing data are imputed, we measure in three ways the impact of imputing the missing data. We measure the changes in the statistical properties of the data, the impact on a two-variable classifier, and the impact on a logistic regression (LR) system. This is done by creating a number of test sets. We take a complete set and remove specifically chosen variables. The variables are chosen according to their mutual correlations and importance, as discussed in section 3.2. The missing values of these variables are thereafter imputed and they are also *guessed* as uniform random numbers. This yields three sets: a complete set, a set with missing data imputed, and a set with random guesses. Analysis of the sets and exposure of the classifier and the LR system to the sets yield results which are compared to test for the impact of missing data imputation.

3.1 Data Evaluation and Preprocessing

The data set used is based on a national human immunodeficiency virus (HIV) and syphilis seroprevalence survey of women attending antenatal clinics in South Africa, and is taken from the study performed in 2001 [10]. The data consist of survey information from 16 743 pregnant women (however, just under 12 000 instances are deemed viable, indicating the significance of missing data in a real world study). The variables contained in the data set include: province (location); age; education; gravidity; parity; father's age; HIV status; rapid plasma reagin (RPR) test status and race as indicated in table 1. Gravidity refers to the number of times a woman has been pregnant, and parity to the number of times the woman has given birth. Father's age indicates the age of the father responsible for the current pregnancy. Education is specified as 0 (no education); 1 - 12 (for grades 1 through to 12); and 13 (tertiary education). An individual's location is categorised through the province variable into one of the nine South African provinces; and race categorised in to one of six race categories [11]. The data are preprocessed to acceptable ranges and are normalised. Categorical data is binary encoded to prevent interference with the machine learning paradigms.

3.2 Feature Selection and Multiple Missing Values

The RF algorithm allows for feature selection by providing estimates of the importance of each variable in the data set in predicting a given output (a missing value) [6]. On predicting variables in the given data set, it is found, as expected, that variables for which effective estimates are obtained when data are missing have high correlations with other known variables in the set.

Table 1. Variable importance for imputing each potentially missing variable

Variable	High Importance Variables	Low Importance Variables
Age	Father's Age; Parity	HIV Status; RPR
Education	All approximately of equal importance (Age slightly dominant)	
Gravidity	Parity	All others
Parity	Gravidity	All others
Father's Age	Age	All others
HIV Status	Education; Race	All others
RPR Status	Province	All others
Province	Race; Parity	Age; Education
Race	Province	All others

If we are to test the impact of two or more missing variables, there are a large number of permutations that require testing. However, feature selection allows us to select and test combinations of variables which have been selected according to their mutual correlations. This allows for a more meaningful analysis to be performed, since variables with high correlations can be specifically and purposefully selected to best test the impact of missing data imputation. In this paper, therefore, we test highly correlated variables to obtain a worst-case impact for this data set. This analysis also offers insight into why certain variables are certainly removable when predicting other certain variables. The analysis also offers a rationale for performing fewer tests on multiple missing variables. The insight gained through this analysis is applied in the impact assessment of section 4. Table 1 presents the notable importance of the other variables for each examined variable. The predictions are generally logically sensible, for example, gravidity is highly correlated with parity, and age with the father's age. It is interesting to note that education has all other variables listed as equally important. The learning paradigm performs relatively badly in predicting education [4]. This means that in imputing education, other variables are more likely equally non-important, offering an explanation as to why education is a difficult variable to impute.

4 Impact and Sensitivity Assessment

The impact of estimating the missing data is evaluated within this section by evaluating three aspects: the statistical impact on the data, the impact on HIV classification, and the impact on a decision making system (a logistic regression system). This assessment gives an overall picture, since it offers insight into the effects of imputation on the data internally (statistical assessment [12]), and on the effects the imputation has on two systems dependent on the imputation: a classifier [3,13] and a decision making system. Study variables are selected based on their mutual correlations as discussed in section 3.2, and based on the prediction performance of the RF predictor. Note that for missing variable(s), two sets are defined, one which has the variable(s) imputed with RFs (*sets RFx*) and one which has the variable(s) uniformly randomly assigned (*sets Rx*). The randomly defined sets act as an experiment control to ensure that the imputed results presented are not spurious, and to show that the statistical properties and the tested cascaded systems are in fact sensitive to the data presented. Note that when these sets are used in conjunction with an HIV classifier or decision making system, as in sections 4.2 and 4.3, HIV data is not used as an input to impute the missing data. Using the variable selection technique discussed in section 3.2 the following sets are defined:

- The original complete target data set (*set T*),
- a single missing variable with average prediction performance - age (*sets RF1A & R1A*),
- a single missing variable with poor prediction performance - education (*sets RF1B & R1B*) - (important for HIV prediction as per table 1),
- a single missing variable with good prediction performance - gravidity (*sets RF1C & R1C*),
- two missing variables which are of high mutual importance (as per table 1) - age and father's age (*sets RF2A & R2A*),
- four missing variables - age, education, father's age, gravidity (*set RF4A & R4A*).

The three studies were also performed on three missing variables - age, education and father's age - the results of which are similar, but are not indicated here in detail. Some of the evaluation techniques include the goodness of fit measures in terms of the KS test [12] and the Mahalanobis Distance (the mean distance is taken) [14]. These give a statistical measure of the similarity between the data sets, and are regarded as a good measure of the fit between results. The mean squared error (MSE) offers a relative indication of the difference between data sets. If N is the entire data set, T_i represents the ith target value and P_i represents the ith predicted value, the MSE is calculated as follows:

$$MSE = \frac{1}{N}\sum_{i=1}^{N}\left(T_i - P_i\right)^2 . \quad (1)$$

4.1 Statistical Impact

The statistical impact on the data is quantified through a number of statistical measures. Table 2 presents statistical results for the missing variable age. Note the similarity between the statistical properties of the original set and of the imputed set, and the deviation in the properties for the set where values are randomly assigned to age. This is particularly evident with the large change in variance and in the combined MSE. Similar results are observed for the study on the variables education and gravidity. Also note the similarity in linear correlation for the original set and the imputed set, and, as expected, the great dissimilarity in linear correlation between the original set and the randomly assigned set. A quantile-quantile plot (QQ plot) allows one to view the deviation in

Table 2. Statistical impact for single missing variable: age

Measure	Age (years)		
	Set T	*Set RF1A*	*Set R1A*
Mean	25.00	25.3	31.2
1st Quartile	20	21	22
Median	24	25	31
3rd Quartile	29	29	41
Standard Deviation	6.3	5.4	11.0
Variance	40.0	29.0	120.3
Combined MSE	-	10.5	195.5
Mean Mahalanobis Distance	-	0.73	3.96
Linear Correlation (with Target Set)	-	85.92 %	2.01 %
Maximum Percentage Deviation	-	84.2 %	163.2 %

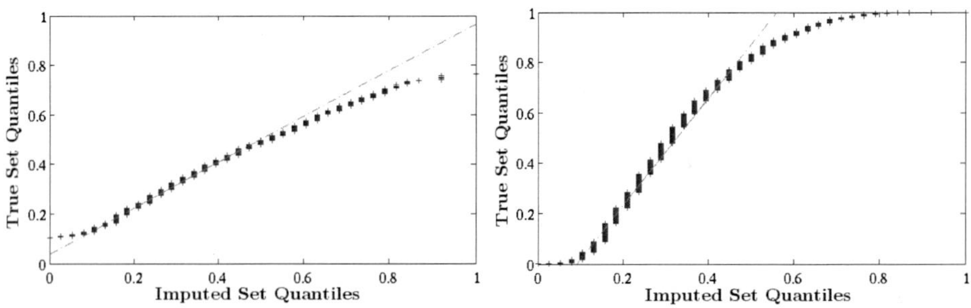

Fig. 1. QQ Plots of the target data set (T) of the age variable with the RF imputed set RF1A (*left*) and with the set with age randomly assigned values R1A (*right*)

the distributions of a given variable [12]. The extent of deviation from a straight line indicates distribution deviation. Figure 1 presents QQ plots first, for the real set (T) and the RF imputed variable (RF1A) and second, for the real set (T) and randomly guessed values (R1A). Note the former plot is fairly linear, with the interpolated line intercepts close to the origin and to the co-ordinates $(1, 1)$, indicating a good distribution match, whereas the latter plot is fairly non-linear.

4.2 Impact on HIV Classification

HIV status of an individual is one of the variables included in the data set, as discussed in section 3.1. Thus, we can attempt to classify the HIV status of an individual by training an RF classifier and then removing the HIV status of the individual in a test set and providing the classifier with the remaining (known) variables. We define for classification the categories in table 3 [13]. From the definitions, we can define evaluation metrics in order to quantify the performance of the classifiers. First, the accuracy of the classifier is defined as $\frac{TN+TP}{TN+TP+FN+FP}$. [13]. *Sensitivity* allows one to assess how well the classifier can recognise positive samples, and is measured as $\frac{TP}{TP+FN}$. *Specificity* measures how

Table 3. Errors for binary classification [13]

	Predicted Negative	Predicted Positive
Actual Negative	True Negative (TN)	False Positive (FP)
Actual Positive	False Negative (FN)	True Positive (TP)

Table 4. HIV classification results for data sets with one, two and four imputed variable(s)

Set	T	RF1A	R1A	RF1B	R1B	RF1C	R1C	RF2A	R2A	RF4A	R4A
TP	837	936	506	926	768	839	310	947	308	1122	109
TN	2986	2787	3057	2766	2463	2955	4040	2391	3589	1825	4358
FP	487	388	1577	1868	2171	1679	594	2243	1045	2809	276
FN	1648	1847	818	398	556	485	1014	377	1016	202	1215
Accuracy (%)	64.2	62.5	59.8	61.2	54.2	63.7	73.0	56.0	65.4	49.5	75.0
Sensitivity (%)	63.2	70.7	38.2	79.9	58.0	63.4	23.4	71.5	23.3	84.7	8.2
Specificity (%)	85.9	87.8	78.9	87.4	81.6	85.9	79.9	86.4	77.9	90.0	78.2
F Measure	0.44	0.46	0.29	0.45	0.36	0.44	0.28	0.42	0.23	0.43	0.13

well the classifier recognises samples as negative, and is evaluated as $\frac{TN}{(TN+FN)}$. *Precision* is a measure of the percentage of samples correctly specified as positive, $\frac{TP}{TP+FP}$ [15]. Note that *recall (Re)* is the same measure as *sensitivity* [16]. The *F measure* is used to assess a system when a single number is preferred, and is given by $\frac{2 \times Precision \times Recall}{Precision + Recall}$ [16]. It is important to note is that the performance of the classifier is not of much interest, but the *relative* performance of the classifier given imputed data is of interest.

In order to assess the impact of imputed data on HIV classification, a number of imputed data sets and data sets with randomly assigned values are propagated through an RF classifier. Table 4 compares these results with that of the classification results of the target (complete set). Evident from this table is that the classifier, though indicated to be of average performance generally from the F measure, shows resilience and almost immunity to the sets with estimated data, especially with 1 or 2 imputed variables. Note that the performance is not degraded when the missing variable is education, which has high importance for HIV prediction as indicated by table 1. The effects of the random data sets are evident, with F measures dropping into the 0.2 range. One can observe this result by comparing the F measure results across the table. The experimental control of using randomly assigned variables ensures that the variables do impact on the classifier, and thus indicates that the experimental results are not spurious.

4.3 Impact on Decision Making System

A logistic regression (LR) decision making system is designed, which, based on input, computes the probability that the output variable belongs to a given set [17]. The output variable is chosen to be HIV status, and we thus obtain probability of an individual's membership to an HIV positive class or an HIV

Table 5. Measures of the result of logistic regression analysis for the original data set and sets with one, two and four imputed variable(s) for comparative purposes

Set	T	RF1A	R1A	RF2A	R2A	RF4A	R4A
1st Quartile (%)	18.7	18.8	19.1	19.7	19.1	20.6	23.9
Median (%)	24.6	24.8	26.2	26.2	26.2	26.8	34.4
3rd Quartile (%)	27.9	27.9	32.8	27.9	32.5	27.9	44.1
Mean (%)	22.6	22.6	25.3	23.1	25.3	23.4	33.4
Variance (%)	89.7	89.8	130.8	88.5	128.1	92.4	230.4
Linear correlation (%)	-	98.9	87.7	96.4	87.3	95.3	77.8
KS test	-	0.02	0.21	0.12	0.21	0.18	0.49
Mean squared error	-	1.8	38.3	6.7	37.9	9.23	214.3

negative class. The original set T is propagated through the LR system, and a set of probabilities that individuals are HIV positive is obtained. Thereafter, the sets with various imputed variables are propagated through the regressor, to yield a set of probabilities that individuals (with imputed demographics) are HIV positive. The probabilities resulting from the original set (T) are compared with the probabilities resulting from the imputed sets (RF1A, etc.). The probabilities are expressed as percentages, and where tests of fit are involved (e.g. the KS test) the relevant result data are compared with the results from the original set T. The results are fairly similar for the various different single imputed variable sets (note that only age is presented here). Results in table 5 present statistical differences in the regressor outputs due to set T, and due to the sets with one, two and four imputed variables.

Once again, the results indicated in table 5 predict a fair amount of immunity to imputed data on the probabilities given by the LR analysis. The original data set and the set with one, two and even four imputed values do not significantly change the predictions of the LR. This is emphasized in figure 2 which indicates the QQ plot of the two probability distributions from the LR analysis: the prediction from the LR analysis given the original data is plotted against the

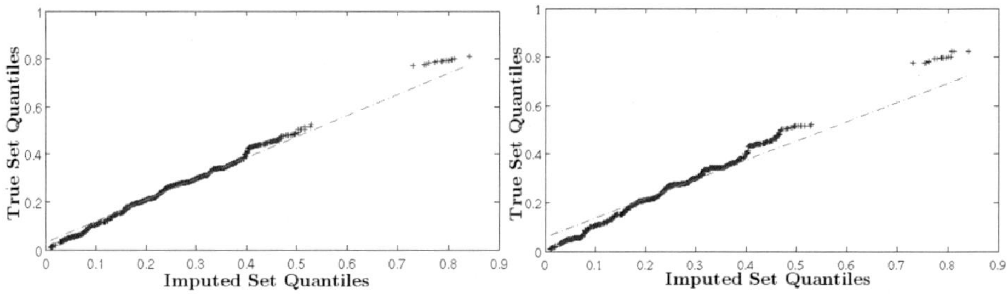

Fig. 2. QQ plot of result from logistic regression probability analysis on HIV status for the true set (T) and for the set with 2 imputed variables (*RF2A*) (left); and QQ plot of result from logistic regression probability analysis on the HIV status for the true set (T) and the set with 4 imputed variables (*RF4A*) (right)

prediction for the 2 imputed variable set; and the prediction from the LR analysis given the original data is plotted against the prediction for the 4 imputed variable set. Note that the plots are fairly linear, indicating high similarity between probability results from the LR. As the number of imputed variables increases, the correlation decreases. This indicates that the number of imputed variables does, as expected, have an effect on the results. Note that the randomly generated sets (R1A, R2A and R4A) indicate significant deviation on propagation through the LR analysis. These sets cause the variance in the data to increase by a significant amount. These indicators show that the LR analysis is sensitive to the data that are tested, thus validating the experiment.

5 Discussion

Through the imputation of missing data and the study of the impact this has on the statistical nature of the data and on the examined systems, it is notable that the missing data imputation does not significantly negatively impact on the statistical properties of the data and on classifiers and decision-making systems reliant on the imputed data. A considerable impact on society can be noted. A decision-based system for preliminary HIV classification in a health-care or study context is invaluable. This has further consequences in fields such as risk analysis, for example, and a number of data mining applications. Furthermore, missing data which renders potentially useful information in a given study meaningless can, through appropriate imputation, be recovered as meaningful information. This can help in studies to uncover the statistical trends that are said to be lost and discarded through the removal of missing entries (in, say, a complete-case method [1,18]). Decisions can therefore be made with fair confidence on instances which were previously unusable due to missing information.

6 Conclusion

It is useful to measure the impact of imputing missing data since a number of systems may be reliant on the provided estimations. RFs are chosen for imputation due to their superiority in both prediction accuracy and in computation time taken. Through the use of survey data of results, data sets for impact analysis are generated with one, two, three and four imputed variables each, and these sets are used for evaluating impact. Impact is determined in three ways: through evaluating statistical deviations of the imputed variables relative to the true values; through the performance of an HIV classifier; and through a logistic regression analysis for probability prediction. Results indicate that for this data set the statistical properties and the decision making systems are in fact rather immune to the imputation of missing data, when an adequate imputation technique is used. A great impact is observed for missing data which are randomly assigned values, while the impact is considerably decreased when the data are imputed rather than simply "guessed". This validates appropriate imputation techniques in providing a viable means to solving the problem of missing data.

Furthermore, systems which are reliant on predicting the missing data seem to operate similarly to when the data are known. These results imply that using appropriate imputation techniques on instances with missing information allows for decision based systems to make informed decisions that may have been previously impossible to make or may have been prone to error.

References

1. Ssali, G., Marwala, T.: Computational Intelligence and Decision Trees for Missing Data Estimation. In: Proceedings of the International Joint Conference on Neural Networks, WCCI 2008, IJCNN, pp. 201–207. IEEE, Los Alamitos (2008)
2. Horton, N.J., Kleinman, K.P.: Much Ado About Nothing: A Comparison of Missing Data Methods and Software to Fit Incomplete Data Regression Models. The American Statistician 61(1), 79–90 (2007)
3. Markey, M.K., Tourassi, G.D., Margolis, M., DeLong, D.M.: Impact of Missing Data in Evaluating Artificial Neural Networks Trained on Complete Data. In: Computers in Biology and Medicine, vol. 36, pp. 517–525. Elsevier, Amsterdam (2006)
4. Pantanowitz, A., Marwala, T.: Missing Data Imputation Through the Use of the Random Forest Algorithm. In: Advances in Intelligent and Soft Computing - IWACI 2009. Springer, Heidelberg (to appear, 2009)
5. Biau, G., Devroye, L., Lugosi, G.: Consistency of Random Forests and Other Averaging Classifiers. Journal of Machine Learning Research 9, 2015–2033 (2008)
6. Breiman, L., Cutler, A.: Random Forests. Department of Statistics, University of California, Berkeley (2004)
7. Brence, J.R., Brown, D.E.: Improving the Robust Random Forest Regression Algorithm (2006)
8. Ho, T.K.: Random Decision Forests. In: ICDAR 1995: Proceedings of the Third International Conference on Document Analysis and Recognition 1 (1995)
9. Qi, Y., Klein-Seetharaman, J., Bar-Joseph, Z.: Random Forest Similarity for Protein-Protein Interaction Prediction from Multiple Sources. In: Pacific Symposium on Biocomputing, vol. 10, pp. 531–542 (2005)
10. Ntsaluba, A.: Summary Report: National HIV and Syphilis Sero-prevalence Survey of Women Attending Public Antenatal Clinics in South Africa, Department of Health, South African Government (2001)
11. Mistry, J., Nelwamondo, F.V., Marwala, T.: Investigation of Autoencoder Neural Network Accuracy for Computational Intelligence Methods to Estimate Missing Data. In: IASTED International Conference on Modelling and Simulation (2008)
12. Gibbons, J.D., Chakraborti, S.: Nonparametric Statistical Inference, 4th edn. CRC Press, Boca Raton (2003)
13. Betechuoh, B.L., Marwala, T., Tettey, T.: Autoencoder Networks for HIV Classification. Current Science 91(11), 1467–1473 (2006)
14. Mahalanobis, P.C.: Generalized Distance in Statistics. In: Proceedings of the National Institute of Science of India 12 (1936)
15. Vazirgiannis, M., Halkidi, M., Gunopulos, D.: Uncertainty Handling and Quality Assessment in Data Mining. Springer, Heidelberg (2003)
16. Ye, N.: The Handbook of Data Mining, 1st edn. Routledge Taylor & Francis Group, Abington (2003)
17. Pampel, F.C.: Logistic Regression - A Primer. SAGE, Thousand Oaks (2000)
18. Fogarty, D.J.: Multiple Imputation as a Missing Data Approach to Reject Inference on Consumer Credit Scoring. Intersat 41(9) (2006)

Discovery of Significant Classification Rules from Incrementally Inducted Decision Tree Ensemble for Diagnosis of Disease

Minghao Piao, Jong Bum Lee, Khalid E.K. Saeed, and Keun Ho Ryu

Database/Bioinformatics Lab, Chungbuk National University,
361-763 Cheongju, Korea
{bluemhp,jongbumlee,abolkog,khryu}@dblab.chungbuk.ac.kr

Abstract. Previous studies show that using significant classification rules to accomplish the classification task is suitable for bio-medical research. Discovery of many significant rules could be performed by using ensemble methods in decision tree induction. However, those traditional approaches are not useful for incremental task. In this paper, we use an ensemble method named Cascading and Sharing to derive many significant classification rules from incrementally inducted decision tree and improve the classifiers accuracy.

Keywords: Classification rules, Incremental tree induction, Ensemble method, Cascading and Sharing.

1 Introduction

Decision trees are commonly used for gaining information for the purpose of decision making. For inductive learning, decision tree is attractive for 3 reasons: (1) Decision tree is a good generalization for unobserved instance, only if the instances are described in terms of features that are correlated with the target concept. (2) The methods are efficient in computation that is proportional to the number of observed training instances. (3) The result of decision tree provides a representation of the concept that explainable to human.

The basis of many existing decision trees is Hunt's algorithm and developed trees are ID3 [1], C4.5 [2], and C5.0 [3]. Tree induction of those non-incremental approaches is based on a consistent set of labeled examples and the process for inducing a decision tree is quite inexpensive because exactly one tree is generated and maps a single batch of examples to a particular tree. ID4 [4], ID5R [5] and ITI [6] are incremental approaches and the results are as correct as non-incremental approaches do.

For improving the classifier's accuracy, ensemble methods are used to construct a set of base classifiers from training data set and perform the classification work by voting on the predictions made by each classifier. The ensemble of classifiers can be constructed in many ways [7] and most widely used is by manipulating the training set like Bagging and boosting. Three interesting observations are described in [8] based on the study of many ensemble methods: (1) Many ensembles constructed by

R. Huang et al. (Eds.): ADMA 2009, LNAI 5678, pp. 587–594, 2009.

the Boosting method were singletons. Due to this constraint, deriving classification rules have a limitation: decision trees are not encouraged to derive many significant rules and they are mutually exclusive and covering the entire of training samples exactly only once. (2) Many top-ranked features possess similar discriminating merits with little difference for classification. This indicates that it is worthwhile to employ different top-ranked features as the root nodes for building multiple decision trees. (3) Fragmentation problem is the another problem does those ensemble methods have: as less and less training data are used to search for root nodes of sub-trees.

Base on those observations, we need a method that can break the singleton coverage constraint and solve the fragmentation problem. Also, the method should be possible to deal with incrementally collected large data set and handle the data set with growing models and guarantee for the accuracy.

2 Related Works

In this section, we will describe some non-incremental and incremental decision tree induction. Also, we will illustrate some widely used ensemble method and the use in incremental induction task.

2.1 Non-incremental Decision Induction Trees

One approach to the induction task would be ID3 which is a simple decision tree learning algorithm developed by Ross Quinlan. The construction of ID3 is based on information theory and constructs the tree by employing a top-down, greedy search strategy through the given sets to test each attribute at every tree node. In order to select an attribute that is most useful for classifying an instance, it uses information gain [1], [10]. Suppose there is a given an example data set *S*, the entropy of *S* could be derived from the Equation 1:

$$Entropy(S) = -P(p)\log_2 P(p) - P(n)\log_2 P(n) \tag{1}$$

Where *P(p)* is the proportion of *positive* examples in *S* and *P(n)* is the proportion of negative examples is *S*. Then the information gain is as shown in Equation 2:

$$InformationGain(S,A) = Entropy(S) - \sum_{i=1}^{V} \frac{S_V}{S} \times Entropy(S_V) \tag{2}$$

Where *A* is an attribute and the value of *i* from *1* to *V* is the domain of *A*, and S_V is the number of records which contains the value *V*.

The decision tree algorithm C4.5 [2] is developed from ID3 in the following ways: Handling missing data, handling continuous data, and pruning, generating rules, and splitting. For splitting purpose, C4.5 uses the Gain Ratio instead of Information Gain. C4.5 uses the largest Gain Ratio that ensures a larger than average information gain. Given a data set *D*, and it is split into s new subsets $S = \{D_1, D_2, \dots, D_s\}$:

$$GainRatio(D,S) = \frac{Gain(D,S)}{SplitINFO} \tag{3}$$

$$splitINFO = -\sum_{i=1}^{s} \frac{D_i}{D} \log_2 \frac{D_i}{D} \tag{4}$$

C5.0 (called See 5 on Windows) is a commercial version of C4.5 now widely used in many data mining packages such as *Clementine* and *RuleQuest*. It is targeted toward use with large datasets. The decision tree induction is close to the C4.5, but the rule generation is different. However, the precise algorithms used for C5.0 are not public. One major improvement to the accuracy of C5.0 is based on boosting and it does improve the accuracy. Results show that C5.0 improves on memory usage by about 90 percent, runs between 5.7 and 240 times faster than C4.5, the error rate has been shown to be less than half of that found with C4.5 on some data sets, and produces more accurate rules [3].

2.2 Incremental Decision Induction Trees

An incremental classifier can be characterized as ID3 compatible if it constructs almost similar decision tree produced by ID3 using all the training set. This strategy is maintained by classifiers such as ID4, ID5, ID5R and ITI. ID4 was the first ID3 variant to construct the incremental learning.

ID4 applies the ID3 in an incremental manner to allow objects to be presented one at a time. The heart of this modification lies in a series of tables located at each potential decision tree root. Each table consists of entries for the values of all untested attributes and summarizes the number of positive and negative instances with each value. As a new instance is add into the tree, the positive and negative count for each attribute value is incremented and those count are used to compute the *E-score* for a possible test attribute at a node. Each decision node contains an attribute that has the lowest *E-score* and if the attribute does not contains the lowest *E-score*, then the attribute is replaced by a non-test attribute with lowest *E-score* and sub-trees below the decision node are discarded. ID4 builds the same tree as the basic ID3 algorithm, when there is an attribute at each node that is the best among other attributes.

ID5 expanded this idea by selecting the most suitable attribute for a node, while a new instance is processed, and restructuring the tree, so that this attribute is pulled-up from the leaves towards that node. This is achieved by suitable tree manipulations that allow the counters to be recalculated without examining the past instances.

ID5R is a successor of the ID5 algorithm. When have to change the test attribute at a decision node, instead of discarding the sub-trees, ID5R uses a *pull-up* process to restructure the tree and retains the training instances in the tree. This *pull-up* process only recalculates the positive and negative counts of training instances during the manipulation. An ID5R tree is defined as: A **leaf node (*answer node*)** contains a class name and the set of instance descriptions at the node belonging to the class. A **non-leaf node (*decision node*)** contains an attribute test with a branch to another decision tree for each possible value of the attribute, and a set of non-test attribute at the node. Each test or non-test attribute is combined with positive and negative counts for each possible value. When classifying an instance, the tree is traversed from the root node until a node is reached that contains the all instances from same class. At that point the class label for the instance is assigned either the node is a leaf or non-leaf node.

The basic algorithm of ITI follows the ID5R, but adds the ability to handle numeric variable, instances with missing values, and inconsistent training instances, also handle multiple classes, not just two. Updating the tree, ITI uses two steps: incorporating an instance into the tree and restructuring the tree as necessary so that each decision node contains the best test. When picking a best attribute it uses the Gain Ratio which is described in C4.5. A table of frequency counts for each class and value combination is kept at a decision node and used for ensuring a best test at a decision node and for tree revision.

2.3 Ensemble Methods

Bagging and boosting are first approach they construct multiple base trees, each time using a bootstrapped replicate of the original training data. Bagging [11] is a method for generating multiple decision trees and using these trees to get an aggregated predictor. The multiple decision trees are formed by bootstrap aggregating which repeatedly samples from a data set and the sampling is done with replacement. It is that some instances may appear several times in the same training set, while others may be omitted from the training set. Unlike bagging, boosting [12] assigns a weight to each training example and may adaptively change the weight at the end of each boosting round.

However, bagging and boosting are difficult to be used in incremental induction tree process because of the expensive cost and they have to manipulate the training data. In CS4 algorithm [8], [13], instead of manipulating the training data, it keeps the original data unchanged and but change the tree learning phase. It forces the top-ranked features to be the root of tree and remain nodes are constructed as C4.5. This is called tree cascading, and for classification, the algorithm combines those tree committees and shares the rules in the committee in a weighted manner. Together, the cascading idea guides the construction of tree committees while the sharing idea aggregates the discrimination powers made by each individual decision tree.

3 Incrementally Inducted Decision Tree Ensemble

In bio-medical research mining area, the useful diagnostic or prognostic knowledge from the result is very important. Only explainable result could be analyzed and easy to understand for the application to the bio-medical and diagnosis of a disease [14], [15]. Among many classification approaches, using classification rules derived from the decision tree induction may helpful to perform this work and previous studies show that it is powerful. We define a rule as a set of conjunctive conditions with a predictive term. The general form of rules is presented as*: IF* $condition_1$ *&* $condition_2$ *& ... &* $condition_m$*, THEN a predictive term.* The predictive term in a rule refers to a single class label. For useful clinical diagnosis purpose, using those rules we can address issues in understanding the mechanism of a disease and improve the discriminating power of the rules. A *significant rule* is one with a largest coverage which the coverage satisfies a given threshold. For example, the given threshold is *60%*, if one rule's coverage is larger than *60%* then it is called significant rule.

However, using traditional ensemble method to build and refine the tree committee and derive significant classification rules is still impossible. So, we introduce a new incremental decision learning algorithm which uses the skeleton of ITI and accepts the cascading and sharing ensemble method of CS4 to break the constraint of singleton classification rules by producing many significant rules from the committees of decision trees and combine those rules discriminating power to accomplish the prediction process. We call this algorithm as *ICS4* and the main steps of the process could be 3 as shown in three functions below:

```
incremental_update(node, training_example)
{
    add_training_example_to_tree(node, training_example)
    { Add examples to tree using tree revision; }
    ensure_best_test(node)
    { Ensure each node has desired test ; }
    sign_class_label_for_test_example(test_example)
    {
        if there are test examples
            for each kth top-ranked tests
          //except the first best test
                force the test to installed at root node
                tor remaining nodes
                    ensure_best_test(node);
         from each constructed decision tree
            Derive significant classification rules;
    }
}
```

Algorithm 1. The skeleton of ICS4

Details of *add_training_example_to_tree* and *ensure_best_test* function are shown in [6]. The third function *sign_class_label_for_test_example* only works at the point when there are test examples or unknown instances that need to be assign a class label. For constructing tree committees, there are two options: construct at the point when we need to perform classification or start from the beginning of tree induction and using incremental manner to construct them. However, at the point when the tree committees are constructed, the top-ranked features used are same for both two strategies because the used examples are no difference. It means that using incremental manner to construct the tree committees is just wasting time and storage. After derived those rules, we use the aggregate score to perform the prediction task. The classification score [9] for a specific class, say class *C*, is calculated as:

$$Score^{C}(T) = \sum_{i=1}^{K_C} Coverage(rule^{C}{}_{i}) \tag{5}$$

Here, K_C denotes the number of rules in the class *C*, $rule^{C}{}_{i}$ denotes *i*th rules in class *C*, if the score for one class *C* is larger than other classes, then the class label for the instance *T* is assigned as *C*.

4 Experiments and Results

Breast cancer is a cancer that starts in the cells of the breast in women and men. Worldwide, breast cancer is the second most common type of cancer after lung cancer and the fifth most common cause of cancer death. Breast cancer is about 100 times as frequent among women as among men, but survival rates are equal in both sexes. In this section, we report empirical behavior of the algorithm discussed above by using the data source named Wisconsin Breast Cancer Dataset which is taken from the University of California at Irvine (UCI) machine learning repository [16]. This dataset consists of 569 instances and have 32 features. Here, we only use 20 top-ranked features to construct tree committees.

Table 1. Attribute information

Feature Name	Description
ID number	*Not used for training*
Diagnosis	*M = malignant, B = benign*
3-32 features are: Ten real-valued features are computed for each cell nucleus	
radius	*mean of distances from center to points on the perimeter*
texture	*standard deviation of gray-scale values*
perimeter	
area	
smoothness	*local variation in radius lengths*
compactness	*perimeter^2 / area - 1.0*
concavity	*severity of concave portions of the contour*
concave points	*number of concave portions of the contour*
fractal dimension	*"coastline approximation" - 1*
symmetry	

The mean, standard error, and "worst" or largest (mean of the three largest values) of these features were computed and resulting in 30 features. For instance, field 3 is Mean Radius, field 13 is Radius SE, and field 23 is Worst Radius.

In the following tables, for example, that *50 vs. 50* means that the example data set is divided into 50% of training and 50% of test.

Table 2. Confusion Matrix

50 vs. 50	Predicted	
Actual	Benign	Malignant
Benign	169	10
Malignant	6	100

70 vs. 30	Predicted	
Actual	Benign	Malignant
Benign	114	5
Malignant	4	66

80 vs. 20	Predicted	
Actual	Benign	Malignant
Benign	88	1
Malignant	7	46

All	Predicted	
Actual	Benign	Malignant
Benign	207	5
Malignant	4	353

Table 3. Detailed accuracy by class

	FP rate	Precision	Recall	F-measure	Class
50 vs. 50	0.057	0966	0944	0955	Malignant
	0.056	0909	0943	0926	Benign
80 vs. 20	0132	0926	0989	0957	Malignant
	0011	0979	0868	092	Benign
70 vs. 30	0057	0966	0958	0962	Malignant
	0042	093	0943	0936	Benign
All	0011	0981	0976	0979	Malignant
	0024	0986	0989	0987	Benign

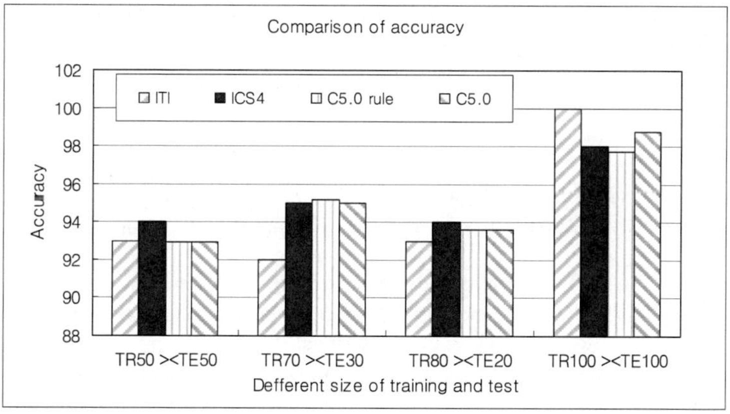

Fig. 1. Comparison of accuracy

At *Figure 1*, ITI and ICS4 are tested in incremental mode, and C5.0 is tested in batch mode (non-incremental) for both rule based and tree based classifier. As shown in above results, the ICS4 algorithms can achieve high performances on different manipulation of the example data and with different types of decision tree learning algorithms. It means that using the Cascading and Sharing method in incremental induction tree to derive significant rules could provide competitive accuracy to incremental induction algorithm and even non-incremental approaches when tested on consistent size of example data. Consider the execution time and storage mechanism, because it just constructs the tree committees on the point of beginning of test or there are new unknown instances, so it can finish the work on acceptable time as ITI.

5 Conclusion

In this paper, base on well-accepted design goals we introduced an approach for discovering many significant classification rules from incrementally induced decision tree by using the Cascading and Sharing ensemble method. Tree induction offers a highly practical method for generalizing from examples whose class label is known and the average cost of constructing the tree committee is much lower than the cost of building a new decision tree committee. For testing the performance of ensembles of incremental tree induction, we used the example data which is about diagnosis of Wisconsin Breast Cancer Dataset. All 31 features are used without variable selection and the threshold of the feature choice was given as 20 to construct the 20 number of trees by forcing 20 top-ranked features iteratively as the root node of a new tree. The first tree is constructed in incremental induction manner, and others are constructed when there are instances that need to be assign a class label. The results show that this new approach is suitable for the bio-medical research.

Acknowledgment. This work is supported by Korea Science and Engineering Foundation (KOSEF) grant funded by the Korea government (MOST) (R01-2007-000-10926-0) and supported by the Korea Research Foundation Grant funded by the Korean Government(MOEHRD) (The Regional Research Universities Program / Chungbuk BIT Research-Oriented University Consortium).

References

1. Quinlan, J.R.: Induction of Decision Trees. Machine Learning, 81–106 (1986)
2. Quinlan, J.R.: C4.5: Programs for Machine Learning. Morgan Kaufmann, San Francisco (1993)
3. RuleQues Research Data Mining Tools, `http://www.rulequest.com/`
4. Schlimmer, J.C., Fisher, D.: A case study of incremental concept induction. In: Proceedings of the Fifth National Conference on Artificial Intelligence, pp. 496–501 (1986)
5. Utgoff, P.E.: Incremental Induction of decision trees. Machine Learning, 161–186 (1989)
6. Utgoff, P.E., Berkman, N.C., Clouse, J.A.: Decision Tree Induction Based on Efficient Tree Restructuring. Machine Learning, 5–44 (1997)
7. Tan, P.N., Steinbach, M., Kumar, V.: Ensemble methods. In: Introduction to data mining, pp. 278–280. Addison Wesley, Reading (2006)
8. Li, J.Y., Liu, H.A., Ng, S.-K., Wong, L.: See-Kiong Ng, Limsoon Wong: Discovery of significant rules for classifying cancer diagnosis data. Bioinformatics 19, 93–102 (2003)
9. Utgoff, P.E.: Decision Tree Induction Based on Efficient Tree Restructuring. Technical report, University of Massachusetts (1994)
10. Tan, P.N., Steinbach, M., Kumar, V.: Decision tree induction. In: Introduction to data mining, pp. 150–172. Addison Wesley, Reading (2006)
11. Breiman, L.: Bagging predictors. Machine Learning 24, 123–140 (1996)
12. Freund, Y., Schapire, R.E.: Experiments with a New Boosting Algorithm. In: The Thirteenth International Conference on Machine Learning, pp. 148–156 (1996)
13. Li, J., Liu, H.: Ensembles of cascading trees. In: Third IEEE international conference on data mining, pp. 585–588 (2003)
14. Lee, H.G., Noh, K.Y., Ryu, K.H.: Mining Biosignal Data: Coronary Artery Disease Diagnosis using Linear and Nonlinear Features of HRV. In: PAKDD 2007 Workshop, BioDM 2007. LNCS. Springer, Heidelberg (2007)
15. Ryu, K.H., Kim, W.S., Lee, H.G.: A Data Mining Approach and Framework of Intelligent Diagnosis System for Coronary Artery Disease Prediction. In: The 4th Korea-Japan Int'l Database Workshop 2008 (2008)
16. UCI Machine Learning Repository, `http://archive.ics.uci.edu/ml/datasets.html`

Application of the Cross-Entropy Method to Dual Lagrange Support Vector Machine

Budi Santosa

Department of Industrial Engineering, Institut Teknologi Sepuluh Nopember Surabaya
60111 Kampus ITS Surabaya, Indonesia
budi_s@ie.its.ac.id

Abstract. In this paper, cross entropy method is used for solving dual Lagrange support vector machine (SVM). Cross entropy (CE) method is a new practical approach which is widely used in some applications such as combinatorial optimization, learning algorithm and simulation. Our approach refers to Kernel Adatron which is solving dual Lagrange SVM using gradient ascent method. Hereby, the cross entropy method is applied to solve dual Lagrange SVM optimization problem to find the optimal or at least near optimal Lagrange multipliers as a solution. As known, the standard SVM with quadratic programming solver suffers from high computational time. Some real world datasets are used to test the algorithms and compare to the existing approach in terms of computation time and accuracy. Our approach is fast and produce good results in terms of generalization error.

Keywords: Cross entropy, generalization error, kernel adatron, Lagrange, Support vector machine, computation time.

1 Introduction

Support Vector Machine (SVM) is an algorithm which is commonly used in data mining for both classification and regression tasks. Basically, SVM is based on the following idea: input points are mapped to a high dimensional feature space where a linear separating hyperplane can be found. To find a separating hyperplane, SVM works by choosing one that maximizes the distance from the closest patterns of two different classes. This is achieved by formulating the problem into a quadratic programming problem which is then usually solved with optimization routines. This step is computationally intensive, required high computing time as the problem getting large. SVM has a proven impressive performance on a number of real world problems.

The Cross Entropy (CE) method is one of the most significant developments in stochastic optimization and simulation in recent years. Firstly, CE was proposed as an adaptive algorithm for rare-event simulation [9]. It was soon realized that the underlying ideas of CE had wider range of application than just rare-event simulation. In the next development, CE could also be applied for solving combinatorial, multi-extremal optimization and machine learning problems [10,2,7].

In this paper, cross entropy method is applied to solve dual Lagrange SVM optimization problem to find the optimal or at least near optimal solution, Lagrange multipliers.

R. Huang et al. (Eds.): ADMA 2009, LNAI 5678, pp. 595–602, 2009.

To judge our approach, the resulting algorithm is applied to some real world datasets with binary output label such as Breast Cancer, WDBC, Pima Indian, Sonar, Bupa and Ionosphere. The experiments shows promising results in terms of computational time and generalization error compared to the standard quadratic programming SVM.

This paper is organized as follows. The second section reviews SVMs and Kernel Adatron (KA) which is the basis of the proposed approach. Section 3 recapitulates the CE method. In section 4, we describe our proposed algorithm. Section 5 explaines the experimental setting and section 6 discusses the results. In section 7, we conclude the results of this research.

2 Support Vector Machines (SVM)

SVM can be explained as follows. Consider a classification problem with two classes of points. The SVM formulation can be written as follows [3],

$$\begin{aligned} &\min_{w,b,\eta} C\sum_{i=1}^{m} \eta_i + \tfrac{1}{2} \| w \|^2 \\ &st \quad y_i (wx_i + b) + \eta_i \geq 1, \eta_i \geq 0, i = 1, \ldots m \end{aligned} \tag{1}$$

where C is a parameter to be chosen by the user. A larger C corresponds to assigning a larger penalty to errors. Introducing positive Lagrange multipliers, α_i, to the inequality constraints in model (1) we obtain the following dual formulation to be minimized:

$$\begin{aligned} &L(\alpha) = \frac{1}{2} \sum_{i=1}^{m} \sum_{j=1}^{m} y_j \alpha_i \alpha_j K(x_i, x_j) - \sum_{i=1}^{m} \alpha_i \\ &\text{st} \sum_{i=1}^{\ell} y_i \alpha_i = 0 \\ &0 \leq \alpha_i \leq C \qquad i = 1, \ldots m \end{aligned} \tag{2}$$

where m is the number of pattern used to train our model. The resulting decision function is

$$f(x) = sign(\sum_{i \in SV} y_i \alpha_i K(x, x_i) + b$$

where α is the solution resulting from Lagrange problem in 2, and b is bias. Solving the problem sometime require high computing time especially for large scale problem. We also need a QP solver for solving this problem. Some algorithms were proposed to cope with this high computing time problem. Hsu and Lin [4] proposed simple decomposition method for bound constrained SVM. Plat proposed SMO to fasten the training phase of SVM [8], Joachim [5] proposed svm light to solve large scale SVMs and many others works had been proposed. Kecman et al. [6] observed the equality between Kernel Adatron and SMO. One of the previous works, what proposed by Frieβ et al. [3] is the simplest one. They used Kernel Adatron (KA) to solve the dual Lagrange SVM problem which basically adopt from gradient ascent method. By introducing kernels into the algorithm it is possible to maximize the margin in the feature space which is equivalent to nonlinear decision boundaries in the input space. The result is a fast robust and extremely simple procedure which implements the same ideas and principles as SV machines at much smaller cost. The kernel *K(x, x')* can be any kernel function. In this paper RBF kernel function is used for doing all experiments.

3 The Cross Entropy Method

Cross entropy is a quite new approach in optimization and learning algorithm. Rubinstein and Kroese. [10], provides complete description on cross entropy method. The basic idea of the CE method is to transform the original (combinatorial) optimization problem to an associated stochastic optimization problem, and then to handle the stochastic problem efficiently by an adaptive sampling algorithm. Through this process one constructs a random sequence of solutions which converges (probabilistically) to the optimal or at least a reasonable solution. Once the associated stochastic optimization is defined, the CE method follows these two phases:

1. Generation of a sample of random data (trajectories, vectors, etc.) according to a specified random mechanism.
2. Update of the parameters of the random mechanism, on the basis of the data, in order to produce a "better" sample in the next iteration.

CE method now can be presented as follows. Suppose we wish to minimize some cost function S(z) over all z in some set Z. Let us denote the minimum value of S by γ^*, thus

$$\gamma^* = \min_{z \in Z} S(z) \tag{3}$$

We randomize our deterministic problem by defining a family of auxiliary pdfs $\{f(\cdot; v), v \in V\}$ on Z and we associate with Eq. (3) the following estimation problem for a given scalar γ:

$P_u(S(Z) \le \gamma) = E_u[I_{\{S(Z)\le\gamma\}}]$ where *u* is some known (initial) parameter. We consider the event "cost is low" to be the rare event $I\{S(Z)\le\gamma\}$ of interest. To estimate this event, the CE method generates a sequence of tuples $\{(\hat{\gamma}_t, \hat{v}_t)\}$, that converge (with high probability) to a small neighborhood of the optimal tuple (γ^*, v^*), where γ^* is the solution of the program (3) and v^* is a pdf that emphasizes values in Z with a low cost. We note that typically the optimal v^* is degenerated as it concentrates on the optimal solution (or a small neighborhood thereof). Let ρ denote the fraction of the best samples used to find the threshold γ. The process that is based on sampled data is termed the stochastic counterpart since it is based on stochastic samples of data. The number of samples in each stage of the stochastic counterpart is denoted by N, which is a predefined parameter. The following is a standard CE procedure for minimization borrowed from Rubinstein and Kroese [10].

We initialize by setting $\hat{v}_0 = v_0 = u$ and choose a not very small ρ, say $10^{-2} \le \rho$. We then proceed iteratively as follows:

1. Adaptive updating of γ_t.

For a fixed v_{t-1}, let γ_t be a $\rho 100\%$-percentile of S(Z) under v_{t-1}. That is, γ_t satisfies $P_{vt-1}(S(Z) \le \gamma_t) \ge \rho$ and $P_{vt-1}(S(Z) \ge \gamma_t) \ge 1-\rho$ where $Z \sim f(\cdot; v_{t-1})$. A simple estimator $\hat{\gamma}_t$ of γ_t can be obtained by taking a random sample Z(1), . . . , Z(N) from the pdf $f(\cdot;v_{t-1})$, calculating the performances $S(Z(\ell))$ for all ℓ, ordering them from smallest to biggest as $S(1) \leq \ldots \leq S(N)$ and finally evaluating the ρ100% sample percentile as $\hat{\gamma}_t = S_{(\lceil \rho N \rceil)}$.

2. Adaptive updating of v_t. For a fixed γ_t and v_{t-1}, derive v_t from the solution of the program

$$\max_v D(v) = \max_v \mathrm{E}_{vt-1} I_{\{S(Z) \leq \gamma t\}} \log f(Z; v) \tag{4}$$

The stochastic counterpart of (4) is as follows: for fixed $\hat{\gamma}_t$ and $\hat{v}_{t-1}$, derive $\hat{v}_t$ from the following program:

$$\max_v \hat{D}(v) = \max_v \frac{1}{N} \sum_{\ell=1}^{N} I_{\{S(Z^{\ell}) \leq \hat{\gamma} t\}} \log f(Z^{(\ell)}; v) \tag{5}$$

We note that if f belongs to the Natural Exponential Family (e.g., Gaussian, Bernoulli), then Eq. (5) has a closed form solution (see [10]). In this paper we will assume that f belongs to a Gaussian family. In our case $Z \in \{0,C\}^n$ and v is an n dimensional vector of numbers between 0 and 1, where C is constant defined by users. The constant C is upper bound of Lagrange multiplier αs which we seek for. The update formula of the kth element in v (Eq. (5)) in this case simply becomes:

$$\hat{v}_t(k) = \frac{\sum_{\ell=1}^{N} I_{\{S(Z^{\ell}) \leq \hat{\gamma} t\}} I_{\{Z_k^{\ell}=1\}}}{\sum_{\ell=1}^{N} I_{\{S(Z^{\ell}) \leq \hat{\gamma} t\}}}$$

This formula has the interpretation that it counts how many times a value of 1 (in $I_{\{Z(\ell)_k = 1\}}$) led to a significant result (matches with the indicator $I_{\{S(Z(\ell)) \leq \hat{\gamma}_t\}}$), how many times a value of 0 led to a significant result, and normalize the value of the parameter accordingly. Instead of the updating the parameter vector v directly via the solution of Eq. (5) we use the following smoothed version

$$\hat{v}_t = \beta \hat{v}_t + (1 - \beta)\, \hat{v}_{t-1}, \tag{6}$$

where $\hat{v}_t$ is the parameter vector obtained from the solution of (5), and β is a smoothing parameter, with $0.7 < \beta < 1$.

Recently, CE had been applied in credit risk assessment problems for commercial banks [12]. Numerical experiments have shown that the cross entropy method has a strong capability to identify the credit risk and it is a good tool for credit risk early warning system. Mannor et al. [7] used cross entropy method to solve support vector machines. Different from other approaches, they use the number of support vectors (the "L_0 norm") as a regularizing term instead of the L_1 or L_2 norms. The experimental results of the method produces generalization errors that are similar to SVM, while using a considerably smaller number of support vectors. Kroese et al. [2] applied the cross-entropy (CE) method to clustering and vector quantization problems.

4 The Proposed Approach

Adopting the idea of gradient ascent method on dual Lagrange SVM as applied in Kernel Adatron, the proposed approach, named CE-SVM, can be explained as follows. The method starts with generate N (number of samples) Lagrange multiplier, α and generate μ and σ values with the size N. Lagrange multiplier α represents Z and parameters μ and σ functioned as v in Section 3. This vector α then inputted to the score function L as denoted in equation 2. L is functioned as *S(z)* in Section 3. The values of L are sorted descendently. *Ne* elite samples selected from the Ne smallest values of L. From these L values, the corresponding α can be identified. If of these α values <=0, as formulated in the second constraint in 2, set α_i=0. If not, then we have to choose min(alpha,C). From the Ne α values, take the average and standard deviation and use smooth updated procedure to compute parameters μ and σ. Use this μ and σ to update α values in the next iteration. The same procedures are repeated until the parameter σ reach specified value epsilon. Fig 1 shows the pseudocode of CE-SVM algorithm.

```
Input:  pattern X, label Y, kernel function, kernel
      parameter, C, Number of sample N and Number of
      clitc sample Ne
Assign beta=1;
Compute kernel Matrix K
Compute H as y(i) * y(j) * K
Generate  initial random vector μ between [0,1],with size
1 x b (b is number of data points)
  Intialize vector σ=1 with size 1 x b
  while absolute value of (sum(σ)) > epsilon do,
    For each data point,
    assign α=mu+σ*normally distribution random number
    For all sample points N
      For all α do
            If α <=0 then set α =0
            Else
            If α >C then set α =C
            EndIf
      Endfor
    For all sample points N
      Compute L(α) = 1/2 Σ_{i=1}^{m} Σ_{j=1}^{m} y_j α_i α_j K(x_i, x_j) − Σ_{i=1}^{m} α_i
      Sort L in descending order
      Select α corresponding to Ne sample points with
      lowest L, and compute the average and standard
      deviation of these α values, denote as ᾱ and σ̄
      Update μ using μ = beta*ᾱ+(1-beta)*μ;
      Update σ using σ = beta*σ̄+(1-beta)*σ;
  Endfor
  Output: α= μ
```

Fig. 1. Pseudocode for CE-SVM algorithm

In this paper the following classifier is used

$$f(x) = sign\,(\sum_{i \in SV} y_i \alpha_i K(x, x_i))$$

where α obtained from implementing algorithm CE-SVM which are nonzero (denoted as support vector, SV). Note that the bias *b* is not used in the classifier.

5 Experiments

In the experiments, the CE based Lagrange SVM (CE-SVM) algorithm was compared to the standard SVM algorithm and Kernel Adatron on six real world of two class problem datasets taken from the UCI repository [17]. For each data set, it is splitted into two samples: training-testing sets where the ratio is 70:30. To have valid results, for each data set, we ran 10 experiments with 10 different pairs of training-testing sample on CE-SVM and SVM. For both SVM and CE-SVM, the same RBF kernel and parameter values, C and σ were used. The generalization error is computed by taking the average of 10 misclassification values. The computational time is computed with the same manner. The experiments were done on core duo processor 2.1 GHz and 2GB RAM.

Table 1. Six real world data sets

No.	Name	# Features	# Patterns
1	Breast cancer	9	683
2	Ionosphere	34	351
3	Pima	8	768
4	Wdbc	30	569
5	Bupa	6	345
6	Sonar	60	208

6 Results and Discussion

The experimental results are summarized in Tabel 2. We ran on six datasets with the same C and σ values for both SVM and CE-SVM. The results show, in general CE-SVM produced the comparable accuracy even better than SVM and Kernel Adatron (KA) yet with faster computational time than SVM. Our purpose of this paper is to tackle the high computing time of using standard SVM. Since we do not need to solve the optimization problem analytically, the computational time can be reduced significantly. Different from what Mannor et al. did [11], which more focused on reducing the number of support vectors through L_0-norm SVM formulation, in this paper we concern more on the computational time. Using standar SVM with L_2-norm, the computation is highly intensive since we have to solve the problem through a quadratic programming. Using CE-SVM, we do not require to utilize any optimization solver.

The main part of the algorithm is how to update αs using specific machanisme through parameters μ and σ by utilizing Gaussian distribution. The updating mechanism mostly affected by the values of fitness function and the corresponing α values. In this context, σ which is used in updating αs is different with those used in RBF kernel. As mentioned in [6], in the classifier function *f(x)* we did not employ bias or set bias to 0. Based on the experiments, the most significant parameter in finding the α values is the kernel parameter σ. Though not shown in this paper, the experimental results indicate that the value of C did not affect much on the accuracy.

Table 2. Comparison of Misclassification Error with Gaussian/rbf kernel

No	Data	Method	Parameter C and σ for RBF kernel	Misclassification average (%)	Computation Time Average (CPU)
1	Breast cancer	SVM	C=.1, σ=1	5.39	15.1
		CE-SVM		**0.49**	**0.21**
		KA		0.49	11.8
2	WDBC	SVM	C=.1, σ=1	8.77	11.9
		CE-SVM		**8.77**	**0.0**
		KA		8.77	0.0
3	Pima	SVM	C=.5, σ=20	24.94	24.01
		CE-SVM		25.97	**0.47**
		KA		**25.11**	0.7
4	Ionosphere	SVM	C= .5 , σ=1	1.90	5.19
		CE-SVM		1.90	0.2
		KA		**1.90**	**0.15**
5	Bupa	SVM	C=.1, σ=10	32	3.16
		CE-SVM		**31**	**0.0**
		KA		**31**	**0.0**
6	Sonar	SVM	C=.1, σ=.5	30.65	1.1
		CE-SVM		**30.65**	**0.0**
		KA		**30.65**	**0.0**

Preliminary results of using polynomail kernel functions without preprocessing the input patterns, indicated poor accuracy. Normalizing the input patterns improved the accuracy signifantly. This finding confirms the investigation in [1]. Because of limited space, the rerults of using polynomial kernel are not presented.

7 Conclusions

We have presented an algorithm cross entropy method to solve dual Lagrange SVM problem. The main advantage of applying CE on SVM is the generalization performance is comparable to SVM while the computational time to find the classifier is significantly lower. The method does not require any optimization routine to find optimal or near optimal solutions, αs. Testing on six real world datasets prove that the proposed method shows promising results. More investigation by applying other kernels rather than RBF and other datasets might reveal interesting insight on the application of CE on SVM.

Acknowledgments. This research is supported by Higher Education Directorate of Indonesia Education Ministry 2008.

References

1. Chin, K.K.: Support Vector Machines applied to Speech Pattern Classification, Dissertation, Darwin College, University of Cambridge (1998)
2. Kroese, D.P., Rubinstein, R.Y., Taimre, T.: Application of the Cross-Entropy Method to Clustering and Vector Quantization. Journal of Machine Learning Research (2004)
3. Frieb, T.T., Christianini, N., Campbell, C.: The Kernel Adatron Algorithm: a Fast and Simple Learning Procedure for Support Vector Machines. In: Shavlik, J. (ed.) Proceedings of the 15th International Conference on Machine Learning, pp. 188–196. Morgan Kaufmann, San Francisco (1998)
4. Hsu, C.W., Lin, C.J.: A Simple Decomposition Method for Support Vector Machines 46(1-3), 291–314 (2002)
5. Joachims, T.: Making large-scale support vector machine learning practical. In: Advances in kernel methods: support vector learning. MIT Press, Cambridge (1999)
6. Kecman, V., Vogt, M., Huang, T.M.: On the Equality of Kernel AdaTron and Sequential Minimal Optimization in Classification and Regression Tasks and Alike Algorithms for Kernel Machines. In: ESANN 2003 proceedings, Belgium, pp. 215–222. d-side publi. (2003)
7. Mannor, S., Peleg, D., Rubinstein, R.Y.: The cross entropy method for classification. In: Proceedings of the 22nd International Conference on Machine Learning, Bonn, Germany (2005)
8. Platt, J.C.: Fast training of support vector machines using sequential minimal optimization. In: Advances in kernel methods: support vector learning, pp. 185–208. MIT Press, Cambridge (1999)
9. Rubinstein, R.Y.: Optimization of computer simulation models with rare events. European Journal of Operations Research 99, 89–112 (1997)
10. Rubinstein, R., Kroese, D.: The cross-entropy method: A unified approach to combinatorial optimization, Monte-Carlo simulation, and machine-learning. Springer, Heidelberg (2004)
11. UCI Repository (2009), `http://www.ics.uci.edu/~mlearn/mlrepository.html`
12. Zhou, H., Wang, J., Qiu, Y.: Application of the Cross Entropy Method to the Credit Risk Assessment in an Early Warning System. In: International Symposiums on Information Processing (2008)

A Predictive Analysis on Medical Data Based on Outlier Detection Method Using Non-Reduct Computation

Faizah Shaari[1,*], Azuraliza Abu Bakar[2], and Abdul Razak Hamdan[2]

[1] Polytechnic Sultan Idris Shah, Polytechnics Malaysia, Ministry of Higher Education, Sungai Lang, 45100 Sungai Air Tawar, Selangor DE
faizahshaari@gmail.com
[2] Center of Artificial Intelligence Technology, Faculty of Information Science and Technology, National University of Malaysia, 43600 Bangi, Selangor DE
{aab,arh}@ftsm.ukm.my

Abstract. In this research, a new method to predict and diagnose medical dataset is discovered based on outlier mining method using Rough Sets Theory (RST). The RST is used to generate medical rules, while outliers are detected from the rules to diagnose the abnormal data. In detecting outliers, a computation of set of attributes or known as Non-Reduct is proposed by proposing two new formula of Indiscernibility Matrix Modula(iDMM D) and Indiscernibility Function Modulo (iDMFM D) based on RST. The results show that the proposed method is a fast detection method with lower detection rate. In conclusion, the computation of the Non-Reduct is expected to give medical knowledge that able to predict abnormality in dataset that could be used in medical analysis.

Keywords: Non-Reduct, abnormal, outliers, non-interesting, redundant, superfluous, rare.

1 Introduction

In medical databases or datasets, relationships and patterns within this data could provide new medical knowledge. Analysis of medical data is often concerned with treatment of incomplete knowledge, with management of inconsistent pieces of information and with manipulation of various levels of representation of data. According to [1], the most commonly intelligent techniques used for predictive analysis in medical are neural network, bayesian classifier, genetic algorithm, decision trees and fuzzy theory. Rough sets theory (RST) has been proposed by [1] to discover the knowledge of breast cancer data using classification approach.

In this paper, RST is used to diagnose medical information based on outlier mining approach. The outlier detection method based on RST is expected to be able to predict the symptom of sickness of a patient from the medical data. A new method is

* Corresponding author.

R. Huang et al. (Eds.): ADMA 2009, LNAI 5678, pp. 603–610, 2009.

proposed for detecting outlier in medical datasets by discovering the concept of Non-Reduct from the RST approach. In computing Non-Reduct, a new concept is hereby defined in calculating the Indiscernibility Matrix Modulo(iDMM *D*) and Indiscernibility Function Modulo(iDFM *D*). The foundation of these concepts can be found in [14,15]. The organization of this paper is as follows. Section 2 discusses the related works in outlier detection methods, section 3 introduces the concept of RST and the computation of the new proposed Non-Reduct. Section 4 describes the experimental design and results tested upon four medical datasets. The effectiveness of performance is measured by comparing the three methods using Detection rate and F-Measure. The conclusion of this paper is discussed in section 5.

2 Outlier Detection Methods

Finding outliers in large dataset has drawn increasing attention among researchers [2-13]. Although many techniques have been proven useful and effective in detecting outlier pattern, the following problems which occurred remain for further explorations among the data mining researchers. As data change its size and dimension, it is found that most algorithms developed faced the problems of handling the in-scalability of the dataset. The curse of dimensionality has caused the using of distances of points inappropriate to discover outliers in high dimensional space [5]. The concept of locality [6], [7] becomes difficult as data become sparse in high dimensional datasets.

The projection in lower density [6] fails to detect outliers in different projections. The clustered-based is found as a method which detect outliers as by product which does not able to interpret the abnormality of the outliers detected [7]. Although the problems of inefficiencies can be improved by hybriding [2] two techniques or more, yet this method is still in study and further research are indeed needed until today. The Frequent Pattern method [8] utilized the frequent patterns in different subspaces to define outliers in high dimensional space however the detection process is time consuming and computationally expensive. In Local Search Algorithm(LSA) [12], the detection of outliers proves able to be detected from feasible solution based on Optimization approach, however the process is reported by He et al. [11] as time consuming on very large datasets. In comparison, the Greedy Algorithm [11], is found faster in order of magnitude than the previous LSA method. One interesting paper is by [13] whom proposed Outlier detection based on Rough Sets Theory. The outlier detection is based on approximation theory where outliers are detected in the boundary region. Outliers are objects which are categorized with certainty as belonging or not to x, where x is a subset of the universe and any equivalence relation on the universe. The computation of outliers is based on exceptional degree of minimal exceptional sets. On the hand the new proposed method which is also based in Rough Sets Theory uses the computation of Non-Reduct.

3 Rough Sets Theory

Rough sets offers two different kinds of knowledge representations called information system and decision system. An information system (IS) is the most basic kind of knowledge that consist of a set of objects where each object is a collection of attributes values. A DS, $\mathcal{A}$=(*U, A, {d}*) is an IS for which the attributes are separated into disjoint sets of condition attributes *A* and decision attributes $d(A \cap \{d\} = \phi)$. The properties of the decision classes are the particular interest. The discernibility matrix modulo *D* (DMM *D*) of $\mathcal{A}$, M_B^d, is $m_B^d(i, j)$ that is the set of attributes that discerns between objects x_i and x_j and also discerns the decision attributes δ where $\delta(x_i) \neq \delta(x_j)$. The discernibility function matrix modulo *f* (DFM *D*) can be calculated from M_B^d. For an object $x_i \in U$, the *object related* discernibility function modulo decision *f* is defined as $f_B^d(x_i) = \underset{x_j \in U}{\wedge\vee} m_B^d(i, j)$, where $m_B^d(i, j)$ is the element matrix of row *i* and column *j* [15]. Reduct is an important part of an IS which can discern all objects that are discernible by the original IS.

3.1 Outlier Detection Method Based on Computation of Reduct in RST

In the following sub-sections, the new concept of Non-Reduct is introduced.

The Concept of Non-Reduct

The concept of Non-Reduct introduced in this section is originated from the concept of Reduct. In RST, a DS is similar to an IS, but a distinction is made between the condition and the decision attributes. A simple DS with distribution of equivalence classes is as shown in Table 1 [15].

Table 1. Example of Equivalence Class of a Decision System (DS)

Class	a	B	C	Decision	Num. of Objects
E1	1	2	3	1	50
E2	1	2	1	2	5
E3	2	2	3	2	30
E4	2	3	3	2	10
E5,1	3	5	1	3	4
E5,2	3	5	1	4	1

As mentioned in section 3, the discernibility function *f* which determined Reduct is computed from the process of DMM *D* and DFM *D*. Similarly, Non-Reduct can be computed using a new formulation of iDMM *D* and iDFM *D*. Below topics explain and the creation of iDMM *D*, iDFM and the Non-Reduct.

Indiscernibility Matrix Modulo Decision (iDMM *D*)

To compute Non-Reduct, the calculation of iDMM *D* is to find a set of attributes from every pair of equivalence classes which are indiscern in attribute values as well as indiscern in the decision attributes from the matrix. In obtaining the set of attributes as mentioned above, iDMM *D* is hereby defined as in Definition 1 below:

Definition 1

In analogy to the definition of DMM *D* as in section 3, the *indiscernibility matrix modulo D* of $\mathcal{A}$, M'^{d}_{B}, is defined as follows where $m'^{d}_{B}(i, j)$ is the set of attributes that indiscerns between objects x_i and x_j and also indiscerns the decision attributes δ where $\delta(x_i) = \delta(x_j)$ where $1 < i, j < n = |U / IND(B)|$ as shown in Eq.(1) below.

$$m'^{d}_{B}(i, j) = \{ a \in B : a(x_i) = a(x_j) \} \tag{1}$$

Table 2 illustrates the iDMM *D* from a decision system $\mathcal{A}$. The simplification of the disjunction and conjunction of the matrix gives the Indiscernibility function modulo *D* (iDFM), f' as shown in the rightmost column in the table. In next topic, the new iDFM *D* is described.

Table 2. Indiscernibility Matrix Modulo(iDMM) from Decision System, $\mathcal{A}$

	E1	E2	E3	E4	E5	*f'*
E1	{}	{}	{}	{}	{}	-
E2	{}	{}	{b}	{}	{}	b
E3	{}	{b}	{}	{a,c}	{}	(a,b)(a,c)
E4	{}	{}	{a,c}	{}	{}	(a,c)
E5	{}	{}	{}	{}	{}	-

Indiscerniblity Function Modulo *D*(iDFM *D*)
In analogy to the definition of DFM *D* as in the section 3, iDFM *D* is defined as in Definition 2 below:

Definition 2

Let $\mathcal{A}=(U, A, \{d\})$ be an IS, $B \subseteq A$ a set of attributes, M'^{d}_{B} is the indiscernibility matrix modulo decision for that DS. For an object $x_i \in U$, the *object related* indiscernibility function modulo decision *f'* is defined as in Eq(2) below:

$$f'^{d}_{B}(x_i) = \wedge\vee_{x_j \in U} m'^{d}_{B}(i, j) \tag{2}$$

where $m'^{d}_{B}(i, j)$ is the element matrix of row *i* and column *j*.

The following topic presents the definition of Non-Reduct.

Definition of Non-Reduct
Reduct is used as defined in section 3. Let us define the very intuitive definition of Non-Reduct as follows:

Definition 3. (Non-Reduct)

Given $\mathcal{A} = (U,A)$, let $B \subseteq A$, let Reduct of B is a set of attributes $B' \subseteq B$ such that all attributes $a \in B\text{-}B'$ are dispensable, and $IND(B') = IND(B)$. A Non-Reduct of B is defined as, there exist a set of attributes $B\text{-}B' \subseteq B$, such that all attributes $a \in B'$ are indispensable, and $IND(B\text{-}B') \neq IND(B)$. The set of Non-Reduct of B is denoted *Non-Red(B)*.

3.2 Detection of Outliers

The approach as in He et al. [10] to identify and detect outlier is followed. An outlier factor value is defined as number of support of all rules in each object over all number of rules generated from Non-Reduct. A Rough Set Outlier Factor value is proposed for every equivalence class as Definition 4 below:

Definition 4. (RSetOF - Rough Set Outlier Factor)

Let $DS = \{E_1, E_2, \ldots, E_n\}$ be a decision system(DS) consists of set of equivalence classes with decision rules described by $\alpha \to \beta$ from Non-Reduct. The term support($\alpha.\beta$) is support for the decision rules $\alpha \to \beta$ that is the total number of rules in each equivalence class of a DS that are being described by α with decision β. Given a threshold known as minimum support or *minsupport*, [15] the set of all decision rules described by $\alpha \to \beta$ from Non-Reduct obtained in a DS, which satisfy the threshold value, is denoted as: *RSet*(*DS*, *minsupport*). For each equivalence class *E*, the *Rough Set Outlier Factor of E* is defined as in Eq. (3) below:

$$RSetOF(E) = \frac{\sum_{\alpha.\beta} \text{support}\left(\alpha.\beta\right)}{\left\| RSet\left(DS, \text{minsupport}\right) \right\|}, \text{ where} \tag{3}$$

$$\alpha.\beta \subseteq \text{E} \;\text{ and } \alpha.\beta \in RSet(DS, minsupport)$$

The formula in Eq.(3) above is defined as follows: if an equivalence class *E* contains more support in the rules, its *RSetOF* value will be large, which indicates that the object is unlikely to be an outlier. In contrast, equivalence class with small support in its *RSetOF* values is likely to be outlier. Two benchmark measurements to detect outliers are applied which follows the work by He et al. [10] & Hawkins et al. [9]. The *top-ratio* measure is number of equivalence classes specified as top-*n* outliers to that of number of equivalence classes in the dataset. The small the value of top-*n* ratio indicates the shorter time in searching of outliers, hence indicates fast speed in detection of outliers. The fast speed detection is referred as low detection rate. Therefore, the speed of detection translates time that is represented by detection rate. The second measure, the *coverage ratio* is defined as number of detected rare classes to that of the number of rare classes in the dataset. The following section 4, describes and explains the experimental design and the results.

4 Experimental and Results

The experimental design is described by explaining and describing the medical datasets chosen for the experiment. In the following subsection 4.1, all datasets are being described and prepared. In subsection 4.2, the experimental results are presented.

4.1 Medical Data Descriptions and Preparation

Four medical datasets from Machine Learning Repository [16] are chosen for testing the effectiveness of the proposed model. These datasets are selected with the idea that they are suitable in predicting abnormality of the rare cases in medical datasets. Each dataset describes the profiles of patients with symptoms of sickness or being healthy.

Table 3. The list of five datasets obtained from UCI Machine Learning Repository and their class distribution. The datasets are in common and rare distributions and for each casethe percentages of records and in equivalence class are shown.

Dataset	Case (Classes)	Class Codes	Records %	Dim.	Eq.Class %	Eq. Class	Total
BRE	Common	2	91.93	10	84.46	212	251
	Rare	4	8.07		15.53	39	
LYM	Common	2,3	95.94	19	95.94	142	148
	Rare	1,4	4.24		4.24	6	
CLV	Common	0,1,2,3	95.55	14	95.54	279	292
	Rare	4	4.45		4.45	13	
HDE	Common	1	93.70	14	94.96	131	139
	Rare	2	6.25		6.47	8	

4.2 Experiment Results

The evaluation of performance is conducted by comparing the RSetAlg method with two other methods referred as FindFPOF and GreedyAlg chosen from the literature review. In this work, the notion detection rate is the preference used to explain the detection of outliers based on top ratio with 100% CR determined. The results from the experiments conducted shows that the RSetAlg method has better detection rate compared to both the FindFPOF and GreedyAlg methods, when tested upon the four datasets. The 100% high score from the total datasets indicates that RSetAlg has a lower detection rate at small percentages of top ratio. The results show that RSetAlg method able to diagnose any symptom of sickness of a patient from the medical information kept in a dataset. The lower detection rate indicates fast detection that may help medical officer to react at fast speed towards patients' survivals.

Table 4. Results Analysis: Comparison of three methods based on Detection Rate

Outlier at 100% CR	**Detection Rate**		
	FindFPOF	RSetAlg	GreedyAlg
BRE	14.00%	8.07%	12.00%
LYM	8.11%	4.05%	4.73%
CLV	53.64%	4.30%	59.60%
HDE	11.88%	6.25%	15.63%

The F-measure is adopted as another metric in evaluating the performance of the three mentioned methods. The measure should be able to give prominent results for positive prediction in the mining of rare cases problem or class imbalanced where outliers can be detected. As shown in Table 5, the results show that the performance of RSetAlg is encouraging for the four (4) datasets when compared to FindFPOF and GreedyAlg. The results show that RSetAlg method able to predict abnormal behaviour in a patient based on the information found in the medical dataset. The high score prediction, promisingly help medical officer to take positive actions on the patient having discover the symptom of sickness obtained based on the medical dataset.

Table 5. Results Analysis: Comparison of three methods based on F-Measure

Outlier at 100% CR	**F- Measure**		
Dataset	FindFPOF	RSetAlg	GreedyAlg
BRE	0.8478	1	0.6170085
LYM	0.6666	1	0.7500699
CLV	0.1665	1	0.1004547
HDE	0.7064	0.8421	0.4496092

5 Conclusion

The proposed RSetAlg method has shown high performance in prediction of rare cases. The results obtained suggest that the proposed method RSetAlg is a competitive method with lower detection rate, hence can be considered as desirable and effective in predictive modeling in mining rare cases. A predictive medical analysis can be produced from the valid and promising method of outliers detection. The capabilities allow advantages in medical diagnosis as the detection rate helps in giving fast speed in diagnosing symptom of sickness of a patient with abnormal data detected. While the highly positive prediction of abnormal cases of a patient give correct decision taken towards patient.

References

1. Hassanien, A.E., Ali, J.M.H.: Rough Sets Approach for Generation of Classification Rules of Breast Cancer Data. Informatica 15(1), 23–28 (2004)
2. Hodge, V.J., Austin, J.: A Survey of Outlier Detection Methodologies. Artificial Intellingence Review (22), 85–126 (2004)
3. Knorr, E.M., Ng, R.T.: Algorithms for Mining Distanced-Based Outliers in Large Datasets. In: 24th VLDB Conference, New York, pp. 392–403 (1998)
4. Breunig, M.M., Kriegel, H.P., Ng, R.T., Sander, J.: OPTICS-OF: Identifying Local Outliers. In: SIGMOD US, pp. 93–104 (2000)
5. Chiu, A.L.-m., Fu, A.W.-c.: Enhancements on Local Outlier Detection. In: Seventh International Database Engineering and Applications Symposium (IDEAS 2003) (2003)
6. Aggarwal, C.C., Yu, P.S.: Outlier Detection for High Dimensional Data. In: SIGMOD 2001, Santa Barbara, pp. 37–46 (2001)

7. He, Z., Xu, X., Deng, S.: Discovering Cluster Based Local Outliers. Pattern Recognition Letters (2003)
8. He, Z., Huang, J., Xu, Z.X., Shengchun, D.: A Frequent Pattern Discovery Method for Outlier Detection. In: Li, Q., Wang, G., Feng, L. (eds.) WAIM 2004. LNCS, vol. 3129, pp. 726–732. Springer, Heidelberg (2004)
9. Hawkins, S., He, H., Williams, G.J., Baxter, R.A.: Outlier detection using replicator neural networks. In: Kambayashi, Y., Winiwarter, W., Arikawa, M. (eds.) DaWaK 2002, vol. 2454, pp. 170–180. Springer, Heidelberg (2002)
10. Williams, G., Baxter, R., He, H., Hawkins, S., Gu, L.: A Comparative Study of RNN for Outlier Detection in Data Mining. In: 2nd IEEE Inf. ICDM 2002, Japan, pp. 709–712 (2002)
11. He, Z., Deng, S., Xu, X., Huang, J.: A Fast Greedy Algorithm for Outlier Mining. In: Ng, W.-K., Kitsuregawa, M., Li, J., Chang, K. (eds.) PAKDD 2006. LNCS, vol. 3918, pp. 567–576. Springer, Heidelberg (2006)
12. He, Z., Deng, S., Xu, X.: An Optimization Model for Outlier Detection in Categorical Data. In: Huang, D.-S., Zhang, X.-P., Huang, G.-B. (eds.) ICIC 2005. LNCS, vol. 3644, pp. 400–409. Springer, Heidelberg (2005)
13. Jiang, F., Sui, Y., Cao, C.: Outlier Detection using Rough Sets Theory. In: Ślęzak, D., Yao, J., Peters, J.F., Ziarko, W.P., Hu, X. (eds.) RSFDGrC 2005. LNCS, vol. 3642, pp. 79–87. Springer, Heidelberg (2005)
14. Mollestad, T., Komorowski, J.: A Rough Set Framework of Prepositional for Mining Default Rules. Springer, Heidelberg (1998)
15. Bakar, A.A., Sulaiman, M.N., Othman, M., Selamat, M.H.: IP Algorithms in Compact Rough Classification Modeling. Intelligent Data Analysis 5(5), 419–429
16. Murphy, M.P.: UCI Machine Learning Repository (online) (retrieved) (1995), `http://www.ics.uci.edu/~mlearn/MLRepository.html` (March 1, 2005)

VisNetMiner: An Integration Tool for Visualization and Analysis of Networks

Chuan Shi, Dan Zhou, Bin Wu, and Jian Liu

Beijing Key Laboratory of Intelligent Telecommunications Software and Multimedia
Beijing University of Posts and Telecommunications,
100876 Beijing
{shichuan,wubin}@bupt.edu.cn, zhoudanqw@gmail.com,
bluemoqi@yahoo.com.cn

Abstract. As the social network analysis has gradually been applied in many scientific fields, a tool for visualizing and analyzing large-scale networks is urgently needed. VisNetMiner is an integration tool for visualization and analysis of networks, which aims at integrating data preprocessing, modeling, analysis and visualization. This tool can analyze different kinds of data and builds a clear architecture for network analysis and visualization. Moreover, VisNetMiner has the following characteristics: 1) organizing the structural data with a uniform and flexible multi-dimensional data model; 2) providing comprehensive methods for network analysis and visualization; 3) assisting users manage the analytic processes with the workflow management. The experiments verify the effectiveness of VisNetMiner with an academic collaboration network.

1 Introduction

From the Internet to the WWW, from the author collaboration network to many relation networks for economy and society, it clearly shows that people have been living in a world which consists of various complex networks [1]. The structure of complex network is usually very complicated and difficult to understand. If we only use tables or texts to indicate networks, the implied information will not be accessed well. Undoubtedly, visualization is the best method to show complex networks intuitively [2]. For the moment, many scientific researchers need combine technologies of visualization and theories of complex network to analyze some actual phenomena. However, many experts in other fields are not familiar with information technology and thus unable to concentrate their energies on solving their problems. So it is very important to develop a visual analytic tool.

There have been several tools for complex network analysis and visualization such as ArnetMiner [3], CFinder [4]. However, these tools usually face the following problems: 1) it is difficult to define a uniform data model to analyze different kinds of data; 2) the functions of analysis and visualization for networks are not comprehensive; 3) the operations in these systems may be complicated for users with little computer experiences due to absence of necessary wizards.

R. Huang et al. (Eds.): ADMA 2009, LNAI 5678, pp. 611–618, 2009.

In this paper, we develop a Tool for Visualizing Network and Mining (called VisNetMiner). This tool proposes a uniform multi-dimensional data model to support the multi-dimensional analysis of networks. It integrates powerful analytic methods and visualization technologies. In addition, the workflow management is used for providing an integrated analytic process.

2 Related Work

A number of tools have been designed for the social network analysis and visualization. For example, ArnetMiner [3], PaperLens [5], and CiteSpace [6] aim at data analysis and visualization in academic networks. These systems provide powerful statistical analysis and search services on authors, papers and conferences. However, it is very difficult to use them to analyze data in other fields.

Some tools focus on special analysis functions. For example, C-Group [7] is a visual analytic tool for pairwise analysis of dynamic group membership over time, and CFinder [4] is a fast program locating and visualizing overlapping, densely interconnected groups of nodes in undirected graphs.

Besides, some other tools can not only analyze networks in various fields, but also provide powerful capabilities of data analysis and visualization. The Network Workbench (NWB) [8] develops a large-scale network analysis, modeling, and visualization cyberinfrastructure for biomedical, social science, and physics researches. The Information Visualization CyberInfrastructure (IVC) [9] is a toolkit for information visualization, which includes many basic tools such as Prefuse [10]. Although these tools are comprehensive and powerful, the independent analytic functions cause the absence of integrated analytic process.

3 System Architecture

In this section, we will present the architecture of VisNetMiner. To make VisNetMiner more flexible, reusable and understandable, we divide the framework into different layers based on their functions. As illustrated in Fig. 1, the system mainly consists of four layers.

1. **Data Source Layer.** This system can handle different kinds of data sources including database, text file and GraphXML. A uniform access interface shields the details of the underlying data.
2. **Data Model Layer.** Data modeling has two steps: 1) data preprocessing is executed to collect and clean data; 2) network extraction transforms different data into the multi-dimensional model. In addition, this layer includes another important module: data configuration management. It is responsible for managing the configuration information of data preprocessing and network extraction.
3. **Visual Analysis Layer.** This Layer aims at analysis and visualization of multi-dimensional data model. On the one hand, there are three analysis modules: multi-dimensional analysis, statistical analysis, and social network analysis. On the other hand, there are three visualization modules: chat visualization, network visualization, and visualization filter. The analysis and visualization are

correlative dependence and interplay. The analysis provides data for visualization and inversely visualization guides the further analysis. Similarly, this layer also includes visual analysis configuration management.

4. **User Access Layer.** In order to improve flexibility of system, we design C/S and B/S model. The C/S model is implemented with more efficiency, but the B/S model is relatively flexible and can be easily modified.

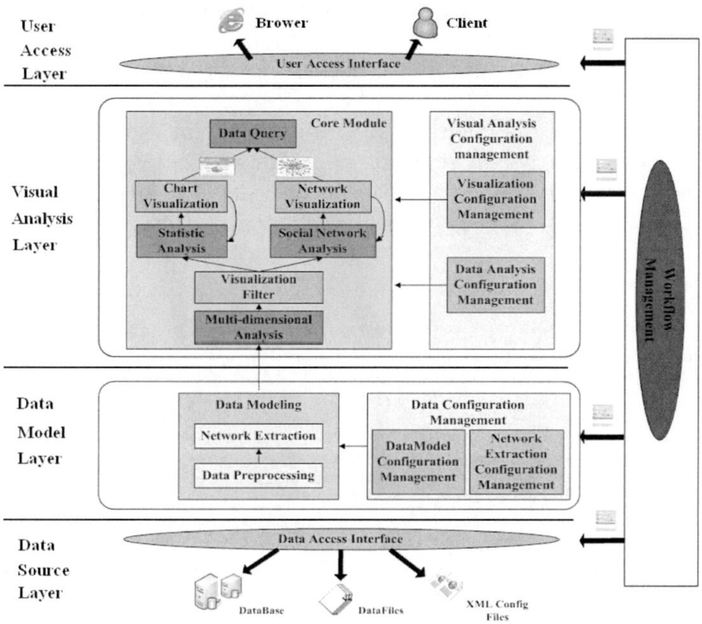

Fig. 1. The architecture of VisNetMiner

On the right part of Fig. 1, the workflow management supports different layers as an integrated tool. It is mainly responsible for coordinating data processing, analysis and visualization in an analytic process.

4 Design and Implementation of VisNetMiner

4.1 Data Model

Traditionally, the information of network is usually represented based on some simple data tables. However, this method isn't sufficient to represent the whole information of networks. We propose a uniform multi-dimensional data model for analyzing networks from different perspectives of entities and their relationships. VisNetMiner concentrates on analyzing relationships among entities instead of analyzing the entities from different dimensions as in data warehouse. Fig. 2 takes an example to illustrate our data model. The original data sources include comprehensive information of authors and papers. They have been changed into two entity dimension tables and two relationship measurement tables. We can select different relationships under different dimensions of entities to do multi-dimensional analysis. For example, when we choose EntityTable2 and Relation2, the entity is representative of province and the relationship represents two provinces have some authors with the same published papers. We can also choose more than one relationship measurements to analyze the network under a dimension of entities, in which two entities have multiple relationships. In addition, the network extraction configuration management module provides the configuration wizards for users to define extraction rules. According to these rules, users can get the appropriate multi-dimensional data model which accords with their analytic demands. After that, the data model is stored by the data model configuration management module.

Here we summarized four contribution of this multi-dimensional data model as follows: 1) this model remains the semantic information of network; 2) it can shield the diversity and complexity of data; 3) it also provides a solid foundation for multi-dimensional analysis; 4) it speeds up the data query and analysis.

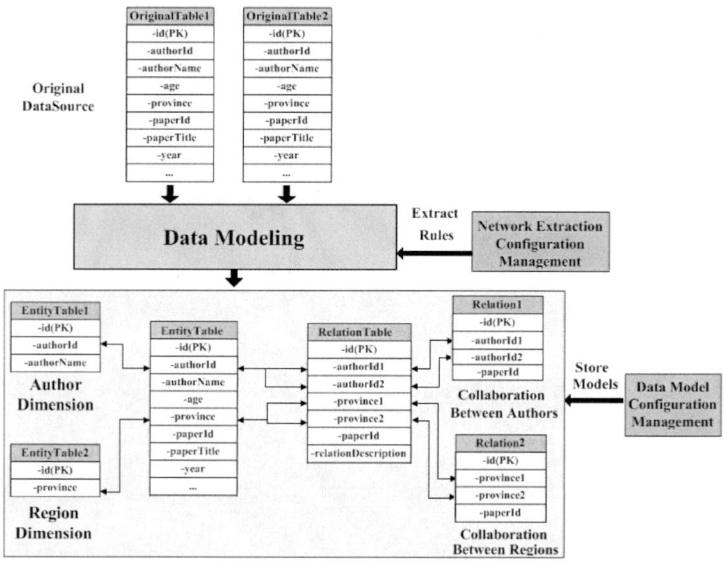

Fig. 2. Multi-dimensional data model. The entity table (EntityTable) and the relation table (RelationTable) describe the base information of entities and relationships which are extracted from OriginalTable1 and OriginalTable2. EntityTable1 and EntityTable2 which are derived from EntityTable describe two dimensions of author and region. Derived from RelationTable, Relation1 and Relation2 represent different relationships under different dimensions of entities.

4.2 Visualization

After the data processing, users can visualize the networks. To make the visualization functions more flexible, VisNetMiner has adopted many technologies such as Prefuse [10] and JUNG 2.0 [11]. Our system can display many kinds of networks, such as the undirected graph, the multi-model graph etc. Additionally, JFreeChat class library is used to display pie, bar and line charts for the statistic analysis. This system also has many convenient functions. For example, the multi-windows are realized for displaying multi-graphs; the transformation of graphs among different dimensions can be conveniently done by mouse operations.

4.3 Data Analysis

4.3.1 Statistic Analysis

Statistic analysis mainly focuses on providing users with the statistic information of the target data. There are two kinds of indicators needing statistic analysis: 1) the attributes of entities and relations; 2) static characteristics of networks such as clustering coefficient, degree distribution etc. These statistic results can be visualized by charts.

4.3.2 Social Network Analysis

Social network analysis is the study of social networks to understand their structure and behavior.We will briefly introduce three social network analysis methods used in our system.

1. **Static Characteristics Analysis.** There are many static characteristics such as the average path length, clustering coefficient, connected components etc. Different static characteristics represent different meanings. For example, the average path length is used to describe the separability of the network.
2. **Community Detection.** Revealing the community structures does much favors to understand the structures and characteristics of networks. Many algorithms have been employed by our system, such as GN [12] and K-Clique-Community Detection Algorithm.
3. **Egocentric Network.** As the scale of networks is getting larger and larger, reducing size of graphs becomes very important. The egocentric network analysis technology which examines only a node's immediate neighbors and associated interconnections can help users to focus on the related information of this node in network.

4.4 Workflow Management

Many network analyses have the similar procedures. These procedures include data preprocessing, network extraction, data analysis and visualization. They can be represented as a configurable workflow module and do cooperative work. The workflow management is responsible for coordinating these configurable modules and provides users with a clear view of network analysis. It also provides UI to make users easily operate. Users can simply modify the configuration files to change the internal states of these processes.

5 Case Study

In this section, we will use VisNetMiner to explore academic collaboration networks and get the collaboration patterns between authors. The data set used is from the records of papers in "Chinese Journal of Virology" spanning from 1999 to 2004.

5.1 Building Data Model

The multi-dimensional data model in this case study has been illustrated in Fig. 2. From OriginalTable1 and OriginalTable2, we want to analyze the co-author network first. Then, we will construct the collaboration network between regions which is called co-region network for short. The evolution of co-author network will be analyzed finally.

5.2 Analyzing Co-author Network

As Fig. 3 (A) shows, there are 2083 authors and 12279 co-author times in this co-author network. The network is too large, so we only select the largest connected subgraph to analyze. And then we use the GN algorithm to detect communities in this subgraph. As shown in Fig. 3 (B), there are 32 communities in which the largest community has more than 60 members, and the smallest one just has less than 10 members. Through analyzing the practical situation, we can find that many members in a community are from the same region or adjacent regions. For example, most of authors in community 1 come from mid-China. In further analysis, almost all authors in community 2 come from Zhejiang University. In this case, we can see that regional collaboration happens frequently.

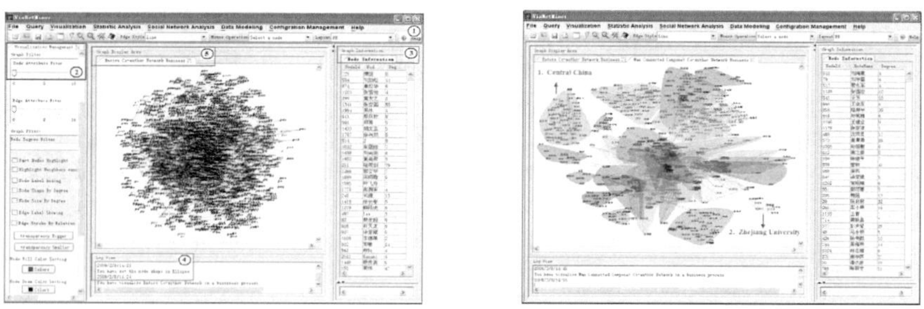

(A) (B)

Fig. 3. (A) The entire co-author network. (B) The community structure in the largest connected subgraph.

5.3 Analyzing Co-region Network

As shown in Fig. 4 (A), we can see that nodes in this graph almost include all provinces of China. In addition, more collaboration usually occurs in the interior but less between countries. Statistic analysis shows that the top four regions with high collaboration times are Beijing, northeast, Guangdong and Zhejiang. These regions communicate ideas closely not only in the interior but also with other regions. For example, Beijing has 3045 collaborations in the interior and 4845 with other regions. As a consequence, we can find that the regions, such as Beijing, may be the backbones with more innovative technologies; inversely, the regions such as Qinghai with few collaboration times seem to be on the fringe of virology research.

5.4 Analyzing Evolution of Co-author Network

In this section, we will reveal the evolution of co-author network from year 1999 to 2004. As shown in Fig. 4 (B), we can draw the following conclusions: 1) the number of authors and collaboration times are both enlarging; 2) as years pass by, the cooperation becomes much closer so that some connected subgraphs have the tendency to combine; 3) in further analysis, the collaboration papers which are signed with green labels are related with the topic of SARS. It may be inferred that social events could influence the rise and fall of the research area.

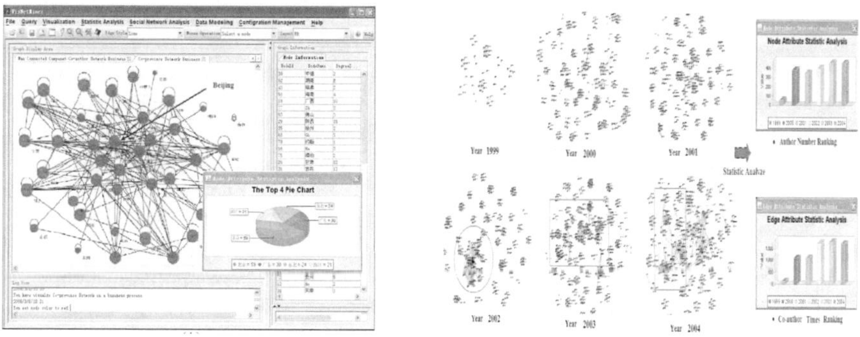

(A) (B)

Fig. 4. (A) Co-region network. (B) The evolution of co-author network.

6 Conclusions

This paper presents a Tool for Visualizing Network and Mining, called VisNetMiner. VisNetMiner has a flexible architecture. A multi-dimensional data model is proposed to enrich the analytic functions. Powerful data processing, analysis and visualization functions are included in the system as an integrated workflow. The paper has taken an academic collaboration network to illustrate the application of the tool.

Many work are desire to be done for completing the system further in the future. A better data model can be designed to support the quick query. The multi-dimensional cube for graph can be constructed to improve the analytic functions.

Acknowledgments. This work is supported by the National Natural Science Foundation of China under Grant No.60402011, 60805041 and the National Key Technology R&D Program of China under Grant No.2006BAH03B05. It is also supported by IBM China Research Laboratory, the Specialized Research Fund for the Joint laboratory between Beijing University of Posts and Communications and IBM China Research Laboratory (Project No.JTP200806014-3).

References

1. Wang, X., Li, X., Chen, G.: The Theory and Application of Complex Network, pp. 1–17. Tsinghua University Press (2006)
2. Wang, B., Wu, W., Xu, C., Wu, B.: A Survey on Visualization of Complex Network. Computer Science 34(04) (2007)

3. Tang, J., Zhang, J., Yao, L., Li, J.: ArnetMiner: Extraction and Mining of Academic Social Networks. In: Proceeding of the 17th International Conference on World Wide Web, pp. 1193–1194 (2008)
4. Adamcsek, B., Palla, G., Farkas, I.J., Derényi, I., Vicsek, T.: CFinder: Locating Cliques and Overlapping Modules in Biological Networks. Bioinformatics 22(08), 1021–1023 (2006)
5. Bongshin, L., Mary, C., George, R., Bederson, B.B.: Understanding Research Trends in Conferences using PaperLens. In: CHI 2005 extended abstracts on Human factors in computing systems, pp. 1969–1972 (2005)
6. Chen, C.: CiteSpace II: Detecting and Visualizing Emerging Trends and Transient Patterns in Scientific Literature. Journal of the American Society for Information Science and Technology 57(3), 359–377 (2006)
7. Hyunmo, K., Lise, G., Lisa, S.: C-Group: A Visual Analytic Tool for Pairwise Analysis of Dynamic Group Membership. In: IEEE Symposium on Visual Analytics Science and Technology, pp. 211–212 (2007)
8. Pullen, J.M.: The Network Workbench: Network Simulation Software for Academic Investigation of Internet Concepts. The International Journal of Computer and Telecommunications Networking 32(03), 365–378 (2000)
9. Herr, B.W., Huang, W., Penumarthy, S., Borner, K.: Designing Highly Flexible and Usable Cyberinfrastructures for Convergence. Annals of the New York Academy of Sciences 1093(1), 161–179 (2006)
10. Prefuse | Interactive Information Visualization Toolkit, `http://prefuse.org`
11. JUNG - the Java Universal Network/Graph Framework, `http://jung.sourceforge.net`
12. Girvan, M., Newman, M.E.J.: Community structure in social and biological networks. PNAS 99(12), 7821–7826 (2002)

Anomaly Detection Using Time Index Differences of Identical Symbols with and without Training Data

Stefan Jan Skudlarek and Hirosuke Yamamoto

Graduate School Of Frontier Sciences, University of Tokyo,
5-1-5 Kashiwa-no-ha, Kashiwa, Chiba, 277-8561 Japan
skudlarek@it.k.u-tokyo.ac.jp, hirosuke@ieee.org

Abstract. Anomaly detection or novelty detection has emerged as a powerful tool for masquerade detection during the past decade. However, the strong dependence of previous methods on uncontaminated training data is a matter of concern. We introduce a novel masquerade detection algorithm based on a statistical test for system parameter drift of time series data. The approach presented may exploit attack-free training data if provided, but is not dependent on it. It transforms the string of commands into a symbol sequence, respectively using the average time index difference of symbols identical to the symbol found at a particular index for anomaly detection. We evaluated the method using the standard data set provided by Schonlau et al., both including and excluding the use of training data. We report the results achieved with and without training data, and compare them to the results attained by several conventional methods using training data.

Keywords: Anomaly Detection, Masquerade Detection, Training Data, Unsupervised Anomaly Detection, Time Series Analysis.

1 Introduction

Novelty Detection has emerged as a new concept of system monitoring during the last decade. It defines an error as a deviation from the normal system behavior previously observed. Because this test does not require any knowledge of the output data probability distribution of an erroneous system state, contrary to the optimum test given by the Neymann-Pearson Lemma[1], errors previously unknown may be detected.

Novelty detection methods based on statistical classification may be divided into two large groups. The first group uses the most recent L observations, $x_{n-L},\ldots,x_{n-1}$, to calculate the probability of the present data x_n[2]. Contrary, the second group employs supervised learning prior to use, requiring training data guaranteed to have been generated by a normal system in order to build a statistical profile of the normal system output[3].

Regarding the field of network intrusion detection, numerous methods applying novelty detection for processing traffic data generated by a network have been proposed. Because attack-free training data as required by the second group mentioned above is difficult to acquire, several methods for detecting and removing attacks buried

R. Huang et al. (Eds.): ADMA 2009, LNAI 5678, pp. 619–626, 2009.

in training data have been devised. Such approaches are usually referred to as *unsupervised anomaly detection*.

Previous efforts of unsupervised anomaly detection concentrated on detecting attacks buried inside regular network traffic[4][5]. However, in case of masquerade detection, which presents a very special case of network intrusion detection, network traffic data will show no anomalies. Masquerade detection relies solely on the pattern analysis of command line input or the sequence of system calls generated by a user who performed a regular login, as may be retrieved from an audit log.

Although the general idea had been around for some time, the first systematic comparison of several statistical classifiers processing the command line input for the purpose of masquerade detection by means of anomaly detection was published in 2001[6], at the same time introducing the data set commonly used for evaluation of the various new detection methods[7]-[10] proposed thereafter. Without exception, those methods adopted the original supposition of the availability of attack-free training data. Although two methods were proposed regarding the problem of unsupervised anomaly detection using sequences of system calls[11][12], the adaption to the analysis of command line input is hampered by the fact that the efficiency of both methods relies on the visibility of any system call, which strictly limits the number of possible symbol sequences.

Our method solves the problem of both supervised and unsupervised anomaly detection with respect to command line input while avoiding the setbacks of many methods relying on attack-free training data: too many parameters (which often have to be set using the trial-and-error method) and high computational cost. We subsequently show that if certain premises regarding command generation and structure hold, we may exploit time index differences of identical commands within the overall log data sequence, modifying a method for testing real valued data[13], in order to detect data generated by a masquerade attack with high efficiency.

2 Theoretical Background and Statement of Algorithms

2.1 Average Index Difference of a Symbol

We are given a sequence of n commands. It may or may not contain a portion of commands generated by a masquerader. The commands may be thought of as symbols x drawn from a finite alphabet, with the alphabet size Z being determined by the number of different commands observed. The generation of intrusions may be interpreted as a change of the internal state of the symbol source, which may switch from 'normal' to 'intrusion' or vice versa.

We subsequently list some premises made with respect to symbol generation:

1. Stationary Ergodic Symbol Generation:

   The symbols are supposed to be output by a stationary ergodic source. A symbol x observed at index j may either have been generated according to

$$P_X^{\mathrm{n}}(x) \overset{\mathrm{def}}{=} P(X = x | \mathrm{normal}) \tag{1}$$

or

$$P_X^{\mathrm{i}}(x) \overset{\mathrm{def}}{=} P(X = x | \mathrm{intrusion}), \tag{2}$$

with $x \in \{1,...,Z\}$. Neither probability is known to us.

2. Intrusion Data Generation:
Intrusion data, if existent, is generated as a single continuous symbol subsequence of unknown length l_{i} within the overall sequence of length n, which is assumed to be bounded by $b \le l_{\mathrm{i}} \le \frac{n}{3}$. $b << n$ is henceforth referred to as the minimum intrusion length or block length, and is supposed to be known.

Let $\Delta(x)$ stand for the interval between two consecutive occurrences of $x \in \{1,...,Z\}$ (see Fig. 1 a)). The expected value of $\Delta(x)$ in either intrusion or normal state of the source is given by $\mathrm{E}(\Delta(x)) = \frac{1}{P_X(x)}$ from Kac's Lemma[1] .

The average index difference $T_j(x)$ of the symbol x found at index j is defined as the average over the respective index differences of the symbol x found at index j and all symbols of identical value x within the whole of the sequence of length n. Using the notation depicted by Fig. 1 a), $T_j(x)$ is written as follows:

$$T_j(x) \overset{\mathrm{def}}{=} \frac{\sum_{o=1}^{C_{\mathrm{b}}} (j - j_o^{\mathrm{b}}) + \sum_{q=1}^{C_{\mathrm{a}}} (j_q^{\mathrm{a}} - j)}{C_b + C_a}. \tag{3}$$

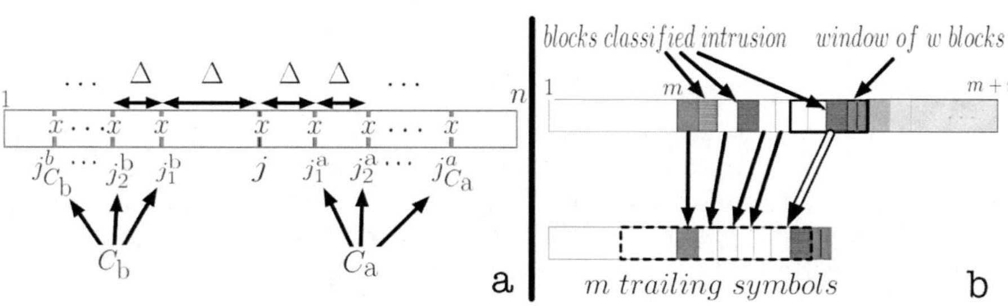

Fig. 1. a) Index notation of identical symbols x. b) Online detection using m symbols.

2.2 Expected Value of the Average Index Difference

If no intrusion is present, we only have to consider $P_X^{\mathrm{n}}(x) \forall x \in \{1,...,Z\}$, $\forall j \in \{1,...,n\}$. The expected values of C_{b} and C_{a} are the product of the length of the subsequence below/above j and $P_X^{\mathrm{n}}(x)$. Using these expected values, as well as Kac`s Lemma, on (3), the expected value of the average index difference can be approximated by

$$\mathrm{E}(T_j(x)) \approx \frac{1}{n \cdot P_X^{\mathrm{n}}(x)} \left(\sum_{o=1}^{j \cdot P_X^{\mathrm{n}}(x)} \frac{o}{P_X^{\mathrm{n}}(x)} + \sum_{q=1}^{(n-j) \cdot P_X^{\mathrm{n}}(x)} \frac{q}{P_X^{\mathrm{n}}(x)} \right) = \\ = \left(\frac{j}{2} \cdot \frac{j \cdot P_X^{\mathrm{n}}(x)}{n \cdot P_X^{\mathrm{n}}(x)} \right) + \left(\frac{(n-j)}{2} \cdot \frac{(n-j) \cdot P_X^{\mathrm{n}}(x)}{n \cdot P_X^{\mathrm{n}}(x)} \right) + \frac{1}{2 \cdot P_X^{\mathrm{n}}(x)} \tag{4}$$

$$\mathrm{E}(T_j(x)) \approx \frac{j^2 + (n-j)^2}{2 \cdot n} = \left(\left(\frac{j}{n} - 1 \right) \cdot j \right) + \frac{n}{2} \geq \frac{n}{4} . \tag{5}$$

The first two terms of (4) show the effect of taking the average of the index differences. The expected value of the average index difference mainly consists of the weighted sum of the differences between the index j and centroid indices of closed areas of time-invariant probability distribution *below and above* the index j. The weights consist of the average percentage of symbols contributed by the respective area. The third term of (4) reminds us that in case of very small generation probability compared to sequence length n, above approximation formula becomes futile.

Table 1. Bounds of the Expected Value of $T_j(x)$ in Case of One Intrusion

	$P_X^{\mathrm{n}}(x) << P_X^{\mathrm{i}}(x)$	$P_X^{\mathrm{n}}(x) \approx P_X^{\mathrm{i}}(x)$	$P_X^{\mathrm{n}}(x) >> P_X^{\mathrm{i}}(x)$
j inside	$\mathrm{E}(T_j(x)) \leq \frac{l_i}{2}$	$\mathrm{E}(T_j(x)) \geq \frac{n}{4}$	$\mathrm{E}(T_j(x)) \geq \frac{n+l_i}{4}$
j outside	$\mathrm{E}(T_j(x)) \geq \frac{l_i}{2}$	$\mathrm{E}(T_j(x)) \geq \frac{n}{4}$	$\mathrm{E}(T_j(x)) \geq \frac{n-l_i}{4}$

Using an approach similar to the case of no intrusion, the expected value of the average index difference in case of one intrusion may be calculated both for the case of j located inside and for the case of j located outside the intrusion, respectively considering three different pairings of the generation probabilities. The bounds of the expected value of the average index difference thus deduced are shown in Table 1, and will be used later for explaining the parameter selection of the algorithm.

2.3 Statement of Algorithm of Masquerade Detection without Training Data

By assigning an individual integer code number to any command, a sequence of n commands is transformed into a symbol sequence. A subsequence is classified as generated by intrusion if the percentage of symbols showing an average index difference below a certain limit τ_{th} surpasses a certain percentage threshold c_{th}. The approach is based on the assumption that the overall sequence of n commands exhibits a single symbol subsequence of a certain length l_{i} generated by intrusion. Thus, the symbols characteristic of the intrusion (i.e. commands frequently used by the intruder *but not* by the regular user) will show a smaller average index difference, because the majority of the identical symbols will be located close to the symbol compared to the overall sequence length n.

A subsequence of length b, which is called a block hereafter, is classified using the subsequent algorithm:

1. average index difference and percentage calculation:
   Calculate the percentage c of the b symbols featuring an average index difference of $T_j(x) \leq \tau_{\mathrm{th}}$ $(0 \leq \tau_{\mathrm{th}} \leq n)$. The average index difference of singular symbols is supposed to be zero.

2. percentage evaluation:
   If $c \geq \tau_{\mathrm{th}}$ $(0 \leq c_{\mathrm{th}} \leq 1)$, the block is classified to have been generated by an intrusion.

After stating the algorithm, we move on to examining the setting of the parameter τ_{th}. The task of the step employing τ_{th} is to search for symbols that were generated by an intrusion *and* are characteristic of the intrusion when compared to the normal output ($P_X^{\mathrm{n}}(x) << P_X^{\mathrm{i}}(x)$).

As for the case of no intrusion and one intrusion, using the upper and lower bounds given by Table 1 and (5), we deduce we have to search for symbols x satisfying

$$T_j(x) \leq \frac{l_{\mathrm{i}}}{2} \text{ and } T_j(x) \leq \frac{n}{4}, \tag{6}$$

while excluding symbols satisfying

$$T_j(x) \geq \frac{n - l_{\mathrm{i}}}{4} \text{ or } T_j(x) \geq \frac{l_{\mathrm{i}}}{2}. \tag{7}$$

The two value ranges given by (6) and (7) will not overlap as long as the intrusion length satisfies $l_{\mathrm{i}} \leq \frac{n}{3}$. Therefore, the optimum parameter setting is

$$\tau_{\mathrm{th}} = \frac{l_{\mathrm{i}}}{2} \leq \frac{n}{6}. \tag{8}$$

Because, apart from the bounds given before, the intrusion length l_{i} is unknown to us, we have to choose the setting of τ_{th} used by the algorithm based on a theoretical evaluation of the impact of possible misestimations $\tau_{\mathrm{th}} \neq \frac{l_{\mathrm{i}}}{2}$. If the primary goal is to completely detect any possible intrusion of length $l_{\mathrm{i}} \leq \frac{n}{3}$, deliberately overestimating the intrusion length l_{i} by choosing a setting close to

$$\tau_{\mathrm{th}} = \frac{n}{6} \tag{9}$$

is preferable. On the contrary, if the primary goal is to minimize the misclassification rate of normal blocks, choosing a substantially lower threshold is reasonable.

2.4 Extension of Algorithm to Masquerade Detection Using Training Data

We also applied our method to the classic masquerade detection scenario featuring the usage of m symbols of clean training data. Supposing online detection, we only used the most recently generated m symbols not classified intrusion for evaluation of the current block.

Figure 1 b) shows the approach using the present data block as well as the m most recently generated symbols, which consist of symbols *not classified intrusion outside a window of w blocks* and all symbols inside the same window besides the current block.

The most significant change compared to the detection without training data is the fact that now the misclassification of one block may influence the classification of subsequent blocks (progressive error accumulation).

In contrast to the default threshold of masquerade detection without training data given by (9), the threshold used for masquerade detection with training data is optimized to $\tau_{th} = \frac{l_i}{2} \leq \frac{w \cdot b}{2}$. This is possible because the training data enables us to limit the position and number of possible intrusion blocks to the window of length w. Even if some intrusion blocks might escape detection, the constant outdating of blocks will limit the impact.

2.5 Computational Complexity

The computational complexity depends on the test data size n, the block length b, the alphabet size Z, and the training data size m.

The computational cost of the algorithm using no training data can be upper bounded by a function of order

$$O(Z \cdot n)\,. \tag{10}$$

The algorithm using training data for online detection requires processing power upper bounded by a function of order

$$O\left(Z \cdot \left(m + n + \frac{m \cdot n}{b^2}\right) + n\right). \tag{11}$$

3 Evaluation of the Methods Presented

Applying our method to the common data set[6], we use $m = 5{,}000$, $n = 10{,}000$ and $b = 100$. For evaluation of the results achieved, we present the ROC curves returned, also plotting the results of two cutting-edge training data-based statistical classifiers for comparison[8][9].

The performance of detection without training data shows a strong dependence on the choice of the parameter τ_{th}. Fig. 2 a) - f) shows the results for descending τ_{th}, with the curve created by respectively raising c_{th} from zero to 100%. The results for $c_{th} = 20\%$ illustrate the correctness of the theoretical considerations regarding the choice of τ_{th} presented in section 2.3. The setting of $\tau_{th} \approx 1{,}700$ offers a high detection rate at the prize of a substantial misclassification rate of the normal blocks. Settings of $\tau_{th} \leq 1{,}000$ minimize the misclassification rate, but the rate of detected intrusion blocks will drop steadily, with an especially sharp decline for any setting below half of the expected intrusion length l_i($\tau_{th} < \frac{E(l_i)}{2} = \frac{5 \cdot 100}{2} = 250$).

As for results achieved using training data, Fig. 2 d) shows the return for a setting $w = 4$ and $\tau_{th} = \frac{w \cdot b}{2} = 200$. The method performs well, coming close to the previous optimum results.

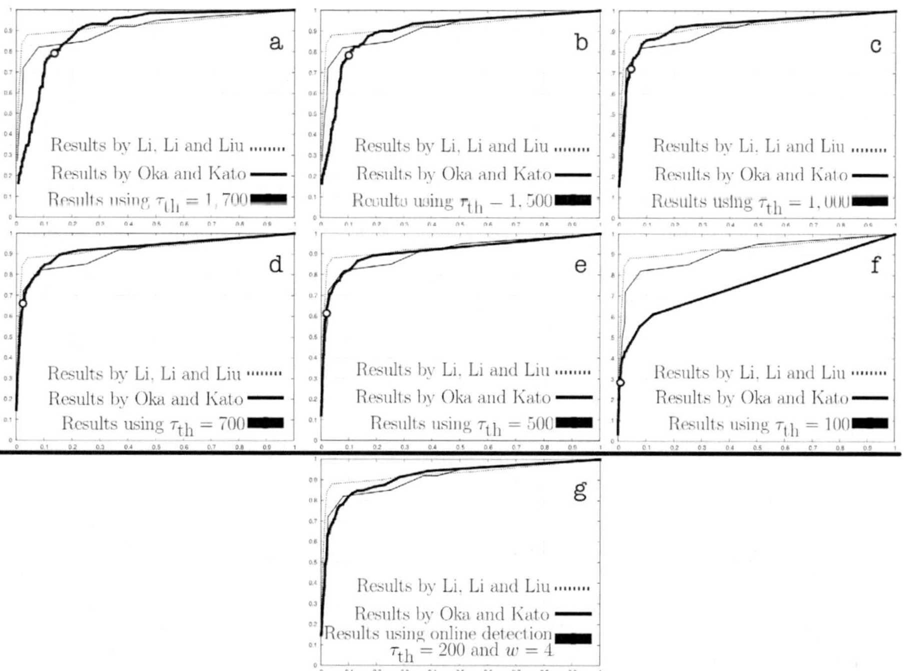

Fig. 2. a) - f) Results without training data for (a) $\tau_{th} = 1{,}700$, (b) $\tau_{th} = 1{,}500$, (c) $\tau_{th} = 1{,}000$, (d) $\tau_{th} = 700$, (e) $\tau_{th} = 500$, (f) $\tau_{th} = 100$. The round mark respectively shows the point within the ROC curve returned by $c_{th} = 20\%$. g) Results of online detection using the $m = 5{,}000$ most recent symbols.

4 Conclusion

The method of average index differences introduced offers three important advantages over conventional methods: The independence from training data, the low computational complexity, and the theoretical foundation of parameter selection.

However, the detection of two separated intrusion block sequences, the theoretical estimation of the impact of misclassification in case of online detection, as well as the problem of soft decision on the average index difference(instead of the present hard decision using τ_{th}) require further examination.

References

1. Cover, T., Thomas, J.: Elements of Information Theory. Wiley & Sons, Chichester (2006)
2. Yamanishi, K., Takeuchi, J.: A Unifying Framework for Detecting Outliers and Change Points From Time Series. IEEE Transactions on Knowledge and Data Engineering 18(I. 4), 482–492 (2006)
3. Clifton, et al.: Combined Support Vector Novelty Detection for Multi-channel Combustion Data. In: IEEE International Conference on Networking, Sensing and Control, pp. 495–500 (2007)
4. Zhang, J., Zulkernine, M.: Anomaly Based Network Intrusion Detection with Unsupervised Outlier Detection. In: IEEE International Conference on Communications, pp. 2388–2393 (2006)
5. Kwitt, R., Hofmann, U.: Unsupervised Anomaly Detection in Network Traffic by Means of Robust PCA. In: IEEE International Multi-Conference on Computing in the Global Information Technology, pp. 10–13 (2007)
6. Schonlau, M., DuMouchel, W., Ju, W., Karr, A., Theus, M., Vardi, Y.: Computer intrusion: Detecting masquerades. Statistical Science 16(1), 58–74 (2001)
7. Wang, K., Stolfo, S.: One Class Training for Masquerade Detection. In: ICDM Workshop on Data Mining for Computer Security, pp. 1–10 (2003)
8. Li, Z., Li, Z., Liu, B.: Masquerade Detection System Based on Correlation Eigen Matrix and Support Vector Machine. In: CIS Conference, pp. 625–628 (2006)
9. Oka, M., Kato, K.: Anomaly Detection Using Integration Model of Vector Space and Network Representation. Information Processing Society of Japan Digital Courier 3, 269–279 (2007)
10. Yamanishi, K., Maruyama, Y.: Dynamic Model Selection with its Applications to Novelty Detection. IEEE Transactions on Information Theory 53(I. 6), 2180–2189 (2007)
11. Eskin, E., Arnold, A., Prerau, M., Portnoy, M., Stolfo, S.: A Geometric Framework for Unsupervised Anomaly Detection: Detecting Intrusions in Unlabeled Data. In: Applications of Data Mining in Computer Security, ch. 4. Kluwer, Dordrecht (2002)
12. Tandon, G., Chan, P., Mitra, D.: Data Cleaning and Enriched Representations for Anomaly Detection in System Calls. In: Machine Learning and Data Mining for Computer Security - Methods and Applications, pp. 137–156. Springer, Heidelberg (2006)
13. Kennel, M.: Statistical Test for Dynamical Nonstationarity in Observed Time-Series Data. Physical Review E 56, 316–321 (1997)

An Outlier Detection Algorithm Based on Arbitrary Shape Clustering

Xiaoke Su[1], Yang Lan[2], Renxia Wan[1], and Yuming Qin[1]

[1] College of Information Science and Technology,
Donghua University,
201620 Shanghai

[2] School of Computer and Information Technology, Xinyang Normal University,
464000 Xinyang, Henan

suxiaoke07@126.com, lyanang@126.com, wrx1022@mail.dhu.edu.cn,
yuming@dhu.edu.cn

Abstract. Outlier detection is an important branch in data mining field. It provides new methods for analyzing all kinds of massive, complex data with noise. In this paper, an outlier detection algorithm is presented by introducing the arbitrary shape clustering approach and discussing the concept of abnormal cluster. The algorithm firstly partitions the dataset into several clusters by proposed clustering approach. Outliers are then detected from the cluster set according to the abnormal cluster concept. Moreover, by introducing inter-cluster dissimilarity measure, the proposed algorithm gains a good performance on the mixed data. The experimental results on the real-life datasets show our approach outperform the existing methods on identifying meaningful and interesting outliers.

Keywords: Outlier detection, arbitrary shape clustering, the mixed data, dissimilarity measure, data mining.

1 Introduction

Outlier detection has a wide range of applications such as credit card, insurance, tax fraud detection, intrusion detection and many other areas. It aims at finding the observation deviating so much from other observations as to arouse suspicions that the observation was generated by a different mechanism [1]. The traditional approaches of outlier detection can be classified as either distribution-based, distance-based, density-based or clustering-based [2].

This paper proposes an outlier detection algorithm based on clustering. The algorithm firstly looks for the arbitrary shape clusters in the mixed data. By this means the natural clusters can be found. It regards the small-scale cluster as a candidate abnormal cluster, and then determines its final end-result by computing the number of its near neighbors. If the number is less than the specified threshold value then the cluster is taken as a true abnormal cluster. It is more reasonable regarding the small-scale clusters as the candidate abnormal clusters in term of their sizes. Every object in the true abnormal cluster is considered as an outlier. The proposed method achieves both higher detection rate and lower false positive rate than previous methods.

R. Huang et al. (Eds.): ADMA 2009, LNAI 5678, pp. 627–635, 2009.

The rest of this paper is organized as follows. The related works are introduced in section 2. In Section 3, some definitions used in the paper are formalized. Section 4 introduces the clustering-based outlier detection algorithm. The experimental results are reported in section 5 and a section of concluding remarks follows.

2 Related Works

The problem of outlier detection in categorical data is formally defined as an optimization problem from a global viewpoint in [3]. Moreover, a local-search heuristic algorithm for efficiently finding feasible solutions is presented. In [4], a very fast greedy algorithm for mining outliers under the same optimization model is presented. In [5], the authors present concrete cooperation between automatic algorithms, interactive algorithms and visualization tools: the evolutionary algorithm is used for obtaining optimal dimension subsets which represent the original dataset without loosing information for unsupervised mode (clustering or outlier detection).

Clustering and outlier detection are closely related. From the viewpoint of the clustering, outliers are objects not locating in any cluster while in the context of outlier detection they may represent outliers. Some clustering algorithms, such as DBSCAN, BIRCH, ROCK, WaveCluster can handle outlier to some extent. In the context of outlier detection, clustering has attracted interests from researchers.

Several clustering-based outlier detection approaches are proposed in [6-8]. They firstly divide the initial dataset into a set of individual clusters by using some clustering algorithms and then detect outliers on the basis of clustering. However, a large part of these approaches have their limitations. [6] and [8] are limited to some special cases. They can handle only the continuous data and the categorical data respectively. The clustering algorithm in [7] is also not reasonable due to its regarding small clusters as abnormal clusters, instead, the algorithm may erroneously partition a non-convex cluster into some small clusters which are finally regarded as abnormal clusters.

3 Notation and Definition

Given the cluster C_1 and the cluster C_2, the object p and the object q of dataset D which is featured by m attributes (m_c categorical and m_n continuous), $m = m_c + m_n$. D_i is the i th categorical attribute. $C \mid D_i$ denotes the projection of C on D_i. Suppose the number of $C_1 \mid D_i$ is $o_{C_1|D_i}$, $0 \le o_{C_1|D_i} \le n_{c_1}$. Similarly the number of $C_2 \mid D_i$ is $o_{C_2|D_i}$, $0 \le o_{C_2|D_i} \le n_{c_2}$ and the number of $(C_1 \cup C_2) \mid D_i$ is $o_{(C_1 \cup C_2)|D_i}$, $0 \le o_{(C_1 \cup C_2)|D_i} \le (o_{C_1|D_i} + o_{C_2|D_i})$.

We modify the cluster feature in [9] and [10] to denote a cluster.

Definition 1. *CF*

For a cluster C, the cluster feature CF is defined as: $CF = (n_C, AbsFreq, Cen)$. n_C is the number of objects in C and $AbsFreq$ is the absolute frequency of the categorical attribute value, $AbsFreq=(AbsFreq_1, AbsFreq_2,..., AbsFreq_{m_c})$. $AbsFreq_i = \{(a, AbsFreq_{C|D_i}(a)) \mid a \in D_i\}$, $1 \le i \le m_c$. Cen is the centroid of the continuous attribute, $Cen = (c_1, c_2, \cdots, c_{m_n})$.

CF is used for denoting the cluster characteristic information. It satisfies additivity property. If the clusters C_1 and C_2 are merged the new cluster is denoted as C_1. C_1 is updated as follows: $n_{C_1} = n_{C_1} + n_{C_2}$. For the categorical attribute D_i, if there exists the same value a_i, the absolute frequency of a_i is $AbsFreq_{C_2|D_i}(a_i)$ in $AbsFreq_i^{(2)}$, after merging the absolute frequency in C_1 becomes

$$AbsFreq_{C_1|D_i}(a_i) = AbsFreq_{C_1|D_i}(a_i) + AbsFreq_{C_2|D_i}(a_i). \tag{1}$$

Otherwise, the attribute value a_i and the corresponding absolute frequency is directly added to $AbsFreq_i^{(1)}$. For the i th continuous attribute the centroid becomes

$$c_i^{(1)} - \frac{c_i^{(1)} \times n_{C_1} + c_i^{(2)} \times n_{C_2}}{n_{C_1} + n_{C_2}}. \tag{2}$$

Definition 2. Absolute frequency

For the cluster C, a is a value of the categorical attribute D_i, the absolute frequency of a in C with respect to D_i is defined as

$$AbsFreq_{C|D_i}(a) = o_{(C|D_i=a)}, 0 \le AbsFreq_{C|D_i}(a) \le o_{C|D_i}. \tag{3}$$

According to the definition 2 the relative frequency of a in C with respect to D_i can be denoted as

$$\mathrm{Re}\,lFreq_{C|D_i}(a) = AbsFreq_{C|D_i}(a) / n_C, 0 \le \mathrm{Re}\,lFreq_{C|D_i}(a) \le 1. \tag{4}$$

Definition 3. The distance between the cluster C_1 and the cluster C_2

The distance is defined as $d(C_1, C_2) = \sum_{i=1}^{m} dif(C_i^{(1)}, C_i^{(2)}) / m$ where $dif(C_i^{(1)}, C_i^{(2)})$ is the dissimilarity measure between C_1 and C_2 on the i th attribute. For the categorical attribute D_i,

$$dif(C_i^{(1)}, C_i^{(2)}) = \sum_{a \in (C_1 \cup C_2)|D_i} \frac{|\mathrm{Re}\,lFreq_{C_1|D_i}(a) - \mathrm{Re}\,lFreq_{C_2|D_i}(a)|}{(\mathrm{Re}\,lFreq_{C_1|D_i}(a) + \mathrm{Re}\,lFreq_{C_2|D_i}(a)) \cdot o_{(C_1 \cup C_2)|D_i}}. \tag{5}$$

If $a \notin C_2 \mid D_i$, $\mathrm{Re}\,lFreq_{C_2|D_i}(a) = 0$. Otherwise, if $a \notin C_1 \mid D_i$ $\mathrm{Re}\,lFreq_{C_1|D_i}(a) = 0$. Therefore:

$$dif(C_i^{(1)},C_i^{(2)}) = 2 - \frac{o_{C_1|D_i} + o_{C_2|D_i}}{o_{(C_1 \cup C_2)|D_i}} + \sum_{a \in (C_1 \cap C_2)|D_i} \frac{|\,RelFreq_{C_1|D_i}(a) - RelFreq_{C_2|D_i}(a)\,|}{(RelFreq_{C_1|D_i}(a) + RelFreq_{C_2|D_i}(a)) \cdot o_{(C_1 \cup C_2)|D_i}} . \quad (6)$$

While for the continuous attribute $dif(C_i^{(1)},C_i^{(2)}) = \left|c_i^{(1)} - c_i^{(2)}\right|$.

Specially, when an object is taken as a cluster and the distance between p and C can be computed that is $d(p,C) = \sum_{i=1}^{m} dif(p_i,C_i)/m$ where $dif(p_i,C_i)$ is the distance between the object p and the cluster C on the i th attribute. For D_i,

$$dif(p_i,C_i) = \left(\frac{|1 - RelFreq_{C|D_i}(p_i)|}{1 + RelFreq_{C|D_i}(p_i)} + o_{(p \cup C)|D_i} - 1\right) / o_{(p \cup C)|D_i} . \quad (7)$$

While for the continuous attribute $dif(p_i,C_i) = \left|p_i - c_i\right|$.

Definition 4. Merged clusters

If the distance between C_1 and C_2 satisfies $d(C_1,C_2) \le r_{(C_1)} + r_{(C_2)}$ C_1 and C_2 are called as the merged clusters. $r_{(C)}$ is the radius of the cluster C that is the distance from the farthest object q in C to the cluster C itself. It is denoted as $r_{(C)} = \max(d(q,C)), \forall q \in C$.

The clusters satisfying the definition 4 can be merged. The merging process is operated according to the additivity property of the cluster feature CF.

Definition 5. Near neighbor of object p

Given threshold value v, if the distance between p and q satisfies $d(p,q) \le v$ q is called as a near neighbor of p.

If the common objects number between the set of the near neighbors of p and the cluster C is k, then the similarity degree between p and C is denoted as $Sim(p,C) = k$.

The definition 5 can be extended to the near neighbor of cluster. Given threshold value nr, if the distance between C_2 and C_1 satisfies $d(C_1,C_2) \le nr$, C_2 is regarded as a near neighbor of C_1.

Definition 6. Candidate abnormal cluster

If the number of objects in C is less than the given threshold value $\min nc$ the cluster C is called as a candidate abnormal cluster.

Definition 7. Abnormal cluster

If the near neighbors number of the candidate abnormal cluster C is less than the given threshold value nn, C is regarded as an abnormal cluster.

The abnormal cluster is taken as a whole in which every object is viewed as an outlier.

4 The Outlier Detection Algorithm

4.1 Overview

Clustering process can be divided into the merging, adding and updating. For an object p it either is absorbed by a cluster, or creates a new cluster including itself. The first preference is to absorb the object into an existing cluster. We first compute the distance between p and every existing object in order to find all the near neighbors of p. If there exist the near neighbors of p and the clusters containing the near neighbors exist then merging is performed. We compute the similarity degree between p and all the cluster containing the near neighbors of p. The object is added to the most similar cluster. If there does not exist the near neighbor of p in the existing clusters, p is viewed as a singleton which is finally taken as a new cluster.

Outlier detection is performed on the basis of the clustering. All the clusters are sorted according to their magnitudes. The small clusters are regarded as the candidate abnormal clusters. The near neighbors of every candidate abnormal cluster are computed. If the number of the near neighbors satisfies the definition 7 that is the outlier requirement, the cluster is the true abnormal cluster.

4.2 The Proposed Algorithm

The algorithm uses the most similar principle to divide dataset into different clusters firstly. The outlier detection is performed on the basis of the clustering result. It is described as follows.

Step 1: Initialize an empty set, and read a new object.

Step 2: Create a cluster with the object.

Step 3: If the dataset is empty, go to step 7, else read a new object p and look for all the near neighbors of p.

Step 4: Compute the number of the clusters containing the near neighbors of p which is denoted as l. If $l = 0$ there does not exist the near neighbor of p then go to step 2.

Step 5: Otherwise determine the merged clusters among the l clusters. Merge all the merged clusters and modify the corresponding CF.

Step 6: Add p into the most similar cluster and update the cluster feature CF. Go to step 3.

Step 7: All the clusters are sorted according to their respective magnitude. The small clusters are taken as the candidate abnormal clusters.

Step 8: Compute the near neighbors number of every candidate abnormal cluster.

Step 9: If the number is less than the threshold value nn then the cluster is taken as the true abnormal cluster and every object in it is regarded as an outlier.

Step 10: Stop.

5 Empirical Results

5.1 Experimental Setting and Evaluation

In order to test the quality of the proposed algorithm, we ran our algorithm on the real-life datasets. The experimental results demonstrate the effectiveness of our method against other methods. All the experiments are performed on a 2.2GHz Intel Pentium IV processor computer with 512MB memory, running on Windows XP professional. Our algorithm is implemented in VC6.0.

The Clustering Result. The clustering accuracy [11] is used as a measure of a clustering result. It is defined as $\varphi = \sum_{i=1}^{c} n_i / N$ where n_i is the number of object occurring in both the i th cluster and its corresponding class and N is the number of objects in the dataset. c is the resultant number of clustering.

The Outlier Detection Result. We use the parameters such as detection rate (DR) and false positive rate (FR) [10] to measure the performance of the outlier detection. The detection rate is defined as the ratio of the detected outliers to the total outliers, and the false positive rate is defined as the ratio of the normal records detected as the outliers to total normal records.

5.2 The Real-Life Datasets

We use the datasets from the machine-learning databases of the UCI repository to test the validity of the algorithm [12].

Lymphography Dataset. Lymphography dataset has 148 records with 18 categorical attributes. The records have been divided into four classes: class one (with two records) and class four (with four records) are the rare class, and the records in the rare class are regarded as outlier.

Table 1. Lymphography outlier detection result

v	accuracy	c	$minnc$	nr	nn	DR	FR
0.1	100%	145	6	0.32	4	100%	7.75%
0.16	99.32%	128	6	0.25	4	100%	53.52%
0.16	99.32%	128	6	0.3	4	100%	12.68%
0.16	99.32%	128	6	0.32	2	100%	2.82%
0.16	99.32%	128	6	0.32	4	100%	7.04%
0.16	99.32%	128	6	0.34	4	83.33%	2.11%
0.2	75%	59	6	0.32	2	100%	9.86%

The experimental results in Table 1 are explained as follows.

We fixed $v = 0.16$, $nn = 4$, $\min nc = 6$ and changed nr , the result is showed from line 2 to line 4 of Table 1. When nr value increases the possibility of the cluster becoming the candidate outlier descends. DR and FR decrease correspondingly.

We fixed $v = 0.16$, $nr = 0.32$, $\min nc = 6$ and changed nn, the result is showed line 4 and line 5 of Table 1. When nn value increases the restriction on the number of the cluster near neighbors is stricter. DR and FR increase correspondingly.

Breast Cancer Dataset. The Wisconsin breast cancer dataset has 699 instances with 9 continuous attributes. Each record is labeled as benign (458 records) or malignant (241 records). We remove some malignant records to form a very unbalanced distribution. The resultant dataset had 39 (8%) malignant records and 444 (92%) benign records.

The corresponding clustering and detection results are given in Table 2. From the Table 2, it can be seen that as the v value increases the number of clusters decreases remarkably. Clustering is used for picking up speed. Most normal records are filtered via clustering.

Table 2. Breast cancer outlier detection result

v	accuracy	*c*	*minnc*	*nr*	*nn*	*DR*	*FR*
0.09	99.79%	49	40	0.08	3	97.44%	3.15%
0.1	99.79%	45	40	0.1	2	100%	2.93%
0.11	99.59%	43	40	0.08	3	100%	2.93%
0.12	99.17%	35	40	0.15	3	97.44%	2.70%
0.13	98.55%	26	40	0.15	3	94.87%	2.48%
0.15	92.55%	4	40	0.1	3	7.69%	0

Table 3 shows the best performance comparison between our method and [8, 10]. The result shows our method is superior to [8, 10].

Table 3. The contrast results in Breast cancer

Ref.	*FR*	*DR*
[8]	5.63%	100%
[10]	24.10%	100%
Our method	2.93%	100%

Kddcup99 Dataset. The dataset contains 41 attributes with 34 continuous and seven categorical attributes. It is not suitable for outlier detection based on the whole dataset. We randomly choose 38841 normal records and 1618 attack records.

Table 4. Kddcup99 outlier detection result

v	accuracy	*c*	*minnc*	*nr*	*nn*	*DR*	*FR*
0.03	99.95%	112	1600	0.03	3	99.01%	2.07%
0.035	99.95%	75	1600	0.04	3	98.95%	1.90%
0.035	99.95%	75	1600	0.035	3	98.95%	1.90%
0.04	99.94%	66	1600	0.03	3	98.83%	1.88%
0.04	99.94%	66	1600	0.03	2	98.83%	1.88%
0.06	97.14%	25	1600	0.05	3	28.41%	0.08%

Table 4 shows the detection results. Table 5 shows the performance comparison between our method and [7, 10, 13]. From the experimental results, we can see that our method outperforms [7, 10, 13] and achieves both higher DR and lower FR .

Table 5. The contrast results in Kddcup99

Ref.	*FR*	*DR*
[7]	8.14%	88%
[13]	10%	93%
[10]	1.30%	98.65%
Our method	2.07%	99.01%

6 Conclusion

Outlier detection is an important task for many applications. In this paper we present a clustering-based outlier detection algorithm. It is based on the assumption that the number of normal records vastly outnumbers that of outliers and records with the same classification are close to each other, and records in different clusters are far apart. Outlier detection is performed on the basis of clustering which can find the arbitrary shape clusters. Furthermore, a new inter-cluster dissimilarity measure handling the mixed data is also introduced. The experimental results show our algorithm outperformed some other methods. However when computing the similarity degree between the new object and the existing clusters and looking for the most similar cluster much operations are performed. They are used for computing the distance between the new object and every existing object. It is not applicable for the large-scale and high dimensional dataset. How to improve the time performance and make the detecting process more efficient will be our future work.

References

1. Hawkins, D.: Identification of Outliers. Chapman and Hall, London (1980)
2. Patcha, A., Park, J.M.: An Overview of Anomaly Detection Techniques: Existing Solutions and Latest Technological Trends. Computer Networks, 3448–3470 (2007)
3. He, Z., Deng, S., Xu, X.: An Optimization Model for Outlier Detection in Categorical Data. In: Huang, D.-S., Zhang, X.-P., Huang, G.-B. (eds.) ICIC 2005, Part I. LNCS, vol. 3644, pp. 400–409. Springer, Heidelberg (2005)
4. He, Z., Deng, S., Xu, X., Huang, Z.: A Fast Greedy Algorithm for Outlier Mining. In: Ng, W.-K., Kitsuregawa, M., Li, J., Chang, K. (eds.) PAKDD 2006. LNCS (LNAI), vol. 3918, pp. 567–576. Springer, Heidelberg (2006)
5. Boudjeloud, L., Poulet, F.: Visual Interactive Evolutionary Algorithm for High Dimensional Data Clustering and Outlier Detection. In: Ho, T.-B., Cheung, D., Liu, H. (eds.) PAKDD 2005. LNCS (LNAI), vol. 3518, pp. 426–431. Springer, Heidelberg (2005)
6. Jiang, M., Tseng, S., Su, C.: Two-phase Clustering Process for Outliers Detection. In: Computational Statistics and Data Analysis, pp. 351–382 (2001)

7. Portnoy, L., Eskin, E., Stolfo, S.: Intrusion Detection with Unlabeled Data Using Clustering. In: ACM Workshop on Data Mining Applied to Security, Philadelphia, PA, pp. 5–8 (2001)
8. He, Z., Xu, X., Deng, S.: Discovering Cluster-based Local Outliers. Pattern Recognition Letters, 1651–1660 (2003)
9. Zhang, T., Ramakrishnan, R., Livny, M.: BIRCH: an Efficient Data Clustering Method for very Large Databases. In: ACM SIGMOD Int. Conf. on Management of Data, Montreal, pp. 103–114 (1996)
10. Jiang, S., Song, X.: A Clustering-based Method for Unsupervised Intrusion Detections. Pattern Recognition Letters, 802–810 (2006)
11. Huang, Z.: Extensions to the K-means Algorithm for Clustering Large Data Sets with Categorical Values. In: Data Mining and Knowledge Discovery, pp. 283–304 (1998)
12. UCI Machine Learning Repository, `http://www.ics.uci.edu/~mlearn/MLRepository.html`
13. Eskin, E., Arnold, A., Prerau, M.: A Geometric Framework for Unsupervised Anomaly Detection: Detecting Intrusions in Unlabeled Data. In: Applications of Data Mining in Computer Security. Advances In Information Security. Kluwer Academic Publishers, Boston (2002)

A Theory of Kernel Extreme Energy Difference for Feature Extraction of EEG Signals

Shiliang Sun and Jinbo Li

Department of Computer Science and Technology,
East China Normal University, Shanghai 200241, China
slsun@cs.ecnu.edu.cn

Abstract. Energy features are of great importance for electroencephalogram (EEG) signal classification in the application of brain-computere interfaces (BCI). In this paper, we address the problem of extracting energy features of EEG signals in a feature space induced by kernel functions. The devised method, based on a recently proposed technique extreme energy difference (EED), is called kernel extreme energy difference (KEED). This paper derives solutions which optimize the KEED criterion by the method of Lagrange multipliers.

Keywords: brain-computer interface (BCI), extreme energy difference (EED), EEG signal classification, feature extraction, kernel machine.

1 Introduction

A brain-computer interface (BCI) is a communication and control system which directly depends on brain activities rather than the brain's normal output of peripheral nerves and muscles [1]. At present, the main impetus to BCI research is the expectation that BCI technology will benefit those with severe neuromuscular disabilities that prevent them from using conventional augmentative communication and control devices. Researchers hope that BCI technology can generate a new channel for people suffering from severe motor disabilities but cognitively intact to send messages and commands to the external world.

So far, study on BCI systems has mainly involved using surface electrodes to record electroencephalogram (EEG) signals which reflect changes in neural mass activities associated with various mental processes. The reason of using surface electrodes is that this kind of recording is relatively convenient, harmless and inexpensive compared with other methods [2,3].

The feasibility of using EEG signals for communication and control is largely dependent on the extent to which they can be reliably recognized. Extracting effective EEG signal features is a very important component in the process of increasing the classification accuracy of an EEG-based BCI. This paper focuses on EEG signal feature extraction using machine learning methodologies. In particular, we aim at energy feature extraction in terms of spatial filtering.

Spatial filters try to derive EEG signal features by manipulating recordings from different and often adjacent electrodes so as to concentrate on activities of

R. Huang et al. (Eds.): ADMA 2009, LNAI 5678, pp. 636–643, 2009.

a particular spatial distribution. There are several spatial filters commonly used, such as bipolar derivation, Laplacian derivation, common average reference, principal component analysis (PCA), independent component analysis (ICA), and common spatial patterns (CSP) [4]. Owing to the lack of an explicit discriminative objective function, the significance and potential of the above mentioned spatial filters can not be understood intuitively. Li and Sun recently presented an extreme energy difference (EED) method, which has a desirable explicit discriminative objective function [5]. During the whole optimizing process, an eigenvalue decomposition of a discrepancy matrix between two covariance matrices respectively belonging to two different classes is conducted. However, the performance of EED can be limited in the sense of its linearity. In fact, any linear algorithm which can be carried out in terms of dot products can be made nonlinear by substituting a chosen kernel [6]. Based on this principle, in this paper we incorporate EED with the kernel trick and put forward a nonlinear extension of the EED method called kernel extreme energy difference (KEED).

The rest of this paper is organized as follows. Section 2 briefly reviews the EED criterion and the energy concept. Section 3 presents the discriminative KEED criterion for energy feature extraction. Finally, conclusions and future work are given in Section 4.

2 EED

Physiologically, EEG waves reflect brain activities generated by some inherent signal sources underneath the surface of the brain cortex. The process of spatial filtering is supposed to recover these latent sources. The variances of the EEG signal sources extracted from each category of brain activities can be regarded as the energy features [7].

In this section, we briefly review the discriminative EED criterion [5] for energy feature extraction of two classes denoted A and B. This method is based on the EEG covariances C_A and C_B from class A and B respectively.

2.1 Feature Extraction of One Source

Denote an observed EEG sample as an $N \times T$ matrix X, where N is the number of recording electrodes and T is the number of points recorded. In other words, one EEG sample can be seen as a distribution of T points in the N-dimensional Euclidean space. Without loss of generality suppose X is normalized and thus energy difference in varying recording time is eliminated [5]. The mean value of each EEG sample is usually taken to be zero as a result of bandpass filters.

The estimation for the covariance of one EEG sample can be written as $C = XX'$ with scale $1/T$ omitted. The covariance of a specific class is computed as the average of all single covariances to get a more accurate and stable estimate.

Assume only one latent signal source from each class is to be recovered. For an EEG sample X, the spatially filtered signal with a spatial filter denoted $\phi_{(N\times 1)}$ will be $\phi' X$. The signal energy after filtering can be represented by the sample variance as $(\phi' X)(\phi' X)' = \phi' C \phi$ where the multiplicative factor $1/T$ is omitted. The discriminative EED criterion is defined as follows [5]:

$$J(\phi) = \phi' C_A \phi - \phi' C_B \phi \,. \tag{1}$$

For classification, by optimizing (1) using the method of Lagrange multipliers with constraint $\phi'\phi = 1$, two optimal spatial filters ϕ^*_{max} and ϕ^*_{min} which satisfy $\phi^*_{max} = argmax J(\phi)$ and $\phi^*_{min} = argmin J(\phi)$ can be obtained. It comes out that the optimal spatial filters ϕ^*_{max} and ϕ^*_{min} are two eigenvectors respectively corresponding to the minimal and maximal eigenvalues of the matrix $(C_B - C_A)$ [5]. The energy feature of a new EEG sample will contain two parts which are respectively the energy values of the sample spatially filtered by ϕ^*_{max} and ϕ^*_{min}.

2.2 Feature Extraction of Multiple Sources

Suppose there are m sources to be extracted from one class of brain activities. Then the m spatial filters for extracting these sources constitute a spatial filter bank $\Phi = [\phi_1, \phi_2, \ldots, \phi_m]$. Since the trace of a covariance matrix represents the sum of signal energy from all the principal direction, it can be used to extend the discriminative criterion (1) to feature extraction of multiple sources. The objective function now becomes [5]:

$$J(\Phi) = tr(\Phi' C_A \Phi - \Phi' C_B \Phi), \tag{2}$$

where $tr(\cdot)$ denotes the trace of a matrix.

By the method of Lagrange multipliers, the optimal filter banks Φ^*_{max} or Φ^*_{min} are solved, which maximize or minimize the energy difference respectively. The solution Φ^*_{max} consists of m eigenvectors corresponding to the m minimal eigenvalues of the matrix $(C_B - C_A)$, while Φ^*_{min} is made up of m eigenvectors whose corresponding eigenvalues are maximal. Each column in Φ^*_{max} and Φ^*_{min} is normalized to have unit length when constructing filters.

3 KEED

The EED method is useful for EEG signal classification since it can guide the derivation of linear spatial filters through which energy features can be simply distilled. However, linear spatial filters are certainly not complex enough for real EEG signals because some signal distributions may have a latent nonlinear structure. The nonlinear combination of signals generated by different electrodes may provide more discriminative features. Therefore, developing nonlinear spatial filters becomes a very important task.

We attempt to combine the EED method with the kernel trick. The kernel trick attracts many attentions in recent years and a great number of powerful kernel-based learning machines have been proposed, e.g., support vector machines (SVMs) [8] and kernel Fisher discriminant (KFD) [9].

A kernel is a function κ that for all $x, z \in \mathbb{R}^N$ satisfies $\kappa(x,z) = (\Psi(x) \cdot \Psi(z))$, where Ψ is a mapping from $\mathbb{R}^N$ to an inner product feature space $\mathscr{F}$. The kernel trick used in this paper is that we first map samples of input space $\mathbb{R}^N$ into a certain high dimensional feature space $\mathscr{F}$ using a nonlinear map

$$\Psi : \mathbb{R}^N \rightarrow \mathscr{F}, x \mapsto \Psi(x), \tag{3}$$

and then pursue linear spatial filters in that feature space. It evaluates the inner product between the images of two inputs in a feature space using a kernel function [6].

Let us start with a general notion of the learning problem considered in this paper. Define $X_{A_u} = \{x_1^{A_u}, \ldots, x_R^{A_u}\}$ $(u = 1, \ldots, T_A)$ and $X_{B_v} = \{x_1^{B_v}, \ldots, x_R^{B_v}\}$ $(v = 1, \ldots, T_B)$ to be samples respectively belonging to class A and B with corresponding sample number T_A and T_B. An observed EEG sample is considered as R N-dimensional snapshots. The gross snapshots is defined as $\mathcal{X} = \{X_{A_u}\} \bigcup \{X_{B_v}\} = \{x_1, \ldots, x_p\}$ with $p = R \times (T_A + T_B)$. The signal covariances C_A^Ψ, C_B^Ψ of class A and B in the kernel space can be written as

$$C_A^\Psi = (1/T_A) \sum_{u=1}^{T_A} C_{A_u}^\Psi, \; C_B^\Psi = (1/T_B) \sum_{v=1}^{T_B} C_{B_v}^\Psi, \tag{4}$$

where $C_{A_u}^\Psi$ and $C_{B_v}^\Psi$ are signal covariances from X_{A_u} and X_{B_v}. In terms of one-sample mean $m_j^\Psi = (1/R) \sum_{i=1}^R \Psi(x_i^j)$ $(j = \{A_u, B_v\})$, $C_{A_u}^\Psi$ and $C_{B_v}^\Psi$ can be represented as

$$\begin{aligned} C_{A_u}^\Psi &= (1/R) \sum_{i=1}^R (\Psi(x_i^{A_u}) - m_{A_u}^\Psi)(\Psi(x_i^{A_u}) - m_{A_u}^\Psi)', \\ C_{B_v}^\Psi &= (1/R) \sum_{i=1}^R (\Psi(x_i^{B_v}) - m_{B_v}^\Psi)(\Psi(x_i^{B_v}) - m_{B_v}^\Psi)'. \end{aligned} \tag{5}$$

3.1 Feature Extraction of One Source

To find the linear spatial filter in $\mathscr{F}$ we need to maximize or minimize

$$J(w) = w' C_A^\Psi w - w' C_B^\Psi w, \tag{6}$$

where $w \in \mathscr{F}$. From the theory of reproducing kernels we know that any solution $w \in \mathscr{F}$ must lie in the span of all training samples in $\mathscr{F}$. Thereby, we can find an expansion for w in the form

$$w = \sum_{i=1}^p \alpha_i \Psi(x_i), \tag{7}$$

where $\alpha_i (i = 1, \ldots, p)$ are coefficients. From equation (4), we have

$$w' C_A^\Psi w = (1/T_A) \sum_{u=1}^{T_A} w' C_{A_u}^\Psi w,$$
$$w' C_B^\Psi w = (1/T_B) \sum_{v=1}^{T_B} w' C_{B_v}^\Psi w. \tag{8}$$

Combining equations (5) and (7), we get

$$\begin{aligned} w' C_{A_u}^\Psi w &= \frac{1}{R} w' \sum_{i=1}^{R} (\Psi(x_i^{A_u}) - m_{A_u}^\Psi)(\Psi(x_i^{A_u}) - m_{A_u}^\Psi)' w \\ &= \frac{1}{R} w' \sum_{i=1}^{R} (\Psi(x_i^{A_u}) \Psi'(x_i^{A_u}) - m_{A_u}^\Psi (m_{A_u}^\Psi)') w = \frac{1}{R} \alpha' N_{A_u} \alpha, \\ w' C_{B_v}^\Psi w &= \frac{1}{R} \alpha' N_{B_v} \alpha, \end{aligned} \tag{9}$$

where $\alpha' = [\alpha_1, \ldots, \alpha_p], N_S = K_S(I - \mathbf{1}_R)K'_S, (K_S)_{p \times R} (S = \{A_u, B_v\})$ is the kernel matrix for class A or class B with $(K_S)_{ij} = k(x_i, x_j^S) = (\Psi(x_i) \cdot \Psi(x_j^S))$, I is the identity matrix and $\mathbf{1}_R$ is the matrix with all entries $1/R$.

From equations (8) and (9), objective function $J(w)$ can be reformulated as

$$J(\alpha) = \frac{1}{R \times T_A} \alpha' \sum_{u=1}^{T_A} N_{A_u} \alpha - \frac{1}{R \times T_B} \alpha' \sum_{v=1}^{T_B} N_{B_v} \alpha. \tag{10}$$

This problem can be solved by the method of Lagrange multipliers with constraint $w'w = 1$ as EED described in Section 2.1.

By virtue of equation (7), the constraint can be translated into

$$w'w = \sum_{i,j=1}^{p} \alpha_j \alpha_i (\Psi(x_i) \cdot \Psi(x_j)) = \alpha' \widetilde{K} \alpha = 1, \tag{11}$$

where $\widetilde{K}_{ij} = (\Psi(x_i) \cdot \Psi(x_j))(i, j = 1, \ldots, p)$. Thus, the Lagrangian can be stated as

$$\mathcal{L}(\alpha, \beta) = J(\alpha) + \beta(\alpha' \widetilde{K} \alpha - 1) = \frac{1}{R} \alpha' (N_A - N_B) \alpha + \beta(\alpha' \widetilde{K} \alpha - 1),$$

where β is the Lagrange multiplier, and

$$N_A = (1/T_A) \sum_{u=1}^{T_A} N_{A_u}, \; N_B = (1/T_B) \sum_{v=1}^{T_B} N_{B_v}. \tag{12}$$

Now we calculate the derivative of $\mathcal{L}(\alpha, \beta)$ with respect to α and β, and let them equal to zero.

$$\begin{cases} \frac{\partial \mathcal{L}}{\partial \alpha} = 0 \\ \\ \frac{\partial \mathcal{L}}{\partial \beta} = 0 \end{cases} \Rightarrow \begin{cases} \frac{1}{R}(N_B - N_A)\alpha = \beta \widetilde{K} \alpha \\ \\ \alpha' \widetilde{K} \alpha = 1 \end{cases} \tag{13}$$

Clearly, we can find that β is the eigenvalue of matrix $(1/R)\widetilde{K}^{-1}(N_B - N_A)$ and α is the corresponding eigenvector.

Substituting equation (13) into equation (6), we obtain $J(w) = -\beta$, and therefore

$$\begin{cases} J_{\max}(w) = -\beta_{\min}, \\ J_{\min}(w) = -\beta_{\max}, \end{cases} \tag{14}$$

where $\beta_{\min} = \beta_1 \leq \beta_2 \leq \ldots \leq \beta_p = \beta_{\max}$ are eigenvalues of matrix $(1/R)\widetilde{K}^{-1}$ $(N_B - N_A)$. The optimal spatial filters w_{max} and w_{min} can be computed in terms of equation (7).

A new EEG sample $Y = [y_1, \ldots, y_R]$ after filtering by w will become

$$\begin{aligned}(w \cdot \Psi(Y)) &= \sum_{i=1}^{p} \alpha_i \Psi(x_i)'[\Psi(y_1), \ldots, \Psi(y_R)] \\ &= \sum_{i=1}^{p} \alpha_i ((\Psi(x_i) \cdot \Psi(y_1)), \ldots, (\Psi(x_i) \cdot \Psi(y_R))) \\ &= \sum_{i=1}^{p} \alpha_i (k(x_i, y_1), \ldots, k(x_i, y_R)).\end{aligned}$$

The extracted energy feature is the variance of $w'\Psi(Y)$. Thus, the energy feature vector of the new sample Y includes two entries which are respectively the variances of $w'_{max}\Psi(Y)$ and $w'_{min}\Psi(Y)$.

3.2 Feature Extraction of Multiple Sources

Suppose there are m sources to be extracted from one class of brain activity pattern. Then the m spatial filters can constitute a spatial filter bank $W = [w_1, \ldots, w_m](w_i \in \mathscr{F}, i = 1, \ldots, m)$. As a nonlinear generalization of (2), we define the discriminative criterion of KEED for seeking optimal nonlinear spatial filters as

$$J(W) = tr(W'C_A^{\Psi}W - W'C_B^{\Psi}W). \tag{15}$$

Similarly, like equation (7), w_j can be found using coefficient vectors α_j.

$$w_j = \sum_{i=1}^{p} \alpha_i^j \Psi(x_i), \tag{16}$$

where $\alpha_j = [\alpha_1^j, \ldots, \alpha_p^j]'(j = 1, \ldots, m)$.

Using (16) and a similar transformation as in (8), (9) and (12), the formulation of (15) can be rewritten as

$$\begin{aligned}J(\alpha_i) &= tr(\frac{1}{R}\begin{bmatrix} \alpha_1'(N_A - N_B)\alpha_1 & \cdots & \alpha_1'(N_A - N_B)\alpha_m \\ \vdots & \ddots & \vdots \\ \alpha_m'(N_A - N_B)\alpha_1 & \cdots & \alpha_m'(N_A - N_B)\alpha_m \end{bmatrix}) \\ &= \frac{1}{R}\sum_{i=1}^{m} \alpha_i'(N_A - N_B)\alpha_i.\end{aligned} \tag{17}$$

We constrain $W'W = I$ to find the filter W that maximizes or minimizes $J(W)$. The constraint amounts to normalizing W by

$$W'W = (w_1, w_2, \ldots, w_m)'(w_1, w_2, \ldots, w_m) = \begin{bmatrix} w_1'w_1 & \cdots & w_1'w_m \\ \vdots & \ddots & \vdots \\ w_m'w_1 & \cdots & w_m'w_m \end{bmatrix} = I. \quad (18)$$

For easy computation, here we loose this constraint and only consider diagonal constraints. That is

$$w_i'w_i = \alpha_i'\widetilde{K}\alpha_i = 1. \quad (19)$$

We use the method of Lagrange multipliers with (19) to optimize (17). The corresponding Lagrangian is

$$\mathcal{L}(\alpha_i, \beta_i) = J(\alpha_i) + \sum_{i=1}^{m} \beta_i(\alpha_i'\widetilde{K}\alpha_i - 1), \quad (20)$$

where β_i is the Lagrange multiplier.

By taking derivatives and letting them equal to zero, we get

$$\begin{cases} \frac{\partial \mathcal{L}}{\partial \alpha_i} = 0 \\ \frac{\partial \mathcal{L}}{\partial \beta_i} = 0 \end{cases} \Rightarrow \begin{cases} \frac{1}{R}(N_B - N_A)\alpha_i = \beta_i \widetilde{K}\alpha_i \\ \alpha_i'\widetilde{K}\alpha_i = 1 \end{cases} \quad (21)$$

From equation (21), we find that β_i is the eigenvalue of matrix $(1/R)\widetilde{K}^{-1}(N_B - N_A)$ and α_i is the corresponding eigenvector. Substituting equation (21) into (15), we get $J(W) = -\sum_{i=1}^{m}\beta_i$, and therefore

$$\begin{cases} J_{\max}(\Phi) = -\sum_{j=1}^{m} \beta_j \\ J_{\min}(\Phi) = -\sum_{j=N-m+1}^{N} \beta_j \end{cases} \quad (22)$$

where $\beta_{\min} = \beta_1 \le \beta_2 \le \ldots \le \beta_p = \beta_{\max}$ are the eigenvalues of matrix $(1/R)\widetilde{K}^{-1}(N_B - N_A)$.

Therefore, the optimal kernel spatial filter banks W_{max} and W_{min} which optimize (15) can be obtained in light of (16). We can see clearly that when $m = 1$, objective function (15) degenerates to function (6) with the same solution.

For a new EEG sample Y consisting of R snapshots $y_j (i = 1, \ldots, R)$, the filtered signal by the nonlinear spatial filter W will be $[(W'\Psi(y_1)), \ldots, (W'\Psi(y_R))]$.

$$\begin{aligned} (W \cdot \Psi(y_j)) &= [w_1, \ldots, w_m]'\Psi(y_j) = [(w_1 \cdot \Psi(y_j)), \ldots, (w_m \cdot \Psi(y_j))]' \\ &= [\sum_{i=1}^{p} \alpha_i^1 k(x_i, y_j), \ldots, \sum_{i=1}^{p} \alpha_i^m k(x_i, y_j)]'. \end{aligned}$$

For better classification of EEG signals, in fact, a $2m$-dimensional filtered signal is formed by $W_{max}'\Psi(Y)$ and $W_{min}'\Psi(Y)$. The energy features can be taken as the variances of these sources.

In summary, energy feature extraction of EEG signals by KEED takes four necessary steps: (1) compute matrices N_A, N_B and $\widetilde{K}$, (2) compute eigenvectors $(1/R)\widetilde{K}^{-1}(N_B - N_A)$ to create a filter, (3) compute a new filtered EEG sample by spatial filters, and (4) compute the variance on each dimension of the filtered sample to get energy feature for further classification of EEG signals.

4 Conclusion and Future Work

In the present study, KEED, a nonlinear energy feature extractor for EEG signal classification has been proposed as an extension of the linear EED method. The detailed process in finding solutions which optimize the KEED criterion is also given.

Theoretical derivation of the KEED method is the contribution of this paper. Experiments with various kernel functions to evaluate the performance of this method are equally important. This work will be considered in the future.

Acknowledgments. This work was supported in part by the National Natural Science Foundation of China (Project 60703005), Shanghai Educational Development Foundation (Project 2007CG30), and Shanghai Rising-Star Program.

References

1. Wolpaw, J.R., Birbaumer, N., McFarland, D.J., Pfurtscheller, G., Vaughan, T.M.: Brain-Computer Interfaces for Communication and Control. Clin. Neurophysiol. 113, 767–791 (2002)
2. Müller-Gerking, J., Pfurtscheller, G., Flyvbjerg, H.: Designing Optimal Spatial Filters for Single-Trial EEG Classification in a Movement Task. Clin. Neurophysiol. 110, 787–798 (1999)
3. Curran, E.A., Stokes, M.J.: Learning to Control Brain Activity: A Review of the Production and Control of EEG Components for Driving Brain-Computer Interface (BCI) Systems. Brain Cogn. 51, 326–336 (2003)
4. Sun, S., Zhang, C.: Adaptive Feature Extraction for EEG Signal Classification. Med. Biol. Eng. Comput. 44, 931–935 (2006)
5. Li, J., Sun, S.: Energy Feature Extraction of EEG Signals and a Case Study. In: Proc. Int. Joint Conf. Neural Networks, pp. 2367–2371 (2008)
6. Shawe-Taylor, J., Cristianini, N.: Kernel Methods for Pattern Analysis. Cambridge University Press, Cambridge (2004)
7. Sun, S.: The Extreme Energy Ratio Criterion for EEG Feature Extraction. In: Kůrková, V., Neruda, R., Koutník, J. (eds.) ICANN 2008, Part II. LNCS, vol. 5164, pp. 919–928. Springer, Heidelberg (2008)
8. Vapnik, V.N.: The Nature of Statistical Learning Theory. Springer, New York (1995)
9. Müller, K.R., Mika, S., Rätsch, G., Schölkopf, B., Weston, J.: Fisher Discriminant Analysis with Kernels. In: Proc. IEEE Int. Workshop Neural Netw. Signal Process, pp. 41–48 (1999)

Semantic Based Text Classification of Patent Documents to a User-Defined Taxonomy

Ashish Sureka, Pranav Prabhakar Mirajkar, Prasanna Nagesh Teli, Girish Agarwal, and Sumit Kumar Bose

Software Engineering and Technology Labs (SETLabs), Infosys Technologies Ltd., India
{Ashish_Sureka,Pranav_Mirajkar,Prasanna_Teli, Girish_Agarwal,Sumit_Bose}@infosys.com

Abstract. We present a generic approach for semantic based classification of text documents to pre-defined categories. The proposed technique is applied to the domain of patent analytics for the purpose of classifying a collection of patent documents to one or many nodes in a user-defined taxonomy. The proposed approach is a multi-step process consisting of noun extraction, word sense disambiguation, semantic relatedness computation between pair of words using WordNet and confidence score computation. The proposed algorithm resulted in good accuracy on experimental dataset and can be easily adapted and customized to other domains other the patent landscape analysis domain discussed in this paper.

Keywords: Sematic-based text classification, patent mining, natural language processing.

1 Introduction

Patent landscape analysis is performed by patent analysts, researchers, scientists, patent attorneys and decision makers to gain insight about technology trends, to perform competitor's or an organizations own patent portfolio analysis and to come up with research and product development strategies. One of the tasks performed while doing patent landscape analysis on a patent dataset is to map each patent to one or multiple nodes in a user-defined taxonomy. The analysts creates a taxonomy (a view or a perspective) based on his business objectives and given a patent document collection (for example a collection of all patents granted or filed by a competitor), the analysts classifies each patent document to one or more nodes or classes within the taxonomy. The result of this classification is then interpreted, visualized & used to make intelligent decisions.

Patent are assigned an International Patent Classification (IPC) number which is an International standard. IPC is maintained by the World Intellectual Property Organization (WIPO). However, the IPC classification number may not be much useful in situations where classifications specific to an organization's business objectives (organization or domain specific taxonomy) or an in-depth

R. Huang et al. (Eds.): ADMA 2009, LNAI 5678, pp. 644–651, 2009.

classification on a focused domain is required. IPC contains thousands of categories and subcategories covering a very large number of topics under which all human inventions can be classified. When an organization specific or a domain specific landscape analysis needs to be performed, a patent analyst first creates a taxonomy defining a technology landscape specific to his needs. The next task is to assign each patent document (from a set of patent documents on which landscape analysis needs to be performed) to one or many nodes within the taxonomy. We present an algorithm to automate this task.

1.1 Related Work and Contribution of This Paper

Patent document classification poses certain challenges (such as varying size of the documents, vocabulary & terminologies used, patent document structure, patent classification hierarchies and taxonomies) which need to be considered while designing a classifier specifically for patent domain [9]. As a result of domain specific challenges, generic text classification tools (for example algorithms & statistical models for classifying news articles) cannot be used directly (or may not result in good accuracy) for the problem of patent document classification. Based on our literature survey, most of the patent classification algorithms leverage machine learning based techniques [6], [7]. Also, there are some papers on patent classification that utilizes citation information for performing patent classification [5],[10]. Algorithms for patent classification have also been proposed for different languages such as English, Japanese and German [8].

Most of the patent classification algorithms are based on machine learning techniques. However, machine learning requires large quantities of training data for the purpose building a model. As a result of this, machine learning based algorithms are not suitable for the problem that we are trying to address in this paper. This is because pre-annotated or pre-classified data may not be available for the various classes or nodes within the user defined taxonomy. The precondition for the problem that we are addressing is that there is no training data or pre-classified data that is available. Also, creating hand-crafted classification rules is a labor intensive, non-scalable and tedious approach. Thus, another requirement for the classification system was an approach that does not rely on hand-engineering rules (for example if-else expressions to check presence of some words and phrases) by a domain expert. The above requirements motivated us to look in the direction of applying semantic based text classification techniques.

Applying semantic based text classification techniques to the problem of classifying patent documents to a user-defined taxonomy is an unexplored or relatively unexplored area (to the best of our knowledge). The contribution of this paper is a novel approach for semantic-based text classification algorithm to solve the problem at hand. The building blocks of our approach are: noun-phrase extraction, word-sense disambiguation, usage of the popular WordNet English lexical database and usage of algorithms for computing semantic relatedness between two words using Wordnet.

2 Solution Approach

There are two inputs to our system: one input is the list of patent abstracts and titles (concatenated as a single text) and the other input is a list of classes each represented by a list of keywords. We use only the title and patent abstract for the purpose of performing classification and do not make use of other information like detailed specifications and claims. This is also a general practice followed by patent analysts. The detailed specifications and claims are referred only in cases where there is a doubt. Normally, the title of the patent and abstracts are enough for performing a first level landscape analysis.

The list of key-words belonging to a single class can be regarded as a bag-of-words (BOW). We also represent the input text that needs to be classified (patent title & abstract) as bag-of-words by extracting all the nouns from input text. Hence, the bag-of-words describes both the entities (patent abstract or the class) so that the semantic similarity between two pair of entities can now be computed using just their respective BOW representations.

Step 1: Noun Extraction (NE)
The first step consists of extracting all the nouns from a patent title & abstract. The raw textual data (free-form text) is first passed through a part-of-speech tagger for extracting singular and plural nouns. We used the Stanford log-linear part-of-speech tagger [1]. We extract just the nouns and ignore other parts of speech. We capture all the occurrences of a particular noun in a sentence and each occurrence is retained for further processing in the text processing pipeline. This is because, the frequency of the occurrence of a word is also important in computing the similarity score. The input to the noun extraction module is a list of all the patent title & abstracts (as a single text document) that needs to be classified and the output is a list of nouns belonging to each of the patent abstracts. All the later computations are done using the list of nouns only and the free-form abstract is not used. Automatic noun extraction step is performed only on the input text (one of the inputs to the system) as the key-words defining the various classes (the other input to the system) are manually created by the patent analyst.

Step 2: Word Sense Disambiguation (WSD)
The second step consists of finding the intended sense for each of the key-words describing a class and each noun extracted from the patent abstracts. The extracted nouns can have different meanings when used in different contexts and hence we perform a WSD (Word Sense Disambiguation) task to identify the intended meaning of a given target word based on the context (surrounding or neighboring words). For word sense disambiguation, we use the "WordToSet" Perl package made available by the University of Minnesota SenseRelate project [2]. We pass two inputs parameters to the PERL script: a target word to which the sense needs to be assigned and a list of context words to be used for the purpose of disambiguating the target word. The target word is assigned the sense which is found to be the most related to its neighboring words or the context.

In our experiments, we use one word as the target word (a noun) and all the remaining words (nouns) as the context for the target word. The program also takes as an input parameter, the name of a "WordNet::Similarity" measure. The default value is "WordNet::Similarity::lesk" and for the purpose of our experiments we used the default setting for this parameter. Word sense disambiguation is performed for each noun in the abstract and each key-word representing a class.

Step 3: Semantic Relatedness Computation (SRC)
The next step consists of computing the semantic relatedness between all the patent abstracts and all the pre-defined classes. This task can be broken up into the following sub-tasks:

1. Compute the semantic relatedness between each word in the bag-of-words representing the patent abstracts with each word in the bag-of-words representing the classes.
2. Compute the semantic relatedness between one patent abstract and one class using results from previous step.
3. Assign the patent abstract to the most probable class.

In this step, we describe the procedure to perform the first sub-task i.e. to compute the semantic relatedness between any two words (the precondition is that the senses have been identified). The other two sub-tasks are described in the next step (Step 4). We use the "WordNet::Similarity" Perl modules for computing the semantic relatedness between two words. The similarity computation between pair of concepts (or word senses) leverages the WordNet lexical database. The "WordNet::Similarity" Perl modules supports a variety of semantic similarity and relatednes measures: Resnik, Lin, Jiang-Conrath, Leacock-Chodorow, Hirst-St.Onge, Wu-Palmer, Banerjee-Pedersen, and Patwardhan-Pedersen. The package provides six measures of semantic similarity and three measures of semantic relatedness [3]. In our experiments, we use the Perl module that implements the method proposed by Hirst-St.Onge for computing semantic relatedness of word senses. We chose the "Hirst and St-Onge" method for the purpose of our experiments as it computes semantic "relatedness" which is more general than computing semantic "similarity". Semantic relatedness is not only limited to considering "is-a" type of relations but includes other types of relations such as "has-part", "is-made-of", and "is-an-attribute-of".

Step 4: Aggregate score computation and class assignment
Computing semantic related between a text document and class is a simple aggregation of semantic relatedness scores between all word pairs and then normalizing it. The final score will be a value between 0 and 16 as Hirst and St-Onge method returns a numeric score between 0 and 16. This value can be normalized to a scale of 0 to 1 or 0 to 100. The next task after computing individual scores between each text document and class is to output the top N classes (top N guesses) for each text document and also output a confidence factor denoting how confident the system is in making its prediction. We compute the confidence

Table 1. List of patents, USPTO classes and keywords used for experiments

Category	Patent Numbers	Keywords
Apparel-Glove (2-159)	6962064, 7210171, 7213419, 7409724, 7434422	Glove, apparel, hand, arm, covering, garment, worn, textile, knit, fabric
Apparel Cap (2-195.1)	6910226, 7062793, 7152250, 7278173, 7454799	Cap, head, crown, visor, covering, hat, hair
Bottle-Cap (215-200)	3980117, 4262813, 5692629, 6426046, 6831552	Cap, bottle, plug, container, closure, vessel, seal, jar

factor for a particular class as a percentage difference in normalized score between that particular class and class having score just below it.

Worked-out example

In this section, we explain the working of the proposed algorithm using a small dataset. In the next section, we present our experimental results on a much larger input dataset and with more number of categories. We downloaded a total of 15 patents from the USPTO (United States Patent and Trademark Office). The 15 downloaded patents belonged to 3 classes. We prepared a list of key-words for each of the 3 classes. The key-words were manually selected from the class and sub-class definitions provided on the USPTO website. Table 1 lists the patent numbers of the documents that we downloaded, the class numbers and title of the 3 classes, keywords chosen for each of the classes and the mapping between the class numbers and the patent numbers. The 3 classes that we selected to test our hypothesis were chosen in such a way that there is some semantic overlap across classes. For example, Apparel-Glove and Apparel-Cap have a commonality that both the classes belong to apparel. Apparel-Cap and Bottle-Cap have the commonality that patent belongings to these categories serve a common function of providing a "cap" (one category for apparel and the other for bottle).

After running the simulations we achieved an overall accuracy of 93.33%. There was just one misclassification: a patent abstract belonging to bottle-cap category got classified into apparel-glove category. However, when we looked at the score for the misclassified patent (patent number 6426046), we found that it's semantic relatedness score for apparel-glove (score of 5.6 on a scale of 0 to 100) and bottle-cap (score of 5.437 on a scale of 0 to 100) was quite close. The correct category was the second best guess and close to the first guess. The score of patent number 6426046 for apparel-cap was 2.9228. Consider one classification case where patent number 5692629 (a bottle-cap patent) got classified into its correct category. We use this case to motivate the advantages of using WorldNet for computing semantic relatedness [2],[3],[4]. The reason for patent number 5692629 getting categorized into the correct class was because of high semantic relatedness score between two pair of words (one from key-word list and other from patent document) such as cap & surface, bottle & container, bottle & surface, plug & closure, plug & surface, container & closure, vessel & container, vessel & bottle, jar & container and jar & closure.

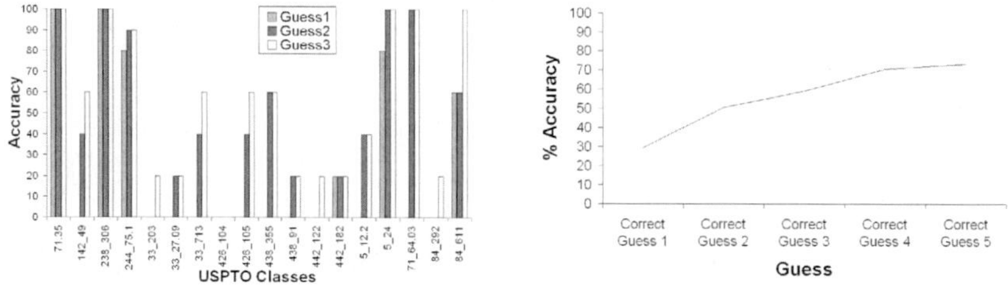

(a) Accuracy results for individual categories

(b) Accuracy results for the algorithm for top 5 guesses

Fig. 1. Accuracy results

3 Experimental Setup and Results

We downloaded a total of 105 patents (from USPTO website) distributed across a total of 18 distinct categories (10 classes plus 8 sub-classes). We downloaded 5-10 most recent patents (as of March 01st 2009) in each of the listed category. We selected categories such that diversity as well as semantic overlap is present. We selected patents belonging to diverse classes such as Bed, Wood turning, aeronautics, food, semiconductor devices, fabric, geometrical instruments, fertilizers, music and railways. Also, some categories are chosen such that they share some communality with other categories. We downloaded 5 patents each from “rotary seats” and “knockdown sofa”, both belonging to the “Bed” class. Similarly, we downloaded 5 patents each from “Metal” and “Rhythm” category, both belonging to the “Music” class. The keywords for each of the category were extracted from the class and sub-class definition provided on USPTO website. We applied the algorithm on the experimental dataset to compute the semantic relatedness score of each patent for each of the 18 categories. The number of execution runs for the algorithm was: 105*18 = 1890. The algorithm outputs the top 5 guesses for each of the patent (this is configurable but for this work we set the value of top N guesses as 5). Each guess is one of the category IDs (USPTO Class & Sub-class). Figure 1(b) shows accuracy results across for the first five guesses. The accuracy achieved for the first 5 guesses are: 29.52, 50.48, 59.05, 70.48 and 73.33 respectively. We notice that almost 30% of the time Guess 1 is correct which increases to 50% if we take into account the second guess also. Also it should be noted that, if a patent is classified into a same parent class but under different sub-class, it is considered as misclassification (we don’t count a correct patent classification but incorrect child classification as a hit).

Our next analysis consists of plotting accuracy results for each of the individual categories separately and draws conclusions. Figure 1(a) shows accuracy results of all 18 categories for the first 3 guesses. We noticed that the accuracy for certain classes are very high whereas for some classes the accuracy are low. For example, the accuracy of the category ID 71.35 for the first guess itself is 100%.

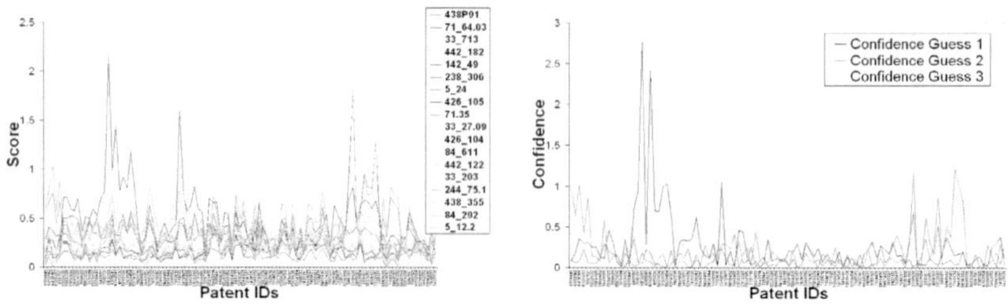

(a) Semantic relatedness score for all the categories

(b) Semantic relatedness score for only the top 3 guesses

Fig. 2. Plot of semantic relatedness score for each of the 105 patents

This means that all the patents belonging to that class are correctly classified in the first guess itself. We observe that for 2 of the 18 classes, the algorithm is able to predict the correct class with 100% accuracy in the first guess itself. According to Figure 1(b), we observe that for 4 classes the algorithm is able to achieve 100% accuracy in including the first two guesses and for 5 classes the algorithm is able to achieve 100% accuracy in including the first three guesses. For 6 classes we are able to get 90% or more accuracy in first 3 guesses. We noticed that for some of the categories the accuracy of the algorithm is less than 60% after 3 guesses also.

Figure 2(a) shows a plot of semantic similarity score (Y-axis) between each patent and each class (X-axis). Patents on X-Axis are arranged according to their classes i.e. all patents belonging to the same class are grouped together. The graph provides useful insights on the working of the algorithm. We observe few instances of spikes or crest in Figure 2(a). For example, for class 238-306, we see a clearly visible spike, which means that all patents in this range got a comparatively much higher score for class 238-306. One interpretation from Figure 2(a) is that for a group of patents belonging to the same class having no such easily differentiable spike or crests means that the algorithm is giving very close scores for all the patent classes. Our conclusion from Figure 2(a) is that the algorithm is highly confident in assigning patents to its correct class for certain categories whereas for patents belonging to some classes, the scores between correct class and incorrect class are quite close (even though the final class assignment is correct). The graph in Figure 2(b) is similar to graph in Figure 2(a) except that the semantic relatedness score for all 18 categories are plotted in Figure 2(a) whereas Figure 2(b) just focuses on the top 3 guesses. We notice presence of few instances of crest and spikes in the graph for Guess 1 as well as Guess 2. We observe that in few cases, the crests corresponding to second guess is more apparent as compared to Guess 1 and Guess 3. This means that for such instances Guess 2 category is clearly distinguished with respect to other categories.

4 Conclusion

We present a generic, adaptable and a scalable approach for semantic based classification of text documents to pre-defined categories or a user-defined ontology. We apply the proposed algorithm to the domain of patent landscape analysis and present experimental results. We conclude that semantic based text classification can be used to classify patent documents to a user defined taxanomy (for patent landscape analysis) in situations where hand-crafted rule based approach and machine learning based approach cannnot be applied.

References

1. Toutanova, K., Klein, D., Manning, C., Singer, Y.: Feature-Rich Part-of-Speech Tagging with a Cyclic Dependency Network. In: Conference of the North American Chapter of the Association For Computational Linguistics on Human Language Technology, Edmonton, vol. 1, pp. 252–259 (2003)
2. Patwardhan, S., Banerjee, S., Pedersen, T.: SenseRelate: TargetWord-A Generalized Framework for Word Sense Disambiguation. In: Annual Meeting of the Association for Computational Linguistics on Interactive poster and demonstration sessions, Ann Arbor, pp. 73–76 (2005)
3. Pedersen, T., Patwardhan, S., Michelizzi, J.: WordNet: Similarity - Measuring the Relatedness of Concepts. In: Nineteenth National Conference on Artificial Intelligence, San Jose, pp. 1024–1025 (2004)
4. Richardson, R., Smeaton, A.F., Murphy, J.: Using WordNet as a knowledge base for measuring semantic similarity between words. In: Artificial Intelligence and Cognitive Science Conference, Dublin (1994)
5. Li, X., Chen, H., Zhang, Z., Li, J.: Automatic Patent Classification using Citation Network Information: An Experimental Study in Nanotechnology. In: The 7th ACM/IEEE-CS Joint Conference on Digital Libraries, Vancouver, pp. 419–427 (2007)
6. Chu, X., Ma, C., Li, J., Lu, B., Utiyama, M., Isahara, H.: Large-scale Patent Classification with Min-max Modular Support Vector Machines. In: International Joint Conference on Neural Networks, pp. 3973–3980 (2008)
7. Koster, C.H.A., Seutter, M., Beney, J.G.: Multi-Classification of Patent Applications with Winnow. In: Broy, M., Zamulin, A.V. (eds.) PSI 2003. LNCS, vol. 2890, pp. 546–555. Springer, Heidelberg (2004)
8. Fall, C.J., Benzineb, K., Guyot, J., Trcsvri, A., Fivet, P.: Computer-Assisted Categorization of Patent Documents in the International Patent Classification. In: International Chemical Information Conference, Nmes (2003)
9. Larkey, L.S.: Some Issues in the Automatic Classification of U.S. patents. In: AAAI 1998 working notes (1998)
10. Varma, K.I., Mirajkar, P.P., Sureka, A.: An Algorithm for Classifying Articles and Patent Documents Using Link Structure. In: The Ninth International Conference on Web-Age Information Management, pp. 203–210. IEEE Press, Zhangjiajie (2008)

Mining Compressed Repetitive Gapped Sequential Patterns Efficiently

Yongxin Tong, Li Zhao, Dan Yu, Shilong Ma, Zhiyuan Cheng, and Ke Xu

State Key Lab. of Software Development Environment, Beihang University,
100191 Beijing
{yxtong,lzh,yudan,slma,zycheng,kexu}@nlsde.buaa.edu.cn

Abstract. Mining frequent sequential patterns from sequence databases has been a central research topic in data mining and various efficient mining sequential patterns algorithms have been proposed and studied. Recently, a novel sequential pattern mining research, called mining repetitive gapped subsequences, has attracted the attention of many researchers. However, the number of repetitive gapped subsequences generated by even these closed mining algorithms may be too large to understand for users, especially when support threshold is low. In this paper, we propose the problem of how to compress repetitive gapped sequential patterns. A novel distance measure of repetitive gapped sequential patterns and an efficient representative pattern checking scheme, δ-dominate sequential pattern checking are proposed. We also develop an efficient algorithm, CRGSgrow (Compressing Repetitive Gapped Sequential pattern grow), including an efficient pruning strategy, SyncScan. An empirical study with both real and synthetic data sets clearly shows that the CRGSgrow has good performance.

Keywords: repetitive gapped sequential pattern, compressing frequent patterns.

1 Introduction

Sequential pattern mining has been a central data mining research topic in broad applications, including analysis of web log, analysis of frequent sequential patterns in DNA and protein sequence, API specification mining from open source repositories, and so on. So far many efficient sequential mining algorithms have been proposed for solving various of real problems, such as the general sequential pattern mining[2, 8, 14], frequent episode mining[7], closed sequential pattern mining[10, 13], maximal sequential pattern mining[6], constraint-based sequential pattern mining[9], etc..

In recent years, some studies have focused on a novel problem of sequential pattern mining, mining repetitive gapped sequential patterns [3, 5]. By gapped sequential patterns it means a sequential pattern, which appears in a sequence in a sequence database, possibly with gaps between two successive events. In addition, for brevity, we use the term sequential pattern instead of gapped sequential patterns in this paper. Although a few approaches have been proposed to solve how to mine all or closed repetitive sequential patterns [3], they cannot avoid an explosive number of output frequent repetitive sequential patterns for Apriori property. Hence, we should find a

R. Huang et al. (Eds.): ADMA 2009, LNAI 5678, pp. 652–660, 2009.

solution to compress the result set of repetitive sequential patterns with a smaller number of representative repetitive sequential patterns.

Recently the problem of compressing frequent itemset has been studied [1, 11, 12], and some algorithms, such as RPglobal and RPlocal[11], have been presented. So, Can we compress repetitive sequential patterns by extension of the approach of compressing frequent itemset? Unfortunately, the answer cannot be so optimistic owing to the two following reasons. Firstly, there is not the one-step algorithm like RPlocal in mining repetitive sequential patterns with very small probability, since RPlocal is not able to solve event order in a sequence. Secondly, although RPglobal algorithm can be applied to this problem, RPglobal consume high computational costs.

In this paper, we propose and study the problem of compressing repetitive gapped sequential patterns. Inspired by the ideas of summarizing frequent itemsets[11, 12], Firstly, to obtain the high-quality compression, we propose a novel distance to measure the quality which shows the similarity between sequential patterns. Secondly, according to the distance threshold given by users, we define δ-sequence cover in order to choose representative repetitive sequential patterns. Finally, to discover minimize the number of representative repetitive sequential pattern, we develop an algorithm, CRGSgrow, including an efficient pruning strategy, *SyncScan*, and an efficient representative pattern checking scheme, *δ-dominate sequential pattern checking*. Empirical results with both real and synthetic data sets prove that the CRGSgrow algorithm can compress repetitive gapped sequential patterns efficiently.

The rest of the paper is organized as follows. The problem formulation will be introduced in Section 2. Section 3 focuses on an efficient and effective algorithm, CRGSgrow. We discuss our experimental results in Section 4, give our conclusion in Section 5, respectively.

2 Problem Formulation

In this section, we formally define the problem of compressing repetitive gapped sequential patterns. Firstly, we define a new distance measure based on Jaccard distance of repetitive gapped sequential patterns. Secondly, we put forward some concepts, such as δ- sequence cover and δ- dominate sequential pattern. Finally, we show the problem of compressing repetitive gapped sequential patterns is equivalent to minimal set-covering problem which is a well-known NP-Hard problem.

To measure the similarity between two repetitive gapped sequential patterns, we need a reasonable measurement criterion. Hence, we propose a novel Jaccard distance for measuring the similarity between repetitive sequential patterns.

Definition 1 (Distance between two repetitive gapped sequential patterns). Let P_1 and P_2 be two repetitive gapped sequential patterns. The distance is defined as:

$$D(P_1, P_2) = 1 - \frac{\text{min-ins}\{S(P_1) \cap S(P_2)\}}{\text{max-ins}\{S(P_1) \cup S(P_2)\}} \quad (1)$$

Where S(P) is the set of sequences in the given sequence database which contains the sequential pattern P, min-ins$\{S(P_1)\bigcap S(P_2)\}$ is the least support between P_1 and P_2 in every sequence including them, and max-ins$\{S(P_1)\bigcap S(P_2)\}$ is the most support between P_1 and P_2 in every sequence including them.

Inspired by the concept of δ-cover in [11], we use similar definition of δ-sequence cover to formulate the above intuition. Note δ is a threshold of distance between two repetitive sequential patterns specified by users and $\delta \in [0,1]$.

Definition 2 (δ-sequence cover). A repetitive sequential pattern P is δ-sequence covered by another repetitive sequential pattern RP if $P \subseteq RP$ and $D(P,RP) \le \delta$ ($\delta \in [0,1]$).

According to the definition of δ-sequence cover, we can simplify the above definition of distance between two repetitive gapped sequential patterns:

$$D(P,RP) = 1 - \frac{\text{min-ins}\{S(P)\bigcap S(RP)\}}{\text{max-ins}\{S(P)\bigcup S(RP)\}} = 1 - \frac{\sup(RP)}{\sup(P)} \qquad (2)$$

Since checking δ-sequence cover between any two sequential patterns will spend much time in computing, we will introduce some novel properties about compressing repetitive sequential pattern to speed up the checking δ-sequence cover.

Definition 3 (min sequence cover). Given a set of repetitive sequential patterns S, min sequence cover of SP (MSC(SP) in short), a repetitive sequential pattern in S, is define as follows:

$$\text{MSC(SP)}\begin{cases} \min\{\text{D(SP,SP}_i)|\forall \text{SP}_i \in \text{S,SP} \subseteq \text{SP}_i\} & \exists \text{SP}_i \in \text{S} \\ +\infty & \forall \text{SP}_i \notin \text{S} \end{cases} \qquad (3)$$

Definition 4 (δ-dominate sequential pattern). Given a set of repetitive sequential patterns S, SP is a repetitive sequential patterns in S. SP is a δ-dominate sequential pattern in S, if MSC(SP)>δ. Equivalently, SP can not be δ-sequence covered by any repetitive sequential patterns in S, if MSC(SP)>δ.

Theorem 1. Given a set of repetitive sequential patterns S, and any set of representative repetitive sequential patterns, RS, which can δ sequence cover S. Then, RS must contain all δ-dominate sequential patterns in S.

Proof: If DP is any δ-dominate sequential pattern in S, based on the definition of δ-dominate sequential pattern, the MSC(DP) in S must larger than δ. Thus, DP can not be δ sequence covered by any repetitive sequential pattern in S. So, DP must be a representative sequential pattern in any set of representative sequential patterns RS that can δ sequence cover S. □

Definition 5(Repetitive Gapped Sequential patterns Compression). Given a sequence database SeqDB, a minimum support min_sup and distance threshold δ, the compressing repetitive gapped sequential patterns is to find a set of representative repetitive gapped sequential pattern RRGS, such that for each frequent repetitive gapped sequential pattern P (w.r.t min_sup), there exits a representative repetitive gapped sequential pattern $RP \in RRGS$ (w.r.t min_sup) which δ-sequence cover P, and the |RRGS|, the size of set of representative repetitive gapped sequential pattern RRGS, is minimized.

Theorem 2. The problem of compressing repetitive gapped sequential patterns is NP-Hard.

Proof: Proved in [11] □

3 Compressing Repetitive Gapped Sequential Patterns Algorithm

In this section, we elaborate an algorithm, CRGSgrow, for compressing repetitive gapped sequential patterns. The CRGSgrow adopts a two-step approach: in the first step, we obtain all closed repetitive sequential patterns as the candidate set of representative repetitive sequential patterns, and at the same time get all δ-dominate sequential patterns; in the second step, we only find the remaining the representative patterns from the candidate set. We firstly introduce the design and implementation of the CRGSgrow, and then analyze the time complexity of all our algorithms.

Algorithm 1: CRGSgrow

Input: sequence database **SeqDB**={ $S_1, S_2, ..., S_n$ }; threshold **min_sup**; a distance threshold δ
Output: A set of representative repetitive sequential patterns
1: $E \leftarrow$ all frequent 1-sequential patterns in SeqDB; Cover $\leftarrow \varnothing$; Covered $\leftarrow \varnothing$;
2: **for each** $e \in E$ **do**
3: **if** e is visited **then** continue;
4: $P \leftarrow e$; $I \leftarrow \{(i, <l>) \mid \text{for some } i, S_i[l] = e\}$;
5: $P' \leftarrow \varnothing$; $I' \leftarrow \varnothing$
6: SyncScan (SeqDB, P, I, P', I', Cover, Covered);
7: Cover $\leftarrow$ Compress (Cover, Covered);
8: **return** Cover;

In algorithm1, the CRGSgrow traverses the pattern space in a depth-fisrt way. In the process of SyncScan, the Cover and Covered sets will be also updated continuously. Especially, all δ-dominate sequential patterns will be obtained in this process. At last, Compress (Cover, Covered) will finish all the compression work, and will be shown in algorithm 2

Algorithm 2: Compress

```
Input: δ-dominate sequential patterns set Cover; other closed frequent sequential patterns set Covered
Output: A set of compressed repetitive sequential patterns set
for each RP in Cover do
   for each P in Covered do
      if RP can δ sequence cover P then
         Put P into Set T;
  T̄ ← Cover - T
for each SP in T̄ do
   for each CR in Covered-Cover do
      if CR can δ sequence cover SP then
         Put SP into Set(CR);
While T̄ ≠ ∅ do
   select a sequential pattern CR which can maximize | Set(CR) |
   for each SP∈ Set(CR) do
         Remove SP from T̄ and other Set(CR′) (CR′ ∈ Covered-Cover)
return Cover
```

According to the algorithm 2, it is an important problem how to obtain δ-dominate sequential patterns efficiently. In the following, we will propose an algorithm called SyncScan to solve this problem in algorithm 3. The process of searching δ-dominate sequential patterns will be conducted together with the process of closed sequential patterns mining in this algorithm. Meanwhile, a reasonable pruning strategy will be applied to improve the efficiency of the algorithm. In addition, in Subroutine Check (P), we will firstly check if the current pattern can be pruned using LBCheck[3]. Then, if the current pattern is closed, we will do the δ-dominate checking with the method, DomCCheck(P).

Algorithm 3: SyncScan

```
Input: sequence database SeqDB={ S_1, S_2, ..., S_n }; threshold min_sup; Pattern P= e_1, e_2, ...e_{j-1} ; P' = e_1, e_2, ...e_{j-1} or ∅ ; leftmost support set I of P in SeqDB; semi-left support set I' of pattern P' in SeqDB; δ-dominate sequential patterns set, Cover; a set of other closed repetitive sequential patterns, Covered
Output: Cove; Covered
Check(P, I, Cover, Covered);
if P' ≠ ∅ then
  P' ← {P − e_1} ∪ P' ; Check(P', I', Cover, Covered);
ev ← the second event of P;
for each e ∈ α do
```

```
      if length of P=2 and all ev occurs in the same sequence with P then
          P' ← ev ;  I' ← {(i,<l>) | for some i, S_i[l]=ev except instances in P};
          obtaining two non-overlapping instance sets of I+ and I+' with e;
          SyncScan(SeqDB, P∘e , I+, P'∘e , I+');
      else P' ← ∅ , I' ← ∅ ;
          obtaining the non-overlapping instance sets of I+ with e;
          SyncScan(SeqDB, P∘e , I+, P'∘e , I+');
Subroutine Check(P)
Input: sequence database SeqDB; Pattern P δ-dominate sequential patterns set, Cover; a set of other closed repetitive sequential patterns, Covered
Output: Cover, Covered
      if |I| ≥ min_sup && LBCheckprune(P) and DomCCheck(P) ≠ nclosed then
      if DomCCheck(P) = δ-dominate then
          Cover ← Cover ∪ P ;
      else Covered ← Covered ∪ P ;
```

To sum up, we will take an example to further explain the pruning strategy above.

Example 1: Table 1 shows a sequence database $SeqDB = \{S_1, S_2\}$. We will compute sup (ABC) and sup(BC) simultaneously in the way of algorithm 3. The complete search space of computing support of each repetitive sequential pattern forms the lexicographic sequence tree shown in Figure1. Thus, we can compute supports of both ABC and BC in the same process, since BC is a subsequence of ABC. We will explain each step as follows:

Table 1. An Example of Sequence Database

Sequence_ID	e_1	e_2	e_3	e_4	e_5	e_6	e_7	e_8
S_1	A	B	B	C	B	A	C	B
S_2	B	A	C	B	A	C	B	C

1) Find a support set I^A of A. I^A is the set of all instances of A and is shown the rectangle labeled I^A.

2) Find a support set I^{AB} of 'AB'. Meanwhile, we also start to compute the support of the patterns with the prefix of 'B', which are not contained in instances of the pattern 'AB' in the sequence database. Those instances of patterns prefixed with 'B' included in the instances of patterns prefixed with 'AB' will not be searched again.These patterns will be pruned and marked as the dotted rectangle of I^B in Figure 1.

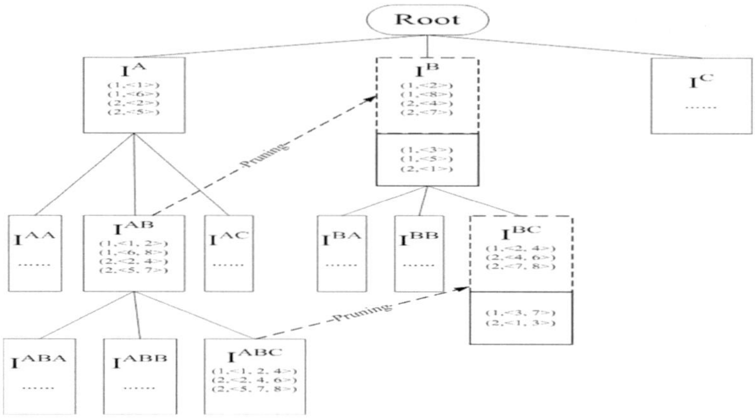

Fig. 1. SyncScan Pruning Strategy

3) Find a support set I^{ABC} of ABC. Similar to step 2, for there is no 'C' to be extended for (1, <6, 8>) in S_1, we stop extending (1, <6, 8>). For (1, <1, 2>) in S_1, it can be extended to (1, <1, 2, 4>). Then we will continue searching in S_2. In the end, we finish computing sup (ABC) =3 and sup(BC)=5 in the same process.

4 Empirical Results

In this section, we report a systematic performance study on both real data sets and synthetic data sets. All of our experiments were performed on a Lenovo ThinkPad T60 with Intel 4200 CPU, 1GB memory and Windows XP professional installed. Algorithms were implemented in Microsoft Visual C++ V6.0. In the experiments, we compared CRGSgrow with a fast closed repetitive gapped sequential pattern mining algorithm CloGSgrow[3] and another compressing frequent patterns algorithm RPglobal[11], using both real data sets and synthetic data sets.

The first data set, *Gazelle*, contains 29,369 web click-stream sequences from customers and 1423 distinct items, which has been a benchmark dataset used by past studies on sequential pattern mining. Although the dataset is sparse since the average sequence length is only, there are a large number of long sequences (the maximum length is 651), where a sequential pattern may repeat many times. More detailed information about this data set can be found in [4].

The second data set, *TCAS* dataset, is a set of software traces collected from Traffic alert and Collision Avoidance System. The dataset contains 1578 sequences and 75 distinct items. The average sequence length of the dataset is 36 and maximum sequence length is 70. More information about this data set can be found in [5].

The third data set, *D5C20N10S20*, is a synthetic set generated by IBM sequence data generator [2]. The data generator requests a set of parameters, D, C, N and S, corresponding to the number of sequences, the average sequence length, the number of distinct items, and the maximum sequence length respectively.

We carry out our experiments to compare three algorithms in the above three dataset mainly on the compression quality and running time. Moreover, we vary the support threshold and fix $\delta = 0.2$ (it is a reasonably good compression quality).

In the experiments of compression quality, as the figures 2-4 shown, we have the following observations: firstly, the number of representative repetitive gapped sequential patterns by CRGSgrow is a little more than the number of patterns generated by RPglobal, and the number of patterns outputted by CRGSgrow is about one-quarter of that closed repetitive gapped sequential patterns mined by CloGSgrow; secondly, in the algorithm of CRGSgrow, we can obtain the δ -dominate sequential patterns which include the most of representative repetitive sequential patterns. In addition, to verify the effectiveness of δ -dominate sequential patterns we proposed in our work, we also make the experiments on the number of δ -dominate sequential patterns which are showed as pink line in figure 2-4. Moreover, if an algorithm cannot finish within 60 minutes, we do not show the results. In the experiments of running time, as the figures 5-7 shown, the running time of CRGSgrow is much less than the time of RPglobal, and is very close to the time of CloGSgrow.

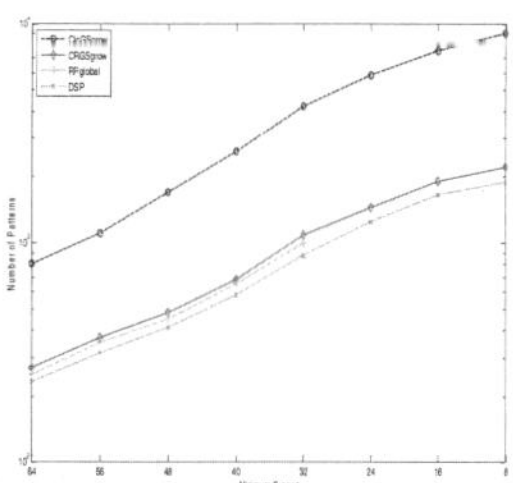

Fig. 2. Num of Patterns in Gazelle

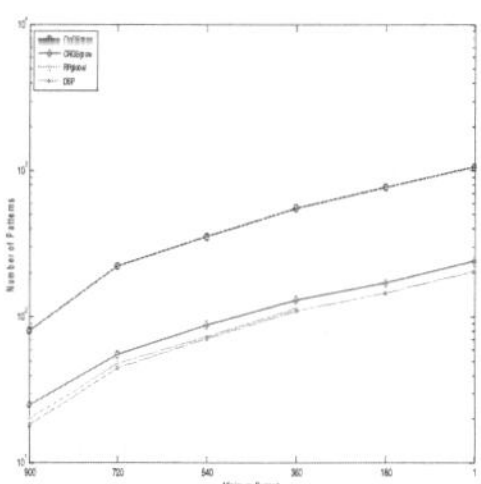

Fig. 3. Num of Patterns in TCAS

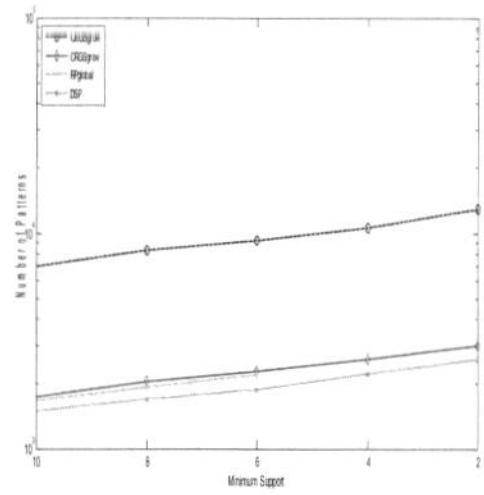

Fig. 4. Num of Patterns in D5C20N10S20

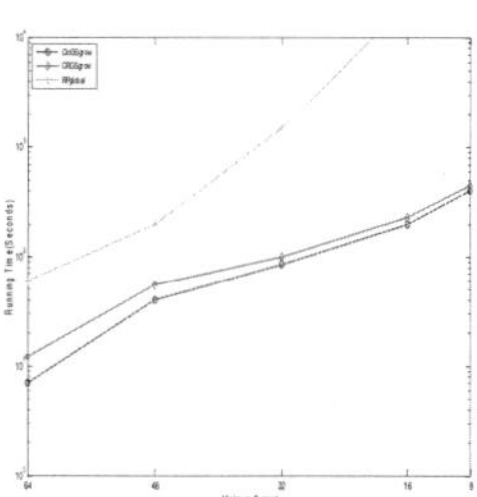

Fig. 5. Running Time in Gazelle

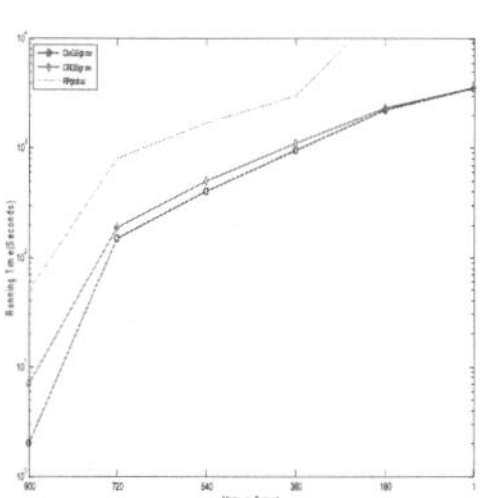

Fig. 6. Running Time in TCAS

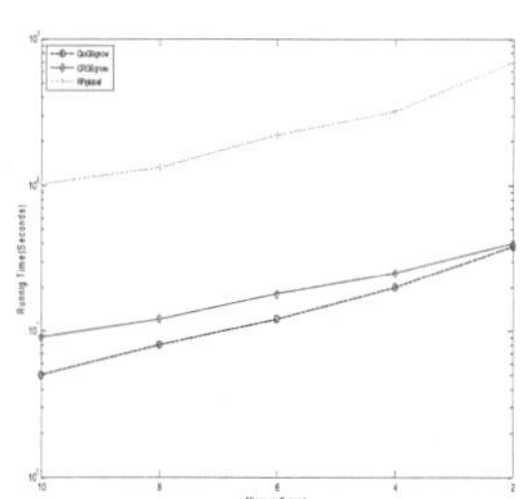

Fig. 7. Running Time in D5C20N10S20

5 Conclusion

This paper studies how to effectively and efficiently compress repetitive gapped sequential patterns from sequence database. To the best of our knowledge, the problem of compressing the repetitive gapped sequential patterns has not been well studied in existing work.

In this paper, we propose and study the problem of compressing repetitive gapped sequential patterns. Inspired by the ideas of compressing frequent itemsets, Firstly, to obtain the high-quality compression, we design a novel distance to measure the quality which shows the similarity between repetitive sequential patterns. Secondly, according to the distance threshold given by users, we define δ -sequence cover in order to choose representative repetitive sequential patterns. Finally, since the problem of compressing repetitive gapped sequential patterns is equivalent to minimizing the number of representative repetitive sequential patterns, we develop an algorithm, CRGSgrow, including an efficient pruning strategy, *SyncScan*, and an efficient representative pattern checking scheme, δ *-dominate sequential pattern checking*. Empirical results prove that the algorithm CRGSgrow can obtain a good compressing quality efficiently.

References

1. Afrati, F., Gionis, A., Mannila, H.: Approximating a Collection of Frequent Sets. In: Proceedings of the 10th ACM SIGKDD International Conference on Knowledge Discovery and Data Mining, pp. 12–19 (2004)
2. Agrawal, R., Srikant, R.: Mining Sequential Patterns. In: Proceedings of the 11th IEEE International Conference on Data Engineering, pp. 3–14 (1995)
3. Ding, B., Lo, D., Han, J., Khoo, S.-C.: Efficient mining of closed repetitive gapped subsequences from a sequence database. In: Proceeding of the 25th IEEE International Conference on Data Engineering, pp. 1024–1035 (2009)
4. Kohavi, R., Brodley, C., Frasca, B., Mason, L., Zheng, Z.: KDD Cup 2000 Organizers' Report: Peeling the Onion. SIGKDD Explorations 2, 86–98 (2000)
5. Lo, D., Khoo, S.-C., Liu, C.: Efficient mining of iterative patterns for software specification discovery. In: Proceedings of the 13th ACM SIGKDD International Conference on Knowledge Discovery and Data Mining, pp. 460–469 (2007)
6. Luo, C., Chung, S.M.: A scalable algorithm for mining maximal frequent sequences using a sample. Knowledge and Information System 15, 149–179 (2008)
7. Pei, J., Han, J., Mortazavi-Asl, B., Pinto, H., Chen, Q., Dayal, U., Hsu, M.: PrefixSpan: Mining Sequential Patterns Efficiently by Prefix-Projected Pattern Growth. In: Proceedings of the 17th IEEE International Conference on Data Engineering, pp. 215–224 (2001)
8. Tong, Y., Zhao, L., Yu, D., Ma, S., Xu, K.: Mining compressed repetitive gapped sequential patterns efficiently. Technical Report, NLSDE, Beihang University (2009), `http://arxiv.org/abs/0906.0885`
9. Xin, D., Han, J., Yan, X., Cheng, H.: Mining Compressed Frequent-Pattern Sets. In: Proceedings of International Conference on Very Large Data Bases, pp. 709–720 (2005)
10. Yan, X., Han, J., Afshar, R.: CloSpan: Mining Closed Sequential Patterns in Large Datasets. In: Proceddings of SIAM International Conference on Data Mining, pp. 166–177 (2003)
11. Yan, X., Cheng, H., Han, J., Xin, D.: Summarizing Itemset Patterns: A Profile-Based Approach. In: Proceedings of the 11th ACM SIGKDD International Conference on Knowledge Discovery and Data Mining, pp. 314–323 (2005)

Mining Candlesticks Patterns on Stock Series: A Fuzzy Logic Approach

Mario Linares Vásquez[1], Fabio Augusto González Osorio[2], and Diego Fernando Hernández Losada[3]

[1] Economic Optimization Research Group. National University of Colombia, Bogotá
[2] Intelligent Systems Research Lab. National University of Colombia, Bogotá
[3] Economic Optimization Research Group. National University of Colombia, Bogotá
{mlinaresv,fagonzalezo,dfhernandezl}@unal.edu.co

Abstract. Candlesticks is a technique used in financial time series in order to forecast future market performance. With candlesticks patterns, traders build active trading strategies in order to buy, sell or hold securities. The process is based on a preliminary stage which consists in identifying individual basic shapes on time series. Identifying candlesticks basic shapes is easy for a human, but recognizing complex patterns is hard because a lot of data is available. In this paper a data mining model for building active trading strategies (using candlesticks assumptions) is proposed looking for frequent itemsets on symbolic stocks series. Model validation is achieved with real data from New York Stock Exchange.

Keywords: Financial market analysis, fuzzy classification, candlesticks, patterns, active trading strategies.

1 Introduction

Stocks forecasting is a growing research field in financial computing. Its foundation is to predict the future market performance using available information. According with the information used in the process, there are two approaches used in the forecasting task: *Technical Analysis and Fundamental Analysis.* In Technical Analysis, historical stock prices are used for the process. So, the aim is to predict futures prices, tendencies and season patterns, through charts and technical indicators. This approach is based on three premises: *price discounts everything, price moves in trends. and history repeats itself.* In Fundamental Analysis, the purpose is to identify the economic forces that cause market action; all the information available is necessary for the analysis (fundamentals, prices, indices, news, financial statements).

In the financial computing case, researches have been oriented to forecasting using prices and transformations of this prices as inputs to the models. Several methods have been used in the process such as econometric models [1], neural networks, support vector machines, evolutionary computation and temporal data mining. Neural networks is the most prolific of these, but financial computing models are driven to predict individual prices and indices as in [2] and [3].

R. Huang et al. (Eds.): ADMA 2009, LNAI 5678, pp. 661–670, 2009.

Candlesticks is a graphical tool used to identify patterns in stocks charts. Patterns are compositions of candlesticks basic shapes, and are associated with a market state. In spite of candlesticks are widely used for technical analysis, the research effort to exploit this technique in computational models has been very poor. With candlesticks patterns and the assumption that it works, the investor can make beliefs about future states of market in order to forecast and build active trading strategies. If the objective is to implement a computational model to build active strategies assuming that candlesticks works, two stages have to be implemented: candlesticks recognition and trading strategies design. So, this paper presents a model to identify candlesticks on real series using fuzzy classification, and a design trading strategies process using a catalog of patterns extracted from real data.

The paper is organized as follows. Section 2 is a summary of candlestick concepts. In Section 3 the model for designing active trading strategies is presented. In Section 4 experiments and results with real data from NYSE are presented; experiments include validation for classification and stocks selection with proposed model. Finally, Section 5 draws conclusions and further work.

2 Candlestick Patterns

Charts and technical indicators are tools used in technical analysis to forecast prices and market performance over time [4]. Charts are the graphical representation of stock prices (open, close, high and low), while technical indicators are "measures" of market behavior. One of these charts is *Candlesticks*.

With candlesticks charts, stocks prices are drawn in one shape with a body and two shadows. Candlestick body is a rectangle that represents the difference between the open and close daily prices. When close price is higher than open the body is white, and in other case the body is black. The lines above and below the body are called shadows. In candlestick charts, the relation among prices are recognized through different basic candlesticks and the historical relation is represented with candles configurations (candle patterns). A combination of body and shadows magnitudes represent a candlestick type, and each one has a meaning or representation of market psychology. Basic candlesticks are (Figure 1) Marubozu (1), Long days (2), Long lower shadow (3), Shaven Head (4), Hanging Man (5), Hammer (6), Spinning Top (7), Inverted Hammer (8), Shooting Star (9), Shaven Bottom (10), Long upper shadow (11) and Dojis (12)[1]. Candlestick patterns are configurations of sequential candlesticks. These patterns represent the market performance that reflects the agents behavior and traders mentality. Therefore, the patterns are used to recognize reversal points or trends continuation. In a reversal point the trend is broken and it takes the reversal way, for example a bullish market changes into a bearish market and vice versa. When the trend carry on the same way, the point is a trend continuation. A full catalogue of candlesticks and patterns is in [4] and [9]. If a pattern represents

[1] Doji candlestick has four types: neutral, long legged, Dragonfly doji, Gravestone doji.

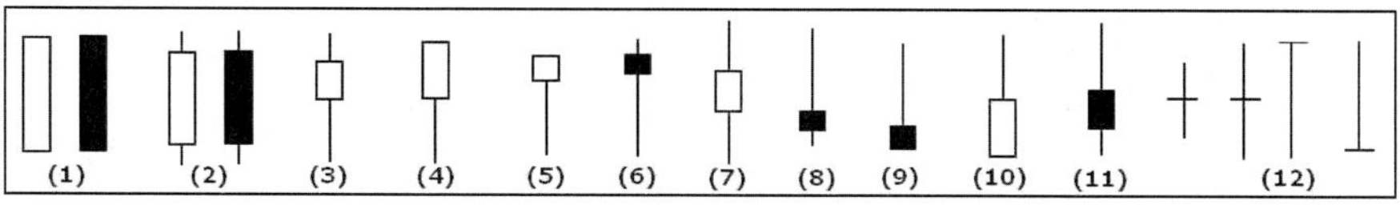

Fig. 1. Candlesticks - Basic shapes

a reversal point is catalogued as a reversal pattern, and if it represents a trend continuation then is catalogued as a continuation pattern.

3 Trading Strategies Model

Technical analysis is an approach to develop trading strategies using historical data, and candlesticks is a tool used to identify signals (reversal or continuation points) on the market in order to buy or selling financial instruments. Under the assumption that candlestick technique works, this paper presents a model based on candlesticks patterns to select stocks with bullish market perspective. In this model, candlesticks shapes are identified with a fuzzy classifier and then patterns are organized as sequences of basic shapes using a prefix tree. Patterns in prefix tree are used to suggest trading strategies according with nature states of each pattern.

3.1 The Fuzzy Classification Model

The human process of identifying candlestick is a visual process of patterns recognition, which is driven by linguistic rules expressed in natural language. These rules define magnitude relations between candlestick body and shadows) and have been built by the observation of lots of stock data. An example of these rules is *if upper shadow is very short or short, body is short and lower shadow is long, then candlestick is a hammer*. So, the conceptual candlestick universe is qualitative. In order to implement a computational model to classify candlesticks, is necessary a technique that handles qualitative information expressed as linguistic rules. In this way fuzzy logic is selected by his power to handle with linguistic rules. The process of identifying candlesticks is a classification task. Components of fuzzy classification system are described below.

System classification input are time series for each stock price. In order to use this data in a fuzzy fashion, crisp time series must be transformed to a natural representation that will be used in fuzzy rules. These rules are evaluated with candlestick magnitudes. Transformations of stock prices to obtain candlesticks magnitudes are defined as follows:

$$body = \frac{|open - close|}{|high - low|}, \qquad upShadow = \begin{cases} \frac{|high - open|}{|high - low|}, & open > close \\ \frac{|high - close|}{|high - low|}, & open \leq close \end{cases},$$
$$lowShadow = \begin{cases} \frac{|close - low|}{|high - low|}, & open > close \\ \frac{|open - low|}{|high - low|}, & open \leq close \end{cases}. \tag{1}$$

A candlestick is represented like a tuple $C = \{body, upShadow, lowShadow\}$. Each tuple must be evaluated with the fuzzy sets defined for the problem. Fuzzy sets used to describe magnitudes of candlesticks parts are *very short, short, medium, long and very long*:

$$\mu_{veryshort}(x) = \begin{cases} 1 - 20x, & x \le 0.05 \\ 0, & x > 0.05 \end{cases}, \qquad \mu_{short}(x) = \begin{cases} 10x, & x \le 0.1 \\ 1, & 0.1 < x \le 0.3 \\ 2.5 - 5x, & 0.3 < x \le 0.4 \\ 0, & x > 0.4 \end{cases},$$

$$\mu_{medium}(x) = \begin{cases} 0, & x \le 0.3, x > 0.7 \\ 10x - 3, & 0.3 < x \le 0.4 \\ 1, & 0.4 < x \le 0.6 \\ 7 - 10x, & 0.6 < x \le 0.7 \end{cases}, \mu_{long}(x) = \begin{cases} 0, & x \le 0.5 \\ 5x - 2.5, & 0.5 < x \le 0.7 \\ 1, & 0.7 < x \le 0.9 \\ 1 - 10x, & x > 0.9 \end{cases},$$

$$\mu_{verylong}(x) = \begin{cases} 0, & x \le 0.9 \\ 10x - 9, & x > 0.9 \end{cases}. \tag{2}$$

For each candlestick shape (Figure 1) is defined a fuzzy rule and each rule represents a class for the process. Fuzzy rules general expression is: *IF* x *is* A^i *and* y *is* B^i *THEN* $z = c^i$, with A and B fuzzy sets, and c a crisp value(fuzzy singleton) for each $i = 1, 2, .., 12$. This kind of rules are used in the *simplified fuzzy reasoning method* [7]. In this method, consequent part is a class assignation. Min operator is used for rules antecedent evaluation. Fuzzy Rules used to describe each candlestick shape are:

1. IF *body* is *verylong* and *upShadow* is *veryshort* and *lowShadow* is *veryshort* THEN c = M (MARUBOZU)
2. . IF *body* is *long* and *upShadow* is *short* and *lowShadow* is *short* THEN c = L (LONG)
3. IF *body* is *medium* and *upShadow* is *short* and *lowShadow* is *medium* THEN c = LLS (LONG-LOWER-SHADOW)
4. IF (*body* is *long* and *upShadow* is *veryshort* and *lowShadow* is *short*) or (*body* is *medium* and *upShadow* is *veryshort* and *lowShadow* is *medium*) THEN c = SH (SHAVEN-HEAD)
5. IF *body* is *short* and *upShadow* is *veryshort* and *lowShadow* is *long* THEN c = HM (HANGING-MAN)
6. IF *body* is *short* and *upShadow* is *short* and *lowShadow* is *long* THEN c = H (HAMMER)
7. IF (*body* is *short* and *upShadow* is *medium* and *lowShadow* is *medium*) or (*body* is *short* and *upShadow* is *short* and *lowShadow* is *medium*) or (*body* is *short* and *upShadow* is *medium* and *lowShadow* is *short*) THEN c = ST (SPINNING-TOP)
8. IF *body* is *short* and *upShadow* is *long* and *lowShadow* is *short* THEN c = IH (INVERTED-HAMMER)
9. IF *body* is *short* and *upShadow* is *long* and *lowShadow* is *veryshort* THEN c = SS (SHOOTING-STAR)
10. IF (*body* is *long* and *upShadow* is *short* and *lowShadow* is *veryshort*) or (*body* is *medium* and *upShadow* is *medium* and *lowShadow* is *veryshort*) THEN c = SB(SHAVEN-BOTTOM)

11. IF *body* is *medium* and *upShadow* is *medium* and *lowShadow* is *short* THEN c = 11 (LONG-UPPER-SHADOW)
12. IF *body* is *veryshort* THEN c = D (DOJI)

A tuple C_i is evaluated with all the fuzzy rules, so the rule with higher antecedent defines the class for the tuple. If antecedent value for each rule is equals to zero, then UNDEFINED class is assigned to the tuple.

3.2 Patterns Catalogue Model

With fuzzy classification, stock series are transformed in symbolic series which are easier for interpretation. Symbolic representation is a sequence of characters where each one represents a candlestick shape. Real candlesticks recognition process includes additional information such as color and relative position for consecutive shapes. Pattern in Table 1 has a black long, a white hammer, and a black Marubozu; Table 1 present the pattern and different representations. Extensions used in the model are:

- **Color:** if difference between open and close prices is greater or equals to zero then the body is black (B), in other case body is white(W).
- **Relative position:** position of two consecutive shapes is defined through location of last shape, according with regions defined by characteristic points of the first shape; these points are high, $max(open, close)$, average price, $min(open, close)$, low, and the regions defined by these points are A,B,C,D, E,F (Figure 2). So the relative position is the range of regions a shape is overlapping with respect to its previous shape.

The proposed model for extracting patterns consist in building a catalogue of patterns using historical data. With this purpose, a pattern is defined by two elements: the pattern sequence and the nature state (Figure 3). The former is the symbolic representation of candlesticks shapes and the later is the state defined

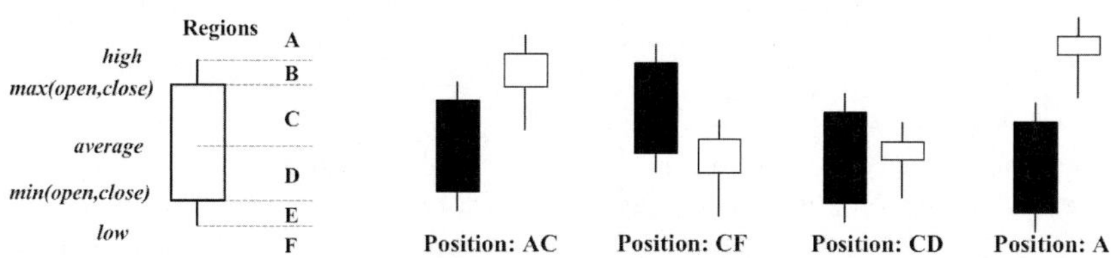

Fig. 2. Relative position

Table 1. Symbolic representations for candlesticks

Pattern	Shapes	Shapes + color	Shapes + position	Shapes + color + position
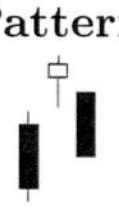	L, H, M	L-B, H-W, M-B	L, H-A, M-EF	L-B, H-W-A, M-B-EF

by the tendencies after and before the sequence. Nature states are continuation-bull, continuation-bear, reversal-bull and reversal-bear, and tendencies (past and future) for each sequence are defined with a linear regression on the averages prices of candlesticks of tendencies period. If the β coefficient of regression is greater than zero then the tendency is *bull*, if β is lesser than zero then tendency is *bear*, in other case tendency is *side*.

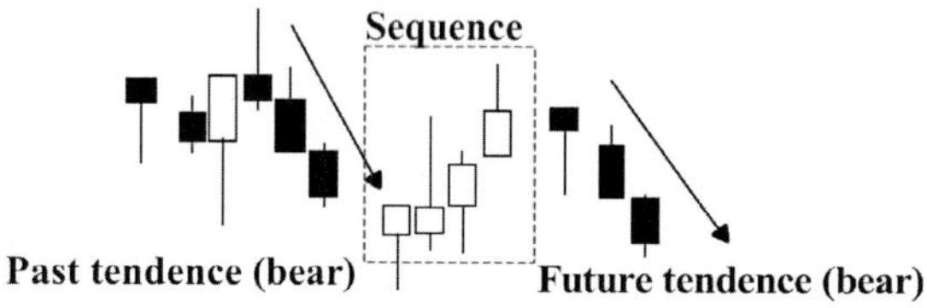

Fig. 3. Pattern

Patterns catalogue is a prefix tree which contains all sequences in the dataset and states associated to each sequence. Prefix tree structure is used to organize patterns in a frequent itemsets approach. A path in the tree is a shapes sequence and leaves are the states of that sequence. Each leaf contains how many times the sequence has appeared in the dataset with that state. In Figure 4, the pattern [L, HM] is illustrated with 50 occurrences for continuation bear and 10 occurrences for continuation bull.The catalogue represents a dictionary of historical patterns for a dataset. So, with a catalogue, traders can evaluate a pattern for a day and know historical behavior; with that knowledge, decision makers can buy or hold those securities with bullish perspective, and sell those one with bearish perspective. The algorithm for the proposed model is:

1. Codify stocks with fuzzy classifier to get symbolic representation. User parameters for codification are: stocks universe, pattern size, tendencies size, extensions for the representation and analysis day. Analysis day is the day selected for the user to building strategies. So, stock series instances are selected since first stock trading day to analysis day(not inclusive).
2. Build prefix tree (patterns catalog), extracting patterns from symbolic series. Patterns are extracted using codification parameters.
3. For each stock in the universe, extract sequences which end at analysis day and past tendencies associated to each sequence.
4. Look for sequences of previous step in prefix tree and get leaves of those sequences related to past tendency. For example if past tendency is bear and sequence is a path in the tree, extract only *cont. bear* and *rev. bear* leaves; if past tendency is bull and sequence is a path in the tree, extract only *cont. bull* and *rev. bull* leaves.
5. Select stocks with bullish perspective using an user selection threshold[2]. That is, select a stock if probability of the bullish state (continuation bull or reversal bear) is greater than user threshold. For example, a stock sequence has a bear past tendency, and in prefix tree has 80 times for continuation bear and 20 times for reversal bear; if user threshold is 0.6 then stock is not selected for buy because the bullish state of the sequence has a occurrence probability of 0.2.

[2] User selection threshold is a number $h \in [0, 1]$.

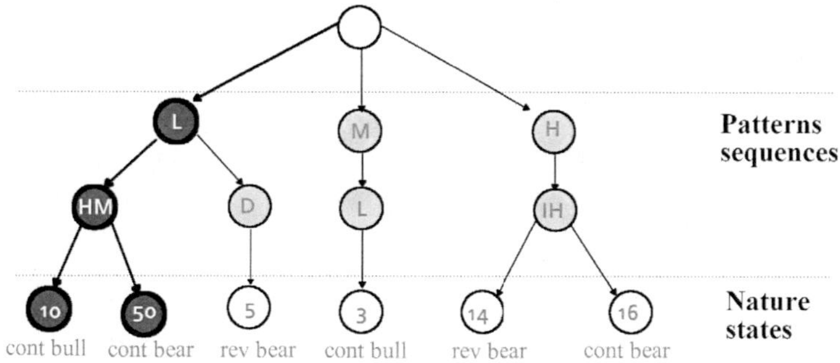

Fig. 4. Prefix tree (catalogue pattern)

4 Experiments and Results

In order to validate the model, a dataset composed with real time stocks series of NYSE (New York Stock Exchange) is used in experimentation. Dataset has 30 time series of NYSE companies, each one with the four prices (open, high, low, close). Time series were downloaded from yahoo finance web site (http://finance.yahoo.com) selecting prices since January 2 of 1962 to September 20 of 2007. The experiments realized are described as follows:

- **Experiment No 1:** the objective is to measure the performance classifying the full data set with two rules sets. The first set includes rules for marubozu, long, spinning top and doji candlesticks. The second rules set includes rules for all the candlesticks displayed in Figure 1. With this experiment, the results will show if the short rules set is enough to identify the candlesticks in the dataset.
- **Experiment No 2:** the objective with this experiment is classify a subset of the whole dataset, and compare the labels generated with the model versus labels generated manually on the same subset. The subset selected is composed with the first 5000 instances of Apple prices.
- **Experiment No 3:** the objective with this experiment is validate the model as a binary classification process. Stocks selection in this model is a binary classification because it suggests to buy stocks with bullish perspective, or to sell stocks with bearish perspective. So, the experiment consist of validate classification with real behavior of stocks since 02/01/2006(dd/mm/yyyy) to 04/01/2009 using several model parameter combinations.

4.1 Experiment Results

Table 2 shows classification percentage for each class using the two rules set in Experiment 1. With Exp. 1 is concluded that for a good classification is necessary use the full rules set. In this case the short rules set assigns 44.86% of the instances to the UNDEFINED class, while the full rules set assigns 1.58% of the dataset to the UNDEFINED class. Experiment 2 shows that 92.3% of selected

Table 2. Experiment No 1 Results

Class	Short rules set	Full rules set
MARUBOZU	14097 (6.56%)	13308 (6.20%)
LONG	38912 (18.12%)	31252 (14.55%)
LONG LOWER SHADOW	-	13365 (6.22%)
SHAVEN HEAD	-	32866 (15.30%)
HANGING MAN	-	7203 (3.35%)
HAMMER	-	5736 (2.67%)
SPINNING TOP	49726 (23.15%)	36811 (17.14%)
INVERTED HAMMER	-	5275 (2.46%)
SHOOTING STAR	-	6447 (3.00%)
SHAVEN BOTTOM	-	30788 (14.33%)
LONG UPPER SHADOW	-	11081 (5.16%)
DOJI	17840 (8.31%)	17257 (8.03%)
UNDEFINED	94206 (43.86%)	3392 (1.58%)

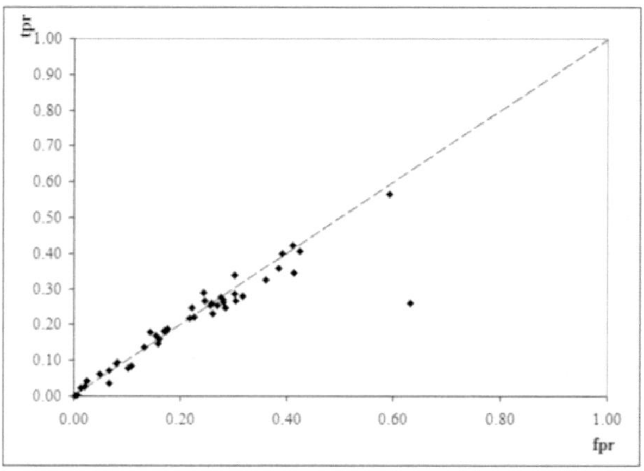

Fig. 5. TPR vs FPR - Stocks selection experiments

dataset is classified correctly, confirming the performance of model proposed. In stocks selection experiment, a binary classification process is realized. Positive class is bullish stock and negative class is bearish stock. Real classes for stocks are determined with behavior of next day; if difference between close prices is positive or equals to zero stock is bullish, in other case is bearish. Values used in confusion matrix to validate the model are True Positive Rate(stocks correctly classified as bullish), False Positive Rate(stocks wrong classified as bullish), True Negative Rate(stocks correctly classified as bearish) and False Negative Rate(stocks wrong classified as bearish).

Figures 5 and 6 show error rate and TPR vs FPR points for 52 combinations of parameters for stocks selection model. In Figure 5 a dashed line is drawn to identify model performance versus a random classifier. Each point in Figure 5 represents performance of a parameters combination for the model. Parameters

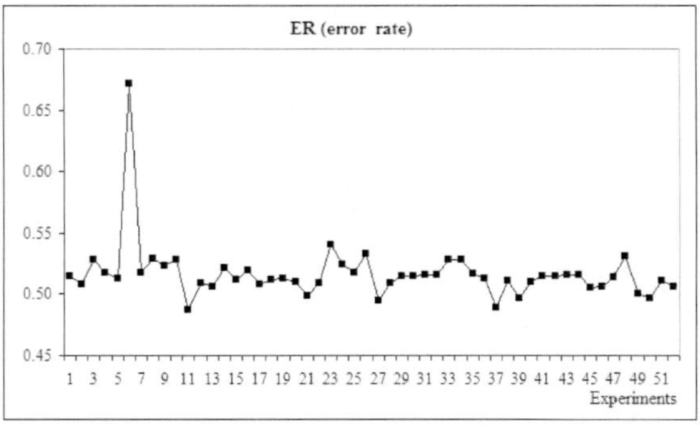

Fig. 6. Error Rate - Stocks selection experiments

combination include: pattern size (2,3,4), tendencies size(5,10,15), representation extensions (the four possible representations) and selection threshold (0.55, 0.6).

5 Conclusions and Further Work

The importance and contribution of proposed model is that it can be used as foundation to implement models to recognize candlesticks patterns using data mining techniques. Experiments show how the fuzzy classifier has a good performance. In the stocks selection case, results show how the selection based on candlesticks has a performance similar to a random classifier. Values used for selection threshold and results in Figure 5 suggest than patterns in catalogue do not have a representative nature state and do not provide additional information for support the decision process. With the results for this model, a conclusion is that models based on candlesticks for stocks selection in trading strategies do not provide useful information for design trading strategies. There are two possible reasons for this; the first one is that representation proposed in the model is not appropriate for the process and do not model the reality on candlesticks; the second is that candlesticks does not work. But for decide this, is necessary a statistical validation in order to conclude if candlesticks work or not.

In this way, two research lines for further work are proposed . The first one includes make experiments with greater datasets and tune fuzzy sets and rules using an evolutionary approach as proposed in [8]. The second research line is to extend the model for making automatic pattern recognition for candlesticks validation (validate patterns reported in [4] and [9]), and perhaps discover and report new patterns through mining lots of data in stocks time series.

References

1. Tsay, R.: Analysis of Financial Time Series. Wiley Interscience, New York (2005)
2. Yao, J., Tan, C.L., Poh, H.L.: Neural networks for technical analysis: A study on KLCI. Journal of theoretical and applied finance 2, 221–241 (1999)

3. Enke, D., Thawornwong, S.: The use of data mining and neural networks for forecasting stock market returns. Expert Systems with Applications 29, 927–940 (2005)
4. Murphy, J.: Technical Analysis of the Financial Markets. New York Institute of Finance, New York (1999)
5. Linares, M., Hernández, D., González, F.: Exploiting stock data: a survey of state of the art computational techniques aimed at producing beliefs regarding investment portfolios. Engineering and Research Journal 28, 105–116 (2008)
6. Clements, M.: Forecasting economic and financial time-series with non-linear models. Journal of Forecasting 20, 169–183 (2004)
7. Tanaka, H.: An introduction to fuzzy logic for practical applications. Springer, New York (1997)
8. Gomez, J., Garcia, A., Silva, S.: Cofre: A fuzzy rule coevolutionary approach for multiclass classification problems. In: IEEE Congress on Evolutionary Computation, pp. 1637–1644. IEEE Press, New York (2005)
9. Nison, S.: Japanese Candlestick Charting Techniques. Prentice Hall Press, New York (2001)
10. Lee, C., Liu, A., Chen, W.: Pattern Discovery of Fuzzy Time Series for Financial Prediction. IEEE Transactions on Knowledge and Data Engineering 18, 613–624 (2006)

JCCM: Joint Cluster Communities on Attribute and Relationship Data in Social Networks*

Li Wan[1,2], Jianxin Liao[1,2], Chun Wang[1,2], and Xiaomin Zhu[1,2]

[1] State Key Laboratory of Networking and Switching Technology,
Beijing University of Posts and Telecommunications,
100876 Beijing
[2] EBUPT Information Technology Co., Ltd,
100083 Beijing
{wanli,liaojianxin,wangchun,zhuxiaomin}@ebupt.com

Abstract. JCCM (Joint Clustering Coefficient Method) algorithm was proposed to identify communities which are cohesive on both attribute and relationship data in social networks. JCCM is a two-step algorithm: In the first step, it clusters tightly cohesive cliques as community cores and we proposed a heuristic method to identify community cores with a probabilistic guarantee to find out all community cores. In the second step, JCCM assigns the community cores and peripheral actors into different communities in a top-down manner resulting in a dendrogram and the final clustering is determined by our objective function, namely Joint Clustering Coefficient (JCC). To consider the power of actors in different roles in community identification, we defined two regimes of communities, namely "union" and "autarchy". Experiments demonstrated that JCCM performs better than existing algorithms and confirmed that attribute and relationship data indeed contain complementary information which helps to identify communities.

Keywords: Graph clustering, Community identification, Social network.

1 Introduction

In many applications, attribute and relationship data are both available, carrying complementary information. Formally we call the datasets contain both attributes and relationships informative graphs (see Def.3.4). Social networks could be exactly represented by informative graphs. Vertices of informative graphs represent actors with attribute value in social networks and edges of informative graphs represent social relationships among actors in social networks.

Identifying communities in social networks is an interesting and challenging problem in social network analysis. Obviously both attribute and relationship data are

* This work was jointly supported by: (1) National Science Fund for Distinguished Young Scholars (No. 60525110); (2) National 973 Program (No. 2007CB307100, 2007CB307103); (3) Development Fund Project for Electronic and Information Industry (Mobile Service and Application System Based on 3G).

R. Huang et al. (Eds.): ADMA 2009, LNAI 5678, pp. 671–679, 2009.

necessary for describing the underlying patterns of communities. A community is naturally defined by a group of actors who share similar interests (attribute data) and whose total number of internal links is greater than the total number of external links (relationship data). So if we represent a social network by an informative graph, community identification is to discover clusters which are cohesive on both attribute and relationship data [1].

While almost all the existing literatures only focus on indentifying community on relationship data solely. The existing algorithm of identifying community falls into four classes: 1) divisive algorithm find disjoint community [2,8,9]; 2) divisive algorithm find joint community [5]; 3) agglomerative algorithm find disjoint community [6] and 4) agglomerative algorithm find joint community [3,4,10,11].

This paper proposed the JCCM (Joint Clustering Coefficient Method) algorithm which combines agglomerative and divisive methodology to identify overlapping community in informative graphs. JCCM could also work on social networks only having relationship data, just by supposing that all the actors in the social network have the same attributes values. We make the following contributions:

1) We proposed a two-step algorithm JCCM to identify overlapping community in informative graphs (see Def.3.4). In the first step (an agglomerative step), the proposed algorithm utilizes a heuristic approach to cluster tightly overlapped cliques which form the community cores (see section4.1).
2) We defined two regimes of community, namely "union community" and "autarchy community". The overlapping patterns of cliques in community cores decide the regimes of communities.
3) We proposed a Joint Clustering Coefficient (JCC) to qualify community clustering by both considering attribute and relationship data. and defined different role-based similarity functions to measure the distance between actors and community cores inheriting different regimes.
4) Experiments demonstrate that JCCM performs better than the clustering algorithms which only consider relationship or attribute data.

The remainder of this paper is organized as follows: after the related work Section, the problem is formulated in Section 3. Section 4 details the JCCM algorithms. Section 5 describes experiments. Conclusions and future works are contained in Section 6.

2 Related Works

Ester et al. studied a general joint cluster model which considers both attribute and relationship data naturally and simultaneously in [7]. In the paper, the Connected k-Center (CkC) problem is proposed, which is to group data objects into k clusters with the goal of minimizing the maximum distance of any data object to its corresponding cluster center, meanwhile satisfying an internal connectedness constraint on the relationship data. Due to the NP-hardness of CkC, a heuristic algorithm is introduced. Different from the CkC model, the joint cluster model, Connected X Clusters, discussed in [1] focuses on the problem of automatically identifying the appropriate number of clusters. Ref. [1] proposes a JointClust algorithm to identify disjoint communities in informative graphs.

Li et al. [13] form overlapping clusters using both the structure of the network and the content of vertices and edges. There are two phase in the algorithm, the first one finds densely connected "community cores", which are defined as densely overlapped cliques. In the second phase, further triangles and edges whose content (assessed using keywords) is similar to that of the core are attached to the community cores.

Unfortunately, the algorithms proposed in Ref. [1] and [7] could not find overlapping community. Ref. [13] just empirically defines "densely overlapped cliques".The proposed JCCM algorithm identifies all the overlapping communities by considering relationship data and attribute data simultaneously without any a-priori knowledge. Not as JointCluster algorithm [1] could only work on informative graphs, our algorithm could work both on informative graphs and graphs without vertices attributes.

3 Preliminaries

Definition 3.1 (Clique adjacent parameter) Two maximal cliques with size L and S are r-adjacent, if they share O vertices, $r = \frac{(S-1)}{(L+S-O)}$, supposing $L \geq S$ .It's clear that $0 \leq r \leq 1$ and if $O = S-1$,r could get its largest value.

Definition 3.2 (Edge Clustering Coefficient) Clustering coefficient of an edge e_{uv} is defined as $C(e_{uv}) = \frac{|\Gamma(u) \bigcap \Gamma(v)|}{|\Gamma(u) \bigcup \Gamma(v)|}$, $\Gamma(u)$ and $\Gamma(v)$ denote the sets of neighboring vertices of u and v respectively. This coefficient measures the ratio of u and v occurring in the same cliques.

Definition 3.3 (Clique Centrality). An actor's clique centrality is defined as the number of cliques which include it. Large clique centrality means the given actor joins many overlapping cliques in social networks. In a way, clique centrality indicates actors' leadership in social networks.

Definition 3.4 (Informative graph). Given a set of n data objects $O = \{o_1, ..., o_n\}$. Let $A(o_i)$ be the attribute values of o_i and $R(o_i, o_j) = 1$ iff there is a relationship between o_i and o_j. An informative graph is a graph G = (V, E, w_v) with following properties:

1) Each vertex $v_i \in$ V corresponds to a uniquely defined data object $o_i \in O$ such that w(v_i) = A(o_i).

2) There is an edge between v_i and v_j iff $R(o_i, o_j) = 1$.

Definition 3.5 (Cluster graph). Let G = (V, E, w_v) be an informative graph. Let { V_1, ..., V_k }, k $\geq$ 2 be a partition of V , i.e., V = $V_1 \cup ... \cup V_k$ and, $V_i \bigcap V_j$j = ∅ ∀1 ≤ i < j ≤k. A graph CG is called cluster graph, if CG = (V_c , E_c),where

$V_c = \{ V_1, \ldots V_k \}$ and $E_c = \{\{ V_i, V_j \} \mid \exists i \in V_i, j \in V_j, (i, j) \in E \}$ and V_i is internally connected in G.

LEMMA 3.1. Minimum edge connectivity

Let graph Q = (V, E) be a graph contains two r-adjacent cliques, and $0.5 \le r \le 1$, $|V| = n > 3$. The edge connectivity of Q cannot be smaller than $\left\lfloor \frac{n}{2} \right\rfloor$, namely, we have to remove at least $\left\lfloor \frac{n}{2} \right\rfloor$ edges in Q to split the two cohesive cliques into two separated sub-graphs.

The proof of the lemma is detailed in [14]. Based on this lemma, we theoretically defined "tightly" cohesive cliques as cliques whose adjacent parameter is larger than 0.5 (i.e. r>0.5).

4 Finding Communities

In a way, prominent properties of actors in a community indicate the property of the community. As shown in Fig.1, if actors in a community have almost the same degree centrality, which means there is not an explicit "leader" who relatively interacts to more actors in the community than others. We call the community inherits this regime "Union Community". In reverse, we call the communities including explicit leaders "Autarchy Community".

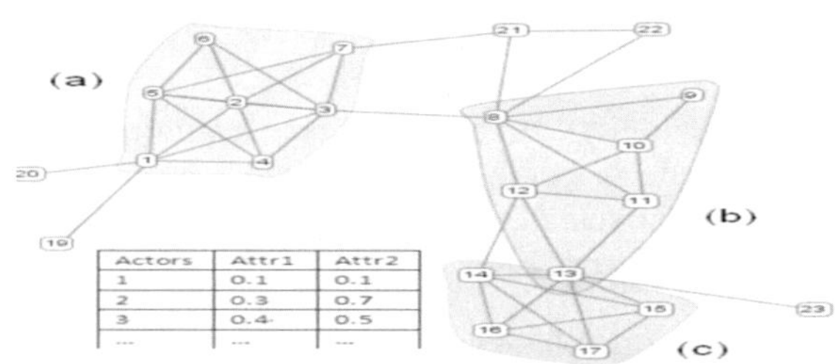

Fig. 1. A informative and cluster graph of three communities. The table contains the attribute data of every actor. The actors circled by polygons are community cores, the others are peripheral actors.

4.1 Community Core

Community cores are the densely overlapping cliques, and the overlapping parameter r must be larger than 0.5(see def. 3.1). The regimes of community are separated by whether there are explicit leaders in communities.

Unfortunately, generating all maximal cliques in a graph is an NP-Hard problem [15]. So we proposed a heuristic method to find community cores. JCCM randomly initialize some vertices in graph as centroids, then assigns vertices to the initial centroids to form cliques. After getting the maximal cliques contain centroids, JCCM merges the maximal cliques and neighboring vertices of these cliques into community cores under the constraint of overlapping parameter r. To guarantee properly clustering every community in the graph, we must guarantee that there would be at least one initial centroid is placed in each true cluster.

Theorem 4.1. Given a graph G(V,E), Let $V_1,\ldots,V_k$ be the partition of V and $|V_i|$ the number of vertices contained in V_i. The probability that we draw exactly one vertex of each community V_i in k trails is between $(\frac{m}{|V|})^k \times k!$ and $\frac{k!}{k^k}$ (where m is the minimum size of a community) .

To illustrate the probabilities in Theorem4.1, we give the following example. Let n = 2000, the minimum size m = 100 and k = 20, then $p = 2.4\times10^{-8}$. Thus, the chance of placing one of the initial centroid in each cluster is very small. In order to guarantee p to a certain value, we draw a larger number of initial centroids than the maximal possible number of clusters (i.e. s>k).

Theorem 4.2. Given a graph G(V,E), Let $V_1,\ldots,V_k$ be the partition of V and $|V_i|$ the number of vertices contained in V_i. Let |V|=n, m is the minimum size of a community and $k = \left\lceil \frac{n}{m} \right\rceil$, In order to choose at least one vertex from each community with a probability of p, we need to draw at least $s = \left\lceil k \ln \frac{k}{-\ln p} \right\rceil$ initial centroids.

Table 1. Number of centroid required for n=2000 and m=100

P	0.9	0.95	0.99	0.999
S	105	120	153	199

The values in Table 1 refer to the same example as described before (n = 2000 and m =100). As described in Theorem4.2, s is the number of initial centroids we have to place in order to guarantee the responding confidence p .

4.2 Joint Clustering Coefficient

Dealing with attribute data, the Silhouette Coefficient [16] is a measurement, which qualifies the clustering quality independent of the number of clusters. By combining a connection constraint to Silhouette Coefficient, we get Joint Clustering Coefficient as follows:

(Joint Clustering Coefficient). Given a graph G = (V, E) and cluster graph CG=(Vc, Ec), the Joint Clustering Coefficient (JCC) is $s = \frac{1}{|V|}\sum_{i\in V}\frac{b(i)-a(i)}{\max\{b(i),a(i)\}}$, where a is the cluster to which i was assigned, a(i) denotes the distance from i to the center of cluster a, b(i) is the average distance from i to all neighboring clusters of "a" in CG. a(i) and b(i) are both calculated by the following function f.

$$f(a,i)=\begin{cases}\dfrac{1}{|a_{uc}|}\sum_{j\in a_{uc}}\sum_{\{u,v\}\in S(i,j)}\dfrac{d(i,j)}{w_{uv}} & (1)\\ \dfrac{1}{|L(a_{ac})|}\sum_{j\in L(a_{ac})}\sum_{\{u,v\}\in S(i,j)}\dfrac{d(i,j)}{w_{uv}} & (2)\\ \dfrac{1}{|C(a_{ac})|}\sum_{j\in C(a_{ac})}\sum_{\{u,v\}\in S(i,j)}\dfrac{d(i,j)}{w_{uv}} & (3)\end{cases}$$

$f(a,i)$ is the similarity function to calculate the distance from a vertex i to the center of a cluster "a". The distance between vertices and cluster centers is defined by both relationship and attribute data. $d(i,j)$ denotes the Euclidean distance in attribute space, namely the similarity in attribute value between i and j. w_{uv} denotes the edge clustering coefficient (see Def.3.2) of edge (u, v) and $S(i,j)$ denotes the shortest path from actor i to j. Equation (1) computes the distance from an actor i to an "union community core" a, where a_u denotes the set of actors in a to which actor i connects. Equation (2) and (3) compute the distance from i to an "autarchy community core".

4.3 JCCM Algorithm

JCCM is an agglomerative and divisive algorithm to find overlapping communities. In first phase, the number of initial centroids is determined by theorem 4.2.

JCCM Algorithm
INPUT: Informative graph G(V, E), p(the probability of guarantee to find all true communities), r(the threshold of clique overlapping parameter), m(the minimum size of communities)
OUTPUT:overlapping communities in graph G(V, E)

1: \\ FIRST PHASE: Finding community cores
2: randomly initialize k vertices as centroids, the value of k is determined by theorem 4.2
3: find cliques contain centroids and extend them into community cores under the constraint of parameter r
4: ClusterGraph=constructCG(community core, G)
5: \\ SECONDE PHASE: Maximizing JCC
6: while the set of edges in "ClusterGraph" is not null
7: delete a randomly selected edge of "ClusterGraph"
8: calculate JCC and reserved it with the corresponding community division
9: goto line 6
10: output the community division corresponds to the largest JCC

The computational complexity of the first phase is $O(s\times M\times D^2\times n^2)$, where n is the number of vertices in graph G, s is the number of initial centroids, M is the largest number of cliques a community core contains and D is the largest degree in graph G. Since $n \gg s,M,D$, we consider s, M, D as constants, therefore the runtime of the first phase is $O(n^2)$. The computational complexity of splitting cluster graph in the second phase is $O(n^3)$. So the runtime of JCCM is $O(n^3)$.

5 Experiments

We compare JCCM with CPM[1] [3] , CONGA [5] and X-mean[2]. [17] algorithms on both informative graphs and graphs without attribute data.The results in Table3 and 4 are gotten by JCCM under the setting that r = 0.5 and m=100.

We test all algorithms on real-world co-authorship networks and use F-measure to numerically measure them.

F-measure provides a value between precision and recall, closer to the lower of them.

- *Recall*: the fraction of vertex pairs labeled with the same labels which are also clustered in the same community.
- *Precision*: the fraction of vertex pairs which are clustered in the same community which are also labeled with the same labels.
- $F\text{-}measure = \dfrac{2 \times recall \times precision}{recall + precision}$

The co-authorship networks were generated based on the two well-known scientific literature digital databases citeseer[3] and DBLP[4]. We chose papers written between 2000-2004, 2002-2006 and 2003-2007, belonging to three different research areas: Theory, Machine Learning, and Databases & Data Mining to construct the co-authorship networks. We employed the term frequency inverse-document frequency [18] to the most important keywords occurring in the abstracts of the corresponding papers and attached them as attribute data to each author. The co-authorship was used to generate the relationship data. Authors were labeled with the majority areas of their papers.

Table 3 shows that JCCM performs better than X-Mean and CPM on community identification in informative graph. This indicates that considering attributes and connecting structure simultaneously is necessary. Table 4 shows JCCM beats CPM and

Table 2. Description of datasets

Dataset	Number of authors	Number of keywords	P[5]
Dataset1	1942	603	3.2%
Dataset2	2103	720	4.0%
Dataset3	3198	887	5.5%

[1] CPM is implemented in CFinder (version-2.0-beta3).

[2] http://sourceforge.net/projects/weka/

[3] http://citeseer.ist.psu.edu/

[4] http://www.informatik.uni-trier.de/ ley/db/

[5] The proportion of actors appear in several communities.

Table 3. Results on informative graphs

Datasets	Algorithm	Recall	Precision	F-measure	Cluster number
Dataset1	X-Mean	0.48	0.96	0.64	4
	CPM	0.71	0.52	0.60	6
	JCCM	0.73	0.96	0.82	3
Dataset2	X-Mean	0.92	0.44	0.60	6
	CPM	0.36	0.89	0.52	15
	JCCM	0.91	0.54	0.68	5
Dataset3	X-Mean	0.82	0.48	0.60	10
	CPM	0.73	0.44	0.54	17
	JCCM	0.90	0.51	0.66	11

Table 4. Results on co-authorship networks without considering attribute data

Datasets	Algorithm	Recall	Precision	F-measure	Cluster Number
Dataset1	CONGA	0.68	0.96	0.80	3
	CPM	0.71	0.52	0.60	6
	JCCM	0.72	0.84	0.78	3
Dataset2	CONGA	0.88	0.46	0.60	5
	CPM	0.36	0.89	0.52	15
	JCCM	0.85	0.48	0.61	5
Dataset3	CONGA	0.73	0.42	0.53	11
	CPM	0.73	0.44	0.54	17
	JCCM	0.75	0.44	0.55	10

performs better than CONGA in most of the time. This indicates that introducing the concept of community's regimes and considering actors in different roles respectively help to identify communities.

6 Conclusion and Future Work

In this paper, we have investigated community identification algorithm on informative graphs. Our algorithm JCCM combines the agglomerative and divisive methodology to find overlapping community in informative graphs. We also defined two community regimes, i.e. autarchy and union, and proposed several different principles to compute JCC for different community regimes. Experiments on both of the informative graph and relationship-only social networks indicate that JCCM is competitive in existing community identification algorithms.

Future works focus on clustering community in dynamic informative graphs.

References

1. Moser, F., Ge, R., Ester, M.: Joint Cluster Analysis of Attribute and Relationship Data Without A-Priori Specification of the Number of Clusters. In: KDD 2007, San Jose, California, USA (2007)
2. Newman, M.E.J., Girvan, M.: Finding and evaluating community structure in networks. Phys. Rev. E 69(026113), 56–68 (2004)
3. Palla, G., Derényi, I., Farkas, I., Vicsek, T.: Uncovering the overlapping community structure of complex networks in nature and society. Nature 435, 814–818 (2005)
4. Derényi, I., Palla, G., Vicsek, T.: Clique Percolation in Random Networks. In: PRL 1994, vol. 160202, pp. 76–85 (2005)
5. Gregory, S.: An Algorithm to Find Overlapping Community Structure in Networks. In: Kok, J.N., Koronacki, J., Lopez de Mantaras, R., Matwin, S., Mladenič, D., Skowron, A. (eds.) PKDD 2007. LNCS, vol. 4702, pp. 91–102. Springer, Heidelberg (2007)
6. Newman, M.E.J.: Fast algorithm for detecting community structure in networks. Phys. Rev. E 69, 066133 (2004)
7. Ester, M., Ge, R., Gao, B.J., Hu, Z., Ben-Moshe, B.: Joint cluster analysis of attribute data and relationship data: the connected k-center problem. In: SDM (2006)
8. Tong, H., Papadimitriou, S., et al.: Colibri: Fast Mining of Large Static and Dynamic Graphs. In: Proceedings of KDD (2008)
9. Sun, J., Xie, Y., Zhang, H., Faloutsos, C.: Less is More: Sparse Graph Mining with Compact Matrix Decomposition. Statistical Analysis and Data Mining 1(1), 6–22 (2007)
10. Zai'ane, O.R., Chen, J., Goebel, R.: Mining Research Communities in Bibliographical Data. In: Advances in Web Mining and Web Usage Analysis: 9th International Workshop on Knowledge Discovery on the Web, WebKDD 2007, and 1st International Workshop on Social Networks (2007)
11. Palla, G., Barabási, A.-L., Vicsek, T.: Quantifying social group evolution. Nature, 446–664 (2007); Adamcsek, B., Palla, G., Farkas, I., Derényi, I., Vicsek, T.: CFinder: locating cliques and overlapping modules in biological networks. Bioinformatics 22, 1021–1023 (2006)
12. Li, X., Liu, B., Yu, P.S.: Discovering overlapping communities of named entities. In: Fürnkranz, J., Scheffer, T., Spiliopoulou, M. (eds.) PKDD 2006. LNCS, vol. 4213, pp. 593–600. Springer, Heidelberg (2006)
13. Zeng, Z., Wang, J., Zhou, L., Karypis, G.: Out-of-Core Coherent Closed Quasi-Clique. Mining from Large Dense Graph Databases. ACM Transactions on Database Systems 32(2), Article 13 (2007)
14. Wan, L., Liao, J., Zhu, X.: CDPM: Finding and evaluating community structure in social networks. In: Tang, C., Ling, C.X., Zhou, X., Cercone, N.J., Li, X. (eds.) ADMA 2008. LNCS, vol. 5139, pp. 620–627. Springer, Heidelberg (2008)
15. Kaufman, L., Rousseeuw, P.: Finding Groups in Data: An Introduction to Cluster Analysis. Wiley, New York (1990)
16. Pelleg, D., Moore, A.: X-means: Extending k-means with efficient estimation of the number of clusters. In: ICML, San Francisco (2000)
17. Salton, G., McGill, M.J.: Introduction to modern information retrieval. McGraw-Hill, New York (1983)

Similarity Evaluation of XML Documents Based on Weighted Element Tree Model*

Chenying Wang, Xiaojie Yuan, Hua Ning, and Xin Lian

Department of Computer Science and Technology, Nankai University,
300071 Tianjin
wangchenying@dbis.nankai.edu.cn

Abstract. The logical presentation model of XML data is the basis of XML data management. After introducing XML tree models and frequent pattern models, in this paper we have proposed a novel Weighted Element Tree Model (WETM) for measuring the structural similarity of XML documents. This model is a concise form of XML tree models, so the efficiency of the operation on this model is higher than XML tree models. And comparing with frequent pattern models, the WETM enhances the expression ability of structural information of sub trees, which can appreciate the accuracy of similarity evaluation. Moreover, in order to compare the performance of the proposed evaluation algorithm, it is applied to XML documents clustering. The experimental results show that our algorithm is superior to the algorithms based on tree models or frequent pattern models.

Keywords: XML, similarity evaluation, clustering, element tree.

1 Introduction

With the growing popularity of XML for data representation and exchange, large numbers of XML documents have emerged. There is a pressing need for discovering valuable information from the massive documents. XML data mining is an important application of knowledge discovery technology, and similarity computation plays a crucial role in XML data mining.

XML documents' mining consists of contextual mining and structural mining. The structure is an important characteristic of XML document. Structural mining could excavate the knowledge among the structures of XML documents, and facilitate XML data extraction, integration and other applications. Classification and clustering are commonly used methods in XML structural mining. The similarity between XML documents is the basis of classification and clustering, and it is a crucial factor of the mining result.

In the XML arena, similarity evaluation between documents has been receiving a lot of attention and many approaches have been developed. At present, there are two main categories of approaches, namely approaches based on XML tree models and approaches based on frequent path models. Modeling XML documents as trees is an

* The work described in this paper is supported by National 863 Plans Projects, China (Project No.: 2009AA01Z152).

ordinary representation. Such as DOM [1], which is a W3C proposed standard, presents XML documents as a hierarchy tree of node objects. A common way of evaluating the similarity of two trees is by measuring their tree edit distance [2], and there are three classical algorithms of tree edit distance: Selkow [3], Chawathe [4] and Dalamagas [5]. The tree models contain all information of the XML documents and the tree edit operations are time-consuming operations, thus the efficiency of the evaluation algorithms based on tree models is considered as a challenge work. Frequent patterns [6] are another category of important models transformed from tree models, such as frequent paths model [7], frequent subtree model [8] and element sequence pattern [9]. The evaluation algorithms based on frequent patterns can evaluate the similarity between XML documents efficiently, but it is difficult to find out all the frequent patterns, and the accuracy of the evaluated similarity is not satisfied because they miss lots of structural information.

In this paper we have proposed a novel model named Weighted Subtree Expression Model (WETM), the model treats the elements as the center, the subtrees as the main part, and the weight of subtree as the connection among elements, and then we give a similarity evaluation algorithm based on the model. In order to verify the performance of this algorithm, we apply the algorithm and other two algorithms, namely Tree Edit Distance (Dalamagas [5]) and PBClustering [7], for XML documents clustering. The results of the experiment show that the similarity evaluation algorithm based on WETM is superior to Tree Edit Distance algorithm based on tree model and PBClustering algorithm based on frequent path model considering either the processing cost or the clustering accuracy.

2 Weighted Element Tree Model (WETM)

The logical XML document model is the basis for processing XML documents. Different data models can be used in different research fields. In order to facilitate the structural similarity evaluation of XML documents, the data model should be concise, and it should contain enough structural information at the same time. Based on the study of the previous data models, we propose a novel data model, called WETM, for measuring the structural similarity of XML documents.

In an XML document, elements, attributes and text nodes are the three most common types of nodes. Since we focus on the structural similarity, and content similarity can be done a good job by the traditional information retrieval technology, text nodes

Table 1. The notations used in this paper

Number	Notation	Description
1	D	An XML document or an XML document tree
2	e	An element node of an XML document
3	E	The element nodes set of an XML document
4	$\pi(e)$	The child nodes set of the element e
5	$\varphi(e)$	The parent node of the element e
6	$\lvert e\rvert$	The level of the element e

can be disregarded. Moreover, because attributes can be seen as a particular case of elements, we can only consider elements disregarding attributes also. The notations used in this paper are shown in Table 1.

The process of similarity evaluation between two documents is to measure the common and different features between the documents. Since common and different features at higher levels in the labeled tree of the documents are more relevant than the ones deeply nested in the tree, we assign a different weight to elements at different levels of the labeled tree. If element e is the root node of the document tree, then $|e|=0$; otherwise, $|e|=|\varphi(e)|+1$. Let $w(e)=\lambda^{-|e|}$ denote the weight of element e, where λ is the factor of relevance of a level with respect to the underlying level, and $\lambda>1$. In general, λ is assigned to 2.

Definition 1 (Weighted Element Tree Model). *Given an XML document D. For each* $e\in E$*, the weighted element tree of e is a 3 tuple* $\gamma=(e, \pi(e), w(e))$*, where e is the parent element,* $\pi(e)$ *is the child elements set of e,* $w(e)$ *is the weight of e ; Then the structural information of document D can be presented as* $\Gamma=\{\gamma|\gamma=(e, \pi(e), w(e)), e\in E\}$.

Given an element tree $\gamma=(e, \pi(e), w(e))$, if $\pi(e) = \Phi$, γ is referred as a *trivial element tree*; Otherwise, it is referred as a *non-trivial element tree*. Note that, the trivial element trees are ignored in the following sections without special pointed out.

3 Similarity Evaluation

The similarity of an XML document D_1 with respect to another document D_2 is a conditional probability of D_1 assuming that D_2 is given. Let $Sim(D_1, D_2)$ denote the similarity of D_1 with respect to D_1, then we can easily obtain that: $Sim(D_1, D_2) = P(D_1| D_2)\in[0,1]$. Since $Sim(D_1, D_2)$ is a conditional probability, the formula $Sim(D_1, D_2) = Sim(D_2, D_1)$ is not always true.

3.1 Tag Similarity

Elements are the base unit of an XML document and XML is self-descriptive, thus it is significant to consider the element tag similarity for measuring the document similarity. An XML document could be same with respect to another except for its tags. Moreover, the two tags can be different but can still represent the same information: they can be synonyms (e.g., movie and film), or syntactically similar according to a string edit distance [11] (e.g., happy and hsppy).

Given two elements e and e_0, and let s and s_0 are their tags, in order to evaluate the similarity of e with respect to e_0, referred as $Sim(e, e_0)$, the following rules are applied: Firstly, if s is same to s_0, we define $Sim(e, e_0)=1$; Secondly, if s and s_0 are synonyms, we define $Sim(e, e_0)=\xi$, where ξ is the factor for measuring synonymous tags and it is assigned to 0.8 here. In this paper, it is according to WordNet [12] to determine whether two words are synonyms or not. Thirdly, in order to compute the syntactical similarity of s with respect to s_0, let k denote the edit distance of s and s_0, which is defined as the minimum number of point mutations required to change s into s_0, where a point mutation is one of: change a letter, insert a letter or delete a letter. If

$1 - \dfrac{k}{max(s.Length,\ s_0.Length)} > \delta$, then $Sim(e, e_0) = 1 - \dfrac{k}{max(s.Length,\ s_0.Length)}$, where δ threshold value of syntactical similarity and it is assigned to 0.4. Here, function *max* returns the maximal value of the input values. Otherwise, s and s_0 are not syntactically similar. Finally, s is not similar to s_0 and $Sim(e, e_0)=0$. Now we give the formula for calculating the tag similarity of elements:

$$Sim(e, e_0) = \begin{cases} 1 & s \text{ is same to } s_0 \\ 0.8 & s \text{ and } s_0 \text{ are synonyms} \\ 1 - \dfrac{k}{max(s.Length,\ s_0.Length)} & s \text{ is syntactically similar to } s_0 \\ 0 & \text{otherwise} \end{cases} . \quad (1)$$

3.2 Similarity of Weighted Element Tree

By Definition 1, the structural information of an XML document can be presented by Γ which consists of weighted element trees, thus we discuss the similarity of a weighted element tree γ with respect to another γ_0 firstly.

Definition 2 (Matched Vector of $\gamma|\gamma_0$). *Given an weighted element tree $\gamma= (e, \pi(e), w(e))$ and another weighted element tree $\gamma= (e_0, \pi(e_0), w(e_0))$, let a matched vector $\vec{e}=(c, i, d)$ describe the similarity of $\gamma|\gamma_0$, referred as $\vec{e}=V(\gamma|\gamma_0)$, where e is the root element of γ; c is the number of elements appearing in both the $\pi(e)$ and $\pi(e_0)$, referred to as common elements, i.e., $c=|\pi(e)\cap\pi(e_0)|$; i is the number of elements appearing in the $\pi(e)$ but not in the $\pi(e_0)$, referred to as inserted elements, i.e., $i=|\pi(e)- \pi(e_0)|$; d is the number of elements appearing in the $\pi(e_0)$ but not in the $\pi(e)$, referred to as deleted elements, i.e., $d=|\pi(e_0)- \pi(e)|$.*

In order to measure the affect factor of the common, inserted and deleted elements to the similarity, we introduce a factor vector $\vec{\sigma} = (\alpha, \beta, \theta)$, where α, β and θ are all greater than 0. Given function *abs* returns the absolute value of the input value, then we can define the structural similarity γ with respect to another γ_0 as follows[1]:

$$Sim(\gamma, \gamma_0)=(1-\frac{abs(|e| - |e_0|)}{max(|e|, |e_0|)}) \times \frac{\lambda \times Sim(e, e_0)+\alpha \times c}{\lambda+\vec{\sigma}\times\vec{e}^{\mathrm{T}}} . \quad (2)$$

Where λ is the factor of relevance of a level with respect to the underlying level defined in Section 2, and $Sim(e, e_0)$ is the tag similarity.

3.3 Similarity of XML Documents

Definition 3 (Similarity Matrix of $\Gamma|\Gamma_0$). *Given two XML documents D and D_0 and their structural information $\Gamma= \{\gamma_1, \gamma_2, ..., \gamma_m\}$ and $\Gamma_0= \{\gamma_1', \gamma_2', ..., \gamma_n'\}$, then we can defined the Similarity Matrix of Γ with respect to Γ_0 as follows:*

---

[1] The vector $\vec{e}^{\mathrm{T}}$ represents the transpose vector of $\vec{e}$. Similarly, the matrix M^{T} represents the transpose matrix of M.

$$M(\Gamma|\Gamma_0) = \begin{pmatrix} s_{11} & s_{12} & \cdots & s_{1n} \\ s_{21} & s_{22} & \cdots & s_{2n} \\ \cdots & \cdots & \cdots & \cdots \\ s_{m1} & s_{m2} & \cdots & s_{mn} \end{pmatrix} \quad \text{where } s_{ij}=Sim(\gamma_i, \gamma_j') \,. \tag{3}$$

For each $\gamma_i \in \Gamma$, let $t_i=Sim(e_i, e_k')$ and $\vec{e_i} = V(\gamma_i|\gamma_k')$, where t_i denotes the tag similarity of e_i and e_k', e_i is the root element of γ_i, $e_k'$is the root element of γ_k', γ_k' satisfies that $s_{ik} =max(s_{i1}, s_{i2}, ..., s_{in})$, $1 \leqslant i \leqslant m$ and $1 \leqslant k \leqslant n$. Then we can define:

(1). The tag similarity vector as $\vec{t} = (t_1, t_2, ..., t_m)$;

(2). The *matched matrix* as $MM_{3m}=[\vec{e_1}^{\,T}, \vec{e_2}^{\,T}, \cdots, \vec{e_m}^{\,T}]$;

(3). The transpose matrix of MM_{3m} as $MM^T_{3m} =[\vec{c}^{\,T}, \vec{i}^{\,T}, \vec{d}^{\,T}$, where $\vec{c}$, $\vec{i}$ and $\vec{d}$ are the vectors consisted of common elements, insert elements and deleted elements respectively.

In order to evaluate the factor of the different levels, we introduced the vector of elements weight as $\vec{w} =(w(e_1), w(e_2), \cdots, w(e_m)) = (\lambda^{-|e_1|}, \lambda^{-|e_2|}, \cdots, \lambda^{-|e_m|})$. Now we can evaluate the similarity of D with respect to D_0:

$$Sim(D, D_0) = \frac{\lambda \times \vec{t} \times \vec{w}^T + \alpha \times \vec{c} \times \vec{w}^T}{\lambda \times \sum_{i=1}^{m} w(e_i) + \vec{\sigma} \times (MM_{3m} \times \vec{w}^T)} \,. \tag{4}$$

4 Experiments

4.1 Experiment Setup

To compare the performance of the proposed WETM with that of other related works, we have implemented the Tree Edit Distance algorithm, referred as TED, which estimates the document similarity using the tree model as well as PBClustering algorithm with the document similarity estimated using the frequent path model. All the algorithms were implemented in C# and all the experiments are carried out on a 2.8 GHz Pentium processor with 1 GB RAM running Windows 2003. The experimental results of WETM are assuming that the affect factors of common elements, inserted elements and deleted elements are the same.

In order to apply the algorithms to similarity-based clustering for performance evaluation, two kinds of datasets were used, namely ACM Sigmod Record [13] and Shakespeare [14]. In order to test the sensitivity of the algorithms on datasets with different sizes, we prepared a number of sub datasets shown in Table 2 for our experiments. After preparing the datasets, all the documents no matter from ACM Sigmod or Shakespeare are belong to one of the categories from C_1 to C_6.

Table 2. Datasets used in our experiments

Category	Datasets	Sources	Documents#
C_1	*IndexTermsPage*	Sigmod-1999	920
C_2	*OrdinaryIssuePage*	Sigmod-1999	51
C_3	*ProceedingsPage*	Sigmod-1999	17
C_4	*OrdinaryIssuePage*	Sigmod-2002	30
C_5	*ProceedingsPage*	Sigmod-2002	16
C_6	*Shakespeare*	Shakespeare	48

4.2 Processing Cost

In this experiment, we compare the processing cost of the three algorithms to the number of documents, which varies from 0.5K to 3K documents. All the documents used in this experiment are from ACM Sigmod Record selected in random, the size ranged from 10KB to 30KB.

In order to compare processing cost, the algorithms are divided into two stages. One is the building model stage, which is the preprocessing stage including to read the documents and turn them into the logical data model. The other is the evaluation stage, which is similarity evaluation stage including to evaluate the similarity of any pair of documents.

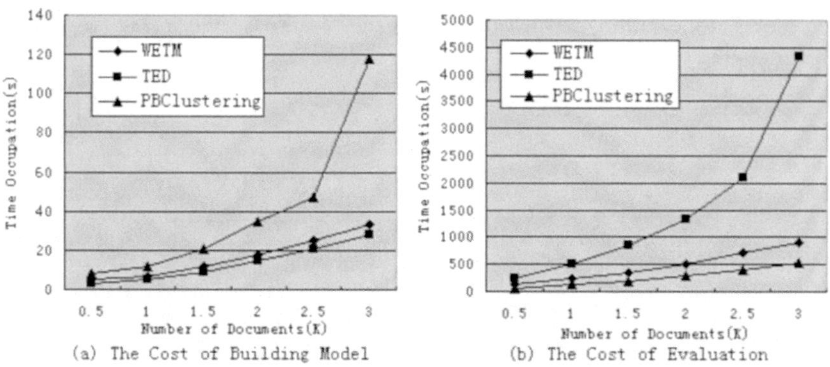

Fig. 1. The Comparison of Processing Cost

Fig.2 shows the comparison of processing cost of the three algorithms. From Fig.3.(a), it can be seen that the building model cost of PBClustering steep rises along the increasing number of documents, which is reasonable because PBClustering should find out the frequent paths for any single document from the paths of the whole document set, while the other two algorithms build the data model for a single document independently. On the one side, since the tree edit operations are time-consuming operations that is mentioned before, the evaluation cost of TED is certainly very high; on the other side, the evaluation of the other two algorithms is quiet simple. Fig.4.(b) just verify this situation.

From the above analysis and experimental results, it can be seen that WETM inherits the convenience of tree models to build logical data model and brings down the evaluation cost close to that of the frequent path modes at the same time.

4.3 Clustering Accuracy

The similarity that can accurately describe the structural relationship between XML documents can be estimated by different evaluation algorithms, and it can benefit all the existing similarity-based clustering algorithms. Without loss of generality, the K-means algorithm was chosen in this paper as the clustering algorithm.

Among the different quality measures for clustering accuracy, we used one of the most common ones the precision and recall rates as clustering accuracy measure. The measure assumes that a cluster is the result of a query and a category is the desired set of the documents for the query. Let C and C_i, where $1 \leqslant i \leqslant 6$ in our experiments, denote the cluster set and category set respectively, then the precision and recall rates of that cluster-category pair can be computed as follows:

$$Precision = \frac{|C \cap C_i|}{|C|} \quad Recall = \frac{|C \cap C_i|}{|C_i|} \ . \tag{5}$$

where $|C \cap C_i|$ is the number of the documents of category C_i falling into the cluster C, $|C|$ is the number of documents in the cluster C, and $|C_i|$ is the number of documents in the category C_i.

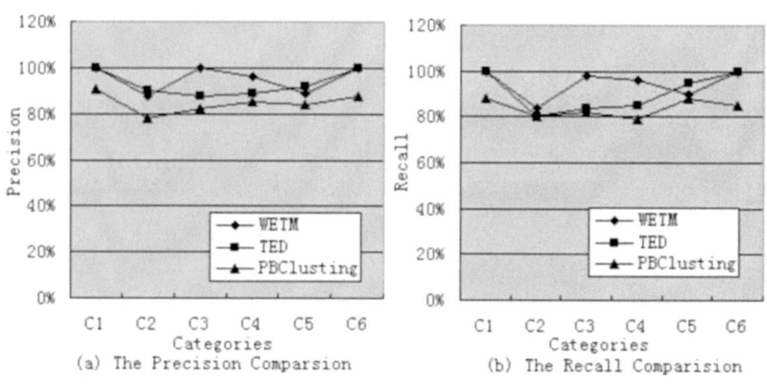

Fig. 2. The Comparison of Clustering Accuracy

According to Fig.2 shown the comparison of clustering accuracy of the three algorithms, the clustering accuracy based on the proposed WETM is found to be significantly better than that based on the others. Note that, the category which contains only 16 documents C_5 is a particular case, because it is not used to evaluate the clustering accuracy but to evaluate the adaptability of the algorithms in the experiments.

5 Conclusion

In this paper we have proposed a novel model named Weighted Element Tree Model, which inherits the convenience of tree models to build logical data model and brings down the evaluation cost closing to that of the frequent path modes at the same time. The clustering accuracy using similarity evaluation algorithm based on the proposed WETM is found to be significantly better than that based on the others. Future works focus on supporting the semantic of element tags and the content of text node for they can be phrase and sentence with particular semantics, and extensional applications of the similarity evaluation.

References

1. W3C: Recommendation, Document Object Model (DOM) Level 3 Core Specification (2004), http://www.w3.org/TR/DOM-Level-3-Core/
2. Bille, P.: A survey on tree edit distance and related problem. Theoretical Computer Science 337, 217–239 (2005)
3. Selkow, S.M.: The tree-to-tree edit problem. Information Processing Letter 6, 184–186 (1997)
4. Chawathe, S.S.: Comparing Hierarchical Data in External Memory. In: Proceedings of the 25th VLDB, pp. 90–101 (1999)
5. Dalamagas, T., Cheng, T., Winkel, K., Sellis, T.K.: A Methodoloay for Clustering XML Documents by Structure. Information Systems 31(3), 187–228 (2006)
6. Nayak, R., Iryadi, W.: XML schema clustering with semantic and hierarchical similarity measures. Knowledge-Based Systems archive 20(4), 336–349 (2007)
7. Leung, H.-p., Chung, F.-l., et al.: XML Document clustering using Common XPath. In: Proc. of the Internation Workshop on Challenges in Web Information Retrieval and Integration, pp. 91–96 (2005)
8. Hwang, J.H., Gu, M.S.: Clustering XML Documents Based on the Weight of Frequent Structures. In: Proc. of the 2007 International Conference on Convergence Information Technology, pp. 845–849 (2007)
9. Zhang, H., Yuan, X., et al.: Similarity Computation for XML Documents by element sequence patterns. In: Proc. of the 10th Asia-Pacific Web conference, pp. 227–232 (2008)
10. XML Path Language (XPath) 2.0. (2007), http://www.w3.org/TR/xpath20/
11. Rice, S.V., Bunke, H., Nartker, T.A.: Classes of cost functions for string edit distance. Algorithmica 18(2), 271–280 (1997)
12. WordNet, http://wordnet.princeton.edu/
13. SIGMOD Record Datasets (2007), http://www.sigmod.org/record/xml/
14. Shakespeare Dataset (1999), http://www.ibiblio.org/bosak/xml/eg/

Quantitative Comparison of Similarity Measure and Entropy for Fuzzy Sets

Hongmei Wang, Sanghyuk Lee*, and Jaehyung Kim

School of Mechatronics, Changwon National University
Sarim-dong, Changwon, Gyeongnam, Korea
iwanghongmei99@163.com, {leehyuk,hyung}@changwon.ac.kr

Abstract. Comparison and data analysis to the similarity measures and entropy for fuzzy sets are studied. The distance proportional value between the fuzzy set and the corresponding crisp set is represented as fuzzy entropy. We also verified that the sum of the similarity measure and the entropy between fuzzy set and the corresponding crisp set constitutes the total information. Finally, we derive a similarity measure from entropy with the help of total information property, and illustrate a simple example that the maximum similarity measure can be obtained using a minimum entropy formulation.

Keywords: Similarity measure, distance measure, fuzzy entropy.

1 Introduction

Analysis of data certainty and uncertainty is essential to process data mining, pattern classification or clustering, and discriminating data. Basically, well known distance measure such as Hamming distance can be used to design certainty and uncertainty measure commonly. To analyze the data, it is often useful to consider a data set as a fuzzy set with a degree of membership. Hence fuzzy entropy and similarity analyses have been emphasized for studying the uncertainty and certainty information of fuzzy sets [1-7].

The characterization and quantification of fuzziness needed in the management of uncertainty in the modeling and system designs. The entropy of a fuzzy set is called as the measure of its fuzziness by previous researchers [1-4]. The degree of similarity between two or more data sets can be used in the fields of decision making, pattern classification, etc., [5-7]. Thus far, numerous researchers have carried out research on deriving similarity measures [8,9]. Similarity measures based on the distance measure are applicable to general fuzzy membership functions, including nonconvex fuzzy membership functions [9]. Two measures, entropy and similarity, represent the uncertainty and similarity with respect to the corresponding crisp set, respectively. For data set, it is interesting to study the relation between entropy and similarity measure. The correlation between entropy and similarity for fuzzy sets has been presented as the physical view [10]. Liu also proposed a relation between distance and similarity measures; in his paper, the sum of distance and similarity constitutes the total

---

* Corresponding Author.

R. Huang et al. (Eds.): ADMA 2009, LNAI 5678, pp. 688–695, 2009.

information [2]. In this paper, we analyze the relationship between the entropy and similarity measures for fuzzy sets, and compute the quantitative amount of corresponding measures. First, fuzzy entropy and similarity measures are derived by the distance measure. Discussion with entropy and similarity for data has been followed. With the proposed fuzzy entropy and similarity measure, the property that the total information comprises the similarity measure and entropy measure is verified. With the total information property, similarity measure can be obtained through fuzzy entropy. Illustrative example help to understand total information property between fuzzy entropy and similarity as the uncertainty and certainty measure of data.

In the following section, the relationship between entropy and similarity for a fuzzy set is discussed. In Section 3, the procedure for obtaining the similarity measure from the fuzzy entropy is derived. Furthermore, the computational example is illustrated. The conclusions are stated in Section 4.

2 Fuzzy Entropy and Similarity Measure

Every data set has uncertainty in its data group, and it is illustrated as the membership functions for fuzzy set. Data uncertainties are often measured by fuzzy entropy, explicit fuzzy entropy construction is also proposed by numerous researchers [9]. Fuzzy entropy of fuzzy set means that fuzzy set contains how much uncertainty with respect to the corresponding crisp set. Then, what is the data certainty with respect to the deterministic data? This data certainty can be obtained through similarity measure. We analyze the relation between fuzzy entropy and similarity as the quantitative comparison.

Two comparative sets are considered, one is a fuzzy set and the other is the corresponding crisp set. The fuzzy membership function pair is illustrated in Fig. 1, crisp set A_{near} represents the crisp set "near" to the fuzzy set A.

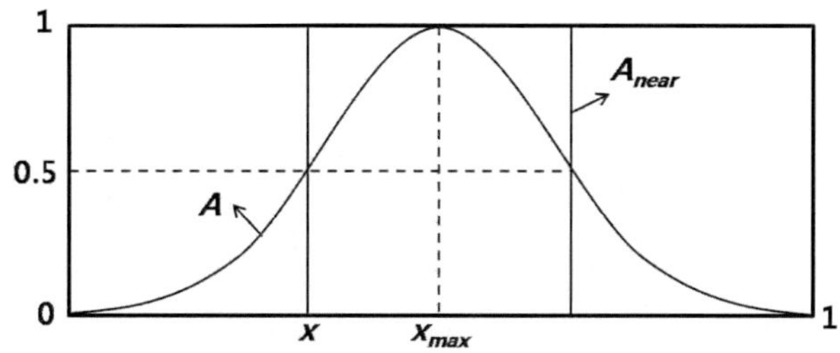

Fig. 1. Membership functions of fuzzy set A and crisp set $A_{near} = A_{0.5}$

A_{near} can be assigned by various variable. For example, the value of crisp set $A_{0.5}$ is one when $\mu_A(x) \geq 0.5$, and is zero otherwise. Here, A_{far} is the complement of A_{near}, i.e., $A^C{}_{near} = A_{far}$. In our previous result, the fuzzy entropy of fuzzy set A with respect to A_{near} is represented as follows [9]:

$$e(A, A_{near}) = d(A \cap A_{near}, [1]_X) + d(A \cup A_{near}, [0]_X) - 1 \quad (1)$$

where $d(A \cap A_{near}, [1]_X) = \frac{1}{n}\sum_{i=1}^{n} | \mu_{A \cap A_{near}}(x_i) - 1 |$ is satisfied. $A \cap A_{near}$ and $A \cup A_{near}$ are the minimum and maximum value between A and A_{near}, respectively. $[0]_X$ and $[1]_X$ are the fuzzy sets in which the value of the membership functions are zero and one, respectively, for the universe of discourse. d satisfies Hamming distance measure. Eq. (1) is not the normal fuzzy entropy. The normal fuzzy entropy can be obtained by multiplying the right-hand side of Eq. (1) by two, which satisfies maximal fuzzy entropy is one. Numerous fuzzy entropies can be presented by the fuzzy entropy definition [2]. Here we introduce same result of (1) as follows.

$$e(A, A_{near}) = d(A, A \cap A_{near}) + d(A_{near}, A \cap A_{near}) \tag{2}$$

The fuzzy entropies in Eqs. (1) and (2) satisfy for all value of crisp set A_{near}. Hence, $A_{0.1}$ and $A_{0.5}$ or some other $A_{0.X}$ can be satisfied. Now, it is interesting to search for what value of $A_{0.X}$ makes maximum or minimum value of entropy.

Eqs. (1) and (2) are rewritten as follows:

$$e(A, A_{near}) = 2\int_0^x \mu_A(x)dx + 2\int_x^{x_{\max}} 1 - \mu_A(x)dx\,. \tag{3}$$

Let $\frac{d}{dx}M_A(x) = \mu_A(x)$; $e(A, A_{near})$ has been shown to be

$$e(A, A_{near}) = 2M_A(x)\,|_0^x + 2(x_{\max} - x) - 2M_A(x)\,|_x^{x_{\max}}\,.$$

The maxima or minima are obtained by differentiation:

$$\frac{d}{dx}e(A, A_{near}) = 2\mu_A(x) - 2 + 2\mu_A(x)\,.$$

Hence, it is clear that the point x satisfying $\frac{d}{dx}e(A, A_{near}) = 0$ is the critical point for the crisp set. This point is given by $\mu_A(x) = 1/2$, i.e., $A_{near} = A_{0.5}$. The fuzzy entropy between A and $A_{0.5}$ has a minimum value because $e(A)$ attains maxima when the corresponding crisp sets are $A_{0.0}$ and $A_{x_{\max}}$. Hence, for a convex and symmetric fuzzy set, the minimum entropy of the fuzzy set is equal to that of the crisp set $A_{0.5}$. This indicates that the corresponding crisp set that has the least uncertainty or the greatest similarity with the fuzzy set is $A_{0.5}$.

All the studies on similarity measures deal with derivations of similarity measures and applications in the distance-measure-based computation of the degree of similarity. Liu has also proposed an axiomatic definition of the similarity measure [2]. The similarity measure $\forall A, B \in F(X)$ and $\forall D \in P(X)$ has the following four properties:

(S1) $s(A,B)=s(B,A)$, $\forall A,B\in F(X)$

(S2) $s(D,D^c)=0$, $\forall D\in P(X)$

(S3) $s(C,C)=\max_{A,B\in F} s(A,B)$, $\forall C\in F(X)$

(S4) $\forall A,B,C\in F(X)$, if $A\subset B\subset C$, then $s(A,B)\geq s(A,C)$ and $s(B,C)\geq s(A,C)$

where $F(X)$ denotes a fuzzy set, and $P(X)$ is a crisp set.

In our previous studies, other similarity measures between two arbitrary fuzzy sets are proposed as follows [9]:

For any two sets $A,B\in F(X)$,

$$s(A,B)=1-d(A\cap B^C,[0]_X)-d(A\cup B^C,[1]_X) \tag{4}$$

and

$$s(A,B)=2-d(A\cap B,[1]_X)-d\left(A\cup B,[0]_X\right) \tag{5}$$

are similarity measures between set A and set B.

The proposed similarity measure between A and A_{near} is presented in Theorem 2.1. The usefulness of this measure is verified through a proof of this theorem.

Theorem 2.1. $\forall A\in F(X)$ and the crisp set A_{near} in Fig. 1,

$$s(A,A_{near})=d(A\cap A_{near},[0]_X)+d(A\cup A_{near},[1]_X) \tag{6}$$

is a similarity measure.

Proof. (S1) follows from Eq. (6), and for crisp set D, it is clear that $s(D,D^C)=0$. Hence, (S2) is satisfied. (S3) indicates that the similarity measure of two identical fuzzy sets $s(C,C)$ attains the maximum value among various similarity measures with different fuzzy sets A and B since $d(C\cap C,[0]_X)+d(C\cup C,[1]_X)$ represents the entire region in Fig. 1. Finally, from $d(A\cap A_{1near},[0]_X)\geq d(A\cap A_{2near},[0]_X)$ and $d(A\cup A_{1near},[1]_X)\geq d(A\cup A_{2near},[1]_X)$, $A\subset A_{1near}\subset A_{2near}$; it follows that

$$\begin{aligned} s(A,A_{1near}) &= d(A\cap A_{1near},[0]_X)+d(A\cup A_{1near},[1]_X)\\ &\geq d(A\cap A_{2near},[0]_X)+d(A\cup A_{2near},[1]_X)\\ &= s(A,A_{2near}) \end{aligned}$$

Similarly, $s(A_{1near},A_{2near})\geq s(A,A_{2near})$ is satisfied by the inclusion properties $d(A_{1near}\cap A_{2near},[0]_X)\geq d(A\cap A_{2near},[0]_X)$ and $d(A_{1near}\cup A_{2near},[1]_X)\geq d(A\cup A_{2near},[1]_X)$. ■

The similarity in Eq. (6) represents the areas shared by two membership functions. In Eqs. (4) and (5), fuzzy set B can be replaced by A_{near}. In addition to those in Eqs. (4) and (5), numerous similarity measures that satisfy the definition of a similarity measure can be derived. From Fig. 1, the relationship between data similarity and entropy for fuzzy set A with respect to A_{near} can be determined on the basis of the total area. The total area is one (universe of discourse $\times$ maximum membership value $= 1\times1=1$); it represents the total amount of information. Hence, the total information comprises the similarity measure and entropy measure, as shown in the following equation:

$$s(A, A_{near}) + e(A, A_{near}) = 1 \tag{7}$$

With the similarity measure in Eq. (5) and the total information expression in Eq. (7), we obtain the following proposition:

Proposition 2.1. In Eq. (7), $e(A, A_{near})$ follows from the similarity measure in Eq. (5):

$$e(A, A_{near}) = 1 - s(A, A_{near}) = d(A \cap A_{near}, [1]_X) + d(A \cup A_{near}, [0]_X) - 1$$

The above fuzzy entropy is identical to that in Eq. (1). The property given by Eq. (7) is also formulated as follows:

Theorem 2.2. The total information about fuzzy set A and the corresponding crisp set A_{near},

$$\begin{aligned} & s(A, A_{near}) + e(A, A_{near}) \\ & = d(A \cap A_{near}, [0]_X) + d(A \cup A_{near}, [1]_X) \\ & \quad + d(A \cap A_{near}, [1]_X) + d(A \cup A_{near}, [0]_X) - 1 \end{aligned}$$

equals one.

Proof. It is clear that the sum of the similarity measure and fuzzy entropy equals one, which is the total area in Fig. 1. Furthermore, it is also satisfied by computation,

$$d(A \cap A_{near}, [0]_X) + d(A \cap A_{near}, [1]_X) = 1 \text{ and}$$
$$d(A \cup A_{near}, [1]_X) + d(A \cup A_{near}, [0]_X) = 1.$$

Hence, $s(A, A_{near}) + e(A, A_{near}) = 1 + 1 - 1 = 1$ is satisfied. ■

Now, it is clear that the total information about fuzzy set A comprises similarity and entropy measures with respect to the corresponding crisp set.

3 Similarity Measure Derivation from Entropy

In this section, with the property of Theorem 2.2 similarity measure derivation with entropy is carried out. Entropy derivation from similarity is also possible. This conversion makes possible measure formulation from complementary measure. With consideration of previous similarity measure (5), fuzzy entropy (1) has been obtained. It is also possible to obtain another similarity measure using fuzzy entropy different from that in Eq. (2). The proposed fuzzy entropy is developed by using the Hamming distances between a fuzzy set and the corresponding crisp set. The following result clearly follows from Fig. 1. Eq. (2) represents the difference between A and the corresponding crisp set A_{near}. From Theorem 2.2, the following similarity measure that satisfies Eq. (7) follows:

$$s(A, A_{near}) = 1 - d(A, A \cap A_{near}) - d(A_{near}, A \cap A_{near}) \tag{8}$$

Here, it is interesting to determine whether Eq. (8) satisfies the conditions for a similarity measure.

Proof. (S1) follows from Eq. (8). Furthermore, $s(D, D^C) = 1 - d(D, D \cap D^C)$ $- d(D^C, D \cap D^C)$ is zero because $d(D, D \cap D^C) + d(D^C, D \cap D^C)$ satisfies $d(D, [0]_X) + d(D^C, [0]_X) = 1$. Hence, (S2) is satisfied. (S3) is also satisfied since $d(C, C \cap C) + d(C, C \cap C) = 0$; hence, it follows that $s(C, C)$ is a maximum. Finally,

$$1 - d(A, A \cap B) - d(B, A \cap B) \geq 1 - d(A, A \cap C) - d(C, A \cap C)$$

because $d(A, A \cap B) = d(A, A \cap C)$ and $d(B, A \cap B) \leq d(C, A \cap C)$ are satisfied for $A \subset B \subset C$. The inequality $s(B, C) \geq s(A, C)$ is also satisfied in a similar manner. ■

Similarity based on fuzzy entropy has the same structure designed from similarity definition. Following corollary insist that two similarity measures has the same structure even though they are derived from different ways.

Corollary 3.1. Proposed similarity measures (6) and (8) are equal.

$$d(A \cap A_{near}, [0]_X) + d(A \cup A_{near}, [1]_X) = 1 - d(A, A \cap A_{near}) - d(A_{near}, A \cap A_{near}).$$

This equality can be verified easily by analyzing Fig 1.

Now, by using Eq. (8), we obtain the maximum similarity measure for the fuzzy set. In our previous result, the minimum fuzzy entropy could be obtained when we considered the entropy between the fuzzy sets A and $A_{0.5}$. Hence, it is obvious that the obtained similarity

$$s(A, A_{0.5}) = 1 - d(A, A \cap A_{0.5}) - d(A_{0.5}, A \cap A_{0.5}) \tag{9}$$

represents the maximum similarity measure.

Computation between Similarity measure and Fuzzy entropy : Let us consider the next fuzzy set with membership function $A = \{x, \mu_A(x)\}$:

{(0.1,0.2), (0.2,0.4), (0.3,0.7), (0.4,0.9), (0.5,1), (0.6,0.9), (0.7, 0.7), (0.8,0.4), (0.9,0.2), (1,0)}.

The fuzzy entropy and similarity measures are calculated using Eqs. (2) and (9) are given in Table 1.

Table 1. Similarity and Entropy value between fuzzy set and corresponding crisp set

Similarity measure	Measure value	Fuzzy entropy	Entropy value
$s(A, A_{0.1})$	0.64	$e(A, A_{0.1})$	0.36
$s(A, A_{0.3})$	0.76	$e(A, A_{0.3})$	0.24
$s(A, A_{0.5})$	0.80	$e(A, A_{0.5})$	0.20
$s(A, A_{0.8})$	0.72	$e(A, A_{0.8})$	0.28
$s(A, A_{0.95})$	0.56	$e(A, A_{0.95})$	0.44

The similarity measure for $s(A, A_{0.5})$ is calculated by using the following equation:

$$s(A, A_{0.5}) = 1 - 1/10(0.2+0.4+0.4+0.2)\ - 1/10(0.3+0.1+0.1+0.3) = 0.8\,.$$

Fuzzy entropy $e(A, A_{0.5})$ is also calculated by

$$e(A, A_{0.5}) = 1/10(0.2+0.4+0.4+0.2) + 1/10(0.3+0.1+0.1+0.3) = 0.2\,.$$

The remaining similarity measures and fuzzy entropies are calculated in a similar manner.

4 Conclusions

Quantification of fuzzy entropy and similarity for fuzzy sets were studied. Fuzzy entropies for fuzzy sets were developed by considering the crisp set "near" the fuzzy set. The minimum entropy can be obtained by calculating area, and it satisfies when the crisp set is $A_{near} = A_{0.5}$. The similarity measure between the fuzzy set and the corresponding crisp set is also derived using the distance measure. The property that the sum of fuzzy entropy and the similarity measure between fuzzy set and corresponding crisp set is derived as a constant value. It is proved that the fuzzy entropy and similarity measure values constitute whole area of information.

Acknowledgments. The authors of this paper were partly supported by the Second Stage of Brain Korea 21 Projects.

References

1. Pal, N.R., Pal, S.K.: Object-background segmentation using new definitions of entropy. In: IEEE Proc., vol. 36, pp. 284–295 (1989)
2. Xuecheng, L.: Entropy, distance measure and similarity measure of fuzzy sets and their relations. Fuzzy Sets and Systems 52, 305–318 (1992)
3. Bhandari, D., Pal, N.R.: Some new information measure of fuzzy sets. Inform. Sci. 67, 209–228 (1993)

4. Ghosh, A.: Use of fuzziness measure in layered networks for object extraction: a generalization. Fuzzy Sets and Systems 72, 331–348 (1995)
5. Rébillé, Y.: Decision making over necessity measures through the Choquet integral criterion. Fuzzy Sets and Systems 157(23), 3025–3039 (2006)
6. Kang, W.S., Choi, J.Y.: Domain density description for multiclass pattern classification with reduced computational load. Pattern Recognition 41(6), 1997–2009 (2008)
7. Shih, F.Y., Zhang, K.: A distance-based separator representation for pattern classification. Image and Vision Computing 26(5), 667–672 (2008)
8. Chen, S.J., Chen, S.M.: Fuzzy risk analysis based on similarity measures of generalized fuzzy numbers. IEEE Trans. on Fuzzy Systems 11(1), 45–56 (2003)
9. Lee, S.H., Kim, J.M., Choi, Y.K.: Similarity measure construction using fuzzy entropy and distance measure. In: Huang, D.-S., Li, K., Irwin, G.W. (eds.) ICIC 2006. LNCS (LNAI), vol. 4114, pp. 952–958. Springer, Heidelberg (2006)
10. Lin, S.K.: Gibbs Paradox and the Concepts of Information, Symmetry, Similarity and Their Relationship. Entropy 10, 15 (2008)

Investigation of Damage Identification of 16Mn Steel Based on Artificial Neural Networks and Data Fusion Techniques in Tensile Test

Hongwei Wang*, Hongyun Luo*, Zhiyuan Han, and Qunpeng Zhong

Key Laboratory of Aerospace Materials and Performance (Ministry of Education),
School of Materials Science and Engineering, Beihang University,
100191 Beijing
Tel.: +86 010 82339085; Fax: +86 010 82337108
wanghongwei-1978@163.com, Luo7128@163.com, hzy19851227@163.com,
zhangpengfei_416@126.com

Abstract. This paper proposes a damage identification method based on back propagation neural network (BPNN) and dempster-shafer (D-S) evidence theory to analyze the acoustic emission (AE) data of 16Mn steel in tensile test. Firstly, the AE feature parameters of each sensor in 16Mn steel tensile test are extracted. Secondly, BPNNs matching sensor number are trained and tested by the selected features of the AE data, and the initial damage decision is made by each BPNN. Lastly, the outputs of each BPNN are combined by D-S evidence theory to obtain the finally damage identification of 16Mn steel in tensile test. The experimental results show that the damage identification method based on BPNN and D-S evidence theory can improve damage identification accuracy in comparison with BPNN alone and decrease the effect of the environment noise.

Keywords: Acoustic emission, Damage identification, Back propagation neural network, Dempster-shafer evidence theory.

1 Introduction

It is well known that 16Mn steel, as loaded member of many structures, has been always applied to bridges, pressure vessels and so on. These structures under stress tend to generate and grow some deforming and damaging which would cause reduction of the stiffness, strength and toughness as well as the remaining life of 16Mn steel. Therefore, in order to estimate the damage, it is important to develop healthy monitor and identify the damage modes of 16Mn steel.

Acoustic emission (AE) technology was selected to monitor the deform and damage, because that acoustic emission has been developed as an effective nondestructive technique for monitoring the deforming and damaging behavior of materials under stress [1], and it has been used in a variety of metal structures, such as pressure vessels, storage tanks, railroad tank cars, manlift booms, and bridges [2, 3].

* Corresponding author.

R. Huang et al. (Eds.): ADMA 2009, LNAI 5678, pp. 696–703, 2009.

But analyzing AE data depends to a large extent on the experience and ability of analyst. So the application of artificial neural networks (ANN) to AE technique has been growing rapidly [4, 5]. An important class of neural network is back propagation neural network (BPNN) [6], which can solve complex problems with multiple layers and use the delta rule for training to reduce the error rapidly [7], Recently, this neural network has been applied to solve pattern identification, classification, and optimizing problems [8]. Therefore, BPNN may be used to identify the damage modes of 16Mn steel in AE test. However, the uncertainties that caused by measurement noise, back propagation neural network model error or environmental changes can impede reliable identification of damage.

In recent years, the information fusion technique has attracted increasing attention to fault diagnosis and structural damage detection because that it can extract the information from different sources and integrate this information into a consistent, accurate and intelligible data set [9]. As one of information fusion techniques, the dempster-shafer (D-S) evidence theory is widely used for dealing with uncertain information. The D-S evidence theory can integrate multi-sensor information including vibration, sound, pressure and temperature in fault diagnosis and structural damage detection [10, 11]. Therefore, the D-S evidence theory can be applied to analyze the acoustic emission data because that many types of sensors were used during AE monitoring of 16Mn steel tensile test.

This paper proposes a damage identification method of 16Mn steel based on back propagation neural network and D-S evidence theory in AE tensile test. Firstly, the AE feature parameters of each sensor are extracted. Secondly, the initial damage decision is made by each back propagation neural network that trained by the extracted data of each sensor respectively. Lastly, the outputs of each back propagation neural network are combined by D-S evidence theory to decrease the effect of the environment noise and enhance the BPNN identification accuracy.

2 Establishment of Damage Identification Model of 16Mn Steel Based on Back Propagation Neural Network and Dempster-Shafer Evidence Theory

2.1 Back Propagation Neural Network of AE

Data fusion can be seen non-linear reasoning process of information space under certain conditions, while the characteristic of artificial neural networks has non-linear process ability, can perform prediction problems and solve modes identification, classification, and optimizing problems, therefore, artificial neural networks is a appropriate method for declaration level fusion. In this research, back propagation neural network (BPNN) is selected to identify damage modes of 16Mn steel in AE tensile test.

Assume that the training input lay is I and the training output lay is O in the AE back propagation neural network, then

$$I = \{AE\ parameter\ i \mid i = 1, 2, \cdots, p\},\quad O = \{Output\ \ neuron\ \ j \mid j = 1, 2, \cdots, q\}$$

Where, p is the numbers of AE parameters, and q is the numbers of output neuron of AE back propagation neural network.

Based on the back propagation neural network of AE, the test output lay O' and mean square error E can be obtained. Therefore, O' can be described as:

$$O' = \{ f(A_1) \ \ f(A_2) \ \ \cdots \ \ f(A_q) \}$$

Where, $f(A_1)$, $f(A_2)$ and $f(A_q)$ are the actual value of first neuron, the actual value of second neuron and the actual value of qth neuron that obtained by AE back propagation neural network respectively.

E can be calculated in the following manner:

$$E = \frac{1}{2}\sum_{i=1}^{q}(f(A_i) - F(A_i))^2 \tag{1}$$

Where $F(A_i)$ is the desired value of ith neuron.

2.2 Basic Probability Assignment (bpa) Based on the Output of Back Propagation Neural Network

Basic probability assignment (bpa) of the dempster-shafer evidence theory of this damage identification model was presented by the output of AE back propagation neural network, and mean square error of back propagation neural network training was applied as uncertainty. Therefore, bpa can be defined as:

$$m(A_i) = \frac{f(A_i)}{\sum_{i=1}^{q} f(A_i) + E} \tag{2}$$

Where $m(A_i)$ is bpa of the ith output neuron, $f(A_i)$ and E are the actual value and mean square error that obtained by AE back propagation neural network respectively.

Uncertainty was presented by mean square error of AE back propagation neural network [12], namely

$$m(\theta) = E \tag{3}$$

2.3 The Damage Identification Model of 16Mn Steel Based on Back Propagation Neural Network and Dempster-Shafer Evidence Theory

Based on the above considerations, damage identification model of 16Mn steel based on back propagation neural network and dempster-shafer evidence theory can be established. The mainly process can be drawn as follows: First of all, four AE sensors collect the information of 16Mn steel in tensile test. Secondly, the back propagation neural network output the diagnosed damage results based on the information of AE four sensors. Then, bpas is presented by the output of back propagation neural

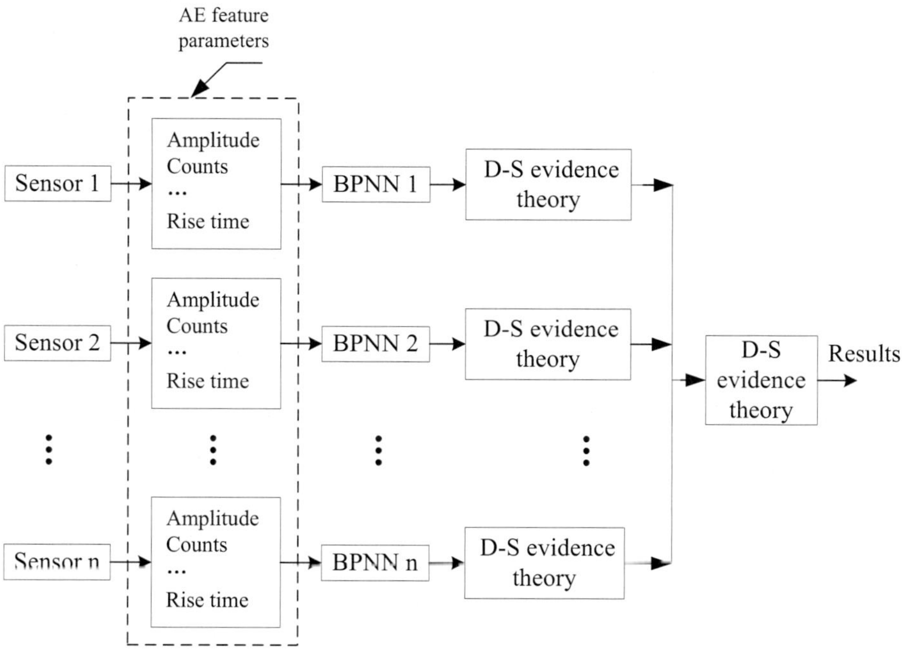

Fig. 1. The damage identification model based on backpropagation neural network and dempster-shafer evidence theory

network. Finally, the dempster-shafer evidence theory is used with bpa s to identify the damage modes of 16Mn steel in tensile test. The damage identification model based on back propagation neural network and dempster-shafer evidence theory is shown in Fig. 1.

3 Experimental Procedure

Acoustic emission measurement instrument manufactured by Physical Acoustic Corporation (PAC) was used in this research, the settings of AE instrument as follows: 1) Test threshold = 40 dB; 2) Peak definition time (PDT) = 300 μs; 3) Hit definition time (HDT) = 600 μs; 4) Hit lockout time (HLT) = 1000 μs.

The two R15 sensors (sensor 1 and sensor 4), two WD sensors (sensor 2 and sensor 3) and four preamplifiers with 40dB gain (2/4/6) which manufactured by Physical Acoustic Corporation were used in the experiment. Those sensors were attached on the specimen by a ringshaped magnet. In addition, vaseline was used at the interface between the sensors and the specimen surface to obtain proper signals.

The specimen dimension of this 16Mn steel tensile test is described in the Fig. 2. In this research, three same specimens ($lh3-1$, $lh3-2$, $lh3-3$) were tested under the condition of 1.5 $mm/\min$ rate of extension.

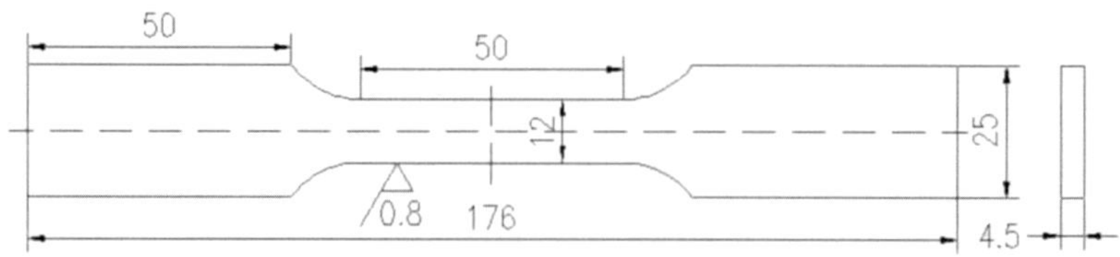

Fig. 2. Specimen dimension

4 Results and Discussion

4.1 The Results of Back Propagation Neural Network

Before using the damage identification model of 16Mn steel based on back propagation neural network and dempster-shafer evidence theory, we assumed that

$$I = \{AE\ parameter\ i \mid i = 1, 2, 3, 4, 5\}$$

Namely, $I = \{amplitude, counts, energy, duration, rise\ time\}$

$$\theta = \{damage\ phase\ j \mid j = 1, 2, 3, 4\}$$

Namely,
$\theta = \{$elastic deformation, yield deformation, strain hardening, necking deformation$\}$

AE back propagation neural network was developed based on the above assumption and the acoustic emission signature analysis during 16Mn steel damage in tensile test. This model was 3-layer neural networks, input layer was amplitude, counts, energy, duration and rise time, output layer was four damage modes which is elastic deformation (code 1000), yield deformation (code 0100), strain hardening (code 0010) and necking deformation (code 0001), and selected Tansig-Logsig as transfer function. The main parameter of AE back propagation neural network model of 16Mn steel in tensile test is shown in Table 1.

Table 1. The main parameter of AE back propagation neural network model

Input layer	Hidden layer	Output layer	Other parameter

Amplitude
Counts
Energy
Duration time
Rise time
	Based on the training results	(1000)
		(0100)
		(0010)
		(0001)
			net.trainparam.lr=0.05;
			net.trainParam.epochs=3000;
			net.trainParam.goal=0.03.

A dataset including 84 data obtained from acoustic emission experiment were used for back propagation neural network. From these data, 48 data (12 data each damage modes) were used for training the network, and the remaining 36 data (9 data each damage modes) were used as the test dataset. Due to the paper limitation, only the data of specimen $lh3-1$ were displayed in this paper, that is to say, 16 data (4 data each damage modes) of each sensor signals were used for training the network, and the remaining 12 data (3 data each damage modes) of that were used as the test dataset. The number of neurons on the hidden layer was adjusted according to the training results and was finally ascertained 45, here, mean square error equals to approximately 0.04. With the trained back propagation neural network, test data was calculated and the actual output of neural network was obtained. The contrast of the desired output and the actual output of AE back propagation neural network model of sensor 1 information about specimen $lh3-1$ is given in Table 2.

Table 2. The contrast of the desired output and the actual output of AE back propagation neural network model of sensor 1 information about specimen $lh3-1$

The actual output of back propagation neural network				The desired output			
$y(A_1)$	$y(A_2)$	$y(A_3)$	$y(A_4)$	A_1	A_2	A_3	A_4
0.9839	0.0918	0.0255	0.0011	1	0	0	0
0.9725	0.0755	0.0172	0.0016	1	0	0	0
0.9496	0.0932	0.0141	0.0017	1	0	0	0
0.8124	0.2129	0.0185	0.0014	0	1	0	0
0.1854	0.8179	0.0416	0.0017	0	1	0	0
0.0629	0.8687	0.1046	0.0022	0	1	0	0
0.0068	0.06	0.902	0.0078	0	0	1	0
0.003	0.0188	0.9088	0.0506	0	0	1	0
0.0011	0.0207	0.665	0.3212	0	0	1	0
0.0005	0.0236	0.2888	0.7252	0	0	0	1
0.0005	0.0236	0.2888	0.7252	0	0	0	1
0.0005	0.0238	0.2762	0.7399	0	0	0	1

It might be found that the most of actual output was close to the desired output based on this AE back propagation neural network model. In this works, there are two falsely estimation in 36 data, therefore, we applied the dempster-shafer evidence theory to combine the AE back propagation neural network results of four sensors to decline error.

4.2 The Results Using Dempster-Shafer Evidence Theory

According to the output of AE back propagation neural network, bpa s were obtained with Eq. (2), bpa s of sensor 1 information about specimen $lh3-1$ are shown in Table 3. The combined bpa s, Bel s (belief function) and Pl s (plausibility function) of four sensors information are shown in Table 4.

It can be found that the damage identification correctness by AE back propagation neural network and sempster-shafer theory model was higher than that of AE back propagation neural network model. In this works, there is no falsely estimation in 36 Pending data.

Table 3. The bpa s based on the output of AE back propagation neural network (specimen $lh3-1$, sensor 1)

$m(A_1)$	$m(A_2)$	$m(A_3)$	$m(A_4)$	$m(\theta)$
0.8888	0.0829	0.023	0.001	0.0042
0.9087	0.0705	0.0161	0.0015	0.0032
0.8922	0.0876	0.0132	0.0016	0.0054
0.4821	0.1263	0.011	0.0008	0.3798
0.1715	0.7565	0.0385	0.0016	0.032
0.0597	0.8238	0.0992	0.0021	0.0152
0.0069	0.061	0.9174	0.0079	0.0068
0.003	0.0191	0.9209	0.0513	0.0057
0.001	0.0185	0.5959	0.2878	0.0967
0.0004	0.0211	0.2584	0.6488	0.0713
0.0004	0.0211	0.2584	0.6488	0.0713
0.8888	0.0829	0.023	0.001	0.0042

Table 4. The bpa s, Bel s and Pl s of sensor 1, 2, 3 and 4 information (specimen $lh3-1$)

	$m(A_1)$	$m(A_2)$	$m(A_3)$	$m(A_4)$	$m(\theta)$	Diagnosis result
bpa	0.9993	0	0	0	0	Elastic
$[Bel,Pl]$	[0.9993, 1]	[0, 0.0007]	[0, 0.0007]	[0, 0.0007]	---	deformation
bpa	0	1	0	0	0	Yield
$[Bel,Pl]$	[0, 0]	[1, 1]	[0, 0]	[0, 0]	---	deformation
bpa	0	0	0.9998	0	0	Strain
$[Bel,Pl]$	[0, 0.0002]	[0, 0.0002]	[0.9998, 1]	[0, 0.0002]	---	hardening
bpa	0	0	0.0006	0.9991	0	Necking
$[Bel,Pl]$	[0, 0.0003]	[0, 0.0003]	[0.0006, .0009]	[0.9991, 0.9994]	---	deformation

It is indicated that damage identification model of 16Mn steel based on AE back propagation neural network and dempster-shafer evidence theory works well, the uncertainty of the system decreases, and the accuracy of specimen damage modes identification increases. But this model need more research and more test.

5 Conclusions

Based on the present test results, analysis and discussions, the preliminary conclusions may be drawn as follows:

(1) Acoustic emission is one of the reliable methods to monitor the damage of 16Mn in tensile test.

(2) It is feasible to identify the damage modes of 16Mn steel based on back propagation neural network and dempster-shafer evidence theory in tensile test.

Firstly, based on the acoustic emission signature analysis during material damage, 3-layer back propagation neural networks model with Tansig-Logsig transfer function was developed. Amplitude, counts, energy, duration and rise time were selected as input neurons, and elastic deformation, yield deformation, strain hardening and necking deformation were selected as output neurons.

Secondly, bpa s and belief function based on the output of AE back propagation neural network were obtained.

Finally, four sensors data was fused with Dempster's combination rule to obtain the finally diagnosis of 16Mn steel damage modes in tensile test.

The results shows that damage identification model of 16Mn steel based on back propagation neural network and dempster-shafer evidence theory works well, this model can increase the reliability and correctness of the diagnosis results, and can decrease the effect of the environment noise.

References

1. Roberts, T.M., Talebzadeh, M.: Acoustic Emission Monitoring of Fatigue Crack Propagation. Journal of Constructional Steel Research 59, 695–712 (2003)
2. Boogaard, J., Van Dijk, G.M.: Acoustic Emission a Review. In: 12th World Conference on Non-Destructive Testing, pp. 429–434. Elsevier Science Publishers B.V., Amsterdam (1989)
3. Jayakumar, T., Mukhopadhyay, C.K., Venugopal, S., et al.: A Review of the Application of Acoustic Emission Techniques for Monitoring Forming and Grinding Processes. Journal of Materials Processing Technology 159, 48–61 (2005)
4. Cheng, R., Gent, T.T., Kato, H., Takayama, Y.: AE behaviors evaluation with bp neural network. Computers & Industrial Engineering 31(5), 867–v871 (1996)
5. Grabec, I., Sachse, W., Govekar, E.: Solving ae problems by a neural network. In: Acoustic emission: current practice and future direction, ASTM STP 1077, pp. 165–182 (1991)
6. Rumelhart, D., Hinton, G.E., Williams, R.J.: Learning internal representations by error propagation. In: Rumelhart, D., McClel1nnd, J. (eds.) Parallel Distributed Processing, vol. 1, section 1.2. MIT Press, Cambridge (1986)
7. Fausett, L.: Fundamentals of Neural Networks: Architectures, Algorithms, and Applications. Prentice Hall, Englewood Cliffs (1994)
8. Weiwei, S., Yiming, C.: Global Asymptotic Stability Analysis for Neutral Stochastic Neural Networks with Time-Varying Delays. Communications in Nonlinear Science and Numerical Simulation 14, 1576–1581 (2009)
9. Hall, D.L.: Mathematical Techniques in Multi-Sensor Data Fusion. Artech House, Boston (1992)
10. Basir, O., Yuan, X.H.: Engine fault diagnosis based on multi-sensor information fusion using dempster-shafer evidence theory. In: Information Fusion, pp. 1–8 (2005)
11. Yang, B.S., Kim, K.J.: Application of Dempster-Shafer theory in fault diagnosis of induction motors using vibration and current signals. Mechanical Systems and Signal Processing 20, 403–420 (2006)
12. Xie, M.: Application of Decision Level Information Fusion for the Fault Diagnosis of Suction Fan. Nanchang University (2007)

OFFD: Optimal Flexible Frequency Discretization for Naïve Bayes Classification

Song Wang[1], Fan Min[1,3], Zhihai Wang[2], and Tianyu Cao[1]

[1] Department of Computer Science, The University of Vermont, 05405 Burlington, VT, USA
[2] School of Computer Science and Information Technology, Beijing Jiaotong University, 100044 Beijing, China
[3] Department of Computer Science and Engineering, University of Electronic Science and Technology of China 610054 Chengdu, Sichuan, China
{swang1,fmin,tcao}@cems.uvm.edu, zhhwang@bjtu.edu.cn

Abstract. Incremental Flexible Frequency Discretization (IFFD) is a recently proposed discretization approach for Naïve Bayes (NB). IFFD performs satisfactory by setting the minimal interval frequency for discretized intervals as a fixed number. In this paper, we first argue that this setting cannot guarantee optimal classification performance in terms of classification error. We observed empirically that an optimal minimal interval frequency existed for each dataset. We thus proposed a sequential search and wrapper based incremental discretization method for NB: named Optimal Flexible Frequency Discretization (OFFD). Experiments were conducted on 17 datasets from UCI machine learning repository and performance was compared between NB trained on the data discretized by OFFD, IFFD, PKID, and FFD respectively. Results show that OFFD works better than these alternatives for NB. Experiments between NB discretized on the data with OFFD and C4.5 showed that our new method outperforms C4.5 on most of the datasets we have tested.

1 Introduction

Naïve Bayes (NB) has been widely deployed in real world applications due to its simplicity and computational efficiency. It assumes that attributes are conditional independent given the class label. Though this simple assumption is violated in most of the real world applications, NB still performs comparable to the state-of-art classifiers such as C4.5. When dealing with numeric attributes, two general mechanisms are used for NB. One is to let NB deal with the data directly by assuming that values of the numeric attributes follow a normal distribution. However, in this way the performance of NB cannot be guaranteed. The other is to employ a preprocessing technique called discretization.

Discretization transforms each numeric attribute *X* into a nominal attribute *X**, and maps the numeric attribute domain into a nominal one. Dougherty et al. [4] stated that "the classification performance tends to be better when numeric attributes are discretized that when they are assumed to follow a normal distribution."

R. Huang et al. (Eds.): ADMA 2009, LNAI 5678, pp. 704–712, 2009.

Two terminologies are widely used in NB discretization, which are interval frequency (the number of training instances in one interval) and interval number (the number of discretized intervals produced by a specific discretization algorithm) [11]. Yang et. al [11] proposed the proportion k-interval discretization technique (PKID). PKID works based on the fact that there is a tradeoff between interval number, interval frequency and the bias, variance component in the classification error decomposition [4]. Also, "large interval frequency incurs low variance but high bias whereas large interval number produces low bias and high variance" [10]. However, PKID does not work well with small data sets, which have at most 1000 instances. In [13], Ying and Webb proposed another technique called Fixed Frequency Discretization (FFD). FFD discretizes the training instances into a set of intervals which contain approximately the same number of *k* instances, where *k* is a pre-specified parameter. Note that FFD is different from Equal Frequency Discretization [1], as in FFD the interval frequency is fixed for each interval without considering the number of training instances. The larger the training data size is, the larger the number of intervals is produced. However, the interval frequency will not change too much.

PKID and FFD are not incremental in nature. Once a new training instance comes, it needs to do the discretization from the scratch, which is rather inefficient, especially for large datasets. To the best of our knowledge, Incremental Flexible Frequency Discretization (IFFD) [8] is the first incremental discretization technique proposed for NB. IFFD sets the interval frequency ranging from *minBinSize* to *maxBinSize* instead of single value *k*. The number *minBinSize* and *maxBinSize* stand for the minimal and maximal interval frequency.

In this paper, we first argue that setting the *minBinSize* as a fixed number does not guarantee that the classification performance of NB is optimal. We also found out that there existed an optimal *minBinSize* for each dataset based on empirical observations. Finally, we propose a new supervised incremental discretization method: Optimal Flexible Frequency Discretization (OFFD) using a sequential search and wrapper based heuristic [5]. We did our experiments on 17 UCI data sets [2] in WEKA [9]. Experiments show that the classification accuracy of NB trained with OFFD is better than NB trained with PKID, FFD, and IFFD. We also compared the classification accuracy of NB trained with OFFD with that of C4.5; our new method defeated C4.5 in most of the domains we have tested.

The rest of the paper is organized as follows: Section 2 reviews previous discretization methods proposed in the context of NB. Details of the OFFD are then given in Sections 3. Section 4 presents experimental results. Finally, Section 5 concludes the paper.

2 Related Discretization Methods for Naïve Bayes Classification

2.1 Proportional K-Interval Discretization

Proportional K-Interval Discretization (PKID) was proposed in [11]. It is designed to reduce the bias and variance of the classification error by adjusting the interval number and interval frequency at the same time. Through out this paper, we denote the interval number as t, the interval frequency as s, the size of training data as N.

PKID discretizes the training data according to the following equations:

$$\begin{cases} s \times t = N; \\ s = t \end{cases} \quad (1)$$

In this way, the training instances are discretized into $\sqrt{N}$ intervals with each interval containing approximately $\sqrt{N}$ number of instances. With the increase of training data size, the interval number and frequency will increase correspondingly. As for large datasets, variance contributes more than bias to the classification error, it is expected that PKID works well in this situation. However, for small datasets, as the number of instances in each interval is also small, therefore PKID incurs high classification bias, whereas under this scenario, bias weights more than variance in the classification error decomposition [10]. Hence, it is difficult for PKID to get reliable statistical probability estimation for small datasets and the performance of NB trained on these datasets using PKID is not satisfactory. In general, PKID will only perform well with numeric attributes with many identical values or small value ranges, henceforth produces much fewer intervals than $\sqrt{N}$. Otherwise, PKID tends to produce larger number of intervals when the datasets size increases.

2.2 Fixed Frequency Discretization

Fixed Frequency Discretization (FFD) is another useful discretization method proposed for NB. FFD adjusts the bias and variance of classification error by setting fixed interval frequency and adapting the interval number according to the increase of training data size. FFD sets the interval frequency at a fixed number, current implementation of FFD set the interval frequency s as 30 [12]. Thus for small datasets, FFD can guarantee that the statistics obtained on the intervals are reliable, thus reduces the bias of probability estimation. For large datasets, with the increase of training data size, the number of intervals will increase, thus reduces the variance of probability estimation. Therefore, the overall classification performance of NB trained with FFD will increase.

Note that PKID and FFD are all unsupervised discretization according to [6] [7]. Also, PKID and FFD are not incremental in nature because they are sensitive to data size change. Once a new instance comes, they need to do discretization from scratch. However, as NB supports incremental learning, it will be upgrade the performance of NB if proper incremental discretization methods for NB are proposed, especially when dealing with large datasets.

3 OFFD: Optimal Flexible Frequency Discretization

In this section, we will describe in detail about our new approach: Optimal Flexible Frequency Discretization (OFFD).

3.1 IFFD Recaptured

Incremental Flexible Frequency discretization (IFFD) was proposed in [8]. The idea of IFFD is that in order not to discretize from scratch when the training data arrives incrementally; it maintains corresponding statistics for each discretized interval. IFFD works as follows: it sets the interval frequency flexibly between pre-specified *minBinSize* and *maxBinSize*, in the current implementation of IFFD, *minBinSize* is set to 30, and *maxBinSize* is set to be twice of *minBinSize*. An interval accepts instances until it reach the upper bound of the interval frequency, if the number of instances in one interval exceeds the upper bound *maxBinSize*, it will call a split procedure to produce two new intervals with interval frequency of at least *minBinSize*, meanwhile, it will update the statistics for those intervals whose values has been affected by the split. IFFD is reported to perform efficiently in [8]. However, we argue that setting the *minBinSize* as 30 for all datasets does not guarantee that the classification performance is optimal in terms of classification error.

3.2 OFFD: Optimal Flexible Frequency Discretization

Our new method OFFD is based on the following claim: there exists an optimal *minBinSize* for the discretization intervals for each numeric attribute as the values of numeric attributes has some distribution. Though such a cumulative distribution does not necessarily be Gaussian distribution, if we could approximate the distribution using the optimal minimal discretization interval frequency (*minBinSize*), it will in turn benefit the classification performance. It is nontrivial to show that such an optimal interval frequency exists theoretically because we know very little about the data distribution, especially for unseen data. Instead, we believe that such an optimal interval frequency does exist and use experiments to justify it.

OFFD works as follows: instead of setting the *minBinSize* as 30 for all the data sets, we set a search space for the optimal *minBinSize* ranging from 1 to 40. The reason for selecting 40 as the upper bound is that for large datasets, 40 works well as stated in [8]. It is also reasonable to set the upper bound as $\sqrt{N}$, but the performance of OFFD when setting upper bound as $\sqrt{N}$ is not good, we decided to use 40 instead. OFFD works in rounds by testing each *minBinSize* values, in each round, we do a sequential search on these range of values and set the current value as *minBinSize* and discretize the data using IFFD based on the current *minBinSize* value, we record the classification error for each round, if the classification error reduces once a *minBinSize* is set, we will update the *minBinSize*, this search process is terminated until all values ranging from 1 to 40 have been searched or the classification error no longer reduces. The pseudo-code of OFFD is listed in Algorithm 1. In OFFD, we also set the *maxBinSize* as twice of *minBinSize*. cutPoints is the set of cut points of discretization intervals. *counter* is the conditional probability table of the classifier. IFFD will update the cutPoints and counter according to the new attribute value V. classLabel is the class label of V. curCfErr denotes the current classification error.

Note the OFFD is a sequential search and wrapper based supervised approach, the search efficiency for optimal *minBinSize* is still efficient in the context of incremental learning, we use the wrapper approach as the heuristic to shrink our search space. Therefore, the efficiency of OFFD is comparable to that of IFFD.

Algorithm 1. Pseudo-code for OFFD

```
Input: cutPoints, counter, V, classLabel, a range of values from 1 to 40
Output: The set of optimal Discretized Intervals
Method:
Do a sequential search from [1, 40], set the current value as minBinSize,
Initialize curCfErr=100% ;
while TRUE do
  Test whether V is greater than the last cut point
  if V is larger than the last cut point then
    Insert V into the last interval;
  update the corresponding interval frequency;
    Record changed interval;
  else
    //This block is adapted from [8]
    for (j=0; j<size-1; j++) do
      if (V ≤ cutPoints[j]) then
        insert V into interval[j];
        intFre[j]++;
        counter[j][classLabel]++;
        chaInt=j; //record the interval which has been changed
        break;
      end if
    end for
  end if
  if (intFre[chaInt] > minBinSize *2) then
    get new cut points;
    insert the new cut points into cutPoints;
    calculate counter[c1] and counter[c2]; //update contingency table
  end if
  Record the current classification error and current minBinSize.
  Set curCfErr as current classification error.
  if (curCfErr does not change) then
    break;
  end if
  Get the next sequential value as new minBinSize;
end while
return the optimal minBinSize and the set of discretized intervals;
```

4 Experiment Design and Result Analysis

In this section, we will justify our claim on the existence of optimal *minBinSize* and evaluate our new discretization method OFFD for NB with other alternatives, including PKID, FFD, and IFFD.

4.1 Experiment Design

We did our experiments on 17 datasets from UCI machine learning repository [2], including 9 large datasets and 8 small ones. Datasets information is summarized in Table 1. Size denotes the number of instances in a dataset, Qa. is the number of numeric attributes, No. is the number of nominal attributes, and C means the number of different class values. We listed the empirical result for the existence of optimal *minBinSize* for each dataset in Figure 1.

Table 1. Data Set Information

ID	DataSet	Size	Qa	No.	C	ID	DataSet	Size	Qa	No	C
1	Sonar	208	60	0	2	9	balance-scale	625	4	0	3
2	hepatitis	155	14	5	2	10	shuttle	58000	9	0	7
3	Crx	690	6	9	27	11	mfeat-mor	2000	47	0	10
4	vehicle	846	18	0	4	12	volcanoes	1520	3	0	4
5	german	1000	7	13	2	13	satellite	6435	36	0	6
6	vowelcontext	990	10	3	11	14	page-blocks	5473	10	0	5
7	cleveland	303	6	7	2	15	hypothyroid2	3772	6	26	4
8	hungarian	284	5	8	2	16	cmc	1473	2	7	3

Table 2. Classification Accuracy Comparison on Different Discretization Methods

DataSet	OFFDNB	IFFDNB	FFDNB	PKIDNB	C4.5
Sonar	**100**	82.2115	88.4615	67.7885	71.6346
Hepatitis	**92.2581**	82.5806	87.7419	84.5161	80.2721
Cleveland	**90.099**	85.8086	84.1584	82.5083	78.2178
Hungarian	**89.1156**	85.034	85.034	83.6735	80.2721
Crx	**91.5942**	87.8261	88.6957	77.6812	85.5072
balance-scale	**92.16**	92.16	92.16	90.4	78.08
Vehicle	**82.6241**	70.2128	72.5768	45.5083	73.5225
German	**87.3**	78	79.9	75.4	74.1
Vowel	**100**	89.798	94.0404	62.9293	78.0808
page-blocks	94.6282	93.9521	94.7926	90.846	**96.9121**
mfeat-mor	**87.15**	73.4	74.4	69.35	72.1
Shuttle	99.4672	99.4345	99.4328	92.9793	**99.9655**
Adult	**86.6877**	84.1776	84.4662	83.2501	86.0509
Cmc	**53.9036**	53.2926	53.3605	50.4413	51.1881
Volcanoes	**93.8158**	69.8684	70.9868	63.6184	66.9079
Satellite	83.5276	83.2479	83.3877	82.3776	**86.5113**
hypothyroid2	98.4093	97.9056	98.0647	95.3871	**99.4698**

As the error rate change with *minBinSize* change for the dataset Adult is special, we list the information about Adult separately: the size of Adult is 48842, with 6 numeric attributes, 8 nominal attributes and 2 distinct class values. For IFFD, we set the *minBinSize* as 30, for FFD, the parameter *k* as 30. For both NB and C4.5 classification, we did 6-fold cross-validation to test the classification accuracy. For IFFD and OFFD, data are fed in an incremental manner. For PKID, FFD, and C4.5, we feed the classifier with all the data at training time. The classification accuracy of NB with these approaches and C4.5 on those 17 datasets is summarized in Table 2. We denote NB with OFFD as OFFDNB, NB with PKID as PKIDNB, NB with FFD and IFFD as FFDNB and IFFDNB respectively.

4.2 Result Analysis

The classification error of NB trained on OFFD and corresponding *minBinSize* ranging from 1 to 40 is shown in Figure 1. It is easy to conclude that an optimal *minBinSize* exists for each dataset from Figure 1.

Figure 1 (a) shows that the error rate of OFFD is minimal when *minBinSize* = 1 on 6 small datasets, i.e., *sonar, hepatitis, crx, vehicle, german* and *vowel-context*. With the increase of *minBinSize*, the performance tends to be worse. For example, when *minBinSize* = 1, for dataset *vowel-context*, the error rate is 0. While when *minBinSize* = 40, the error rate is 0.142. Empirically for these datasets, the best choice is to perform NB directly without discretization.

Figure 1 (b) also illustrates the performance OFFD on the other 3 small datasets. The algorithm performs the best for *cleveland* while *minBinSize* = 2. Even if we set *minBinSize* = 1, the accuracy is only 0.0033 worse, which is not significant. Also, for dataset *hungarian*, the best setting is *minBinSize* = 3. Comparing to this setting, when *minBinSize* = 1 the accuracy is only 0.0034 worse. With the increase of *minBinSize* over the optimal setting, the accuracy continues to be worse. Experimental results on *balance-scale* is somewhat strange, the performance is not influenced by the setting of *minBinSize*. One conclusion could be drawn up to now is that for small datasets, discretization is generally not necessary, or can even degrade the performance of NB.

As depicted in Figure 1(c) (d) and (e), the results of *shuttle* come to the same conclusion with that of small datasets. That is, no discretization is needed. This holds regardless of the fact that there are 58,000 instances in the dataset. The results of *mfeat-mor*, *volcanoes*, *satellite* and *adult* have the same and even stronger trend as that of *shuttle*.

The result of NB on *page-blocks* shows that the optimal setting of *minBinSize* is 8. With the setting of *minBinSize* farther to the optimal one, the error rate almost continues to increase. Therefore, 8 is a stable optimal setting for *minBinSize*. The error rate of NB on the *hypothyroid2* dataset is similar with that of *page-blocks*. However, the trend toward the optimal setting of *minBinSize* is not so strong. 15 and 23 are two settings that produce the same error rate. One reason lies in that the error rate is already very low, thus some unimportant factors may cause “significant" difference. We also observe that there is some rational setting of *minBinSize*. The error rate of NB on *cmc* shows that the error rate drops as *minBinSize* approaches 40, therefore we should set *minBinSize* $\leq$ 40 to obtain the optimal setting. From Figure 1 and the above analysis, we can conclude empirically an optimal *minBinSize* for each dataset exists. Meanwhile, this observation also provides us a technique to decide whether we need to do discretization for a given dataset or not.

Table 2 indicates that the classification performance of NB with OFFD is much better than that of NB with IFFD, FFD, and PKID. NB with OFFD outperforms NB with PKID and NB with IFFD on all 17 datasets we have tested. Also, the classification accuracy of NB with OFFD is better than that of NB with FFD on 15 datasets. Although for *page-blocks* NB with FFD is better, on average, NB with OFFD works better. Table 2 also shows that the classification performance of NB with OFFD is better than that of C4.5 on most domains (those numbers in bold means that the classification accuracy of NB with OFFD defeats that of C4.5). The reason is that NB with OFFD used a wrapper based approach and tried to improve the classification performance of NB as much as possible.

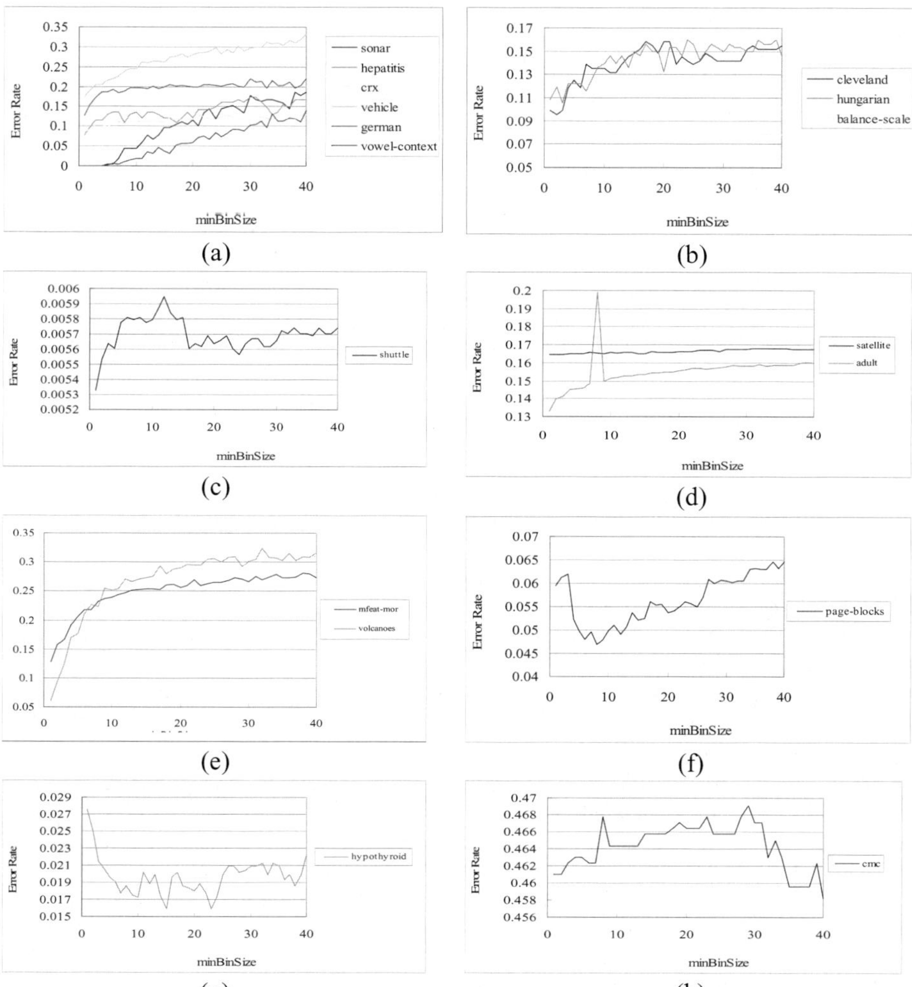

Fig. 1. Error Rate of NB with *minBinSize* Change

5 Conclusions

In this paper, we empirically found out that an optimal interval frequency exists for the datasets we have tested and proposed a sequential search and wrapper based incremental discretization approach (OFFD) for NB. Experimental results show that our new approach outperforms its peers in most cases and it can also make the classification accuracy of NB scalable to that of C4.5 in most of the domains we have tested. However, we did not provide a sound theoretical analysis on why the optimal interval frequency exists; it will be very useful if we can find a complete proof of such a fact. Also, it would be fruitful if we could know more about the data distribution and use such domain knowledge to guide the discretization process.

Acknowledgements

This research is partially supported by National Natural Science Foundation of China, Project No. 60673089. The authors would like to thank Dr. Jingli Lu for providing the codes of IFFD.

References

1. Catlett, J.: On changing continuous attributes into ordered discrete attributes. In: Kodratoff, Y. (ed.) EWSL 1991. LNCS, vol. 482, pp. 164–178. Springer, Heidelberg (1991)
2. Blake, C.L., Merz, C.: UCI machine learning repository. University of California, Irvine (1998)
3. Domingos, P., Pazzani, M.J.: On the optimality of the simple Bayesian classifier under zero-one loss. Machine Learning 29(2-3), 103–130 (1997)
4. Dougherty, J., Kohavi, R., Sahami, M.: Supervised and unsupervised discretization of continuous features. In: ICML, pp. 194–202 (1995)
5. Kohavi, R., John, G.H.: Wrappers for feature subset selection. Artificial Intelligence 97(1-2), 273–324 (1997)
6. Kohavi, R., Sahami, M.: Error-based and entropy-based discretization of continuous features. In: KDD, pp. 114–119 (1996)
7. Liu, H., Hussain, F., Tan, C.L., Dash, M.: Discretization: An enabling technique. Data Mining & Knowledge Discovery 6(4), 393–423 (2002)
8. Lu, J., Yang, Y., Webb, G.I.: Incremental discretization for naïve-bayes classifier. In: Li, X., Zaïane, O.R., Li, Z.-h. (eds.) ADMA 2006. LNCS, vol. 4093, pp. 223–238. Springer, Heidelberg (2006)
9. Witten, I., Frank, E.: Data mining: Practical machine learning tools and techniques, 2nd edn. Morgan Kaufmann, San Francisco (2005)
10. Yang, Y., Webb, G.I.: Discretization for naive-bayes learning: Managing discretization bias and variance. Machine Learning 74(1), 39–74 (2009)
11. Yang, Y., Webb, G.I.: Proportional k-interval discretization for naïve-bayes classifiers. In: Flach, P.A., De Raedt, L. (eds.) ECML 2001. LNCS, vol. 2167, pp. 564–575. Springer, Heidelberg (2001)
12. Yang, Y., Webb, G.I.: A comparative study of discretization methods for naïve-bayes classifier. In: PAKW, pp. 159–173 (2002)
13. Yang, Y., Webb, G.I.: On why discretization works for naive-bayes classifiers. In: Australian Conference on Artificial Intelligence, pp. 440–452 (2003)

Handling Class Imbalance Problems via Weighted BP Algorithm

Xiaoqin Wang and Huaxiang Zhang

College of Information Science and Engineering, Shandong Normal University, 250014 Jinan, Shandong
piao_123qin@163.com

Abstract. When using Neural Networks (NN) to handle class imbalance problems, there exists a fact that minority class makes less contribution to the error function than the majority class, so the network learned prefers to recognizing majority class data which we pay less attention to. This paper proposes a novel algorithm WNN (Weighted NN) to solve this problem using a newly defined error function in BP (BP) algorithm. Experimental results executed on 20 UCI datasets show that the approach can effectively enhance the recognition rate of minority class data.

Keywords: Imbalance dataset, Neural Network, BP algorithm, weight.

1 Introduction

At present, growing interest is being devoted to the class imbalance classification problem. This increasing interest results in two workshops being held, one by AAAI (American Association for Artificial Intelligence) in 2000, and the other by ICML (International Conference on Machine Learning) in 2003. In 2004 SIGKDD Exploration also published one special issue about class imbalance problem [1]. This problem is mainly caused by imbalance data, in which one class contains much fewer instances than others. Imbalance data is encountered in a large number of domains, such as fraud detection [2], medical diagnosis [3] and text classification [4]. It will result in negative effects on the performance of traditional learning methods which assume a balanced class distribution. Class imbalance problem is regarded as one of the big hotspot of future machine learning.

When facing imbalance data, traditional learning algorithms tend to produce high predictive accuracy for majority class data but poor predictive accuracy for minority class data which people pay much more attention to. That is mainly because traditional classifiers seek predictive accuracy over the whole dataset [1]. They are not suitable to cope with imbalanced learning tasks [5, 6] since they tend to assign all instances to majority class which is usually the less important. The cost of misclassifying minority class data is tremendous.

To address this problem, several techniques are proposed. They can summarized into two groups, one is by changing the class distribution of the training set before building classifiers, the other is by improving learning algorithms to make them

R. Huang et al. (Eds.): ADMA 2009, LNAI 5678, pp. 713–720, 2009.

adaptable to imbalance dataset. The first group mainly involves the following approaches: (a) over-sampling, methods that balance the training data by increasing the minority class data randomly. A popular approach is SMOTE [7], it synthetically over-samples the minority class. The disadvantage of this kind of approaches is they may cause overtraining and increase the computing complexity; (b) under-sampling. In this kind of methods, the training dataset is made class balance by deleting a number of majority class data randomly. With ignoring part of the majority class data, some important information vanishes too; (c) over-sampling and under-sampling combination. Experiments have shown that the combination of the two did not obtain significant improvement; (d) cluster based sampling, methods in which the representative examples are randomly sampled from clusters [8]; (e) constructing a learning algorithm that is intrinsically insensitive to class distribution in the training set. A representation is SHRINK algorithm, which finds only rules that best summarize the positive instances of the minority class, but makes use of the rules from the negative instances of the majority class [9].

The second group mainly contains cost-sensitive learning and Boosting methods. One method of cost-sensitive learning improves the prediction accuracy by adjusting the cost (weight) matrices for each class [10]. MetaCost is another method for making a classifier cost-sensitive and the representative approach of Boosting is AdaBoost algorithm.

When using approaches mentioned above to cope with class imbalance problems, researchers usually like to choose decision trees as the base classifiers, in deed, a much better alternative is neural networks (NN). NN is an important research field in machine learning; it provides a robust approach to approximating real-valued, discrete-valued, and vector-valued target functions. BP is one of the most important methods of NN. When applying it to deal with class imbalance problems, the majority class data contribute more to the error function than the minority class data. This leads to the low classification precision of the minority class data. To solve this problem, the paper proposes a novel algorithm WNN using weighted BP. This method weights each instance according to the size of the class the instance belongs to, i.e., give minority class a bigger weight and majority class a smaller weight in order to balance their influence to the error function. Experiments on 20 UCI datasets show that this approach can improve the recognition rate of the minority class data in imbalance dataset.

2 BP Algorithm

BP algorithm [11] is one of the most practical parts of NN which was developed at the beginning of 80s, it settled the multilayer networks learning problem which perception can't cope with. For multilayer networks that interlinked by a set of fixed units, BP algorithm is used to learn weights of the network. It employs gradient descent to attempt to minimize the squared error between the network output values and the target values. For a NN with multiple output units, its error function E is given as

$$E(\vec{w}) \equiv \frac{1}{2}\sum_{d \in D}\sum_{k \in outputs}(t_{kd} - o_{kd})^2 \tag{1}$$

Where, D is the training dataset, $outputs$ is the set of the network's output units, t_{kd} and o_{kd} is the output values associate with the training example d and the kth output unit.

BP algorithm commonly uses the sigmoid unit as the base of constructing multi-layer networks, and the sigmoid unit computes its output as

$$o = \sigma(\vec{w} \bullet \vec{x}) \text{ , where } \sigma(y) = \frac{1}{1+e^{-y}} \tag{2}$$

BP algorithm searches a large hypothesis space defined by all the possible weight values for all the units in the network, and stops until termination conditions are satisfied. Details are given in [11].

3 Weighted BP Algorithm

When training a NN, there exits a default hypothesis that the number of instances each class contains is balanced, namely, the contribution of each class to the error function (loss function) is identical. But to imbalance dataset, the instance number of the majority class is far greater than that of the minority class, thus the majority class makes more contribution to the training error function than minority class. The final hypothesis converged by BP is propitious to the recognition of the majority class data but not the recognition of the minority class data.

For the above reasons, a variation of BP named WNN is developed to deal with imbalance data recognition problems. In this method, we redefine the error function as:

$$E_w(\vec{w}) \equiv \frac{1}{2}\sum_{d \in D}\sum_{k \in outputs}[(t_{kd} - o_{kd}) * weight_d]^2 \tag{3}$$

Where, D is the training dataset, $outputs$ is the set of the network's output units, t_{kd} and o_{kd} is the output values associate with training example d and the kth output unit, $weight_d$ is the weight of instance d .

Equation (3) can also be written as:

$$E_w(\vec{w}) = \frac{1}{2}\sum_{d \in D}[\sum_{k \in outputs}(weight_d \cdot t_{kd} - weight_d \cdot o_{kd})^2] = \frac{1}{2}\sum_{d \in D}\sum_{k \in outputs}(t'_{kd} - o'_{kd})^2 \tag{4}$$

Where, $weight_d \cdot t_{kd} = t'_{kd}$ and $weight_d \cdot o_{kd} = o'_{kd}$

Hertz [13] had proved that with Gradient Descent rule Equation (4) can converge toward the minimum error hypothesis, and similar conclusions of equation (3) can be made.

We set the weight for each instance according to the distribution of the training dataset, i.e., weight each example based on the class it belongs to and the number of instances the class contains. Details of algorithm WNN are described in Table 1.

Table 1. WNN algorithm

input : training examples training_dataset , learning rate η , a multilayer feed forward network

output : NN that can correctly classify instances

process:

For the training data, count the number n of instances in minority class and the number m of instances in majority class;

Calculate the weight for each instance d :

if $d \in majorityclass$, $weight_d = (double)n/(double)m$, else $weight_d = 1.0$;

initialize all network weight as a small random number;

while(termination condition is not satisfied){

for each $d \in D$:

(a) Import instance d to network, compute output o_u of every output units u : $o_u(\vec{x}) = \sigma(\vec{w_u} \bullet \vec{x})$, and propagate the errors backward through the network;

(b) For each output unit k of the network, compute its error Err_k

$$Err_k \leftarrow o_k(1-o_k)(t_k - o_k) \bullet weight_d \tag{5}$$

(c) For every hidden units h of the network, compute its error Err_h

$$Err_h \leftarrow o_h(1-o_h) \sum_{k \in outputs} w_{kh} Err_k \tag{6}$$

(d) Compute the increment of weights $\Delta w_{ji}(t) = \eta Err_j x_{ji} + \alpha \Delta w_{ji}(t-1)$ (7)

(e) Update every weight of network $w_{ji} \leftarrow w_{ji} + \Delta w_{ji}(t)$ (8)}

(Note: x_{ji} is the i th input of unit j , w_{ji} is the weight relevant to the i th input of unit j , $weight_d$ is the weight of instance d , α is the momentum.)

In the algorithm, we weight the target and output values associated with each output unit and training example. From Table 1 we can see, the error Err_k of output unit k is similar to that in the original BP algorithm but multiplied by the weight of instance d $weight_d$. Hidden layers' outputs error Err_h is calculated similarly. We introduces a momentum term α ($0 \le \alpha < 1$), and it determines the influence of the prior weights to current weights.

Weight of each class changes along with the unbalance degree of dataset in this method, so the algorithm is also suit for balanced dataset. To balanced dataset, instances in each class is even, therefore majority class' weight n/m is close to 1, and equivalent to that of the minority class, so their contributions to the error function are equivalent, and the performance of the classifier would not be affected.

4 Experiments

4.1 Experiments Design

The network we design in this paper is a multilayer feed forward network containing two layers of sigmoid units. Node number of hidden layer is the mean of attribute number and class number. Node number of input layer equals to the number of class of dataset. The basic algorithm we choose is the BP algorithm. Experiments are finished on 20 UCI datasets, and the learning rate η is dissimilar to different dataset, on balance-sc, ionosphere, Anneal, weather and bridge-ver η =0.1; on cmc, dermatol, haberman, solar-flare, tic-tac-toe, hepatitis and heart-h η =0.01; on the rest datasets η =0.001.The momentum α =0.1.The value of the two parameters are obtained through many experiments and choose the best results. The proportion of minority class in each dataset is showed in Table 2.

In this experiment, we compare the performance of the following four algorithms: AdaBoostM1, Original BP algorithm (BP), WNN and an algorithm CascadeBag that reduces the imbalance degree of dataset by cascade structure combining undersampling. The basic classifier of AdaBoostM1 and CascadeBag is decision tree.

4.2 Evaluation Criterion of Classifier Performance [12]

We choose 4 measures that commonly used in class imbalance field as the evaluation criterions: recall, precision, F-value and AUC. AUC is the underneath area of ROC. The bigger AUC is, the better the algorithm is. According to Table 3, the first 3 measures are defined as:

$$recall = \frac{TP}{TP + FN} \tag{9}$$

$$precision = \frac{TP}{TP + FP} \tag{10}$$

$$F-value = \frac{(1+\beta^2)\cdot recall \cdot precision}{\beta^2 \cdot recall + precision}$$, β (usually set to 1) is a factor (11)

Table 2. Proportion of minority class in each dataset

dataset	balance-sc	breast-canc	car	cmc	colic
%	17.01	29.72	3.76	22.62	36.97
dataset	dermatol	haberman	ionosphere	lung-cancer	solar-flare
%	5.46	26.47	35.90	28.13	2.17
dataset	weather	tic-tac-toe	credit-g	vote	Anneal
%	35.71	34.66	30.00	38.62	0.891
dataset	breast-w	pima-diab	hepatitis	bridge-ver	heart-h
%	34.48	34.90	20.65	9.35	36.05

Table 3. Confusion matrix

	Predicted as positive (minority class)	Predicted as negative (majority class)
Actually positive (minority class)	True positive(TP)	False negative(FN)
Actually negative (majority class)	False positive(FP)	True negative(TN)

4.3 Experimental Results and Analysis

The experimental results are shown in Table 4 and Table 5, where bold fonts are the best of each measure in each dataset.

We make the following observations from Table 4 and 5:

(1) The influence of dataset's imbalance degree to classifier is not absolute, for example, AdaBoost have excellent performance on dataset vote and Breast-w, both its F-value exceeds 0.9.
(2) AdaBoost algorithm and BP algorithm focus more on precision, in our experiments, AdaBoost algorithm has the highest precision on 5 datasets, and BP has highest precision on 8 datasets. However, they both obtain poor recall, that mainly because the two original algorithms do not change the distribution of imbalance datasets, and with high recognition rate on majority class, they mistakenly take minority class as majority class easily, what could cause high FN.
(3) On 13 datasets, F-value of BP is higher than AdaBoost, which proves that without handling, performance of NN at imbalance dataset is better than decision tree, because NN owns advantages such as anti-jamming and high robustness.
(4) CascadeBag algorithm first adopts the mode of combining cascade structure and under-sampling method to reduce the imbalance degree of dataset, and then use the resample technique of Bagging to train classifier. It can obtain high recall, but low precision relatively.
(5) From Table 4 and Table 5 we can find, the WNN method we proposed obtains highest recall on 15 datasets, highest F-value on 18 datasets, highest AUC on 13 datasets compare with other three algorithms, and highest value on the average of every criterion on 20 UCI datasets. That mainly because this algorithm first change the distribution of dataset by weighting examples to increase the influence of minority in the training process, a result of this process is the reducing of FN, which represent the part of minority class data that been mistakenly classified as majority class, then combine the advantage of anti-jamming and high robustness of NN, finally enhance classify precision of minority class.

Table 4. Results on 20 datasets

(balance-sc)	pre	rec	F-value	AUC	(breast-can)	pre	rec	F-value	AUC
AdaBoost	0	0	0	0.794		0.500	0.424	0.459	0.697
CascadeBag	0	0	0	0.632		0.360	0.576	0.443	0.607
BP	***0.531***	0.694	0.602	0.929		***0.750***	0.035	0.067	***0.697***
WNN	0.45	***0.918***	***0.604***	***0.941***		0.491	***0.624***	***0.549***	0.694
(car)	pre	rec	F-value	AUC	(cmc)	pre	rec	F-value	AUC
AdaBoost	0	0	0	0		0	0	0	0.535
CascadeBag	0.620	0.877	0.726	0.990		***0.477***	0.405	0.438	***0.741***
BP	0	0	0	0.994		0.422	0.408	0.415	0.706
WNN	***0.970***	***0.985***	***0.977***	***1***		0.325	***0.865***	***0.473***	0.734
(colic)	pre	rec	F-value	AUC	(dermatol)	pre	rec	F-value	AUC
AdaBoost	0.756	0.728	0.742	0.867		0	0	0	0.640
CascadeBag	0.777	0.794	0.785	***0.902***		0.944	0.850	0.895	0.999
BP	***0.825***	0.765	***0.794***	0.882		0.056	0.050	0.053	0.492
WNN	0.768	***0.801***	0.784	0.883		***1***	***1***	***1***	***1***
(haberman)	pre	rec	F-value	AUC	(ionosphere)	pre	rec	F-value	AUC
AdaBoost	***0.500***	0.259	0.341	0.637		***0.970***	0.770	0.858	0.944
CascadeBag	0.355	0.543	0.429	0.605		0.853	***0.873***	0.863	***0.956***
BP	0	0	0	0.606		0.952	0.794	0.866	0.919
WNN	0.400	***0.568***	***0.469***	***0.638***		0.945	0.825	***0.881***	0.919
(lung-canc)	pre	rec	F-value	AUC	(solar-flare)	pre	rec	F-value	AUC
AdaBoost	***0.667***	0.444	0.533	0.717		***0.500***	0.143	***0.222***	0.596
CascadeBag	0.462	***0.667***	0.545	0.676		0.049	0.429	0.088	***0.791***
BP	0.429	0.333	0.375	0.691		0	0	0	1
WNN	0.556	0.556	***0.556***	***0.749***		0.114	***0.714***	0.196	0.782
(weather)	pre	rec	F-value	AUC	(tic-tac-toe)	pre	rec	F-value	AUC
AdaBoost	0.333	0.200	0.250	.0400		0.655	0.440	0.526	0.794
CascadeBag	0	0	0	0.267		0.782	0.928	0.848	0.973
NeuralNet	***0.667***	0.400	0.500	0.733		0.966	0.955	0.961	0.996
WNN	0.600	***0.600***	***0.600***	***0.756***		***0.981***	***0.955***	***0.968***	***0.996***
(credit-g)	pre	rec	F-value	AUC	(pima-diab)	pre	rec	F-value	AUC
AdaBoost	0.485	0.270	0.347	0.723		0.658	0.552	0.600	0.801
CascadeBag	0.492	***0.743***	0.592	0.779		0.614	***0.772***	0.684	0.832
BP	***0.602***	0.490	0.540	***0.784***		***0.707***	0.586	0.641	0.833
WNN	0.528	0.733	***0.614***	0.781		0.625	0.757	***0.685***	***0.834***
(hepatitis)	pre	rec	F-value	AUC	(vote)	pre	rec	F-value	AUC
AdaBoost	0.586	0.531	0.557	0.851		***0.930***	0.952	0.941	0.992
CascadeBag	0.490	0.750	0.593	0.861		0.920	0.964	0.942	0.989
BP	***0.621***	0.563	0.590	0.83		0.930	0.946	0.938	0.993
WNN	0.545	***0.750***	***0.632***	***0.867***		0.926	***0.970***	***0.948***	***0.993***
(heart-h)	pre	rec	F-value	AUC	(Anneal)	pre	rec	F-value	AUC
AdaBoost	***0.736***	0.604	0.663	0.891		0	0	0	0.83
CascadeBag	0.664	0.821	0.734	0.881		0.500	0.625	0.556	0.981
BP	0.733	0.726	0.730	0.873		0.897	0.875	0.886	***0.968***
WNN	0.690	***0.821***	***0.750***	***0.911***		***1***	***0.875***	***0.933***	0.966
(bridge-ver)	pre	rec	F-v	AUC	(breast-w)	pre	rec	F-v	AUC
AdaBoost	0	0	0	0.580		0.929	0.921	0.925	0.989
CascadeBag	0	0	0	0.499		0.918	***0.979***	0.948	0.989
BP	***0.638***	0.560	0.596	0.858		***0.953***	0.934	0.943	0.993
WNN	0.636	***0.700***	***0.667***	***0.868***		0.951	0.959	***0.955***	***0.994***

Table 5. Average of each measure on 20 UCI datasets

	pre	rec	F-value	AUC
AdaBoost	0.460	0.362	0.398	0.696
CascadeBag	0.493	0.611	0.535	0.661
BP	0.584	0.506	0.525	0.839
WNN	***0.675***	***0.799***	***0.712***	***0.865***

5 Conclusions

This paper proposes an improved BP algorithm WNN by giving different training data different weight according to the class it belongs to. Experiments on 20 UCI datasets show that this method improves the capability of NN in handing imbalanced dataset. Because weights of instances change with the dataset's imbalance degree, so this algorithm is also suitable to handling class balance problems.

Acknowledgments. This work is partially supported by the Science and Technology Projects of Shandong Province, China (No.2007ZZ17, Y2007G16, 2008GG10001015).

References

1. Chen, M.-C., Chen, L.-S.: An information granulation based data mining approach for classifying imbalanced data. Information Sciences 178, 3214–3227 (2008)
2. Fawcett, R.E., Provost, F.: Adaptive fraud detection. Data Mining and Knowledge Discovery 3(1), 291–316 (1997)
3. Japkowicz, N., Stephen, S.: The class imbalance problem: a systematic study. Intelligent Data Analysis 6(5), 429–450 (2002)
4. Liu, Y., Lob, H.T., Sun, A.: Imbalanced text classification: A term weighting approach. Expert Systems with Applications 36, 690–701 (2009)
5. Guo, H., Viktor, H.L.: Learning from imbalanced data sets with boosting and data generation: the DataBoost-IM approach. SIGKDD Explorations 6, 30–39 (2004)
6. Batista, G., Prati, R.C., Monard, M.C.: A study of the behavior of several methods for balancing machine learning training data. SIGKDD Explorations 6, 20–29 (2004)
7. Li, C.: Classifying Imbalanced Data Using A Bagging Ensemble Variation (BEV). In: Proceedings of the International Conference (2006)
8. Altincay, H., Ergun, C.: Clustering based under-sampling for improving speaker verification decisions using AdaBoost. In: Fred, A., Caelli, T.M., Duin, R.P.W., Campilho, A.C., de Ridder, D. (eds.) SSPR&SPR 2004. LNCS, vol. 3138, pp. 698–706. Springer, Heidelberg (2004)
9. Kubat, M., Holte, R., Matwin, S.: Machine Learning for the Detection on Oil Spills in Radar Image. Machine Learning 30, 195–215 (1998)
10. Cristianini, N., Shawe-Taylor, J.: An Introduction to Support Vector Machines and Other Kernel-Based Learning Methods. Cambridge University Press, Cambridge (2000)
11. Mitchell, T.M.: Machine Learning, 1st edn., pp. 60–94 (2006)
12. Witten, I.H., Frank, E.: Data Mining Practical Machine Learning Tools and Techniques, 2nd edn., pp. 110–116 (2006)
13. Hertz, J., Krogh, A., Palmer, R.G.: Introduction to the theory of neural computation. Addison Wesley, Reading (1991)

Orthogonal Centroid Locally Linear Embedding for Classification

Yong Wang, Yonggang Hu, and Yi Wu

Department of Mathematics and Systems Science, National University of Defense Technology, 410073 Changsha, Hunan
yongwang82@gmail.com

Abstract. The locally linear embedding (LLE) algorithm is an unsupervised technique for nonlinear dimensionality reduction which can represent the underlying manifold as well as possible. While in classification, data label information is available and our main purpose changes to represent class separability as well as possible. To the end of classification, we propose a new supervised variant of LLE, called orthogonal centroid locally linear embedding (OCLLE) algorithm in this paper. It uses class membership information to map overlapping high-dimensional data into disjoint clusters in the embedded space. Experiments show that very promising results are yielded by this variant.

1 Introduction

Dimensionality reduction can obtain compact representations of the original high-dimensional data and curb the effect known as "the curse of dimensionality", thus it's an important and essential preprocessing operation to deal with multidimensional data in classification. Traditional methods to perform dimensionality reduction are mainly linear, such as principal component analysis (PCA). However, linear dimensionality reduction methods fail to unfold a nonlinear manifold, which we often confront with when the data are sampled from real-world, while trying to preserve its structure in a low-dimensional space. Therefore, several methods to reduce the dimensionality of the original data nonlinearly have been proposed, such as multi-dimensional scaling (MDS), locally linear embedding (LLE) etc.

Nevertheless, all methods mentioned above belong to the unsupervised feature extraction algorithms, which are mostly intended for data mining and visualization when no prior knowledge about data is available. While prior knowledge, such as data label information, is usually available in practice especially in classification problem. This means that we can employ prior information about the original data to supervise the implement of dimensionality reduction. Linear discriminant analysis (LDA) [1] is such a linear supervised feature extraction algorithm. It maps the data points belonging to different classes as far from each other as possible, while requiring the data points of the same class to be close to each other. The orthogonal centroid method (OCM) [2], which is based on QR decomposition, is another linear supervised algorithm. To supervise the implement of dimensionality reduction nonlinearly, a supervised extension of LLE for classification, called supervised locally linear embedding (SLLE) algorithm, have been independently proposed by Okun et al. [3] and Ridder

R. Huang et al. (Eds.): ADMA 2009, LNAI 5678, pp. 721–728, 2009.

et al. [4]. This method uses a control parameter α to modify the first step of the conventional LLE by changing the elements of pre-calculated data distance matrix, while leaving the other two steps unchanged. However, this extension has certain limitations. One limitation is that the optimal value of the control parameter is difficult to determine. In many practical applications, extra experiments are usually designed to determine the optimal value of α [4], which is very expensive and time-consuming; or, it is set by experience [5][6]. Both should be avoided as best as we can. Moreover, like LLE, the optimal embedding dimensionality of SLLE (except for 1-SLLE [3]) is not determined which restricts its applications to some extent.

In this paper, we propose a new supervised variant of LLE, named orthogonal centroid locally linear embedding (OCLLE) algorithm for classification. Based on class membership information, OCLLE uses the orthogonal centroid transform to map overlapping data into disjoint clusters. The proposed orthogonal centroid transform can either be used to modify the first step of the original LLE algorithm which selects neighbors of each data, or be used to change the embedded coordinates obtained by LLE to separate different classes. Furthermore, no additional parameters are needed and the optimal embedding dimensionality is selected as less by one than the number of classes, which are beneficial to practical applications. Experimental results show that this variant can yield very promising classification results.

The rest of this paper is organized as follows: Section 2 reviews the steps and principle of LLE and SLLE algorithms. Section 3 introduces the orthogonal centroid locally linear embedding algorithm while Section 4 presents the experimental results and compares them with LLE and SLLE. Finally, a conclusion is given in Section 5.

2 LLE and SLLE Algorithms

The locally linear embedding (LLE) algorithm [7] is an unsupervised algorithm for nonlinear dimensionality reduction which can obtain low-dimensional, neighborhood preserving embeddings of high-dimensional data. The supervised locally linear embedding (SLLE) algorithm [3][4] is a supervised extension of LLE, which uses class membership information to form the neighborhood of each data point.

2.1 LLE Algorithm

The LLE algorithm's input data consist of N real-valued vectors x_i (or points), each of dimensionality D, assembled in a matrix X of size $D \times N$. Its output is a matrix Y consisting of N columns representing points y_i, $y_i \in \mathbb{R}^d, i = 1, \cdots, N$, where $d \ll D$ and y_i corresponds to x_i. The algorithm has the following three sequential steps:

Step 1. K nearest neighbor points are found for each individual point x_i.

Step 2. Compute the linear coefficients (i.e., weights w_{ij}) that best reconstruct the data point x_i from its neighbors. The optimal weights w_{ij} are found by minimize the following reconstruction cost function:

$$\varepsilon(W) = \sum_{i=1}^{N} \left\| x_i - \sum_{j=1}^{N} w_{ij} x_j \right\|^2 , \tag{1}$$

subject to two constraints: i.e., $w_{ij} = 0$ if x_i and x_j are not neighbors, and $\sum_{j=1}^{N} w_{ij} = 1$.

Step 3. The final step of the algorithm computes embedded coordinates y_i in the low-dimensional space for each x_i. Projections are found by minimizing the embedding cost function for the fixed weights:

$$\Phi(Y) = \sum_{i=1}^{N} \left\| y_i - \sum_{j=1}^{N} w_{ij} y_j \right\|^2 = tr(YMY^{\mathrm{T}}) \ , \tag{2}$$

where $M = (I - W)^{\mathrm{T}}(I - W)$ is a sparse, symmetric, and semipositive definite matrix.

The details and proofs of the standard LLE algorithm can be found in Ref. [8].

2.2 SLLE Algorithm

The main idea of the SLLE algorithm is to find a mapping separating within-class structure from between-class structure. This can be achieved by adding distance between samples in different classes, while leaving them unchanged if samples come from the same class:

$$\Delta' = \Delta + \alpha \max(\Delta)\Lambda, \alpha \in [0,1] \ , \tag{3}$$

where Δ is the distance matrix of the original data, $\max(\Delta)$ is its largest value, and Δ' is the distance matrix integrating with class membership information. If x_i and x_j belong to different classes, then $\Lambda_{ij} = 1$, otherwise $\Lambda_{ij} = 0$. In Eq. (3), $\alpha \in [0,1]$ controls the amount that class membership information should be taken into account. After the neighbors of each data point are selected using Δ', then the following two steps of LLE are implemented to obtain the embedded coordinates.

Since SLLE is a supervised nonlinear algorithm which can deal with data sampled from nonlinear manifold and obtain better recognition results than LLE, SLLE is used widely in classification [3][4][5][6]. However, it also has certain limitations. First, though Ridder et al. [4] had pointed that the best performance of SLLE can be obtained when the control parameter α is between 0 and 1, the problem is which value is the best? In many practical applications, extra experiments are usually designed to determine the optimal value of α [4], which is very time-consuming; or, it is set by experience [5][6]. Both should be avoided as best as we can. Second, just like LLE, the optimal embedding dimensionality of SLLE (except for *1*-SLLE [3]) is not determined which restricts its applications to some extent.

3 Orthogonal Centroid Locally Linear Embedding Algorithm

SLLE uses class membership information to change the elements of pre-calculated data distance matrix. Then, SLLE selects neighbors of each data point from the changed distance matrix while leaving the coordinate of each data point unchanged. However, our method not only separates data points belonging to different classes, but also changes their coordinates. Moreover, it can either be used to modify the first step of LLE or be used to disjoin the embedded coordinates of different classes.

Suppose there are N points $z_i, i=1,\cdots,N$ in L-dimensional space, which come from m different classes. Let $Z_j, j=1,\cdots,m$ denotes the matrix that contains the points with class label j. Then, all Z_j are concatenated into a single matrix Z. To roughly represent each class's character, its centroid coordinate is a good choice [2].

Definition 1. Suppose $a_1, a_2, \cdots, a_n \in \mathbb{R}^{k\times 1}$ are all the data points in the same class. Then their *centroid coordinate* is defined as:

$$c = 1/n \cdot \sum_{i=1}^{n} a_i = 1/n \cdot Ae \ , \tag{4}$$

where $A = [a_1 \ a_2 \ \cdots \ a_n] \in \mathbb{R}^{k\times n}$ and $e = [1\,1 \cdots 1]^{\mathrm{T}} \in \mathbb{R}^{n\times 1}$.

Since the data matrix Z is partitioned into m different classes, we can define its *centroid matrix* as:

$$C = [c_1 \ c_2 \ \cdots \ c_m] \in \mathbb{R}^{L\times m} \ , \tag{5}$$

where the ith column is the centroid coordinate of the ith class in L-dimensional space. It can also be represented as:

$$C = ZH \ , \tag{6}$$

where $H \in \mathbb{R}^{N\times m}$ is a *grouping matrix* as:

$$H = F \cdot (diag(diag(F^{\mathrm{T}}F)))^{-1} \ , \tag{7}$$

and $F \in \mathbb{R}^{N\times m}$ with the element:

$$F(i,j) = \begin{cases} 1 & if \ z_i \ is \ one \ column \ of \ Z_j \\ 0 & otherwise \end{cases} . \tag{8}$$

Thus, each column of the centroid matrix stands for the mean value of its corresponding class and includes some class information which can be used to separate each class. One way to achieve this purpose is to orthogonalize the centroid matrix first and then all the data points belonging to the same class have the same transform as their centroid coordinate do.

There is no loss in generality in considering c_1 as the origin of the L-dimensional space, then $C = [c_1 \ c_2 \ \cdots \ c_m]$ transforms to $C^* = [0 \ c_2 - c_1 \ \cdots \ c_m - c_1]$. The reason for this transform is that with one class's centroid coordinate lies on the origin, we can make others orthogonalized reciprocally. Now, we focus our attention on the matrix $C_{new} = [c_2 - c_1 \ \cdots \ c_m - c_1] \in \mathbb{R}^{L\times(m-1)}$. It is well known that when C_{new} has full rank column, the QR decomposition of C_{new} gives an orthogonal matrix $Q \in \mathbb{R}^{L\times L}$ and an upper triangular matrix $R \in \mathbb{R}^{(m-1)\times(m-1)}$ such that

$$C_{new} = Q \begin{pmatrix} R \\ 0 \end{pmatrix} . \tag{9}$$

With $Q = [Q_m \ Q_n]$, $Q_m \in R^{L\times(m-1)}$ and $Q_n \in \mathbb{R}^{L\times(L-m+1)}$, we have $C_{new} = Q_m R$. Note that, Q_m can stand for the orthonormal form of C_{new}, and then we define the matrix

$$Q^* = sqrt(\max(N, \max(\Delta^*))) \cdot [0\ Q_m] \in \mathbb{R}^{L \times m}, \tag{10}$$

as *orthogonal centroid matrix*. In Eq. (10), $\max(\Delta^*)$ is the largest distance between any two points in Z. The scalar $sqrt(\max(N, \max(\Delta^*)))$ is used to guarantee that each class can be separated as well as possible. As we can see that each column of the orthogonal centroid matrix can be obtained through certain translations of the corresponding column of the centroid matrix in *L*-dimensional space. Note that, each column of the orthogonal centroid matrix stands for the centre position of its corresponding class. To keep the structure of each class, every points belonging to the same class should have the same translations as their centroid coordinate transform to the corresponding orthogonal centroid coordinate. Thus, *Z* is transformed to

$$Z^* = Z - (C - Q^*)F^{\mathrm{T}}, \tag{11}$$

which is called *orthogonal centroid data matrix,* and this transform from the data matrix to the orthogonal centroid data matrix is defined as *orthogonal centroid transform*. Thus, one of the classes is centered on the origin, while others are centered on different points which are orthogonal via the orthogonal centroid transform.

Note that, no additional parameters are needed in the above operation. What we really need is just a transform of the given data matrix, so the increased computational requirement is much lower. Moreover, our method can either be used to modify the first step of LLE which selects neighbors of each data, or be used to change the embedded coordinates obtained by LLE to separate different classes. For the first case, when the training data set is given, we can obtain its data matrix and its orthogonal centroid data matrix using class membership information. Subsequently, *K* neighbors of each point are selected and the last two steps of LLE algorithm are implemented to obtain embedded coordinates. In fact, this case usually selects the neighbors of a data point only from the same class. Mapped classes are separated due to the constraint $1/N \bullet \sum_{i=1}^{N} y_i y_i^{\mathrm{T}} = I_{d\times d}$, and all samples in a certain class are mapped onto a single point in $\mathbb{R}^{m-1}$ just as *1*-SLLE do [9] since they have the same principle. We define this variant as *orthogonal centroid locally linear embedding 1 algorithm* (abbreviated as OCLLE*1*). However, the original geometric structure of the training data set is destroyed in the embedded space by this way. Moreover, some points which can reconstruct one of the data point well in the original space do not include in its neighbors any longer since they belong to different classes. Therefore, the *orthogonal centroid locally linear embedding 2 algorithm* (abbreviated as OCLLE*2*) performs LLE algorithm on the original training data set first, and then the embedded coordinates are managed using the orthogonal centroid transform. Because of orthogonality of class centre, when there are *m* different classes, all the data embed into $m-1$ dimensional space can get well separability under this algorithm. Thus, the optimal embedding dimensionality in OCLLE is less by one than the number of classes in both variants. After training, we use a simple linear generalization method (see Ref. [8]) to obtain the coordinate of a new sample in the test data set.

4 Experiments

In this Section, we investigate the performance of the proposed method on several data sets from UCI repository [10].

4.1 Iris Data Set

The Iris data set contains 150 samples composed of 4 measurements. There are three different types of iris and each class has 50 samples. In this experiment, this data set is randomly split 10 times into training and test sets containing 80% and 20% of the data samples, correspondingly. The data sets are projected onto two-dimensional space. Nearest neighbors K for each data point is set to 12 and the results implemented by α-SLLE ($\alpha = 0.03$) and *1*-SLLE are also given. Fig.1 shows the two-dimensional projections of the training set obtained by each algorithm. We can see that each certain class is mapped onto a single point by *1*-SLLE and OCLLE*1*, while some of the class structures are retained by α-SLLE with within-class dispersion reduced compared to LLE. However, OCLLE*2* keeps the within-class dispersion, but

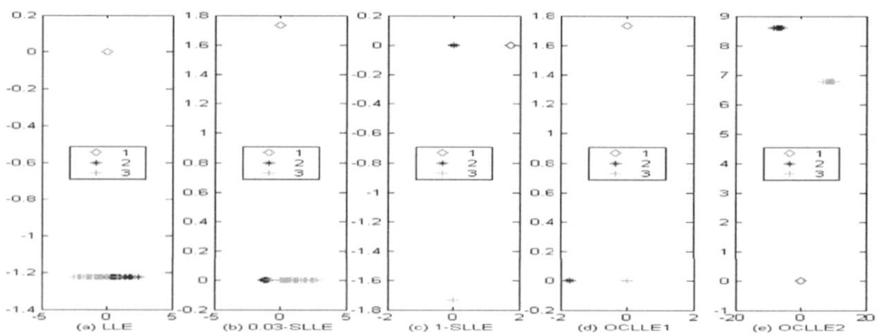

Fig. 1. Two-dimensional projections of the training set obtained by different algorithms

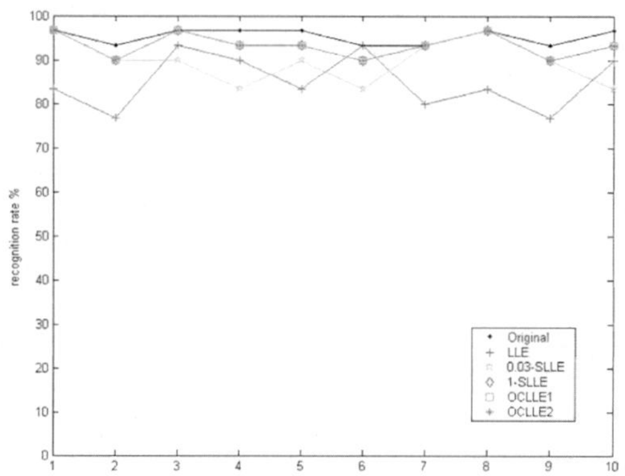

Fig. 2. Recognition rates on different data sets with six algorithms

separates each class compared to LLE. Although, internal structure of each class is (partially) lost during mapping by the SLLE algorithm and the OCLLE algorithm, they are easy to be distinguished in the two-dimensional space.

Fig.2 shows the recognition rates on different data sets selected randomly 10 times. Obviously, OCLLE*1* performs as well as *1*-SLLE algorithm since they have the same principle. Moreover, to this data set, OCLLE*2* gets the same recognition rates as OCLLE*1*. However, they all generally lead to better classification performance than LLE while *0.03*-SLLE outperforms LLE sometimes.

4.2 Multiple Features Data Set

To illustrate the effectiveness of our method on large high-dimensional data set, we perform experiments on the Multiple Features data set. This data set consists of features of 2,000 handwritten numerals ('0'--'9') images, each of 649 dimensions. There are 10 pattern classes and each class has 200 samples. Each class is randomly split 20 times into training and test sets containing 180 and 20 data samples, correspondingly. Nearest neighbors K for each data point is set to 15 in this experiment and nearest neighbor classifier is used. Table 1 presents the average accuracy of recognition on the test set with the standard deviation. The results confirm that OCLLE generally leads to better classification performance and higher stability. This is to be expected, as OCLLE can extract nonlinear manifolds in a supervised way and map overlapping classes into disjoint clusters.

Table 1. Average accuracy (in %) with the standard deviation

algorithm	average accuracy	standard deviation
LLE	86.5	2.4
0.05-SLLE	91.5	3.2
OCLLE1	95.6	1.8
OCLLE2	96.0	0.9

5 Conclusion

In this paper, we propose a new supervised extension of LLE, called orthogonal centroid locally linear embedding (OCLLE) algorithm for classification. It employs class membership information to supervise the implement of dimensionality reduction and obtain disjoint clusters of the overlapping high-dimensional training data set in the embedded space. It uses orthogonal centroid transform to modify the first step of LLE which selects neighbors of each data, or to change the embedded coordinates obtained by LLE to separate different classes. Compared with SLLE, OCLLE does not need any additional parameters and its optimal embedding dimensionality is determinate which is less by one than the number of classes. Thus, OCLLE is time-saving and beneficial to applications. The experimental results indicate that OCLLE is a promising method for classification even when followed by simple classifiers.

Our future work will try to combine OCLLE with other classifiers to get more successful classification results. Other possible research directions include comparing the classification performance of this method to a number of data sets varying in the number of samples, dimensions and classes.

Acknowledgements. The research described in this paper has been supported by the National Natural Science Foundation of China (Grant No.60673090).

References

1. Belhumeur, P.N., Hespanha, J.P., Kriegman, D.J.: Eigenfaces vs.Fisherfaces: Recognition using Class Specific Linear Projection. IEEE Trans. Pattern Anal. Mach. Intell. 19, 711–720 (1997)
2. Park, H., Jeon, M., Rosen, J.B.: Lower Dimensional Representation of Text Data Based on Centroids and Least Squares. BIT Numerical Math. 43, 427–448 (2003)
3. Okun, O., Kouropteva, O., Pietikainen, M.: Supervised Locally Linear Embedding Algorithm. In: Tenth Finnish Artificial Intelligence Conference, pp. 50–61 (2002)
4. de Ridder, D., Duin, R.P.W.: Locally Linear Embedding for Classification. Technical Report PH-2002-01, Pattern Recognition Group, Department of Imaging Science and Technology, Delft University of Technology, Delft, The Netherlands (2002)
5. Liang, D., Yang, J., Zheng, Z.L., Chang, Y.C.: A Facial Expression Recognition System based on Supervised Locally Linear Embedding. Pattern Recogn. Lett. 26, 2374–2389 (2005)
6. Wang, M., Yang, J., Xu, Z.J., Chou, K.C.: SLLE for Predicting Membrane Protein Types. J. Theor. Biol. 232, 7–15 (2005)
7. Roweis, S.T., Saul, L.K.: Nonlinear Dimensionality Reduction by Locally Linear Embedding. Science 290(5500), 2323–2326 (2000)
8. Saul, L.K., Roweis, S.T.: Think Globally, Fit Locally: Unsupervised Learning of Low Dimensional Manifolds. J. Mach. Learning Res. 4, 119–155 (2003)
9. de Ridder, D., Kouropteva, O., Okun, O., Pietikainen, M., Duin, R.: Supervised Locally Linear Embedding. In: Thirteenth International Conference on Artificial Neural Networks, pp. 333–341 (2003)
10. Hettich, S., Blake, C., Merz, C.: UCI repository of machine learning databases. University of California, Irvine, Dept. of Information and Computer Sciences (1998), http://www.ics.uci.edu/_mlearn/MLRepository.html

CCBitmaps: A Space-Time Efficient Index Structure for OLAP*

Weiji Xiao and Jianqing Xi

School of Computer Science & Engineering, South China University of Technology, 510000 Guangzhou, Guangdong
weiji_xiao@163.com, jianqingxi@163.com

Abstract. Quotient Cube is a classic algorithm to build data cube for OLAP. QC-Tree is based on Quotient Cube that can compress Cube further and accelerate query. This paper proposes a new index structure CC-Bitmap for OLAP which needs less storage space and improves queries performance significantly against QC-Trees, avoiding large memory and time consumption costs meanwhile. A new intersection algorithm CCListsIntersection is presented in this paper, which improves queries efficiency highly against traditional methods. Finally, detailed experiment evaluation is given to show that CC-Bitmap is a Space-Time efficient index structure for OLAP.

Keyword: Quotient Cube, Closed Cube, QC-Trees, CC-Bitmaps, CCListsIntersection.

1 Instruction

With the development of OLAP technology, many methods on constructing data cube have been proposed, such as Condensed Cube[1], Dwarf[2] and Quotient Cube[3]. Condensed Cube and Dwarf compress data cube by sharing tuples, while Quotient Cube constructs data cube by equivalence classes and semantical relationship of roll-up and drill-down. Compared with Quotient Cube, Closed Cube[4] requires less storage space and computing time as only upper bounds of cube is reserved. However, semantics of roll-up and drill-down is lost. The main compression ideas of QC-Tree[5] is prefix sharing. First, it numbers and sorts temp classes produced by Quotient Cube, and then constructs a complete tree after all temp classes have been sorted. So that its space inefficient, and the construction time is long.

The main contributions of this paper are summarized as below:

(i) A CC-Bitmaps construction algorithm, which can reduce storage space to QC-trees, is proposed. And the time and memory needed are also less than QC-Trees.

(ii) A new intersection algorithm CCListsIntersection is presented, which can speed up the queries.

(iii) A CC-Bitmaps query algorithm is given and is proved to be more effective than QC-Trees.

* The research is supported by the Science & Technology Project of Guangdong Province(NO.2006B80407001) and International Cooperation Project (NO. 0711420500008).

R. Huang et al. (Eds.): ADMA 2009, LNAI 5678, pp. 729–735, 2009.

2 CC-Bitmaps

2.1 Construction Algorithm of CC-Bitmaps

There are two steps in the construction algorithm of CC-Bitmaps, the first one is to set up inverted index[6], and the second is to set up bitmap index.

2.1.1 Bitmap Index

Definition 1(Base Upperbound Tuple, BUT). In the upper bounds set of closed cube, BUT is the upper bound $(v_1,...,v_i,...,v_n)(1 \le i \le n)$ with n as number of dimensions and $\forall v_i \ne *$.

Definition 2 (Query Bitmap, QBitmap). Let $q=(v_1,...,v_k,...v_n)(1 \le k \le n)$ be a query, the QBitmap of q is a bitmap vector $(B_1,...,B_k,...,B_n)(1 \le k \le n)$ with $B_k = 0$ if $v_k = *$ and $B_k = 1$ otherwise.

Definition 3 (CC-Bitmaps). CC-Bitmaps is a table of Closed Cube, the first line means all possible QBitmap which corresponding to all 2^n cuboids below (n is the number of dimensions) and sorted by ascending. The first column of the table means all BUTs in Closed Cube. The middle part of the table is filled by the aggregate values of the upper bounds produced by Closed Cube.

Table 1 shows a simple base table Sales with three dimensions and one measure, SUM as aggregate function, Table 2 is the CC-Bitmaps constructing on Table 1.

Table 1. Base Table Sales

TID	**Store(S)**	**Customer(C)**	**Product(P)**	**Price(M)**
1	S1	C2	P2	$70
2	S1	C3	P1	$40
3	S2	C1	P1	$90
4	S2	C1	P2	$50

Table 2. CC-Bitmaps for Sales

QBitmap	000	001	010	011	100	101	110	111
cuboid / aggregate value / BUT	ALL	P	C	CP	S	SP	SC	SCP
(S1,C2,P2)	$250	$120			$110			$70
(S1,C3,P1)		$130						$40
(S2,C1,P2)							$140	$50
(S2,C1,P1)								$90

2.1.2 Construction Algorithm

Algorithm 1.[Construct CC-Bitmaps]
Input: base table B.
Output:CC-Bitmaps for closed cube of B.
Method:
1. //declare some data structures
  aggsVector;//store aggregate values
  BUTVector; //store all BUTs
  ubBitmapVector;//store QBitmaps about all upperbounds
2. b = (* ,..., *);
3. call CCDFS(b, B, 0);//(*, … ,*)has no lattice child;
4. for all values of each dimension in BUTVector,build inverted indexes and store to files.When building inverted indexes,the RLE compression algorithm is optional.
5. sort all bitmaps belongs to each BUT by ascending,the same to the corresponding aggregate values in the same order;
6. set all the sorted bitmaps of all BUTs except the bitmap of each BUT itself to chars one by one continuously,then store all chars to a file.
7. store all sorted aggregate values of all BUTs one by one to a file and record writing file positions for each BUT simultaneously and store all positions to a file.
8. if(CCListsIntersection)//whether to use the CCListsIntersection Algorithm is optional
    store all BUTs to a file;
9. return;

Fig. 1. CC-Bitmaps Construction Algorithm

Function *CCDFS*(c[], B_c,dim)
//c=a cell+its TID(store in the last position of array c):B_c is the corresponding partition of the base table;
1. for each 0≤d<dimNum do
    if B_c[d] has unique value then
        closedTuple[d] = B_c[d];
    else
        closedTuple[d] = ALL;
2. if there is some j<dim s.t. c[j]=ALL and closedTuple[j]≠ALL
    return;//such a bound has been examined before
3. //find the minimal TID in this Partition
  minTID = the minimal TID in all cells included in B_c
4. //compute the aggregate values of cell c
  values = aggregate(B_c);
5. //set the upper bound bitmap
  for each 0≤d<dimNum do
    if closedTuple[d]≠ ALL, set bitmap in position d to 1;//namely its QBitmap
6. store aggregate values, closedTuple(only when it is a BUT) and bitmap to aggsVector,BUTVector and ubBitmapVector according to the minTID, respectively.
7. for each dim<j<dimNum s.t. closedTuple[j]==ALL do
    for each value x in dimension j of partitionB_c
        let closedTuple[j] = x;
        form partition B_d;
        if B_d is not empty,call CCDFS(closedTuple,B_d, j);
8. return;

Fig. 2. CCDFS Function

Figure 1 is the construction algorithm of CC-Bitmaps. CCDFS is the procedure to create upper bounds of data cube and information needed by CC-Bitmaps, as illustrated in Figure 2.

2.2 Query Algorithm of CC-Bitmaps

Definition 4 (Super Bitmap, SBitmap). Given a QBitmap($B_1,...,B_k,...,B_n$), its SBitmap ($B_1',...,B_k',...,B_n'$) is $B_k'=1$ if $B_k=1$ and there exists $B_i'=1$ when $B_i=0 (1 \le i \le n)$.

Definition 5 (first meet BUT). The first meet BUT is gotten by intersection algorithm in top-down order when scanning all BUTs of CC-Bitmaps table according to all instantiated values of query q.

2.2.1 CCListsIntersection Algorithm

Let $q=(v_1,...,v_k,...v_n)(1 \le k \le n)$ be a point query, for $\forall v_k \ne *$, let its TID list be L_k. Then all TID lists is ($L_1...L_k$) and all the sizes of TID lists are ($L_{s1}...L_{sk}$). Figure 3 gives the detailed algorithm of CCListsIntersection. For convenience, the intersection algorithm is named as int getQueryTID(int q[]).

```
Algorithm 2. [CCListsIntersection]
Input: q[] = (v1..,vk,...vn)(1≤ k ≤n)
Output: the minimal TID satisfy q or -1 if not find
1. //declare some important data structures
   tidListsVec(2);//two list vectors to store TIDs data and map to each other when dimension changed.
   flag = 0;      //the value is 0 or 1 to indicate which tidListsVec being used.
   pos = -1;      //find the query dimension position which TID List size is minimal.
   size = the max size of an integer;    //record the minimal TID List size.
2. for each  0≤ d <dimNum do
        if (q[d]) //q[d] ≠*
            if(Lsd <size) size = Lsd ;
                 pos = d;
3. if(pos == -1) return -1;  //query fails
4. tidListsVec[flag] = the minimal TID List;
   for each  0≤ d <dimNum do
        if(d≠pos && q[d])
            for each 0≤ j <size do//BUTVec stores all BUTs by row reading from file saved in Algorithm 1.
                if(BUTVec[tidListsVec[flag]][d]==q[d])
                      map tidListsVec[flag]][d] to tidListsVec[!flag];
            if(tidListsVec[!flag] is empty) return -1;//query fails
            clear tidListsVec[flag];
            flag = !flag;
   if(tidListsVec[flag] is not empty)
            return tidListsVec[flag][0]; //return the minimal TID which satisfies the query q,namely first meet BUT
```

Fig. 3. CCListsIntersection Algorithm

2.2.2 CC-Bitmaps Query

Figure 4 gives the query algorithm of CC-Bitmaps.

```
Algorithm 3. [CC-Bitmaps Query]
Input:q[] = (v1.., vk,... vn)(1 ≤ k ≤n)
Output:query result aggregate values.
1. for each   0≤d<dimNum do
        if(q[d])
            set the position d of QBitmap to 1;
3. if(QBitmap all positions are 0)
        return the first aggregate values stored in the beginning of the aggregate file;
2. qryTID = getQueryTID(q[]);
3. if(qryTID == -1) return NULL;//query fails
4.if(QBitmap all positions are 1)
      return the aggregate values of BUT itself;
5. read all bitmaps belong to this BUT (QBit1,...,QBitm) to an array QBit[];//all bitmaps have been sorted.
6. for each   0≤ j <QBit.size do
        if(QBit[j] ≥ QBitmap)//if QBitmap exits,it will be =,otherwise means the first SBitmap > QBitmap
            return the aggregate values corresponding to the QBit[j];
7. return the aggregate values of the BUT itself;//BUT is covered by query q directly.
```

Fig. 4. CC-Bitmaps Query Algorithm

3 Experimental Evaluation

All experiments are running on HP-BL35P blade server, AMD Processor 1.8GHz × 2 with memory DDR 1G, Windows Server 2003, Microsoft Visual C++ 6.0.

The real data sets we use are Foodmart Sales and ocean weather data set[7]. And we use Data Generator presented in reference [8] to create two synthetic data sets with 9 and 15 dimensions, and name them DS9 and DS15 respectively. To each test data set, we generate 1000, 5000, 10000 random queries, respectively.

We name the version using RLE, ListsIntersection algorithms and all TID lists loaded into memory only when required by the query as CC-BitmapsRL. Name the version using RLE, ListsIntersection algorithms and all TID lists loaded into memory when the program initializing as CC-BitmapsRLM. And the version using CCListsIntersection algorithm (that not using RLE compression algorithm) and all TID lists and BUTs loaded into memory when the program initializing as CC-BitmapsCCLM.

3.1 Storage Space Comparison

As the storage capacity needed by CC-BitmapsRL and CC-BitmapsRLM are the same, CC-BitmapsRL is selected as representative when storage space is compared. In order to compare quantity relationship of storage space consumption, Table 3 gives size

Table 3. Size Ratios and Construction Time

Algorithms	Datasets(size ratio/construction time)			
	Sales	weather	DS9	DS15
QC-Trees	41.7%/9.782s	67.1%/3957.610s	41.0%/8.266s	23.0%/170.157s
Closed Cube	42.0%/5.328s	35.8%/8.056s	16.3%/4.438s	12.8%/41.047s
CC-BitmapsRL	25%/5.983s	9.9%/8.264s	7.8%/4.695s	3.5%/41.375s
CC-BitmapsCCLM	27.7%/5.999s	14.5%/8.562s	9.6%/4.766s	3.9%/41.234s

ratios of other algorithms comparing with Quotient Cube as a baseline. From the four data sets in Table 3, we can work out that storage consumption of the two versions of CC-Bitmaps is only 14.8% of QC-Trees at the best case, and only 66.4% at the worst case.

3.2 Query Performance Comparison

3.2.1 Real Data Sets

Figures 5 and 6 give the performance comparisons of queries on Sales and weather data sets, respectively. We can see CCListsIntersection algorithm has larger performance improvement against ListsIntersection, the improvement is 2 to 4 times on data set Sales, and is 12 times on data set weather.

Observing queries on Sales, the query time of CC-Bitmaps is half of QC-Trees, and the speed of BitmapsRLM is close to speed of QC-Trees. Observing queries on weather, the query performance on various versions of CC-Bitmaps performed very well. The query performance presents several to ten times improvement to QC-Trees.

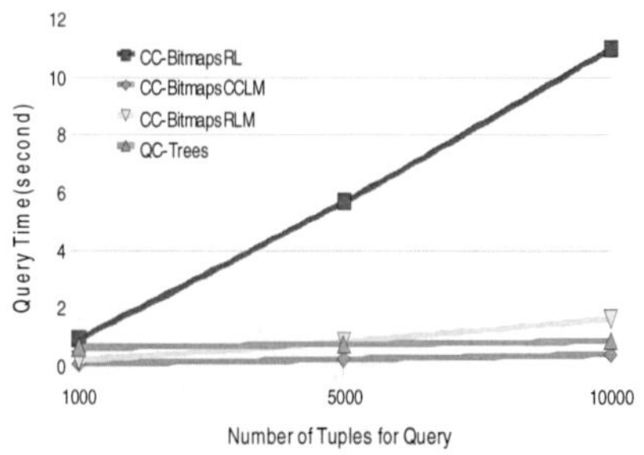

Fig. 5. Query Performance on Sales

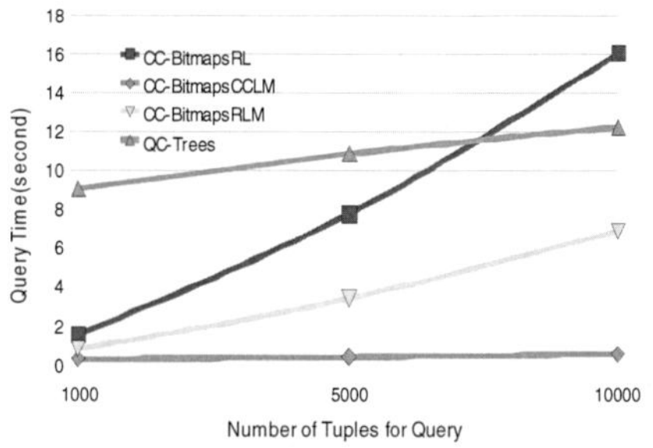

Fig. 6. Query Performance on weather

3.2.2 Synthetic Data Sets

Figures 7 and 8 present the queries performance of the 3 algorithms on DS9 and DS15 data sets. For observing the queries performance of various memory algorithms much better, the CC-BitmapsRL algorithm is removed.

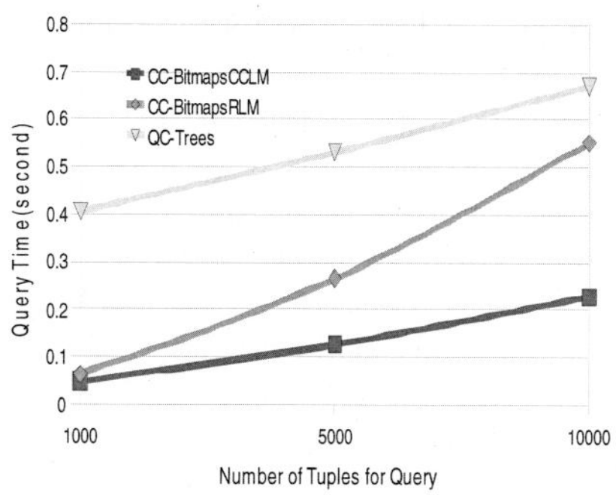

Fig. 7. Query Performance on DS9

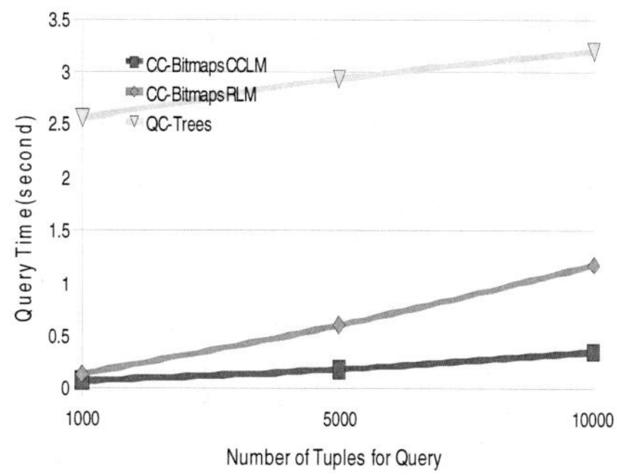

Fig. 8. Query Performance on DS15

4 Conclusion

From analysis above, we can draw some conclusions as follows. 1) RLE compression algorithm is effective, and with the increasing size of base table the compression ratio has the trend to increase. 2) The intersection algorithm CCListsIntersection proposed in this paper is obviously superior to the traditional algorithm ListsIntersection. 3) The construction of CC-Bitmaps is very simple, and the time, space consumption of constructing are greatly reduced comparing to the QC-Trees. 4) The memory version of CC-BitmapsCCLM is almostly better than QC-Trees. BitmapsRLM is better than QC-Trees in most cases.

In one word, CC-Bitmaps is a Space-Time efficient index structure for OLAP. In near future, we will study and implement the update of CC-Bitmaps.

References

1. Wang, W., Lu, H.J., Feng, J.L., Yu, J.X.: Condensed cube: An effective approach to reducing data cube size. In: Proc. of the 18th Int'l Conf. on Data Engineering, pp. 155–165. IEEE Computer Society, San Jose (2002)
2. Sismanis, Y., Deligiannakis, A., Roussopoulos, N., Kotidis, Y.: Dwarf: Shrinking the PetaCube. In: Franklin, M.J., Moon, B., Ailamaki, A. (eds.) Proc. of the 2002 ACM SIGMOD Int'l Conf. on Management of Data, pp. 464–475. ACM Press, Madison (2002)
3. Lakshmanan, L.V.S., Pei, J., Han, J.W.: Quotient cube: How to summarize the semantics of a data cube. In: Bressan, S., Chaudhri, A.B., Lee, M.L., Yu, J.X., Lacroix, Z. (eds.) Proc. of the 23rd Int'l Conf. on Very Large Data Bases, pp. 778–789. Morgan Kaufmann, Hong Kong (2002)
4. Li, S.E., Wang, S.: Research on closed data cube technology. Journal of Software 15(8), 1165–1171 (2004)
5. Lakshmanan, L.V.S., Pei, J., Zhao, Y.: QC-Trees: An Efficient Summary Structure for Semantic OLAP. In: Proc. of the ACM SIGMOD Int. Conf. on Management of Data, pp. 64–75 (2003)
6. Zhao, Y.: Quotient Cube and QC-Tree: Efficient. Summarizations for Semantic OLAP. Ph.D thesis. The University of British Columbia (2003)
7. Hahn, C., Warren, S., London, J.: Edited synoptic cloud reports from ships and land stations over the globe (1996), http://cdiac.esd.ornl.gov/cdiac/ndps/ndp026b.html
8. http://www.codeplex.com/datagenerator

Rewriting XPath Expressions Depending on Path Summary

Xiaoshuang Xu[1,2], Yucai Feng[2], Feng Wang[1], and Yingbiao Zhou[2]

[1] School of Educational Sci. & Tech., Huanggang Normal Univ, 438000 Huanggang, Hubei

[2] School of Computer, Huazhong Univ. of Sci. & Tech, 430070 Wuhan, Hubei

{xxsh99,mwfwf}@hgnu.edu.cn, {Fengyc,Zhouyb}@hust.edu.cn

Abstract. Optimizing XPath expressions before query execution is an important measure for efficient improvement of the holistic performance. After path summary is derived from a XML document, we equivalently transfer a path expression into the ones without descendent axes or wildcards. To rewrite a linear path expression, its analytical path set is calculated in polynomial time under path summary. For complex expressions, we disclose path restraints among nodes in its tree pattern under path summary, and develop a rewriting algorithm to completely eliminate its descendent and wildcard features. Thereof, path summary provides an excellent basis for rewriting XPath expressions. The experimental results show that our algorithm can rewrite XPath expressions effectively.

Keywords: XML, XPath, structural join, tree pattern, rewriting.

1 Introduction

As a simple XML query language, XPath expressions make an effective way of navigating an XML document and extracting information from them. Our discussions focus on $XP^{\{/,*,//,[]\}}$ expressions[1] used frequently in practice. Many previous works on path index[2] and structural join[3,4] have sped efficient evaluation of XPath queries.

More recently, the problem of rewriting XPath expressions to reduce '//' and '*' is attracting more attention. Sometime minimization of XPath expressions [5,6] partly decreases '//' and '*'. Given a DTD, P.T.Wood proposed that wildcards could be removed from $XP^{\{/,//,*\}}$ expressions[7] in polynomial time. Authors in [8] employ tree automaton from a unambiguous DTD to eliminate '//' and '*' in XPath expressions. As stated in [9], since a recursive DTD has the capability of infinitely expressing $XP^{\{/\}}$ paths, it is impossible that descendant axes and wildcards are completely reduced from $XP^{\{/,//,*\}}$ expressions under the DTD.

In practice, an arbitrary XPath expression is always used to query a certain XML document. No matter how large the document is, its different paths are not only limited but also far less than total elements. Thereof, path information in the document should be considered during rewriting expressions.

In this paper, the problem of rewriting an XPath expression $p \in XP^{\{/,//,*,[]\}}$ is to find an algorithm that, given path summary[10] coming from an XML document D,

R. Huang et al. (Eds.): ADMA 2009, LNAI 5678, pp. 736–744, 2009.

computes XPath expression $q_1, q_2, \ldots, q_n \in XP^{\{/,[]\}}$ such that query result by p is equivalent to the union of that by q_i over D ($i \in [1,n]$). To the best of our knowledge, this paper is the first that addresses the problem of rewriting XPath expressions depending on path summary to completely eliminate '//' and '*'. The contributions of the work reported here can be summarized as follows:

(1) We propose analytical path set for a linear expression under path summary coming from the XML document.
(2) We define valid paths of every node in a tree pattern. Meanwhile, we present calculation of valid paths according to path restraints in the context of the tree pattern.
(3) We describe path join operation among valid paths and develop a novel rewriting algorithm to completely eliminate wildcards and descendant axes. The experimental results demonstrate the effectiveness of our algorithm.

2 Preliminaries

Because the work of Gerome Miklau and Dan Suciu is notable exception [1], we directly cite several notions from their classic work with no explanation in this section. These notions will be used throughout the rest of the paper.

If t is a tree, we denote its node set, edge set, and root node as $V(t)$, $E(t)$ and $Root(t)$ respectively. Since an $XP^{\{/,*,//,[]\}}$ expressions is often modeled as a tree pattern belonged to $P^{\{[],*,//\}}$ equivalently, we don't distinguish the two concepts in this paper. As stated in [1], let $p \in P^{\{/,*,//\}}$, $q \in P^{\{/\}}$ be two linear tree patterns, q is contained in p ($p \sqsupseteq q$) if and only if there exists an embedding f from p to q.

Definition 1. *Given linear tree pattern $p \in P^{\{/,*,//\}}$, the path length of p, denoted as $\|p\|$, is the number of nodes in p. Let $q \in P^{\{/,*,//\}}$ be a linear tree pattern, if there exists an embedding f from p to q, denoted as $f{:}p \to q$, then $f(p)$ is the embedded sub-tree pattern of q.*

Definition 2. *Let $p \in P^{\{/,*,//\}}$, $q \in P^{\{/\}}$ be two linear tree patterns. If there exists an embedding from p to q such that both the head node and the tail node of p are mapped to those of q respectively, then q is implied in p, denoted as $p \models q$.*

In Fig. 1(a), $\|p\|=3$,$\|q_1\|=4$ and $p \models q$. In Fig. 1(b), q_1 is not implied in p.

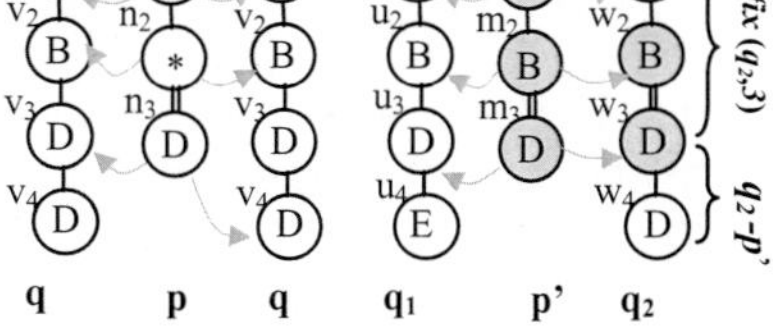

Fig. 1. $q \sqsupseteq p, q \models p$ and $p' = \text{Prefix}(q_2, 3)$

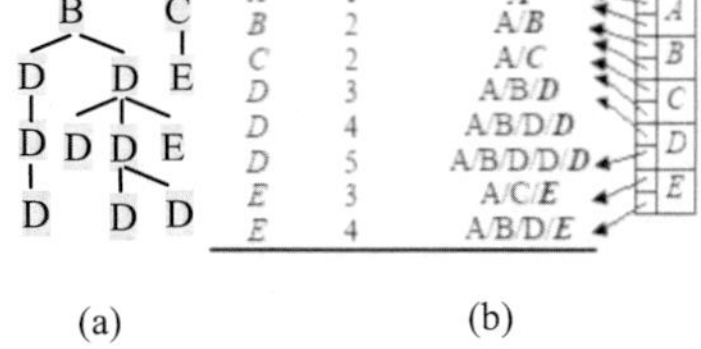

Fig. 2. An XML tree D and its S(D)

Definition 3. *Let p, q∈*$P^{\{/,*,//\}}$ *be two linear tree patterns. If p equals to a sub-tree pattern which begins from the root to the n-th node in q. then q is prefixed with p, denoted as p=Prefix(q,n), and q-p is a sub-tree pattern which begins from the n-th node to the tail one in q.*

In Fig. 1(b), *p'=Prefix(*q_2*,3)* and q_2*-p'=D/D*.

Lemma 1. *Given p∈*$P^{\{/,*,//\}}$ *and q∈*$P^{\{/\}}$*, p ⊨ q if and only if there exists embedding e:p→q such that e(p)=q.*

3 Analytical Path Sets for Linear Expressions

In practice, most XML documents have a great deal of repetitive linear paths whose tree patterns belong to subclass $P^{\{/\}}$. Obviously, the number of all different linear paths is far less than that of total elements. the number of all different labels is far less than that of total linear paths.

Definition 4. *Given an XML document D , table PT (Label, Len, Expr) records all different linear paths belonged to* $XP^{\{/\}}$*. For an arbitrary tuple t∈PT, t.Expr represents an unique linear path, t.Len=||t.Expr||, and t.Label is the last label of t.Expr. Summary path w.r.t D , denoted as S(D), consists of all paths in field Expr.*

During travelling the XML document by depth-first order(See Fig.2(a)), we record all different paths into Table PT orderly.

Definition 5. *Given XML document D and pattern p∈*$P^{\{/,*,//\}}$*, Analytical Path Set for p w.r.t D is* A*(p,D)={q|q∈S(D)∧p ⊨ q},where S(D) is path summary of D.*

Definition 5 shows a rewriting method for linear expressions to eliminate '//' and '*'. For example, let *p=A//*//D*. From *S(D)* in Fig.2(b), linear expression *A/B/D*, *A/B/D/D* and *A/B/D/D/D* are implied in *p* by definition 2, so A*(p,D)={A/B/D, A/B/D/D, A/B/D/D/D}*. Since *S(D)* is a finite set, A*(p,D)* is also finite.

4 Valid Paths for Nodes

In order to rewriting an XPath expression completely, it is an important step to disclose path restraints in the context of its tree pattern. For any node in the pattern, we present it must accept path restraints from itself, its parent and children.

4.1 Path Restraints for Nodes

Definition 6. *Let P be a tree pattern. For n∈V(P), n.Path is the root-to-self path.*

For example, in Fig.3(a), since n_1 is the root node in pattern *P*, n_1*.Path=A*. For node n_5, n_5.Path =*A//*//D*. As a leaf node, n_6*.Path=A//*//D/D*.

Definition 7. *Given XML document D and its path summary S(D), let P∈*$P^{\{/,[\ ],*,//\}}$*, Q∈*$P^{\{/,[\ ]\}}$ *be two tree patterns . If there exist an embedding f:P→Q such that for each n∈V(P), f(n.Path)∈S(D), then f(n.Path)∈*$P^{\{/\}}$ *is one of valid paths for n. We denote* E *(n.Path) as the set of valid paths for n, so f(n.Path)∈* E *(n.Path).*

Fig. 3(a) shows embedding f:P→Q, where P=A[/C/E]//*[//D/D][//E]∈ $P^{\{/,[],*,//\}}$, Q = A[/C/E]/B/D[/D/D][/E]∈ $P^{\{/,[]\}}$. For any n∈V(P), f(n.Path)=f(n).Path by definition 1 and 6. Since n_6.Path=A//*//D/D, f(n_6.Path)=f(n_6).Path=v_7.Path=A/B/D/D/D. Similarly, f(n_1.Path)=A, f(n_4.Path)=A/B/D, and f(n_5.Path)=A/B/D/D. Because all paths as above are in S(D), A∈E(n_1.Path), A/B/D∈ E(n_4.Path), A/B/D/D∈ E(n_5.Path) and A/B/D/D/D∈E(n_6.Path).

From Fig. 3(b), there exist A/B∈E(n_4.Path), A/B/D∈E(n_5.Path) and A/B/D/D∈ E(n_6.Path).

Lemma 2. *Given XML document D and pattern P* $\in P^{\{/,[\ ],*,//\}}$. *For each n∈V(P),* E *(n.Path)⊆*A*(n.Path).*

We obtain this result by Definition 5 and 7. Lemma 2 gives the first constraint condition for valid paths.

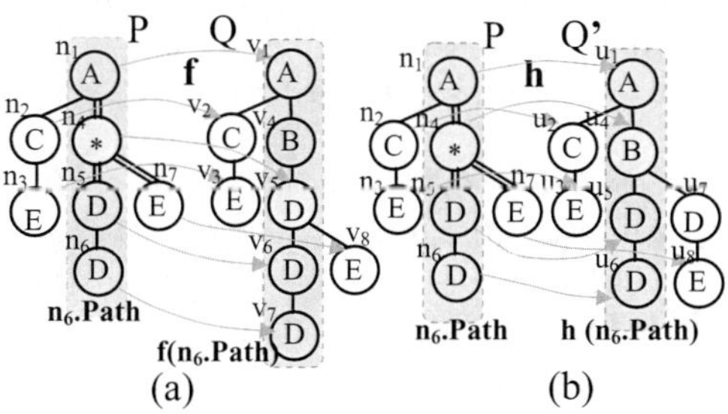

Fig. 3. f: P→Q and h: P→Q'

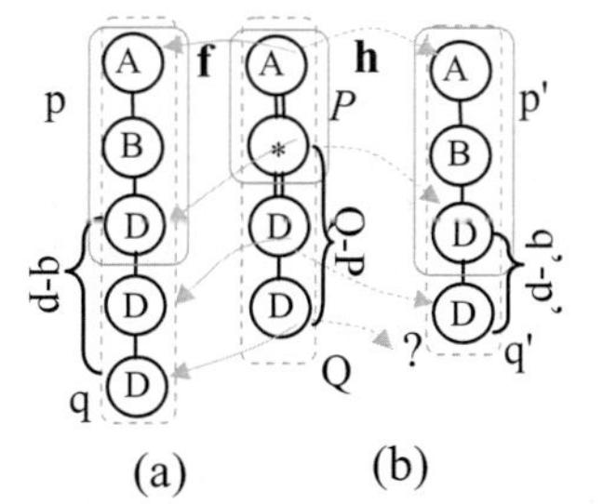

Fig. 4. f: Q→q and h: P' →p'

Lemma 3. *Given P, Q*$\in P^{\{/,*,//\}}$ *and p, q*$\in P^{\{/\}}$ *satisfying P=prefix(Q,||P||), p=prefix(q,||p||), P ⊨ p and Q ⊨ q. There exist embedding f:Q→q such that f(Q)=q and f(P)=p if and only if f(Q-P)=q-p.*

In Fig. 4(a), both *P* and *Q* share an embedding *f* such that *p* ⊨ *P* and *q* ⊨ *Q* with no conflict by Lemma 3.

Definition 8. *Given XML document D and pattern P* $\in P^{\{/,[\ ],*,//\}}$, *let n be a non-root node and n' be its parent (i.e. <n',n >∈E(p)). We define path join as follows:*

$$n'.Path \bowtie n.Path=\{(p,q) \mid \exists p\in E(n'.Path), \exists q\in E(n.Path)\ (p=prefix(q,||p||)) \rightarrow \exists f: Q\rightarrow q\ (f(Q-P)=q-p\ )\}.$$

In Fig. 3, *<n_5 ,n_6>∈E(p), A/B/D/D∈*E*(n_5.Path), A/B/D /D/D∈*E*(n_6.Path).* By Lemma 3, *(A/B/D/D, A/B/D/D/D) ∈ n_5.Path⋈n_6.Path.* On the other hand, although *A/B/D∈* E*(n_5.Path)* and *A/B/D/D/D∈*E*(n_6.Path), (A/B/D, A/B/D/D/D)∉n_5.Path ⋈ n_6.Path.* Lemma 3 assures a pair of linear paths for two nodes with the parent-child relationship in the pattern can be embedded to a certain path belonged to $P^{\{/\}}$ with no conflict.

For any node in the pattern, its valid paths must accept path restraints from its parent and children. Thereof, it is necessary to disclose path-restraints by the context of the pattern.

Lemma 4. *Given XML document D and pattern* $P \in P^{\{/,[\ ],*,//\}}$, *the following properties are satisfied:*

(1) If non-leaf node n has a child ň, then $E\ (n.Path) \subseteq \{p \mid \exists p \in S(D), \exists q \in E\ (\check{n}.Path)\ (p=prefix(q,\|p\|)) \rightarrow \exists f{:}\check{n}.Path \rightarrow q\ (f(\check{n}.Path - n.Path) = q-p)\}$;

(2) If non-root node n has a parent n', then $E\ (n.Path) \subseteq \{q \mid \exists q \in S(D), \exists p \in E\ (n'.Path)\ (p=prefix(q,\|p\|)) \rightarrow \exists f{:}n.Path \rightarrow q\ (f(n.Path - n'.Path) = q-p)\}$.

Lemma 2, 4 show all constraint conditions for valid paths.

4.2 Calculation of Valid Paths

By Lemma 2 and 4, we can calculate the set of valid paths for each node in the pattern. Main thought is presented in Algorithm 1, where function *ValidPath(P)* works as follows: Firstly, the set of valid paths for each leaf node is assigned with its Analytical Path Set. Secondly, function *PreparePath(node n)* finds all possible valid paths for other nodes by proposition (1) in Lemma 4 from bottom up. Finally, function *VerifyPath(node n)* verifies valid paths for each node by proposition (2) in Lemma 4 from top down.

In Algorithm 1, there are three the self-explanatory functions. *ParentNode(n)* returns the parent of node *n*; *ChildNode(n,i)* returns *i*-th child of node *n*; *ChildNum(n)* returns the number of *n*'s children. By Algorithm 1, the valid paths for each node in pattern *P* in Fig. 3 is presented with Graph *G* in Fig.5.

Theorem 1. *Given XML document D and pattern* $P \in P^{\{/,[\ ],*,//\}}$, *Algorithm 1 correctly returns all valid paths of each node in P.*

```
PreparePath (node n)
  if n is a leaf node return;
  for i=1 to ChildNum(n) do
      ň←ChildNode(n,i);
      FindValidPath (ň);
  end for
  E(n.Path) ←S(D);
  for i=1 to ChildNum(n) do
      ň←ChildNode(n,i);
      for each q∈ E(ň.Path) do
    E(n.Path)←E(n.Path) ∩
        {p|∃p∈S(D)(p=prefix(q,||p||))→
      ∃f:ň.Path→q(f(ň.Path-n.Path)=q-p)}
      end for
  end for

VerifyPath (node n)
  if n is root node return;
  n'←ParentNode(n);
  for each  p∈E (n'.Path) do
      E (n.Path)←E (n.Path)∩
       {q|∃q∈S(D)(p=prefix(q,||p||))→
      ∃f:n.Path→q(f(n.Path-n'.Path)=q-p)};
  end for
  for i=1 to ChildNum(n) do
      ň←ChildNode(n,i);
  VerifyValidPath (ň)
  end for

ValidPath (Pattern P)
  for each leaf node n in P
  E(n.Path)←A(n.Path,D);
  end for
  PreparePath (Root(P));
  VerifyPath (Root(P));
```

Algorithm 1. Calculation of Valid Paths

In order to prove Theorem 1, we definition path-join graph as follows:

Definition 9. *Path-join Graph of pattern P is defined as* $G=(V(G),E(G))$*, where* $V(G)=\bigcup_{n_i\in V(P)} \mathrm{E}\,(n_i.Path)$ *and* $E(G)=\bigcup_{<n', n>\in E(P)} \{<p, q>\mid (p, q)\in n'.Path \bowtie n.Path\}$.

Path-join Graph of P in Fig. 3 is showed $G=(V(G),E(G))$ in Fig.5. $V(G)$ presents all valid paths and $E(G)$ does all path joins.

Proof for Theorem 2. For each node n_i, we select one element from $\mathrm{E}\,(n_i.Path)$ to build subtree $t=(V(t),E(t))$ such that $|V(t)\cap \mathrm{E}\,(n_i.Path)|=1$, $V(t)\subseteq V(G)$ and $E(t)\subseteq E(G)$. Subtree t is similar to pattern P structurally and corresponding to an unique pattern belonged to $P^{\{/,[]\}}$ by following steps:

Step 1: If node p is the root in subtree t, create the tree pattern of p;
Step 2: If node p is a non-root node, extend its parent pattern to that of p;
Step 3: Repeat Step 2 until all nodes are accessed.

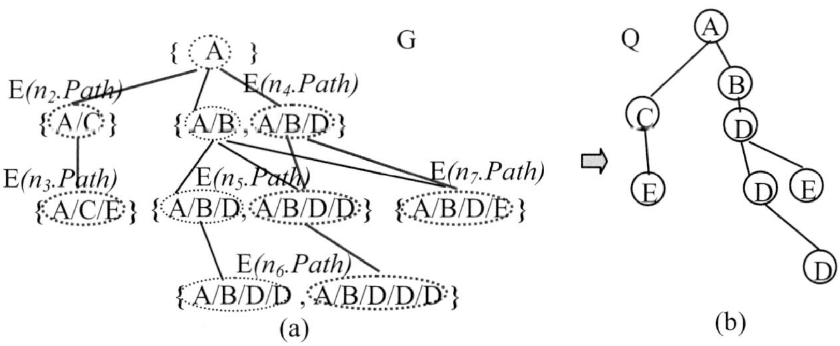

Fig. 5. Path-join Graph G, Subtree t and Pattern Q

In Fig.5(a), using blue edges and nodes, we build sub-tree t, which is corresponding to $Q\in P^{\{/,[]\}}$ (See Fig.5(b)). According to Fig.3(a), there is an embedding $f{:}P\rightarrow Q$.

Let $f=f_0{:}n_0.Path\rightarrow u_0 \cup (\bigcup_{n_i\in V(T)} f_i{:}n_i.Path\text{-}n_i'.Path\rightarrow u_i\text{-}u_i')$, where n_i is non-root node and n_0 is the root node, f is an embedding from P to Q. By Definition 7, u_i is one of valid paths for n_i. Thereof, valid paths for each node is sound in Algorithm 1.

Line 22-23 in Algorithm 1 assure that valid paths for each leaf node do not miss by Lemma 2. Lemma 4 assure valid paths for other nodes do not miss. Thereof, valid paths for each node is complete in Algorithm 1.

Theorem 2. *Given XML document D and pattern* $P\in P^{\{/,[\,],*,//\}}$*, Algorithm 1 has worst-case CPU cost linear to* $|V(P)||S(D)|^2$*, where* $|S(D)|$ *is the count of elements in* $S(D)$*.*

5 The Rewriting Algorithm

In this section, we present an algorithm of rewriting patterns based on valid paths and their path joins. The algorithm can naturally processing tree patterns with branch([]), child edge(/), descendant edge(//), and wildcard(*) synthetically.

When Path-join Graph G of pattern P has been worked out, the rewriting task is to extract all sub-trees such that for each node $n_i\in V(P)$, $|V(t)\cap \mathrm{E}(n_i.Path)|=1$. The main thought is represented in Algorithm 2. Since transferring these sub-trees to patterns belonged to $P^{\{/,[]\}}$ easily, we omit the procedure in Algorithm 2. According to Algorithm 2, pattern *P* in Fig.3 have three rewriting results as following: *A[/C/E]/B/D[/D/D][/E]*, *A[/C/E]/B[/D/D/D][/D/E]*, and *A[/C/E]/B[/D/D][/D/E]*.

Since Algorithm 2 uses path summary to compute all possible tree patterns , each of which is belonged to subclass $P^{\{/,[]\}}$ and contained in $P \in P^{\{/,[],*,//\}}$,the union of query result by these tree patterns over the XML document must be equivalent to that by *P*.

```
MatchPattern(Node n, int i )
    p←E (n_i'.Path)[i]; T[n] ←p;
    bMatch←true;
    for k=1 to ChildNum(n) do
        ň←ChildNodes(n,k);
        bfound←false;
            for j=1 to |E (ň.Path)|
                q←E (ň.Path)[j];r←false;
            if (p,q) ∈n.Path ⋈ ň.Path
            then r←MatchPattern(ň,j);
            else continue;
        if r then T[ň] ←q;
        if (ň = N )∧r then output T;
      bfound ←bfound ∨r;
    end for
    bMatch←bMatch ∧ bfound;

Globe Variable:
 node N, node R

RewritePattern(Pattern P)
N←the last node
R←Root(P)
    for i=1to |E (R.Path)|
MatchPattern(R, i );
end if;
```

Algorithm 2. Rewrite Patterns

6 Experimental Evaluation

This section presents experimental results on rewriting performance implemented using C++. All experiments were run on a 2.1G AMD4200 processor with 1GB of main memory and 160GB quota of disk space, running windows XP system.

In following experiments, we focused on 116M XMark[11] with many repetitive and recursive paths. Our experiments consist of ten twig queries showed in Table 1.

Although descendant axes and wildcards bring about flexibly matching, the valid paths, which are selected by the contexts from P1-P10, are not too many(See Table 2). Like P1,P4 and P6, our algorithm works out an unique expression belong to $XP^{\{/,[]\}}$ by the context of their tree patterns. Owing to there exist a few similar paths, several results are obtained after rewriting P2, P3 and P5. The XMark has so many recursive paths that the count of final rewriting results from P7-P10 arranges from 9 to 81.

According to the experimental results, we conclude that our algorithm can effectively rewrite XPath expressions under path summary to completely eliminate descendant and wildcard features.

Table 1. Ten XPath Expressions

No.	XPath Expression
P1	site/*/*/name
P2	site//name
P3	site/*//*//*/name
P4	site/*/*/person [/homepage]/name
P5	site/regions/*/*/*/text[/emph][/keyword]
P6	site//person[//homepage]//name
P7	site//category[//emph][//keyword]
P8	site//*/closed_auction//text[//keyword]
P9	site//*[/closed_auction]//text[//keyword]
P10	site//description//text[/*//bold][//keyword]

Table 2. Rewriting Properties

No.	|V(G)|	|E(G)|	#
P1	4	3	1
P2	8	7	7
P3	23	22	6
P4	6	5	1
P5	38	37	6
P6	4	3	1
P7	20	19	81
P8	9	13	9
P9	9	14	9
P10	81	72	54

7 Related Work

The rewriting problem has been first discussed in traditional client-server databases[12]. There has been several research approaches to rewrite XML query in a host of previous works. To answer XML queries using RDBMS, Authors in [13] propose rewriting path queries over DTDs to SQL. Many works like to emphasize rewriting XPath expressions on their semantic structure. The minimization of XPath queries has been studied in [5,6,7].To reduce non-deterministic features in XPath queries, document [8] uses graph schemas and DTDs. These measures difficulty deal with '*' and '//' when nested paths occur in XML documents.

8 Conclusions

In this paper, we rewrite an XPath expression into a aeries of ones belonged to subclass $XP^{\{/,[]\}}$ depending on path summary coming from the XML document. Since '//' and '*' in XPath expressions can be completely eliminate after rewriting, our work provide an effective pre-processing way during answering XPath queries with '*' and '//'.

Acknowledgments. The first author was supported by the National High-Tech Research and Development Plan of China under Grant No. 2005AA4Z3030.

References

1. Miklau, G., Suciu, D.: Containment and equivalence for a fragment of XPath. Journal of the ACM, 2–45 (2004)
2. Chen, Y., Zheng, Y., Davidson, S.B.: BLAS: An Efficient XPath Processing System. In: SIGMOD 2004, pp. 47–58 (2004)
3. Bruno, N., Koudas, N., Srivastava, D.: Holistic Twig Joins: Optimal XML Pattern Matching. In: SIGMOD 2002, pp. 310–321 (2002)
4. Lu, T., et al.: From Region Encoding To Extended Dewey: On Efficient Processing of XML Twig Pattern Matching. In: Proceedings of VLDB, pp. 193–204 (2005)
5. Flesca, S., Furfaro, F., Masciari, E.: On the minimization of XPath queries. In: VLDB (2003)

6. Amer-Yahia, S., Cho, S.R., et al.: Minimization of Tree Pattern Queries. In: SIGMOD 2001, pp. 497–508 (2001)
7. Wood, P.T.: Minimizing Simple XPath Expressions. In: Proc of Intl. Conf. on WebDB, pp. 13–18 (2001)
8. Chidlovskii, B.: Using regular tree automata as XML schemas. In: Proc. of the IEEE Advances in Digital Libraries, Washington, pp. 89–104 (2000)
9. Yang, W.D., Wang, Q.M., et al.: Complex Twig Pattern Query Processing over XML Streams (in Chinese). Journal of Software 16(2), 223–232 (2005)
10. Moro, M.M., Vagena, Z., et al.: XML Structural Summaries. In: Proc. of Intl. Conf. on VLDB, Auckland, pp. 1524–1525 (2008)
11. Schmidt, A.R.: XMark: an XML benchmark project, `http://monetdb.cwi.nl/xml/index.html`
12. Halevy, A.Y.: Answering Queries Using Views: A Survey. VLDB Journal 10(4), 270–294 (2001)
13. Fan, W., Yu, J.X., et al.: Query translation from XPath to SQL in the presence of recursive DTDs. In: VLDB (2005)

Combining Statistical Machine Learning Models to Extract Keywords from Chinese Documents

Chengzhi Zhang[1,2]

[1] Department of Information Management, Nanjing University of Science & Technology, 210093 Nanjing, Jiangsu

[2] Institute of Scientific & Technical Information of China, 100038 Beijing

zhangchz@itic.ac.cn

Abstract. Keywords are subset of words or phrases from a document that can describe the meaning of the document. Many text mining applications can take advantage from it. Unfortunately, a large portion of documents still do not have keywords assigned. On the other hand, manual assignment of high quality keywords is time-consuming, and error prone. Therefore, most algorithms and systems aimed to help people perform automatic keywords extraction have been proposed. However, most methods of automatic keyword extraction cannot use the features of documents effectively. A method which integrates the statistical machine learning models is proposed in this paper. This method extracts keyword from Chinese documents through voting of multiple keywords extraction models. Experimental results show that the proposed method based on ensemble leaning outperforms other methods according to F_1 measurement. Moreover, the keywords extraction model based on ensemble learning with the weighted voting outperforms the model without the weighted voting.

Keywords: keywords extraction, ensemble Learning, statistical machine learning, text mining.

1 Introduction

Automatic keyword extraction (AKE) is a technology to identify a small set of words or segments that are meaningful and representative. In library and information science, automatic keyword extraction is usually called automatic indexing. Since keyword is the smallest unit which can express the meaning of document, many text mining applications can take advantage of it, e.g. automatic summarization, automatic classification, automatic clustering, etc. Therefore, keywords extraction can be considered as the core technology of all automatic processing for documents. However, a large number of documents including web pages do not have keywords. At the same time, manual assignment of high quality keywords is costly and time-consuming, and error prone. Therefore, automatic keywords extraction is a technology worthy of researching.

Currently, most methods of automatic keywords extraction cannot use the features of documents effectively. The statistical machine learning models including support vector machine, conditional random fields, can use the features of documents more

R. Huang et al. (Eds.): ADMA 2009, LNAI 5678, pp. 745–754, 2009.

sufficiently and effectively. At the same time, the automatic indexing models performance varies in the task of automatic indexing. If we combine these models to index the documents by ensemble learning, the performance of indexing can be improved. In order to improve the performance of keywords extraction, a method which integrates the statistical machine learning models and ensemble learning method is proposed in this paper. This method extracts keywords from the documents through voting of multiple keywords extraction models. Moreover, we also evaluate the performance of the keywords extraction model with the weighted voting and the model without the weighted voting.

The rest of this paper is organized as follows. The next section reviews some related work on keyword extraction. In section 3, a detailed description of the proposed approach is presented. Subsequently in section 4, the authors report experiments results that evaluate the proposed approach. The paper is concluded with summary and future work directions.

2 Related Works

In the task of keyword extraction, words occurred in the document are analyzed to identify apparently significant ones, on the basis of properties such as frequency and length. Existing methods about automatic keywords extraction can be divided into four categories, i.e. simple statistics, linguistics, machine learning and other approaches.

Statistics Approaches are simple and do not need the training data. The statistics information of the words can be used to identify the keywords in the document. The statistics methods include N-Gram [1], word frequency [2], TF*IDF [3], words co-occurrence [4], PAT-tree [5], etc.

Linguistics Approaches use the linguistics feature of the words mainly, sentences and document. The linguistics approach includes the lexical analysis [6], syntactic analysis [7], semantic analysis [8], discourse analysis [9] [10], and so on.

Keyword extraction can be seen as supervised learning from the examples. Machine learning approach employs the extracted keywords from training documents to learn a model and applies the model to find keywords from new documents. This approach includes NB [11], Maximum entropy model, SVM [8], CRF [12], Bagging [7], etc. Some keyword extraction tools, e.g. GenEx [13], KEA [14], have been developed.

Other Approaches about keyword extraction mainly combine the methods mentioned above or use some heuristic knowledge in the task of keyword extraction, such as the position, length, layout feature of the words, html tags [15], etc.

In 2003, keywords extraction based on bagging algorithm was first proposed by Hulth [7]. The method based on ensemble learning is still a direction in the task of automatic keywords extraction so far.

3 Keyword Extraction Based on Ensemble Learning

In this section, an overview of ensemble learning method and some descriptions of the general ensemble learning algorithms in the task of classification are provided.

Then, some base classifiers based on statistical machine learning models and two baseline models are given. Finally, these models are combined to extract keywords from Chinese documents according to ensemble learning methods.

3.1 Overview of Ensemble Learning in the Classification

Ensemble learning tries to solve the same problem through multiple learning models and it can improve the generalization ability of learning models. Therefore, the machine leaning community has attached a great importance to ensemble learning, and Dietterich regards it as the top of the four major directions of research on machine learning. Ensemble learning is also known as integrated learning. In the classification, a group of base classifiers according to the training data are built, and they are used to classify the objects through voting of each classifier [16]. For example, Schapire & Singer applied Boosting algorithm to classify documents. They found that Boosting algorithm is more effective than or at least as well as common classification techniques (e.g. Rocchio) in the task of classification [17]. Moreover, Weiss successfully achieved a high precision in classification based on a small dictionary by ensemble learning [18].

3.2 Description of General Ensemble Learning Algorithms

Figure 1 shows the logical view of ensemble learning method. We can use following methods to build a combination of base classifier [16].

- Processing training data sets, such as Bagging and Boosting;
- Processing the input features, this method is more effective if data set has a large number of redundant features;
- Processing category label, this method is suitable for the case that has enough categories;
- Processing learning algorithm, this method can get different models in the same training data by using different algorithms.

The following are some descriptions of typical methods based on ensemble learning [16], e.g. Bagging, AdaBoost, integrated algorithm based on combination of different learning algorithms.

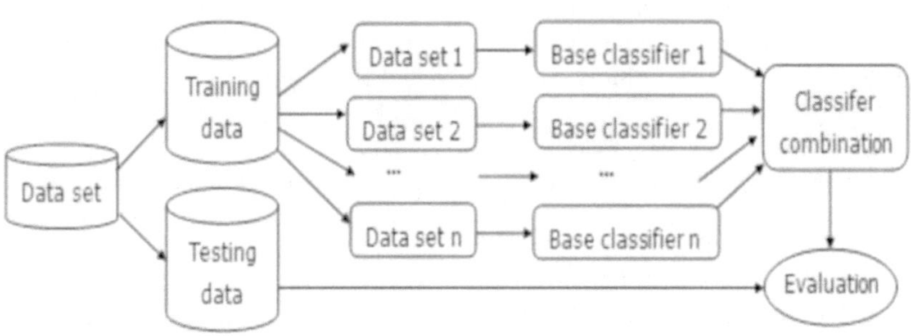

Fig. 1. Logical view of ensemble learning method

Bagging algorithm, which is relatively simple, samples repeatedly from the data set on basis of a uniform probability distribution. AdaBoost algorithm is a method which weighs prediction value of every base classifier instead of a majority vote, and this makes it possible for AdaBoost to punish the models that has a low precision [16].

It is a common method to build a combination of base classifiers through getting different models by conducting different algorithms on the same training set. Figure 2 shows this integrated algorithm. As mentioned previously, this algorithm is used to extract keywords from Chinese documents based on ensemble learning in this paper.

Algorithm: Integrated algorithm based on different learning algorithms

Input: training set $D=\{(x_1, y_1),\ldots,(x_m, y_m)\}$, where $x_i \in \mathbf{X}$, $y_i \in \mathbf{Y}$, $\mathbf{T}$ represents testing set and $\mathbf{P_i}$ represents precision of each base classifier;

Output: ensemble learning machine $C^*(x)$

Steps: first set the number of classifiers, let's say k

For i=1 to k {

  Build training set D_i based on D;

  Build base classifier C_i based on D_i }

For each record of testing data $x \in \mathbf{T}$

$$C^*(x) = \mathrm{Vote}\left(\frac{P_1}{\sum_{i=1}^{k} P_i} * C_1(x), \frac{P_2}{\sum_{i=1}^{k} P_i} * C_2(x), \ldots, \frac{P_k}{\sum_{i=1}^{k} P_i} * C_k(x)\right)$$

Fig. 2. Integrated algorithm based on different learning algorithms [16]

3.3 Base Classifier for Keywords Extraction

The task of automatic keyword extraction can be viewed as a classification problem. In this section, four base classifiers base on statistical machine learning used for automatic keywords extraction are provided, i.e. Conditional Random Fields (*CRF*), support vector machines (*SVM*), multiple linear regression (denoted as *MLR*), logistic regression (denoted as *Logit*). Moreover, two baseline models based on heuristic approaches is introduced, namely, BaseLine1 and BaseLine2.

(1) CRF Classifier

Conditional Random Fields model is a new probabilistic model for segmenting and labeling sequence data [19]. CRF model is an undirected graphical model that encodes a conditional probability distribution with a given set of features. The main advantage of CRF model comes from that it can relax the assumption of conditional independence of the observed data often used in generative approaches, an assumption that might be too restrictive for a considerable number of object classes. Additionally, CRF avoids the label bias problem. In 2008, Zhang & Wang used CRF model for automatic keyword extraction [12].

(2) SVM Classifier

Support Vector Machine was proposed by Vapnik in 1995 to solve the problem with recognition of binary classification pattern [20]. In 2004, Zeng used SVM model to extract the significant phrases in the task of search results clustering [21]. Two years later, Zhang proposed SVM-based automatic keywords extraction model [8]. This model is also used in this paper.

(3) MLR Classifier

Linear regression is the simplest form of regression. Zeng used multiple linear regression models for the extraction of the significant phrases. They found that MLR model achieved better results in solving the problem of significant phrases extraction [21].

(4) Logit Classifier

When the dependent variable is a binary type, Logistic regression model is more suitable for predicting samples label. Zeng also used Logistic regression model to extract the significant phrases and found this model also achieved good results [21].

(5) BaseLine1 Classifier

In BasaLine1 model (abbreviated as *BL1*), we use the normalized word frequency (TF), normalized inverse document frequency (IDF) and the length of words as a factor for weight computation, and weight formula is as follows:

$$Score = TF * IDF * Len \tag{1}$$

Where, Len is normalized by the maximum length of words in the document.

(6) BaseLine2 Classifier

BaseLine2 model (abbreviated as *BL2*) is different from BL1 model. It is added the position of the first appearance of the word (*Dep*) as another factor except for TF*IDF and the length of words. The weight formula is as follows:

$$Score = TF * IDF * Len * length\ (d)/Dep \tag{2}$$

Where, length (d) is the length of the document.

When the models above are used to automatic extract keywords from Chinese documents, a series of pre-processes are prepared, e.g. segmenting the Chinese sentence into Chinese words, tagging the Part-Of-Speech of the Chinese words, features' weight computation and sorting, etc. More detailed description of these steps as well as training and testing of the models can be found in [12].

3.4 Automatic Keywords Extraction Method Based on Ensemble Learning

In this paper, we use those automatic keyword extraction models such as *BaseLine1*, *BaseLine2*, *MLR*, *Logit*, *SVM* and *CRF*, as base classifiers and combine them to extract keywords from Chinese documents based on ensemble learning method. In this section, we mark these automatic keywords extraction methods based on ensemble learning as *Ens*.

According to whether base classifier use precision of keywords extraction as the weigh, the keywords extraction model based on ensemble learning can be divided into two categories, e.g. one with the weighted voting and the other without the weighted voting.

(1) Keywords Extraction Model without the Weighted Voting
In the process of voting, this model gives same weight to each base classifier, and gets a majority voting result as the final keywords extraction result according to the formula as follows:

$$C^*(x) = Vote\,(C_1(x),\ C_2(x), \ldots,\ C_k\,(x)) \tag{3}$$

(2) Keywords Extraction Model with the Weighted Voting
Figure 2 shows descriptions of the ensemble learning method with the weighted voting. In the process of voting, this model gives different weights to each base classifier, and gets the classification result according to the formula as follows [16]:

$$C^*(x) = Vote\,\left(\frac{P_1}{\sum_{i=1}^{k} P_i} * C_1(x),\ \frac{P_2}{\sum_{i=1}^{k} P_i} * C_2(x),\ \ldots,\ \frac{P_k}{\sum_{i=1}^{k} P_i} * C_k(x)\right) \tag{4}$$

Where k is the number of base classifiers and the precision of each classifier $\mathbf{P_i}$ turns to $P_i \Big/ \sum_{i=1}^{k} P_i$ ($i \in [1, k]$) after normalization.

4 Experimental Results

In this section we combine the six keywords extraction models described above to extract keywords from Chinese documents based on ensemble learning. Then we analyze the experiment results.

4.1 Data Sets and Evaluation Measures

(1) Data Sets
In this study, we collect documents from database of 'Information Center for Social Sciences of RUC' (http://ww.zlzx.org). We randomly chose 600 academic documents in the field of economics from the database. Each document includes the title, abstract, keywords, full-text, heading of paragraph or sections, boundaries information of paragraphs or sections, references, etc.

(2) Evaluation Measures
In the evaluation, there are two types of words or phrases in manual assignment of keywords, which are keywords and non-keywords assigned by humans. On the other

hand, there are two types of words or phrases in automatic keyword extraction, i.e. keywords and non-keywords extracted by keyword extraction system. Table 1 shows the contingence table on the result of keywords extraction and manual assignment keywords.

From all experiments on keyword extraction, we conducted evaluations according to the general measuring method used in the Information retrieval evaluation, i.e. precision (P), recall (R) and F_1-Measure. The evaluation measures are defined as follows:

$$P=a/(a+b) \tag{5}$$

$$R=a/(a+c) \tag{6}$$

$$F_1(P,R)=2PR/(P+R) \tag{7}$$

Where, a, b, c and d denote number of instances. In this paper, we get the evaluation results by using 10-fold cross-validation.

4.2 Analysis of Experimental Results

We use the data set in section 4.1 for 10-fold cross-validation and combine general and domain-specific dictionary to extract features. The general dictionary directly uses *PKU dictionary*[1] as segmenting dictionary while the domain-specific dictionary is a collection of the keywords in a specific domain. Table 2 shows us the precision of each base classifier.

Table 2. Precision of each base classifier in automatic keywords extraction

	BaseLine1	BaseLine2	MLR	Logit	SVM	CRF
Precision of base classifier	0.1188	0.171	0.2135	0.2116	0.5140	0.4702
Normalized Precision of base classifier	0.0699	0.1010	0.1256	0.1245	0.3024	0.2766

Figure 3 shows the result of a sample record. The keywords of this document are extracted based on ensemble learning without the weighted voting.

Table 3 shows the precision, recall and F_1-Measure of the eight keywords extraction methods including methods based on ensemble learning. In table 3, *Ens* means the keywords extraction model without the weighted voting, and *Ens-W* with the weighted voting based on the normalized precision of each classifier in table 2.

From table 3, we know that the automatic keywords extraction method based on ensemble learning is above average in both precision and recall and also have the largest F_1-Measure. According to the F_1 value, the keywords extraction model based on ensemble learning outperforms the other seven models, which shows its effectiveness.

[1] http://ccl.pku.edu.cn/doubtfire/Course/Chinese%20Information%20Processing/Source_Code/Chapter_8/Lexicon_full_2000.zip

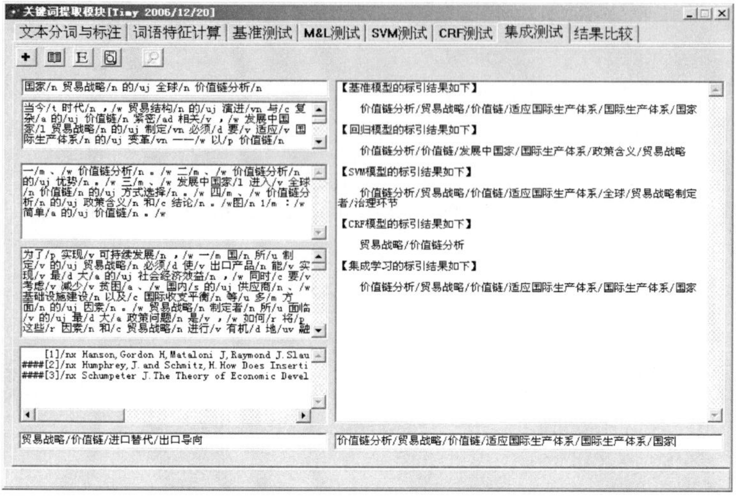

Fig. 3. Result sample of the keywords extraction model without the weighted voting

Table 3. Performance of eight keywords extraction methods

Model	P	R	F_1
BL1	0.1188	0.2284	0.1563
BL2	0.1717	0.3302	0.2260
MLR	0.2135	0.3519	0.2657
Logit	0.2116	0.3488	0.2634
SVM	0.5140	0.1698	0.2552
CRF	0.4702	0.2438	0.3209
Ens	0.4520	0.3097	0.3676
Ens-W	0.4701	0.3178	0.3792

In the keywords extraction based on ensemble learning, the model with the weighted voting outperforms the model without the weighted voting, which means the former is more effective. We will consider other factors and optimize the weighing method to improve the quality of keywords extraction in the future work.

5 Conclusion and Future Works

In this paper we extract keywords from Chinese documents based on ensemble learning by using the integrated algorithm. Experimental results show that the keywords extraction model based on ensemble learning can improve F_1 value. Moreover, the model with the weighted voting outperforms the model without the weighted voting.

The future works include three aspects. First, experiments will be extended to different combination ways of classifiers or on different data scales. We will also use selective ensemble learning [22] in the task of keyword extraction. Second, we will make a further step to optimize CRF-based automatic keywords extraction. For example, use the results of different keywords extraction models as the input of CRF, use CRF-based integrated automatic keywords extraction, etc. Finally, we plan to find out a more reasonable way such as considering precision and recall in the process of voting. For example, we will use F_1 value as the weight.

Acknowledgments. This research was partially supported by National Key Project of Scientific and Technical Supporting Programs (No.2006BAH03B02), Project of the Education Ministry's Humanities and Social Science funded by Ministry of Education of China (No.08JC870007), Youth Research Support Fund funded by Nanjing University of Science & Technology (No.JGQN0701), Scientific Research Foundation funded by Nanjing University of Science & Technology (No.AB41123).

References

1. Cohen, J.D.: Highlights: Language and Domain-independent Automatic Indexing Terms for Abstracting. Journal of the American Society for Information Science 46(3), 162–174 (1995)
2. Luhn, H.P.: A Statistical Approach to Mechanized Encoding and Searching of Literary Information. IBM Journal of Research and Development 1(4), 309–317 (1957)
3. Salton, G., Yang, C.S., Yu, C.T.: A Theory of Term Importance in Automatic Text Analysis. Journal of the American society for Information Science 26(1), 33–44 (1975)
4. Matsuo, Y., Ishizuka, M.: Keyword Extraction from a Single Document Using Word Co-occurrence Statistical Information. International Journal on Artificial Intelligence Tools 13(1), 157–169 (2004)
5. Chien, L.F.: PAT-tree-based Keyword Extraction for Chinese Information Retrieval. In: Proceedings of the 20th Annual International ACM SIGIR Conference on Research and Development in Information Retrieval (SIGIR1997), Philadelphia, PA, USA, pp. 50–59 (1997)
6. Ercan, G., Cicekli, I.: Using Lexical Chains for Keyword Extraction. Information Processing and Management 43(6), 1705–1714 (2007)
7. Hulth, A.: Improved Automatic Keyword Extraction Given More Linguistic Knowledge. In: Proceedings of the 2003 Conference on Empirical Methods in Natural Language Processing, Sapporo, Japan, pp. 216–223 (2003)
8. Zhang, K., Xu, H., Tang, J., Li, J.Z.: Keyword Extraction Using Support Vector Machine. In: Proceedings of the Seventh International Conference on Web-Age Information Management (WAIM 2006), Hong Kong, China, pp. 85–96 (2006)
9. Dennis, S.F.: The Design and Testing of a Fully Automatic Indexing-searching System for Documents Consisting of Expository Text. In: Schecter, G. (ed.) Information Retrieval: a Critical Review, pp. 67–94. Thompson Book Company, Washington (1967)
10. Salton, G., Buckley, C.: Automatic Text Structuring and Retrieval –Experiments in Automatic Encyclopedia Searching. In: Proceedings of the Fourteenth SIGIR Conference, pp. 21–30. ACM, New York (1991)
11. Frank, E., Paynter, G.W., Witten, I.H.: Domain-Specific Keyphrase Extraction. In: Proceedings of the 16th International Joint Conference on Artificial Intelligence, Stockholm, Sweden, pp. 668–673 (1999)
12. Zhang, C.Z., Wang, H.L., Liu, Y., Wu, D., Liao, Y., Wang, B.: Automatic Keyword Extraction from Documents Using Conditional Random Fields. Journal of Computational Information Systems 4(3), 1169–1180 (2008)

13. Turney, P.D.: Learning to Extract Keyphrases from Text. NRC Technical Report ERB-1057, National Research Council, Canada, pp. 1-43 (1999)
14. Witten, I.H., Paynter, G.W., Frank, E., Gutwin, C., Nevill-Manning, C.G.: KEA: Practical Automatic Keyphrase Extraction. In: Proceedings of the 4th ACM Conference on Digital Library (DL 1999), Berkeley, CA, USA, pp. 254–255 (1999)
15. Keith Humphreys, J.B.: Phraserate: An Html Keyphrase Extractor. Technical Report, University of California, Riverside, pp. 1–16 (2002)
16. Tan, P., Steinbach, M., Kumar, V.: Introduction to Data Mining, pp. 276–290. Addison-Wesley, Boston (2006)
17. Schapire, R.E., Singer, Y.: BoosTexter: a Boosting-based System for Text Categorization. Machine Learning 39(2-3), 135–168 (2000)
18. Weiss, S.M., Apte, C., Damerau, F.J., Johnson, D.E., Oles, F.J., Goetz, T., Hampp, T.: Maximizing Text-mining Performance. IEEE Intelligent Systems 14(4), 63–69 (1999)
19. Lafferty, J., McCallum, A., Pereira, F.: Conditional Random Fields: Probabilistic Models for Segmenting and Labeling Sequence Data. In: Proceedings of the 18th International Conference on Machine Learning (ICML 2001), Williamstown, MA, USA, pp. 282–289 (2001)
20. Vapnik, V.: The Nature of Statistical Learning Theory, pp. 1–175. Springer, New York (1995)
21. Zeng, H.J., He, Q., Chen, Z., Ma, W.Y., Ma, J.: Learning to Cluster Web Search Results. In: Proceedings of 27th Annual International Conference on Research and Development in Information Retrieval (SIGIR 2004), Sheffield, UK, pp. 210–217 (2004)
22. Zhou, Z.H., Wu, J.X., Tang, W.: Ensembling Neural Networks: Many Could Be Better Than All. Artificial Intelligence 137(1-2), 239–263 (2002)

Privacy-Preserving Distributed *k*-Nearest Neighbor Mining on Horizontally Partitioned Multi-Party Data

Feng Zhang[1,2], Gansen Zhao[1], and Tingyan Xing[3]

[1] School of Software, Sun Yat-sen University,
510275 Guangzhou, Guangdong

[2] Guangdong Key Laboratory of Information Security Technology, Sun Yat-sen University,
510275 Guangzhou, Guangdong

[3] School of Information Engineering,China University of Geosciences,
100083 Beijing

zhfeng@mail.sysu.edu.cn

Abstract. *k*-Nearest Neighbor (*k*-NN) mining aims to retrieve the *k* most similar objects to the query objects. It can be incorporated into many data mining algorithms, such as outlier detection, clustering, and *k*-NN classification. Privacy-preserving distributed *k*-NN is developed to address the issue while preserving the participants' privacy. Several two-party privacy-preserving *k*-NN mining protocols on horizontally partitioned data had been proposed, but they fail to deal with the privacy issue when the number of the participating parties is greater than two. This paper proposes a set of protocols that can address the privacy issue when there are more than two participants. The protocols are devised with the probabilistic public-key cryptosystem and the communicative cryptosystem as the core privacy-preserving infrastructure. The protocols' security is proved based on the Secure Multi-party Computation theory.

Keywords: Privacy-reserving Data Mining, *k*-NN Mining; Probabilistic Cryptosystem, Communicative Cryptosystem, Secure Multi-party Computation.

1 Introduction

The wide spreading of data mining applications reveal many new challenges. Privacy-preserving in data mining is one of the challenges [1]. It is common that data-owners are concerned with the misuse of their data, and do not want any sensitive information to be mined out. With such privacy concerns, data owners are not willing to provide their data to participate in data mining.

Privacy Preserving Data Mining (PPDM) makes tradeoffs between data mining contributions and data privacy. It aims at carrying out data mining precisely and efficiently while protecting data privacy to a certain extent. Privacy preserving data mining has achieved significant results in many data mining areas [1].

Techniques employed in data mining can be roughly classified as two categories, the Cryptography-based Technique and the Randomized Perturbation Technique (RPT). The former is primarily applied to distributed data storage scenario, and the latter is generally used in centralized data storage scenario.

R. Huang et al. (Eds.): ADMA 2009, LNAI 5678, pp. 755–762, 2009.

In the case of the Cryptography-based PPDM research, some privacy-preserving protocols for *k*-NN mining (The readers can refer to definition 1 for the detailed definitions of *k*-NN mining and privacy-preserving *k*-NN mining) based on horizontally partitioned multi-party data have been developed [2,3]. However, these protocols are only capable of deal with two-party case and the security is only proved for each of the parts. To the best of our knowledge, no protocols have been proposed as of the time of writing to deal with the privacy issue of *k*-NN mining when the number of the participants is greater than two. The paper follows the previous work [2,3] and designs several protocols to address this issue.

The rest of the paper is organized as follows. Section 2 presents the privacy-preserving *k*-NN problem definitions. Section 3 explains the basic concepts and related security infrastructures. Section 4 proposes a set of protocols to tackle the issue of having more than two participants in *k*-NN data mining. The security proofs of the proposed protocols are shown in Section 5. The performance of the proposed protocols is discussed in Section 6. Section 7 concludes the work with discussion on potential future work.

2 Problem Definitions

Let the resulting *k*-NN set be $N_k(q)$ and the set of all objects be *D*. *k*-NN mining can be formally defined as below [3]:

Definition 1. (*k*-NN mining). Given a data set of object *D*, a query object $q \in D$ and a query parameter *k*, *k*-NN mining returns the set $N_k(q) \subseteq D$ of size *k* such that:

$$\forall o \in N_k(q), \forall o' \in D : \forall o' \notin N_k(q) \Rightarrow d(o,q) \leq d(o',q) \quad (1)$$

In a horizontally partitioned distributed data setting, it is assumed that there are *s* distinct objects in a data set *D*, and they are distributed over *N* $(N>2)$ sites. All objects hold the same dimensions. The *w* dimensions correspond to *w* attributes, $A_1, A_2, \ldots, A_w$. Let attribute set $A=(A_1, A_2, \ldots, A_w)$. Site *i* $(i=0,1,\ldots,N-1)$ collects s_i objects, where $\sum_{i=0}^{N-1} s_i = s$. The s_i objects constitute the set D_i, satisfying $\bigcup_{i=0}^{N-1} D_i = D$. And for any $i \neq j$, $D_i \cap D_j = \phi$. The *k*-NN query returns the set $N_k(q) \subseteq \bigcup_{i=0}^{N-1} D_i$ of size *k* such that:

$$\forall o \in N_k(q), \forall o' \in D_i (1 \leq i \leq m) : \forall o' \notin N_k(q) \Rightarrow d(o,q) \leq d(o',q) \quad (2)$$

To ensure that the *k*-NN mining is privacy-preserving, the proposed protocols must be able to compute the nearest neighbors without revealing private information about the other parties' input, except that can be computed with the respective input and output of the computation.

3 Preliminaries

3.1 Security in Semi-honest Model

The main idea of the definition is that under the semi-honest model, a protocol is said to securely (privately) compute *f* if the information obtained by every participating party after carrying out the protocol, can be polynomially simulated through the participating party's input and output.

Due to space limitation, interested readers are suggested to refer to [4] for the definition of Secure Multi-party Computation.

3.2 The Probabilistic Public-Key Cryptosystems

A Probabilistic Cryptosystem is a Public Key Cryptosystem. In probabilistic encryption, the encryption results are probabilistic instead of deterministic. The same plaintext may map to two different ciphertexts at two different probabilistic encryption processes:

$$C_1 = E_k(M), C_2 = E_k(M), C_3 = E_k(M), \ldots, C_n = E_k(M). \tag{3}$$

The Paillier Cryptosystem provides significant facilities to guarantee the security of the designed protocols [5].

3.3 The Commutative Cryptosystems

A cryptosystem system is called a commutative cryptosystem if the following three properties satisfied: (1) The ciphertext of a compositely commutative encryption is identical regardless of the encryption order; (2) Two different plain messages will never result in the same encrypted messages; (3) The encryption is secure. In particular, given a plain message x and its randomly-chosen-way encryption $f_e(x)$, for a new plain message y, an adversary cannot distinguish in polynomial time between $f_e(y)$ and a randomly chosen value z from the domain of encrypted messages. Interested readers can refer to [6, 7] for more details.

4 Secure *k*-NN Mining Protocols

Without loss of generality, the following protocols only consider how to mine the *k*-NN of the object $O_i=(x_{i1},x_{i2},\ldots,x_{iw})$ at site 0. The protocols also observe the fact that O_i's *k*-NN may locate at site 0 as well or at any other sites. The Euclidian Distance is used in measuring the distance between objects in the protocols.

Given two objects $O_i=(x_{i1},x_{i2},\ldots,x_{iw})$ and $O_j=(x_{j1},x_{j2},\ldots,x_{jw})$, the first problem is to compute their scalar product in additively split form:

$$c = O_i \bullet O_j = \sum_{r=1}^{w} (x_{ir} \times x_{jr}) \tag{4}$$

For example, suppose Alice has the object O_i and Bob has the object O_j. The goal is for Alice to obtain a z' and Bob a z'' such that $z' + z'' = O_i \cdot O_j$. This makes possible the computation, in additively split form, of the square of Euclidean distance (Formula 5) between x and y.

$$dist(O_i,O_j) = \sum_{r=1}^{w} (x_{ir} - x_{jr})^2 = \sum_{r=1}^{w} (x_{ir})^2 + \sum_{r=1}^{w} (x_{jr})^2 - 2 \times \sum_{r=1}^{w} (x_{ir} x_{jr}). \tag{5}$$

Since Alice can compute $\sum_{r=1}^{w}(x_{ir})^2$, Bob can compute $\sum_{r=1}^{w}(x_{jr})^2$, it is only necessary to securely compute $2\times\sum_{r=1}^{w}(x_{ir}x_{jr})$ to compute the distance, which is the core work of the Secure Scalar Product Protocol (SSP)[8].

4.1 Secure Scalar Product Protocol

Protocol 1: Private Homomorphic SSP Protocol	
Private Input: Alice has object O_i, and Bob has object O_j	
Private Output: Alice gets n_A, Bob gets n_B, $n_A+n_B\equiv O_i\bullet O_j \bmod m$	
1	**Setup phase. Alice does:**
1.1	**Generate** two big primes p, q , and g B;
1.2	**Compute** $n = pq$ and $\lambda=lcm(p-1,q-1)$;
1.3	**Generate** the pair (n, g) as the public key, λ as the private key;
1.4	**Send** public key (n,g) to Bob through secure channel.
2	**Alice does :**
2.1	**Compute** $\varepsilon(O_i)=(\varepsilon(x_{i1}),\varepsilon(x_{i2}),...,\varepsilon(x_{iw}))$ and send it to bob;
3	**Bob does:**
3.1	**Randomly choose** r from Z_n^* and **compute:**
3.2	$w=\varepsilon(O_i\bullet O_j)=(\varepsilon(x_{i1}))^{j1}\times(\varepsilon(x_{i2}))^{j2}\times.....\times(\varepsilon(x_{iw}))^{jw}\times r^n \bmod n^2$;
3.3	**Generate** a random plaintext n_B and a random number r' from Z_n^* , **Compute** $w'=w\times\varepsilon(-n_B)$; **Send** w' to Alice
4	**Alice does:**
4.1	**Decipher** w', **getting** $n_A=O_i\bullet O_j-n_B$;

4.2 The Secure Set Union Protocol

Protocol 2: Set Union Protocols based on protocol 1	
Input: Each site i $(i=0,1,\dots,N-1)$ collects s_i objects, $\sum_{i=0}^{N-1}s_i=s$. The s_i objects constitute the set L_i, satisfying $\bigcup_{i=0}^{N-1}L_i=L$	
Output: Generate union $L=\bigcup_{i=0}^{N-1}L_i$ without revealing which elements belongs to which site	
1	**For** each site $i=0,1,\dots,N-1$ **do** // *Encrypt L_i at site i, getting* L_{e_i}
1.1	**Generate** $(e_i,d_i), e_i\in_r E, d_i\in_r D$; $L_{e_i}=\phi$;
1.2	**For each** $x\in L_i$ **do** $L_{e_i}=L_{e_i}\cup f_{e_i}(x)$ **End for**
1.3	**End for**
2	**For** round $j=0$ to $N-2$ **do //Encrypt** L_{e_i} **by all sites**
2.1	**If** j==0 **then** Each site i sends permuted L_{e_i} to site $(i+1) \bmod N$;
2.2	**Else**

Protocol 2: Set Union Protocols based on protocol 1	
2.3	Each site i encrypts all elements in $L_{e_{(i-j)\,mod\,N}}$ with e_i, permutes, and sends it to *site (i+1) mod N*;
2.4	**End if**
2.5	**End for**
2.6	Each site i encrypts all elements in $L_{e_{(i+1)\,mod\,N}}$ with e_i;
3	**For each site *i=2,3,…,N-1* do // Merge L_{e_i} , getting *L***
3.1	Send $L_{e_{(i+1)\,mod\,N}}$ to site *1*;
3.2	**End for**
3.3	**At site *1*: Generate** $L = \bigcup_{i=1}^{N-1} L_{e_{(i+1)\,mod\,N}}$, permute and send it to site 0;
3.4	**At site *0*: Generate** $L = L \cup L_1 = \bigcup_{i=0}^{N-1} L_{ei}$;
4	**For** *i=0* to *N-1* **do //Decryption *L***
4.1	**Site** *i* decrypts all elements in *L* using d_i and sends permuted *L* to site *(i+1) mod N*;
4.2	**End for**
4.3	Site *0* broadcasts *L* to all other sites.

4.3 The *k*-NN Mining Protocol based on Horizontally Partitioned Data

Protocol 3: Mining the *k*-NN of object $O_i=(x_{i1},x_{i2},\ldots,x_{iw})$ (at site 0)	
Input: Each site *i (i=0,1,…,N-1)* collects s_i objects, $\sum_{i=0}^{N-1} s_i = s$. The s_i objects constitute the set D_i, satisfying $\bigcup_{i=0}^{N-1} D_i = D$	
Output: Global *k*-NN located at each site *i*;	
1	**At site 0:**
1.1	**Compute** $\sum_{r=1}^{w} (x_{ir})^2$
2	**For each site *u=1,2,…,N-1***
2.1	**Set up** public key and private key (similar to 1.1-1.3 of protocol 1);
2.2	**Send** public key *(n,g)* to site *0* through secure channel.
2.3	**For** each object O_j in D_u
2.4	**Execute** Protocol 1 with O_i and O_j, site *0* obtaining random number n_A and site *u* obtaining random number n_B, $n_A + n_B = O_i \bullet O_j$;
2.5	**At site *0*: compute** $p_0 = \varepsilon(\sum_{r=1}^{w} (x_{ir})^2 + n_A)$, and send it to site *u*;
2.6	**At site *u*: compute** $p_0 \times \varepsilon(\sum_{r=1}^{w} (x_{jr})^2 + n_B)$, and decipher it, getting: $\sum_{r=1}^{w} (x_{ir})^2 + \sum_{r=1}^{w} (x_{jr})^2 + n_A + n_B = dist(O_i, O_j)$.
2.7	**End for**
2.8	**End for**
3	**For each site *u=0,1,…,N-1* do**
3.1	**Compute** the local minimum *k* distances at site *u*;
3.2	**Constitute** set L_u using the *k* distances.
3.3	**End for**

Protocol 3: Mining the k-NN of object $O_i=(x_{i1},x_{i2},\ldots,x_{iw})$ (at site 0)	
4	**Execute protocol 2** with the input L_u $(u=0,1,\ldots,N\text{-}1)$, **securely getting** the union $L=\bigcup_{u=0}^{N-1} L_u$ **at site 0**.
5	**At site 0**
5.1	**Sort** the elements in L and **locate** the kth minimum element (distance) $dist_k$;;
5.2	**Broadcast** $dist_k$ to all other sites;
6	**For each site** $u=0,1,\ldots,N\text{-}1$ **do**
6.1	The objects at site u whose distance with O_i is greater than $dist_k$ constitute the global k-NN located at site u;
6.2	**End for**

5 Security Proof of the Protocols

The security of the two protocols proposed in section 4 can be proved as follows.

Theorem 1. If all involved parties satisfy the semi-honest assumption, protocol 1 is secure.

Proof. Since private data transferring only takes place at step 2 and step 3 of protocol 1, we only need to prove the security of the 2 steps.

Step 2: Since Alice does not receive any private data, it can simulate its view by running the step on its own input.

Step 3: Bob receives ciphertext $\varepsilon(O_i)=(\varepsilon(x_{i1}),\varepsilon(x_{i2}),\ldots,\varepsilon(x_{iw}))$, which can be simulated by randomly choosing w numbers r_j $(j=1,\ldots,w)$ from $[0,n)$, constituting $(r_1, r_2, \ldots, r_w)$. As both the real view and the simulated view come from the same distribution, they are computationally indistinguishable.

Step 4: w' may be simulated by the way of randomly choosing r from $[0,n)$.

Theorem 2. If all participating parties satisfy the semi-honest model assumption, protocol 2 is secure.

Proof. Protocol 2 is similar to protocol 1 in [9]. Interested readers may refer to [9] for the security proof.

Theorem 3. If all participating parties satisfy the semi-honest model assumption, protocol 3 is secure.

Proof. Data privacy is only likely to be breached in step 2.3-2.7 and step 4. Their securities are obvious. Due to space limitation, the complete security proofs are skipped here.

6 Computation Complexity and Communication Costs

Suppose that the commutative encryption/decryption time cost is T_c; the Probabilistic encryption/decryption time cost is T_p; transferring plain message/ciphertext requires B bites; transferring public key requires P bites. The computation complexity (CComp) and communication costs (CCost) of protocol 1 are shown in Table 1.

Table 1. Computation Complexity and Communication Costs of Protocol 1

Step	CComp	CCost
1	*Const*	$O(P)$
2	$O(w \times T_p)$	$O(w \times B)$
3	$O(T_p)$	$O(B)$
4	$O(T_p)$	*N/A*

$Comp_{p1}$ is used to denote the total computation complexity, and $Cost_{p1}$ to denote the total communication cost of protocol 1. The CComp and the CCost of protocol 2 are shown in Table 2.

Table 2. Computation Complexity and Communication Costs of Protocol 2

Step	CComp	CCost
1	$O(s \times T_c)$	*N/A*
2	$O((N\text{-}1) \times s \times T_c)$	$O((N\text{-}1) \times s \times w \times B)$
3	$O(N)$	$O((N\text{-}1) \times s \times w \times B)$
4	$O(N \times s \times T_c)$	*Ignored*

$Comp_{p2}$ is used to denote the total CComp, and $Cost_{p2}$ to denote the total CCost of protocol 2. The CComp and the CCost of protocol 3 are shown in Table 3.

Table 3. Computation Complexity and Communication Costs of Protocol 3

Step	CComp	CCost
1	$O(w)$	*N/A*
2	$O((N\text{-}1) \times Comp_{p1})$	$O((N\text{-}1) \times Cost_{p1})$
3	$O(s^2)$	*N/A*
4	$Comp_{p2}$	$Cost_{p2}$
5	$O(s^2)$	*const*
6	$O(s)$	*N/A*

7 Conclusion

The main contribution of this paper is three folded. Firstly, we identify the incapability of existing protocols in handling the privacy issue of the scenarios when there are more than two participants in privacy-preserving k-NN mining. Secondly, we propose two protocols to tacked the above identified issue Thirdly, we briefly show the security of the proposed protocols.

The proposed protocols are not without their limitations. The proposed protocols rely on the semi-honest model, which may not be true in some scenarios The issue of designing privacy-preserving k-NN mining protocol, as well as other actual multi-party computation protocols under malicious models are major challenges in SMC applications. This challenge, along with how to design a protocol further reducing the time and communication cost, are our future concerns.

Acknowledgments. The work is supported the 2008 Sun Yat-sen University Science Foundation under Grant No. 1131014.

References

1. Vaidya, J., Clifton, C., Zhu, M.: Privacy Preserving Data Mining (Advances in Information Security). Springer, New York (2005)
2. Shaneck, M., Kim, Y., Kumar, V.: Privacy Preserving Nearest Neighbor Search. In: 6th IEEE International Conference on Data Mining Workshops, pp. 541–545 (2006)
3. Qi, Y., Atallah, M.J.: Efficient Privacy-Preserving k-Nearest Neighbor Search. In: 28th International Conference on Distributed Computing Systems, pp. 311–319 (2008)
4. Goldreich, O.: Foundations of Cryptography. Basic Applications, vol. 2. Press Syndicate of the University of Cambridge, Cambridge (2004)
5. Paillier, P.: Public-key Cryptosystems based on Composite Degree Residuosity Classes. In: Stern, J. (ed.) EUROCRYPT 1999. LNCS, vol. 1592, pp. 223–238. Springer, Heidelberg (1999)
6. Boneh, D.: The Decision Diffie-Hellman Problem. In: Buhler, J.P. (ed.) ANTS 1998. LNCS, vol. 1423, pp. 48–63. Springer, Heidelberg (1998)
7. Agrawal, R., Evfimievski, A., Srikant, R.: Information Sharing Across Private Databases. In: The 2003 ACM SIGMOD Int'l Conf. Management of Data, pp. 86–97 (2003)
8. Goethals, B., Laur, S., Lipmaa, H., Mielikainen, T.: On Private Scalar Product Computation for Privacy-preserving Data Mining. In: 7th Annual International Conference in Information Security and Cryptology, pp. 104–120 (2004)
9. Kantarcioglu, M., Clifton, C.: Privacy-preserving Distributed Mining of Association Rules on Horizontally Partitioned Data. IEEE Trans. Knowledge and Data Engineering 9(16), 1026–1037 (2004)

Alleviating Cold-Start Problem by Using Implicit Feedback

Lei Zhang, Xiang-Wu Meng, Jun-Liang Chen, Si-Cheng Xiong, and Kun Duan

State Key Laboratory of Network and Switching Technology,
Beijing University of Posts and Communications,
100876 Beijing
jhon565@yahoo.com.cn

Abstract. Collaborative Filter (CF) methods supply favorably personalized predictions relying on adequate data from users. But the ratings, of new users or about new items are not always available and CF can't make a precise recommendation in this case. In our paper, we present our consideration on alleviating cold-start problem by using users' implicit feedback data, which is not the same as the traditional methods which focus completely on the sparse data. To exploit the significance of users' implicit feedback for alleviating cold-start problem, we present two independent strategies—the neural network-based M1 method and the collaboration-based M2 method, by which the significance of users' implicit feedback for cold-start recommendation has been preliminarily demonstrated.

Keywords: cold-start recommendations, implicit feedback, collaborative filtering.

1 Background and Related Work

Among the several novel recommendation methods, CF recommendation [1][2] is proved to be the most promising technologies. Even though CF is hard to beat by other technologies once abundant rating data is available, it can not supply satisfied results when rating data is sparse, for instance, the new users or new items in system. Recommendation with scarce user ratings is defined as cold-start recommendation. Actually, Cold-start problem is intractable and researchers didn't supply any convergence measures of the cold-start recommendation their system generated [3]. A new similarity method [4] to alleviate the cold-start problem is presented by Ahn, which clearly outperforms the classical COS or COR method when the data is very sparse. A Bayes-based technique [5] is proposed by Andrew for the purpose of solving the cold-start problem. Even though some ideas are brought forward, the effects are not as good as expected. Actually, the current methods are mainly focusing on the very finite data itself but the initial ratings are so little that even though all the information is extracted, it is far from enough to reflect one user's tastes for satisfied recommendation, setting aside the fact that it's really hard to distinguish which information is effective or not when data is sparse. In another aspect, using implicit feedback—Clickthrough data, to enhance Information Retrieve precision is an effective way to enhance IR performance, which is a rewarding reference for our work. Actually, from

R. Huang et al. (Eds.): ADMA 2009, LNAI 5678, pp. 763–771, 2009.

a more practical point, new users in recommendation system generally don't like to give too many ratings, because it is too boring to constantly find what movies they've watched from a large amount of films. As a more natural way and before giving too many ratings, one (new) user would prefer to have a browse to movies presented—a public recommendation or some new movies, etc., may be proper, to decide which movie(s) is/are the best he wants to watch now, or give check mark(s) for the movie(s) he'll purchase later, etc. We collect these so called Wishlist data (the favored data, and certainly, the data browsed by user but not interested in by him should also be recorded) implemented in http://movielens.umn.edu, which is very valuable to overcome the scarcity of useful information in initial stage and acquired naturally just like getting Clickthrough data in search engine utility.

In our paper, we present the preliminary exploration for alleviating cold-start problem by using user's implicit feedback data, which is not the same as the traditional methods which focus completely on the sparse data. We present two independent strategies to exploit the significance of users' implicit feedback for cold-start problem, the neural network-based M1 method and the collaboration-based M2 method. Specifically, we use M1 to learn the feedback data itself, to mine users' preferences to such factors as item slots, etc. from the relatively "superiority" or "inferiority". For M2 method, we make the basic but effective transformation for the available data, by which the similar information will be skillfully extracted from the implicit feedback and item ratings which are of no comparability originally. In most cases, M2, belonging to collaborative filtering category, will be more effective for item recommendation than M1 which belongs to the content-based analysis category and the significance of users' implicit feedback for cold-start recommendation has been preliminary demonstrated.

2 The Neural Network-Based M1 Method

2.1 Preliminary Knowledge

As mentioned above, we divide items browsed by user U into two sets, W_u and R_u namely, the same idea as Joachims [6]. Data in W_u is taken from Wishlist of U, with R_u as a set of data browsed but outside W_u. We define set $<_r^u$ as below:

$$<_r^u = \{ < x_p, x_q > \mid \forall x_q \in W_u, \forall x_p \in R_u \} \quad (1)$$

where $< x_p, x_q > \in <_r^u$ means U prefers x_q to x_p. Then the problem becomes learning $U's$ preferences using $<_r^u$. We take an intuitive and basic idea from Burges [7] to utilize the partially ordered pairs. That is, suppose $< x_p, x_q > \in <_r^u$ and network outputs for x_p and x_q are o_p and o_q, then the larger is $o_p - o_q$, the larger the cost should be made for modification, with one element in pair as the reference.

Now we first take a look at the process of one user choosing fruit in supermarket. Taking apple purchase for example, different users have different preferences, some paying their first attention to apple type, with color& size as the second, some focusing on color & size, but neglecting apple type, etc. For specific user V , he chooses apples in the following way. Focusing mainly on apple type, he may first select Fuji, and then with his second attention to color & size, he may choose the big and red ones from Fuji apples. Lastly, as for the producing area, maybe Australia is better, but other area is acceptable either, for he doesn't care much about this attribute. Apples rated highly by V should have a good type and color &size, to get high ranking. In other words, only the apples, each with a synthesized high score or high ranking based on users' preferences toward attributes, will be selected at last.

Therefore, motivated by content-based methods and much like the above decision-making process, to finally decide which ones are the best, we give the preference model of user U towards movies using the following definition.

Let $\overline{f} =< c_1, c_2, \ldots, c_m >$ be the attribute/slot vector for items—movies, for example, and $\forall c_k \in \overline{f}$, $\upsilon_{k,\sigma_e} = \{v_{k1,\sigma_e}, v_{k2,\sigma_e}, \ldots, v_{kz,\sigma_e}\}$ be feature set of item σ_e corresponding to attribute c_k , and then we can get preference degree p of U towards σ_e by formula (2).

$$p_{\sigma_e} = \sum_{k=1}^{m} w_k \sum_{s=1}^{z} \zeta_{ks,\sigma_e} \tag{2}$$

where $w_k \in \overline{\omega} = < w_1, w_2, \ldots, w_m >$ is the interest degree of U towards corresponding attribute in $\overline{f}$, ζ_{ks,σ_e} is the preference corresponding to $v_{ks,\sigma_e} \in \upsilon_{k,\sigma_e}$. In the following, based on BP neural network, we'll take advantage of classical steepest gradient descent rule to effectively modify corresponding weights (preferences) such as w_k and ζ_{ks,σ_e} . Actually, by adding up the preference degrees (weights) of all features in one movie to present target user's preferences towards this movie, many content-based recommendation methods modify the feature weights (preference degrees) using some simple strategies——for every movie, enlarge or narrow each feature weight using the same factor each time (that is, suppose constant α is the proportion, for every modification of each movie, each feature weight of the movie will be multiplied by α or each one of the movie will be divided by α), or just add (or subtract) each feature's tf-idf value to (from) the previous weight of this feature each time. Intuitively, comparing with these methods, our neural network method based on preference model has its effectiveness and we'll present the explicit explication on our M1 method in the coming section.

2.2 The Neural Network-Based M1 Method

With the preference model presented in section 2.1, every feature of attribute c_k is chosen as one input node of kth node in hidden layer and hidden layer nodes basically denote item attributes. Then obviously, taking the forward prop results which are good presentation of rating outputs, the key to this problem lies in back-prop process (exploring solution space), to convert feedback data to user preferences for better rating predictions. Actually, with preference model especially aiming at this specific scene, the mentioned architecture is light-weighted but very effective comparing with a general 2-layer network which just inputs all features of items to each node of hidden layer with generally empirical but not fully qualified hidden node number and inevitably posing the learning complexity with intricate characteristics unsuitable to the relatively sparse data in cold-start stage. With corresponding results of forward prop well presenting rating outputs and by taking advantage of back-prop relative error to constantly modify acquired tastes, it is possible for us to make full of user's implicit feedback by converting feedback date into preference weight vector, the valuable data for further prediction. Here, we firstly present network output of forward prop for item σ_e in formula (3).

$$o_{\sigma_e} = f(\sum_{j=1}^{s^1} w_{1,j}^2 \cdot g(\sum_{i=1}^{s^0} w_{j,i}^1 \cdot a_{j,i}^{0,e} + b_j^1) + b_1^2) \equiv o_e \tag{3}$$

where $f(\cdot), g(\cdot)$ is the transfer function, s^1 and s^0 in our model are the number of item attributes and the number of features of corresponding attribute, and for the offsets b, weight w or output a (a gets its nonzero value, as 1 only if corresponding feature of attribute occurs correspondingly in item σ_e), the upper (left) index indicates the node layer, the lower (left) one of w or b indicates the corresponding node, the subscript i of $w_{j,i}^1$ denotes the ith input, and the lower indices of a denote output $a_{j,i}^{0,e}$ (of $0th$ layer) is the ith input of jth node (of $1st$ layer), with the superscript e denoting that σ_e is the eth one in item set.

Then, denoting the hth and eth training items as σ_h and σ_e, with corresponding net outputs as $o_h and o_e$, we define one judgment function $\phi(\cdot)$ as follows.

$$\phi(\cdot) = \tau \cdot (o_h - o_e) \tag{4}$$

where $\tau = \begin{cases} 1 & \sigma_h \in W_u, \sigma_e \in R_u \\ -1 & \sigma_h \in R_u, \sigma_e \in W_u \\ 0 & otherwise \end{cases}$, and once $\phi(\cdot) < 0$, back-prop occurs. Note that no ideal output will be supplied in the actual situation, so the classical neural network isn't suitable to be applied here directly. However, taking advantages of the reversion punishment idea mentioned in section 3.1, the deviation could be measured by the distance (the monotonic increasing distance function, denoted as $\hat{F}$) between o_h and o_e , the larger is $\hat{F}$, the larger the compensation once disorder occurs, which ensures the proper direction for the following searching.

Here, we define transfer function of output layer as linearity function and the one of hidden layer as sigmoid function, which satisfies our preference-based model properties naturally, with the biggest curve slopes about origin, a gradually slope decreasing but still keeping the function monotonic increasing and the function value no more than 1. Consequently and more explicitly, we present the BP rule according with our model in the following formula (5), (6).

$$\begin{aligned}\Delta w^2_{1,j} =& -\eta_2 \Big(f\big(\sum_{j=1}^{s^1} w^2_{1,j}\cdot g(\sum_{i=1}^{s^0} w^1_{j,i}\cdot a^{0,h}_{j,i}+b^1_j)+b^2_1\big)-f\big(\sum_{j=1}^{s^1} w^2_{1,j}\cdot g(\sum_{i=1}^{s^0} w^1_{j,i}\cdot a^{0,e}_{j,i}+b^1_j)+b^2_1\big)\Big)\\ &\Big(f'\big(\sum_{j=1}^{s^1} w^2_{1,j}\cdot g(\sum_{i=1}^{s^0} w^1_{j,i}\cdot a^{0,h}_{j,i}+b^1_j)+b^2_1\big)\cdot g(\sum_{i=1}^{s^0} w^1_{j,i}\cdot a^{0,h}_{j,i}+b^1_j)-f'\big(\sum_{j=1}^{s^1} w^2_{1,j}\cdot g(\sum_{i=1}^{s^0} w^1_{j,i}\cdot a^{0,e}_{j,i}+b^1_j)+b^2_1\big)\cdot\\ &g(\sum_{i=1}^{s^0} w^1_{j,i}\cdot a^{0,e}_{j,i}+b^1_j)\Big)=-\eta_2\Big(\big(\sum_{j=1}^{s^1} w^2_{1,j}\cdot g(\sum_{i=1}^{s^0} w^1_{j,i}\cdot a^{0,h}_{j,i}+b^1_j)+b^2_1\big)-\big(\sum_{j=1}^{s^1} w^2_{1,j}\cdot g(\sum_{i=1}^{s^0} w^1_{j,i}\cdot a^{0,e}_{j,i}+b^1_j)+b^2_1\big)\Big)\\ &\Big(g(\sum_{i=1}^{s^0} w^1_{j,i}\cdot a^{0,h}_{j,i}+b^1_j)-g(\sum_{i=1}^{s^0} w^1_{j,i}\cdot a^{0,e}_{j,i}+b^1_j)\Big)\end{aligned} \tag{5}$$

$$\begin{aligned}\Delta w^1_{j,i} =& -\eta_1\Big(\big(\sum_{j=1}^{s^1} w^2_{1,j}\cdot g(\sum_{i=1}^{s^0} w^1_{j,i}\cdot a^{0,h}_{j,i}+b^1_j)+b^2_1\big)-\big(\sum_{j=1}^{s^1} w^2_{1,j}\cdot g(\sum_{i=1}^{s^0} w^1_{j,i}\cdot a^{0,e}_{j,i}+b^1_j)+b^2_1\big)\Big)\\ &\left(\frac{a^{0,h}_{j,i} w^2_{1,j}\exp(-(\sum_{i=1}^{s^0} w^1_{j,i}\cdot a^{0,h}_{j,i}+b^1_j))}{(1+\exp(-(\sum_{i=1}^{s^0} w^1_{j,i}\cdot a^{0,h}_{j,i}+b^1_j)))^2}-\frac{a^{0,e}_{j,i} w^2_{1,j}\exp(-(\sum_{i=1}^{s^0} w^1_{j,i}\cdot a^{0,e}_{j,i}+b^1_j))}{(1+\exp(-(\sum_{i=1}^{s^0} w^1_{j,i}\cdot a^{0,e}_{j,i}+b^1_j)))^2}\right)\end{aligned} \tag{6}$$

3 The Collaboration-Based M2 Method

In this section, we'll present the collaboration-based M2 method to learn users' implicit feedback. For traditional recommendations, users give their ratings (e.g. 1-5) and CF methods could work well to calculate the similarities between users and then present rating prediction. However, we can't use CF method directly here because the new users' implicit feedback is the "relative superiority or inferiority", and we can't make a straightforward versus between feedback data and users' ratings. Here, we make the basic but effective transformation for the available data, by which the similarities will be skillfully drawn from the implicit feedback and item ratings which are of no comparability originally. And then, the rating prediction will be performed.

With the definition in part 2, suppose that new user U is the target user, and then we get set $<_r^u = \{<\sigma_p, \sigma_q> | \sigma_q \in W_u, \sigma_p \in R_u\}$ according to U 's feedback and $<\sigma_p, \sigma_q> \in <_r^u$ which means U prefer σ_q to σ_p. For user V (V is any user of recommendation system but not a new user, and we can make full use of his relatively adequate ratings), suppose his rating set is R^v and the item set is ϖ^v, with each $\sigma_e \in \varpi^v$ corresponding to $r_e^v \in R^v$. Firstly, according to V 's rating, we build set $<_r^v = \{<\sigma_f, \sigma_g> | r_f^v < r_g^v, \sigma_f \in \varpi^v, \sigma_g \in \varpi^v\}$. Next, we perform the following operations on $<_r^u$ and $<_r^v$ to build new sets $<_r^{'u}$ and $<_r^{'v}$.

Set $<_r^{'u}$ is defined as following. $\forall <\sigma_{p'}, \sigma_{q'}> \in <_r^{'u}$, there is $(<\sigma_{p'}, \sigma_{q'}> \in <_r^u) \wedge ((<\sigma_{p'}, \sigma_{q'}> \in <_r^v) \vee (<\sigma_{q'}, \sigma_{p'}> \in <_r^v))$. Correspondingly, set $<_r^{'v}$ is defined as follows. $\forall <\sigma_{f'}, \sigma_{g'}> \in <_r^{'v}$, there is $(<\sigma_{f'}, \sigma_{g'}> \in <_r^v) \wedge ((<\sigma_{f'}, \sigma_{g'}> \in <_r^u) \vee (<\sigma_{g'}, \sigma_{f'}> \in <_r^u))$.

In the following part, according to $<_r^{'u} \subset <_r^u$, we get U 's $|<_r^{'u}|$-dimension multi-tuple $\nabla^u = (<\sigma_{a'}, \sigma_{b'}>, \ldots, <\sigma_{p'}, \sigma_{q'}>, \ldots, <\sigma_{c'}, \sigma_{d'}>)$ which satisfies that for each element $<\sigma_{p'}, \sigma_{q'}> \in \nabla^u$, there is $<\sigma_{p'}, \sigma_{q'}> \in <_r^{'u}$. Taking target user U 's preferences as basic reference, we define function $\iota^u(\cdot)$ as below. $\forall <\sigma_{p'}, \sigma_{q'}> \in \nabla^u$, there is $\iota^u(<\sigma_{p'}, \sigma_{q'}>) \equiv 1$. Then, we get the ordered multi-tuple $\angle^u = (1, \ldots, 1, \ldots, 1)$, that is, let $<\sigma_{p'}, \sigma_{q'}> \in <_r^{'u}$ be the kth element in ∇^u, then the kth element in $\angle^u$ gets its value by $\iota^u(<\sigma_{p'}, \sigma_{q'}>) = 1$. Next, we build user V 's ordered multi-tuple $\angle^v$ as below. For each element $<\sigma_{p'}, \sigma_{q'}> \in <_r^{'u}$, suppose $<\sigma_{p'}, \sigma_{q'}>$ is the kth element in ∇^u, and then the value of kth element in $\angle^v$ can be acquired by the following function $\iota^v(\cdot)$ (7).

$$\iota^v(<\sigma_{p'}, \sigma_{q'}>) = \begin{cases} 1, & <\sigma_{p'}, \sigma_{q'}> \in <_r^{'v} \\ 0, & <\sigma_{p'}, \sigma_{q'}> \notin <_r^{'v} \end{cases} \quad (7)$$

Now we have finished the basic data transformation, after which we can effectively explore the users' similarities using target user's implicit feedback and the others' actual ratings which are of no comparability originally.

With the multi-tuples obtained above, the similarity between target user U and any user V can be measured using cosine vector-based approach [8] as below. For target user U, with his top-N similar neighbors $\{V_1, V_2, ..., V_N\}$ in hand, we could get the rating prediction of item σ_h by the memory-based formula [9].

$$sim(U,V) = \frac{\angle^u \cdot \angle^v}{|\angle^u||\angle^v|} \tag{8}$$

$$r_h^u = \frac{\sum_{i=1}^{N} sim(U,V_i) \times r_h^{v_i}}{\sum_{i=1}^{N} |sim(U,V_i)|} \tag{9}$$

4 Experimental Evaluations

4.1 Data Set and Experimental Setup

We choose Movielens (http://www.movielens.umn.edu) data set for our experimental evaluations. Moreover, we use data from IMDb (The Internet Movie Database, http://www.imdb.com) to get features for movies in Movielens.

We take *pairwise accuracy* [10] as evaluation metric. For each user U, we define two sets X and Y below.

$X = \{<S_p, S_q> | S_p \in H, S_q \in H, r_{s_p} < r_{s_q}\}$, $Y = \{<S_e, S_f> | S_e \in H, S_f \in H, r'_{s_e} < r'_{s_f}\}$,

where H is the set of movies used for test, r_{s_p} and r_{s_q} are human ratings of movie S_p and S_q, while r'_{s_e} and r'_{s_f} are predicted ratings of movie S_e, S_f, and then the *pairwise accuracy* can be given in formula (10).

$$pairwise\ accuracy = \frac{|X \cap Y|}{|X|} \tag{10}$$

We select 50 users whose rating numbers are larger than 200 as the target users with the capacity of supplying abundant partially ordered pairs. For each target user U, suppose his whole rating number is Z and we randomly select M ratings to build set $<_r^u = \{<S_c, S_d> | (S_c, S_d \in T_M^u) \wedge (r_{S_d} >= 4) \wedge (r_{S_c} \leq 2)\}$, where T_M^u is the set of M ratings and r_{S_c} and r_{S_d} are U 's ratings towards S_c and S_d respectively. The left $(Z - M)$ ratings are used for his test set and the actual training set is the subset of $<_r^u$. For M1, suppose the maximal pair number (for one user) used for training is 300, and then we set ANN scale (input node number) according to the number of the whole features of movies, each movie occurring in the 300 pairs of that user.

4.2 Experimental Results

In practice, one user will browse the movie candidates at a very high speed when selecting the ones he is interested in. Suppose user U select 2 movies from 100 candidates in his whole browsing process, then the number of partially ordered pair will be N =198. If user U repeats the similar operations, the pair number will be larger than the above-mentioned. In our experiment, we set the number of partially ordered pairs as N =0, N =20, N =100 and N =300 (M can change) to present the *pairwise accuracy* values of M1 and M2 method in table 1. In comparison, the *pairwise accuracy* of classical CF method (when the rating number is no more than 20) is shown in table 2.

Table 1. The pairwise accuracy of M1 and M2 method with different parameter values

accuracy	$N = 0$	$N = 20$	$N = 100$	$N = 300$
M1	0.4	0.49711	0.53498	0.54753
M2	0.4	0.69122	0.70247	0.70878

Table 2. The pairwise accuracy of User-based CF method

accuracy	$K=0$	$K=2$	$K=5$	$K=10$	$K=15$	$K=20$
CF	0.4	0.4475	0.58533	0.65083	0.67025	0.68549

It's not hard to see that the *pairwise accuracy* of completely random prediction is 40%. Obviously, M2 has made a better performance than CF method. Without any actual ratings, the *pairwise accuracy* of M2 gets constant improvement as N increases. Specially, when the pair number is relatively limit initially, M2 accuracy gets major enhancement as the pair number increases. Then, with further increase of the pair number, the accuracy rise slows down. For example, as the pairs increase from 100 to 300, the accuracy is basically maintained at the scope of 0.70-0.71. M2 has made a better performance than CF method when ratings of CF are relatively sparse, and it is not hard to see that the rational utility of implicit feedback is very helpful to alleviate cold-start problem. As for M1, *pairwise accuracy* holds the trend of constant rise with the increase of pair number. However, even thought the pair number has reach 300, M1 accuracy still stays at 0.54735. Therefore, M1 has a positive influence on cold-start recommendation, but the effect is not obvious enough. M1 belongs to the scope of content-based analysis and it has obvious performance gap with M2 belonging to the scope of collaboration method with relatively abundant pairs. Actually, in the non-cold-start stage, CF precision is much higher than that of content-based method for current progress. Up until now we have finished our main consideration on implicit feedback for cold-start problem. In the following part, we'll present simple discussion on new item problem. With few users' ratings to one new item, CF methods have difficulties in recommendation. Adopting the strategy of content-based analysis, M1 can obtain better performance than M2 in the situation. We can suppose one of the utility scenes of M1 as follows. We want to get the prediction of a new movie for the user who joins in the system newly. With very few ratings of the new item available, we can relatively effective to make sure whether the movie is worthy to be recommended by M1 method by using feedback data. With 300 pairs in hand, for example, the prediction accuracy can reach 0.54753, which is much higher than random. After all, the content-based method will not suffer from the new item problem.

5 Conclusions

In this paper, we presented our considerations on alleviating cold-start problem by making full use of implicit feedback data, rather than omitting them, even through the ratings are far from enough in the beginning like traditional methods. By the presented strategies—M1 and M2 method, we has preliminarily demonstrated the significance of users' implicit feedback on cold-start recommendation and we believe that using implicit feedback data is one promising research direction for improving cold-start recommendation.

Acknowledgement. The presented work was supported by the National Natural Science Foundation of China under Grant No. 60872051.

References

1. Resnick, P., Iacovou, N., Suchak, M., Bergstrom, P., Riedl, J.: GroupLens: An open Architecture for Collaborative Filtering of Netnews. In: CSCW 1994, pp. 175–186. Chapel Hill (1994)
2. Nakamura, A., Abe, N.: Collaborative Filtering Using Weighted Majority Prediction Algorithms. In: 15th Int'l Conf. Machine Learning, San Francisco, pp. 395–403 (1998)
3. Salter, J., Antonopoulos, N.: CinemaScreen recommender agent: combining collaborative and content-based filtering. J. IEEE Intelligent Systems 21(1), 35–41 (2006)
4. Ahn, H.J.: A new similarity measure for collaborative filtering to alleviate the new user cold-starting problem. J. Information Sciences 178(1), 37–51 (2008)
5. Schein, A.I., Popescul, A., Ungar, L.H., Pennock, D.M.: Methods and Metrics for Cold-Start Recommendations. In: 25th annual international ACM SIGIR conference on Research and development in information retrieval, New York, pp. 253–260 (2002)
6. Joachims, T.: Optimizing search engines using clickthrough data. In: The SIGIR Workshop on Mathematical/Formal methods in Information Retrieval, New York, pp. 133–142 (2002)
7. Burges, C., Shaked, T., Renshaw, T., Lazier, A., Deeds, M., Hamilton, N., Hullender, G.: Learning to Rank using Gradient Descent. J. Computer and System Sciences 50, 32–40 (1995)
8. Breese, J.S., Heckerman, D., Kadie, C.: Empirical analysis of predictive algorithms for collaborative filtering. In: 14th Conference on Uncertainty in ArtificialIntelligence, Stockholm, pp. 43–52 (1998)
9. Delgado, J., Ishii, N.: Memory-Based Weighted-Majority Prediction for Recommender Systems. In: ACM SIGIR 1999 Workshop Recommender Systems: Algorithms and Evaluation (1999)
10. Richardson, M., Prakash, A., Bill, M.: Beyond PageRank: machine learning for static ranking. In: 15th Int'l Conf. WWW 2006, New York, pp. 705–715 (2006)

Learning from Video Game: A Study of Video Game Play on Problem-Solving

Xue-Min Zhang*, Zijiao Shen, Xin Luo, Chunhui Su, and Jiaqi Wang

School of Psychology and Beijing Key Lab of Applied Experimental Psychology,
State Key Lab of Cognitive and Neuroscience and Learning, Beijing Normal University,
100875 Beijing
xmzhang@bnu.edu.cn

Abstract. This research was intended to explore the influence of video game experience on problem representation, efficiency of strategies, meta cognition, and quality of mental models during solving the problems that encounter in a new game. This experiment asked all the participants to keep thinking aloud during the 20-minite process of playing, and their verbal report was analyzed to study the problem solving. The results indicate a significant influence of computer games on game players' performance in an unfamiliar game. The most frequently referred comment is direct strategy, next are game rules and cues, monitoring and game progress. Expert players performed better than novice players on problem representation, efficiency of strategies, and meta cognition.

Keywords: Video Game Play, Expert players, Novice players, Problem solving.

1 Introduction

1.1 The Computer Game and Related Learning Studies

With the development of internet technology, computer video game as a new type of entertainment and competition has been acctracting more and more attention. According to the quarterly reports of video game market in China, there were 27 million of video game players in China in 2006, twenty-nine percent of whom played over 60 hours each month.

As many adolescents spend a lot of time playing video games on the computer and even become addicted to them, parents and school teachers start to worry about the negative influence on those adolescents. People held that playing too much computer games can not only cause depression and violent behavior in adolescents, but also do harm to their physiological, cognitive and social development and students who play video games are no longer interested in their school study any more(Li, Zhou, and Zhu; 2007).

Since video game is becoming more and more popular, it is impossible to treat it as some kind of evil thing and get rid of it from our life. Some work(e.g., Wu and Li;

* Corresponding Author.

R. Huang et al. (Eds.): ADMA 2009, LNAI 5678, pp. 772–779, 2009.

2008) had been done to find out the value of video game, such as the learning factors involved in it. Actually, the computer video games were described by Klabbers (1999) as actions involving various technical knowledge and skills. Generally, Playing video games provides the possibility to improve cognitive abilities of players, because game-playing is the process of solving problems by using some cues and following specific rules (Zhang and Lei; 2006).

1.2 Studies on Problem Solving and Cognitive Skills

In 1950s, the theory of information processing made further progress in the realm of problem solving(Newell, Shaw and Simon,1957). Anderson (1992) divided the development of cognitive skill into three stages: the declarative stage, the knowledge compilation stage and the procedural stage. Glaser, Lesgold, and Lajoie (1985) proposed 6 dimensions of cognitive skills. (1) the structure and organization of knowledge; (2) the depth of problem representation; (3) the quality of mental models; (4) the efficiency of procedures or methods; (5) the automaticity of cognitive performance; (6) The meta-cognitive skills. Experts were better in planning, monitoring, and self regulation.Compared with traditional evaluation methodology of behaviorism, Glaser's theory could access the cognitive skills from the aspect of individual development. That's why we choose Glaser's evaluation methodology.

1.3 Further Studies Based on the Learning Video from Game

In the field of computer video game, the related research on problem-solving and cognitive skills was mainly focused on the so called "impasse-negotiation". Cognitive psychologists, such as VanLehn (1994, 2003) and his colleagues, through studying the differences among computer video game players on cognitive skills, suggested that in the process of problem-solving, impasse cognition could help to advance individual study. In few studies on the difference of strategies using between expert and novice players, Hong and Liu (2003) found out that expert players manifested greater analogical thinking than novice players who manifested more trial and error over the game playBlumberg and Sokol L.M. (2004) divided the strategies that players used while playing computer video games into the internal-based strategies and the external-based strategies. Blumberg etc. (2008) made use of the think-aloud method to investigate the different performances of expert players and novice players while playing computer video games. Findings turned out that when encountered with the impasse in the video game, expert players made significantly greater reference to insight and strategies than novice players in their verbal report. Over the course of the game play, both expert players and novice players were making more and more reference to the insight and strategies in their verbal report. This paper is a model of both quantitative and qualitative research using the think-aloud method to investigate the game play process.

According to the analysis above, we carried out this research also using the think-aloud paradigm and re-designed the related data classification standard based on Glaser's theories on the evaluation of cognitive skills.

1.4 The Hypothesis of Present Study

We regarded playing computer video games as a process of problem-solving for game players who involved with the usage of different cognitive skills and strategies. Then, based on Glaser's theories on the evaluation of cognitive skills, we could investigate into the influence of computer games on game players' cognitive skills from the developmental point of view.

Specifically, we drew up to use the think-aloud method and retrospective interview, combining the quantitative and qualitative analysis, to find out the influence of computer video games on participants' cognitive-skill development, including problem representation, efficiency of strategies, meta cognition, and quality of mental models etc.. We coded and made analysis of the verbal report of participants' to compare the differences between expert players and novice players on the items of the above-referred cognitive skills.

On the whole, the research assumption of this experiment was: during playing, compared with novice players, expert players could make problem representation according to the essence characteristic of problems, use strategies more accurately, directly and effectively and thus, could manifest to be more well-planed, self-monitored and more effective on self-regulation and finally, could construct more complicated mental models and use them to guide their game behaviors.

2 Method

2.1 Participants

Participants included 9 expert and 11 novice adult video game players. At the beginning of the experiment, all potential participants were asked to complete a questionnaire to identify their experience with video games (9 expert players who presently played video games more than once a week and at least one half hour one time; 11 novice players who played video games less than once a week and less ten minutes one time, This criteria had been used in prior studies by Blumberg in 1998, 2000 and 2008). The age of all participants ranged from 19 to 22 years of age. All participants got payment after the experiment was finished.

2.2 Procedure

There were two phases in the experiment: the first phrase was training of thinking aloud and the second phrase was formal experiment of playing video game with thinking aloud while playing. During the first phase of the study, participants were trained to ''think aloud'' while solving a basic algebra problem. The ''think aloud'' procedure is a methodological technique (Chi, 1997; Van Someren, Barnard, & Sandberg, 1994) for elucidating problem-solving strategies used in such diverse settings in related historical documents (Wineburg, 1991, 1998; Hong & Liu, 2003) and video game study (Blumberg et al., 2008). During the experimental phase, each participant individually played a non-violent adventure game for 20 minutes developed by Sega in the early 1990s called "**Sonic the Hedgehog 2 for Game Gear**" This game contains eight different zones, which are comprised of three acts per zone. On the third act of each zone is a "boss" or difficult obstacle that must be overcome to continue to the next zone.

All the participants' playing processes and their thinking aloud behaviors were audio taped for later transcription, and all verbal comments were coded. The following dependent variables were used to evaluate actual game performance: (1) highest number of levels completed; (2) highest level attained; (3) number of Sonics lost; and (4) number of games started. These variables were recorded by the experimenter for each participant on an individual scoring sheet. After 20 minutes, participants were asked five questions. 1) To classify the problems they met during playing and recall the solutions. 2) To evaluate their planning and monitoring skills. 3) To assess the validity of the strategies. 4) To recall how they regulate their strategies to fit the conditions in the game. 5) To assess their performance as a whole. Finally, they were asked to assess how much they liked the game on a five point scale (1 = disliked very much; 5 = liked very much). Responses to these questions were also audio-taped.

3 Results

The findings presented below primarily involve the game performance of all participants, their comments made during the think-aloud, and the relationship between comments made and game performance.

3.1 The Classification Standard of Verbal Report

Transcripts of all participants' comments during video game play were prepared by the second author and verified by an independent reader using the original audiotapes. Comments were labeled by act. And from Act 1 to the highest Act that the player passed was encoded separately in document format. In accordance with Chi's (1997) method for quantifying verbal reports and Blumberg and her colleague's (2008) study, each transcript was segmented into meaningful units for coding which the smallest possible grouping of words that conveyed meaning in the game context.

A coding scheme was devised to characterize the meaningful unit within these transcripts. The verbal report was divided into five types. (1).***Game oriented*** : This category included reference to game cues and rules, comments that reflect the affection towards the games, and prior experience. These comments were also sub-categorized into ***game rules and cues,*** which meaned reference to specific features of the game including essential game features, such as the landscape. ***game evaluations***, which referred to how much the player was enjoying the game. ***background knowledge***, which reflected prior experience with the game or one played before. (2).***Problem representation :*** This category included reference to the representation of the problems according to the superficial or essential cues. These comments were also sub-categorized into ***superficial representation and Deep representation***, meaning representing problems by the non essential or essential cues while solve the problem。 (3).***Meta cognition:*** This category included reference to plans made during the game, self-monitoring and self-regulation. These comments were categorized as ***short term plan***, ***Long term plan, Monitoring***, ***self Regulation, Game progress***, and ***Unexpected action*** (4).***Strategy oriented*** This category included reference to strategies adopted in the process of playing. These comments were categorized as ***direct strategy***, which means reference to solving a problem directly. ***Indirect strategy***, meaning trying to find a solution through trial and error. (5).***Unable to code,*** which were characterized coding unit that did not fit under any of the above categories and were irrelevant to the goals of the task in present study.

3.2 Game Performance

Participants' performance while playing was characterized by four indicators: scores, number of levels completed, number of Sonics lost, and the game preference.

Table 1. Means (percentage) for video game experience by game performance

Performance	**Expert Players**	**Novice players**	**t**	**p**
score	17734.44	7376.37	2.428	0.026*
number of Sonics lost	4.00	6.09	-2.194	0.042*
number of levels completed	3.56	2.91	1.094	0.289
game preference	3.2	2.91	0.456	0.654

* ***p < 0.05.***

An independent samples t-test of scores of expert and novice video game players yielded a significant difference ($t_{(18)}$=2.428, P<0.05), whereby expert players outperformed novice players. An independent samples t-test of the number of Sonics lost of expert and novice video game players indicates a significant difference ($t_{(18)}$=-2.194, p<0.05), whereby expert players suffer less lost. No significant difference of number of levels completed is found ($t_{(18)}$=1.094, p>0.05). Participants' game preference failed to indicate a significant difference between expert and novice players (t=0.456, p>0.05). Overall, participants rated the game low in likeability

3.3 Comments during Game Play

According to descriptive results in table 2, we found that the most frequently referred comment was direct strategy, next were game rules and clue, monitor and game progress. The least referred comments were deep representation, background knowledge and unable code. In order to find the difference of the five categories which were characterized in the video game play, a series of non-parameter mean test of five categories comment percentage were done (see table 2), which yielded significant difference of game evaluations , deep representation, short term plan, long term plan, regulation and direct strategy . Expert players performed better on game evaluations, direct strategy, long term plan and deep representation than novice players. Novice players performed better on short term plan and regulation. No significant differences were found on other categories.

Table 2. Means (%) and non-parameter mean Mann-Whitney U test of comments

Strategy	Expert Players	Novice players	Means Difference	U-test	P
Game oriented					
game rules and cues	17.17	17.23	0.06	45.5	0.761
game evaluations	2.40	.9182	1.4818	19	0.019*
knowledge	.3667	.9545	-0.5878	46.5	0.797
Problem representation					
superficial	9.589	9.173	0.416	45.5	0.761
deep	1.020	.1000	0.92	26	0.028*
Meta cognition					
short term plan	2.056	8.355	-6.299	2	0.000**
long term plan	2.411	1.082	1.329	27	0.083
monitor	13.16	14.67	-1.51	40.5	0.494
regulation	2.211	4.146	-1.935	27	0.087
Game progress	13.54	12.65	0.89	44	0.676
Unexpected action,	1.644	4.746	-3.102	29	0.117
Strategy oriented					
direct strategy	29.20	13.67	15.53	2	0.000**
indirect strategy	3.989	5.655	-1.666	35.5	0.287
Unable to code	1.300	.8273	0.4727	46.5	0.811

$p < 0.05$., **$p < 0.01$

3.4 The Relationship between Comments and Performance

Participants' game preference failed to indicate a significant relationship with game score($r=0.203$, $p=0.379>0.05$). In consideration of the fact that the number of Sonics lost or levels completed was too limited to avoid range restrictive effect, we adopted game score as the standard to stand for the performance. After transforming the percent of each kind of comments and score attained by players to ranked data, we analyzed Spearman rank correlation coefficients. As shown in the table 4, game evaluations, short term plan, long term plan, direct strategy and self-regulation were significantly related with game performance.

4 Discussion

According to the statistics of comment, most was on direct strategy, next are game rules and cues, monitor and game progress. This was consistent with expectation. So we can say thinking aloud could help players to regulate more effectively and improve the attention paid to the solution of a problem. The least referred comments were related to deep representation, indicating that it was difficult to represent problems by the essential cues and be concentrated on the whole game. This result showed that 98.7% comments of all participants focused on the game task.

According to the performance records, expert players got higher scores and lost less lives than novice players, indicating that they did better in the game. In accord with previous studies, we found even when facing an absolutely unfamiliar game, an expert player's performance was better that a novice player, due to the impact of previous game experience. And the difference in performance proved the classification of subjects was reasonable. When considering the Problem representation, we found no difference in superficial representation between expert and novice players. But a significant different in deep representation shows that expert players did better in seeking solutions from the whole. When encountered some difficulties, they could take the influence of present problem on the whole game into consideration. This was in accord with Glaser's theory claiming that the experts were different with novices in deep representation.

For the meta-cognitive, expert game players made significantly more reference to the long-term plans than novice players did, while novice players reported more short-term plans than expert players. The result was consistent with Glaser's cognitive-skill evaluation theories on the difference of Meta cognition, particular of the planning ability, between experts and novices. We considered the result to be associated with the players' game-playing experiences. Expert players had rich experiences of game-playing and they could generally have a foresight of the game-development trends, thus, could focus more attention on overall objects, to make more long-term plans than novice players. However, with the lack of experience, novice players could not grasp the overall goal of the game and were only able to deal with the current situation and make small-step plans. As a consequence they made more short-term plans than expert players. Expert players used more direct strategies and long-term plans to solve the problem effectively, so they reported less behavior regulation.

As for the strategy, the descriptive statistics showed that novice players reported more about trial and error strategies than expert players, but the result did not reach statistically significant level. However, expert players made significantly more reference to the direct strategy than novice players, which was consistent with Hong and Liu's (2003) study results. We believed that expert players tended to use the direct strategy, because they could well definite the initial state and target state of the problem and could make a relatively straightforward analyze on the plan they use.

According to previous analysis, expert players performed better than novice players in terms of problem representation, strategy effectiveness and meta-cognition. We could conclude that expert players could build more complex mental models than novice players and could use the model to guide their game behavior.

5 Conclusion

From present study, we can conclude: (1).The experience of game playing had a significant impact on players' game performance while playing a novel game. And there was no significantly positive correlation between game performance and players' preference level of the game. (2).Both expert and novice game players made most reference to the direct strategy, game cues or rules and self monitor in their verbal report. (3).Game performance had a significantly positive correlation with players' verbal comments of game evaluations, long-term plans and direct strategies, but a significantly negative correlation with short-term plans and behavior regulations. (4).Expert game players reported significantly more than novice players on direct strategies, game evaluation, deep structural characterization of problems, and long-term plans. But novice players reported significantly more on short-term plans and behavior regulation. That is expert players performed better than novice on problem-solving: problem representation, strategy effectiveness and meta-cognition.

References

1. Anderson, J.R.: Skill Acquisition: Compilation of Weak-Method Problem solution. Psychology Review 94, 192–210 (1987)
2. Anderson, J.R.: Automaticity and the ACT Theory. American Journal of Psychology 105, 165–180 (1992)
3. Blumberg, F.C., Sokol, L.M.: Boys' and Girls' Use of Cognitive Strategy When Learning to Play Video Games. Journal of General Psychology 131(2), 151–158 (2004)
4. Blumberg, F.C., Rosenthal, S.F., Randall, J.D.: Impasse-driven learning in the context of video games. Computers in Human Behavior 24(4), 1530–1541 (2008)
5. Glaser, R., Lesgold, A., Lajoie, S.: Toward A Cognitive Theory for the Measurement of Achievement. In: Ronning, G., et al. (eds.) The Influence of Cognitive Psychology on Testing and Measurement, pp. 41–85. Erlbaum, Hillsdale (1985)
6. Greenfield, P.M., DeWinstanley, P., Kilpatrick, H., Kaye, D.: Action video games and informal education: Effects on strategies for dividing visual attention. Journal of Applied Developmental Psychology 15, 105–123 (1994)
7. Greenfield, P.M., et al.: Cognitive socialization by computer games in two cultures: inductive discovery or mastery of an iconic code Special issue: effects of interactive entertainment technologies on development. In: Interacting with video, Norwood, pp. 141–168 (1996)
8. Lisi, R.D., Wolford, J.L.: Improving children's mental rotation accuracy with computer game playing. The Journal of Genetic Psychology 163(3), 272–282 (2002)
9. Dan, L., Zhihong, Z., Dan, Z.: The Interrelation among Adolescent Behavior Problems, Computer Game Playing and Family Factors. Psychological Science 30(2), 450–453 (2007)
10. Royer, J.M., Cisero, C.A., Carlo, M.S.: Techniques and Procedures for Assessing Cognitive Skills. Review of Educational Research 63, 201–243 (1993)
11. Subrahmanyam, K., Greenfield, P.M.: Effect of video game practice on spatial skills in girls and boys. Journal of Applied Developmental Psychology 15, 13–32 (1994)
12. Sweller, J., Mawer, R.F.: Some examples of Cognitive task analysis with instructional implications. Aptitude, Learning and Instruction 1, 1–2 (1986)
13. VanLehn, K., Siler, S., Murray, C.: Why do only some events cause learning during human tutoring Cognition and Instruction 21, 209–249 (2003)

Image Classification Approach Based on Manifold Learning in Web Image Mining

Rong Zhu[1,2], Min Yao[1], and Yiming Liu[1]

[1] School of Computer Science & Technology, Zhejiang University, 310027 Hangzhou, Zhejiang
[2] School of Math & Information Engineering, Jiaxing University, 314001 Jiaxing, Zhejiang
{zr,myao,liuym}@zju.edu.cn

Abstract. Automatic image classification is a challenging research topic in Web image mining. In this paper, we formulate image classification problem as the calculation of the distance measure between training manifold and test manifold. We propose an improved nonlinear dimensionality reduction algorithm based on neighborhood optimization, not only to decrease feature dimensionality but also to transform the problem from high-dimensional data space into low-dimensional feature space. Considering that the images in most real-world applications have large diversities within category and among categories, we propose a new scheme to construct a set of training manifolds each representing one semantic category and partition each nonlinear manifold into several linear sub-manifolds via region growing. Moreover, to further reduce computational complexity, each sub-manifold is depicted by aggregation center. Experimental results on two Web image sets demonstrate the feasibility and effectiveness of the proposed approach.

Keyword: Web image mining, Data mining, Image classification, Dimensionality reduction, Manifold learning, Distance measure.

1 Introduction

With the rapid advances in computer technology and digital technique, cost reducing of available capturing devices and data storage, the amount of images on Internet has been extremely growing. These Web images of various kinds of real-world scenes, if analyzed, can provide plenty of useful knowledge to Internet users. Nowadays, image mining is rapid gaining attention among researchers in the fields of data mining, information retrieval, and multimedia database due to its ability to extract semantically meaningful information from images that may push various research fields to new frontiers [1]. Therefore, for effectively utilizing the image resources on Internet, efficient image mining systems for Web images are increasingly in demand. Classification is a basis for image mining and its results are very crucial for improving the execution performance of image mining application.

Researchers have made great efforts in developing various approaches to improve the accuracy of image classification. There has been a large quantity of prior work. A

R. Huang et al. (Eds.): ADMA 2009, LNAI 5678, pp. 780–787, 2009.

survey was provided in [2]. Here, we divide image classification approaches into two groups: image space-based approaches [3], [4], [5] and feature space-based approaches [6], [7], [8], [9], [10]. The former approaches use the visual features directly extracted from images in a data space, while the latter ones transform the features into a feature space by certain dimensionality reduction, and then realize classification based on the features in the feature space. That is, the high-dimensional features are mapped into the low-dimensional ones without losing critical image characteristics. Therefore, compared with image space-based approaches, these groups can reduce feature dimensionality and discard redundant information, especially suitable for dealing with the high-dimensional data in applications. Manifold learning is a recently developed technique [11] for dimensionality reduction, which provides a new way for solving the problem of image classification. Among current manifold learning-based methods, locally linear embedding (LLE) [12] is a representative one and has been used in image classification [10], [13]. However, most Web images from search engines (e.g., Yahoo! and Google) and photo management tools (e.g., Flickr and Picasa) are usually taken from real-world scenes with different imaging conditions, so the image data in an image set may be not well distributed. Therefore, if a whole manifold is constructed for all data in the set (only considering the relativity between data but ignoring the discrimination among categories) [10], not only the topological structure of the image data won't be correctly revealed but also the classifier's efficiency will be reduced.

In this paper, we propose a new feature space-based image classification approach, i.e., transforming the solving of image classification problem from a high-dimensional data space into a low-dimensional feature space (i.e., manifold). We propose an improved feature reduction algorithm via neighborhood optimization to reduce feature dimensionality. We also construct a new classifier based on distance measure and simplifying strategy. The proposed approach is evaluated on the task of Web image classification. Experimental results have demonstrated that our approach outperforms the competing ones. The rest of paper is organized as follows. An improved feature reduction algorithm is described in Section 2. In Section 3, we propose a novel classification method based on manifold learning. Experiments are given in Section 4. In the last section, we conclude and discuss possible future work.

2 Feature Reduction Based on GLLE

Inspired by geometric intuition, in this paper, we propose a new feature reduction algorithm based on LLE by using neighborhood optimization, which is called global-based locally linear embedding algorithm (GLLE). In GLLE, a regular neighborhood is constructed by globular radius instead of an irregular neighborhood via neighbor number in LLE. And then GLLE searches the candidate data within the globular neighborhood by using radius increment and the chosen data will be regarded as the nearest neighbors of current data. Therefore, GLLE meets the demand of shape preserving mapping that the nearer neighbors of current data in the high-dimensional data space make more contributions to the data reconstruction on the low-dimensional manifold.

2.1 Construction of Globular Neighborhood

Let $X = \{\vec{x}_1, \vec{x}_2, ..., \vec{x}_N\}$ ($\vec{x}_i = (\vec{x}_i^1, \vec{x}_i^2, ..., \vec{x}_i^D)^T; i = 1,2,..., N$) denotes N data in a data space (R^D) .They can be denoted as $Y = \{\vec{y}_1, \vec{y}_2, ..., \vec{y}_N\}$ on the manifold, where $\vec{y}_i$ $(i = 1,2,..., N)$ is a d -dimensional column vector ($d << D$). We assume that N data are distributed in R^D , taking $\vec{x}_i$ (current data) as globular core and using a given radius r to identify a globe as G , and then G is the globular neighborhood of $\vec{x}_i$. If there have p_i data within the neighborhood, they will be regarded as the nearest neighbors of $\vec{x}_i$ (each data is denoted as $\vec{x}_{ij}$ ($j = 1,2,..., p_i$)). We define the distance measure between $\vec{x}_{ij}$ and $\vec{x}_i$ based on L2 distance [14] as follow:

$$d_{ji} = dist(\vec{x}_{ij}, \vec{x}_i) = \left\| \vec{x}_{ij} - \vec{x}_i \right\|_{L2}, i = 1,2,..., N, j = 1,2,..., p_i . \tag{1}$$

We use N p_i -by-1 distance matrixes: $D_i = [d_{ji}]_{p_i \times 1} (i = 1,2,..., N)$ to express the neighbor number of each current data for convenience. It is worth pointing that different from LLE, the neighbor selection scheme in GLLE only relies on the distance between current data and candidate data, which alleviates its sensitivity to noise. Additionally, for precise similarity measure and short execution time, we introduce a radius increment Δr when implementing GLLE.

2.2 Outline of GLLE

The GLLE algorithm can be summarized as follows:

Step 1: For each $\vec{x}_i$, construct its globular neighborhood via a radius r and calculate d_{ij} between $\vec{x}_i$ and $\vec{x}_{ij}$ (use Eq. (1)). If there have p_i candidate data satisfy the condition: $d_{ij} < r$, these data will be chosen as the nearest neighbors of $\vec{x}_i$.

Step 2: For each $\vec{x}_i$, calculate the reconstruction weight w_{ij} based on its neighbors. Let W ($W = [w_{ij}]_{N \times N}$) denotes the local reconstruction weight matrix, then the reconstruction error function can be minimized as:

$$\varepsilon(W) = \arg\min \sum_{i=1}^{N} \left| \vec{x}_i - \sum_{j=1}^{p_i} w_{ij} \vec{x}_{ij} \right|^2 \quad s.t. \sum_{j=1}^{p_i} w_{ij} = 1 \, if \; \vec{x}_{ij} \in gnei(\vec{x}_i); else \; w_{ij} = 0 , \tag{2}$$

where $gnei(\vec{x}_i)$ denotes the globular neighborhood of $\vec{x}_i$.

Step 3: Calculate the low-dimensional embedding Y based on W and the nearest neighbors of each $\vec{x}_i$. To realize the shape preserving mapping, minimizing the embedding cost function as follow:

$$\varepsilon(Y) = \arg\min \sum_{i=1}^{N} \left| \vec{y}_i - \sum_{j=1}^{p_i} w_{ij} \vec{y}_{ij} \right|^2 \quad s.t. \sum_{i=1}^{N} \vec{y}_i = 0 \, and \sum_{i=1}^{N} \vec{y}_i \vec{y}_{ij}^{\,T} \Big/ N = I, \tag{3}$$

where $\vec{y}_i$ denotes the mapping of $\vec{x}_i$ on the manifold. $\vec{y}_{ij}$ ($j = 1,2,..., p_i$) denotes the nearest neighbors of $\vec{y}_i$. Eq.(3) can be rewritten in the following matrix form:

$$\varepsilon(Y) = \arg\min \sum_{i=1}^{N} \sum_{j=1}^{N} m_{ij} \vec{y}_i^{\,T} \vec{y}_{ij}, \tag{4}$$

where M ($M = [m_{ij}]_{N \times N}$) is a cost matrix, given as $M = (I - W)^T (I - W)$. Finally, by Rayleigh-Ritz theorem, Eq.(4) is performed by finding the eigenvectors with the smallest (nonzero) eigenvalues of M .

3 Image Classification Based on Manifold Learning

In most existing image classification tasks, image data are trained and classified on a few samples. However, in many applications, both training set and test set may be very large, i.e., the classification is performed with a set of known images which are related to several semantic categories and a set of unknown images rather than single image. Enlightened by the work in [15], we formulate the problem of image classification as the calculation of the distance measure between training manifold (learned from training images) and test manifold (learned from test images). Specifically, we assume that the images in each semantic category are distributed on one nonlinear manifold. So, image classification problem can be formulated as the calculation of the distances between several training manifolds and a test manifold.

Let X_{tr} and X_{te} denote a training set and a test set in the data space respectively. After performing the feature reduction based on GLLE, all images in the two sets can be mapped into the low-dimensional manifolds. Let a set of training manifolds are denoted as $\mathrm{M}_{tr1}, \mathrm{M}_{tr2}, ..., \mathrm{M}_{trl}$ and the test manifold is denoted as M_{te}, where l is the number of semantic category. Our classification approach is illustrated as Fig.1.

3.1 Extraction of Sub-manifolds

Since GLLE guarantees that the obtained low-dimensional manifold is locally linear, we assume that a nonlinear manifold can be represented by a set of linear sub-manifolds. Let a data set $Y = \{\vec{y}_1, \vec{y}_2, ..., \vec{y}_N\}$ on a d -dimensional manifold M is given, then $\mathrm{M} = \{S_1, S_2, ..., S_m\}$, $S_i = \{\vec{y}_1^{\,i}, \vec{y}_2^{\,i}, ..., \vec{y}_{N_i}^{\,i}\}$ ($S_i \cap S_j = \phi; i, j = 1,2,..., m; i \neq j$), where m denotes the number of sub-manifolds, N_i denotes the number of the images in S_i . The extraction algorithm can be summarized as follows:

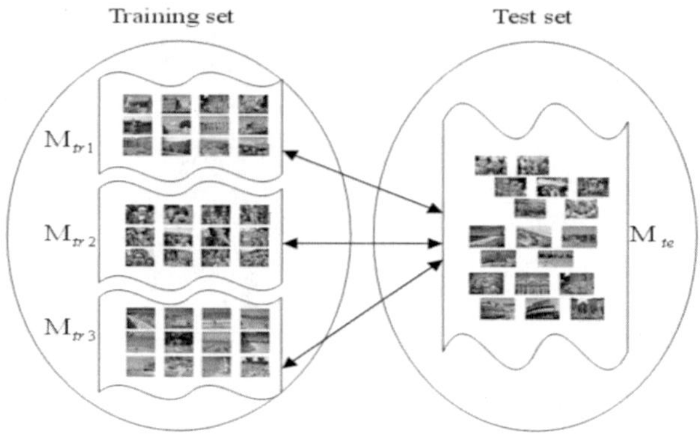

Fig. 1. Image classification based on manifold learning ($l = 3$)

Step 1: Initialize $i = 1$; $S_i = \phi$; $T = \mathrm{M}$,

Step 2: While ($T \neq \phi$)

Step 2.1: Select a data $\vec{y}_1^{\,i}$ from T and let $S_i = S_i + \{\vec{y}_1^{\,i}\}; T = T - \{\vec{y}_1^{\,i}\}$,

Step 2.2: For (each $\vec{y}^{\,i}\,' \in S_i$)

Search $\vec{y}^{\,i}\,'$'s neighbor, denoted as $\vec{y}''$, if ($\vec{y}'' \in T$ and $diff\,(\vec{y}', \vec{y}'') \leq \theta$ [15]), then let $S_i = S_i + \{\vec{y}''\}; T = T - \{\vec{y}''\}$,

Step 2.3: Let $i = i + 1$; $S_i = \phi$, go to Step 2.

Step 3: Finally, a set of sub-manifolds are obtained as $S_1, S_2, ..., S_m$.

3.2 Computation of Aggregation Center

Due to the aggregation capability of similar data in GLLE and the characteristics of linear distribution on the sub-manifolds, we use an aggregation center to represent each sub-manifold based on the principle of clustering method. Its advantages are that: (1) classification results are unaffected by abnormal data, and (2) computation time can be shortened effectively. Let $\vec{c}^{\,i}$ denotes an aggregation center of S_i ($i = 1,2,..., m$). We define a cost function and it can be minimized as follow:

$$J_i = \arg\min \sum_{j=1}^{N_i} \left\| \vec{y}^{\,i}_j - \vec{c}^{\,i} \right\|, \; i = 1,2,..., m \,. \tag{5}$$

By means of matrix norm, Eq.(5) can be solved as follow:

$$J_i = \arg\min \sum_{j=1}^{N_i} trace[(\vec{y}^{\,i}_j - \vec{c}^{\,i})(\vec{y}^{\,i}_j - \vec{c}^{\,i})^T] \,. \tag{6}$$

Then the solving of Eq. (6) is equivalent to that of $\vec{c}^{\,i}$ that satisfies the equation:

$$\vec{c}^{\,i}_{\max} = \arg\max \sum_{j=1}^{N_i} (\vec{y}^{\,i}_j \vec{c}^{\,iT}) \quad s.t. \quad \vec{c}^{\,i}\vec{c}^{\,iT} = 1,\ i = 1,2,\ldots,m \,. \tag{7}$$

Thus, Eq.(7) is performed by finding the eigenvectors with the largest eigenvalues of the matrix $\sum_{j=1}^{N_i} \vec{y}^{\,iT}_j \vec{y}^{\,i}_j$.

3.3 Definition of Classifier

According to the analyses above, the distance measure between a set of training manifold $\mathrm{M}_{tri}(i = 1,2,\ldots,l)$ and a test manifold M_{te} can be defined as:

$$d_i(\mathrm{M}_{tri}, \mathrm{M}_{te}) = \min d_i(S_j, \mathrm{M}_{te}),\ S_j \in \mathrm{M}_{tri},\ i = 1,2,\ldots,l,\ j = 1,2,\ldots,N_i \,. \tag{8}$$

If we use aggregation center to represent each sub-manifold, then $d_i(S_j, \mathrm{M}_{te})$ will be replaced by $d_i(\vec{c}^{\,i}, \mathrm{M}_{te})$. Suppose $Y_{te} = \{\vec{y}_{te1}, \vec{y}_{te2}, \ldots, \vec{y}_{teN}\}$ is the test set embedding on the manifold, two strategies can be applied to deal with these data. The first is "single strategy", i.e., $d_i(\vec{c}^{\,j}, \mathrm{M}_{te})$ is replaced by $d_i(\vec{c}^{\,j}, \vec{y}_{tep})$ $(p=1,2,\ldots,N)$. The second is "peer strategy", i.e., $d_i(\mathrm{M}_{tri}, \mathrm{M}_{te}) = \min d_i(S_j, S_k')$, where $\mathrm{M}_{te} = \{S_1', S_2', \ldots, S_m'\}$, $S_j \in \mathrm{M}_{tri}, S_k' \in \mathrm{M}_{te}$, and $d_i(S_j, S_k')$ can be further reduced to $d_i(\vec{c}^{\,j}, \vec{c}^{\,k}{}')$. To compare these two strategies, "single strategy" achieves more accurate results than "peer strategy" but need longer computation time.

Finally, based on the distance measure between the training manifold and the test manifold, we can define a novel manifold learning-based classifier as follow:

$$\min d_i(\mathrm{M}_{tri}, \mathrm{M}_{te}),\ i = 1,2,\ldots,l \,. \tag{9}$$

4 Experimental Results

We designed two experiments to verify our approach, i.e., image classification based on manifold learning (ICML). All Web images were downloaded from Internet and normalized into fixed size. To avoid the additional classification errors caused by the partition of the test manifold, we applied "single strategy" to handle the test images. Hardware: Pentium PC (CPU 1.86GHZ; Memory 2.0GB); software: MATLAB 7.0.

Experiment 1: Binary Classification
The training set includes 1350 erotic and 1800 normal images and the test set contains 105 erotic and 200 normal images. An 86-dimenasional feature vector was extracted from each image data [16]. For evaluation, we define the correct rate (Correct) as the ration of the number of correctly classified erotic images to the total number of the classified erotic images; the recall rate (Recall) as the ratio of the number of correctly

classified erotic images to the total number of erotic images; the error rate (Error) as the ratio of the number of wrongly classified normal images to the total number of normal images.

As seen from Table 1, since two separate manifolds were established for erotic images and normal images, ICML achieves higher Correct (%) (81.58), higher Recall (%) (88.57) and lower Error (%) (10.05), compared with other three approaches. So ICML shows a good behavior for the binary classification of real-world images.

Experiment 2: Multi-class Classification
The image set includes 2547 images: beach (C1, 582), sunset/sundown (C2, 462), grassland (C3, 504), high building (C4, 480) and rose (C5, 519). The images in each semantic category were randomly divided into two sets (a training set and a test set) according to 2:1 proportion. We use average accurate rate (AAR) for evaluation, which is defined as the ration of the number of correctly classified images to the total number of the images in each semantic category (over five randomly generated test sets). The extracted 40-dimensinal feature vectors include color histogram (24), wavelet texture (3), shape invariant moment (7) and edge histogram (6).

We can see from Table 2 that the AARs of LLE+Center and ICML are both higher than that of LLE+Mean [13]. Since GLLE embodies better stability and stronger anti-noise ability, compared with LLE+Center, ICML yields better performance benefits. Additionally, among three approaches, the classification time of per image in ICML is the shortest, which makes our approach become a strong candidate for the classification tasks that have time requirements, such as Web image classification.

Table 1. Performance comparison among four classification approaches. All images were resized to 200×150 or 150×200. The dimension of feature space was set to 12(see [10]).

Approach	Correct (%)	Recall (%)	Error (%)
LLE+ k-Nearest-Neighbor(KNN)	68.18	71.43	17.50
LLE+ Neural Network(NNET)	56.52	61.90	25.00
LLE+ SVM with RBF Kernel (SVMRBF)	73.91	80.95	15.00
ICML	81.58	88.57	10.05

Table 2. Performance comparison among three classification approaches. All images were resized to 96×64 or 64×96. The dimension of feature space was set to 8(see [10]).

Approach	Average Accurate Rate (%)					Time
	C1	C2	C3	C4	C5	(Second/Per image)
LLE+Mean	60.43	63.34	62.89	70.21	81.64	0.418
LLE+Center	72.12	74.27	72.52	81.01	92.34	0.385
ICML	75.01	78.62	76.16	85.47	96.21	0.342

5 Conclusion and Future Work

This paper has proposed a novel feature space-based image classification approach based on manifold learning, fully considering the quantities and diversities of real-world images. One main contribution is that: the problem of image classification in a high-dimensional data space is transformed into a low-dimensional feature space, and its solution process is viewed as the calculation of the distance measure between manifolds. Experimental results have shown promising performance of our approach in terms of both accuracy and efficiency. In the future, we will intend to test our approach over large test sets via "peer strategy". More discriminative information among categories will be introduced into the cost function of aggregation center. Additional, our approach has also applied into a license plate recognition system and obtained a preliminary result, we will further improve its precision and accuracy.

Acknowledgement. This paper is the partial achievements of project 2008AA01Z132 supported by National Hi-Tech Research & Development Program of China; project 2007C21170 supported by Science & Technology Department of Zhejiang Province.

References

1. Hsu, W., Lee, M.L., Zhang, J.: Image Mining: Trends and Developments. J. Intelligent Information Systems 19(1), 7–23 (2002)
2. Lu, D., Weng, Q.: A Survey of Image Classification Methods and Techniques for Improving Classification Performance. Int. J. Remote Sensing 28(5), 823–870 (2007)
3. Li, J., Wang, J.Z.: Automatic Linguistic Indexing of Pictures by a Statistical Modeling Approach. IEEE Trans. on Pattern Analysis and Machine Intellige 25, 1075–1088 (2003)
4. Meyer-Bäse, A.: Pattern Recognition in Medical Imaging. Academic, New York (2004)
5. Nezamabadi-pour, H., Kabir, E.: Concept Learning by Fuzzy k-NN Classification and Relevance Feedback for Efficient Image Retrieval. Expert Systems with Applications 36(3), 5948–5954 (2009)
6. Borgne, H.L., Guérin-Dugué, A., Antoniadis, A.: Representation of Images for Classfication with Independent Features. Pattern Recognition Letters 25(2), 141–154 (2004)
7. Fortuna, J., Capson, D.: Improved Support Vector Classification Using PCA and ICA Feature Space Modification. Pattern Recognition 37(6), 1117–1129 (2004)
8. Luo, J., Boutell, M.: Natural Scene Classification Using Overcomplete ICA. Pattern Recognition 38(10), 1507–1519 (2005)
9. Lu, C.D., Zhang, C.M., Zhang, T.Y., Zhang, W.: Kernel Based Symmetrical Principal Component Analysis for Face Classification. Neurocomputing 70(4-6), 904–911 (2007)
10. Xu, Z.J., Yang, J., Wang, M.: A New Nonlinear Dimensionality Reduction for Color Image. J. Shanghai Jiaotong University 38(12), 2063–2072 (2004) (in Chinese)
11. Seung, H., Lee, D.: The Manifold Ways of Perception. Science 290(5500), 2268–2269 (2000)
12. Roweis, S.T., Saul, L.K.: Nonlinear Dimensionality Reduction by Locally Linear Embedding. Science 290(5500), 2323–2326 (2000)
13. Yao, L.Q., Tao, Q.: One Kind of Manifold Learning Method for Classification. J. Pattern Recognition and Artificial Intelligence 18(5), 541–545 (2005) (in Chinese)
14. de Juan, C., Bodenheimer, B.: Cartoon Textures. In: ACM SIGGRAPH Symposium on Computer Animation, Grenoble, France, pp. 267–276 (2004)
15. Wang, R.P., Shan, S.G., Chen, X.L., Wen, G.: Manifold-Manifold Distance with Application to Face Recognition Based on Image Set. In: IEEE Conf. on Computer Vision and Pattern Recognition 2008, Anchorage, AK, pp. 23–28 (2008)
16. Jiang, Z.W.: Research on Content-Based Web Image Filter Technology. Ph.D Thesis, Zhejiang University, Zhejiang, China (2007) (in Chinese)

Social Influence and Role Analysis Based on Community Structure in Social Network

Tian Zhu, Bin Wu, and Bai Wang

Beijing Key Laboratory of Intelligent Telecommunications Software and Multimedia,
Beijing University of Posts and Telecommunications,
100876 Beijing

Abstract. Recent graph-theoretic approaches have demonstrated remarkable success for ranking networked entities, including degree, closeness, betweenness, etc. They are mainly considering the local link factors only, while not so much work concentrates on the social influence ranking based on the local structure in social network. In this paper, two new social influence ranking metrics, InnerPagerank and OutterPagerank are proposed based on the concept of modified Pagerank, by considering the community structure knowledge. It is well adapted to direct and weighted networks also. Using the two metrics, we also show how to assign community-based node roles to the nodes, which is an effective supplement for single metric used as social influence measure. Identifying and understanding the node's social influence and role is of tremendous interest from both analysis and application points of view. This method is shown to give rasonable results than previous metrics both on test networks and real networks.

Keywords: social network, community, node role, PageRank.

1 Introduction

Community structure is found to display in many networks after the global properties, such as the average shortest path length between nodes, the clustering coefficient, and other measures of degree-degree correlations. Community structures are often referred as dense connections within them but sparse connections between them. It has been proven to be of interest both in their own right as functional building blocks within networks and for the insights they offer in to the dynamics or modes of fomation of networks ([1]).

There are a number of metrics that can be used to ranking individual nodes in a network, such as degree, closeness, betweenness, etc. These simple counting metrics are attractive, because it is convenient to have a single number that is easy to interpret, particularly with applications to search engines. A node role is a subjective characterization of the part it plays in a network structure, which are highly related to the node's social influence. Knowing the role of a node is important for many link mining applications ([2,3,4]). In this paper, we argue that the utilization of community knowledge in the network with natural communities can lead to a valuable extension of existing social influence ranking problem.

R. Huang et al. (Eds.): ADMA 2009, LNAI 5678, pp. 788–795, 2009.

PageRank ([5,6]) is the well known ranking measure which successfully used in the World Wide Web network. PageRank relies on the 'democratic nature of the Web' by using its topology as an indicator of the score to be attached to any page. PageRank is so useful that we can also apply it to social networks besides World Wide Web network. For example, the PageRank value of an individual in a Telecommunication network might indicates the social influence of the individual. Since there are only a few nodes with high PageRank, the telecom operators can target at these influential people to retain them.

In this paper, we propose the two-dimensional form of PageRank which are InnerPagerank and OutterPagerank based on community structure. We assume that there has some nodes which have high OutterPagerank and low InnerPagerank can act as some kind of bridge between the communities, and also some nodes with low OutterPagerank and high InnerPagerank can take the core role of the community. The contributions of this paper include: (1) new measure for social influence ranking and node role assigning are proposed based on community structure; (2) the new measure is especially adapted to directed and weighted networks; (3) several realistic applications have been performed both on test datasets and real datasets by utilization of the roles induced by the new measure. Obtained results suggest that the new measure is successful in define various roles in directed and weighted networks with natural communities.

2 Community-Based Node Role

In this section we will give the common community discovery methods, and define the two-dimensional community-based PageRank measures. We then introduce the modified PageRank-based node roles. Finally we give the analysis of the algorithm to prove the measures' adaption to large scale networks.

2.1 Community Structure Discovery

After the concept of community been brought out by Newman etc. ([7]), a large volumn of research has been devoted to the development of algorithmic tools for discovering communities ([8,9,10,11]). The widest used method is the Fast GN algorithm, which is based on the idea of modularity (Q). It assumes a high value of Q represents a good community division, and simply optimize Q over all possible divisions thus to find the best one. The modularity function is defined as follows:

$$Q = \frac{1}{2m} \sum_{ij} (A_{ij} - \frac{k_i k_j}{2m}) \delta_{c_i, c_j} , \tag{1}$$

Newman etc. also propose a modified version to tackle the directed network discovering problem, which revised the famous modularity function as follows:

$$Q = \frac{1}{m} \sum_{ij} [A_{ij} - \frac{k_i^{in} k_j^{out}}{m}] \delta_{c_i, c_j} , \tag{2}$$

where A_{ij} is an element of the adjacency matrix, δ_{ij} is the Kronecker delta symbol, k_i is the degree of vertex i and m is the total number of edges in the network.

2.2 Community-Based PageRank

The basic idea of PageRank is that of introducing a notion of page authority. In PageRank, the authority reminds the notion of citation in the scientific literature. In particular, the authority of a page p depends on the number of incoming hyperlinks (number of citations) and on the authority of the page q which cites p with a forward link. Moreover, selective citations from q to p are assumed to provide more contribution to the score of p than uniform citations. Hence, PageRank x_p of p is computed by taking into account the set of pages pa[p] pointing to p. According to Brin and Page ([5]):

$$x_p = d \sum_{q \in pa[p]} \frac{x_q}{h_q} + (1-d) \ , \tag{3}$$

here $d \in (0,1)$ is a dumping factor and h_q is the outdegree of q, that is the number of hyperlinks outcoming from q.

PageRank computes the node influence considering the direct links in the network. The links of a graph can also be labeled, and the relationships between the labels of links provide extra information for graph mining. This is the weighted graph mining problem where each link has an associated weight. This weight could signify the strength of the link for example, each link in the Telecom call graph has an associated number specifying the times or lengthes of communication over the call.

Definition 1. *Weighted PageRank is defined as:*

$$x_p = d \sum_{q \in pa[p]} \frac{w_{qp}}{w_q} x_q + (1-d) \ , \tag{4}$$

here w_{qp} is the weight of link $q \rightarrow p$, w_q is the outweight of node q, which is the sum of wights outcoming from q.

Definition 2. *The InnerPagerank and OutterPagerank of the community G_I are defined respectively as:*

$$x_I^i = d \sum_{\substack{q \in I \\ q \in pa[i]}} \frac{w_{qp}}{w_q} x_q^i + (1-d) \ , \tag{5}$$

$$x_I^o = d \sum_{\substack{q \notin I \\ q \in pa[i]}} \frac{w_{qp}}{w_q} x_q^o \ . \tag{6}$$

According to Definition 1, the Outter PageRank depends linearly on $x_{in(I)}^o$, a vector includes only the external nodes that have links to the community.

Due to the linearity of Eq.(2), PageRank meets the decomposition property.

2.3 Modified-PageRank-Based Node Roles

Our approach is based on the general idea that nodes with the same role should have similar topological properties. The InnerPagerank measures how 'well connected' node i is to other nodes in the community, and the OutterPagerank is the measure of the links distributed among other communities. We hypothesize that the role of a node can be determined, to a great extent, by its InnerPagerank degree and OutterPagerank degree, which define how the node is positioned in its own community and with respect to other communities. The two properties are easily computed once the community of a network are known.Simple calculations suggest that each nodes can be naturally assigned into four roles, as shown in Fig. 1.

There are four roles, core nodes (role R1): if a node has larger social influence within the community; bridge nodes (role R2): if a node has larger social influence out of the community; huge_influential nodes (role R3): if a node has great social influence both inside and outside of the community, which means the node has a great pagerank value and influence widely; normal nodes (role R4): if a node has few social influence both inside and outside of the community, which means the node has a small pagerank value.

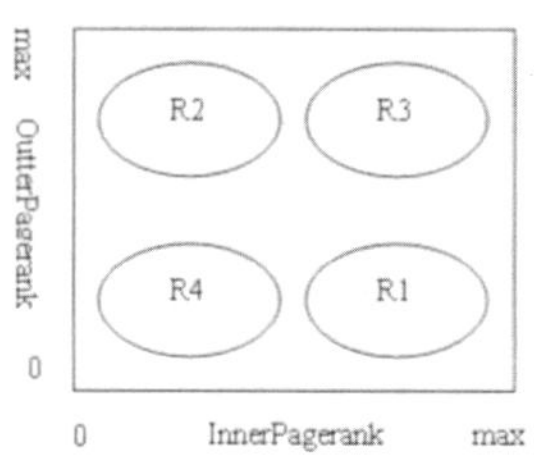

Fig. 1. Node role chart

Table 1. Datasets Used in Our Experiments

Network	karate	enron	T.C.1	T.C.2
V(G)	34	151	222	6971
E(G)	78	2082	9915	39546
L(G)	78	41998	118008	
k_{max}	17	100	222	736
Ave.C.C.	0.5706	0.5349	0.5322	0.1665
Ave.A.C.	-0.4756	-0.0192	-0.0156	-0.1455

2.4 Node Role Analysis

The role detection framework is composed of 4 steps as follows:

Step 1: read network as graph;
Step 2: community discovery using suitable partition algorithms;
Step 3: compute InnerPagerank and OutterPagerank metrics using Algorithm1;
Step 4: obtain node role list.

Algorithm 1 is described as follows, where step 2-5 are forming the inCaller and outCaller lists of each node; step 6-12 are counting the two initial matrix; step 13-23 are counting the InnerPagerank and OutterPagerank value.

Algorithm 1. InnerPagerank and OutterPagerank Algorithm

Input: Graph G with n nodes of the network, community list.
Output: InnerPagerank and OutterPagerank lists.

Procedure InnerPagerank and OutterPagerank
  for every node i $\in$ each community c_k do
    node i,j are connected;
    if i,j are of the same community **then** j $\rightarrow$ inCaller(i);
    if i,j are of the different community **then** j $\rightarrow$ outCaller(i);
  end for
  for every node i **do**
      if j $\in$ inCaller **then** $inM[i][j] = \frac{w_{ji}}{w_j} * d;$
      else inM[i][j] = 0;
      if j $\in$ outCaller **then** $outM[i][j] = \frac{w_{ji}}{w_j} * d;$
      else outM[i][j] = 0;
  end for
  inTP[1..n] = $\frac{1}{n}$, outTP[1..n] = $\frac{1}{n}$;
  do until err < eps
    for every node i $\in$ G **do**
      InnerPagerank[i] = inTP[j] * inM[j][i];
      OutterPagerank[i] = outTP[j] * outM[j][i];
      inTP = InnerPagerank, outTP = OutterPagerank;
      maxErr = max(pagerank[i]-tempprr[i]);
    end for
    err = maxErr;
  end
  InnerPagerank = InnerPagerank + (1 - d);
end Procedure

3 Experimental Evaluation

The purpose of this section is to provide the results of experiments which will demonstrate the distinctiveness and utility of InnerPagerank and OutterPagerank. Specially we show that:

(1) InnerPagerank and OutterPagerank can give two useful roles R1, R2 and R3 which Pagerank can also provide;
(2) the community-based role nodes follow a fairly predictable distribution.

3.1 DataSet Used

In our problem setting, we need the datasets which has natural community structure. We choose a standard directed and weighted dataset: enron data, and two Telecommunication Call networks (T.C.) to illustrate the effectiveness of our measure. To illustrate our measure can also reduced to be used in the undirected

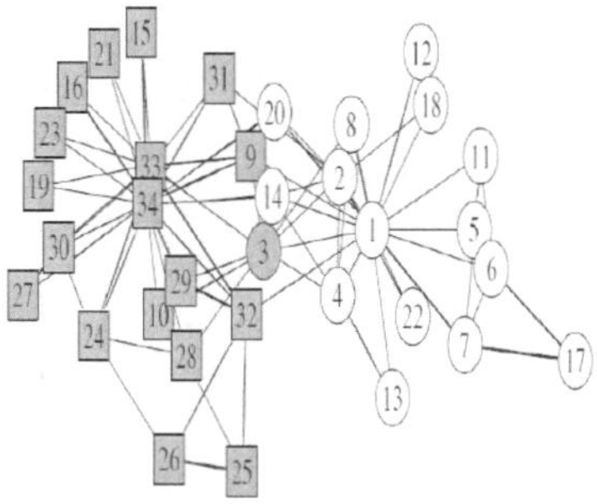

Fig. 2. Karate Club Network

Table 2. Comparison of the metrics

Node	*Degree*	Clut. Coef.	*Close.*	*Between.*	InnerP.	OutterP.
1	2	31	1	1	1	3
2	5	22	9	7	4	4
3	4	27	2	4	6	1
4	6	12	10	15	5	6
5	17	12	20	21	13	12
6	11	17	17	10	13	12
7	11	17	17	10	13	12
8	11	1	14	23	8	6
9	8	17	5	6	10	9
10	23	33	15	20	13	12
11	17	12	20	21	13	12
12	34	33	32	23	13	12
13	23	1	26	23	13	12
14	8	16	5	8	8	5
15	23	1	26	23	13	12
16	23	1	26	23	13	12
17	23	1	34	23	13	12
18	23	1	22	23	13	12
19	23	1	26	23	13	12
20	17	22	8	9	11	11
21	23	1	26	23	13	12
22	23	1	22	23	13	12
23	23	1	26	23	13	12
24	8	21	16	13	13	12
25	17	22	22	18	13	12
26	17	22	22	16	13	12
27	23	1	33	23	13	12
28	11	30	11	12	13	12
29	17	22	13	19	13	12
30	11	12	17	17	13	12
31	11	17	12	14	12	8
32	6	28	4	5	7	9
33	3	29	5	3	3	12
34	1	32	3	2	2	2

and unweighted network, we also tested on the well-known Zachary Karate Club dataset with some changes of our algorithm. Here, we choose the weight value as the times that two nodes contacts.

All experiments are done on a single PC (2.66GHz processor with 2Gbytes of main memory on Win xp OS). Table 1 shows the general description of our datasets.

3.2 Zachary Karate Club

Zachary Karate Club is one of the classic studies in social network analysis. The network is of connections among members of the club based on their social interactions shown in Fig. 2. Table 2 lists the ranked results for degree, cluster coefficient, closeness, betweenness centrality as well as InnerPagerank and OutterPagerank. The degree metric appears to have many common results as InnerPagerank. Closeness and betweenness like degree correlated somewhat with InnerPagerank, which can indicate the influence of a node. Therefore, OutterPagerank is able to identify node with id 3 considering the community knowledge while other metrics can not.

Fig. 3 shows the InnerPagerank and OutterPagerank distribution of the data. The node in the extreme right side is node with id 1, which is the core node of its community. The node on the upside is node 3, which has much connection between the two communities.

3.3 Enron Email Data

The Enron data set is a large scale email collection from a real organization over a period of 3.5 years. For our problem setting this dataset is of particular

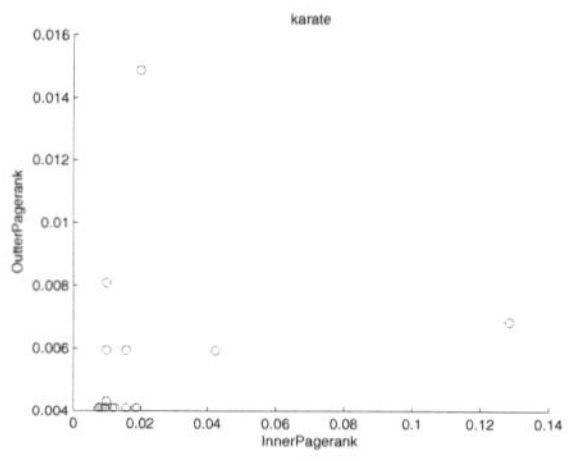

Fig. 3. Zachary Karate Club

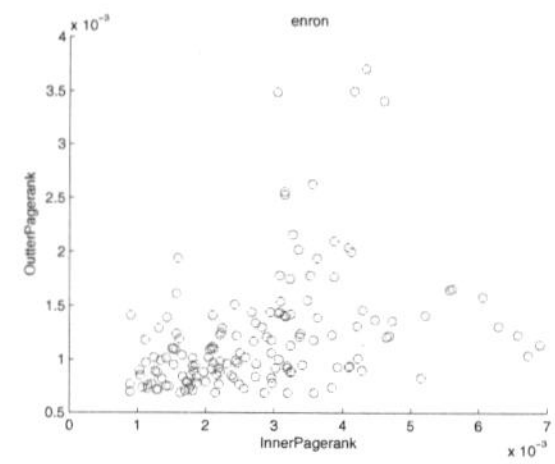

Fig. 4. Enron Email Data

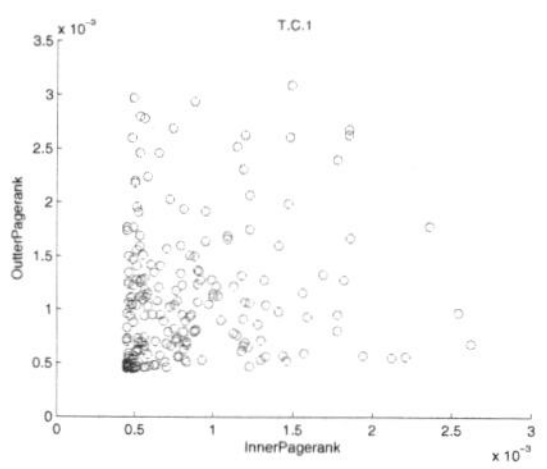

Fig. 5. T.C.1 data

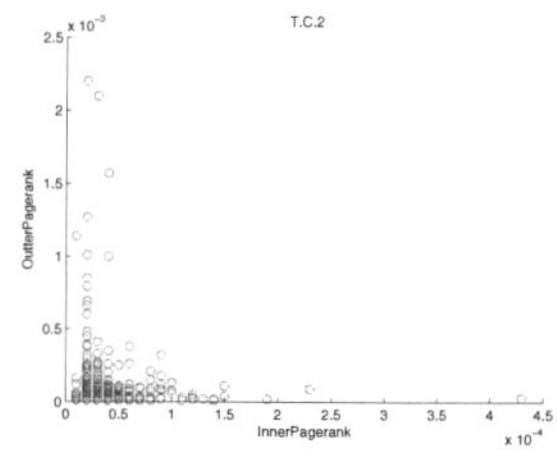

Fig. 6. T.C.2 data

interest and potential value because it enables the examination of interactions and processes within and among the entities of an organization. Fig. 4 shows the InnerPagerank and OutterPagerank distribution based on the group level of Enron data. The nodes on the right side of the chart are most of the presidents in the organization.

3.4 Telecommunication Call Networks

The first telecommunication call network is built from the dataset of a company in one city within a period of six months from a Telecom Operator in China, and the second call network is the dataset of another city in China among three months. We regard each subscriber as a vertex and two vertices will share an edge if the subscribers have once contacted with each other by their mobile phones. Fig. 5 and Fig. 6 shows each network's InnerPagerank and OutterPagerank distribution.

The two charts of the networks show very different distributions because of the first network is of a company, which connected closely and likely to form one single community, whereas the second is of a typical industry customers in one city. Therefore, our problem setting is more inclined to the network with natural communities.

4 Conclusion and Future Work

In this paper, we have identified two social influence ranking metrics, InnerPagerank and OutterPagerank, combines of which can assigning the role of a node in the network. The two ranking metrics are unique among the many available centrality metrics because they use the community information specifically besides the global topological properties. The metrics enable a new way of the use of community structure to understand the interesting information in the network. The combination of InnerPagerank and OutterPagerank also enables the node role assigning procedure to use the direct and weight factors among the network, which leads to a more reasonable result. In our future work, we will discuss the applications of the node role measure, and analysis the method to predict for the potential customer.

Acknowledgement. This work is supported by the National Natural Science Foundation of China under Grant No.60402011; the National Key Technology R&D Program of China under Grant No.2006BAH03B05; it is also supported by IBM China Research Laboratory, the Specialized Research Fund for the Joint laboratory between Beijing University of Posts and Communications and IBM China Research Laboratory (Project No.JTP200806014-3).

References

1. Newman, M.E.J.: The structure and function of complex networks. SIAM Review 45, 167–256 (2003)
2. Scripps, J., Tan, P.N., Esfahanian, A.H.: Node Roles and Community Structure in Networks. In: Joint 9th WEBKDD and 1st SNA-KDD Workshop, San Jose, California, pp. 2–35 (2007)
3. Guimerà, R., Sales-Pardo, M., Amaral, L.A.N.: Classes of complex networks defined by role-to-role connectivity profiles. Nature physics 3(26), 63–69 (2007)
4. Guimerà, R., Amaral, L.A.N.: Cartography of complex networks: modules and universal roles. Journal of Statistical Mechanics: Theory and Experiment (February 1-12, 2005)
5. Brin, S., Page, L.: The anatomy of a large-scale hypertextual Web search engine. In: Proceedings of the 7th World Wide Web Conference (WWW7), Brisbane, Australia (1998)
6. Bianchini, M., Gori, M., Scarselli, F.: Inside PageRank. ACM Transactions on Internet Technology 5(1), 92–128 (2006).
7. Givan, M., Newman, M.E.J.: Community structure in social and biological networks. PNAS 99(12), 7821–7826 (2002)
8. Du, N., Wang, B., Wu, B.: Community detection in complex networks. Journal of Computer Science and Technology 23(4), 672–683 (2008)
9. Newman, M.E.J.: Fast algorithm for detecting community structure in networks. Physical Review E 69, 066133 (2004)
10. Leicht, E.A., Newman, M.E.J.: Community structure in directed networks. Phys. Rev. Lett. 118703(100), Epub. (2008)
11. Karrer, B., Levina, E., Newman, M.E.J.: Robustness of community structure in networks. Physic Review E 77, 046119 (2008)

Feature Selection Method Combined Optimized Document Frequency with Improved RBF Network*

Hao-Dong Zhu[1, 2], Xiang-Hui Zhao[1, 2], Yong Zhong[1, 2]

[1] Chengdu Institute of Computer Application, Chinese Academy of Sciences,
610041 Chengdu, Sichuan
[2] Graduate University of Chinese Academy of Sciences,
100039 Beijing
zhongyong@cimslabsoft.com

Abstract. Feature selection is the core research topic in text categorization. Firstly, it combined word frequency with document frequency and presented an optimized document frequency (ODF) method. Then it proposed an adaptive quantum-behaved particle swarm optimization (AQPSO) algorithm in order to train the central position and width of the basis function adopted in the RBF neural network. Next the weight of the RBF network was computed by means of least-square method (LSM). Finally, a combined feature selection method was provided. The combined feature selection method firstly uses the optimal document frequency method to filter out some terms to reduce the sparsity of feature spaces, and then employs the improved RBF neural network to select more outstanding feature subsets. The experimental results show that the combined method is effective.

Keywords: Feature Selection, Document Frequency, PSO, RBF, LSM.

1 Introduction

In text categorization, texts are usually expressed in vector form, which is characterized by high dimension and sparsity [1]. While in the Chinese text classification, lexical entry is usually used as the smallest independent semantic carrier, so the original feature space may be constituted by all the lexical entry that appear in the text. The total number of the Chinese lexical entry is more than 200,000, which makes its high-dimension and sparsity more obvious, and greatly limits the choice of classification algorithms and reduces its efficiency and accuracy. Therefore, to find an efficient feature selection method, which can be used to reduce feature space dimension, avoid dimension disaster and improve the efficiency and accuracy of text classification, has

* Biography: Hao-Dong ZHU (1980-), male, Ph.D., The member of China Computer Federation, NO: E20-0012527G, main research: Software process technology and method, Text mining; Xiang-Hui ZHAO (1982-), male, Ph.D., The member of China Computer Federation, NO: E20-0012404G, main research: Machine learning, Data Mining, Image analysis and pattern recognition; Yong ZHONG (1966-), male, Tutor and Researcher, main research: Software process technology and method.

R. Huang et al. (Eds.): ADMA 2009, LNAI 5678, pp. 796–803, 2009.

become more important problem that is urgent to be solved in the automatic text categorization [2].

So, a new feature selection method was provided. The new feature selection method firstly uses an optimal document frequency method to filter out some terms to reduce the sparsity of feature spaces, and then employs an improved RBF neural network to select more outstanding feature subsets.

2 Two Classic Feature Selection Methods

The classic feature selection methods of texts include information gain, word frequency, document frequency [3, 4], etc. It simply introduces word frequency and document frequency here, and for the others, please refer to the literature [3, 4].

2.1 Word Frequency

When Word Frequency is used to choose features, this method only considers the times that the features appear in the document set. If the times that a feature appears in the text set reach to a pre-set threshold, it will leave the feature that mentioned, or it will delete it. The disadvantage of this method is that it only selects the word that appears frequently as the feature and neglects the low-frequency words.

2.2 Document Frequency

When Documents Frequency is used to choose features, this method only considers the number of documents which have the features. If the number of documents that a feature appears in the text set reaches a pre-set threshold, it will leave the feature that mentioned, or it will delete it. The disadvantage of this method is that it only considers whether the feature words appear in the document or not and neglects the times that they appear. So the problem appears, that is, if the document frequency of the feature words a and b are the same, then this method holds that the contribution of the two feature words are the same and neglects the times that they appear in the document. However, the common situation is that the words that appear less frequently in the document are the noise words. Therefore, that will make the features chosen by this method not very representative.

3 Optimized Document Frequency

By analyzing the word frequency method (2.1) and the document frequency method (2.2), we find that these two methods are complementary. Therefore, if two methods are combined together, we can have good results. For this reason, we put forward an optimized document frequency---document frequency based on minimum word frequency.

Definition 1. Document Frequency Based on Minimum Word Frequency of the feature f signifies the number of the documents when the times that the feature f appears reach to a certain extent, and is made as DF_n, in it, n means the least times that the feature words appear in the document.

4 Brief Introduction of RBF Neural Network

RBF network is 3-layer forward feed neural network [5,6,7]. It is constituted by input layer, hidden layer and output layer. The output calculating formula of the network output layer is:

$$y_i = \sum_{k=1}^{N} w_{ik} \varphi_k (x, c_k) = \sum_{k=1}^{N} w_{ik} \varphi_k (\| x - c_k \|_2) \tag{1}$$

Here $x \in R^{n \times l}$ is the input vector and $\varphi_k(.)$ is the function got from the positive real numbers set to the real numbers set. As there are many given forms of this function, here Gaussian function is used:

$$\varphi(x) = \exp(-x^2 / \sigma) \tag{2}$$

$\|.\|_2$ signifies Euclidean paradigm, w_{ik} is the weight value of the output layer, N signifies the number of the neurons in the hidden layer and $c_k \in R^{n \times l}$ signifies the RBF network center of the input vector space. For each hidden layer neuron, it needs to calculate the Euclidean distance between the associated centers and the network input. The output of the hidden layer neuron is the nonlinear function of the distance. The total number of the added weight of the hidden layer output is the output of RBF network [5, 6, 7].The number of hidden layer nodes of the RBF network can be determined by SOM (self organization mapping) network, while for the center c_k and the width σ , they are solved by means of the following adaptive quantum particle swarm optimization algorithm.

5 Adaptive Quantum-Behaved Particle Swarm Optimization (AQPSO)

Particle Swarm Optimization (PSO) Algorithm is a kind of random search algorithm based on the swarm intelligence, its particles fly in the classical mechanics space. The flight track is definite, so the search space is limited and is easy to fall into local minimum [8]. In order to avoid this defect, from the point of the quantum mechanics, literature [9] puts forward the particle swarm optimization (QPSO) with the quantum behavior. As the particles that move in the quantum space do not have the definite tracks, which makes particles can randomly search the optimal solution in the whole possible solution space. Therefore, the global search ability of the QPSO algorithm is far better than the PSO algorithm [9]. Owing to that, in the quantum space, the particles' position and velocity cannot be determined, so we can describe the state of particles through the wave function. And by solving the Schrödinger equation, we can get the probability density function of particles appear in the certain dot of the space. Then, by means of Monte Carlo random simulation, we can get the equations of the particles' position in the quantum space, as following [10]:

$$p = a \times p_b(i) + (1-a) \times g_b \tag{3}$$

$$m_b = \frac{1}{N}\sum_{i=1}^{N} p_b(i) \tag{4}$$

$$b = 1 - i_t / i_{t\max} \times 0.5 \tag{5}$$

In the QPSO Algorithm, there is only one coefficient b, and the choice and control of this parameter is very important, it is closely related to the convergence performance of the whole algorithm[10]. The literature [10] has proved that when b is smaller than 1.7, the particles converge and are close to the current best position of particles swarm; when b is bigger than 1.8, the particles diffuse and are away from the current best position of particle swarm. The formula (5) shows that the coefficient b decreases linearly as the evolution generations increase in the process of the particles evolution. In this regard, this paper has improved as following: $h = f_{bfit2} / f_{fitness}(i)$

If h is smaller than 0.5, then b is equal to two multiplies h (6)
If h is bigger than and is equal to 0.5, then b is equal to one plus h (7)

In it, f_{bfit2} is the adaption degree of which the last generation group gets the best position g_b , $f_{fitness}(i)$ is the current adaption degree of the i particle, and h is the ratio of the two that mentioned above, when h is smaller, it means that the particles are away from the current best position of the particles swarm; when h is bigger, it means that the particles are closer to the current best position of the particles swarm. In this paper, we take the value 0.5 of h as a dividing line, if h is smaller than 0.5, it means that the particles are away from the best position g_b of the swarm, the coefficient b of contraction and expansion should be smaller than 1.7 and makes it converges. Therefore, the value of b is set to 2 multiplies h , so that it does not exceed 1. When h is bigger than and is equal to 0.5, it means that the particles are close to the current best position g_b so the value of b is set to one plus h in order to increase its probability of being bigger than 1.8 , and enables it to diffuse as far as possible and expand the search area.

6 Optimized RBF

The specific model of the optimized RBF network is as follows:

Step1: To initialize the particles swarm POP, the best position of each particle is $p_b(0) = \varnothing$, the best position of the particles swarm $g_b = \varnothing$, the adaptation degree of the particles $f_{fitness}(0) = 0$, the optimal adaption degree of the current particles swarm $f_{bfit1} = 0$, the optimal adaption degree of the previous generation of the particles swarm $f_{bfit2} = 0$ and its presupposed precision ε=0.09, and the maximum times of the iterations $t_{t\max} = 600$, t=1;

Step2: Basing on the current position of the particle i (we obtain the center and the width of the network), and combining with the method of the least 2 multiplication (we obtain the connection weight value of the network) , we calculate the adaption degree of the particle i to all the training samples; moreover, we compare the adaption degree of the particles $f_{fitness}(i)$ the adapt ion degree of the whole particles swarm f_{bfit1}. If $f_{fitness}(i) < f_{bfit1}$, then we update the best position $p_b(i)$ of the particle i ;

Step3: To determine whether all the particles have finished the search, if yes, and then go to Step4, otherwise return to Step2;

Step4: To compare the optimal adaption degree f_{bfit1} of the current swarm with the optimal adaption degree f_{bfit2} of the previous generation swarm, if $f_{bfit1} < f_{bfit2}$, then update the optimal position g_b and the optimal adaption degree f_{bfit1} of the particles swarm;

Step5: To determine whether the optimal adaption degree of the particles swarm, that is the smallest E_{MS}, is smaller than the presupposed precisionε. If yes, then go to the step8, otherwise go to step6.

Step6: t=t+1, if $t \geq t_{t\max}$, then go to step8, otherwise return to step7;

Step7: To update the position of each particle according to the formula from (3) to (7) and produce new particles swarm, then return to step2;

Step8: The training of the RBF network is completed, and the best position g_b of the particles swarm is output. In it, $g_b(1:m)$ is corresponding to m optimal data centers of the RBF network, and $g_b(m+1:2\times m)$ is corresponding to m optimal width of the RBF network. Then we calculate the weight value of the network connection by the LMS and build the model of RBF network prediction based on the AQPSO algorithm.

7 Offered Feature Selection Method

The offered feature selection method in this paper has combined the document frequency that is based on the minimum word frequency with the optimization RBF neural network that is based on AQPSO algorithm. The simple process is as follows:

To train the sample to extract and get the original feature set by segmentation and feature, then filter out some terms by means of optimization document frequency (the smallest threshold value of the word frequency is: n = 2, the minimum threshold value of documents is: m = 5, at this time, the feature set is the primary feature set).

The method of using the optimization RBF network to select the optimal features is as follows:

To denote each training sample in the form of vector, and each primary feature (the selected feature by the optimization document frequency) is corresponding to a weight value in this vector. In this paper, we take the times that the feature appears in the text and the total number of the texts that contain the feature in the whole training texts that the text belongs to as the weight value and the vector of all the text (equivalent to the initial particles swarm) as the training samples, the number of neurons of the RBF networks input layer is equal to the number of primary features, and the number of output neurons is equal to the number of the training text types. The number of the hidden layer neurons is relatively constant (to determine it according to the generalization and the training efficiency of the network). After training, there are some hidden layer neurons that the larger weight value corresponds to and the feature that the neurons of the input layer that connect to it signifies is the feature and their combined set is the selected feature subset.

8 Experimental Examples

8.1 Experimental Corpus and Environment

In the aspect of Chinese text classification, after analysis and comparison, the corpus in this paper is from Chinese text classification corpus of Fudan University. This corpus is constructed by the group of natural language processing of the international database center in computer information and the technology department of Fudan University. All documents completely come from the Internet. It can be downloaded on-line. The website is: http://www.nlp.org.cn/categories/default.php? cat_id=16.

This corpus consists of 20 categories. This paper only takes the first 10 categories of partial documents, various category documents distribution are shown in Table 1.

Table 1. Various Categories Documents Distribution

Number	Category	Training documents	Test documents
1	Economy	480	419
2	Sport	584	489
3	Computer	628	591
4	Politics	573	482
5	Agriculture	547	435
6	Environment	405	371
7	Art	510	286
8	Outer space	506	248
9	History	466	468
10	Military	74	75

When it carries out Chinese participle processing, we use ICTCLAS system of the open source project Chinese lexical analysis system of institute of computing Technology of Chinese academy of sciences. The software tool used in experiment is Weka that is a series of machine learning algorithm related to data mining and developed by the New Zealand Waikato University. The website is: http://www.cs.waikato.ac.nz/ml/weka/. The computing equipment used in experiment is MATLAB 7.0.

8.2 The Classifier and the Evaluation Criteria

This experiment aims to compare the influence on the performance of subsequent text classification by the method proposed in this paper with information gain (IG) and statistics (CHI), mutual information (MI). Therefore, this experiment uses the same classifier to classify the text after a variety of feature selection methods. It uses KNN classifier to compare these kinds of feature selection methods (it sets the value of K as 10).In order to appraise the classification performance, it choose classification accuracy rate and recall rate as the evaluation criteria: Accuracy =a/(a+b), Recall=a/(a+c). The a, b and c refer to document number respectively and their meanings are shown in the Table 2.

Table 2. Binary Association tables

	In category	No in category
Judged belongs to the category	a	b
Judged not belong to the category	c	d

8.3 Experimental Results

The experiment is taken 10 times. Final experimental results are the average experimental results of these experimental results, as shown in Fig.1 and Fig. 2. Fig.1 shows contrast results in accuracy rate, and the vertical axis expresses accuracy rate, the horizontal axis expresses category number. Fig.2 shows contrast results in recall rate, and the vertical axis expresses recall rate, the horizontal axis expresses category number.

Fig.1 and Fig.2 summary accuracy rate and recall rate of the four methods in the selected data sets. On the whole, the method in this paper is > IG> CHI> MI.

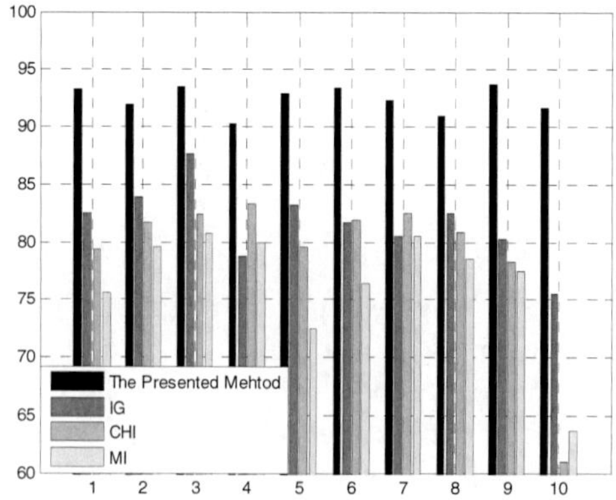

Fig. 1. Contrast Results in Accuracy Rate

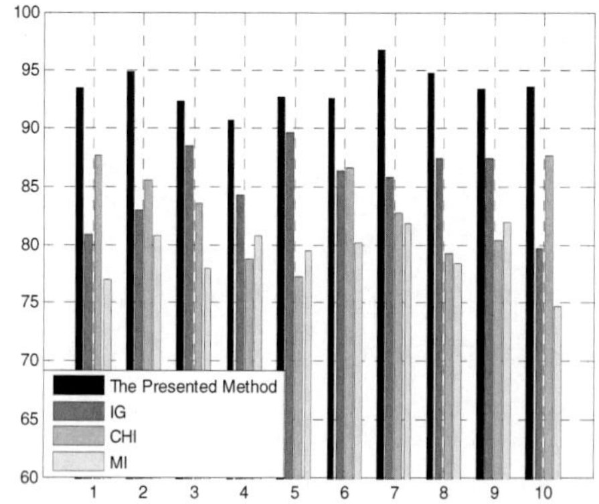

Fig. 2. Contrast Results in Recall Rate

9 Conclusion

This paper firstly discussed two classic feature selection methods and summarized their deficiencies. After that it put forward a method of optimal document frequency and used this method to filter out some terms in order to reduce the sparsity of the texts matrix. Then it proposed an adaptive quantum particle swarm optimization (AQPSO) algorithm in order to train the central position and width of the basis function adopted in the RBF neural network, computed the weight value of the network with least-square method (LSM), and the RBF neural network was improved. Finally, a combined feature selection method was provided. This method not only reduces the dimension of vector space, but also makes the selected feature subset more representative. The experiment shows that the feature selection method in this paper has higher degree of accuracy and recall rate compared with the three classical methods of feature selection, they are "Information Gain", " x^2 statistics" and "Mutual Information", and reduces the complexity of time and space for the follow-up knowledge discovery algorithms, all of which provide with certain practical value in the text classification.

Acknowledgments. This article is supported by a grant from the Science and Technology Plan Projects in Sichuan Province of China (No: 2008GZ0003). Meantime, it is supported by key technologies R & D program of Sichuan province of China (No: 2008SZ0100).

References

1. Delgado, M., Martin-Bautista, M.J., Sanchez, D., Vila, M.A.: Mining text data: special features and patterns. In: Proceedings of ESF Exploratory Workshop, London, UK, pp. 32–38 (2002)
2. Zhu, H.-D., Zhong, Y.: New Feature Selection algorithm based on multiple heuristics. Chinese Journal of Computer Applications 29(3), 848–851 (2009)
3. Friedman, N., Geiger, D., Goldszmidt, M.: Bayesian NetworkClassifiers. Machine Learning 29(2), 131–163 (1997)
4. Zhang, H.-L., Wang, Z.-L.: Automatic text categorization feature selection methods research. Chinese Journal of Computer Engineering and Design 27(20), 3838–3841 (2006)
5. Jiang, H.-G., Wu, G.-F.: RBF Network Prediction Model Based on Artificial Immune Principal. Chinese Journal of Computer Engineering 34(2), 202–205 (2008)
6. Yan, S.-Y., Yu, X.-Y., Zhang, Z.-J.: Subjective evaluation of user interface design using an RBF network. Chinese Journal of Harbin Engineering University 28(10), 1150–1155 (2007)
7. Zang, X.-G., Gong, X.-B., Chang, C.: An Online Training RBF Network Based on Immune System. Acta Electronica Sinica 36(7), 1396–14000 (2008)
8. Liu, X.-Z., Yan, H.-W.: A RBF Neural Network Learning Algorithm Based on Improved PSO. Chinese Journal of Computer Technology and Development 16(2), 185–187 (2006)
9. Chen, W., Feng, B., Sun, J.: Simulation study on the parameters optimization of radial basis function neural network based on QPSO algorithm. Chinese Journal of Computer Applications 26(8), 19–28 (2006)
10. Sun, J., Feng, B., Xu, W.-B.: Particle swarm Optimization with particles having quantum behavior. In: Proceeding of 2004 Congress on Evolutionary Computation, Piscataway CA, pp. 325–330. IEEE Press, Los Alamitos (2004)

Author Index

MIX
Papier | Fördert gute Waldnutzung
FSC® C083411

Kolibri 360 GmbH
Ferdinand-Jühlke-Straße 7
99095 Erfurt, Deutschland
produktsicherheit@kolibri360.de